CLIMATE CHANGE AND TERRESTRIAL CARBON SEQUESTRATION IN CENTRAL ASIA

BALKEMA – Proceedings and Monographs
in Engineering, Water and Earth Sciences

Climate Change and Terrestrial Carbon Sequestration in Central Asia

Editors

R. Lal

The Ohio State University, Carbon Management and Sequestration Center, Columbus, Ohio, USA

M. Suleimenov

International Center for Agriculture Research in Dryland Areas-Central Asia Caucasus, Tashkent, Uzbekistan

B.A. Stewart

Department of Agricultural Sciences, West Texas A&M University, Canyon, Texas, USA

D.O. Hansen

The Ohio State University, International Programs in Agriculture, Columbus, Ohio, USA

P. Doraiswamy

USDA-ARS Hydrology and Remote Sensing Laboratory, Beltsville, Maryland, USA

CRC Press is an imprint of the
Taylor & Francis Group, an **informa** business

CRC Press
Taylor & Francis Group
6000 Broken Sound Parkway NW, Suite 300
Boca Raton, FL 33487-2742

First issued in paperback 2019

Typeset by Charon Tec Ltd (A Macmillan company), Chennai, India

ISBN-13: 978-0-415-42235-2 (hbk)
ISBN-13: 978-0-367-38877-5 (pbk)

British Library Cataloguing in Publication Data
A catalogue record for this book is available from the British Library

Library of Congress Cataloging-in-Publication Data
Climate change and terrestrial carbon sequestration in Central Asia/R. Lal ... [et al.], editors.
p. cm.
Includes bibliographical references and index.
ISBN 978-0-415-42235-2 (hardcover : alk. paper) 1. Carbon sequestration—Asia, Central. 2. Climatic changes—Asia, Central. I. Lal, R.
SD387.C37C58 2007
368.700958—dc22 2007009819

Visit the Taylor & Francis Web site at
http://www.taylorandfrancis.com

and the CRC Press Web site at
http://www.crcpress.com

Table of Contents

Foreword

The countries of Central Asia and the Caucasus (CAC), which were a part of the former Soviet Union, represent a vast area of some 416 million hectares. This area has great potential for carbon sequestration through better land management. Since the CAC countries are now "economies in transition," they offer an opportunity to examine the role of land use change in both reducing poverty and in ensuring long-term sustainability of natural resources.

Agriculture in CAC occupies around 70% of the land area and is characterized by relatively low productivity and increasing land degradation. After independence, all CAC countries dismantled the former large state-run farms and are now facing the challenge of addressing the problems of smaller leased or privately owned land. During this period of transition, crop and livestock production has declined markedly as inputs have become scarcer and unaffordable. Recovery has been slow, putting enormous pressure on rural populations as they struggle to come to terms with these profound changes. Some areas have been abandoned because of land degradation, lack of resources and low returns. Management of the production systems and of the natural resource base that they depend on remains less than satisfactory in many areas.

A Collaborative Research Program for Sustainable Agricultural Production in Central Asia and the Caucasus was established in 1998, involving nine CGIAR Centers and eight CAC NARS, with ICARDA as the lead Center. The Consortium has helped the region to halt the erosion of its genetic resources, organize joint research on improved crop and livestock production, and on efficient control of pests and diseases. Many conservation agriculture technologies including zero tillage, crop diversification and reduction of summer fallow area, as well as improved rangeland management practices, are contributing to improved soil organic carbon maintenance. The Central Asian Countries Initiative on Land Management (CACILM), in which ICARDA has been playing the lead role in a Research Component on Sustainable Land Management, will address carbon sequestration as one of priority issues.

Since much still remains to be done, it was only timely that Ohio State University, jointly with the USDA-ARS, ICARDA and CIMMYT, organized a workshop on Carbon Sequestration in Central Asia on 1–5 November, 2005. The prospects for carbon trading were discussed at the workshop. The trading schemes that emerged could offer the rural poor in CAC the opportunity to generate income and conserve the natural resource base. The challenge will be to stimulate a widespread adoption of the promising resource-conserving practices that were identified at the workshop and to link them to carbon trading possibilities.

We are delighted that the proceedings of this workshop are now available to a wider audience through this publication. We look forward to participating in, and catalyzing further interactions between the national agricultural research systems of the region and international research organizations and other partners globally.

Mahmoud Solh
Director General, ICARDA.

Preface

Climate change and desertification are major global issues of the 21st century. The earth's mean global temperature rose by $0.6 \pm 0.2°C$ during the second half of the 20th century, at a rate of 0.17°C/decade. If the present trend continues, a drastic increase in global temperature is projected by the end of the 21st century and consequences will be a rise in sea level and accelerated meltdown of polar ice sheets. Scientists recently observed that a shelf of floating ice, which was larger than 100-km^2, and which jutted into the Arctic Ocean for 3,000 years from Canada's northernmost shore, broke away in the summer of 2005 because of sharply warming temperatures. Increasing concentrations of CO_2 along with CH_4 and N_20 are several of the causes of the accelerated greenhouse effect. The problem is exacerbated by a rising global demand for energy and a corresponding increase in fossil fuel combustion. The world used 420 Q of energy in 2003. This amount is projected to increase to 470 Q in 2010 and 620 Q in 2025.

Thus, growing and strong interest in renewable energy sources is rightfully justified. For all their potential, however, wind does not always blow nor does the sun always shine when they are most needed. For this reason a reduction of carbon emissions will remain an essential component of any strategy to address global warming.

The U. N. declared 2006 the "Year of Deserts and Desertification". Despite the severity of the problem and the good intentions of all parties concerned, no concrete action was undertaken during 2006 to combat it nor has a sharply focused action plan been designed for the future.

It is in this context that the topic of terrestrial carbon sequestration in Central Asia is extremely relevant and timely. Serious problems of soil and environmental degradation in general, and that in Central Asia in particular, have been exacerbated by the collapse of Soviet Union which helped to coordinate regional use of soil and water resources. Land use change from natural steppe vegetation to agricultural ecosystems also resulted in severe problems of wind and water erosion and desertification in the region. The total desert area in the region is estimated to be about 150 Mha or 37% of the total land area. Most agricultural and range land soils lost 30 to 50% of their soil organic carbon pool, and soils have experienced a corresponding decline in quality. Inappropriate land use, soil mismanagement, and excessive irrigation with high evaporation have caused severe and unprecedented problems of degradation of soil, water, vegetation and other elements of the environment of the region with long-term adverse impacts on agricultural sustainability, environmental quality and economic well being of the region's inhabitants.

A workshop was held at The Ohio State University campus in fall, 2005. It addressed soil and other environmental problems in the Central Asia region. The rationale for organizing the workshop included the following considerations:

Soil degradation: Large areas of arable land in Central Asia are being lost to production as a consequence of inappropriate cropping systems and inappropriate irrigation schemes. Some of it is being transformed into semi-desert conditions with an attendant loss in soil biodiversity, soil organic carbon pool, plant nutrient reserves, and plant available water capacity. Declines in soil quality have severe adverse impacts on net primary productivity (NPP), agronomic sustainability, and water quality.

Loss of Aral Sea and Water Resources: Land us change and expansion of irrigation have had a major negative impact on the hydraulic balance of the region. Two major rivers (Amu-Darya and Syr-Darya) feed the Aral Sea. Overuse of their flowage for irrigation purposes has drastically shrunk the Aral Sea and adversely affected its water quality. This has resulted in a major reduction in the availability of water to sustain human and animal populations in the region and major adverse changes in the surrounding ecoregions.

Water Pollution and Contamination: Excessive water use and indiscriminate use of agricultural chemicals have led to considerable salinization and water logging of soils. Pollution and contamination of water resources are serious problems throughout the region.

Unsustainable Agriculture and Food Insecurity: The severe problems of soil degradation and depletion of water resources in the region threaten the production of food for inhabitants of the region. The problem is exacerbated by the projected climate change which may accentuate the frequency and intensity of extreme events.

Global Climate Change: Depletion of the soil organic carbon pool exacerbates emission of CO_2 into the atmosphere. Soil degradation decreases CH_4 uptake by agricultural soils. Indiscriminate use of nitrogenous fertilizers and water logging also accentuate N_2O emissions from croplands. The projected climate change may be a positive feedback which increases the risk of soil degradation and the rate of decomposition of soil organic matter.

Eminent scholars, who are familiar with these problems in Central Asia, were invited to contribute chapters to this book. Topics and authors were specifically chosen to achieve the following 5 objectives: (a) identify land use and soil/vegetation management strategies that restore degraded soils and ecosystems, enhance soil quality, improve water use efficiency, and sequester carbon in soil biomass; (b) develop strategies to facilitate dialogue among scientists and policy makers so that soil and ecosystem recovery is an integral component of any governmental program to mitigate climate change; (c) encourage dialogue on scientific and technological exchange; (d) create multi-disciplinary teams to facilitate carbon trading in national and international markets; and (e) identify social, economic, and bio-physical factors and processes that restore degraded soils and ecosystems, thus making agriculture a contributor to the solution of the environmental degradation problems in Central Asia.

The 34-chapter volume is a state-of-the-knowledge compendium on terrestrial C sequestration in Central Asia. It is sub-divided into 8 thematic sections. Section A deals with the biophysical environments of the region and consists of 3 chapters: one describing the principal biomes and the other two reviewing the predominant vegetative cover of the region. Section B deals with the water resources of Central Asia. It consists of 4 chapters that address the current water regime, possible impacts on it of climate change, problems caused by water mismanagement, contamination of surface and ground waters by non-point source pollution, and increasing salinization. Section C also consists of 4 chapters in which existing challenges to sustainable agriculture, problems of soil degradation, and the effects of irrigation schemes on secondary salninzation are discussed. Section D consists of 12 chapters that address the principal theme of the book, namely, “soil management and its relationship to carbon dynamics”. Several chapters focus on the impact of tillage methods, soil fertility management, and summer fallowing on soil carbon dynamics, water conservation and agronomic productivity. Section E, contains two chapters that describe the important relationship between forest management and carbon dynamics. Section F also contains two chapters in which economic analyses of land use practices are presented. Materials found in Section G deal with important methodological issues regarding the use of GIS, remote sensing, carbon budgeting and scaling. Section H consists of only one chapter in which knowledge gaps on carbon and climate change are identified and related researchable priorities are recommended.

Organization of the workshop and publication of this volume were possible because of the cooperation and support of sponsoring organizations. The workshop was jointly sponsored by The Ohio State University, the International Center for Agricultural Research in Dry Areas (ICARDA), the United States Department of Agriculture – Agricultural Research Service (USDA-ARS), and the International Maize and Wheat Improvement Center (CIMMYT). Dr. Mekhlis Suleimenov, Assistant Regional Coordinator, of ICARDA-CAC Office in Tashkent, Uzbekistan played an important role in identifying and contacting scientists from the region. All authors are to be thanked for their outstanding efforts to document and present current research information and summary analyses of the current status of accumulated knowledge on these topics. The authors’ contributions help increase general understanding of opportunities and challenges encountered when attempting to enhance terrestrial carbon sequestration in Central Asia, and the potential sink capacity of different biomes through adoption of recommended land use and management practices. Research reported in this volume has advanced the frontiers of soil and environmental science with regards to terrestrial C sequestration, enhancing biomass-productivity, improving soil quality, advancing sustainability and mitigating climate change.

Special thanks are also due to Dr. Bobby A. Stewart. He undertook the most difficult and tedious task of formatting each of the 34 chapters and getting them camera ready. It is a pleasure and honor to work with him. He is a role model regarding dedication, hardwork, sincerity and commitment to excellence. Thanks are also due to Dr. Jerry Ladman of OSU for his support through the CIRIT-Climate Change initiative. Help received from staff of the Carbon Management and Sequestration Center and Ms. Lynn Everett in relation to organizing the workshop is also much appreciated. Preparation of this volume also depended on assistance from many staff of the publisher Taylor & Francis, Leiden, The Netherlands.

15 January 2007
Columbus, OH 43210

R. Lal
Chair, Organizing and Editorial
Committee

Contributors

Bakhyt Akhmedou, Science Systems and Applications, Inc., Lanham, MD, USA.

Khukmatullo Akmadov, Academy of Agricultural Sciences of the Republic of Tajikistan, Dushanbe, Tajikistan.

A. Akramhanov, International Maize and Wheat Improvement Center (CIMMYT), Astana, Kazakhstan.

U. Akramov, Soil Science Institute of Tajikistan, Dushanbe, Tajikistan.

Kanat Akshalov, Barayev Kazakh Research and Production Center of Grain Farming, Shortandy, Kazakhstan.

Evsey Aleksandrovskiy, Research and Production Centre of Ornamental Gardening and Forestry, Tashkent, Uzbekistan.

T.E. Astanakulov, Samarkand Agricultural Institute, Samarkand, Uzbekistan.

Clanton C. Black, Biochemistry and Molecular Biology, University of Georgia, Athens, Georgia, USA.

W. Bleuten, University of Utrecht, Utrecht, The Netherlands.

Warren Busscher, Coastal Plains Soil, Water and Plant Research Center, USDA-ARS, Florence, SC, USA.

V.E. Chub, Uzhydromet, Tashkent, Uzbekistan.

Eddy De Pauw, International Center for Agricultural Research in the Dry Areas (ICARDA), Aleppo, Syria.

Salamat Djumabaeva, Research Institute of Forest and Nut Industry, Bishkek, Kyrgyzstan.

Paul C. Doraiswamy, USDA-ARS Hydrology and Remote Sensing Laboratory, Beltsville, MD, USA.

D. Egamberdiyeva, Tashkent State University of Agriculture, Tashkent, Uzbekistan.

Pirnazar Elmuratov, Tashkent State University of Economics, Tashkent, Uzbekistan.

O. Erokhina, Soil Research Institute, Academgorodok, Almaty, Kazakhstan.

Shinya Funakawa, Graduate School of Agriculture, Kyoto University, Kyoto, Japan.

I.M. Ganiev, Samarkand Agricultural Institute, Samarkand, Uzbekistan.

I. Garfurova, Tashkent State University of Agriculture, Tashkent, Uzbekistan.

B. Haitov, Tashkent State Agrarian University, Tashkent, Uzbekistan.

Nazar Hakimov, Tashkent State University of Economics, Tashkent, Uzbekistan.

Rashid Hakimov, Tashkent State University of Economics, Tashkent, Uzbekistan.

A.X Hamzaev, Samarkand Agricultural Institute, Samarkand, Uzbekistan.

D.O. Hansen, International Programs in Agriculture, The Ohio State University, Columbus, OH, USA.

Sh. T. Holikulov, Samarkland Agricultural Institute, Samarkland, Uzbekistan.

G.A. Ibragimov, Samarkand Agricultural Institute, Samarkand, Uzbekistan.

R.K. Ikramov, Central Asian Research Institute of Irrigation, Tashkent, Uzbekistan.

K.R. Islam, Crops, Soil and Water Resources, Ohio State University South Centers, Piketon, OH, USA.

Temirbulat Jalankuzov, Akademgorodok, Institute of Soil Science, Almaty, Kazakhstan.

Ekaterina Juylova, Department of Desert Ecology and Water Resources Research, Samarkland, Uzbekistan.

M. Karabayev, International Maize and Wheat Improvement Center (CIMMYT), Astana, Kazakhstan.

Elmira Karbozova-Saljnikov, Institute of Soil Science, Belgrade, Serbia and Montenegro.

Abdukhalil Kayimov, Department of Forestry, Tashkent State Agrarian University, Tashkent, Uzbekistan.

Anwar U.H. Khan, University of Agriculture, Faisalabad, Pakistan.

T.M. Khujanazarov, Department of Desert Ecology and Water Resources Research, Samarkland, Uzbekistan.

Rasulmant Khusanov, Research Institute of Market Reforms, Tashkent, Uzbekistan.

S.N. Kirpotin, Tomsk State University, Tomsk, Russia.

Takashi Kosaki, Graduate School of Global Environmental Studies, Kyoto University, Kyoto, Japan.

Muhammad Kosimov, Research Institute of Market Reforms, Tashkent, Uzbekistan.

N.P. Kosych, Institute of Soil Science and Agrochemistry, Novosibirsk, Russia.

Osamu Kozan, University of Yamanashi, Kofu-city Yamanashi, Japan.

Flarida Kozybaeva, Akademgorodok, Institute of Soil Science, Almaty, Kazakhstan.

S.P. Kulizhski, Tomsk State University, Tomsk, Russia.

R. Lal, Carbon Management and Sequestration Center, SENR/OARDC, The Ohio State University, Columbus, OH, USA.

A. Landi, University of Ahwaz, Faculty of Agriculture, Department of Soil Science, Ahwaz, Iran.

E.D. Lapshina, Yugorskiy State University, Khanty-Mansiysk, Russia.

Allan Lines, Department of Agricultural, Environmental, and Development Economics, The Ohio State University, Columbus, OH USA.

Elena Lioubimtseva, Department of Geography and Planning, Grand Valley State University, Allendale, MI, USA.

Larry D. Makus, Professor of Agricultural Economics, University of Idaho, ID, USA

J. Marquand, University of Oxford, Oxford, UK.

Naoko Matsuo, Graduate School of Agriculture, Kyoto University, Kyoto, Japan.

Greg W. McCarty, USDA-ARS Hydrology and Remote Sensing Laboratory, Beltsville, MD, USA.

A.R. Memut, University of Saskatchewan, Department of Soil Science, Saskatoon, Saskatchewan, Canada.

N.P. Mironycheva-Tokareva, Institute of Soil Science and Agrochemistry, Novosibirsk, Russia.
A. Morgounov, International Maize and Wheat Improvement Center (CIMMYT), Ankara, Turkey.

B. Mukanov, Research and Production Centre of Forestry, Shuchinsk, Kazakhstan.

Tolib Mukimov, Department of Desert Ecology and Water Resources Research, Samarkland, Uzbekistan.

N. Muratova, Kazakh Space Research Institute, Almaty City, Kazakhstan.

R. Nasyrov, Soil Research Institute, Academgorodok, Almaty, Kazakhstan.

A.V. Naumov, Institute of Soil Science and Agrochemistry, Novosibirsk, Russia.

Jeff Novak, Coastal Plains Soil, Water and Plant Research Center, USDA-ARS, Florence, SC, USA.

M.A. Oripov, Bukhara State University, Bukhara, Uzbekistan.

T.K. Ortikov, Samarkland Agricultural Institute, Samarkland, Uzbekistan.

Konstantine M. Pachikin, Department of Soils, Institute of Kazakhstan Ministry of Education and Science, Almaty, Kazakhstan.

Mustafa Pala, International Center for Agricultural Research in the Dry Areas, Aleppo, Syria.

Raj Paroda, Program for Central Asia and Caucasus, International Center for Agricultural Research in Dryland Areas, Tashkent, Uzbekistan.

Paul E. Patterson, Extension Professor of Agricultural Economics, University of Idaho, ID, USA.

B. Qushimov, Tashkent State University of Economics, Tashkent Uzbekistan.

D.C. Reicosky, North Central Soil Conservation Research Laboratory, USDA Agricultural Research Service, MN, USA.

I. Rustamova, Tashkent State Agrarian University, Tashkent, Uzbekistan.

John Ryan, International Center for Agricultural Research in the Dry Areas, Aleppo, Syria.

S. Sanginov, Soil Science Institute of Tajikistan, Dushanbe, Tajikistan.

Ashyrmuhammed Saparmyradov, Department of Science and Technologies of the Ministry of Agriculture, Ashgabat, Turkmenistan.

A. Saparov, Soil Science Institute, Academgorodok, Almaty, Kazakhstan.

Ken Sayre, International Maize and Wheat Improvement Center (CIMMYT), El Batan, Texcoco, Mexico.

Sarah E. Schafer, International Programs, The Ohio State University, Columbus, OH, USA.

B.A. Stewart, Department of Agricultural Sciences, West Texas A & M University, Canyon, TX, USA.

Mekhlis Suleimenov, International Center for Agriculture Research in Dryland Areas-Central Asia Caucasus, Tashkent, Uzbekistan.

Beibut Suleymenov, Akademgorodok, Institute of Soil Science, Almaty, Kazakhstan.

U. Sultangazin, Kazakh Space Research Institute, Almaty, Kazakhstan.

Yusuke Takata, Graduate School of Agriculture, Kyoto University, Kyoto, Japan.

A. Terekhov, Kazakh Space Research Institute, Almaty City, Kazakhstan.

Richard J. Thomas, International Center for Agriculture Research in Dryland Areas, Aleppo, Syria.

Kristina Toderich, Department of Desert Ecology and Water Resources Research, Samarkland, Uzbekistan.

Tsuneo Tsukatani, Institute of Economic Research, Kyoto University, Japan.

Muhabbat Turdieva, CWANA Subregional Office for Central Asia, Tashkent, Uzbekistan.

Erik R. Venteris, Ohio Division of Geological Survey, Columbus, OH, USA.

S.N. Vorobiov, Tomsk Regional Environmental Administrative Unit, Tomsk, Russia.

P.C. Wall, International Maize and Wheat Improvement Center (CIMMYT), Harare, Zimbabwe.

Roger A. Williams, School of Environment and Natural Resources, The Ohio State University, Columbus, OH, USA.

Junta Yanai, Graduate School of Agriculture, Kyoto Prefectural University, Kyoto, Japan.

N. Yushenko, Central Kazakh Agricultural Research Institute, Karaganda, Kazakhstan.

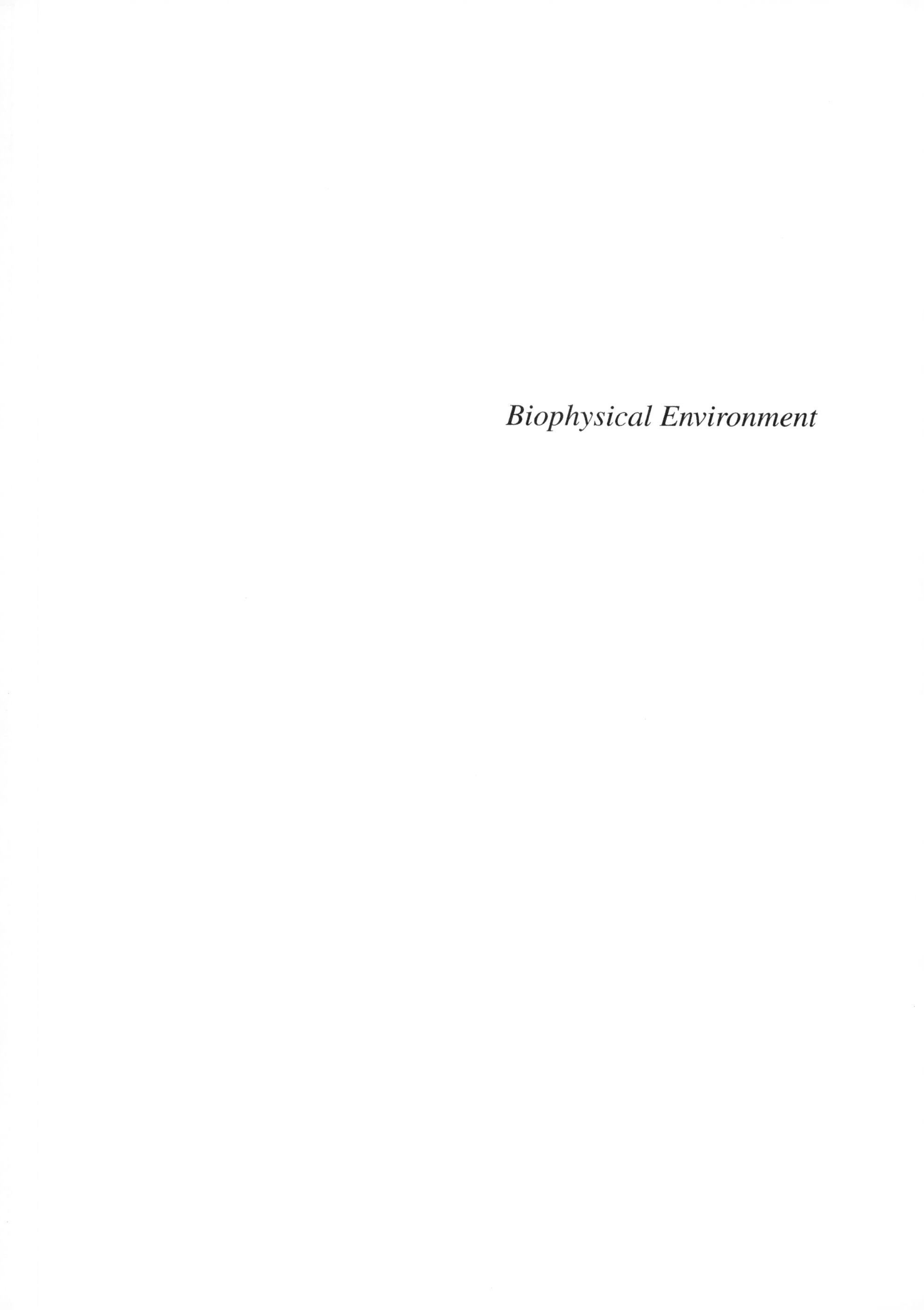

Biophysical Environment

CHAPTER 1

Principal biomes of Central Asia

Eddy De Pauw
International Center for Agricultural Research in the Dry Areas (ICARDA), Aleppo, Syria

1 INTRODUCTION

If we wanted to describe Central Asia in a few lines, the following would do: a huge barren or sparsely vegetated plain, bordered by snow-covered mountains in the south, and broken by large irrigated areas fed by rivers with their headwaters in the mountains.

Correct as this may be as seen from space, the reality on the ground is much more complex. Central Asia is a unique region among the world's agroecosystems. The particular mix of climatic conditions, terrain, soils, vegetation and land use patterns, as well as the major socioeconomic shifts taking place, particularly with respect to land and water use, are not found elsewhere. Despite the apparent (and misleading) monotony of the landscapes in most of the region, there is a surprising diversity in agroecologies. Moreover, it is a region that has witnessed some major environmental catastrophes and degradation of its land and water resources in its recent past, and is particularly vulnerable to the threat of climate change.

Several studies and projects have been undertaken to assess the potential of Central Asia's vast land resources to act as a sink for the greenhouse gas CO_2 and thus to help mitigate the impact of global warming. At the same time, there is concern that poor management of the natural resource base, compounded by ongoing warming trends, may negatively affect the region's capacity to absorb atmospheric carbon.

In dealing with the issue to what extent Central Asia ultimately has the capacity to act as a carbon sink, and particularly on how to manage this carbon store, perhaps insufficient attention has been paid to the diversity of its environments. The sequestration potential is determined by climatic conditions, the kinds of vegetation and land use, the soil types, the land management systems and their interactions. An understanding of the characteristics and dynamics of these factors is essential for guiding carbon sequestration research and management.

It is not as if there is a shortage of information. During the Soviet era there was a major investment in resource monitoring, characterization and mapping summarized in National Atlases, which exist for each country of Central Asia. These maintain rich collections of thematic maps, most of which remain relevant to the present day. However, the use of these paper databases in the post-Soviet era is largely limited to educational purposes, mostly in view of limitations of scale, particularly for large countries, and the difficulty in combining this information with other datasets (De Pauw et al., 2004). What is needed is a new approach, based on the use of geographical information systems (GIS) to consolidate this thematic information into a more integrated spatial database, available to C-sequestration researchers, and update it.

2 METHODS

This paper essentially offers a synthesis of Central Asia's agricultural environments on the basis of secondary sources and new information in the form of maps, some of which were developed using methodologies reported elsewhere. The production of the climatic maps is based on a two-stage approach, in which at first basic climatic variables, obtained from meteorological station data, were interpolated to create 'basic climate surfaces'. These GIS layers were then converted into new layers, 'derived climate surfaces', through either elementary grid calculations in GIS software, or through pre-programmed and more complex models. The specific transformation rules for generating the maps in Figures 4, 6, 7, 10–15 can be found in De Pauw (2001) or De Pauw et al. (2004). The methods used for the production of the Agroecological Zones are reported further on in the paper.

3 GENERAL SETTING

Central Asia occupies an area of 3,994,400 km^2, or about half the area of the United States (Encyclopedia Britannica, 1994). Based on information from the Internet statistical database Geohive, its total population in 2000 was approximately 58 million, the majority of which (about 70%) lives in rural areas or in small towns. Rural population density varies widely, from a high of 200–500 persons km^{-2} in the densely populated Fergana Valley, Uzbekistan, to only 0–2 people km^{-2} in most of 'empty' Kazakhstan (Figure 1). Level of aridity and presence of water resources (see further) were key, but not the only, factors in the initial establishment of settlements and subsequent population growth.

The physical geography of Central Asia (Figure 2) mainly consists of a huge plain, with vast steppe lands in the north in Kazakhstan, and the drainage basin of the Aral Sea in the south. This plain is bordered in the east and south by mountain ranges, which include some of the highest peaks in the world.

Central among the plain features of Central Asia is the fairly monotonous Turanian Plain, which extends northward from the sandy Karakum desert and links up to the West Siberian Plain in Russia through a narrow corridor, the Turgay 'Valley', which is in fact more of a longitudinal depression within the plain rather than a true valley. The Turanian Plain borders the Aral Sea to the west and is located at an absolute elevation of 0–100 m. Further west is the higher-lying (100–200 m) Ustyurt plateau, a plain with low ridges, salt marshes, sinkholes and other karstic features, bordering with dissected edges the Aral Sea, and is formed on gypseous deposits (Gintzburger et al., 2003). The Ustyurt Plateau separates the Turanian Plain from the Caspian Depression, a featureless lowland

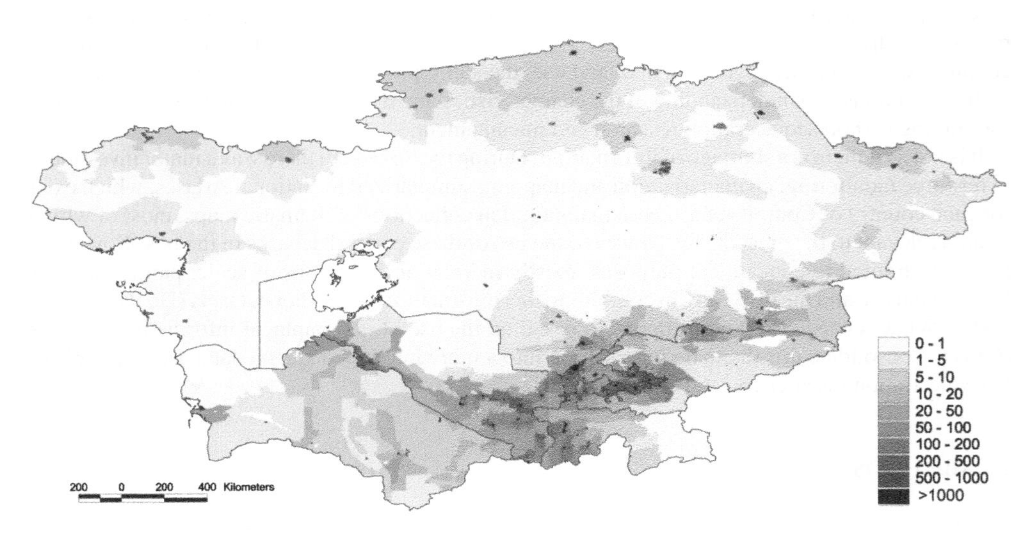

Figure 1. Population density in 2000, people km^{-2} (CIESIN and CIAT, 2005).

descending imperceptibly from 0 m to −30 m at the edge of the Caspian Sea. At its maximum extent, north of the Caspian Sea, this lowland area extends up to 300 km inland.

In the north east of the region, in central Kazakhstan, the landscape rises to a central ridge with west-east orientation and an elevation range of 500–1000 m. The Kazakh Uplands in the middle of this central ridge consist of more hilly landscapes and are in mountainous places.

Breaking further the relative uniformity of the Central Asian plain landscapes are rivers, deserts, several large lakes, and mountains (Figure 2). The main river systems of the Amu-Darya and Syr-Darya with their extended irrigation systems originate in and drain the southern mountain waters into the Aral Sea basin. The huge 840 km-long Karakum Canal, constructed in the early 50s–60s, diverts water from the Amu-Darya river to the agricultural areas west of Ashkhabad and is navigable for about 450 km. Water losses and abstractions for agriculture along this artificial waterway are one of the principal factors for the drying out of the Aral Sea.

The deserts, while plain-like and at fairly low elevations (0–500 m), have their own characteristics. The Karakum ('Black Sand' in Turkic) desert occupies 70% of Turkmenistan and is, with 350,000 km^2 one of the world's largest sand deserts. The Kyzylkum ('Red Sand') desert, with an area of about 300,000 km^2 in both Uzbekistan and Kazakhstan, is also plain-like with isolated bare hills rising up to 900 m. The Muyunkum desert, another sandy desert in southern Kazakhstan, north of the Karatau Range, is about half the size of the Kyzylkum or Karakum deserts. The Betpak-Dala desert (or 'Severnaya Golodnaya Steppe') occupies about 75,000 km^2 east of the Turanian Plain and has many shallow, mostly saline lakes. The western part is clayey, the east is more stony desert.

Lake Balkhash, with a length of 600 km and an extent varying between 15,000–19,000 km^2, depending on precipitation levels in the lake's huge catchment area, is, after the Aral Sea, the second largest water body in Central Asia. It is fairly shallow (10–30 m) and partially saline. The salty Lake Issykul, with a length of 182 km and a width of up to 60 km is the third largest water body, and with a depth of up to 700 m the deepest. It is located at an elevation of 1,600 m.

Central Asia's southeastern and eastern rim consists entirely of mountains, foothills and a few plain depressions. The Tien Shan is major dividing line in Asia, with the highest peaks reaching 7,500 m, from which 'satellite' mountain ranges extend deep into the lowlands. Some of these Tien Shan 'branches' are shown in Figure 1. The topography is one of very steep slopes with narrow intervening valleys, generally trending from east to west. Characteristic for the Tien Shan is a high area under glaciers (about 10,000 km^2) of which 80% in Kyrgyzstan and Kazakhstan. The Karatau Range is the most westward extension of the Tien Shan, reaching 450 km into the Kazakhstan

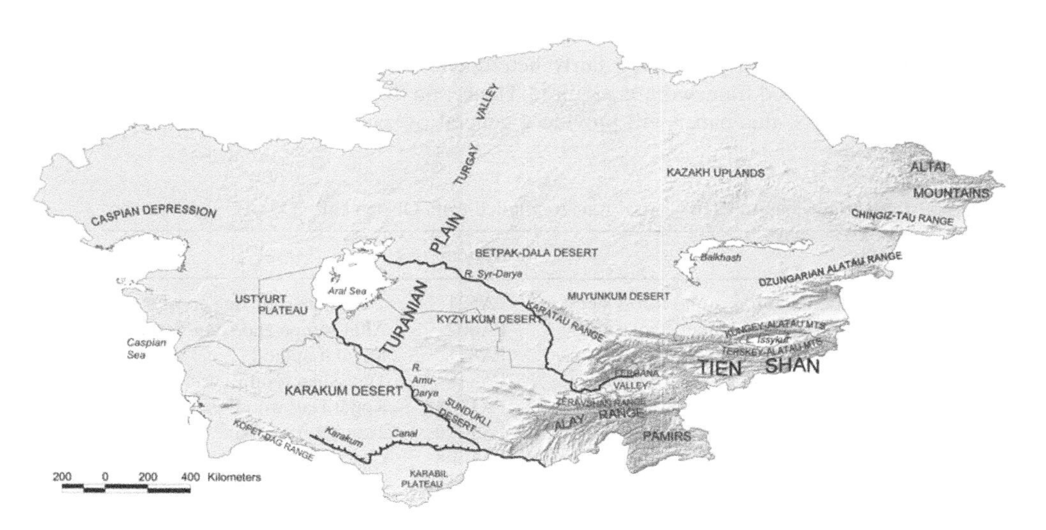

Figure 2. Main physiographical features of Central Asia (Compiled from Encyclopedia Britannica, 1994).

lowlands and rising on average about 1,000–1,500 m above the plain. The Dzungarian Alatau Range, the most northerly westward 'branch' of the Tien Shan in Kazakhstan, rises up to 4,500 m.

Another central knot in the mountain systems of Central Asia are the Pamirs. With their core in the highlands of Tajikistan, they are at the hub of a system of intersecting east-west and north-south ranges. They easily reach 5,000–6,000 m in elevation, with the highest peaks up to 7,500 m. Glaciation in the Pamirs is dominated by the massive Fedchenko Glacier, also the ultimate source of the Amu-Darya river.

In all mountain regions of Central Asia there is a great complexity and diversity of climatic conditions, terrain, vegetation and land use systems, which is best understood in the context of prevailing wind directions, rain exposure or shadow effects, elevation zones and northern or southern slope exposure to radiation.

4 BIOMES OF CENTRAL ASIA

'Biomes', also known as 'major life zones' (Encyclopedia Britannica, 1994) are usually defined as major communities of plants and animals with similar life forms and environmental conditions. With this kind of 'flexible' definition it is obvious that there can be no single system of classification. A well-known system is Bailey's Eco-regions classification (Bailey and Hogg, 1986; Bailey, 1989), which maps for Central Asia 15 Ecoregion Provinces, of which the following three cover 75% of the region:

- Dry steppes of continental climates
- Semi-deserts of continental climates
- Deserts of continental climates.

In addition to its coarse resolution, a disadvantage of this system is that no documentation appears to exist for areas outside the continental USA, hence the exact meaning of the terminology used, although obviously related to climate and vegetation, is not clear.

A more up-to-date, well documented and carefully mapped system is the classification of Terrestrial Ecosystems of the World Wildlife Fund (Olson et al., 2001). This system, designed as a basis to serve conservation of flora, fauna and their habitats, differentiated 867 ecoregions worldwide, of which 20 are in Central Asia (Table 1). Figure 3 shows their location.

Whereas the WWF Ecoregions offers a useful framework for targeting biodiversity and wildlife conservation projects, it is less suitable for carbon sequestration research given its focus on macro-environments that themselves may be fairly heterogeneous. For agricultural management and planning a more detailed framework is required. Hence, on the basis of available global, regional and national datasets, this paper will provide a general overview of the various environmental

Table 1. Terrestrial ecosystems in Central Asia (compiled from Olson et al., 2001).

System	Name	System	Name
A502	Altai montane forest and forest steppe	PA521	Tien Shan montane conifer forests
A801	Alai-Western Tien Shan steppe	PA802	Altai steppe and semi-desert
A806	Emin Valley steppe	PA811	Kazakh upland
A814	Pontic steppe	PA818	Tien Shan foothill arid steppe
A1001	Altai alpine meadow and tundra	PA1008	Kopet Dag woodlands and forest steppe
A1014	Pamir alpine desert and tundra	PA1019	Tien Shan montane steppe and meadow
A1306	Badghyz and Karabil semi-desert	PA1308	Caspian lowland desert
A1310	Central Asian northern desert	PA1311	Central Asian riparian woodlands
A1312	Central Asian southern desert	PA1317	Junggar Basin semi-desert
A1318	Kazakh semi-desert	PA1319	Kopet Dag semi-desert

factors that influence C-sequestration and integrate these in a new GIS-based biophysical framework ('Agroecological Zones') to identify the region's principal agricultural environments, without either oversimplification or unnecessary detail.

5 CLIMATE

The key features of Central Asia's climate are a pronounced degree of aridity, untypical in comparison to the rest of the Eurasian landmass, and large temperature fluctuations, unmitigated by any oceanic influence. The main arid regions of the world are normally found between 15° and 35° in both northern and southern hemisphere in the subsidence branch of the Hadley global circulation belt. The presence of arid regions between elevations 30° N and 50° N, as is the case in Central Asia, is caused by a massive rain-shadow effect due to the high mountains in the southeast and east. These mountains also act as an effective barrier between cold northern Eurasian and warm southern airflows primarily from the Indian Ocean. Warm air can not reach northern Central Asia and Siberia, resulting in strong temperature fluctuations typical of a continental climate (Buslov et al. 2006).

Within this overall setting there is a lot of variability in space and time. The main features and spatial and temporal patterns of Central Asia's different climates are explained in the following sections.

5.1 *Agroclimatic zones*

Using UNESCO's classification system (UNESCO, 1979) and the spatial interpolation methods explained earlier, 21 major agroclimatic zones were identified in Central Asia. Figure 4 shows their distribution and Table 2 presents the characteristics of each zone. Six zones occupy 90% of the region, and two of these occupy two thirds. Hence, although climatically diverse, most of the region has only a few climates, and its climatic diversity finds expression only in relatively small 'niches'.

Climate diagrams for stations that represent the most important agroclimatic zones are illustrated in Figure 5. The location of these stations is indicated by the blue dots in Figure 4. These diagrams show that climatic conditions across Central Asia can vary a lot, but that, with the exception of

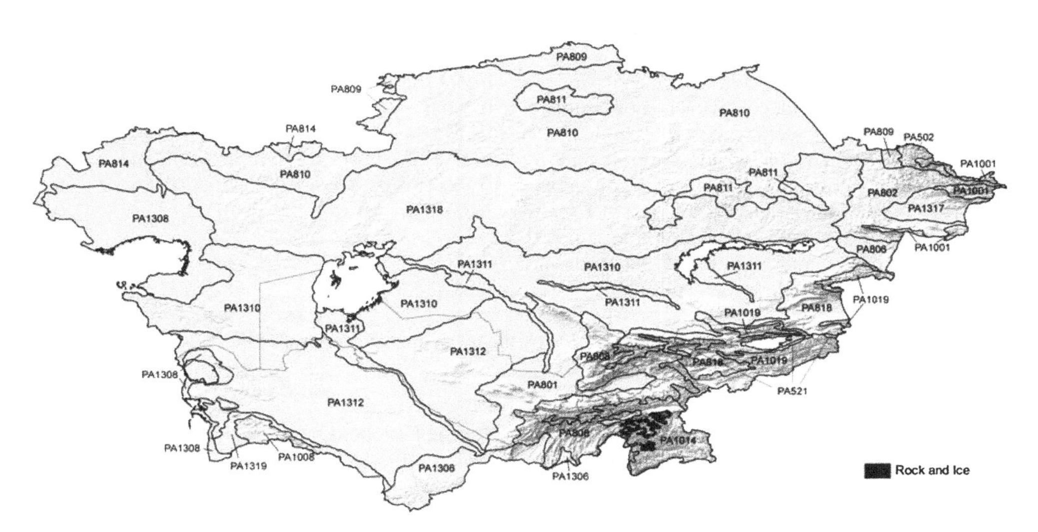

Figure 3. Terrestrial ecoregions of Central Asia (compiled from Olson et al., 2001).

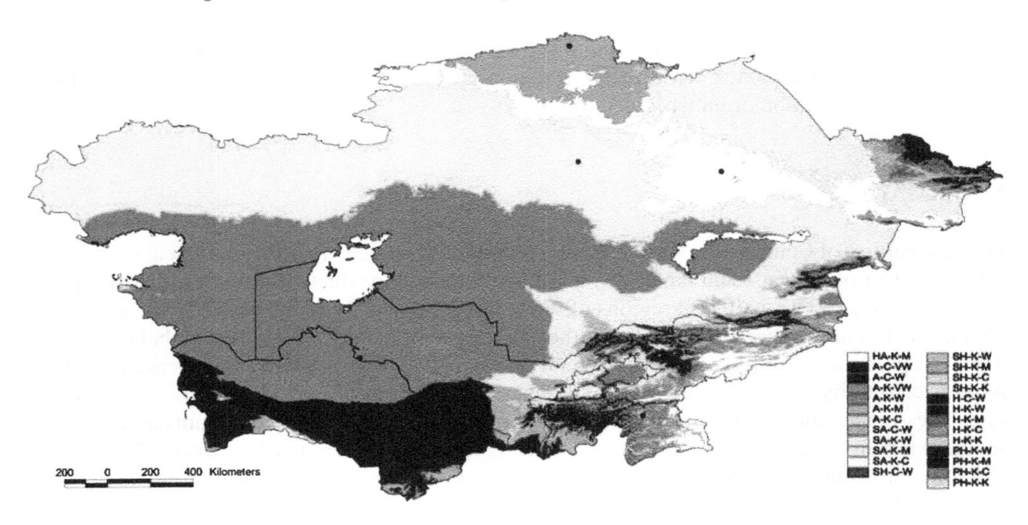

Figure 4. Agroclimatic zones of Central Asia.

Table 2. Agroclimatic zones.

Agroclimatic Zone	Description	Aridity index[1]	Temp. range coldest month	Temp. range warmest month	% of total
SA-K-W	Semi-arid, cold winter, warm summer	0.2–0.5	≤0°C	20°–30°C	37.9
A-K-W	Arid, cold winter, warm summer	0.03–0.2	≤0°C	20°–30°C	30.8
SA-K-M	Semi-arid, cold winter,	0.2–0.5	≤0°C	10°–20°C	6.6
SH-K-M	Sub-humid, cold winter,	0.5–0.75	≤0°C	10°–20°C	5.9
A-C-W	Arid, cool winter, warm summer	0.03–0.2	0°–10°C	20°–30°C	4.9
A-C-VW	Arid, cool winter, very warm summer	0.03–0.2	0°–10°C	>30°C	2.9
PH-K-C	Per-humid, cold winter, cool summer	>1	≤0°C	0°–10°C	2.0
H-K-M	Humid, cold winter, mild summer	0.75–1	≤0°C	10°–20°C	1.6
SA-C-W	Semi-arid, cool winter, warm summer	0.2–0.5	0°–10°C	20°–30°C	1.5
SH-K-W	Sub-humid, cold winter, warm summer	0.5–0.75	≤0°C	20°–30°C	1.4
A-K-VW	Arid, cold winter, very warm summer	0.03–0.2	≤0°C	>30°C	1.2
PH-K-M	Per-humid, cold winter,	>1	≤0°C	10°–20°C	1.2
SH-K-C	Sub-humid, cold winter, cool summer	0.5–0.75	≤0°C	0°–10°C	0.5
SA-K-C	Semi-arid, cold winter, cool summer	0.2–0.5	≤0°C	0°–10°C	0.5
H-K-C	Humid, cold winter, cool summer	0.75–1	≤0°C	0°–10°C	0.5
H-K-W	Humid, cold winter, warm summer	0.75–1	≤0°C	20°–30°C	0.2
SH-C-W	Sub-humid, cold winter, warm summer	0.5–0.75	0°–10°C	20°–30°C	0.1
A-K-M	Arid, cold winter, mild summer	0.03–0.2	≤0°C	10°–20°C	0.1
PH-K-K	Per-humid, cold winter, cold summer	>1	≤0°C	≤0°C	0.1
PH-K-W	Per-humid, cold winter, warm summer	>1	≤0°C	20°–30°C	0.0
A-K-C	Arid, cold winter, cool summer	0.03–0.2	≤0°C	0°–10°C	0.0

[1]The ratio of the mean annual precipitation divided by the mean annual potential evapotranspiration.

mountain peaks with high snowfall (see the diagram for the Fedchenko Glacier station), most of the region has low precipitation. The precipitation patterns tend to shift from a dominance of winter precipitation in the south of the region, to a higher share of summer precipitation in the north.

Temperature conditions are characterized by a high range between summer and winter, reflecting the continental climates, but also by a high diurnal range, particularly in desert and steppe areas.

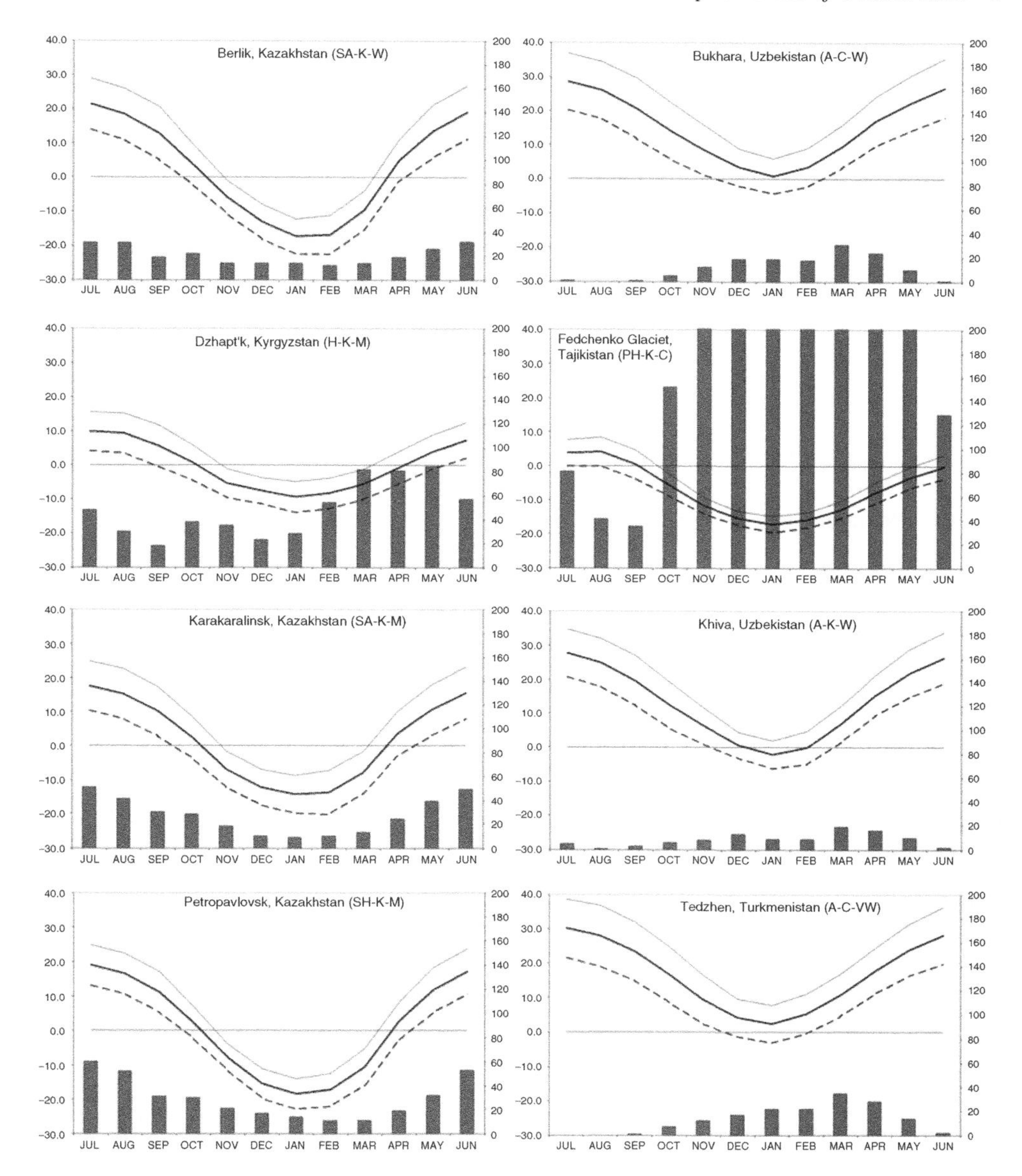

Figure 5. Climatic diagrams for representative stations in Central Asia (left axis and lines: mean, maximum and minimum temperature, °C right axis and shaded blocks: monthly precipitation, mm).

5.2 *Precipitation*

The distribution of precipitation is shown in Figure 6 and summarized in Table 3. Both confirm the high level of aridity, with nearly 90% of the region receiving 400 mm or less and 70% in the 100–300 mm range. Precipitation decreases towards the center, south of Aral Sea, and increases towards the north, east, and southern edges of the region. Rain shadows can be particularly pronounced in valleys that are entirely surrounded by high mountains (e.g., Fergana, Issykul).

The seasonal distribution of precipitation is shown in Figure 7 as the percentage of annual precipitation that falls in each of the four seasons. Winter or spring precipitation is the dominant

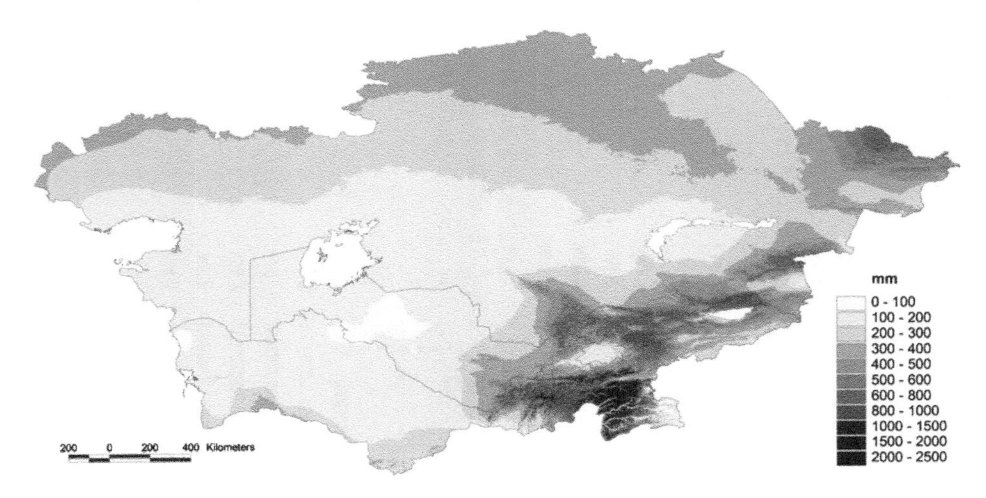

Figure 6. Mean annual temperature.

Table 3. Summary of precipitation levels in Central Asia.

Precipitation class (mm)	% of Asia	Mean (mm)	Standard deviation (mm)
0–100	1.19	266	160
100–200	38.28		
200–300	31.08		
300–400	18.73		
400–500	3.93		
500–600	2.60		
600–800	2.62		
800–1000	0.89		
1000–1500	0.53		
1500–2000	0.14		

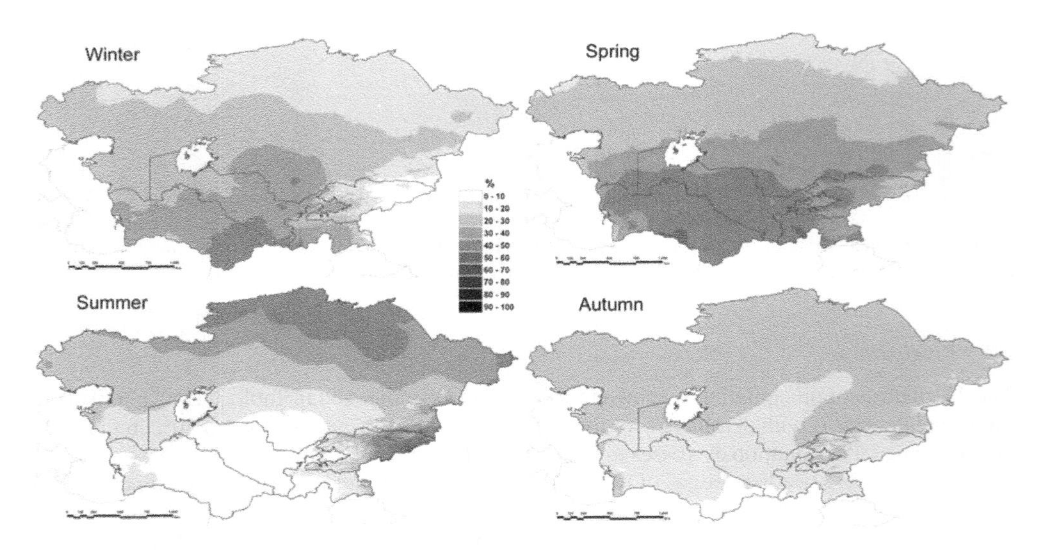

Figure 7. Seasonal distribution of precipitation as a percentage of the annual total.

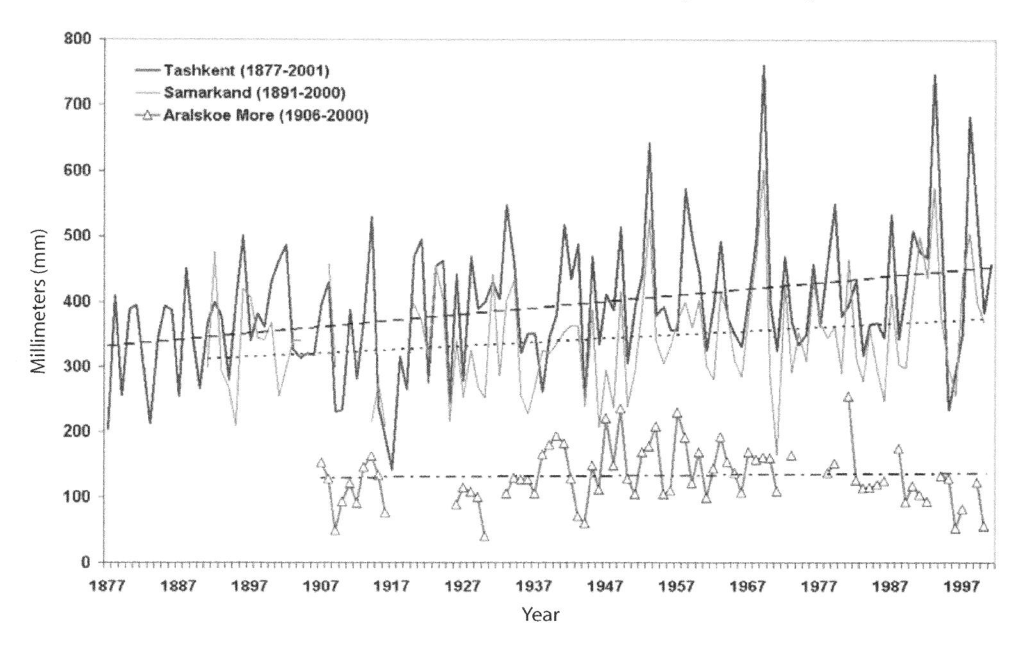

Figure 8. Variability and trends of annual precipitation for Tashkent, Samarkand and Aralskoe More (Williams and Konovalov, 2004).

source of moisture in the south of the region, whereas summer precipitation prevails in the north and the eastern mountains. The overall share of autumn precipitation appears less prominent.

As is typical for arid regions, the temporal variability of precipitation is very high and the vulnerability to drought severe. Figure 8 shows typical examples of inter-annual precipitation variations for three stations with long records and different precipitation levels. Absolute variations, as expressed by the standard deviation, increase with increasing precipitation. Figure 8 also shows the trend lines of precipitation over the observation periods. They are positive for Tashkent, less so for Samarkand and neutral for Aralskoe More.

Figure 9 shows the probability distributions of annual precipitation for the same stations, as approximated by log-normal transforms. With a 42% probability to exceed the mean precipitation, they display the typical negative skewness of precipitation patterns in arid regions, and the associated greater likelihood of precipitation below than above average.

5.3 *Temperature*

Central Asia is generally a cold region. As Table 2 indicates, 90% of Central Asia has cold winters, whereas less than 5% has very warm summers. The inter-annual variation of the mean temperature is quite pronounced, as shown in the climate diagrams of Figure 5 and the maps of Figure 10.

Temperature can be represented as a pool of energy for plant growth and biomass production through the concept of *accumulated heat units* or *growing degree days*, which sum the daily temperature above a threshold (e.g., 0°C) for a specified period (e.g., one year). The map of mean annual heat units accumulated above 0°C (Figure 11) shows that two thirds of the region accumulate less than 4,000 degree-days. This is very low in comparison to tropical regions, which easily reach more than double this amount.

Given the general cold character of the region, it is no surprise that virtually all of Central Asia is affected by frost to some extent (Figure 9). About 50% of the region has 150–210 frost days. Very high frost occurrences (270–365 days) are confined to highest mountain ranges, particularly the Pamir and Tien Shan (Figure 12).

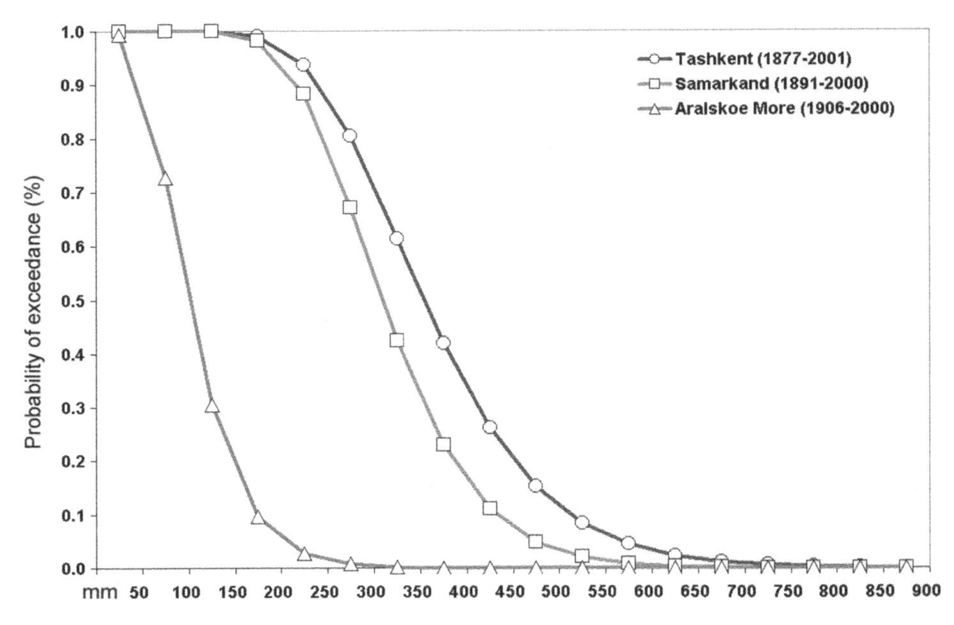

Figure 9. Probability distributions of annual precipitation for Tashkent, Samarkand and Aralskoe More (Williams and Konovalov, 2004).

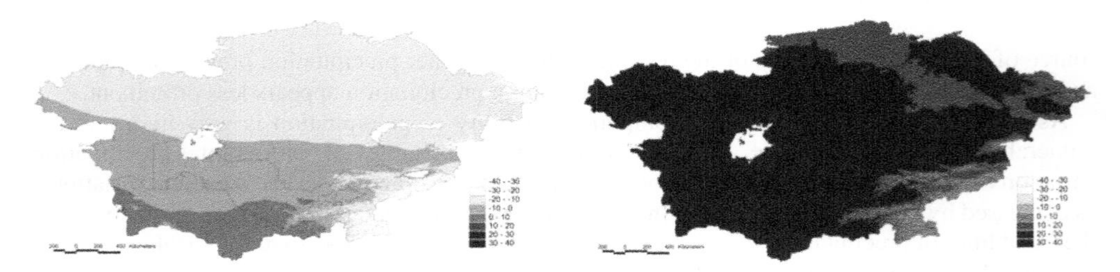

Figure 10. Mean temperature of the coldest (left) and warmest month (right).

Figure 11. Mean annual heat units (AHU) accumulated above 0°C.

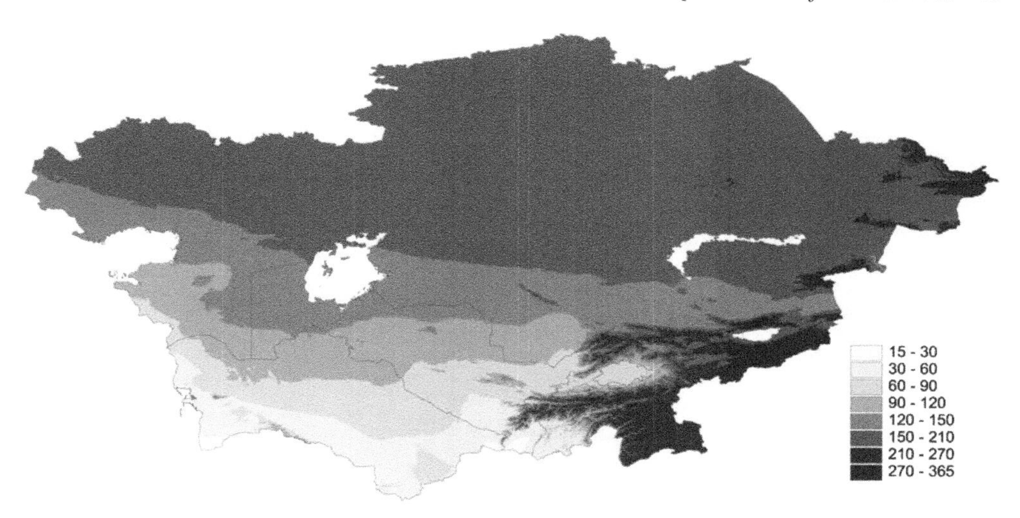

Figure 12. Mean number of frost days.

Table 4. Summary of potential evapotranspiration (PET) levels in Central Asia.

Range	% of Central Asia
400–500	1.59
500–600	2.30
600–800	17.18
800–1000	41.89
1000–1500	31.24
1500–2000	5.80

5.4 *Potential evapotranspiration*

Potential evapotranspiration (PET), a measure of the water demand of live vegetation, is determined principally by air temperature. Other important factors are radiation, humidity and wind speed. As Table 4 indicates, PET is relatively low in Central Asia, with a mean of 993 mm and standard deviation of 265 mm, and this is parallel with the low levels of heating units (see earlier section).

The pattern is determined by latitude, with a notable decrease when going north, but the lowest PET levels are in the mountains in the southeast and east of the region (Figure 13). Considering the balance between precipitation and PET, 97% of Central Asia has a water deficit (i.e., can not meet the full water demand of live vegetation). In about 50% of the region the deficit is in the range 500–1000 mm, which is high. Only 3% has a surplus (mountain tops).

5.5 *Growing periods*

The growing period, as a climatic concept, is the time of the year when neither temperature nor moisture limit biomass production. As explained in earlier sections, both temperature and moisture can be major limiting factors for biological activity in Central Asia. It therefore makes sense to assess them separately as either a temperature-limited or a moisture-limited growing period, and then to combine both into a temperature- and moisture-limited growing period.

The mean temperature-limited growing period (TGP) in Central Asia is 206 days, with a standard deviation of 51 days. About 94% of the region has a TGP of more than 150 days, hence with the

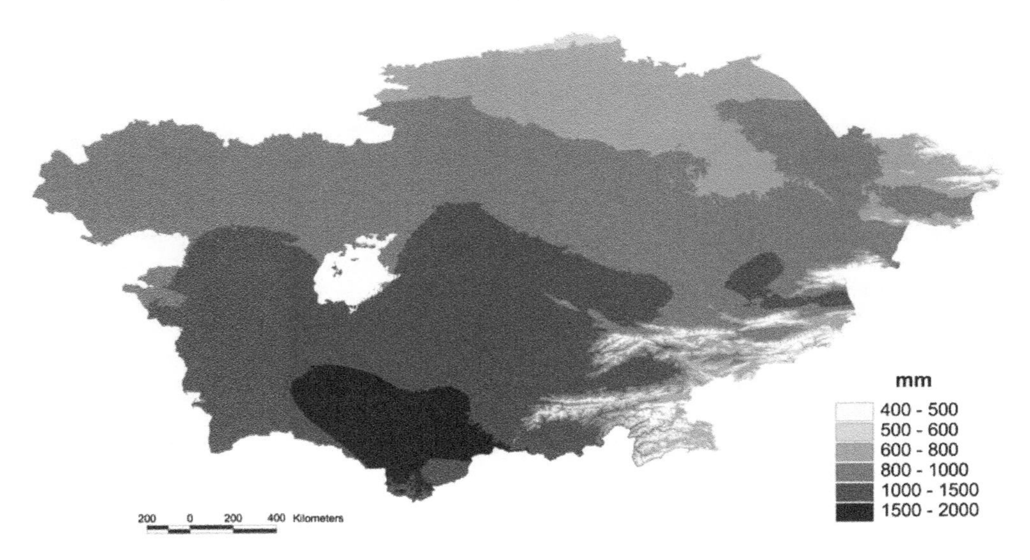

Figure 13. Mean annual potential evapotranspiration (based on Pennman-Monteith method).

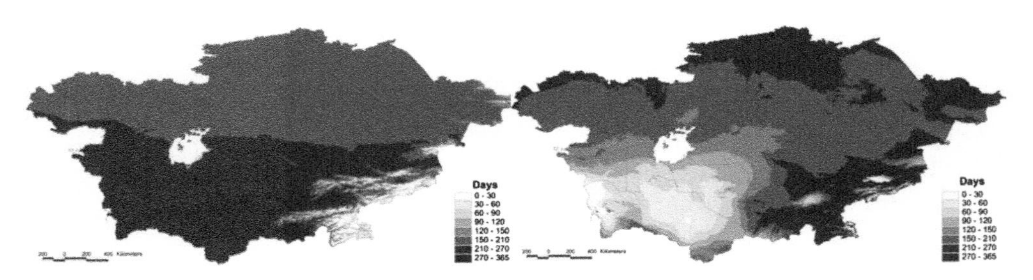

Figure 14. Distribution of the temperature-limited (left) and moisture limited (right) growing period.

exception of high mountains there is no major temperature limitation for crop production. The spatial distribution of the TGP (Figure 14, left) is primarily determined by a latitudinal gradient of increasing growing period duration from north to south, and secondly by a declining growing period with increasing elevation in the high mountains.

The mean moisture-limited growing period (MGP) in Central Asia is 178 days with a standard deviation of 69 days. About 67% of the region has a MGP above 150 days, hence in most of the region there is sufficient moisture for growing crops. The spatial distribution of MGP (Figure 14, right) is foremost characterized by a trend of decreasing duration along a north-south latitudinal gradient, deflected to the west by the presence of the southern and eastern mountain ranges. The lowest values occur in the deserts and steppes south of the Aral Sea, and in intra-montane depressions and linear valleys.

The periods in which TGP and MGP overlap constitute the temperature- and moisture-limited growing period (TMGP). The mean TMGP in Central Asia is only 30 days with a standard deviation of 36 days, and 90% of the region has a TMGP not exceeding 90 days (Figure 15).

These figures point to a severe mismatch between the period in which moisture is available and the period in which temperature is adequate. In much of Central Asia it is too cold at the time that soil moisture is adequate for biomass production. It is one of the reasons why most of Central Asia is rangeland, and not cropland. It also explains why rainfed agriculture occurs only in the north of the region, where there is a larger share of precipitation during the warmer part of the year

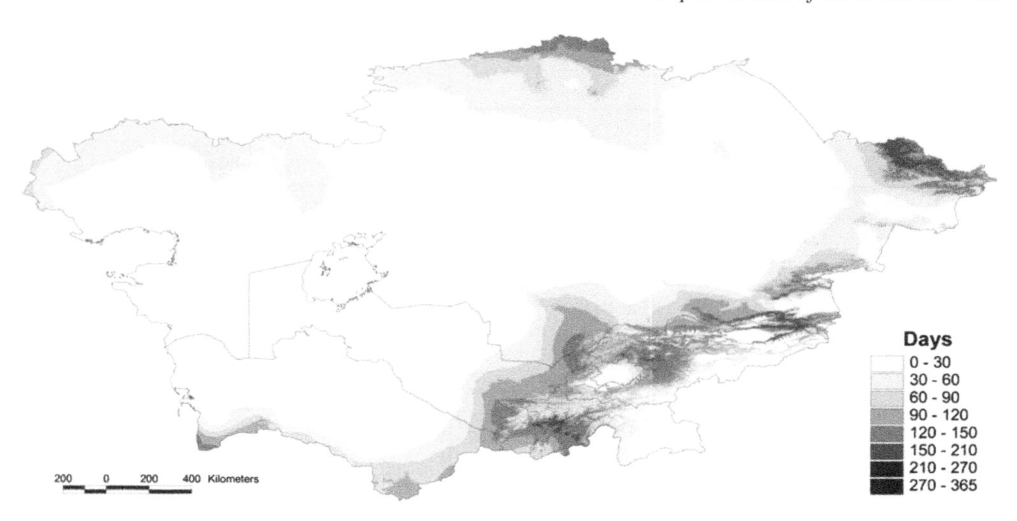

Figure 15. Duration of the growing period, limited by both temperature and moisture.

(Figure 7), and in areas bordering the southern mountains (where both temperature and moisture are adequate during a longer overlapping period).

Where irrigation water is available, areas with very low MGP but high TGP can become highly productive, as is the case in e.g. the Fergana intramontane basin.

6 SOILS

Central Asia has a great diversity in soil types. Eighteen of the 26 Soil Orders according to the 1974 FAO soil classification (FAO-UNESCO, 1974) are represented in Central Asia, distributed over 244 soil associations. Of the Soil Orders, eight occupy 82% of the region (Table 5). In addition, a large proportion of the region (12.6%) is covered by mobile sand and sand sheets, particularly in the sandy deserts of Karakum, Kyzylkum and Myunkum, and south of Lake Balkhash.

The distribution of the main soil types is shown in Figure 16. From north to south the soil patterns follow climatic gradients of decreasing precipitation and increasing temperature.

Above 52°N *Chernozems* are dominant. These soils are deep, have high porosity, high moisture storage capacity, dark and thick topsoils, and high organic matter content, ranging from 4 to 16% (Deckers et al., 1998), and have more fine, clayey textures. They contain free calcium carbonate, but usually not above 2 m depth.

Between 48°N and 52°N the main soils are *Kastanozems* (also called *Chestnut Soils* in Russia), which differ from the *Chernozems* by a lower organic matter content (2–4%) and more shallow topsoils, lighter colors, and generally more loamy textures. They usually contain free carbonates within 1 m depth.

Below 48°N the trend towards declining organic matter content, lighter color and rising carbonates continues as the land becomes drier and warmer, at first with the *Xerosols*, and in the driest parts of the region with the *Yermosols*. Whereas the *Xerosols* are usually strongly enriched in lime in some soil horizons above 1 m depth, the *Yermosols* may also contain substantial quantities of gypsum. Both types of soils are alternatively known in Central Asia as *Sierozems* (Plyusnin, 1960) or '*grey-brown soils*' (Lobova, 1960), which cover a wide range of soils with different morphology and physical characteristics.

A particular feature of the soil patterns in Central Asia is a comparatively high proportion of sodic soils (a.k.a. *Solonetz* and *solonetzic soils*) and saline soils (*Solonchaks*). *Solonetz* are characterized

Table 5. Most common soil types of Central Asia.

Soil classification system			
FAO, 1974	FAO, 1988 WRB, 1998[2]	Soil Taxonomy, 1994	Central concept and extent[1]
Xerosols, Yermosols	Calcisols (Gypsisols)	Calcids (Gypsids)	Soils with high lime concentrations in the subsoil. Gypisols have high subsoil gypsum concentrations (20.8%)
Kastanozems	Kastanozems	Ustolls	Lighter colored soils with fertile, well-structured, organic topsoil, enriched in lime or gypsum (18.4%)
Solonetz	Solonetz	Soils with *natric* horizon	Sodic soils (11.3%)
Lithosols + rock outcrops	Leptosols	Lithic subgroups of different soil orders	Shallow soils and rock outcrops (9.3%)
Solonchaks + salt flats	Solonchaks	Salids	Saline soils (9.3%)
Chernozems	Chernozems	Ustolls	Dark soils, rich in lime with a fertile, well-structured, topsoil rich in organic matter (5.8%)
Gleysols	Gleysols	Soils with *aquic* conditions	Poorly drained soils (4.7%)
Fluvisols	Fluvisols	Fluvents	Soils formed on recent alluvium (2.8%)

[1]Percentages in parentheses refer to the relative amount of land area in each soil type.
[2]World Reference Base for Soil Resources (Deckers et al., 1998).

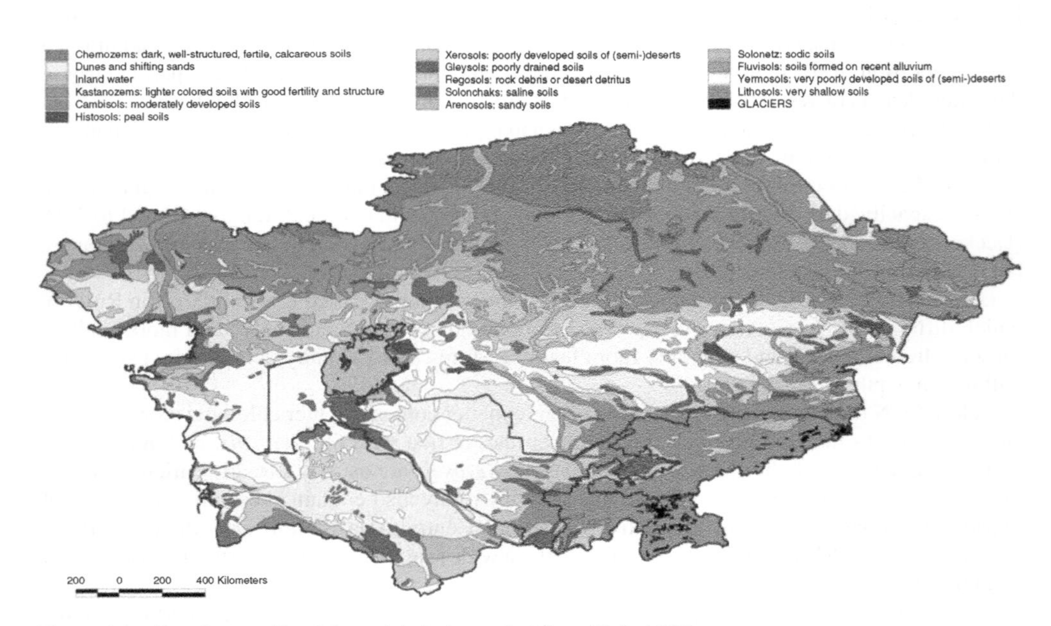

Figure 16. Dominant soils of Central Asia (compiled from FAO, 1995).

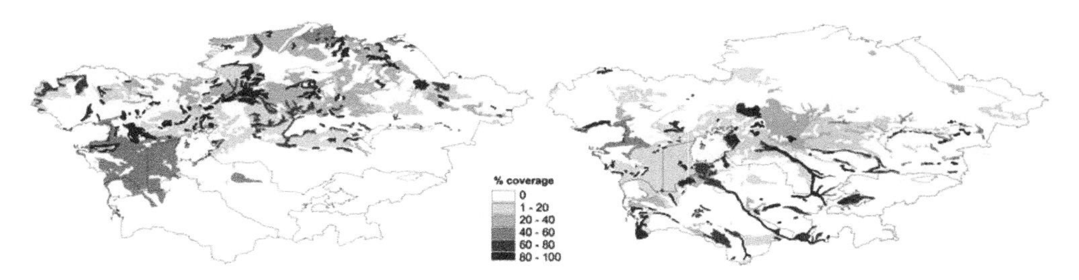

Figure 17. Relative coverage (%) by sodic (left) and saline soils (right) (compiled from FAO, 1995).

by a *natric horizon*, a dense, strongly structured clay horizon with a high proportion of adsorbed sodium and/or magnesium ions (Driessen et al., 2000). They are particularly well represented in the Ustyurt Plateau and in a latitudinal belt north of 45°N, where they are associated with *Chernozems, Kastanozems, Xerosols* and *Solonchaks*, and where, according to Plyusnin (1960), they represent a 'zonal', rather than a site-specific, feature. The unusual importance of these soils in Central Asia is linked to a particular combination of climatic and micro-topographic conditions in the region.

Solubility of common sodium and magnesium compounds in the soil increases sharply as temperature rises from 0°C to 30°C. The dissolution and accumulation of Na/Mg-compounds during the warmer part of the year is followed by much slower leaching during the wet but cold winter season (Driessen et al., 1998).This process of Solonetz formation is promoted by the level, slightly undulating relief, typical for the steppe plains, in with concave parts capture surface water, especially after snowmelt, soils get more humid and often leached, whereas convex parts are much drier (Plyusnin, 1960). The process starts already in the *Chernozem* zone, reaches its maximum importance in the south of the *Kastanozem* zone, weakens in the zone of dry steppes and is virtually absent in the deserts (Figure 17, left).

In Central Asia saline soils are in most cases associated to sodic soils and occur under similar climatic and micro-topographic conditions (Figure 17, left and right). In addition, they are widespread in alluvial plains, where their presence is mostly related to secondary salinization under prolonged irrigation and waterlogging.

To complete the picture, *Lithosols* are the dominant soils in the mountain ranges where they are associated with rock outcrops, whereas in the main alluvial plains and their tributaries *Fluvisols* are prevalent. *Gleysols* (soils with poor drainage) can also be important in depressions of alluvial plains or could result from improper irrigation without drainage.

As much as by the soils that do occur, the region is characterized by the soils that are absent. The FAO Soil Map of the World does not report for Central Asia the following soils, which are widespread in other parts of Eurasia or the Mediterranean region:

- *Andosols*: soils formed on volcanic ash
- *Acrisols*: acid soils with clay-enriched subsoil and good nutrient retention capacity
- *Phaeozems*: strongly colored soils with fertile, well-structured, organic topsoil
- *Luvisols*: non-acid soils with clay-enriched subsoil and good nutrient retention capacity
- *Nitosols*: well drained fine-textured soils with constant clay content in subsoil
- *Podzols*: soils with a subsoil enriched in mobilized iron and organic matter
- *Vertisols*: darker colored cracking clays with (somewhat) deficient drainage

Whereas the absence of *Andosols* is indicative of a lack of volcanic activity in Central Asia, the lack of the other soil types is mainly related to the absence of the climatic conditions needed for the formation of these soils.

The organic matter content of the soils of Central Asia is variable. On the basis of the soil classification units and their relative proportions in the associations of the Soil Map of the World, the Food and Agriculture Organization of the United Nations developed global maps of the distribution of organic carbon (FAO, 1995). Figure 18 is based on these maps and shows the distribution of

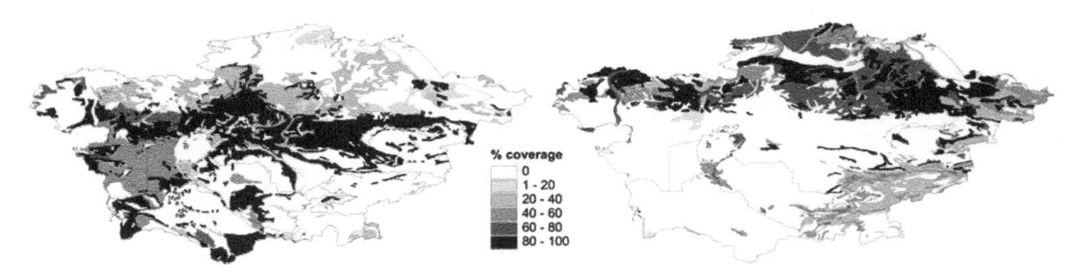

Figure 18. Distribution of soils with low organic matter content (left) and medium-high organic matter content (right) (compiled from FAO, 1995).

soils with very low to low OC-content in the topsoil (less than 0.6% OC) on the left, and medium to high OC-content (more than 1.2% OC) on the right.

In Figure 18 it is easy to recognize the patterns of *Chernozems, Kastanozems, Xerosols* and *Yermosols*, with the associated southward trend of declining organic matter.

7 VEGETATION

Within the overall context of aridity, general coldness and soil patterns of Central Asia, the vegetation shows great diversity. In Uzbekistan alone, Gintzburger et al. (2003) mention the existence of 3750 species of vascular and other plants.

The best way to describe the distribution of the main plant communities is through a phytogeographical framework linked to the main ecological zones. The Vegetation Map of Uzbekistan, (Geography Department of the Academy of Sciences of the Uzbek SSR, 1982), differentiates the following broad ecosystems:

- Riparian vegetation and river banks ('tugai'): communities dominated by either reeds, riparian forest species, halophytes, *Tamarix*, or mixtures
- Vegetation of plains (desert-'chol'): communities dominated by various Haloxylon, Calligonum, Astragalus, Ammodendron, Artemisia, Anabasis, Carex, Salsola and other species
- Vegetation of foothills ('adyr'): communities dominated by various Artemisia, Pistacia, Hordeum, Elytrigia and other species
- Mountain vegetation ('tau'): communities dominated by various fruit trees forests and bushes, Acer and Juniperus species
- Vegetation of high mountains ('alau'): communities dominated by various Prangos, Festuca, Polygonum and other species.

In other countries of Central Asia the same zoning system for regrouping the plant communities is followed.

Despite the high aridity, the 'Chol' vegetation, shows surprising diversity in response to local micro-climatic and soil conditions. Gintzburger et al. (2003) describe in Uzbekistan different plant communities on either sandy desert, 'gypseous' desert, *Solonchak* desert and clay desert.

For a more extensive treatment of vegetation in Central Asia is referred to the relevant literature. Gintzburger et al. (2003), although mostly focused on Uzbekistan, offer a very comprehensive monograph and literature list, whereas the WWF Ecoregion descriptions (Olson et al, 2001) contain excellent introductions to vegetation communities and their associated fauna.

8 LAND USE

The map in Figure 19, derived from interpretation of 1-km resolution satellite imagery for the base year 1993 (Celis et al., 2006a), shows the distribution of the main land use/cover categories.

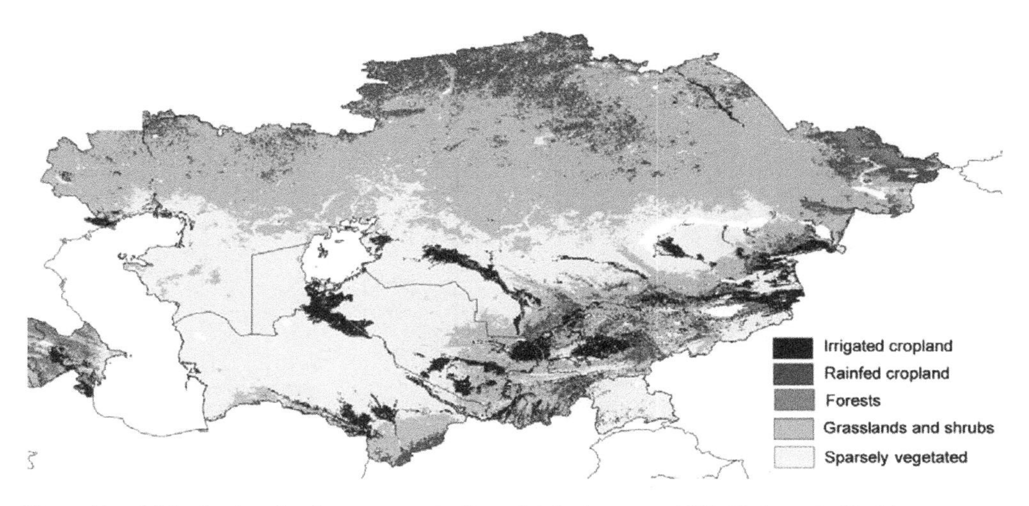

Figure 19. Major land use/land cover types in Central Asia, base year 1993 (Celis et al., 2006a).

Table 6. Summary of land use/land cover in Central Asia.

Land use/land cover	% of region
Irrigated cropland	6.4
Rainfed cropland	8.2
Grasslands and shrubs	48.4
Forests and woodlands	1.1
Sparse vegetation	33.8
Other (water, urban)	2.1

Agriculture is practiced in about 15% of the region. Unsurprisingly, rainfed agriculture is confined to the north of the country and the foothill areas along the southeastern mountains, where both temperature and precipitation allow sufficiently long growing periods (Figure 15). Irrigation occurs along main river courses, the Karakum canal and natural basins, such as the Fergana valley. Most of the region (Table 6) is covered by either a good cover of grasslands and shrubs, or by sparse vegetation, for a total of more than 80%. Forests or woodland barely cover 1% of the region.

This picture of land use patterns is not static. A regional analysis of low-resolution satellite imagery (Celis et al., 2006b) evidenced during the observation period 1982–1999 significant and consistent (i.e. not related to climatic variability) changes in land use/land cover. These changes could be summarized into 4 major trends (Figure 20):

- '*intensification of agriculture*': a consistent upward trend in land use with higher biomass productivity
- '*intensification of natural vegetation*': a similar upward trend for the natural vegetation
- '*retrenchment of agriculture*': a consistent downward trend in land use tending to lower biomass productivity
- '*retrenchment of natural vegetation*': a similar downward trend for the naural vegetation

Comparing Figure 20 with Figure 19, it is clear that the hotspots of change are mainly located at the interface between the grassland-shrub areas and the sparsely vegetated rangelands on the one hand, and between the grassland-shrubs and the rainfed agriculture systems. Within each system individually there is not much change.

Quantifying these changes, it appears that intensification of agriculture occurred on 147,520 km^2, mostly in the form of a change from grasslands/open shrublands into rainfed cultivation. The inverse

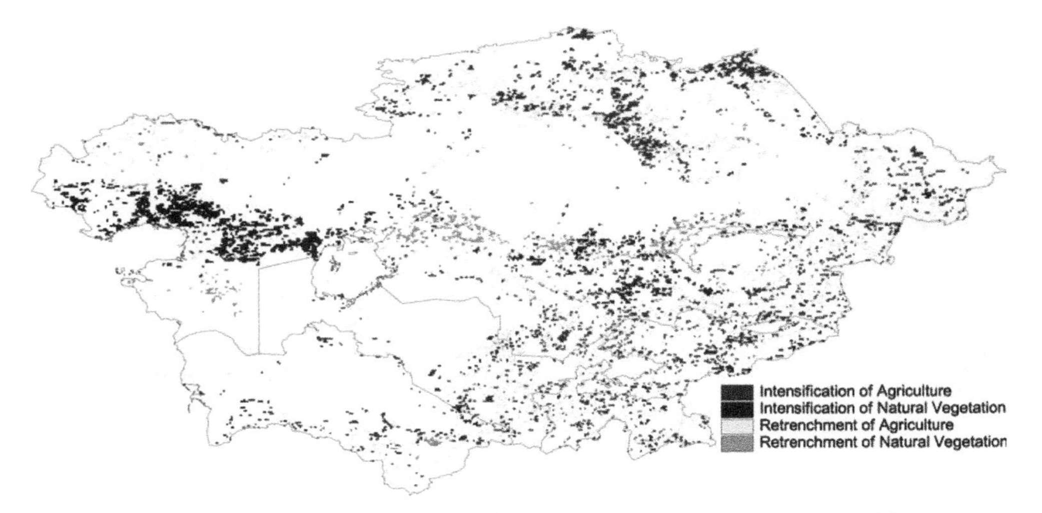

Figure 20. Land use/land cover change trends in Central Asia over the period 1982–1999 (Celis et al., 2006b).

trend, a change from rainfed cultivation to grasslands/open shrublands, occurred on 80,960 km^2 and is the major component of the retrenchment of agriculture (125,248 km^2).

Intensification of natural vegetation occurred on 111,040 km^2, almost all as a result of change from barren/sparsely vegetated areas to grasslands/open shrublands. This intensification trend is very noticeable in Southwest Kazakhstan near the Caspian Sea. There is no clear explanation for this trend in the literature. It is possible that the trend is related to the decline of livestock herds and corresponding decrease in grazing intensity after the collapse of the Soviet Union in the early 1990s, a cause proposed for the NDVI patterns in other parts of Kazakhstan (de Beurs and Henebry, 2004). The inverse trend, a change from grasslands/open shrublands into barren/sparsely vegetated areas, occurred on 46,272 km^2, and is particularly clear in Central Kazakhstan.

9 AGROECOLOGICAL ZONES

In the previous sections it was evidenced that the key environmental drivers that influence carbon sequestration, climate, terrain, soils and land use systems, are subject to considerable spatial and often temporal variability. Hence it is obvious that the interaction of these factors may lead at local level to a wide range of different domains for the management of the natural resource base. To capture the major differences in these factors and link them in a way that management recommendations can be more easily formulated in a spatially explicit manner, a more holistic spatial framework is needed. In this context the term *agroecological zones* is used to demarcate integrated and more or less homogeneous spatial units, in which available water resources, climate, terrain, and soil conditions are combined to create unique environments, associated with distinct farming systems and recommendation domains for land use management. Whereas in the past the manual integration of spatial data from different disciplines was problematic, GIS technology makes this now perfectly practicable.

9.1 *Methodology*

The agroecological zones (AEZ) for Central Asia were generated at 1-km resolution by an overlay procedure in GIS that involves climatic, terrain, soils, land use/land cover themes and the following steps:

- Generating raster surfaces of basic climatic variables through spatial interpolation from station data;

Table 7. Soil management groups of Central Asia.

Soil types by symbols (FAO, 1974)	Soil Management Groups and their extent (%)
B, Bc, Bd, Be, Bh, C, Ch, Ck, Cl, H, K, Kh, Kl, Mo	Agricultural soils (26.6%)
G, Gc, Ge, Gm, J, Jc, Je, Oe	Soils of wetlands, poorly drained areas and floodplains (7.9%)
Qc + DS	Sandy soils, sand sheets and mobile sands (12.9%)
S, Sg, Sm, So, Ws, Z, Zg, Zo, Zt, Zm, St	Sodic and saline soils (21.4%)
I, RK, U	Rock outcrops and shallow soils (9.7%)
X, Xh, Xk, Xl, Xy	Semi-desert soils (12.2%)
Y, Yh, Yk, Yt, Yy	Desert soils (8.6%)
Gx, Rc, Rd	Non-agricultural soils (.3%)
Dd	Soils with high acidity and/or low nutrient status (0.1%)

- Generating a spatial framework of agroclimatic zones (ACZ) by combining the basic climatic surfaces into more integrated variables that provide a better, although simplified, synthesis of climate conditions;
- Simplifying the relevant biophysical themes (agroclimatic zones, land use/land cover, landforms and soils);
- Integrating the simplified frameworks for agroclimatic zones, land use/land cover, landforms and soils by overlaying in GIS;
- Removal of redundancies, inconsistencies, and spurious mapping units;
- Characterization of the spatial units in terms of other relevant themes.

The generation of raster surfaces of basic and derived climatic variables was described in earlier sections. As input theme for climate the 1-km layer of Agroclimatic Zones (Figure 4) was used. The land use/land cover layer developed by Celis et al. (2006a) was used as second overlay theme. The input theme used to represent landforms was a layer consisting of 7 elevation-'relief intensity' classes, based on the classification of the 1-km resolution GTOPO30 digital elevation model (Gesch and Larson, 1996). For the soil theme the 1:5,000,000 scale Digital Soil Map of the World (FAO, 1995) was used, with attribute table to represent the entire soil association.

These themes were then simplified in order to avoid (i) a replication of the single-theme maps, and (ii) unnecessary complexity for the purpose of the AEZ map. Thus the theme 'agroclimatic zones' was condensed from 21 to 9 classes, the land use/land cover theme from 12 to 3 classes, the landform theme from 7 to 3 classes, and the soils theme from 55 FAO Soil Groups and 140 soil associations to 9 'soil management groups' (Table 7).

The creation of new units in GIS through overlaying propagates errors that were already present in the component datasets. Simplification of the input data layers was one way to reduce the errors resulting from overlaying. In addition, special rules were established to remove redundancy or inconsistencies as well as GIS-based automatic procedures to remove 'spurious' mapping units, created by the overlaying process.

A final step was the characterization of the AEZ in terms of additional themes relevant to planning or management. This could be biophysical themes, such as frost occurrence, growing periods, soil management groups, but also socio-economic themes, such as farming systems, and is presented as a summary table. An example is given in Table 8 for the theme 'Number of frost days'.

9.2 *Results*

Following the above methodology 185 AEZ were identified in Central Asia.

Table 8. Percentages of frost day classes within a specific agroecological zone (AEZ).

AEZ	0–1 days	1–30 days	30–60 days	60–90 days	90–120 days	120–150 days	150–180 days	>180 days
160	0	100	0	0	0	0	0	0
251	0	6	11	53	31	0	0	0
351	0	0	0	0	76	23	0	0
206	0	14	36	40	11	0	0	0
120	0	100	0	0	0	0	0	0
306	0	0	0	0	56	43	2	0
456	0	0	0	0	0	37	60	3

Table 9. Five largest agroecological regions (AEZ) in Central Asia.

AEZ	Description
33100	Arid climates with cool or cold winters and warm or very warm summers; winter coldness is an ecological constraint; non-agricultural land use; plains and plateaux; undifferentiated soils
53110	Wemi-arid climates with cool or cold winters and mostly warm summers; winter coldness is an ecological constraint; non-agricultural land use; plains and plateaux; predominantly agricultural soils
53100	As above, but with undifferentiated soils
33170	Arid climates with cool or cold winters and warm or very warm summers; winter coldness is an ecological constraint; non-agricultural land use; plains and plateaux; predominantly desert soils
33140	As above but with predominantly sodic and saline soils

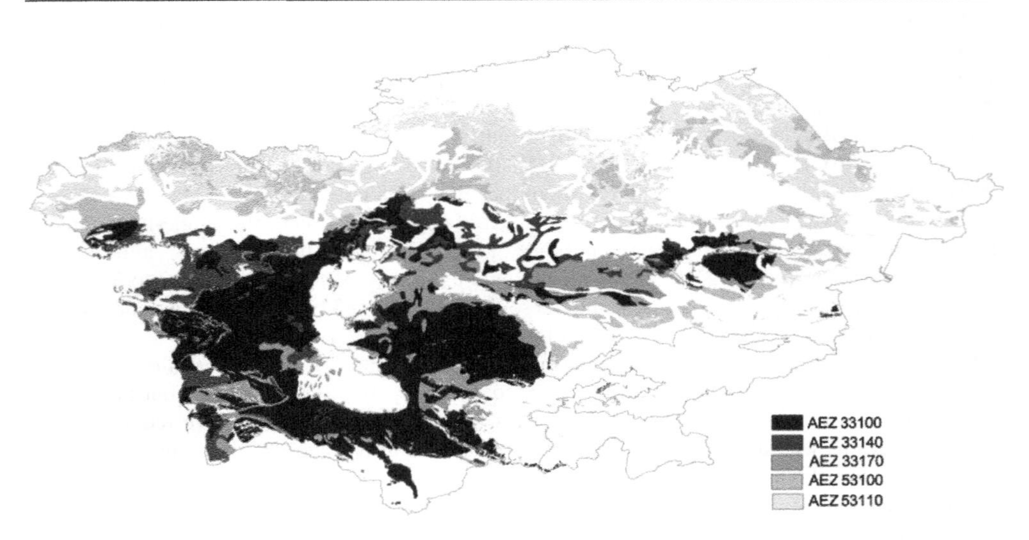

Figure 21. Distribution of the five largest agroecological zones (AEZ) in Central Asia.

Of these, the five largest (Table 9) cover nearly 48% of Central Asia (Figure 21), whereas the 20 largest cover 80% of the region. On the other hand, 50% of all AEZ in Central Asia cover a total of only 1%, evidencing that virtually all of them constitute 'niche' agroecologies, the vast majority located in mountain areas.

10 CONCLUSION

Keeping the soils of Central Asia as a net sink for atmospheric carbon becomes a new goal for agricultural planning, on top of the need to create productive and economic land use systems, improve rural livelihoods and maintain environmental sustainability. In this respect it becomes another optimization challenge in balancing often conflicting interests.

Although it is certainly possible to make generalized management recommendations for the agricultural environments of Central Asia, this paper make it clear that the environmental drivers of land use management are far from uniform across the region. Climatic drivers, particularly those determining moisture supply and temperature stress, vary considerably, not only in space but also in time. Micro-climatic conditions can be very important, particularly, but not exclusively, in mountain areas, and the large number of AEZ with rather small spatial extent points to the existence of 'niche' agroecologies in much of the region.

In order to reduce the great complexity of agricultural environments without oversimplification, this paper proposes an approach for defining agroecological zones using GIS procedures. It is based on the integration of thematic layers related to climatic, landform, land use/land cover, and soil data, and the characterization of the resulting spatial units. The method is useful to define areas that can be considered homogeneous in their biophysical characteristics at a relatively fine resolution ($1-5\ km^2$) and can thus serve as a first basis for area-specific agricultural (research) planning. It is also sufficiently flexible to allow, again using GIS procedures, aggregation into larger spatial units for higher-level planning. The delineation of the zones, while based on objective criteria, is not static and their updating, as could be required due to changes in the more dynamic components, such as land use or even climate change, is perfectly feasible.

REFERENCES

Bailey, R.G. and H.C. Hogg, 1986. A world ecoregions map for resource reporting. Environmental Conservation 13: 195–202.

Bailey, R.G. 1989. Explanatory supplement to Ecoregions Map of the Continents. Environmental Conservation 16: 307–309.

Celis, D., E. De Pauw, and R. Geerken. 2006a. Assessment of land cover/land use in the CWANA region using AVHRR imagery and agroclimatic data. Part 1. Land cover/land use – Base Year 1993. Technical report. ICARDA, Aleppo, Syria, 50 pp.

Celis, D., E. De Pauw, and R. Geerken. 2006b. Assessment of land cover/land use in the CWANA region using AVHRR imagery and agroclimatic data. Part 2. Hot spots of land cover change and drought vulnerability. Technical report. ICARDA, Aleppo, Syria, 58 pp.

Center for International Earth Science Information Network (CIESIN), Columbia University; and Centro Internacional de Agricultura Tropical (CIAT). 2005. Gridded Population of the World Version 3 (GPWv3): Population Density Grids. Palisades, NY: Socioeconomic Data and Applications Center (SEDAC), Columbia University. Available at http://sedac.ciesin.columbia.edu/gpw. (2006/11/11)

Deckers J.A., F.O. Nachtergaele, and O.C. Spaargaren (eds.). 1998. World Reference Base for Soil Resources. Introduction. Acco Leuven, 165 pp.

de Beurs, K.M. and G.M. Henebry. 2004. Land surface phenology, climatic variation, and institutional change: analyzing agricultural land cover change in Kazakhstan. Remote Sensing of Environment 89: 497–509.

De Pauw, E. 2002. An agroecological exploration of the Arabian Peninsula. ICARDA, Aleppo, Syria, 77 pp. ISBN 92-9127-119-5.

De Pauw, E., F. Pertziger, and L. Lebed. 2004. Agroclimatic mapping as a tool for crop diversification in Central Asia and the Caucasus. p. 21–43. In: J. Ryan, P. Vlek, and R. Paroda (eds.). Agriculture in Central Asia: Research for Development. ICARDA, Aleppo, Syria.

Driessen,P., J. Deckers, O. Spaargaren, and F. Nachtergaele. 2001. Lecture Notes on the Major Soils of the World. World Soil Resources Report 94, FAO, Rome. 334 pp.

Encyclopedia Britannica. 1994. The New Encyclopedia Britannica. Chicago.

FAO. 1974. FAO-UNESCO Soil Map of the World. Vol.I: Legend. UNESCO, Paris

FAO. 1988. FAO-UNESCO Soil map of the world: Revised legend. World Soil Resources Report No. 60. Food and Agriculture Organization of the United Nations, Rome, Italy.

FAO-UNESCO. 1995. The Digital Soil Map of the World and Derived Soil Properties. Land and Water Digital Media Series 1. FAO, Rome. CD-ROM.

Geohive. 2006. Geohive Global Statistics. Available at http://www.xist.org/default1.aspx (2006/12/04)

Gesch, D.B., and K.S. Larson. 1996. Techniques for development of global 1-kilometer digital elevation models. (on-line document, URL http://edc.usgs.gov/products/elevation/gtopo30/papers/geschtxt.html) (2007/01/14)

Gintzburger, G., K.N. Toderich, B.K. Mardonov, and M.M. Mahmudov. 2003. Rangelands of the Arid and Semi-Arid Zones in Uzbekistan. CIRAD-ICARDA. ISBN 2-87614-555-3, 426 pp.

Lobova, E.V. 1960. Génèse et classification des sols gris-bruns des déserts de l'URSS. Bulletin AFES, May 269–282.

Olson, D.M., E. Dinerstein, E.D. Wikramanaya, N.E. Neil, D. Burgess, G.V.N. Powell, E.C. Underwood, J.A. D'Amico, I. Itoua, H.E. Strand, J.C. Morrison, C.J. Loucks, T.F. Allnutt, T.H. Ricketts, Y. Kura, J.F. Lamoreux, W.W. Wettengel, P. Hedao, and K.R. Kassem. 2001. Terrestrial Ecoregions of the World: A New Map of Life on Earth. BioScience 51: 933–938 Terrestrial Ecoregion GIS Database available at: http://www.worldwildlife.org/science/data/terreco.cfm (2006/12/04)

Plyusnin, I.I. 1960. Reclamative Soil Science. Foreign Language Publishing House, Moscow. 398 pp.

Soil Survey Staff. 1994. Keys to Soil Taxonomy. United States Department of Agriculture, Soil Conservation Service. Sixth Edition, Washington, DC., USA.

UNESCO, 1979. Map of the world distribution of arid regions. Map at scale 1: 25,000,000 with explanatory note. United Nations Educational, Scientific and Cultural Organization, Paris, 54 pp. ISBN 92-3-101484-6

Williams, M.W. and V.G. Konovalov. 2004. GHCN: Regional Data Base on Climate of Central Asia, Volume 1.0. Available at: snobear.colorado.edu/Markw/Geo**data**/BD_INFOR_E.doc (12/5/2006)

CHAPTER 2

Forests in Central Asia: Current status and constraints

Muhabbat Turdieva
CWANA Subregional Office for Central Asia, Tashkent, Uzbekistan

Evsey Aleksandrovskiy
Research and Production Centre of Ornamental Gardening and Forestry, Tashkent, Uzbekistan

Abdukhalil Kayimov
Department of Forestry,Tashkent State Agrarian University, Tashkent, Uzbekistan

Salamat Djumabaeva
Research Institute of Forest and Nut Industry, Bishkek, Kyrgyzstan

B. Mukanov
Research and Production Centre of Forestry, Shuchinsk, Kazakhstan

Ashyrmuhammed Saparmyradov
Department of Science and Technologies of the Ministry of Agriculture, Ashgabat, Turkmenistan

Khukmatullo Akmadov
Academy of Agricultural Sciences of the Republis of Tajikistan, Dushanbe, Tajikistan

1 INTRODUCTION

Humans have used forest genetic resources for a long time, but without understanding the necessity and methods of their conservation. Thus, over-exploitation and insufficient forest protection, misuse of forest lands, uncontrolled cattle grazing which continues until present, felling trees for firewood, over-harvesting of wild fruits and nuts, forest fires and other factors have adversely impacted forest resources. Consequently, the forest vegetation in Central Asian countries does not form large contiguous areas, and is represented with relatively fragmented forest lands, located far from human settlements, and often, in poorly accessible locations. Forest ecosystems are degraded and many plant and animal species are endangered to a great extent. The ecosystem services of forests are being compromised, many species are losing their natural habitat, and their gene pool is progressively impoverished under the influence of human activities. An important role in conservation of forest biodiversity can be played by a national system of creating protected areas, which include nature reserves ("zapovednik"), national parks, nature refuges ("zakaznik"), and nature monuments. However, the ratio of specially protected areas is on average about 5% in the region and this situation is a major concern for the future of forest resources in Central Asia.

Forests in Central Asia are not only a source of timber but also provide ecosystem services including moderating climate, protecting soils, storing water, etc., which ensure sustainability of forest and agricultural production. Forests in the region are an important source of non-wood forest products for rural communities and forest dwellers living there. The demand of construction industry for timber is mainly met by importation from Russia.

Area under forests in Central Asia has been reduced on average by a factor of 4–5 times since the beginning of the 20th century. Desert and floodplain native forests were especially affected drastically by anthropogenic pressure due to their use for agricultural production (Regional Report "Environment Status in the Aral Sea Basin", 2000). This chapter outlines the need for undertaking urgent activities on conservation of genetic resources of main forests and associated forest tree species.

2 CENTRAL STATUS OF FORESTS IN CENTRAL ASIA

The Region of Central Asia occupies a large territory – about 3.996 million km^2 with a population of 54.6 million. The Region of Central Asia includes Kazakhstan, Kyrgyzstan, Tajikistan, Turkmenistan and Uzbekistan and belongs to the area categorized as sparsely covered with forests. On average, the portion of area covered with forest barely exceeds 5% in the entire region. State forest farms are the main entities in the region, whose main objective is protection and management of forest land in Central Asian countries.

2.1 *Kazakhstan*

Total area of all forest types in Kazakhstan is more than 26.5 million hectares (M ha), including forest covered area of 11.4 M ha. The ratio of forest-covered area in Kazakhstan is very low, about 4.2%. Ratio of plain area is 42.5% of the total area of the country, and is characterized with the least forest-covered area (1.63%). Forests in Kazakhstan occur almost in all climatic zones, but are unevenly distributed throughout the country. Eastern and south-eastern parts of the country are the areas most covered with forests. Diversity of forest tree species in Kazakhstan is quite rich and includes 622 species.

Coniferous forests in the country are represented by pine (*Pinus silvestris, P.sibirica*) woods in north, with fir (*Abies sibirica*), larch (*Larix sibirica*), spruce, and cedar forests in east and south-east. Total area of coniferous forests is 1.69 M ha. Hard-wood forest area is 5.4 M ha and is comprised mainly of oak (*Quercus robur*) and elm (*Ulmus laevis*). Soft-wood forests occupy 1.3 M ha. *Betula* sp. is the main forest forming species in soft-wood forests. Saxaul forest is widely spread in southern part of the country. *Haloxylon* sp. grows in about 5.3 M ha. Brushwood forest area (*Artemisia* sp, *Colligonum* sp.) has formed recently through transfer of areas under bush and degraded pastures in southern part of the country to the forest fund, and has area of 2.52 M ha. An area under wild fruit species is 0.19 M ha in the southern part of the country and mainly comprises *Malus sieversii, Armeniaca vulgaris, Crataegus* and *Berberis* spp.

Forests in Kazakhstan are of great importance to ecosystem services and 96% of them belong to groups I and II of the forest protection system. There are nine nature reserves with total area of 0.87 M ha, five National Parks with area of 1.28 M ha, and 57 natural refuges in Kazakhstan. The total area of specially protected areas is 5.4% of the whole territory of Kazakhstan, where 0.14% of national plant gene pool is conserved (Krever et al., 1998).

2.2 *Kyrgyzstan*

Total area of state forest land is 3.16 M ha in Kyrgyzstan and forest covered area is 0.85 M ha, which comprises 4.25% of the total area of the country (Kolov et al., 2001). All forests in Kyrgyzstan are of particular value. Under the Forest Code of Kyrgyz Republic, these are classified as environment protecting forests and according to composition of dominant species and habitat conditions they are divided into four groups: (a) Spruce, (b) Juniper, (c) Nut-Fruit, and (d) Flood-lands forests.

Total area of spruce forests is 0.124 M ha, including *Picea Shrenkiana* F. et. M. (0.114 M ha), *Abies Semenovii* Fed. (3,600 ha), *Pinus silvestris* L. (2,300 ha), *Larix* sp. (1,600 ha) and *Betula pendula* (5,400 ha). Ratio of total area covered with spruce forests is 2.3% of the total forest area in Kyrgyzstan (Figure 1).

Juniper forests, formed with three juniper species grow in the south of the country in Osh Province in mountainsides of the Altay ridge, at the altitude of 900–3,700 m above sea level (m.a.s.l.). Lower belt of these forests is formed with *Juniperus seravschanica* Kom. at the altitude

Figure 1. Spruce forest in Kyrgyzstan (photo of s. Djumabaeva).

of 900–1,300 m.a.s.l., medium belt of *Juniperus semiglobosa* Rgl. at the altitude of 1,400 to 3,100 m a.s.l., and higher belt of *Juniperus turkestanica* at the altitude of 2,500 to 3,300 m a.s.l. Total area of juniper forests in Kyrgyzstan is 250,300 ha, including cedar forests with area of 90,510 ha.

Despite the fact that conservation of juniper forests is of great importance, their status is gradually becoming worse. Since 1980, the area of juniper forests is reduced to 18% and the area of hessians has increased to 31%. Intensive tree felling in the past, forest fires and excessive cattle grazing have led to this situation. Severe soil erosion is observed on 0.115 M ha in Kyrgyzstan, which was covered with juniper forests in recent past. Natural regeneration of juniper forests is practically stopped due to bad seed germination and cattle grazing. Degradative processes are rapidly accelerating. Desertification of areas covered with juniper forests, replacement of forest plants with desert and semi-desert species, strong mud streams and soil washing are widespread (Kolov et al., 2002).

Natural nut and fruit tree forests, a unique world heritage, are located in Jalalabad and Osh Provinces in Kyrgyzstan, in the mountainsides of Fergana and Tien Shan ridges. As many as 183 tree and shrub species grow in nut and fruit forests there. Nut and fruit tree forest is of special interest and has area of 0.63 M ha, including 0.28 M ha covered with forest. *Juglans regia* L. (35,100 ha), *Pistacia vera* L. (32,600 ha), *Malus domestica* (16,500 ha), *Prunus divaricata* Ldb. (600 ha), *Acer* sp. (27,600 ha), *Picea tienshanika* and *P. shrenkiana* F. et. M. (64,000 ha), *Abies Semenovii* Fed. (1,400 ha), *Juniperus* sp. (47,200 ha), *Crataegus* sp. (3,000 ha) are the main species comprising nut and fruit tree forests. About 2,500 t of walnut, 8,000 t of apples, 1,000 tons of cherry-plum and other wild nuts and fruits are annually harvested in natural forest stands in Kyrgyzstan (Figures 2 and 3). Other forest tree species grow on 113,300 ha (Kolov, 2003).

A national system of specially protected areas in Kyrgyzstan includes six nature reserves with total area of 0.22 M ha and forested area of 20,262 ha, six national parks with total area of 0.22 M ha, 52 nature refuges and 18 nature monuments. The total special protected area is 0.86 M ha, or 4.4% of total area of Kyrgyzstan.

2.3 *Tajikistan*

All forests in Tajikistan are mountainous and belong to the first category of the forest protection system. The total area of forest reserve is 1.8 M ha, of which 0.41 M ha is covered with forest. The ratio of forest covered area is about 3% in Tajikistan.

Figure 2. Walnut natural stands in Southern Kyrgyzstan (photo of S. Djumabaeva).

Figure 3. Pistachio (Pistacia vera L.) natural stands in Tajikistan (photo of L. Nikolyai).

Juniper sp. is the main forest species in Tajikistan and grows at the altitude of 1,500–3,200 m.a.s.l. Area of juniper forests, represented by *Juniper sibirica, J. turestanica, J. seravschanica, J. semiglobosa* and *J. schugnanica* is 0.15 M ha.

Pistachio (*Pistacia vera*) sparse forests grow mainly in the southern part of Tajikistan at the altitude of 600–1,400 m.a.s.l. (Figure 3). Area of pistachio forest in Tajikistan is 78,000 ha.

Walnut (*Juglans regia*) stands occupies area of 8,000 ha and grows in central part of Tajikistan at the altitude of 1,000–2,000 m.a.s.l. (Figure 2), sometimes in mixture with *Malus sieversii, Acer Regelii, Prunus darsvasica. Betula, Populus, Hippophae rhamnoides, Fraxinus* growing along riversides at the altitude of 1,500 m.a.s.l. and occupying an area of 15,000 ha (Novikov, Safarov, 2002). *Populus, Elaeagnus, Tamarix* contribute floodplain forests (tugay) in Tajikistan. In the beginning of 19th century, the area of tugay forest was about 1 M ha, whereas now its area is only 120,000 ha or one-eighth of the original area (Akhmadov and Kasirov, 1999).

There are four nature reserves with area of 0.17 M ha, 13 nature refuges with area of 0.31 M ha, two national parks (2.6 M ha), and 26 nature monuments in Tajikistan.

Figure 4. Juniper forests in Turkmenistan (photo of M. Turdieva).

2.4 *Turkmenistan*

Arid Turkmenistan is one of the forest deficient regions in Central Asia. All forests in Turkmenistan belong to State Forest Fund (forests of state importance and forests in reserved land) and its area is 9.996 M ha. Forest covered area is about 4.13 M ha. Forests in Turkmenistan are distinguished into three types: (a) Desert forests, (b) Mountainous forests, and (c) Native floodplain (tugay) forests.

Mountain forests (mainly juniper and pistachio tree stands) grow on 0.524 M ha in Turkmenistan with forest covered area of 0.15 M ha (Figure 4). Juniper forests comprise of *Juniperus turcomanica* and *J. zeravschanica*. These are fragmented forests and barely form big all-over covered forest areas. Total area of juniper forests is 66,200 ha. Juniper tree stands provide water saving and nature-preserving service to mountain ecosystem of Kopetdag and Kugitang.

Wild populations of *Pistacia vera* L. and *Punica granatum* L. grow in hills in the south of Turkmenistan. Main natural stands of *Pistacia vera* L. grow in Badkhiz, in the border of Turkmenistan, Iran and Afghanistan. Total area of pistachio stands in Turkmenistan is over 0.1 M ha with forest covered area of 36,400 ha. Today, pistachio stands are fragmented and do not form large natural stands, though a dense belt of semi-savannah pistachio groves ranges along the northern mountainside from east-end to central part of Kopetdag ridge in the past.

Total area of desert forests is 9.35 M ha, of which the forest cover occupies 3.96 M ha. *Haloxylon persicum* is the main forest species and covers a large area of 3.66 M ha. *Haloxylon aphyllum* grows on an area of 0.73 M ha. The area covered with *Calligonum* sp. is 41,000 ha of which *Salsola* sp. occupies 4,000 ha (Allamuradov et al., 2005).

Native floodplain forests (tugay) grow in valleys of Amu-Darya, Tejen and Murgab rivers, and mountain rivers of Central and especially South-Western subtropical Kopetdag. Total area of tugay forests is 44,500 ha, of which the forest cover is 26,000 ha. The main forest species of tugay forests in flood-plains of Amudarya river are 'turanga' (*Populus diversifolia*) (4,800 ha) and Russian olive (*Elaeagnus angustifolia*) (1,500 ha). *Glycyrrhiza glabra* is of particular value in tugay forests (Kamakhina, 2005).

There are eight nature reserves with a total area of 0.785 M ha, 14 nature refuges with an area of 1.16 M ha, and 17 nature monuments. Total area of specially protected lands is 1.98 M ha or

Figure 5. Saxaul forest in Uzbekistan.

4.02% of the total area of the country. About 62.6% or 72 species of plants are conserved in these protected areas.

2.5 *Uzbekistan*

Total area of forest land in Uzbekistan is 8.78 M ha, or 19% of the total area (data of 1.01.2006). Uzbekistan is a low-forested country, and the forest covered area is only 2.3 M ha or about 5.1% of the total area of Uzbekistan (Kayimov, 2006). The principal owner of forest lands is Main Forestry Office of Ministry of Agriculture and Water Management of Uzbekistan. The total area of forest fund belonging to these Offices is 8.05 M ha (91.8%) of which 2.37 M ha is covered with forest. Forest fund of the Main Forestry Office is classified into different ecological zones as: (a) Mountainous, (b) Desert, (c) Tugay (native floodplain forests), and (d) Valley zone.

The main part of forests (88%) is located in deserts (total area of 7.02 M ha with forest covered area of 1.95 M ha). Desert forests are formed mainly by *Haloxylon persicum* (1.12 M ha) and *H. aphyllum* (0.27 M ha). *Salsola Paletzkiana, S. Richteri, Calligonum* sp. and *Tamarix* sp. are also widely spread in desert forests (Figure 5).

Ratio of mountains forests is 9% with total area of 0.76 M ha and forest covered area of 0.25 M ha. *Juniperus* sp. (0.2 M ha) and such nut-bearing species as *Juglans regia* L., *Pistacia vera* L. and *Amugdalus* sp. (46,400 ha) are the main forest species in mountainous area. *Malus siversi* L., local species of *Crataegus* sp., *Acer* sp., *Prunus armeniaca*, shrub species of *Berberis* sp., *Rosa* sp., *Hippophae rhamnoides* and others also grow in forests there.

Area of tugay forests is 75,500 ha with forest covered area of 17,800 ha. The area of valley zone forests is 38,000 ha with forest covered area of 8,800 ha. The main forest species in tugay forests are *Populus pruinosa* and *P. diversifolia*, which usually grow together with *Elaeagnus* sp. and *Salix* sp. In contrast, Poplar, elm, sycamore and other fast-growing trees and shrubs are forest spp. in valley zone forests (online data of Main Forestry Office of Ministry of Agriculture and Water Resources of the Republic of Uzbekistan; 2006).

In comparison with 1996, forest area in Uzbekistan was reduced by more than 1 M ha by logging, converting forest lands to arable lands, and increasing effects of soil erosion. Area of tugay forests was reduced to less than one-tenth of the original area. Cutting saxaul, juniper and other forest trees for firewood has led to desertification of 6.5 M ha of land previously covered with vegetation in Uzbekistan.

Table 1. Wood stock of main forest forming species groups in Central Asian countries.

	Wood stock (million m^3)			
Species	Kazakhstan	Kyrgyzstan	Turkmenistan	Uzbekistan
Coniferous tree species	231.0	15.95	1.5	4.135
Soft-wood tree species	121.9	2.42	–	0.845
Saxaul sp.	10.2	–	9.3	12.745
Hard-wood tree species	2.9	–	–	–

National system of specially protected areas in Uzbekistan includes nine nature reserves with total area of 0.22 M ha, two national parks with area of 0.61 M ha, and nine nature refuges with area of 1.22 M ha (National Strategy and Action Plan on Conservation and Sustainable Use of Biodiversity of the Republic of Uzbekistan, 1998).

3 FOREST PRODUCTIVITY

It is very difficult to estimate forest productivity in the countries of the region for a number of reasons. Information is lacking or outdated to make analysis of wood stock by species and age structure. Average forest yield in Central Asia is 82.04 million m^3, of which 60.2% is wood stock of coniferous species, 7.2% is hard-wood, 28.2% is soft-wood and 4.4% comprises other tree species. On average, the per capita forest area is 0.4 ha and per capita wood stock is about 7.0 m^3 in Central Asian region, which are very low.

Distribution of forested area is uneven throughout the region. Main part of forested area and consequently wood stock is in Kazakhstan (366.1 M m^3) (Table 1). Tajikistan is the country with the lowest wood stock (5.0–5.2 M m^3) in the region (Regional report, Environment status of the Aral Sea Basin).

Forests in Central Asian countries mainly (89%) belong to 1st group of Forest Functions Classification and provide ecological, sanitational, recreational and protective functions. Therefore, commercial timber logging in these forests is prohibited. There are forests of 2nd and 3rd groups where timber logging is allowed, but only in Kazakhstan. Forests in the region are considerably affected by such negative factors such as pest outbreaks, forest fires, cutting trees for firewood due to lack of alternative energy sources, overgrazing, climate change, transforming forest lands into arable.

4 MAIN CONSTRAINTS IN FOREST MANAGEMENT

The principal constraints in sustainable management of forests in Central Asian region are:

- Low capacity of the current national forest protection and management systems in the region to ensure effective conservation of forest genetic resources due to lack of resources both financial and physical;
- Lack of awareness on value and importance of forest conservation and sustainable use for national development at national and regional levels. This should be recognized by all groups of stakeholders, including policymakers, planning agencies and local communities;
- Lack of definite prospects for forest enterprises functioning in new economic situations during the phase of economy transition in the region;
- Lack of qualified staff at forest enterprises trained to manage forests in effective way under market driven economy.

For low areas covered with forest, a high level of land reclamation and continuous anthropogenic pressure on forests are required to manage the forests in a sustainable way to ensure their availability for future generations in Central Asia.

REFERENCES

Akhmadov Kh.M. and K.Kh. Kasirov. 1999. Tigrovaya balka nature reserve. Wildlife Conservation [in Russian] 1(12): 20–21.

Allamuradov A., A. Abdurakhimov and A. Kuliev. 2005. Turkmenistan: Policies affecting forest land use and forest products markets – Forest resources assessment for sustainable forest management. Country statement. UNECE/FAO/LCR Workshop: Capacity building in sharing forest and market Information, October 24–28, 2005, Prague and Kristny, Czech Republic [online]. Available from: http://www.fao.org/regiona/seur/events/krstiny

Forestry of Uzbekistan [online]. Available from: http://www.msvx.uz/rus/forest Kamahina G.L. 2005. Flora of Central Kopetdag: In past, nowadays and in future [in Russian]. Ashgabat, Turkmenistan. p. 245

Kayimov A.K. 2006. Conservation of forets genetic resources biodiversity and their rational use in Uzbekistan. Journal Jalal-Abad State University 1:82–85

Kolov O.V., T.S. Musuraliev, Sh.B. Bikirov, V.D. Zamoshnikov, and T.M. Koblitshkaya. 2001. Forest and forest use. p. 103–120. In: Mountains of Kyrgyzstan [in Russian]. Technologiya Press, Bishkek, Kyrgyzstan.

Kolov O.V., Sh.B. Bikirov, N.S. Bikirova, and A.Sh. Bikirova. 2002. Conservation and rational use of forest biodiversity and forest genetic resources of Kyrgyz Republic. pp. 161–166. In: E.D. Shukurov (ed.). Biodiversity of Western Tien-Shan: Status and prospects [in Russian]. Bishkek, Kyrgyzstan.

Kolov O.V. 2003. Current status of nut and fruit forests biodiversity in Southern Kyrgyzstan. p. 157–161. In: I.S. Sodombekov (ed.). Conservation and sustainable use of plant resources. [in Russian]. Bishkek. Kyrgyzstan. pp. 157–161

Krever V., O. Pereladova, M. Williams, and H. Jungius (eds.). 1998. WWF. Biodiversity Conservation in Central Asia: An Analysis of Biodiversity and Current Threats and Initial Investment Portfolio. TEXT Publishers, Moscow, Russia.

National Biodiversity Conservation Strategy and Action Plan of the Republic of Uzbekistan. 1998 [in Russian]. Tashkent, Uzbekistan

Novikov V. and N. Safarov. 2002. Tajikistan:. Environment status. Ecological report. [online]. Available from: http://enrin.grida.no/htmls/tadjik

Regional report: Environment status of the Aral Sea Basin. [online]. Available from: http://enrin.grida.no/aral/aralsea

CHAPTER 3

C3/C4 plants in the vegetation of Central Asia, geographical distribution and environmental adaptation in relation to climate

Kristina Toderich
Department of Desert Ecology and Water Resources Research, Samarkland, Uzbekistan

Clanton C. Black
Biochemistry and Molecular Biology, University of Georgia, Athens, Georgia, USA

Ekaterina Juylova
Department of Desert Ecology and Water Resources Research, Samarkland, Uzbekistan

Osamu Kozan
University of Yamanashi, Kofu-city Yamanashi, Japan

Tolib Mukimov
Department of Desert Ecology and Water Resources Research, Samarkland, Uzbekistan

Naoko Matsuo
Graduate School of Agriculture, Kyoto University, Kyoto, Japan

1 INTRODUCTION

An unparalleled floral diversity is found in the continental deserts of Central Asia that stretch for thousands of miles from Iran and Afghanistan across Uzbekistan, Turkmenistan into Western China and Mongolia. Two major pressures subject Central Asian deserts to powerful forces that lead to desertification, namely, their extremely dynamic environments plus a host of human driven activities. Great natural contrasts exist in these continental environments, e.g., with mostly autumn-winter rains and almost absence of rainfall in summer along with quite high and low annual temperature fluctuations, and complicated desert, mountain, steppe and plain relief. Human activities demand lands, waters, plants, and other natural resources for activities such as domestic livestock grazing, fuels, cropping agriculture, irrigation, dams, reservoirs and canals, mining, drilling, roads, dwellings, and other human infrastructures. Such extreme natural environmental changes coupled with human activities often induce dynamic detrimental environments for the growth and seed reproduction of plants; indeed difficult environments for all biological life forms.

The present paper includes our field and laboratory work describing the natural floral diversity of arid/semiarid zones plus the unusual physiology, biochemistry, and reproductive biology, which allows these plants to grow, reproduce, adapt, and survive, even with the environmental degradation found in Central Asian deserts. However data about the relationships between the occurrence of plants and their spatial distribution with different pathways of photosynthesis and climate; as well as mapping of plant species according the new divisions and detailed climatic data are absent. The Central Asian desert flora of from this point of view is insufficient studied yet.

The objectives of this study were to obtain a complete list of plants with C4 photosynthesis from different desert zones of Central Asia based on results of experimental field studies and on previously published data and to analyze the geographical distribution patterns of C4 Poaceae and Chenopodiaceae species as a function of climatic variables. We also studied climatic and ecological distribution of C4 grasses and chenopods with different photosynthesis biochemistry, mesophyll structures, seed reproduction systems and life forms. Here we demonstrate that C4 grasses and chenopods differ in their relationship to climatic variables and that morphological-anatomical and seed reproduction features are important for surviving in different habitats. This work allows us to suggest ways to identify plant functional types (PFTs) for Asian arid territories. Knowledge about photosynthetic pathways and PFTs can be useful in the evaluation and prediction of vegetation changes during global climatic changes; in the conservation and restoration of natural ecosystems; and in improving pastures in the domestic livestock grazing based society of Central Asian Desert areas. Several recent biological discoveries also are presented regarding plants living in specific environments that can give us better insights into the unique environmental traits found in desert plants. These new findings will be combined with our understandings of how plants interact with their soil and air environments to adapt and cope with the enormous concentration ranges in nutrients, ions, gases, and toxicants which occur naturally or as pollutants. Plants must take up their essential growth minerals, but simultaneously they often take up non-essential ions and pollutants that can be toxic to plants, toxic to animals, and to other organisms that feed on plants. Thus, as sessile organisms, plants have very strong adaptive interactions with their environments which must be known to present sound and balanced desert restoration and management activities. Ultimately we wish to propose reliable scientific and social reasons for remediating, restoring, and balancing the management of deserts as stable productive economic ecosystems for Central Asian citizens.

2 METHODS AND MATERIAL

Photosynthetic types of plants were determined during studies on the basis of Kranz-anatomy microscopy, $\delta^{13}C$ fractionation, and primary $^{14}CO_2$ photosynthetic products and turnover. Many experiments were conducted from 1997 till 2003 during series of field expeditions, mainly in steppe, semidesert, and desert areas of Uzbekistan, Turkmenistan and Mongolia.

The assignment of photosynthetic types and biochemical groups of plants was carried out on the basis of the data obtained from these studies and from available published works. The photosynthetic types of some species, that were not available from experimental studies, were identified on the basis of closely related species and plant taxonomy for grasses (Hatteresley and Watson, 1992) and chenopods (Akhani et al., 1997; Pyankov et al., 1992, 1997). Plant samples (leaves, flowers and fruits) were collected from natural desert environments for isotope studies. Different floral organs of all examined species were immersed in 3% glutaraldehyde in sodium cacodylate phosphate buffer (pH = 7.2) for 3 hours prior to mounting on stubs. Then the material was placed in an Edward freeze-dryer for 24 h-55°C. Two methods of sample preparation were used for SEM, i.e., chemical fixation and freeze-drying. For chemical fixation, material was fixed, post-fixed and dehydrated as described for TEM, critical point dried, putter-coated with gold and observed with the JEOL JSM-T330 scanning electron microscope (Bozolla and Russell, 2002). Experiments were conducted at the laboratory of plant cells structure at the Kyoto University (Japan), Department of Botany of Poznan University (Poland) and Laboratory of Electron Microscopy at the University of Georgia (USA).

Morphology of fruits and seeds were investigated according to the N.N. Kaden et al. (1971) and N.M. Dudik (1971). Anatomical sections of bracts, perianth segments, anthers, embryo and fruits were selective stained with safranin in combination with fast green, haematoxylin or toluidine blue. Samples for anatomical studies of fruit covers were fixed in alcohol:glycerin:water (1:1:1). Sections were stained with methylene blue.

We have used aridity index defined as de Martonne (Encyclopedia of Climatology, 1987), $I = P/T+10$, where P is average annual precipitation, T is average annual temperature. The minimum absolute values of this index correspond to the maximum aridity of climate. Pearson's correlation coefficients were calculated for the relationships between climatic variables and occurrence of percentage of Chenopodiaceae and Poaceae species with different photosynthesis types.

3 CENTRAL ASIAN DESERT CLIMATE AND PLANT DIVERSITY DISTRIBUTION

Biogeografic studies habitually address Central Asian deserts as the territory covering Mongolia (Jigjidsuren and Johson, 2003), the desert in Chinese Inner Mongolia (Petrov, 1973) that is continuously stretching throughout Kyzylkum (Uzbekistan), Karakum (Turkmenistan) towards Shirmohi (Afghanistan) and northern African as a unique link of sandy deserts (Figure 1).

The climate in almost all of mentioned region is characterized by a dry continental with hot, sunny summer and cold or very cold winter with most of annual precipitation (67%) occurring as snow and rain in spring (March–April), and only 23% during September–December (Figure 2). This winter-spring precipitation feature is most characteristics for Middle Asia (Gintzburger et al., 2005), in contrast with Central Asia covering Northwestern China and Mongolia receiving monomodal summer precipitation. The best example is Uzbekistan that is also subject to the cold influence of the Arctic and Siberian fronts that may reach the southern border near Afghanistan. This characterizes the Middle Asian deserts rangelands climate as predominantly continental and subdesertic with an average 42°C continentally Index (M-m) as is a case for Uzbekistan (Gintzburger et al., 2003). The mean annual precipitation (MAP) received as snow and rain on the lowland is between 80–110 mm in the Kyzylkum sandy desert and north of the Karakum desert (Turkmenistan) and the Amudarya River around the Aral Sea. It increases to 200–300 mm between the Kyzylkum desert and west of a line Karshi-Navoi-Nuratau in Uzbekistan. Temperature ranges from 5–10°C or extreme minima −40°C in January as is a case in Mongolia, Karakalpakistan and Usturt Plateau between the Aral and Caspian seas. Average relative air humidity during April to September ranges from 38–63%. Transpiration during this period is about 914–1216 mm. From a bioclimatic point of view, the Irano-Turanian provinces have much colder winters preventing winter plant growth even when soil water is available. Despite of this the Irano-Turanian provinces possesses slightly over 2,680 species, 670 of which are endemic (Le Houuerou, 2005).

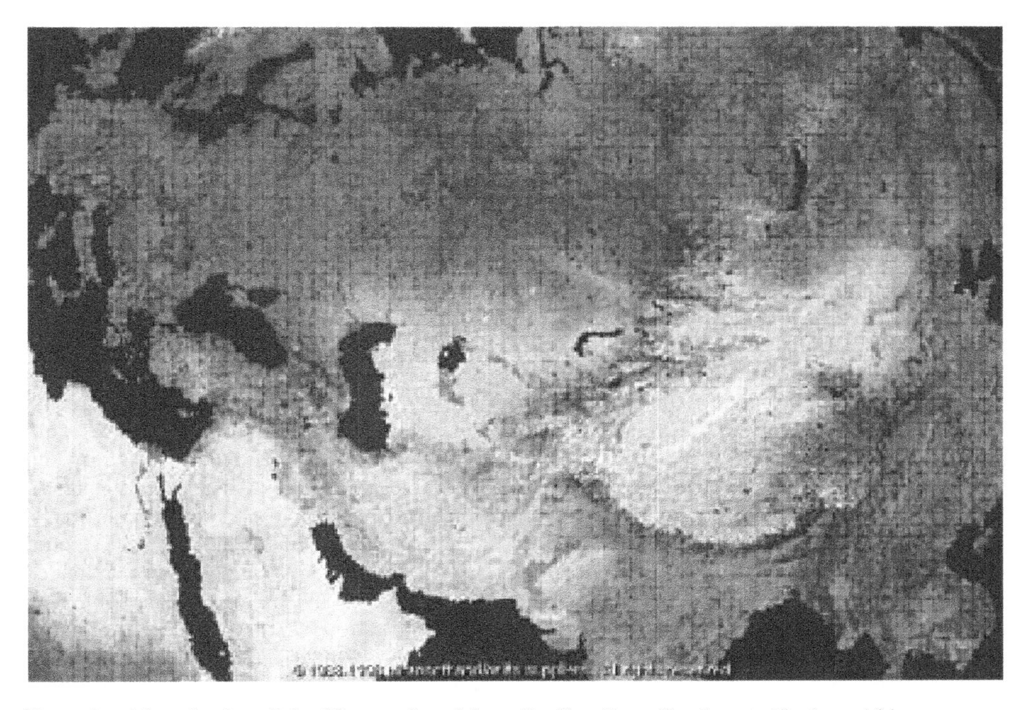

Figure 1. Map of unique link of Deserts from Mongolia, Kyzylkum-Karakum to Northern Africa.

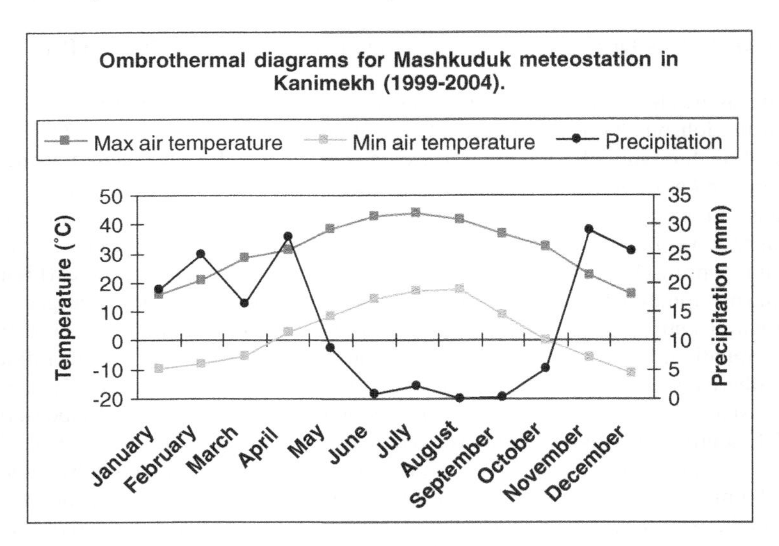

Figure 2. Ombrothermal diagrams for Central Kyzylkums (Uzbekistan).

Asian ecosystems, particularly the continental desert ecosystems, have not been studied in detail in relation to the occurrence of C4 species and their climatic distribution. The desert territories of Central Asia differ markedly from the arid ecosystems of North America, Australia and Europe in their taxonomic groups of C4 species. According our results Chenopodiaceae is the leading family in abundance of C4 species in Asia, and Kranz-grasses have a reduced importance, particularly in arid territories (Pyankov et al., 1986, 1992b; Pyankov and Mokronosov, 1993; Toderich et al., 2001). Several studies have identified many C4 Chenopodiaceae plants from the Near East and Central Asia (Shomer-Ilan et al., 1981; Winter, 1981; Ziegler et al., 1981; Zalenskii and Glagoleva, 1981; Voznesenskaja and Gamaley, 1986; Gamaley et al., 1988; Pyankov et al., 1992a, 1997; Pyankov and Vakhrusheva, 1989; Akhani et al., 1997). The C4 syndrome occurs in about 250 to 300 chenopods that mainly belong to subfamilies Salsoloideae (tribes Sauedeae and Salsoleae) and Chenopodioideae (tribes Camphorosmeae and Atripliceae); but many species are not yet classified. The occurrence of C4 species has been studied in some Asian desert regions, e.g., the Central Karakum desert (Pyankov et al., 1984; Pyankov and Vakhrusheva, 1988), in some parts of Mongolia (Gamaley, 1985; Gamaley and Shirevdamba, 1988), in South Tadjikistan (Pyankov and Molotkovskii, 1993), in some territories of the former Soviet Union (Pyankov and Mokronosov, 1993), the Pamirs mountains (Pyankov and Mokronosov, 1993; Pyankov, 1993), and in Northeast China (Redman et al., 1995) and Central Kyzylkum (Toderich et al., 2006).

One of the plant features uncovered in Central Asia deserts was that globally this is the most Northern and coldest edge of native C4 plant growth. In a global comparison, the desert vegetations of Central Asia have a unique taxonomic composition along with the presence of quite diverse and unusual biochemical, physiological, and structural features. One unusual finding (Pyankov et al., 1986, 2000; Botschantzev, 1969; Gamaley et al., 1988) is the large taxonomic biodiversity and tendency for perennial members of the Chenopodiaceae family to dominate Asian desert vegetation. In Figure 3, we have presented a global taxonomic comparison based on C4 photosynthesis plants. As is shown marked taxa domination of plants by Chenopodiaceae is found in the deserts of Central Asia. In contrast, grass species (Poaceae) dominate in North America, Australia and Europe. Indeed grasses are a minor taxa in Central Asia deserts, i.e., in the Kyzylkum Desert it is remarkable how few native grasses are present as pastoral plants compared to the strongly grass-supported herbivory found in the America's, Africa, Europe, and Australia (Smith et al., 1977; Jones, 1985). As it's shown in Figure 3, C4 members of Polygonaceae are present only in Central Asian deserts (Winter, 1981). Also absent from the deserts of Central Asia, but present in other world deserts, is a large

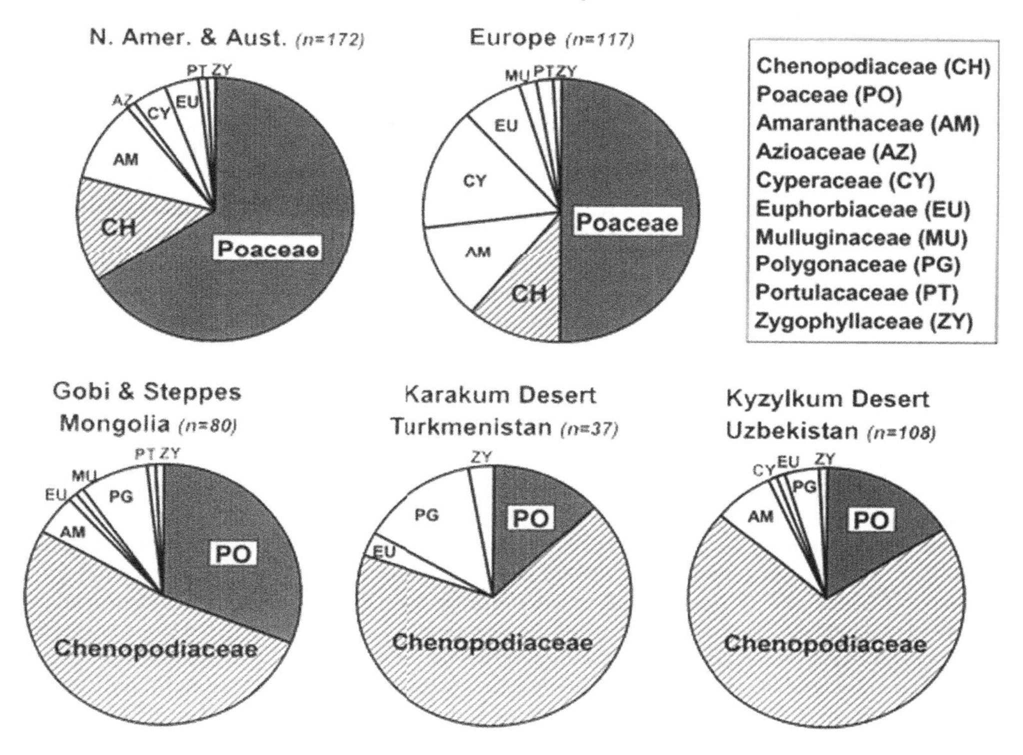

Figure 3. A global comparison of C4 photosynthesis taxonomic diversity versus C4 plant diversity in Central Asian deserts.

group of plants with a powerful water conserving type of physiology and biochemistry called Crassulacean Acid Metabolism (CAM). This is a readily visual difference in plant taxa because the deserts of the America, Africa and Australia contain large numbers of huge, even tree-like, CAM plants such as cactus, Agave, Aloe, and Euphorb's (Black et al., 2003). In fact, for a decade we have searched in Central Asia for CAM plants and CAM is almost absent. However, recently we discovered CAM in one native Asian genus, Orostachys (Oyungerel et al., 2000).

4 TAXONOMIC OCCURRENCE OF C3/C4 SPECIES

According our preliminarily analysis there are 16 families of vascular plants in which C4 photosynthesis occurs. Among them two belongs to subclasses of Monocotyledoneae (Poaceae, Cyperaceae) and 14 are subclasses of Dicotyledoneae (Acanthaceae, Aizoaceae, Amaranthaceae, Asteraceae, Boraginaceae, Caryophyllaceae, Chenopodiaceae, Euphorbiaceae, Nyctaginaceae, Molluginaceae, Polygonaceae, Portulacaceae, Scrophyllariaceae, Zygophyllaceae). According to the latest taxonomic data (Gubanov, 1996) flora of Mongolia consists of 2823 vascular species belonging to 128 families; while the vegetation desert flora of Uzbekistan is estimated as 3100 vascular plants. From our studied more than 96 species with C4 photosynthesis occurs in 8 families: Amaranthaceae, Chenopodiaceae, Euphorbiaceae, Molluginaceae, Polygonaceae, Portulacaceae, Zygophyllaceae and Poaceae.

4.1 *Chenopodiaceae*

C4 photosynthesis species in Chenopodiaceae were found in the subfamily Salsoloideae, tribes Salsoleae and Suaedeae, and in tribes Atripliceae and Camphorosmeae of the subfamily Chenopodioideae (Carolin et al., 1975; Osmond et al., 1980; Winter, 1981; Gamaley, 1985; Pyankov and Vakhrusheva, 1989; Gamaley et al., 1988; Pyankov et al., 1992a, 1997; Akhani et al., 1997). The total chenopod that exhibit C4 photosynthesis comprises about 45% of the total. The largest number of species with the Kranz-syndrome is in the Salsoleae tribe, where 34 of 45 are C4 photosynthesis species (Table 1). Sixteen species have a direct confirmation of C4 photosynthesis based on leaves/bracts/fruits/flowers/anatomy and biochemical tests.

Some experimentally unidentified species belonging to Anabasis, Climacoptera, Nanophyton and Petrosimonia are C4, because these genera are derivatives of Salsola (Botchatnzev, 1969) in which all experimentally studied species demonstrated typical features of C4 photosynthesis (see reference in Table 2). Among 11 Salsola species in the Mongolian flora, and 9 from them were identified as C4 species. Salsola mostly contains C4 plants, but a few are C3 species, mainly shrubby forms (Pyankov et al., 1993, 1997; Akhani et al., 1997). The Kranz-syndrome was found in seven species experimentally (Table 2). Two annual species Salsosa were not identified in regards to photosynthetic type (*S. ikonnikovii* and *S. rosaceae*) that belong to the most young and progressive Salsola section of genus Salsola (Botchantzev, 1969) in which all studied species have C4 photosynthesis (Pyankov et al., 1997; Akhani et al., 1997).

There are two shrubby Salsola species, *S. abrotanoides* and *S. laricifolia*, that are not included in the list with Kranz-syndrome. *S. laricifolia* and *S. montana* according our measurement have a C3-like $\delta^{13}C$ value of $-23.1‰$, while *S. arbusculiformis* keep a tendency to C3-C4 internmediate type of photosynthesis. *S. abrobtanoides* and *S. pachyphylla* were not studied, but can be included as a C3 plant based on it taxonomy. According to Botchantzev (1976), *S. abrobtanoides* belong to section Coccosalsola, subsection Genistoides together with another two *S. webbi* and *S. genistoides*. C3 photosynthesis in *S. webbi* and *S. genistoides* is well documented (Carolin et al., 1975; Winter, 1981; Akhani et al., 1997; Pyankov et al., 1997). Hence it seems likely that *S.abrobtanoides* has a C3 carbon metabolism. C3 photosynthesis in *Sympegma regelii* is well documented. This species has a C3-like centric non-Kranz type named Sympegmoid by Carolin et al. (1975). The vasculature in *Sympegma regelii* is close to Kranz Salsoloid-type, therefore Carolin et al. (1975) suggested the possible origination of Sympegmoid type from Salsloid with a reversion of C4 to C3 photosynthesis. We found a high percentage incorporation of ^{14}C to C3 primary photosynthetic products as in C3 species, but incorporation of label to primary C4 acids varied from 6 to 24% during 10-sec of $^{14}CO_2$ assimilation; somewhat more in comparison with typical C3 species (Black, 1973).

Some genera of Chenopodiaceae that grow in the Central Asian deserts, are known to have both C3 and C4 photosynthesis types, such as Suaeda, Salsola, Kochia, Atriplex, and Bassia. There are ten species of Suaeda amongst the Mongolian flora (Gubanov, 1996). *Suaeda* has species with different photosynthetic types (Carolin et al., 1975; Winter, 1981; Pyankov et al., 1992a; Akhani et al., 1997; Fisher et al., 1997). We included two species in our list of C4 species, *S. arcuata* and *S. heterophylla*, based on previous results of yearly photosynthesis products and measurement of enzyme activities (Pyankov and Vakhrusheva, 1989; Pyankov et al., 1992a). *S. arcuata* has C3-like values of $^{13}C/^{12}C$ fractionation (Winter, 1981; Akhani et al., 1997) which confirmed the recent review of photosynthesis types in Suaeda (Fisher et al., 1997). It seems likely that eight other species, *S. corniculata, S. foliosa, S. glauca, S. kossinskyi, S. linifolia, S. prostrata, S. prezhewalskii* and *S. salsa* are C3. Some of them, *S. corniculata, S. glauca, S. linifolia* and *S. salsa* show C3-like carbon fractionation values (Winter, 1981; Akhanii et al., 1997). In addition, taxonomic analysis of photosynthesis amongst Suaeda species show C3 CO_2 fixation for species belonging to the most primitive sections Chenopodina (syn. Heteriospermae) (*S. corniculata, S. kossinskyi, S. prostrata, S. salsa*), and Schanginia (*S. glauca, S. linifolia*). Some disagreement in identification of photosynthetic types exists, for instance, Gamaley (1985) included *S. corniculata* as a C4 plant, but this may be incorrect species identification. Experimental data are completely absent regarding *S. foliosa* and *S. przhevalskii*, so we have not included them in our list of C4 species.

Table 1. Occurrence, ecological distribution and structural-biochemical features of C4 plants of Central Asian Deserts.

No.	Family, tribe and species[1]	Life form[2]	Habitat[3]	Kranz-anatomy[4]	Biochem. type	$^{13}C/^{12}C$ value	Criteria[5]
Chemopodiaceae Atripliceae							
1	*Atriplex cana C. A. Mey.*	semishrub	6	ATR	NAD-ME	−13.9 (+)	M
2	*A. laevis C. A. Mey.*	annual	6	ATR	NAD-ME	−13.0 (W)	A
3	*A.sibirica L.*	annual	5, 6	ATR	NAD-ME		A, M
4	*A. tatarica L.*	annual	1, 6	ATR	NAD-ME	−13.6 (W)	A, E
Camphorosmeae							
5	*Bassia hyssopifolia (Pal.) O. Kuntze*	annual	6	KOCH	NAD-ME	−13.4 (+)	A, M, F
6	*Camphorosma lessingii Litv.*	semishrub	5, 6	KOCH	NAD-ME	−13.1 (+)	A, E, F
7	*Kochia densiflora Tursz. ex Moq.*	annual	5, 6	KOCH	NAD-ME	no data	A
8	*K. iranica Litv.*	annual	6	KOCH	NAD-ME	−13.9 (W)	A, M
9	*K. krylovii Litv.*	annual	3, 5	KOCH	NAD-ME	no data	*
10	*K. melanoptera Bunge*	annual	6	KOCH	NAD-ME	no data	A
11	*K. prostrata (L.) Shrad.*	semishrub	5, 6	KOCH	NAD-ME	−14.0 (+)	A, M, E, F
12	*K. scoparia (L.) Schrad*	annual	1, 2	KOCH	NAD-ME	−13.4 (A)	A, F, E
13	*Londesia eriantha Fish.* et Mey.	annual	5, 6	KOCH	NAD-ME	no data	A
Suaedeae							
14	*Suaeda acuminate (C.A. Mey.) Moq.*	annual	6	ATR	NAD-ME	−12.8 (W)	A, F, M
15	*S. heterophylla (Kar. Et Kir.) Bunge*	annual	6	SUA	NAD-ME	no data	A, E
Salsoleae							
16	*Anabasis aphylla L.*	semishrub	4–6	SALS	NADP-ME	−14.5 (W)	A, F
17	*A. brevifolia C.A. Mey.*	semishrub	4–6	SALS	NADP-ME	−12.9 (+)	A, M, F
18	*A. elatior C.A. Mey. Schisnhk.*	semishrub	5, 6	SALS	NADP-ME	−13.5 (W)	F
19	*A. eripoda (Shrenk) Benth. Ex Volkens*	semishrub	5	SALS	NADP-ME	−13.0 (A)	A, M, F
20	*A. pelliotti Danguy*	semishrub	5	SALS	NADP-ME	no data	*
21	*A. salsa (C.A. Mey.) Benth. Ex Volkens*	semishrub	4–6	SALS	NADP-ME	−13.2 (+)	F, E
22	*A. truncate (Shrenk)*	semishrub	4–6	SALS	NADP-ME	no data	*
Bunge							
23	*Cimacoptera affinis (C.A. Mey.) Botsch.*	annual	3, 6	SALS	NAD-ME	no data	a*, F*, M
24	*C. subcrassa (M. Pop.) Botsch*	annual	3, 6	SALS	NAD-ME	−13.07 −13.1 (+)	*
25	*Halogeton glomeratus (Bieb.) C.A. Mey.*	annual	3, 6	SALS	NADP-ME	−12.8 (+)	A, E, M
26	*Haloxylon ammodendron (C.A. Mey.) Bunge*	shrub	3, 6	SALS	NADP-ME	−13.2 (+)	A, F, M
27	*Iljinia regelii (Bunge) Korov.*	shrub	3–6	SALS	NADP-ME	−11.9 (+)	A, M, F
28	*Micropeplis arachnoideae (Moq._ Bunge*	annual	3–6	SALS	NADP-ME	−12.6 (+)	A, M,

(Continued)

Table 1. Continued

No.	Family, tribe and species[1]	Life form[2]	Habitat[3]	Kranz-anatomy[4]	Biochem. type	$^{13}C/^{12}C$ value	Criteria[5]
29	*Nanophyton grubovii Pratov*	semishrub	3–6	SALS	NAD-ME	no data	A*, M*, F*
30	*N. Monogolicum Pratov*	semishrub	3–6	SALS	NAD-ME	no data	*
31	*Petrosimonia litwinovii*	annual	3, 6	SALS	NAD-ME	no data	A*, F*, M*
32	*P. sibirica (Pall.) Bunge*	annual	3, 6	SALS	NAD-ME	no data	*
33	*S. arbuscula Pall.*	shrub	3, 5	SALS	NADP-ME	−12.9 (+)	A, F, M, E
34	*S. arbusculiformis Drob.*	shrub	4, 5	no data	C3-C4 intermediate photosynthesis	−26.30 (in leaves)	*
35	*S. drobovii*	shrub	3, 4	SALS	NAD-ME	−25.79 (in fruits cover)	*
36	*S. Richteri Kar. Et Litv*	woody tree-like	3,4	SALS	NAD-ME	−12.28 (in fruits)	*
37	*S. paletzkiana Litv.*	woody tree-like	3,4	SALS	NAD-ME	−13.24	*
38	*S. montana Litv*	shrub	3,4	no data	no data	−26.40 (in flower's perianth); −20.88 (in leaves)	*
39	*Salsola dendroides Pall*	shrubs	3, 4, 6	SALS	NADP-ME	−12.63 (flowers)	*
40	*S. incanescens C.A. Mey*	perennial	3, 4	SALS	no data	−13.67 (fruits)	*
41	*S. foliosa (L) Schrad.*	annual	3, 4	SALS	NAD-ME	−11.9 − 12.0	
42	*S. carinata*	annual	1, 3, 6	SALS	NADP-ME	−13.1 (+)	A, M, E, F
43	*S. orientalis S, G. Gmel.*	semishrub	4, 6	SALS	NAD-ME	−13.5 (+)	A, M, E, F
44	*S. gemmascens*	semishrublet	3, 4	SALS	NADP-ME	−12.9–12.98 (+)	*
45	*S. monoptera Bunge*	annual	3, 5	SALS	NADP-ME	−12.7 (+)	*
46	*S. passeria Bunge*	semishru	4–6	SALS	NAD-ME	−13.0 (+)	*
47	*S. paulsnii Litv.*	annual	3, 6	SALS	NAD-ME	−12.8 (+)	A, M, F, E
48	*S. rosaceae L.*	annual	3, 5	SALS	NADP-ME		*
49	*S. tragus L.*	annual	1, 3, 6	SALS	NAD-ME	−13.0 (+)	A, E
50	*S. pachyphylla Botsch.*	annual	1, 3, 4	SALS	–	−20.8 − 20.9	A, E
51	*S. pestifer A. Nelson*	annual	1, 3, 4, 5	SALS	NAD-ME	−13.42	*
52	*S. aperta Pauls.*	annual	1, 3, 4, 5	SALS	NAD-ME	12.78	*
53	*S. tamariscina*	annual	1, 3, 4, 5	SALS	NAD-ME	−11.93	*
54	*S. tomentosa*	annual	1, 3, 4, 5	SALS	NAD-ME	−12.91 (fruits)	*
55	*S. aucheri (Mog.) Bge*	annual	1, 3, 4, 5	SALS	NAD-ME	−13.25	*
56	*S. sclerantha C.A. Mey*	annual	1, 3, 4, 5	SALS	NAD-ME	−13.4	*
	Poaceae Andropogonae						
57	*Spodiopogon sibiricus Trin.*	perennial	2, 5	PANIC	NADP-ME	−12.3 ®	F, A*
58	*Echinochloa crus-galli (L.) Beauv.*	annual	1, 2	PANIC	no data	−16.2 (S)	A, E
59	*Panicum miliaceum L.*	annual	1, 2	PANIC	NAD-ME	no data	A
60	*Pennisetum centralasiaticum Tzvel.*	perennial	1, 5	PANIC	NADP-ME	no data	A*, F*
61	*Setaria glauca (L.) Beauv.*	annual	1	PANIC	NADP-ME	no data	A, F, E

(Continued)

Table 1. Continued

No.	Family, tribe and species[1]	Life form[2]	Habitat[3]	Kranz-anatomy[4]	Biochem. type	$^{13}C/^{12}C$ value	Criteria[5]
62	*S. pumila (Poiret) Roem. et Schult*	annual	1, 2	PANIC	NADP-ME	no data	*
63	*S. virdis (L.) Beauv.*	annual	1, 2	PANIC	NADP-ME	−13.2 (+)	A, E, F
Aristideae							
64	*Aristida heymannii Regel*	annual	3, 5	ARIST	NADP-ME	−14.6 (+)	A, F, M
Arundinella							
65	*Arundinella anomala Steud.*	perennial	1–2, 5	PANIC	NADP-ME	no data	A
66	*Aeluropus micranterus Tzvel.*	perennial	6	CHLOR	NAD-ME	no data	A, E, F
67	*Chloris virgata Sw.*	annual	1, 4–5	CHLOR	PEP-CK	−14.6 (S)	A, F
68	*Cleistogenes caespitosa Keng.*	perennial	5	CHLOR	no data	no data	*
69	*C. kitagawae Honda*	perennial	5	CHLOR	no data	no data	*
70	*C. foliosa Keng*	perennial	2, 5	CHLOR	no data	no data	*
71	*C. songorica (Roshev.) Ohwi*	perennial	2, 5	CHLOR	no data	−14.7 (+)	A
72	*C. squarrosa (Trin) Keng*	perennial	1, 5	CHLOR	no data	−16.4 ®	A, F
73	*Crypis aculeate (L.) Ait.*	annual	6	CHLOR	no data	no data	F
74	*C. schoenoides (L.) Lam.*	annual	6	CHLOR	no data	no data	F
75	*Enneapogon borealis (Griseb.) Honda*	annual	2, 5	CHLOR	NAD-ME or PEP-CK	−14.7 (+)	A, F
76	*Eragrostis cilianensis (All.) Vign.-Lut.*	annual	1,3	CHLOR	NAD-ME	−15.4 (S)	A, F
77	*E. minor Host*	annual	1–3, 5	CHLOR		−13.7 (+)	A
78	*E. pilosa (L.) Beauv.*	annual	1, 3, 5	CHLOR	NAD-ME	−15.0 (S)	A
79	*Tragus mongolorum Ohwi*	annual	3, 5	CHLOR	NAD-ME	no data	A*, F*
80	*Tripogon chinensis (Franch.) Hack*	annual	5	CHLOR	no data	no data	*
81	*T. purpurascens Duthie*	annual	5	CHLOR	no data	no data	*
Polygonaceae							
82	*Calligonum chi-nurum Ivanova ex Soskov*	shrub	3	SALS	NAD-ME	no data	A*, F*, M
83	*C. gobicum Bunge ex Meissn*	shrub	3	SALS	NAD-ME	no data	*
84	*junceum (Fish. Et Mey.) Litv.*	shrub	3	SALS	NAD-ME	−12.7 (W)	F
85	*C. litvinovii Drob.*	shrub	3	SALS	NAD-ME		*
86	*C. mongolicum Turcz*	shrub	3	SALS	NAD-ME	−12.9 (+)	A
87	*C. pumilum Losinsk.*	shrub	3	SALS	NAD-ME		A
Amaranthaceae							
88	*Amaranthus albus L.*	annual	1	ATR	NAD-ME	no data	A
89	*A. cruentus L.*	annual	1	ATR	NAD-ME	no data	A
90	*A. blitoides S. Wats.*	annual	1	ATR	NAD-ME	no data	A
91	*A. retroflexsus L.*	annual	1	ATR	NAD-ME	−12.6 (+)	A, F
Molluginaceae							
92	*Mollugo cerviana (L.) Ser.*	annual	3	ATR	NAD-ME	no data	A
93	*Portulaca oleaceae L.*	annual	1	ATR	NAD-ME	no data	A

(Continued)

Table 1. Continued

No.	Family, tribe and species[1]	Life form[2]	Habitat[3]	Kranz-anatomy[4]	Biochem. type	$^{13}C/^{12}C$ value	Criteria[5]
Euporbiaceae							
94	*Euphorbia mongolicum Prokh.*	annual	3	ATR	NAD-ME	−13.8 (+)	94
Zygophyllaceae							
95	*Tribulus terrestris L.*	annual	3, 4	ATR	NADP-ME	no data	A, F
Tamaricaceae							
96	*Tamarix hispida*	shrubs	no data	no data	no data	+27.3	no data

[1]List of species, names of species and families by Gubanov (1996), division of Chenopodiaceae by Iljin (1936) and Botchantzev (1969); division of Poaceae by Tszvelev (1987).
[2]Life forms according Grubov (1982).
[3]Habitat: (1), moist, meadow or banks of rivers and lakes, oasises; (2), dry and non-saline, dry medow and steppe; (3), sandy and semisandy deserts; (4), takyrs, week saline clay deserts; (5), rocky surfaces; (6), salt-marshes.
[4]Kranz-anatomy types after Carolin et al. (1973, 1975). ATR – atriplecoid, KOCH – kochioid, SUE – suaedoid, SALS – salsoloid, PANIC – panicoid, ARIST – aristidoid, CHLOR – chloridoid.
[5]Criteria: A, Kranz-anatomy; M, protosynthesis carbon metabolism; E, enzyme activity of C4 cycle; F, carbon isotope fractionation. Asterisk (∗) indicated C4 traits for another species of the same genus.

In the Camphorosmeae tribe, C4 species exist in Camphorosma, Bassia, Kochia and Londesia. Camphorosma lessingi, Londesia eriantha and almost all Kochia species have direct evidences of C4 photosynthesis. *K. krylovii* has no experimental studies, but was included as a C4 plant. All Asian species of Kochia species, like *K. prostrata, K. scoparia, K. iranica* demonstrate C4 photosynthesis (Pyankov and Mokronosov, 1993, Toderich et al., unpublished data); and we agree with Carolin et al. (1975) that Eurasian Kochia species commonly have the Kranz-syndrome. Four of five Atriplex species were classified in the Kranz-anatomy group based on experimental evidence. Mesophyll structure and photosynthesis enzyme activities were studied in *A. tatarica* and *A. sibirica* (syn. *A. centralasiatica*) (Pyankov et al., 1992a,b). *A. cana* is included on the basis leaf anatomy, $\delta^{13}C$ value and primary photosynthetic products (Table 1). There are data showing Kranz-anatomy (Gamaley, Shirevadamba, 1988) and a C4-like $\delta^{13}C$ value (Winter, 1981) in *A. laevis*; but Akhani et al. (1997) found a C3-like $\delta^{13}C$ value and it is a C3 species according Carolin et al. (1975). *A. fera* was not included in our C4 list, because data about its type of carbon metabolism are absent.

Akhani et al. (1997) reported a C4-like $\delta^{13}C$ value in *Axyris amarantoides*, which habitats in Mongolia (Gubanov, 1996). However it seems likely, that this species is C3. Carolin et al. (1975) did not find Kranz-anatomy in *A. amaranthoides* and Axyris is very close to Krashennikovia systematically, all representatives of which are typical xerophytes with a C3 isopalisade leaf structure (Gamaley and Sirevdamba, 1988).

The relationships between systematic groups of C4 Chenopodiaceae and their Kranz-anatomy and biochemical types seem clear now (Carolin et al., 1975; Voznesenskaja and Gamaley, 1986; Pyankov and Vakhrusheva, 1989; Pyankov et al., 1992a, 1997). All species from Atriplex and Suaeda are NAD-ME. The species of Camphorosmeae (Camphorosma, Kochia, Bassia) are NADP-ME. But there are different biochemical pathways of CO_2 metabolism among representatives of Salsoloideae. Annual plants belonging to section Salsola and derivative (Halogeton, Micropeplis), Anabasis and shrubs from section Coccosalsola (*S. arbuscula*) and its derivative (Haloxylon, Ijinia) have NADP-ME biochemical type. Salsola species of Malpigipila section (*S. gemmascens, S. passerina*), derivative species of sections Belantera (Climacoptera), Caroxylon (Nanophyton) and species of Petrosimonia genus likely are all NAD-ME photosynthesis types.

Table 2. Coefficients of correlations between occurrence of C4 species and climatic variables. Correlations >0.50 are significant at $P < 0.05$.

Plant group[1]	Aridity index[2]	Sum of temperature >10°C	Annual precipi-tation (mm)	Annual average temperature (°C)	July average temperature (°C)	July mean minimum temperature (°C)	July mean minimum temperature (°C)
POA+CH, total species	0.21	−0.63	0.26	−0.57	−0.57	−0.53	−0.54
%C4 POA+CH, of total POA+CH species	−0.8	0.92	−0.72	0.92	0.88	0.9	0.84
POA, C3+C4 species	0.52	−0.82	0.53	0.78	−0.77	−0.75	−0.73
POA, C3 species	0.56	−0.86	0.54	−0.83	−0.81	−0.80	−0.78
POA, all C4 species	−0.45	0.47	−0.14	0.55	0.53	0.54	0.60
%all C4 POA of POA total	−0.60	0.90	−0.50	0.90	0.83	0.84	0.86
POA, C4 native number	−0.42	0.44	−0.14	0.52	0.48	0.49	0.57
%C4 native POA of POA number	−0.59	0.87	0.48	0.89	0.8	0.8	0.84
POA, NAD-ME + PEP-CK species	−0.49	0.51	−0.24	0.6	0.56	0.58	0.62
%NAD-ME + PEP-CK species of POA total	−0.62	0.89	−0.53	0.91	0.83	0.84	0.84
POA, NADP-ME species	−0.03	0.04	0.26	0.01	0.10	0.07	0.15
%NADP-ME species number of POA total	−0.44	0.72	−0.31	0.69	0.65	0.63	0.69
CH, C3+C4 species	−0.70	0.40	−0.62	0.47	0.43	0.48	0.40
CH, C3 species	−0.60	0.35	−0.47	0.44	0.40	0.44	0.42
CH, C3 succulent species	−0.80	0.70	−0.70	0.79	0.68	0.69	0.70
CH, %C3 succulent species of CH total	−0.68	0.72	−0.63	0.76	0.65	0.64	0.69
CH, C4 species	−0.71	0.39	−0.68	0.44	0.41	0.46	0.33
%C4 of CH total	−0.74	0.49	−0.64	0.46	0.48	0.51	−0.38
CH, C4+C3 succulent species	−0.81	0.54	−0.74	0.60	0.55	0.59	0.50
%C4+C3 succulents of CH total	−0.97	0.80	−0.82	0.80	0.77	0.78	0.73
CH, C4 shrubby NADP-ME Salsoloid type species	−0.63	0.50	−0.62	0.53	0.49	0.55	0.34
%C4 shrubby NADP-ME Salsoloid type of CH total	−0.69	0.69	−0.75	0.72	0.64	0.67	0.54
CH, C4 annual NADP-ME Salsoloid type species	−0.5	0.27	−0.31	0.28	0.34	0.36	0.34
%of C4 annual NADP-ME Salsoloid type of CH total	−0.16	0.15	0.11	0.07	0.22	0.19	0.23
CH, C4 shrubby NADP-ME Kochioid type spec.	−0.65	0.18	−0.64	0.24	0.16	0.21	0.14
%, NADP-ME shrubby Kochiod type CH total	−0.57	0.17	−0.47	0.15	0.15	0.15	0.13

[1]POA – Poaceae, CH – Chenopodiaceae;

[2]Aridity index de Martonne, $I = P/T + 10$. P – annual pct., T – avg. annual.

The composition of chenopods with different types of CO_2 fixation and structural and biochemical subtypes of C4 also changed along the warming and aridity gradient. The relative abundance of non-succulent C3 chenopod species decreased from cool moist to warm desert regions. In general, the percentage of C4 species from Chenopodiaceae positively correlated with increasing of aridity (or negative with aridity index) in each district (Table 2), $r = -0.74$, $P < 0.05$. Temperature parameters, i.e., the sum of effective temperatures or average year or July temperatures correlated positively, but not significantly, with the relative abundance of C4 Chenopodiaceae.

The chenopods with C4 CO_2 fixation have different in types of Kranz-anatomy, biochemical subtypes, and life forms. As is noted in Table 1, there are 4 types of Kranz anatomy: Atriplecoid, Suaedoid, Kochioid and Salsoloid, and two biochemical subtypes, NAD-ME and NADP-ME. Plants with Salsoloid Kranz-anatomy presented both biochemical variants plus two contrasting groups with different life forms – shrubs and annuals.

We hypothesized that various groups of C4 Chenopodiaceae responded differently to environmental variables. It would be clear to test this hypotesis by analyzing each group of Chenopodiaceae versus climatic variables. However the number of C4 Atriplex and Suaeda species was not enough for analysis. Therefore we studied the climatic patterns in Camphorosmeae and the main groups of Salsoleae that contain the most number of chenopods with the Kranz-syndrome. We found a different composition of groups along the aridity gradient (Table 2; Figure 4 a, b). Large shrubs with a Salsoloid type of assimilation organs and NADP-ME biochemical subtypes, such as Haloxylon, Iljinia, Salsola arbuscula, and semishrubs of genus Anabasis, appeared only in dry and hot districts with aridity index near 10 or less, and habitat mainly in Middle Asian and Gobian deserts. Their relative abundance positively and significantly correlated with aridity ($r = 0.69$), the average year temperature ($r = 0.69$), and negatively with annual precipitation ($r = -0.75$).

The annuals, Salsolas NADP-ME species, C4-herbaceous halophytes known as solyanki are common in all of investigated by us regions. These mostly succulent drought/salt tolerant species form a dominant group in the flora and vegetation of saline/arid environments.

The highest coefficient between climate aridity and C4 percentage were found for a group, combined of C4 NADP-ME shrubs with Salsoloid anatomy and C3 succulents, that consist of Halocnemum, Kalidium, Salicornia, Salsola, Suaeda, Sympegma. The relative abundance for this group of combined C3 and C4 succulents correlated with aridity at the highest value ($r = 0.97$)

The correlation with temperature also was high ($r = 0.80$), but less than aridity. Thus chenopods with C4 photosynthesis types and C3 succulents are favored in areas with hot and dry climate.

Recently it was a surprise to discover that the green leaves of some C4 species express C4 genes while their green cotyledons expressed C3 genes. Clearly some, yet unknown, types of control for gene silencing and expression exists in leaves versus cotyledons of the same plant related to Kranz or non-Kranz cell anatomy and the expression of C4 or C3 photosynthesis genes.

4.2 *Polygonaceae*

Calligonum, mostly tree-like and woody shrubs life forms, is a single genus in Polygonaceae, that contains all C4 species (Winter et al., 1981; Voznesenskaja and Gamaley, 1986). Species of genus Calligonum are mostly distributed on the sandy areas of Central Asian deserts. Only *C. mongolicum* has a broader geographical distribution in Mongolia. All species have a centric type of Kranz-anatomy that is similar in appearance with Salsoloid type (Voznesenskaja and Gamaley, 1986).

Other dicots. C4 species were identified in another five dicotyledonous families (Table 2). There are mainly well-known naturalized species Amaranthus (Amaranthaceae), *Portulaca olearceae* (Portulacaceae), and cosmopolitian species *Mollugo cerviana* (Molluginaceae) and *Tribulus terrestris* (Zygophyllaceae). We found a C4-like $\delta^{13}C$ value in *Euphorbia mongolicum* (Euphorbiaceae), which is included as a C4 species. Voznesenskja, Gameley (1986) reported the C4 syndrome in the Mongolian species *E. humifusa*; but our $\delta^{13}C$ determination of -26.3 is a C3-like value.

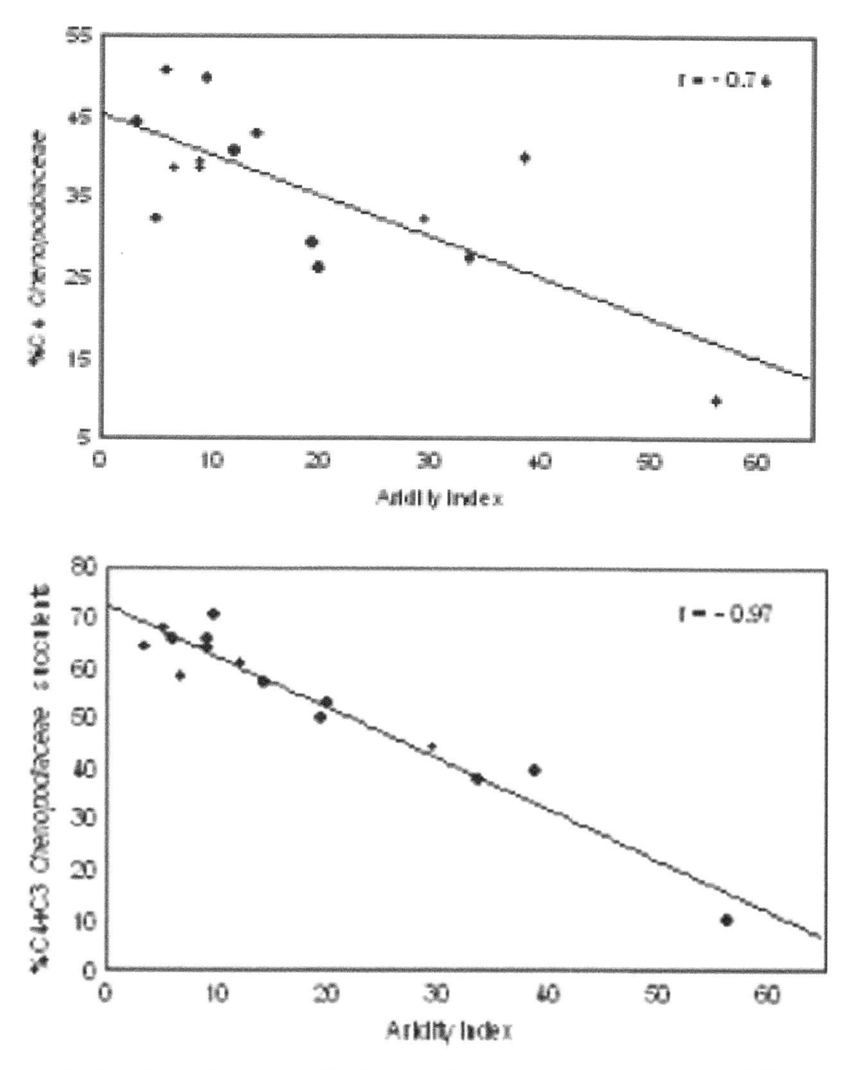

Figure 4. a (top), b (bottom). Ratio of C3/C4 types of photosynthesis occurrence in different taxonomic groups and aridity index.

4.3 *Poaceae*

Poaceae family is in the third place in abundance amongst species in the Mongolian flora (Gubanov, 1996) and Kyzylkum (Tsvelev, 1987). Identification of C4 grass species was on the basis of experimental studies of Kranz-anatomy and $\delta^{13}C$ values plus published data (Table 2). Beside these, there are strong relationship between taxonomic position and photosynthesis types, Kranz-anatomy and biochemical types (Hatteresley and Watson, 1992). Based on these characters we identified 25 grasses with the Kranz-syndrome, i.e., 10% of Poaceae. We found species with the C4 syndrome in five tribes Andropogoneae, Paniceae, Aristideae, Arundinelleae and Cynodonteae. A majority of C4 grasses, 16 species, belong to Cynodonteae and have NAD-ME or PEP-CK types. Species with NADP-ME are comparatively few, and tribes Andropogoneae, Aristideae, and Arundinelleae consist of only a single species. The representatives of Paniceae (Ehinochloa, Panicum and Setraria) are naturalized species and habitat mainly in annually disturbed territories and oasises. Other C4 annuals with aspartate-forming type of metabolism, NAD-ME or PEP-CK, i.e., *Chloris virgata*, Enneapogon and Eragrostis species are cosmopolites, and their appearance in the natural ecosystems very strongly depends on precipitation. Thus, only a few C4 grasses, such as species from Cleistogenes, Aristida, Aeluropus, Tragus and Eragrostis can play outstanding roles in natural ecosystems.

4.4 *Cyperaceae*

We could not identify C4 photosynthesis amongst representatives of the Cyperaceae family, which is the second in the world flora abundance for C4 photosynthesis. In the Mongolian flora, the Cypreaceae family includes 127 species belonging to 27 genera (Gubanov, 1996). Among these only two genera, Cyperus and Eleocharis contain species with Kranz-anatomy (Ueno and Takeda, 1992; Li, 1993). *Cyperus fuscus*, the single species in Cyperus has a C3 type of CO_2 fixation (Li, 1993). Seven species are in Eleocharis genus, and C3 photosynthesis type was previously demonstrated in five species: *E. acicularis, E. mamillata, E. mitrocarpa, E. palustris*, and *E. unglumsis* (Ueno et al., 1989; Ueno and Takeda, 1992). The photosynthesis type is not known for *E. klingei* and *E. meridionalis*; but the occurrence of C4 species among Eleocharis is rare (Ueno and Takeda, 1992); therefore, the presence of C3 photosynthesis in all Eleocharis species is very likely.

5 SEXUAL REPRODUCTION, PHOTOSYNTHETIC PATHWAYS AND BIOCHEMICAL LINKAGE

This insight came from extensive studies regarding their biochemistry, physiology, and cellular structural diversity (Pyankov et al., 1986, 1997, 1999; Butnik et al., 2001; Toderich et al., 2001; Zalenskii and Glagoleva, 1981). Space limits us to a brief summary here of findings about this surprising floral diversity in photosynthetic biochemistry, cellular anatomy, and gene expression. Some key findings include studies on their CO_2 biochemistry, with ^{14}C and $\delta^{13}C$, which show many C4 plants in addition to numerous C3 plants. Their Kranz-type C4 leaf cell anatomy diversity is unusual with several distinctive structural modifications (Zalenskii and Glagoleva, 1981; Pyankov et al., 1999, 2000).

Within C4 leaf biochemistry we know that a specific plant expresses dominance by only one of three known C4-acid decarboxylases (Burris and Black, 1976; Buchanan et al., 2000). In naturally-occurring Central Asian plants, PEP carboxykinase decarboxylating plants apparently are absent; while one decarboxylase, either NAD-malic enzyme or NADP-malic enzyme, is dominant in specific C4 plants (Pyankov et al., 1986). We have noted that there is a strong tendency for NAD-malic enzyme plants to dominate the vegetation growing in saline soils and for NADP-malic enzyme plants to dominate in non-saline soils.

Diversity in anatomy of assimilation organs and their photosynthetic pathway has been marked within representatives of genus Salsola. Two anatomical types, Salsoloid and Sympegmoid (Carolin et al., 1975; Butnik, 1979; Toderich et al., 2002; Freitag et al., 2002) occur in leaves and bracts of species of Salsola. It was also determined that some species with Salsaloid anatomy have NAD-ME C4 photosynthesis, whereas others have NADP-ME C4 subtype (Pyankov et al., 1997). Plants with Sympegmoid anatomy have C3-like C^{13}/C^{12}carbon discrimination values (Akhani et al., 1997; Pyankov et al., 1997). Variations also occur in structural and biochemical features in cotyledons (Pyankov et al., 1999). Two non-Kranz anatomy, Atriplicoid and Salsoloid, are found in Salsola cotyledons (Butnik, 1979). Finally, Kranz-type cotyledons and leaves may or may not contain a hypodermis. The result is a number of unique combinations of structural and biochemical photosynthetic types in leaves and cotyledons in species of Salsoleae. So, multiple origins of C4 photosynthesis as was described for the families of Poaceae, Cyperaceae, Asteraceae and Zygophylaceae appear likely within Chenopodiaceae and the diversity of photosynthetic types and anatomical structures in Salsoleae suggests a dynamic pattern of photosynthetic evolution within this single tribe.

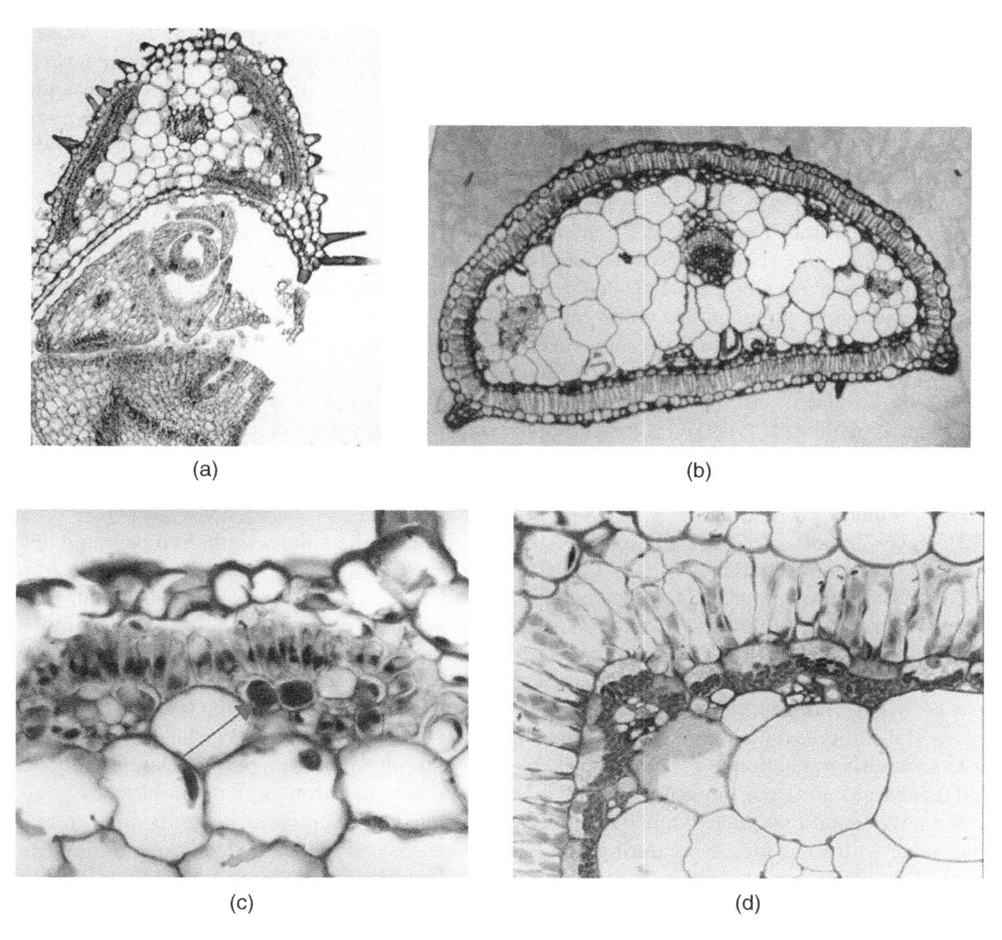

Figure 5. (a) The first stage of differentiation of sclerenchyma of *S. praecox*. (b) Salsaloid bracts anatomy and indefinite Kranz bundle sheath cells of *S. praecox* during budding stage. (c) Cross section through bracts of *S. praecox* during flowering process; Kranz bundle sheath cells is visible. (d) Cross section through bract of *S. praecox* during fruit maturation. Photosynthetic activity of reproductive organs of *Salsola praecox*.

From our phenological observations and experimental results, it seems very likely that structural polymorphism of floral organs and sexual reproduction system in some Asiatic Salsola species are coupled with the diversity of photosynthetic pathways and anatomy of the CO_2 assimilative organs. Great floral micromorphology and anatomical diversity of assimilative and floral like organs (bracts, bracteoles) were described for some annual Salsolas, contrasting by their growing habitats. It was found that *S. pestifer, S. praecox* and *S. paulsenii* are similar in photosynthesis types: C4-Sals (-H) both in leaves, cotyledons and bracts.

Differences were revealed in the anatomy of bracts. All Asiatic annual Salsola species of Section Salsola (Kali) have so-called Salsoloid (Carolin et al., 1975) or crown centric (Voznesenskaya and Gamaley, 1986) Kranz leaf and bracts anatomy. The anatomy of bracts in the different Asiatic species of Salsola we studied in relation to their photosynthetic activity (Figure 5 a, b, c, d). The first features of differentiation of bracts and bracteoles chlorenchyma cells are marked at the early stage of pollen sacs development, reaching a maximum during blooming stage. Cross sections of fruits perianth of many annual Salsola species during its maturity also show an insignificant development of chlorenchyma tissue. The similar situation was described for the species of Section Belanthera. In bracts or fruiting body of this type, chlorenchyma is represented by two layers of green cells positioned around the periphery of the organs: the outer layer is composed of palisade mezophyl cells and the inner layer is composed of palisade mezophyll cells and inner layer of bundle sheath cells. The main vascular bundle with greatly thick-walled in the centre, surrounded by the water-storage tissue, and only small peripheral bundles have contact with chlorenchyma. It is known that all species with Salsaloid Kranz anatomy in photosynthetic organs (irrespective of whether these are leaves, stems, cotyledons or bracts) have C4 type photosynthesis (Pyankov and Vakhrusheva, 1989; Pyankov et al., 1999, 2000). However, chlorenchyma of *S. ruthenica,* consisting of palisade and Kranz cells, is interrupted by longitudinal colenchymatic ridges.

S. arbusculiformis manifest a Sympegmoid leaf and bracts anatomy and non-Kranz bundle sheath cells (Voznesenskaya et al., 2001; Toderich, unpubl. data). Other species of Section Coccasalsola forming a unique plant functional group can be united by Salsoloid (with hypodermis both in leaves and reproductive organs) or a Crownary – centrical Kranz type of photosynthetic cell arrangement (Voznesenskaya and Gamaley, 1986). The Salsoloid type of Kranz assimilation tissues anatomy is always associated with the C4 syndrome (Carolin et al., 1975) that is well supported by, as represented in Table 1, C4-like $^{13}C/^{12}C$ carbon discrimination values in leaves, flowers and fruits with a range of −12.0−14.08. Such similarity of anatomical and biochemical features is well coordinated with developmental stability of reproductive system noted by us for *S. arbuscula, S. richteri* and *S. paletzkiana.*

However, plants of *S. arbusculiformis* from their natural habitats with Sympegmoid leaf and bracts anatomy maintained their C3$^{13}C/^{12}C$ carbon fractionation values in the range from −23.6−26.31 throughout their ontogeny, although significant variation was found within plant organs with 2.69‰ in flowers. In addition *S. arbusculiformis* is characterized by a set of primitive embryological features such as ana-campylotropous, crassinucellate, bitegmic ovule, autogamy (self pollination/fertilization system), narrow specialization of sexual reproductive system that may be an evidence of lower reproductive plant functional activities leads to the lower level of seed set, seed viability and their seed germination energy. It is also known that C3 is the primary type of photosynthesis in relation to C4. Apparently there is a strong connection between structural floral and fruits traits and their physiological and biochemical activity throughout their ontogeny.

We have found that photosynthetic activity of reproductive organs was insignificantly in budding stage with some increasing during flowering processes and gradually decreasing during fruit maturation. From our experimental results, it seems very likely that structural polymorphism of floral organs and sexual reproduction system in Asiatic shrubby Salsola species are coupled with the diversity of photosynthetic pathways and anatomy of the CO_2 assimilative organs. We have also found out that *S. arbusculiformis* manifest a Sympegmoid leaf and bracts anatomy with all structural features for C3-like species. Almost the same tendency was revealed in the flowers and fruits of two other taxonomically close related species of section Arbuscula as *S. montana* and *S. drobovii.* It was found also a positive correlation between structure and biochemistry of *S. arbusculiformis* that shows the bundle sheath cells were found to be Krantz-like with rather numerous chloroplasts, and the walls of these cells were thicker than those of the mezophyll cells. Since *S. arbusculiformis* has two-to-three layers of mezophyll cells, but a much more distinctive chlorenchymatous sheath than found in Sympegmoid species like S. *oreophilia* it was suggested it might be a C3-C4 intermediate photosynthesis (Pyankov et al., 1997; Voznesenskaya, 2001). However in comparing with structure of assimilation tissues and biochemistry of other C3-C4 intermediates there is a little evidence of C4 in *S. arbusculiformis.* The $^{13}C/^{12}C$ carbon discrimination values of *S. arbusculiformis* are less negative than in *S. oreophilia,* indicating that a C4 cycle may make a small contribution to photosynthesis under certain conditions in this species. *S. arbusculiformis* might be an intermediate in the path of evolution from C3 to C4 type of photosynthesis. *S. arbuscula, S. Richteri* and *S. paletzkiana* with some very similar morphological, palynological and embryological features can be united by Salsoloid (with hypodermis both in leaves and reproductive organs) or a Crownary-centrical Kranz type of photosynthetic cell arrangement into unique plant structural functional group. The Salsoloid type of Kranz assimilation tissues anatomy is always associated with the C4 syndrome (Carolin et al., 1975) that is well support by C4 like $^{13}C/^{12}C$ carbon discrimination values in leaves, flowers and fruits with a range of −12.00−14.08. Such similarity of anatomical and biochemical features is well coordinated with developmental stability of reproductive system noted by us for taxonomically close *S. arbuscula, S. richteri* and *S. paletzkiana, S. montana* and *S. drobovii* species of section Arbuscula.

However, plants of *S. arbusculiformis* from their natural habitats with Sympegmoid leaf and bracts anatomy maintained their C3$^{13}C/^{12}C$ carbon fractionation values in the range of −23.6−26.31 throughout their ontogeny, although significant variation was found within plant organs with 2.09‰ in flowers and fruits. It is also known that C3 is the primary type of photosynthesis in relation to C4. Primitive type of phtosynthesis and biochemistry in *S. arbusculiformis* are well correlated with the lower reproductive functional activity due to the presence of primitive embryological features, such as ana-campylotropous, crassinucellate, bitegmic ovule, autogamy (self pollination/fertilization system) and narrow specialization of sexual reproductive organs. Such structural-functional mechanisms of vegetative- reproductive systems in geographically restricted in distribution and reproductively isolated *S. arbusculiformis* might estimate reduction of seed set, seed biomass, as well as weight of seeds, seed viability and rate of their germination. Apparently there is a strong connection between structural floral and fruits traits and their physiological and biochemical activity throughout their ontogeny.

We suggested that the C4-syndrom evolved in the leaves and only later extended to the cotyledons and floral organs. Sex expression of flower' organs have been most marked for *S. ruthenica* (European species) that may be related with the habitats, where this species was recently introduced and naturalized. This is a good evidence of origin and diversification of Salsola from Asia to Europe. Analysis of karyotype revealed that these two species have the same number of chromosome 2n = 18 (Wojnicka-Poltorak et al., 2002).

Annual C4-Asiatic Salsolas, known in pasture economy of Uzbekistan as solyanki, differ from European ones by saltglands/trichomes morphology (shape of their head): mainly clavate or capitate and its density. Abundant papillae prickle hairs and secreted salts secretion between ridges on bracts/bracteoles surfaces of annual Asiatic Salsola are described. Frequently salt glands are globose or club-shaped and readily distinguishable from unicellular papillae and sharp-pointed prickles. These parameters could be used as discriminating characters between different ecological halophytes Salsolas groups. Variation in the indumentum density is believed to be mainly the effect of stress/desert environmental factors and/or even herbivores pressure.

Floral structure diversity (hypoginous disk, stigma, ovule, anthers), pollen grain morphology and fruit anatomy are clearly defined, suggesting their potential value as diagnostic characters concerning taxonomic and evolutionary linkages within Chenopodiaceae, particularly genus Salsola. Remarkable examples of intraspecific variation in pollen grain traits have been described for C/D ratio, thickness of exine, size and numbers of pores. These parameters being more conservative than other flowers and fruits traits were discovered to be highly specialized and support the evident cladistic grouping of some critical species within genus Salsola. Diversity in the anatomy of fruits reflects ways and character of adaptive coevolution of woody Salsola taxons and plays a more significant role in the species identification than other elements of floral organs. For instance at *S. richteri* and *S. paletzkiana* the direction of adaptive specialization to the xeric-arid conditions going towards the intensification of sclerification of fruiting perianth and increasing of size and number of cells layers of pericarp and even embryo tissues. The presence of pigments in the fruits covers, singular hydrocytic cells, partial myxospermy and development of membranous layer in the spermoderma intense the defending function against sun radiation. A fully development of embryo organs and differentiation of their tissues indicate the completely readiness of embryo of Salsola species to the germination. Seed dispersal is manifested by the development of large and wide wings; all elements of fruiting cover and embryo organs of studied species has adaptive value pigmentation, partial myxospermy, thickenings of external walls; membranous and aleironic layers in the spermoderma, intensification of succulence features as a results of well development of aerial parenchyma, abundant of reserve store nutritional substances, which stimulate the good defending of embryo from extreme deserts environments.

We found out that Asiatic Salsola species of section Arbuscula Coccosalsola section with both C4 and C3 photosynthesis types represents a unique example of the evolutionary convergence of ecological, structural, physiological and biochemical traits. The great range of variation, far more marked in ploidity of genome and fruit structures than in floral and pollen morphology explains the high phenotypic plasticity and good adaptation of *S. richteri* and *S. arbuscula* to various geographical and ecological desert habitats. Controversy, *S. paletzkiana* and *S. arbusculiformis* are characterized by narrow structural specialization of reproductive organs, partly seeds to germinate only on the sandy or stony gypsumferous soils that perhaps explained the strict local distribution of this species in the Central Asian Flora (Toderich et al., 2006b).

The embryological data of perennial Asiatic woody Salsola species from their natural habitats presented briefly in this paper lead us to the conclusion that *S. arbusculiformis* by reproductive biology is evidently not allied close to *S. arbuscula, S. richteri* and *S. paletzkiana*, despite these species form a taxonomic separate monotypic subsection Arbuscula in Coccosalsola section (Botschantzev, 1969). This supports a view that many arid plants being morphological and taxonomic closely related develop a specific plant reproductive functional system that ensures them to reproduce and survive in extreme desert conditions. Our anatomical, biochemical and physiological data in regard to the C3 and C4 pathways of photosynthesis in limits of Coccosalsola complex both in leaves and reproductive organs confirm that this delimitation is well founded. Based on morphological features, Akhani (1997) recently mentioned that section Coccosalsola could be most primitive group of Salsola. However high specialized centric type anatomy of mesophyll of cotyledons of *S. arbuscula, S. richteri* and *S. paletzkiana* indicate on their relatively younger phylogenical group. *S. arbusculiformis* manifest a unique combination of structural and biochemical phosynthetic types in leaves and cotyledons within genus Salsola (Pyankov et al., 1997; Voznesenskaya et al., 2001). In addition several aspects of the species' ecology indicate on reproductive assurance in this taxon as an adaptive explanation of selfing control for surviving under extreme desert condition. First, *S. arbusculiformis* typically occurs in small patchy geographical isolated populations or singular individuals, where the availability of outcross mates might be limited. Floral biology of *S. arbusculiformis* supports the hypothesis that mating between close relatives is expected to reduce the number or vigor plants. Second, embryological data showed that *S. arbusculiformis* produces simultaneously selfed, outcrossed and apomictic seeds in the absence of pollinators. This feature perhaps is a main reason of lower viability and germinability of seeds both under field and laboratory condition.

Based on whole plant morphological features, sexual reproduction and functional systems, as well as ecological habitats and geographical distribution of Asiatic Coccosalsola complex, we propose that *S. arbusculiformis* was primitive and primary with an origin in Central Asia.

We supposed that diversities in sexual reproduction mechanisms and CO_2 fixation pathways, e.g., in tree-like Salsola species, also are important factors regarding reproduction and survival under changing desert environments. Structural polymorphism of floral organs is coupled with a diversity of biochemical pathways and anatomy of CO_2 assimilative organs as illustrated with two woody perennial species C4 species with NADP-ME photosynthesis, *S. arbuscula*, and Salsoloid type of Kranz-anatomy both in leaves and reproductive organs form a unique "functional reproductive plant group" that is very resistant to the severe stresses in these arid habitats. Other species, e.g., *S. arbusculiformis*, have both sexual and apomixis systems of reproductive linked with a Sympegmoid-type cell structure and C3-like $\delta^{13}C$ values both in leaves and reproductive organs; which likely relate to their ancient origin from which Salsola species with the C4-syndrome were derived recently. The coexistence of two reproduction fitness components (sexual and apomictic) in *S. arbusculiformis* are structural organ adaptations which may enable this plant to survive and reproduce fruits in rocky niches in the extreme isolation of these continental deserts. However, these highly specialized reproductive systems result in a low seed set plus poor germination and viability. Likely these traits result in *S. arbusculiformis* currently being a very rare plant in the stressed deserts of Central Asia. Based on whole plant morphological features, sexual reproduction and functional systems, as well as ecological habitats and geographical distribution of Asiatic Coccosalsola complex we suppose that *S. arbusculiformis* was primitive and primary with an origin in Central Asia.

6 REDUNDANCY IN ENVIRONMENTAL ADAPTATIONS BY PLANTS

When we consider global vegetation it is clear that green canopies of forests, grassland, and often crops cover much of our lands and act as continuous active green biological interfaces between the atmosphere and other terrestrial life. However the desert land-atmosphere biological interface contains only scattered vegetation, plus unstable sands, rocks, soils, and surface microbes that makes spatial and niche management practices regarding biodiversity-space-harvesting-production relationships in desert ecosystems especially difficult. Also natural and human driven activities such as hooved grazing animals, firewood collection, shifting sands, salinization, industrial actions, pollutants, etc. greatly modify the interface of this difficult green desert lands-atmosphere mosaic.

Plants as autotrophic sessile organisms face a continually changing environments. Even internally higher plants have micro-environmental problems to solve within numerous organs and cell types which continually face changing supplies of, e.g., nutrient ions, sugars, amino acids, gases, light and water. Some major external environmental problems that plants must solve are: (I) their biophysical soil and air environments are continually changing, far beyond normal daily environmental changes; (II) their biological environments of microbes, herbivores, and other creatures change constantly; and (III) humans particularly move and destroy plants and add both beneficial materials and toxic pollutants to plant environments. For example in the Kyzylkum Desert some plants, which we characterize as metallohalophytes (Toderich et al., 2004, 2005a,b, 2006a), are found growing well in either natural or contaminated soils containing salt and metals. But the Kyzylkum vegetation contains only a restricted number of species, mainly from Salsola (both annuals and perennial), tamarix, graminous, which have metal/salts removal abilities and also can survive and reproduce under these contaminated environments. Under such condition some successful species produce large quantities of small, easily dispersed seeds, hence facilitating colonization. From a host of biochemical and physiological studies, it has become clear that plants have multiple often redundant pathways and mechanisms that accomplish the same function or goal. These genetically built-in mechanisms for redundancy in numerous plant functions act as fail-safe mechanisms. Redundancy apparently gives sessile plants two major capabilities: (I) their normal developmental ability to form diverse functions in different types of organs, tissues and cells; and (II) a very powerful means to adapt the functions of these structures to cope with whatever happens in their biophysical and biological environments. The following are some well known plant examples of fail-safe redundancy-type activities acting as strongly functional adaptations to various environmental stresses. As external environmental CO_2 levels vary, the internal CO_2 levels in green photosynthetic tissues can be modified to provide this essential nutrient. Currently we know that specific plants and green organs can express one of 9 or 10 separate CO_2 fixation pathways. Figure 6 represents an example of two plant responses to atmospheric CO_2 changes, showing that C4 tissues are well adapted to air CO_2 levels; while air CO_2 is rate limiting for C3 photosynthesis. Within the interior of green tissues C4 plants can increase their micro-CO_2 levels in bundle sheath cells 50 to 60 times the level of CO_2 in equilibrium with air (Figure 6); or green CAM plants can raise their CO_2 levels during the day to 2 to 4% CO_2, in contrast to air at 0.035% CO_2.

A host of redundancy network type examples of plants continually coping with environmental changes now are well known in plants including: the simultaneous presence of several pigment networks for harvesting light across the entire visible electromagnetic spectrum during photosynthesis; huge numbers of either large or small diameter cells acting as conduits for transporting H_2O; plant roots commonly have both short-lived root hairs plus long-lived mutualistic mycorrhizal fungi for nutrient and water uptake in soils; metabolically, multiple enzymes are present in cells at various steps in glycolysis forming a carbon metabolism network that, e.g., with an immediate environmental stress, a plant can direct large quantities of carbon into defense responses; a number, perhaps a dozen, light sensing phytochromes and phototropins, even chimaera's of these, that trigger different plant responses as light changes; large numbers of plastids and mitochondria in a cell to insure energy capture and supply; extensive gene duplications; a variety of post-transcriptional RNA modifications for either silencing or synthesizing various proteins; and extensive protein post-translational modifications for the utilization, activation or inactivation of specific proteins.

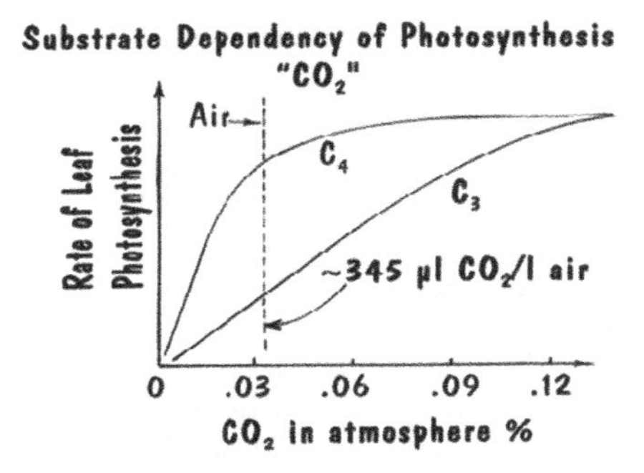

Figure 6. Separate responses of C3 and C4 photosynthesis plants to changing the nutritional levels of CO_2 in air.

A most critical, indeed a universal trait of plants is the ability to begin meristematic development anew, hence to change completely the primary function of a cell or group of cells; even for a plant to regrow completely, reproduce, and survive. Indeed the constant ability to initiate new meristematic activity is an essential and central functional adaptation in plants when environmental stresses occur and/or decline. Each of these redundancies have strong plant biology roles in Central Asia because of widespread environmental problems such as off-seasonal wet and dry soils, salinization, overgrazing herbivory, and pollutants (Toderich et al., 2003).

In addition, our last years of investigation revealed that native desert plants tend to accumulate the highest ion concentrations (primarily sodium, chlorates and oxalates) in the epidermal and subepidermal tissues, including various glandular structures on leaves, bracts/bracteoles and perianth segments. *Tamarix hispida* as a C3 salt exluder hyperhalophyte has a remarkably high Fe and Co levels in the aerial dry matter biomass that it deserves being described as a hyperaccumulator plant. A significant ability for heavy-metal removal has been noted for *Artemisia diffusa, A. turanica, Alhagi pseudoalhagi, Peganum harmala, Haloxylon aphyllum,* as well as annual and perennial species of the genus Salsola and some grasses (Toderich et al., 2002, 2005a, 2006a,b).

It was also noted that the occurrence of calcium oxalate crystals was almost absent in root and stems. An abundance of these crystals was described in the tissue of the seed coats of many xero- and gemi-halophytes (Figure 7 a, b).

Sclerenchyma, vascular and pigmental layers, as is shown in Figure 5 a, c for annual species of section Salsola, consisting of small-sized cubic cells, that partially filled with inclusions of unknown origin. Epidermis can be rather thin-walled (internal walls), but, as a rule, at *S. dendroides, S. incanescens* (section Caroxylon) and *S. iberica* (section Salsola) outer walls are thickened are selected by uniform thickening of all walls epidermis. The evolution of parenchyma goes from radially elongated large thin-walled cells to small sized cells thick-walled izo-diametric shape cells. In the same direction the cells of sclerenchyma from very large radial elongated to small sized izodiametric form are changed. Two directions of adaptation are detected for species of genus Salsola. The large cells parenchyma and sclerenchyma are positively correlated with the development of strong pubescence, whereas at almost not pubescent species (exception short papillae for a few species) the small sized with thickened walls of parenchyma is observed. Increasing of sclerification, availability of pigments and tracheids like cells holding moisture, abundance of crystals in the perianth tissues also promote the protection of embryo from unfavorable conditions.

In addition some naturally highly adapted metallohalophytes develop a cellular mechanism to partition toxic salts into vacuoles or to exclude salt at the root zone so it does not impact on cell metabolism and division, i.e., a high concentration of various ions can accumulate in the vacuoles of bladder-trichomes terminal cells that are frequently developed on the adaxial surface of epidermal cells of leaves or bract/bracteoles (Figure 7 a, b).

The prominent levels of sclerification of perianth segments combined with thickening of pericarp and spermoderma epidermis bearing papillae-shaped proturbences (as in the case indicated in

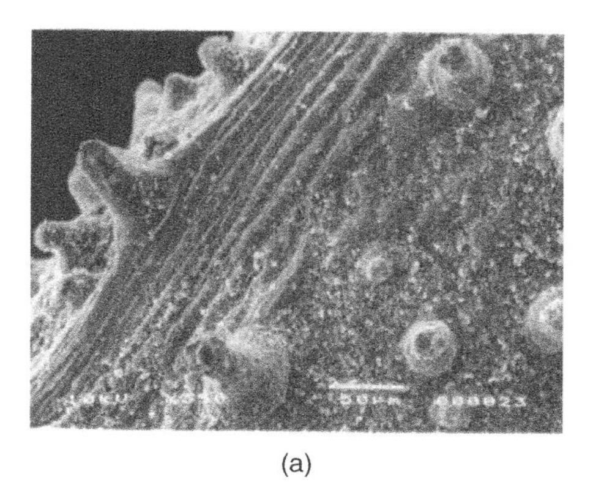

(a)

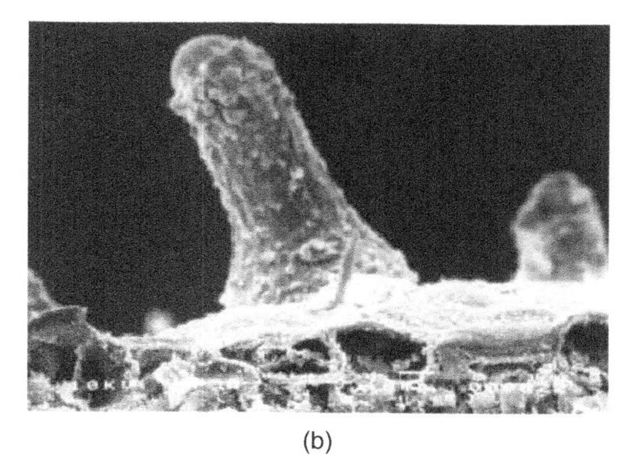

(b)

Figure 7. Scanning micrographs: exclusion effect of contaminants through salt active trichomes/glands that are abundantly developed at the adaxial epidermis of annual Salsola species, growing under contaminated conditions of Kyzylkum sandy desert.

Figure 8 for *Salsola paulsenii*) are related to the defending of embryo against entrance of toxic elements. Diversities in sexual reproduction mechanisms and CO_2 fixation pathways, as were described in this paper for tree-like Salsola species, also are important factors regarding reproduction and survival under saline and technogenic contaminated desert environments.

Most essential plant nutrients come from soil-plant interactions via root and microbial contacts; simultaneously essential nutrient uptake must cope with the presence of any toxicants and non-essential elements in soils. The roles of fungi, bacteria, and other organisms as they interact with plants are crucial but, also due to space limits and the limited information about desert plant/soil organism's interactions, these are not detailed here. Biological lipid bilayer membranes are essentially impermeable to ions, sugars, and polar molecules; hence channels, pumps, diffusion, solution, and mass flow are used to cross biological membranes. For example, the uptake of mineral ions from soils by plant roots occurs through protein-built channels in a biphasic fashion, first with a strong high affinity active carrier mechanism, followed by a slower diffusion uptake. Such active transport channels and pumps are powered, usually by ATP, and may involve an active co-transport with other ions or an exchange with other ions. Figure 9 illustrates processes and ion competition plus their partial independence.

Normally the uptake of K, versus K concentration, is biphasic similar to the Rb curve. Rb is a pollutant non-essential element; K competes with the active uptake process but does not influence diffusion uptake. The same biphasic type uptake occurs with other essential plant mineral nutrients, such as NO_3^-, SO_4^{2-} and Fe; indeed biphasic nutrient uptake also occurs inside of plants with the apoplastic uptake of sugars, ions, and amino acids by cells. Diffusion is a mass action driven process that likely is especially important in metal and salt polluted soils in Central Asian deserts. To remediate desertification the unique plant soil interactions in Central Asian deserts must be known to present reliable restoration management procedures.

7 SOIL ORGANIC CARBON AND SOIL NUTRIENT FLUXES OF CENTRAL ASIA ECOSYSTEMS

The higher grazing pressure of desert/semidesert rangelands and the intensive cultivation of marginal and virgin lands in Central Asia (as were a case with grassland Mirzachul steppe in Uzbekistan and large areas of Kazakhstan steppe) resulted in a substantial declining in soil organic matter (Nasyrov et al., 2004).

During the more than 50 years of cotton monoculture leads to decreasing of in the amount of organic carbon. The same trend, though less pronounced, was observed in the case of cotton in

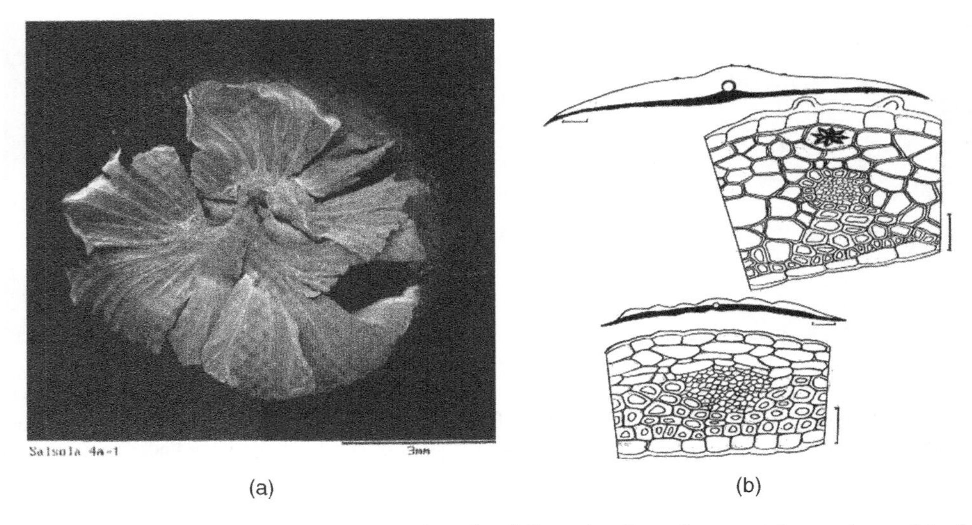

Figure 8. Fruit morphology and anatomy of tepals of *S. paulsenii*; oxalate crystals are observed in the sub-epidermal cells; special layer of tracheids like holding moisture cells is also developed.

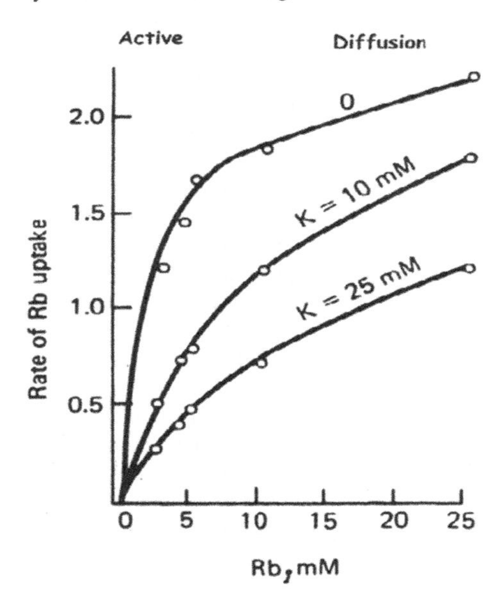

Figure 9. The biphasic uptake of a non-essential element, Rb, and the competition by an essential plant element, K, during the active uptake phase in barley roots. Note the uptake rate of Rb by diffusion is little influenced by K levels; but K very effectively competes with Rb at the active uptake channel.

monoculture with fertilizers. Even crop rotation (3 years of alfalfa and 7 years of cotton) could not maintain soil organic carbon at the initial level (Figure 10).

Desert plant communities with predominance of perennial fodder species in arid zones of Central Asia are effective and singular key for maintaining or increasing soil organic matter at a satisfactory level. The shrubs vegetation in the sandy desert in Karakum (Turkmenistan) and Kyzylkum (Uzbekisan) had the longest growing season at 173 days, followed by 140 days for the arid grassland steppes of Mirzachuli (Kazakhstan) and Kazakhstan, while the Karnabchuli Artemisia ephemeroid semidesert has the shortest vegetation cycle with 111–126 days.

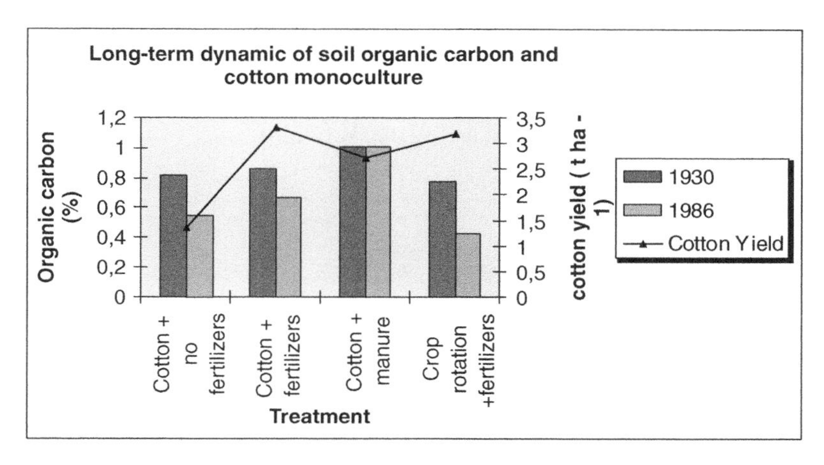

Figure 10. Cotton yield and dynamic soil organic carbon under different cropping system.

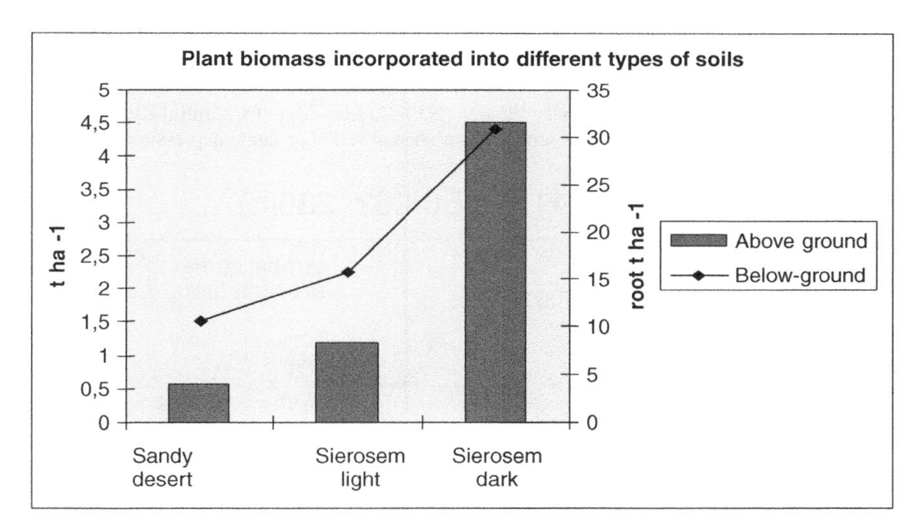

Figure 11. Dynamic of above ground and below ground plant biomass under different types of soils.

As it's shown in Figure 11, the magnitude of biomass both above-ground and below ground at the peak of plant growing seasons was highest on the sierosem dark soils in the grassland steppe and foothills areas, intermediate on light sierosem and gray brown as is the case of Karnabchuli Artemisia ephemeroid semidesert and lowest on the sandy desert soils with predominance of psammophytic woody shrubs type of vegetation of Karakum and Kyzylkum deserts.

A strong correlation between plant growing season, biomass accumulation (both above ground and below ground) and time course of daily net ecosystem CO_2 exchange (NEE, g CO_2 m^{-2} d^{-1}). Total flux of CO_2 for the period of 1998–2000 varied from 659 ± 67 g CO_2 m^{-2} for grassland steppe; 347 ± 178 g CO_2 m^{-2} for Artemisia ephemeroid pastures to 151 ± 121 g CO_2 m^{-2} under condition of shrubby sandy desert (Saliendra et al., 2004). This phenomenon indicates that huge open spaces of rangelands of Central Asia are potential sinks for atmospheric CO_2. Long-term precipitation data from nearby meteorological stations showed that mean annual precipitation was in the limits of 324 at Kazakhstan steppe >237, at Artemisia ephemeroid pasture >113 mm yr^{-1}. The decreasing trend of temperature and annual precipitation mentioned by Saliendra et al. (2004)

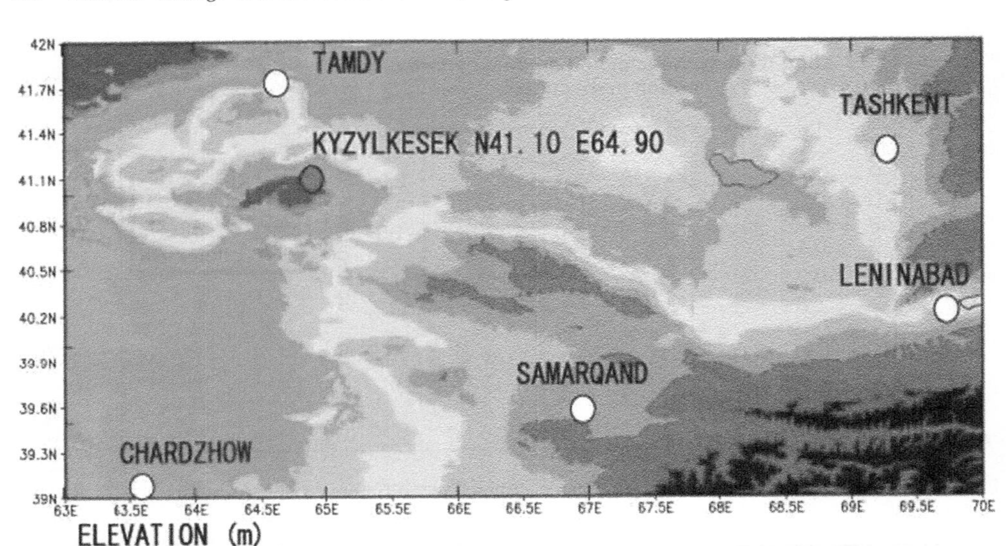

Figure 12. Digital Elevation Map of Kyzylkesek area (N39-42, E63-70) at the Central Kyzylkum – that we have chosen as a target area; this areas represents a basin-shaped valley or saline depression.

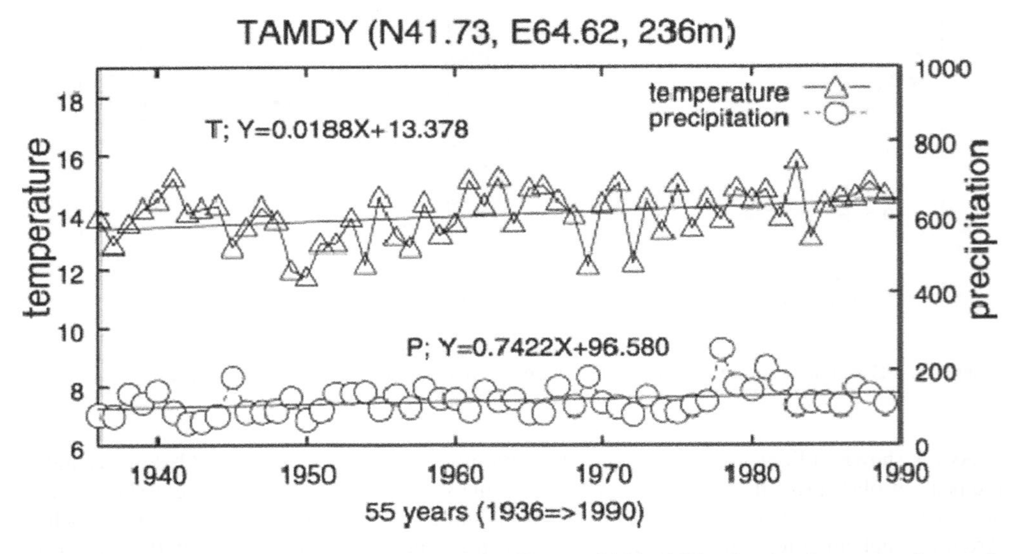

Figure 13. Long term trend of Tamdy station data for 55 years (1936–1990); triangles shows annual mean air temperature and circles shows annual precipitation. Tamdy is the main meteorological station in the Central Kyzylkum of Uzbekistan.

and noted by our preliminarily data for Tamdy (Figures 12 and 13) (shrubby sandy desert vegetation of Central Kyzylkum) correlated well with the decreasing pattern of total NEE for the growing season for all three typical types of rangelands in Central Asian dryland terrestrial ecosystems.

Air temperature and precipitation, as it's shown in Figure 13, have increasing trend. Air temperature increasing rate per 1 year is 0.0188 degree in Celsius (1.034 degree for 55 years). Precipitation increasing rate per year is 0.74 mm (40.82 mm for 55 years).

However the above-mentioned data were obtained for the relatively pristine (virgin) deserts ecosystems. At present time, however, continuously degradation of rangelands as result of overgrazing, up-rooting of trees and shrubs for fuel, extension of irrigated agriculture on pastures of best quality leads to the decreasing of their capacity for fixation of atmospheric CO_2 during the plant growing season, as well as sequestration of carbon in soils. Increasing soil organic matter content represents a key component for more productive rangelands and sustainable agricultural ecosystems development.

8 NEW DISCOVERIES RELATIVE TO DESERT PLANTS

Recently several new biological features of desert plants have been reported that help give us better insights into plant growth and potential management possibilities. First, the roots of desert plants can be populated by previously unrecognized large groups of fungi that I.) are adapted to symbiosis with desert plants and II.) are tolerant of high temperatures. Recently a great fungal diversity was found on a single grass species, with about 49 fungal phylotypes (Vandenkoornhuyse et al., 2002). A further development regarding symbiotic fungi was the detection of a thermal tolerant $\leq$40°C, fungus (Redman et al., 2002). Importantly, the fungus conferred heat resistance to the plant host and both the plant and the fungus were required for high temperature growth (as found in deserts). Unfortunately the fungal diversity on Central Asian desert plants is unknown along with desert microbial and plant thermal tolerances. Secondly, new pathways of photosynthetic CO_2 assimilation have been described in desert plants and in saline-living diatoms. Two Asian plants, *Borszczowia aralocaspica* and *Bienertia cycloptera* in Chenopodiaceae, have been described as conducting a new type of C4 photosynthesis within a single green cell (Freitag et al., 2000; Voznesenskaya et al., 2001; Freitag et al., 2002); i.e., without the well known two green cell types needed by most C4 photosynthesis plants. The physiological and environmental adaptations or advantages in deserts of these new types of plant biochemistry are unknown. Another new discovery is for single cell saline-growing diatoms to first fix bicarbonate into C4 acids, and then to decarboxylate them to release high levels of CO_2 for increasing Rubisco activity and C3 photosynthesis (Reinfelder et al., 2000). This pathway would facilitate photosynthesis in low CO_2 environments such as Asian desert salt marshes and saline lakes.

An unusual night fixation of CO_2, via CAM has been detected in some desert plants that allow them to survive in cold, higher altitude, rocky desert environments. We found that CAM occurred in *Orostachys spinosa* in Mongolia where it is used as a pastoral and herbal medicine plant. Subsequently we learned that Orostachys species are widely adapted for survival and growth in Asian deserts and grow naturally in Russia, Tibet, and much of Central Asia. Orostachys seems to be the coldest adapted CAM plant known, but its special environmental adaptation mechanisms remain to be elucidated.

9 CONCLUSIONS AND DISCUSSIONS

The Central Asian desert vegetation has a unique composition apparently as an evolutionary result of their extremely harsh and dynamic environments and, in recent times, from strongly human driven pressures. These combined pressures have resulted in very difficult biological growth environments. The Chenopodiaceae family dominates their quite diverse flora, specifically by perennials. Surprisingly essentially absent in Central Asian deserts is a large worldwide group of desert plants, called Crassulacean acid metabolism (CAM) plants. Only one small genus Orostachys has been found with CAM. Also this is the Northern most edge of growth for C4 photosynthesis plants that dominate the desert flora. Several new aspects of desert photosynthesis, symbiosis and reproduction have been detected recently, which give clues about the unusual biology needed by plants to successfully grow and produce seeds under harsh desert environments. Some of adaptive features are well presented on seed production, photosynthesis and soil nutrient uptake mechanisms.

The Central Asian Flora, mostly from Uzbekistan and Mongolia was studied for C4 photosynthesis species, their natural geographical occurrence plus their carbon assimilation, structural, and biochemical diversity in relation to climatic characteristics. Mainly families' spectrum of desert flora of mentioned areas was screened for C3/C4 plants by using $^{13}C/^{12}C$ isotope fractionation, determining the early products of $^{14}CO_2$ fixation, microscopy of leaf mesophyll cell anatomy, floral structure and from reported literature data. C4 species being dominated in the cover vegetation of Central Asian Deserts were found among 8 families: Amaranthaceae, Chenopodiaceae, Euphorbiaceae, Molluginaceae, Poaceae, Polygonaceae, Portulacaceae and Zygophyllaceae. Most of the C4 species are in 3 families: Chenopodiceae, Poaceae and Polygonaceae (mostly species of genus Calligonum). C4 Chenopodiaceae species comprise 45% of the total chenopods and are very important ecologically in saline areas and in cold arid deserts likes whole Irano-Turanian lowlands and northern Mongoilian desert ecosystems. C4 grasses include about 10% of the total Poaceae species and these species naturally concentrate in foothills and steppe zones. Naturalized Kranz-grasses of genera Setaria, Aeluropus, Bromus,Eremopyrum, Agropyron, Echinochloa, Eragrostis, Panicum, Pennisetum and Chlori, which occurs in almost all of Irano-Turanian and Mongolias botanical-geographical regions were commonly grown in disturbed areas and in desert oasises. The relationships between the occurrence of C4 plants and their major climatic influences for the main botanical-geographical regions of Central Asian desert showed that the occurrence of C4 species increases with the decreasing of geographical latitude and along the warming North to South gradient; however grasses and chenopods differ in their responses to climate. The abundance of Chenopodiaceae species was closely correlated with aridity index; but the distribution of the C4 grasses was more dependent of temperature. Also, we found a unique distribution of different C4 Chenopodiaceae structural and biochemical subtypes along the aridity gradient. NADP-ME tree-like species with a Salsoloid type of Kranz-anatomy, such as *Haloxylon ammodendron, H. aphyllum, Iljinia regelii, Halothamnus subaphyllus* plus shrubby perennial species of Salsola, (except *Salsola arbusculiformis*) and Anabasis species were the most resistant plants to ecological stress and habitat in extra arid conditions with less than 100–120 mm of annual precipitation. Most of the annual C4 chenopods are halophytes, succulent, and grow well in saline and arid environments in steppe and sandy desert areas. The relative abundance of C3 succulent desert species also increased along the aridity gradient. This was the case of C3 Tamarix species from Tamaricaceae that showed strong ranges of variability of $^{13}C/^{12}C$ value (23.0–27.6). Native C4 grasses were mainly annual and perennial species from the Cynodonteae tribe with NAD-ME and PEP-CK photosynthesis types. Their habitats occurs across of Gobi (Mongolia), Karakum (Turkmenia) & Kyzylkum (Uzbekistan) Deserts, although are most common in steppe zones where they can play a dominant role in grazing ecosystems.

A concept of plant functional reproduction system (PFRS) based on common structural floral, fruits and functional (photosynthetic pathways and biochemistry) features of perennial shrubby and annual species of different taxonomic groups of plants in their desert natural habitats is developed. Our analysis of C4 plant composition in different botanical-geographical regions of Central Asian deserts showed that species with a similar type life forms, reproductive biology, mesophyll anatomy, CO_2 biochemistry are usually characterized by a similar habitation and belong to a similar taxonomical unit. Such similar structural-functional plant groups amongst C4 chenopods have the largest spatial distribution in different areas of Turanian deserts with a much smaller number of species for Mongolia. Shrubs with salsoloid Kranz-anatomy from Chenopodiaceae (Haloxylon, Ilinia, Halothamnus, *Salsola arbuscula, S. montana, S. drobovii, S. paletzkiana, S. richteri*) and Polygonaceae (mostly species of genus Calligonum) occurs in the sandy desert, i.e. the most hot hyperarid deserts in the whole Central Asia. Annuals such as Salsola and derivative genera (Micropeplis, Halogeton, Ghamanthus, Haloharis) with NADP-ME biochemistry are cold-resistant and are distributed in broadest climatic range, including boreal regions. Some of them are weedy plants and distributed worldwide. At the same time, the species of this group are not abundant in true deserts and do not habitat in saline conditions. Annual chenopods with salsoloid and suaedoid mesophyll structure and NAD-ME biochemical type (Climacoptera, Petrosimonia, Suaeda) commonly habitat is saline environment. Plants with kochioid type of Kranz-anatomy all exhibit NADP-ME biochemistry and can occur in a broad range of climatic conditions; but their relative abundance is highest in steppe and semi-desert conditions. In our mind, we have a mosaic distribution of functional plant groups along the main ecological gradients, i.e. temperature, drought and salinity. The correspondence between ecological and structural-biochemical features of these plants is a result of plant evolution and natural selection.

Our studies do indeed illustrate two quite contrasting plant functional groups with both groups involving plants with C4 photosynthesis of various life forms. First, it is the vegetation of the extremely arid, cold and hot Kyzylkum and Karakum Deserts, where the C4 Chenopodiaceae species dominate the vegetation; indeed their shrubby to tree life forms form a forest like woodland. These might be species of Haloxylon, Iljinia, Anabasis, Salsola, Calligonum are either NADP-ME or NAD-ME C4 plants with salsoloid assimilation organ anatomy and have deep rooted tree and shrub life forms combined with succulence of assimilation organs and show strong correlation with aridity. This same plant structural-functional group dominated by C4 Chenopodiaceae is found, with some variations in floral richness, in the desert of Western China, Central Asia, down in Arabian Peninsula, Near East, and North Africa being insignificantly distributed in Europe and American continent.

In contrast we also found in Central Asia the well known appearance of C4 annuals in annually disturbed areas in agricultural crop lands, and around oasis as weedy cosmopolitian species (Black et al., 1968). These species of Amaranthus, Halogeton, Chenopodium, Micropeplis, Salsola, Ehinochloa, Setaria, and grasses of the Cynodonteae tribe, usually have distinct types of grass Kranz or salsoloid anatomy, NADP-ME or NAD-ME, and have a quick growing annual life form and show little correlation with climatic factors. This C4 plant functional groups occurs world wide on regularly distributed lands up to 60–67° North latitude.

By means of SEM analyses it was defined that structure of perianth segments reflects ways and character of adaptive coevolution of these taxons and plays a more significant role in the species identification than other elements of floral organs. The structural polymorphism of floral organs is coupled with the diversity of pathway and anatomy of the CO_2 assimilative organs. We suppose that sexual reproduction mechanism and CO_2 fixation pathway diversities of tree-like Salsola species, detected in this study, are important factors supported by their successful reproduction under desert harsh environments. Species with similar floral developmental features, C4-MADP-ME photosynthesis and Salsoloid type of Kranz-anatomy both in leaves and reproductive organs forming a unique plant functional reproductive group were widely distributed and most resistant plants to various ecological arid habitats. Ana-campylotropous, crassinucellate and butegmic ovule, autogamy (self pollination/fertilization system), narrow specialization of sexual reproductive structures linked with Sympegmoid – leaf anatomy and C3 – like $\delta^{13}C$ values both in leaves and reproductive organs testify the ancient origin of *S. arbusculiformis*, from which Salsola species with C4 – syndrome must have derived recently. Evolutionary changes in the mating system of *S. arbusculiformis* under extreme desert environments through advantages of self-fertilization, reproductive assurance (when pollinators are scarce), geographical restricted distribution (growing preferentially on rocks, crushed stones or other ancient substrates) and reproductive isolation between sympatric species support a strong tendency to inbreeding depression in this taxon. The coexistence of two fitness components (sexual and apomictic) in *S. arbusculiformis* we estimate as a structural adaptive modifications of reproductive organs enable plants to survive and reproduce fruits under extreme continental desert conditions.

Complex ecosystems populations, as in the deserts of Central Asia, can be restored over time by utilizing existing species and by adding new uniquely adapted or transformed species. The stable recovery of ecosystem functions can be considered best from the viewpoint of development over time. Note that includes human activity interventions; even recognizing that the first priorities of local citizens living in deserts are concerns about producing food, animals, and useful home-type products for themselves. Naturally, they are not too concerned with topics such as biodiversity, plant adaptations, or science and technology. For example, at numerous Central Asian desert sites the soils are polluted with metals or toxic organics that plants can cope with, indeed some plants can remove these toxicants to assist in land restoration! Phytoremediation technology is considered a potentially valuable technique for dealing with heavy metals, which are typically the most difficult pollutants to remove from soils. The use of metallohalophytes from the Central Asian flora to reclaim soils could represent both a practical and economically viable strategy. Today there is much hope that plants can be transformed to remove specific pollutants for land cleaning and a host of other plant improvements. Even though the scientific technology for molecularly transforming plants is very well established, unfortunately plants that are well adapted to desert environments have not yet been transformed. Plant transformation knowledge needs to be applied immediately to the special needs of desert-adapted plants in Central Asia. Since these deserts have been exploited, polluted, and are regularly overgrazed, many plants well adapted to natural Central Asian deserts have nearly disappeared and are being very difficult to locate. This shortage of high quality plants greatly hinders work to mass propagate, to restore, and maintain desert plant ecosystems.

By combining our accumulated knowledge of biodiversity, reproductive biology, biochemistry, physiology, and environmental adaptations, recommendations should be developed for more effectively and sustainable managing the deserts of Central Asia; even in the presence of strong anthropogenic pressures. The elaboration of new strategies of proper management and re-vegetation of degraded rangelands and salt-affected contaminated desert pastures should be a prime priority for all Central Asian countries.

REFERENCES

Akhani, H., P. Trimborn, and H. Ziegler. 1997. Photosynthetic pathways in Chenopodiaceae from Africa, Asia and Europe with their ecological, phytogeographical and taxonomical importance. Plant Systematics and Evolution 206:187–221.

Bender, M.M., I. Rouhani, H.M. Vines, and C.C. Black, Jr. 1973. $^{13}C/^{12}C$ ratio changes in Crassulaceae acid metabolism. Plant Physiology 52:427–430.

Berry, J.A. 1993. Global distribution of C4 plants. Carnegie Institution Year Book 92:66–69.

Black, C.C. 1973. Photosynthetic carbon fixation in relation to net CO_2 uptake. Annual Review Plant Physiology 24:253–286.

Black, C.C., T.M. Chen, and R.H. Brown. 1968. A biochemical basis for plant competition. Weed Sci. 17:338–344.

Black, C.C. and C.B. Osmond. 2003. Crassulacean Acid Metabolism Photosynthesis: Working the night shift. Photosynthesis Research 76:329–341.

Botschantzev, V.P. 1969. The genus Salsola: a concise history of its development and dispersal. Botanicheskii Zhurnal 54:989–1001. (in Russian).

Bozolla, J.J. and L.D. Russell. 2002. Electron Microscopy: principles and techniques for biologists, Second edition, Boston, USA. 644 p.

Butnik, A.A., O.A. Ashurmetov, R.N. Nygmanova, and S.A. Paizieva. 2001. Ecological anatomy of desert plants of Middle Asia. Vol. 2. Subshrubs, subshrublets, Tashkent: Fan AS Ruz. 132 p.

Burris, R.H. and C.C. Black. 1976. CO_2 Metabolism and Plant Productivity. Univ. Park Press, Baltimore, MD. 431 p.

Buchanan, B.B., W. Gruissem and R.L. Jones. 2000. Biochemistry and Molecular Biology of Plants. Amer. Soc. of Plant Physiol., Rockville, MD. 1367 p.

Carolin, R.C., S.W.L. Jacobs, and M. Vesk. 1975. Leaf structure in Chenopodiaceae. Botanishe Jahrblcher fur Systematishe Pflanzengeschichte und Pflanzengeographie 95:226–255.

Carolin, R.C., S.W.L. Jacobs, and M. Vesk. 1978. Kranz cells and mesophyll in Chenopodiaceae. Australian Journal Plant Physiology 26:683–698.

Dudik, N.M. 1971. To the methodology of compiling of glossary based on fruits and seeds. p. 20–21. Biological Basis of Improvement of Seed Production and Seed Quality of Introduced Plants. Kiev, Naukova Dumka.

Fisher, D.D., H.J. Schenk, J.A. Thorson, and W.R. Ferren, Jr. 1997. Leaf anatomy and subgeneric affiliations of C3 and C4 species of Suaeda (Chenopodiaceae) in North America. American Journal of Botany 84: 1198–1210.

Freitag, H. and W. Stichler. 2000. A remarkable new leaf type with unusual photosynthetic tissue in a Central Asiatic genus of Chenopodiaceae. Plant Biol. 2:154–160.

Freitag, H. and W. Stichler. 2002. Bienertia cycloptera Bunge ex Boiss., Chenopodiaceae, another C4 plant without Kranz tissues. Plant Biol. 4:121–132.

Gamaley, Y.V. 1985. The variations of the Kranz-anatomy in Goby and Karakum deserts plant. Botanicheskii Zhurnal 70:1302–1314. (in Russian).

Gameley, Y.V. and T. Shirevdamba. 1988. The structure of plants of Trans-Altai Gobi. p. 44–106. In: Y.V. Gamalei, P.D. Gunin, R.V. Kamelin and N.N. Slemnev (eds.), Deserts of Trans-Altai Gobi: Characteristic of Dominant Plants. Nauka, Leningrad.

Grubov, V.I. 1982. Key to the vascular plants of Mongolia. Nauka, Leningrad. In Russian.

Gubanov, I.A. 1996. Conspectus of flora of Outer Mongolia (vascular plants). Valang, Moscow. In Russian.

Gintzburger, G., K.N. Toderich, B.K. Mardonov, and M.M. Makhmudov. 2003. Rangelands of of the arid and semi-arid zones in Uzbekistan. Centre de Cooperation Internationale en Resherche Agronomique pour le Development (CIRAD) , Monpellier, France. 498 p.

Gintzburger, G, S. Saidi and V. Soti. 2004b. Ravnina Rangeland current vegetation conditions and utilization. Final DARCA Report INCO-COPERNICUS. Centre de Cooperation Internationale en Resherche Agronomique pour le Development (CIRAD), Monpellier, France. 202 p.

Gintzburger, G., H.N. Le Houeoru, and K.N. Toderich. 2005a. The steppes of Middle Asia: Post-1991 Agricultural and Rangeland Adjustment. J. Arid Land Research and Management 19:215–239.

Gunin, P.D., E.A. Vostokova, N.I. Dorofeyuk, P.E. Tarasov, and C.C. Black. 1999. Vegetation Dynamics of Mongolia. Kluwer Academic Publishers. 233 p.

Hatteresley, H.A. and L. Watson. 1992. Diversification of photosynthesis. p. 38–116. In G.P.Chapman (ed.), Grass Evolution and Domestication. Cambridge University Press, Cambridge.

Iljin, M.M. 1936. Chenopodiaceae Less. p. 4–272. In: V.L. Komarov and B.K. Shishkin (eds.) Flora of the USSR. Vol.VI, Centrospermae (English translation, 1970, by N.Landau). Israel Program for Scientific Translation, Jerusalem.

Jigiidsuren, S. and D.A. Johnson. 2003. Forage plants in Mongolia. RIAH., Ulanbaator, Mongolia.

Jones, C.A. 1985. C4 Grasses and Cereals. John Wiley & Sons, NY. 419 p.

Kaden, N.N. and S.A. Smirnova. 1971. To the methodology of karpological description. In: Biological Basis of Increasing of Productivity and Quality of Seeds of Introduced Plants. Kiev., Maukova Dumka. 24 p.

Le Houuerou, H.N. 2005. Bioclimatology, phytogeography and diversity of the isoclimatic Mediterranean biome. Copymania Publication, Montpellier, France.

Li, M. 1993. Distribution of C3 and C4 species of Cyperus in Europe. Photosynthetica 28:119–126.

Nasyrov, M., N. Ibragimov, B. Halikov, and J. Ryan. 2004. Soil organic carbon of Central Asia's Agroecosystems. p. 126–139. In: J.Ryan, P. Vlek, and R. Paroda (eds.), Agriculture in Central Asia: Research for Development. ICARDA, Aleppo, Syria.

Oliver, J.E. and R.W. Fairbridge (eds.). 1987. The Encyclopedia of Climatology. Vol. XI. New York, NY.

Osmond, C.B., O. Bjorkman and J. Andersen. 1980. Physiological processes in plant ecology: toward a synthesis with Atriplex.Ecological Studies 36.

Oyungerel, O., T. Tsendeekhuu, and C.C. Black. 2000. Studies of photosynthesis by succulent plants in Mongolia. Ph.D. Thesis (in preparation).

Petrov, M.P. 1973. The deserts of the Globe. Nauka Publishing House, Leningrad, USSR. 414 p. (in Russian).

Pyankov, V.I. 1984. Relationship between primary CO_2 fixation pathways in C4 plants of arid zones as affected by various temperature. Soviet Plant Physiology 31:644–649.

Pyankov, V.I. 1993. C4-species of high mountain desert of Eastern Pamirs. Russian Journal of Ecology 24:156–160.

Pyankov, V.I. and A.T. Mokronosov. 1993. General trends in changes of the Earth's vegetation related to global warming. Russian Journal of Plant Physiology 40:443–458.

Pyankov, V.I. and Y.I. Molotkovskii. 1993. C4-flora of Tiger Valley reserve, South Tadjikistan. Proceedings Acad. Sci. Tadjikistan 35:38–43. (in Russian).

Pyankov, V.I. and D.V. Vakhrusheva. 1989. The pathways of primary CO_2 fixation in Chenopodiaceae C4-plants of Central Asian arid zone. Soviet Plant Physiology 36:178–187.

Pyankov, V.I., D.V. Vakhrusheva, and O.L. Burundukova. 1986. The photosynthetic pathway types in the hot desert of Central Karakum and its ecological importance. Problems of Desert Development N2:45–54. (in Russian).

Pyankov, V.I., D.V. Vakhrusheva, and M. Durikov. 1988. Structural-functional especialities of ecotypes of *Kochia prostrata (L.) Shrad.* in agrocenosis of Central Kara-Kum and Kopetdag. Problems of Desert Development N1:67–73. (in Russian).

Pyankov, V.I., A.N. Kuzmin, E.D. Demidov, and A.I. Maslov. 1992a. Diversity of biochemical pathways of CO_2 fixation in plants of the families Poaceae and Chenopodiaceae from arid zone of Central Asia. Soviet Plant Physiology 39:411–420.

Pyankov, V.I., E.V. Voznesenskaja, A.N. Kuzmin, E.D. Demidov, A.A. Vasiljev, and O.A. Dzubenko. 1992b. C4-photosynthesis in alpine species of the Pamirs. Soviet Plant Physiology 39:421–430.

Pyankov, V.I., E.V. Voznesenskaja, A.V. Kondratschuk, and C.C. Black. 1997. A comparative anatomical and biochemical analysis in Salsola (Chenopodiaceae) species with and without a Kranz type leaf anatomy: a possible reversion of C3 to C4 photosynthesis. American Journal of Botany 84: 597–606.

Pyankov V.I., C.C. Black, E.G. Artyusheva, E.G. Voznesenskaya, M.S.B. Ku, and G.E. Edwards. 1999. Features of photosynthesis in Haloxylon species of Chenopodiaceae that are dominant plants in Central Asian deserts. Plant Cell Physiol. 40:125–134.

Pyankov, V.I., P.D. Gunin, S. Tsoog, and C.C. Black. 2000. C4 plants in the vegetation of Mongolia: their natural occurrence and geographical distribution in relation to climate. Oecologia 123:15–31.

Pyankov V.I., E.V. Voznesenskaya, A.N. Kuzmin, M.S.B. Ku, E. Ganko, V.R. Franceschi, C.C. Black, and G.E. Edwards. 2000. Occurrence of C3 and C4 photosynthesis in cotyledons and leaves of Salsola species (Chenopodiaceae). Photosynthesis Research 63:69–84.

Redman, R.E., L. Yin, and P. Wang. 1995. Photosynthetic pathway types in grassland plant species from Northern China. Photosynthetica 31:251–255.

Redman, R.S., K.B. Sheehan, R.G. Stout, R.J. Rodriguez, and J.M. Henson. 2002. Thermotolerance generated by plant/fungal symbiosis. Science 298:1581p.

Reinfelder, J.R., A.M.L. Kraepiel, and F.M.M. Morel. 2000. Unicellular C4 photosynthesis in a marine diatom. Nature 407:996–999.

Saliendra N.Z., D. Johnson, M. Nasyrov, K. Akshalov, M. Durikov, B. Mardonov, T. Mukimov, T. Gilmanov, and E. Laca. 2004. Daily and growing season fluxes of carbon dioxide in rangelands in Central Asia. p. 140–153. In: J. Ryan, P. Vlek, and R. Paroda (eds.), Agriculture in Central Asia: Research for Development. ICARDA, Aleppo, Syria.

Schulze, E.D., R. Ellis, W. Schulze, and P. Trimborn. 1996. Diversity, metabolic types and $\delta^{13}C$ carbon isotope ratios in the grass flora of Namibia in relation to growth form, precipitation and habitat conditions. Oecologia 106:352–369.

Shomer-Ilan, A.S., A. Nissenbaum, and Y. Waisel. 1981. Photosynthetic pathways and the ecological distribution of the Chenopodiaceae in Israel. Oecologia 48:244–248.

Smith, S.D., R.K. Monson, and J.E. Anderson. 1997. Physiological Ecology of North American Desert Plants. Springer, Berlin.

Toderich, K.N., R.I. Goldshtein, V.B. Aparin, K. Idzikowska, and G.S. Rashidova. 2001. Environmental state and an analysis of phytogenetic resources of halophitic plants for rehabilitation and livestock feeding in arid and sandy deserts of Uzbekistan. 154 p. In: Sustainable Land Use in Deserts. Berlin.

Toderich, K.N., T. Tsukatani, C.C. Black, K. Takabe, and Y. Katayama. 2002. Adaptations of plants to metal/salt contained environments: glandlar structure and salt excretion, KIER Discussion paper, No 552, Kyoto University, Kyoto, Japan. 18 p.

Toderich, K.N., V.B. Aparin, T. Tsukatanu, and A.B. Konkin. 2003. A strategy for land rehabilitation by salt and heavy metal removal using the integration of Asiatic desert plant diversity. Chinese Journal of Arid Land Geography 26:10:150–159.

Toderich, K.N., T. Tsukatani, O.F. Petukhov, V.A. Gruthinov, T. Khujanazarov, and E.A. Juylova. 2004. Risk assessment of Environmental contaminants of Asiatic Deserts Ecosystems in relation to plant distribution and structure. Journal Arid Land Studies 14S:33–36.

Toderich, K.N., T. Tsukatani, E.V. Shuyskaya, T. Khujanazarov, and A.A. Azizov. 2005a. Water quality and livestock waste management in the arid and semiarid zones of Uzbekistan. p. 574–583. In: Proceedings of the University of Obihiro, Japan.

Toderich, K.N., V.V. Li, C.C. Blak, T.R. Yunusov, E.V. Shuiskaya, G. Mardanova, and L.G. Gismatulina. 2005b. Linkage studies of structure, isoenzymatic diversity and some biotechnological procedures for Salsola species under desert saline environments. p. 73–82. In: Biosaline Agriculture and Salinity Tolerance in Plants. Birkhouser Verlop AG Basel – Boston – Berlin.

Toderich, K.N., C.C. Black, A.A. Ashurmetov, U. Japakova, L.G. Gismatulina, and T. Yunusov. 2006a. Salsola spp. as a model for understanding plants reproductive system. p. 204–206. In: Biochemistry and Environmental Physiology in Desert Ecosystems. Baytenovskie Chteniya. Almata.

Toderich, K.N., P.N. Yensen, Y. Katayama, Y. Kawabata, V.A. Grutsinov, G.K. Mardonova, and L.G. Gismatullina. 2006b. Phytoremediation Technologies: using plants to clean up the metal/salts contaminated desert environments. Journal Arid Land Studies 15S:183–186.

Tsvelev, N.N. 1987. The system of grasses (Poaceae) and their evolution. 37th Komarov Lecture, Nauka, Leningrad. (in Russian).

Ueno, O., M. Samejima, and T. Koyama. 1989. Distribution and evolution of C4 syndrome in Eleocharis, a sedge group inhabiting wet and aquatic environment, based on calm anatomy and carbon isotope ratios. Annals of Botany 64:425–438.

Ueno, O. and T. Takeda. 1992. Photosynthetic pathways, ecological characteristics, and the geographical distribution of the Cyperaceae in Japan. Oecologia 89:195–203.

Vandenkoornhuyse P., S.L. Baldauf, C. Leyval, J. Straczek, and J.P.W. Young. 2002. Extensive fungal diversity in plant roots. Science 295:2051.

Voznesenskaja, E.V. and Y.V. Gamaley. 1986. The ultrastuctural characteristics of leaf types with Kranz anatomy. Botanicheskii Zhurnal 71:1291–1307. (in Russian)

Voznesenskaya, E.V., V.R. Franceschi, O. Kiirats, H. Freitag, and G.E. Edwards. 2001. Kranz anatomy is not essential for terrestrial C4 plant photosynthesis. Nature 414:543–546.

Winter, K. 1981. C4 plants of high biomass in arid regions of Asia: occurrence of C4 photosynthesis in Chenopodiaceae and Polygonaceae from the Middle East and USSR. Oecologia 48:100–106.

Wojnicka-Poltorak, A., E. Chudzinska, E. Shuiskaya, H. Barczak, K. Toderich, and W. Prus-Glowacki. 2002. Isoenzymatic and cytological studies of some Asiatic species of genus Salsola. Acta Societatis Botanicorum Poloniae. 71, No 2:115–120.

Zalenskii, O.V. and T.A. Glagoleva. 1981. Pathway of carbon metabolism in halophytic desert species from Chenopodiaceae. Photosynthetica 15:244–255.

Ziegler, H., K.H. Batanouny, N. Sankhla, O.P. Vyas, and W. Stichler. 1981. The photosynthetic pathway types of some desert plants from India, Saudi Arabia, Egypt, and Iraq. Oecologia 48:93–99.

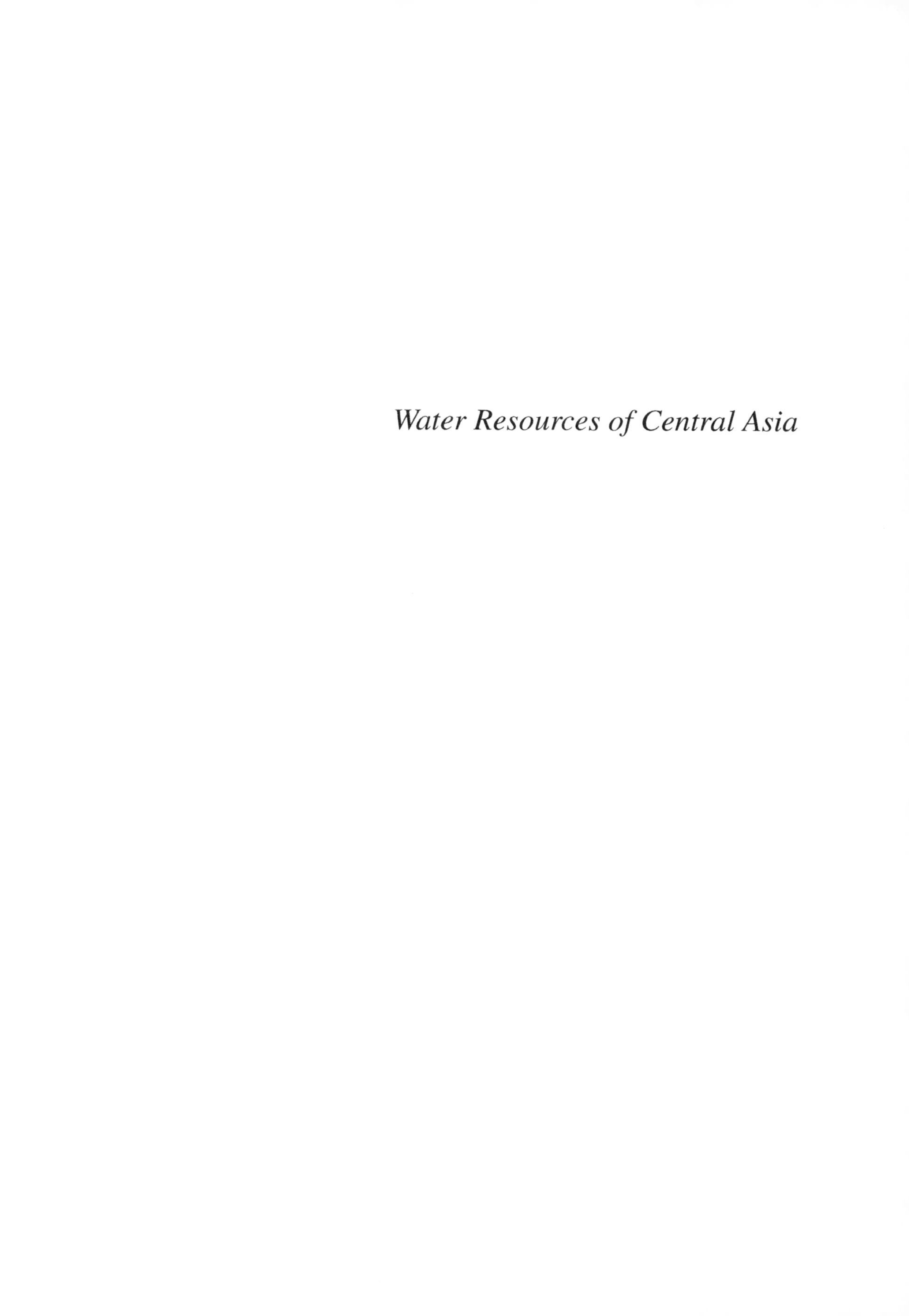

Water Resources of Central Asia

CHAPTER 4

Water resources of the Central Asia under conditions of climate change

V.E. Chub
Uzhydromet, Tashkent, Uzbekistan

1 INTRODUCTION

Most of the Central Asian region is located in areas that feed into the Aral Sea, Balkhash, Issyk-Kul and Karakul lakes basins. The region is part of the Eurasian continent and is located near the northern border of the subtropical zone. It is characterized by a continental climate, uneven distribution of atmospheric precipitation and a peculiar hydrological cycle which is greatly affected by wind movement fluctuations. Most water resources originate in mountainous regions and are concentrated in transboundary rivers whose waters are jointly used by the Central Asian states.

The upper catchments of the watersheds of these rivers are characterized by highly variable hydrometeorological features that are greatly affected by climate fluctuations and changes. These features should be analyzed because they will increase in importance as climate changes that are caused by increases in greenhouse gas concentrations in the atmosphere.

In the past, the climate of the region was quite different from what it experiences today. During the time of the Lyavlyakon pluvial (7–4 thousand years B.C.) current desert areas were characterized by a plains-type climate. Water flows from the Amudarya and Zeravshan rivers were greater than 2.5 times the current flows (Mamedov and Trofimov, 1996). Desertification and its impact on the landscape greatly increased during the end of the third – beginning of the second millennia B.C. According to archeological, geomorphological and much historical data, humidity levels actually increased during the early Middle Ages and during the seventeenth to twentieth centuries.

2 CHARACTERISTICS OF THE HYDROLOGICAL CYCLE

Unidirectional trends are being recorded for a number of important characteristics of the hydrological cycle. They are very evident in changes in regional glaciation data. For example, only in 3–4 cases, based on data associated with 30 years of instrumental observations about the Abramov glacier, show a positive mass balance. Analyses of aerial photos and satellite images made in different years indicate a reduction in the glaciation found in mountainous regions. The glaciation area of Hissar-Alai region was reduced 15.6% during the period 1957–1980, while in Pamir it was reduced by 10.5%. A 7.6% to 10.6% increase in the moraine area covered by glaciers was recorded in Pamir-Alai and a similar increase of from 4.8% to 11% was recorded in Pamir (Tsarev, 1996). Most of the river flow to the Aral Sea basin is being formed by atmospheric precipitation during cold seasons. Its distribution over the area of the flow formation is determined by the peculiarities of synoptic processes, especially regional atmospheric events such as wind movements.

Regularities in the hydrological cycle of the mountainous regions are evident through analyses of changes in specific hydrological elements that depend on the altitude and nature of slopes (Getker, 1981; Kupriyanova, 1969; Chub, 2000; Shetinnikov, 1993). Range ridges have a major impact on wind movements. They cause increases in precipitation layers at similar altitudes as they approach the windward ranges. They cause the opposite effect when moving from these range barriers into the interior of the mountain system. Mean long-term gradient values increase to 20 mm km^{-1} near the ranges of the barriers and decrease to 1–5 mm km^{-1} when moving away from them.

Different types of relationships are formed between precipitation and local altitude depending on the orientation of ranges and their relationship to the main moisture bearing masses. For the windward slopes the relationship is represented by convex shaped, constantly increasing correlations, the steepness of which increases for the favorably oriented ranges. For the leeward slopes and inner mountain screened basins, the correlation relationships take the form of a concave curve (Getker, 1981).

It is possible for prevailing types of synoptic processes to vary such that the same slopes can experience both windward and leeward conditions, thus resulting in high temporal and spatial distributions of precipitation. Indeed, rather significant climate fluctuations were recorded during past meteorological observation periods in Central Asia.

The 1930–1960 period was characterized by a predominance of latitudinal wind circulation while the 1961–1985 period was characterized by a predominance of meridional wind circulation. Many stations recorded high variability and mixed characteristics in the distribution of precipitation were common during the periods of meridional wind circulation. On the other hand, precipitation deficits were common during the prevalence of periods of latitudinal wind circulation. Muminov (1995) has noted the distinguishable "dry" decade from 1941–1950, the moderately dry decades from 1961–1980, and the humid decades from 1951–1960 and from 1981–1990. He also noted the initiation of a marked increase in latitudinal wind circulation trend during the 90's. Seasonal snow cover in the Central Asian mountains is rather dynamic and has a major impact on flora and fauna of the region as well as on economic activities.

Estimations of a linear trend based on observation results from the Chirchik-Akhangaran basin suggest that snow reserves decreased during the period 1935–1990. The decrease was 610 thousand m^3 yr^{-1} for the Pskem river basin in the inner mountain region. For the Akhangaran valley approaching from the west, the decrease is estimated to have been 1.6 million m^3 yr^{-1}; and for the wide Chatkal valley the decrease was estimated to be 5.63 million m^3 yr^{-1} which is about 1.0–1.5% of the natural snow reserves standard (Ivanov Yu et al., 1987) (See Figure 1).

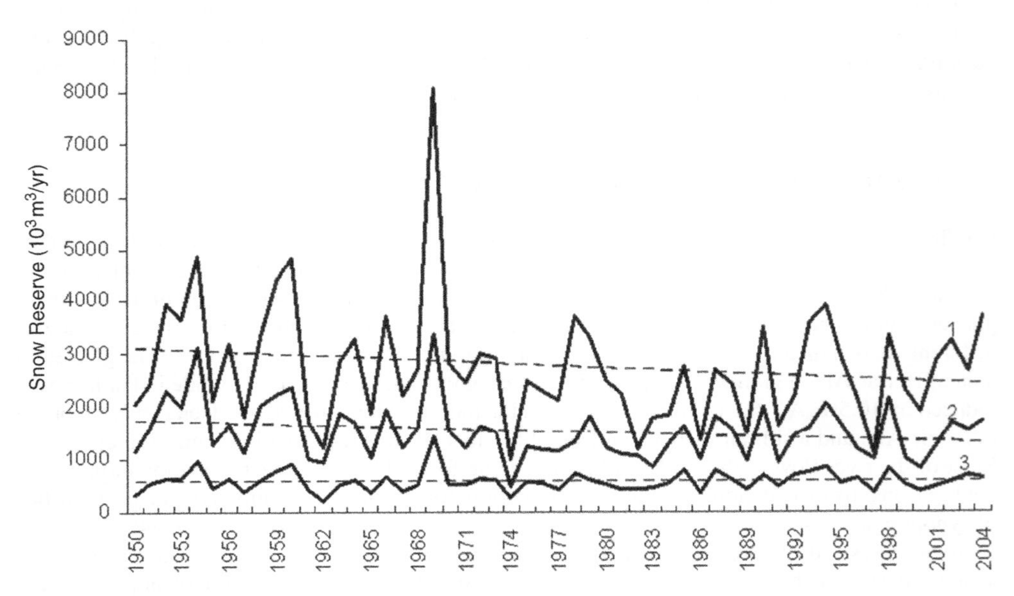

Figure 1. Intrannual and linear trends of the integral snow reserves accumulated by the end of March in the basins of Chatkal (1), and Pskem (2) and Akhangaran rivers.

3 HYDROLOGICAL NETWORK

The hydrological network was most highly developed during the 1970's and 1980's. It is virtually impossible to identify exact relationships between the flow of rivers in the Central Asia mountains and key meteorological factors because of lack of adequate data. This explains why the mean flow calculation for mountain rivers is most frequently estimated by using relationships between the median of the mean long-term flow (Ì) and the mean weighted height (Z) and why hydrometeorological data are used to make these calculations.

There are 89,000 rivers and almost 6,000 lakes in the region (Table 1). The formula most often used to estimate the mean long-term flow in mountains is $M = f(H)$ which is based on the zoning of the land in question.

Central Asia can be divided into two major regions that are distinguished from one another by major hydrographic and geomorphologic features. One is a plains region where almost no river flowage which occupies almost 70% of the area. The other is the mountainous region where all rivers of the region are being formed. It occupies about 30% of the total area. The mountainous region contains the large Amudarya and Syrdarya river basins. Other major river basins are found in the southern and south-eastern parts of Turkmenia. Also among the larger water flowage systems are the Talas, Chu and Aksu river basins and the Issyk-Kul lake basin. These large hydrological units are subsequently subdivided into smaller basins or regions. For example, the Amudarya river basin is divided into the Pyandj, Vakhsh, Kafirnigan, Surkhandarja river basins. The Sherabad, Kashkadarja and Zeravshan rivers are also found in the hydrological region. The Syrdarya river basin is divided into the Naryn, Karadarya, Angren, and Chirchik river basins. It also contains the Keles and Arys rivers, the Southern-Fergana and Northern-Fergana regions, the northern slopes of the Nuratau range and the southwestern slopes of the Karatau range. Turkmenia contains the northeastern slopes of the Kopetdag range and the Badhyz region. The Issyk-Kul lake region is in part defined by rivers on its northern and southern banks.

Almost all of these regions are subdivided into subregions which are used to estimate the relationship between river flowage rates and altitude. The water resources of the Central Asia have been estimated several times by different scientists. All have based their studies on observational data provided by the network of hydrometric measurement stations of the regional hydrometeorological system.

River flowage values were taken at points representing where rivers entered plains regions from their mountain sources. During most years, it is impossible to gauge the rates of water flowage in rivers in the plains regions, since the flowage is practically non-existent. Flowage data calculated by NIGMI of Uzhidromet are presented in Table 2. Omitted from this table are flowage rates for left bank tributaries of Amudarya River which originate in Afghanistan. They are estimated to be 17 $km^3\ yr^{-1}$ (Table 3). Little difference in the estimates of water resources presented by the different authors exists for basins with high density hydrometric networks, principally because the results of reference stations are not distorted by economic activities.

Table 1. The distribution of rivers and lakes on the main regions of Central Asia.

	Rivers		Lakes	
River basin, region	Number	Length, km	Number	Area, km^2
Amudarya	40,999	1,787	2,619	129
Syrdarya	29,790	1,907	1,405	65
Talas	3,632	276	467	23
Chu	5,244	491	506	38
Chinese Aksu	4,495	214	260	4
Issyk-Kul lake	1,976	134	183	20
Turkmenistan	2,972	167	211	42
Total number	89,018	4,978	5,961	321

Table 2. Water resources ($km^3\ yr^{-1}$) of the rivers of Amudarya river basin.

River basins	Mean annual long-term discharge, $m^3\ s^{-1}$	Volume of the annual flow, $km^3\ yr^{-1}$		
		Mean	Maximum	Minimum
Amudarya basin				
Pyandj	1,140	35.9	–	–
Vakhsh	661	20.8	27.6	16.2
Kafirnigan	187	5.9	9.8	4.1
Surkhandarja with Sherabad	127	4.0		
Kashkadarja	49	1.6	2.7	0.9
Zeravshan	169	5.3	6.9	3.8
Total	2,334	73.5	–	–
Syrdarya basin				
Naryn	448	13.8	2.3	0.8
Rivers of Fergana valley	401	12.8		
Rivers of the northern slope of range to the west of Turkistan range to the west from Fergana valley	10	0.3	0.4	0.2
Akhangaran	38	1.2	3.0	0.6
Chirchik	248	7.8	14.1	4.5
Keles	7	0.2	0.5	0.1
Arys	64	2.0	–	–
Rivers of the south-western slope of Karatau range	21	0.7	–	–
Total	1,237	38.8	–	–
Krygyzstan				
Talas	68	2.1	–	–
Chu	137	4.3	–	–
Issyk-Kul lake	118	3.7	–	–
Chinese Aksu	225	7.1	–	–
Total	548	17.2	–	–
Turkmenistan				
Atrek	10	0.3	–	–
Tedjen	27	0.8	0.5	0.1
Murgab	53	1.7	2.6	0.4
Rivers of the northern slopes of Kopetdag range	11	0.3	–	–
Total	101	3.1	–	–
Overall for Aral Sea basin	4219	132.6	–	–

Table 3. Water resources of the left bank tributaries of Amudarya River.

River	Mean annual long-term discharge value, $m^3\ s^{-1}$	Volume of the annual flow, $km^3\ yr^{-1}$
Kokcha	211	6.6
Kunduz	165	5.2
Khulm	2.7	0.1
Balkhab	49.8	1.6
Sarypul	8.0	0.2
Shirintagao	4.1	0.1
Murgab	61.1	1.8
Gerirud	48.5	1.4
Total	550.0	17.0

Table 4. Estimates of levels of water flowage for the different river basins of Central Asia.

River basin	Schults, 1965	Golshakov, 1974	IWP*, 1985	SHI*, 1987	NIGMI*, 1997
Amudarya	79.0		72.8	69.5	73.6
Srdarya	37.8	38.3	36.7	37.0	38.8
Naryn	13.5	13.8		13.8	
Chu and Talas	5.98	6.47			
Issyk-Kul lake	3.62	3.72			

*IWP-Institue of Water Problem, Academy of Science of Russia; SHI-Science Hydrological Institute of Russia Hydrometservices; NIGMI-Hydrometeorological Research Institute of Uzhydromet.

Table 5. Estimates of water flowage in rivers of Central Asia ($km^3 yr^{-1}$).

		Flow which is formed within the states				
State	Area '000 km^2	SHI*, 1967	Kupriyanova, 1969	IWP*, 1985	SH*, 1967	IWP*, 1985
Turkmenistan	488.0	1.0	–	1.1	–	2.9
Uzbekistan	447.4	11.1	10.6	9.5	–	
Tajikistan	143.0	51.2	53.4	47.4	20.0	20.7
Kyrgyzstan	198.5	52.8	49.7	48.7		
Total	1,765.9	126.1	–	106.7	–	23.6

*IWP-Institute of Water Problem, Academy of Science of Russia; SHI-Science Hydrological Institute of Russia Hydrometservices.

Data in Table 4 indicate that discrepancies in the water resources of the rivers of the Fergana valley, as represented by the Chirchik-Akhangaran irrigation area and the Amudarya river basin, are rather high.

Different authors have estimated the water resources of the Amudarya River. Some have taken into account the flow which was formed in Afghanistan territories in doing so, while some have not. The resulting discrepancies can be dealt with by assessing flow commonalities. Data presented suggest that calculations of total and regional water resources are quite inaccurate and may seriously underestimate total water flowage. National hydrometeorological systems should strive to preserve and further develop observational networks in order to improve measurement processes and the quality of resulting data. Greater effort will need to be devoted to estimation of water resources in the Central Asian states.

The relatively low level of reliability of data estimates is illustrated by data found in three published documents that are presented in Table 5. Discrepancies exceed expected error parameters for these data.

Error variance in estimations can be accounted for in part by differences in interpretation of size of particular regions and area elevation. Variation in the calculation of river flowages primarily results from cyclical fluctuations in water availability, flow variability and errors made in adjusting results of measurements at specific measurement sites on the rivers. General trends in variability of water flowage and changes in the availability of water resources can be gauged by analyzing time series data over long observation periods.

Based on available data it can be concluded that the temporal variability of river flowages in Central Asia over the past fifty to seventy years is less than in Kazakhstan, Siberia and the European part of Russia. This is due to the fact that most major rivers flow represents water that results from the melting of permafrost snow and glaciers. This process is more intensive during years that experience high summer temperatures and less intensive during years that experience lower summer temperatures. The latter can guarantee that remaining of snow and ice reserves will persist from year to year. Unlike in these regions, air temperature fluctuations in Central Asia are negligible from year to year.

Basins for most large rivers sprawl out for long distances. Thus fluctuations in water availability are very low and many times are less than for the distant tributaries. Table 2 shows that the flow for the Vakhsh basin during high water years is only 1.7 times more than the flow during low water years. Water flowage is 1.8 times less for the Zeravshan basin during low water years.

Differences in flowage rates for the Pyandj river basin for high and low water years are probably less than 1.6 times. Until now it has been impossible to estimate river flows during the dry years because of very short periods of observation that resulted from the lack of water flowage. Variations in the volume of temporal flows were not obtained for other river basins, such as the Surkhandarja river, the Fergana river valley, the Arys river, the rivers of the south-western slope of Karatau range, the Chu and Talas rivers, the Issyk-Kul lake region, the Kopetdag river and the Atrek river for similar reasons. Reliable detailed estimates of flow variability are difficult to estimate because of the short duration of observation series and because of variations in the timing of measurement observations at the monitoring stations.

Despite that fact that flow formation zones for the different rivers of the region represent a rather limited geographic area, a high level of fluctuation in flowage rates is apparent. This is determined in part by variations in synoptic processes and to a great extent by wind movements in different regions. It is common to record high water levels for rivers with watersheds located at high mean elevations, especially during years characterized by high air temperatures. On the other hand, high water levels are recorded at lower elevations during years characterized by high levels of precipitation. Variations in water flowage are reduced in large river basins, in part by the size of the region affected.

On average, expected variations are observed in annual flow fluctuations in the Amudarya and Syrdarya river basins. These are illustrated by differential integral curves that were calculated using the restored data series in from Kerki and Bekabad flow monitoring stations and taking into account the additional flow from the Chirchik-Akhangaran basin (Figure 2).

The Bekabad flow monitoring station, which is located in Syrdarya river basin, is found at the point where the river exits the Fergana valley. This location accounts for 68% of the basin flow. Flow from the Chirchik and Akhangaran rivers to this basin represents another 21% of the total flowage. Together they account for about 89% of the entire flow of Syrdarya river basin.

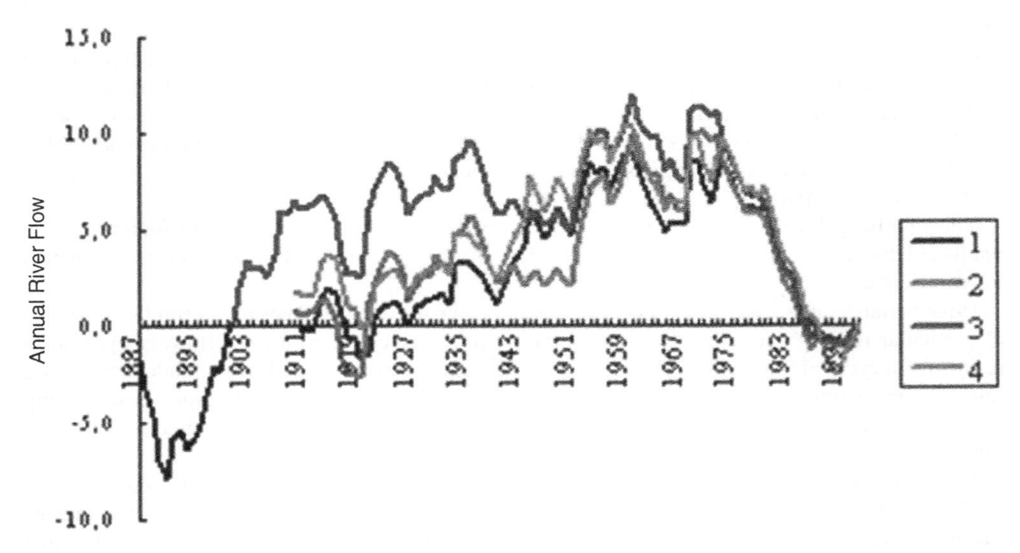

Figure 2. Integral differential curves of the river flow for Amudarya-Kerki town monitoring station (1,), Sydara town – total flow up to Chardara water storage (2,3).

The differential integral flow curves found in Figure 2 show that only one long period of flow increase was recorded during 1891–1960 in the Amudarya basin. Flowage rates tended to decrease from that period to 1991. However, increases after 1991 may represent the onset of a prolonged period of flow increase against a background short periods of decrease. Similar trends were observed, for example, in 1914, 1920, 1934–1940.

Evidently, the flow increase in Syrdarya river basin began about 1891 and continued up to 1973 after which it began to decrease. Periods of the flow decrease against a background of its increase up to 1973 were characterized by somewhat different features. Similar occurrences took place in the Amudarya river basin during the period 1914–1920 and during the period 1924–1944. The flow decrease since 1973 in the Syrdarya river basin has been continuous while periods of increase were observed in the Amudarya river basin after the trend in flow decreases began in 1961.

The common water flowage feature for these two basins is a period of extremely reduced flowage followed by a number of high water years. Water flowage was high in both river basins in 1969. The decrease in flowage in the Amudarya river basin lasted for 17 years from 1969 to 1986. The decrease in flowage in the Syrdarya river basin lasted for 15 years from 1969 to 1982.

4 CONCLUSION

Even the availability of a limited measurement observations over a long period of time and of measurements related to the restoration of the conventionally-natural river regimes makes it possible to use statistical analysis methods and to simulate data series as well as to calculate data for the purposes of forecasting.

REFERENCES

Bolshakov, M. Water resources of the rivers of the Soviet Tien-Shan and methods of their calculation – Frunze, ILIM, 1974. 308 p.

Getker, M. Method of the maximum snow reserves calculation in the mountain and glacial regions for the purpose of creation of maps in the Atlas of the World Snow and Ice Resources Materials of Glaciological Studies. Chronicle. Discussions. Moscow, Akademkniga, 1981. No. 40. p. 133–142.

Chub, V. 2000. Climate change and its impact on the natural resources potential of the Republic of Uzbekistan. Tashkent, Glavgidromet. 252 p.

Climate variability in Central Asia (F. Muminov, ed.), Tashkent, Glavgidromet of Uzbekistan. 1995. 216 p.

Denisov Yu., A. Sergeev, and G. Ibragimova. New approach to the elaboration of the method of the long-term forecasting of mountain rivers flow//Proceedings of SANIGMI (Denisov, Ed.). Tashkent, Glavgidromet of Uzbekistan. 1977. Vol. 153(234). p. 3–10.

Kim I., and T. Spectorman. 1991. Climate change effect on the water availability in rivers and on precipitation regime in the mountain regions of the Central Asia. p. 54–67 In: Ushakov (ed.). Reports presented at XV International Conference on the Meteorology of the Carpathian region. Kiev Academy of Science,

Kupriyanova E., I.Tsigelnaya, M. Ljvovich, and N. Dreyer. 1969. Peculiar features of the water balance in the mountain regions. p. 90–124 In: V. Beltman (ed.). Water balance of USSR and its transformation. Moscow: Nauka.

Mamedov, E. and G. Trofimov. 1996. On the matter of the long-period flow fluctuations of the Central Asian rivers. p. 12–16. In: A. Babaev (ed.). Problem of the desert areas development. Ashgabat, Yilim.

Resources of the surface water of USSR. 1987. Amudarya River. Ivanov Yu (ed.). Vol. 14, No. 3. Leningrad, Gidrometeoizdat. 576 p.

Schults, V. 1965. Rivers of the Central Asia. Leningrad, Gidrometeoizdat. 692 p.

Shetinnikov, A. 1993. Change of the water resources in the glaciers of Pamir-Alai mountains during 1957–1980. Materials of glaciological observations. Chronicle. Discussions. Moscow, Akademkniga 76:83–89.

Tsarev B. 1986. Monitoring of snow cover in mountain territories. Tashkent, Glavgidromet. 76 p.

Water resources of USSR and their use. Leningrad, Gidrometeoizdat, 1987. 302 p.

Water resources and water balance of the territory of USSR. Leningrad, Gidrometeoizdat, 1967. 200 p.

Voropaev, G. and D. Ratkovich. 1985. Problem of the territorial re-distribution of the water resources – Moscow. Publishing House of USSR Ac. Sci. 504 p.

CHAPTER 5

Climate change and water resource alteration in Central Asia: The case of Uzbekistan

Nazar Hakimov
Tashkent State University of Economics, Tashkent, Uzbekistan

Allan Lines
The Ohio State University, Columbus, USA

Pirnazar Elmuratov & Rashid Hakimov
Tashkent State University of Economics, Tashkent, Uzbekistan

1 INTRODUCTION

Uzbekistan is a landlocked country in Central Asia with a total area of 447,400 km^2. It is bordered on the north by Kazakhstan, on the east by the Kyrgyz Republic and Tajikistan, and on the south by Afghanistan and Turkmenistan. Geopolitically, the country is divided into 12 provinces (viloyats) and the autonomous republic of Karakalpakstan, which is located in the far west of the country near the Aral Sea. The country is divided into three distinct agro-ecological zones:

- the desert (Kyzylkum), steppe and semi-arid region – covering 60% of the country and located mainly in the central and western parts of the country;
- the fertile valleys that skirt the Amu Darya and Syr Darya rivers, including the Ferghana valley; and
- the mountainous areas in the east with peaks at about 4,500 m above sea level (Tien Shan and Gissaro-Alay mountain ranges).

2 CLIMATIC CONDITIONS

The climate of Uzbekistan is very continental and arid. It is noted for an abundance of solar radiation, sparse cloud cover, and limited atmospheric precipitation. All of Uzbekistan is characterized by a cold, unstable winter and a dry, hot summer. Winter climate is governed by dry, cold Arctic and Siberian air masses from the north and tropical air masses from the south. In the summer the territory is influenced by local tropical air. The mountains also have an important effect on climatic conditions.

In flat areas of the country, total solar radiation averages from 130 to 160 cal cm^{-2} annually. The actual duration of solar light is between 2,800 and 3,130 hours depending on the latitude and topography of a given region. In the mountains, the duration of solar light is substantially determined by the proximity of the horizon, slope exposition and cloud cover. The average temperature in the plains region increases from the north to the south: in January from −8°C to +3°C. Low temperatures can reach −30°C in the north and −25°C in the south during severe winters. In July, which is the hottest month, the average monthly temperature reaches +32°C with an absolute maximum of +50°C in the Surkhandarya region.

Mean annual precipitation in the plains ranges from 100–150 mm in the desert to 200–400 mm in the foothills. Approximately 60–70% of annual precipitation occurs during the winter and spring months. Autumn precipitation is much less and summer precipitation is essentially insignificant. Droughts in desert areas last 6–7 months and typically occur from May until November. In the foothill areas the drought period is reduced to 4–5 months. For regions in close proximity to foothills, precipitation levels increase to more than 800–900 mm. Distribution of precipitation in the mountains depends on the altitude, relief forms, and slope exposure.

Moisture deficits in the southern part of Uzbekistan are greatest from April to September. The largest oasis area of this region has an ancient irrigated agricultural system and is characterized by a special microclimate. Available water resources in Uzbekistan are formed from renewed surface and underground water of natural origin and from return water of anthropogenic origin.

3 WATER RESOURCES

Uzbekistan is the major water consumer in the Aral Sea Basin. In accordance with interstate agreements, on average 43–52 km^3 of water per year is allotted for use by Uzbekistan from its boundary rivers.About 90% of river flow is formed beyond Uzbekistan's boundaries (UNFCCC, 1999).

Water resources are subdivided into two categories: national and trans-boundary. National water resources include flows of the local rivers and underground and return water formed within a given country. Trans-boundary water resources include rivers, aquifers, and return water that is common to the territory of two or more countries. It also includes water deposits that are connected hydraulically with trans-boundary rivers, or with artificial, anthropogenic, or reservoir resources.

The natural mean annual flow of the rivers for periods during which it has been gauged is 123 m^3 yr^{-1}. This flow is for territories that include 81.5 km^3 in the Amu-Darya basin for the period 1932 to 2003, and 41.6 km3 in the Syr-Darya basin for the period 1926 to 2003.

At present, Uzbekistan uses about 42 km^3 of trans-boundary river flow, 34 km^3 of which is from the Amu Darya and Syr-Darya basins. Surface flow formed in Uzbekistan makes up 11.5 km^3.

The flow of the rivers is characterized by considerable annual and long-term irregularity. In low water years which occur about 90% of the time, the flow decreases by an average of 23 km^3, compared to an average water year. Groups of high water years reoccur at every 6- to 10 year intervals and their duration lasts from 2 to 3 years. Low water periods occur at 4- to 7 year intervals and can last up to 6 years.

The primary concern of the water resource management strategy is to ensure the availability of water resources. It takes into account not only natural water flows, but also their regulation through reservoirs as well as the use of return flow and underground water.

Available water resources in the Aral Sea Basin include the actual long-term regulated flow of Syr-Darya water and the seasonal flow of Amu-Darya water. Total amount available is about 121.69 km^3 yr^{-1}, and has a 90% probability of availability.

The Rogun reservoir on the Vakhsh River was constructed to facilitate the long-term regulation of the flow of Amu-Darya and to make other technical interventions possible in order to obtain up to 133.6 km^3 of available water resources in The Amu-Darya basin. A similar magnitude of availability of water can occur for the Syr-Darya basin. Of this total resource, the share of Uzbekistan's is 72.4 km^3. It includes 61.1 km^3 for irrigation purposes and 11.3 km^3 for non-irrigation uses including communal, industrial, agricultural water supply and other non-irrigation uses.

These water resources also feed the Aral Seaside region and the Aral Sea. According to figures that are part of the water schemes, the Aral Seaside should receive 3.25 km^3 yr^{-1} from the Syr-Darya River and 3.2 km^3 yr^{-1} from The Amu-Darya River. These rivers should also provide 3.2 km^3 yr^{-1} of water for sanitary drawdown; 1 km^3 of water for fishery drawdown and 1 km^3 of water for delta flooding. Thus, the planned supply of water from these two rivers to the Aral Seaside area should be 8.4 km^3.

During the low water years, these parameters are reduced by up to 54.2 km^3. The volume of prescribed flow established by the schemes is also decreased by that much. This deficiency is to be corrected by projected future water management conditions.

Under the current regime of long-term flow regulation of the Syr-Darya River and seasonal flow regulation of the Amu-Darya River, these sources represent up to 11.5 $km^3 yr^{-1}$ (18.4%) of the total water consumption and 457 m^3 of yearly per capita consumption in Uzbekistan.

River surface flows are undergoing significant changes due to anthropogenic impacts. Increases in water intake from the rivers into irrigation canals and subsequent water losses in canals cause a quantitative flow reduction. Further, discharge of collector drainage water worsens its natural mode and quality. Total reserves of underground waters in Uzbekistan sum to 18.9 km^3, including 7.6 km^3 of mineralized water with less than 1 gl^{-1}, and 7.9 km^3 of water with 1–3 gl^{-1} of salinity.

Overall, the water needs of consumers in Uzbekistan are completely met during years of average and increased water probability. Crisis situations arise in low water years, especially for middle and downstream regions of the Syr-Darya and the Amu-Darya water basins. The creation of the Toktogul hydro system for electrical power generation and the regulated use of the Syr-Darya flow for this purpose have aggravated the shortage of water supply for the Syr-Darya and Djizzak Oblasts during the growing season.

Irrigation development has resulted in the formation of artificial lakes in Uzbekistan. They serve as sinks to collect drainage water. The Sinchankul, Dengozkul and Solyenoe Lakes were formed mid-stream in the Kashkadarya and Bukhara Oblasts of the Amu-Darya river basin. Downstream at the Amu-Darya delta, there is a whole system of lakes that was previously fed by Amu-Darya flow. Some of these lakes have subsequently begun to be used as sinks to collect drainage flow. Among these is Lake Sudochye which is well known for the large and diversified bird population that it helps to sustain. In the Syr-Darya river basin, the Arnasayi system of lakes acts as a sink to collect drainage water from irrigated areas. It was formed in the Syr-Darya and Djizzak Oblasts.

Catastrophic winter draw downs from the Chardarya reservoir and its adjoining lake system located in Kazakhstan in previous years resulted in water flow volumes of 35 km^3 where, the water surface is about 2,200 km^2. The Arnasai system of lakes has turned into a profound natural and economic entity that influences four Oblasts of Uzbekistan, namely, Tashkent, Syr-Darya, Djizak and Navoi.

Change in use of water during the last decade not only created the downstream summer water shortages, but also winter water excesses. These winter excesses, however, do not help to solve the Aral Sea problem. In order to prevent downstream flooding the excess water was diverted into an isolated Arnaysay depression, in which several independent lakes were created. These lakes cover a total of 2000 km^2 area and are responsible for raising groundwater tables. As a result, newly formed wide-spread swamps now exist, covering over 20,000 km^2 in area.

Uzbekistan presently has 52 reservoirs with total volume of more than 19.3 km^3. Twenty-one of these reservoirs in the Syr-Darya basin have a total volume of 5 km^3 and 31 reservoirs in the Amu-Darya basin have a total volume of more than 14.3 km^3. There are two large reservoirs in the Syr-Darya river basin, namely the Charvak and Andijan reservoirs. The Charvak reservoir is located on the river Chirchik near Tashkent and holds a total volume about 1.99 km^3. The Andijan reservoir is located on the river Karadarya in Andijan and has a capacity of 1.9 km^3. Oblasts use these reservoirs for many purposes, including irrigation, flood control and electric power generation.

The Tuyamuyun downstream section of the Amu-Darya river basin has the largest multipurpose reservoir which contains 7.8 km^3 of water. This reservoir consists of four separate parts located in Turkmenistan and Uzbekistan. However, the parts located in Uzbekistan are responsible for management of the entire reservoir. The Amu-Darya downstream area, especially in the northern Aral seaside of the Republic of Karakalpakstan, has numerous serious problems with its communal and potable water supply. These problems initially resulted from the drying of the Aral Sea and have subsequently been aggravated by substantial increases in grounded water salinity.

The maximum potential of hydroelectric power generation on the rivers of Uzbekistan is estimated between 85,000 and 90,000 $GWt yr^{-1}$. Of this amount, 15,000 $GWt yr^{-1}$ is economically suitable for development. Actual capacity, based on existing infrastructure, can produce 1,739.0 $GWt yr^{-1}$. In 2000, the production of electricity by hydroelectric power stations was 4.9 billion $KWt hr^{-1}$, which was less than 20% of the average annual production for previous years. This reduced production can be explained in part by the critical low waters in the Aral Sea Basin during 2004.

Total water intake in 2000 was 54.2 km^3, of which 47.8 km^3 of water intake was from surface water and 6.4 km^3 of underground water return flow. Irrigation intake from surface water and underground water amounted to 44.1 km^3 and 3.3 km^3, respectively.

Uzbekistan utilizes over 90% of its available water resources for irrigation purposes. Water resources of the river basin of the Aral Sea belong to Uzbekistan and other nations in the region. They are fully regulated and distributed among water consumer countries, of which Uzbekistan is a primary benefactor. However, Uzbekistan is currently experiencing a large deficit in water resources, the most crucial problem being the Aral Sea crisis.

The level of anthropogenic load on water resources is very high in all parts of Uzbekistan and threatens the production of irrigated crops. There is a widespread problem of inappropriate drainage practices associated with poor irrigation. The degradation of cultivated land is primarily a consequence of increased salinization and susceptibility to all types of soil erosion. Natural pastures are being reduced in size and are progressively being degraded. Biological diversity in crops is also being reduced and the entire ecosystems in the Amu-Darya delta are either being destroyed or are disappearing. Approximately 50% of the country's agricultural land area is prone to degradation of one type or another.

4 CLIMATE CHANGE

Extremes of weather conditions are typical of Uzbekistan's climate, and adversely impact crop yields. Frosts in late spring and early fall, high temperatures and frequent and enduring droughts in summer are of common occurrence. Additionally, increase in population growth results in increased demand for food production. All of these conditions increase the vulnerability of Uzbekistan to climate change. These changes necessitate development and implementation of adaptive measures to address climate change, especially in less favored ecological regions. These measures can substantially reduce the negative consequences of droughts and exert positive socio-economic influences throughout the entire nation. Examples of such measures include current attempts to address the consequences of the Aral Sea crisis (Coop. Prog. on Water and Climate, 2004).

In the late 1950s, the Aral Sea was the Earth's fourth-largest inland body of water with respect to its surface area. In 1960, the mean level of the Aral Sea was 53.4 m, its surface area was 66,000 km^2, and its volume at about 1,090 km^3. There was a flourishing sea fishery industry, based on the exploitation of a variety of commercially valuable species. During the last three decades of the 20th century, the Aral Sea region became a major world-class ecological and socio-economic problem, and became the sixth-largest inland water body (UNU, 1998; Chub et al., 1999).

The problem with focus on assessments of the consequences of climate change is that it is very difficult to distinguish the impact among nations. This requires the investigator to choose among scenarios of greenhouse gas emissions, to determine the time intervals between evaluations, to create regional climatic scenarios, and to choose among distinct models and methods in order to assess actual conditions and temporal changes.

According to the majority of climatic scenarios, Uzbekistan is likely to have as much access to water in the future as it has during the first decade of the 21st century. However, under extremely adverse scenarios, there may be a considerable decline in water drainage which would undoubtedly have a further negative impact on the private sector agriculture in Uzbekistan. In the foreseeable future, the price of water will continue to increase. Water shortages will probably become more frequent; thus events such as the failure of the Amu-Darya to flow to the Aral Sea in 2000 and 2001 are likely to occur with increasing frequency (Chub et al., 1999).

Even if water supplies should remain the same, excessive use of water flow from the major rivers is likely to continue to occur since the number of people who currently reside in the Amu-Darya and Syr-Darya river basins will increase from 40 million in 2006 to a projected 55–60 million in the coming decades.

5 UZBEKISTAN AGRICULTURE

The development strategy for the agricultural sector is designed to ensure food and environmental security, while increasing the efficiency and export potential of domestic production (Djalalov and Chandra, 2006). According to the macroeconomic forecast, the agrarian sector will maintain its leading role in the economy. To meet the strategic goals by 2010, annual growth in agricultural output must be at least 5–6% (UNFCCC, 1999).

As shown in Graph 1, agriculture represented 26.8% of the GDP of Uzbekistan in 2004. Over the last few years, this sector of the economy has experienced continued growth largely due to a conservative approach to use of water resource. Over the same period animal production has also experienced a growth tendency.

The total amount of land used for all types of agricultural production in 2004 was 7,242,000 hectares which represented a slight decrease from 2003. This land use pattern included an increase of 62,700 hectares to produce cotton the previous year and an increase of 69,000 hectares to produce other industrial crops (See Table 1). Land used to produce cereals decreased by 125,000 hectares and that used to produce other grains also decreased by 37,000 hectares. Land used to produce rice decreased by almost half from 119,000 hectares to 65,000 hectares. Land under feed crop production decreased by 35,000 hectares. Production of food crops remained stable or decreased slightly during that period.

The decline in proportion of total area under collective farm production and the increase in the proportion under modern farm production and "peasant" ("dekhkan") farm production continued in 2004. Land under collective farm production in 2004 was 2,970,000 hectares. This signified a decline of 23% from tht under collective farm production in 2003. In contrast, land under modern farm production increased to 3,400,000 hectares, or an increase of 23% over that under modern farm production in 2003. Land area under modern farm cultivation was 36.3% of the total cultivated area. The land area under cultivation on peasant farms increased slightly from 850,000 hectares to 868,000 hectares from 2003 to 2004 (Mumunov and Abdulleav, 1997; Ryan et al., 2004).

These major shifts were mainly due to favorable weather conditions and to a number of other factors including intensified restructuring of unprofitable collective farms into modern farms, the

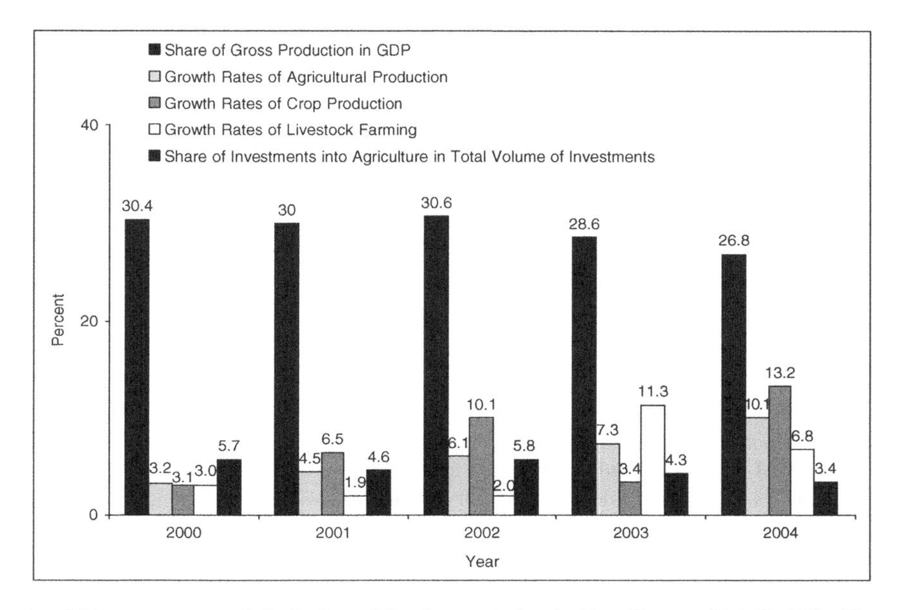

Figure 1. Main macroeconomic indicators of development of agriculture (Source: State Statistics Committee of Uzbekistan, 2004).

Table 1. Structural changes of areas under crop by form of management for 2003–2004.

	2003				2004			
Commodity	AFC[1]	CF[2]	MF[3]	PF[4]	AFC[1]	CF[2]	MF[3]	PF[4]
	thousand hectares							
Cereals	1790	912	680	198	1665	661	805	199
Grain	1506	766	581	159	1469	593	715	161
Rice	119	67	47	5	65	22	40	3
Maize for seed	33	7	7	19	33	5	8	20
Industrial crops	1444	847	590	7	1515	690	815	10
Cotton	1393	820	573	–	1454	662	792	–
Potatoes, melons and gourds	236	47	36	153	226	31	37	158
Potatoes	48	3	2	43	51	3	3	45
Vegetables	145	34	17	94	137	21	19	97
Melons	40	8	17	15	34	5	14	15
Fruits and berries	200	108	27	65	198	97	37	64
Grapes	117	76	10	31	115	67	16	32
Feed crops	315	160	94	61	280	113	103	64
Total area under production	7386	3855	2681	850	7242	2970	3404	868

AFC[1]: All Farm Categories; CF[2]: Collective Farms; MF[3]: Modern Farms; PF[4]: Peasant Farms.
Source: State Statistics Committee of Uzbekistan, 2004.

Table 2. Gross harvests of agricultural products in farms of all categories.

	2003				2004			
Commodity	AFC[1]	CF[2]	MF[3]	PF[4]	AFC[1]	CF[2]	MF[3]	PF[4]
	thousand tons							
Cotton	2802	1741	1061	–	3535	1707	1828	–
Grain	6318	3075	2264	979	6016	2211	2781	1024
Wheat	5624	2805	2030	789	5501	2085	2584	832
Rice	349	162	145	42	186	54	105	27
Potatoes	833	49	33	751	891	45	43	803
Vegetables	3300	626	353	2321	3314	335	384	2575
Melons	585	77	177	331	571	46	174	351
Fruits and berries	764	231	56	477	844	215	109	520
Grapes	400	160	22	218	576	264	56	256

AFC[1]: All Farm Categories; CF[2]: Collective Farms; MF[3]: Modern Farms; PF[4]: Peasant Farms.
Source: Estimates of State Statistics Committee of Uzbekistan., 2004.

implementation of new credit mechanisms for farmers, timely inputs of technology, an adequate supply of mineral fertilizers, improvement of contractual relationships and increased modernization of water resource utilization.

The relationship between crop and animal production was substantially the same in 2003 and 2004. Crop production amounted to 53.7% of total production in 2004 as compared to 51.6% in 2003, while that of livestock farming amounted to 46.3% of total production in 2003 compared to 48.4% in 2004. As shown in Table 2, cotton harvest in 2004 was more than 3.5 million tons due to the increase in area under cultivation, timely agro-technical inputs, repairs of critical irrigation systems, an increase in raw cotton production output on farms and favorable weather conditions. In contrast, there was a decline in the production of grains, rice, melons and gourds in 2004. In comparison with 2003, wheat production declined by 123,000 t in 2004 and rice production declined by 163,000 t over the same time period. Melon production also declined by 14,000 t over

Table 3. Livestock and poultry population in all categories of farms.

	2003				2004			
Commodity	AFC[1]	CF[2]	MF[3]	PF[4]	AFC[1]	CF[2]	MF[3]	PF[4]
	thousand heads							
Cattle	5878	284	291	5303	6231	228	309	5694
Sheep and goats	9927	2460	309	7158	10,558	2449	419	7690
Pigs	88	31	12	45	82	20	14	48
Poultry	17,674	5630	770	11,274	18,725	5139	766	12,820

AFC[1]: All Farm Categories; CF[2]: Collective Farms; MF[3]: Modern Farms; PF[4]: Peasant Farms.
Source: Estimates of State Statistics Committee of Uzbekistan, 2004.

the same period. These production declines were due to the decrease in areas under cultivation of these crops in 2004 as is evidenced by data in Table 1.

In contrast to grain crops, livestock production increased from 2003 to 2004. The number of head of cattle, sheep and goats, and poultry continued to grow. Water-supply for livestock farms and peasant farms also improved. Consequently, there was 6.8% growth rate in livestock production in all types of farms. As is indicated in Table 3 comparatively high growth rates were observed for sheep and goats (6.4%) and poultry (5.9%).

6 CHANGES IN TYPES OF FARMS

Implemented institutional and structural reforms helped farms that were progressively managed to succeed, and led to a decrease in the number of collective farms over this period. Over time 326 collective farms have been reorganized, thus allowing for the establishment of 15,118 modern farms which occupy 769.1 thousand hectares. Among the newly established farms, 10,667 of them, representing 469.9 thousand hectares under cultivation, are specializing in the production of cotton and grains.

The restructuring of unprofitable collective farms into modern farms continued during 2004. The share of collective farms in the total agricultural production decreased from 22.2% in 2003 to 19.9% in 2004.

The number of peasant farms in agricultural production remained stable and their cumulative output also remained stable. However, their share in the total gross production decreased, which was primarily due to the increase in the number of farms.

Economic reforms in agriculture also caused a number of problems. For example, the need for further improvement of the mechanisms for establishing farms became more evident, as was the need for contract discipline, reform of banking-finance institutes, and the creation of service related infrastructure for agricultural producers in the rural areas.

Economic reforms in the agrarian sector intensified further in 2005. The accelerated development of modern farms remains a priority task. The anticipated liquidation of collective farms will progressively lead to the establishment of about 20 thousand new modern farms.

7 SUMMARY AND CONCLUSIONS

A comprehensive review of salient water and climate issues in Central Asia, focusing on Uzbekistan, is found in this chapter. The discussion included consideration of Uzbekistan's agricultural sector since it directly influences water and climate issues in neighboring countries in Central Asia as well as Uzbekistan. Central Asia has a land area of around 3.9 million square kilometers and a population of around 54 million. About 690,000 square kilometers belong to the so called Aral Sea Basin. About 80% of the population of Central Asia depends on the water from the Amu Darya and Syr Darya rivers. Sufficient water exists in Central Asia to meet the needs of the population. However, water resources are used inefficiently because of rapid deterioration of irrigation and sanitation infrastructure and poor administrative allocation systems that do not allocate water equitably and efficiently. Furthermore, people perceive water as a free good; hence the increasing scarcity for an increasing number of people. Indiscriminate use of water for non agricultural purposes, inefficient irrigation practices, excessive use of chemicals for growing cotton and rice crops and lack of adequate drainage have caused extensive water logging, increased salinity and polluted the ground water and drainage inflow to the rivers and the Aral Sea. Water pollution from urban and industrial wastes has further aggravated pollution problems and climate changes.

Data presented in this chapter lead to a single principal conclusion, namely, that all critical issues concerned with water and climate in Central Asia, have the same cause. These problems are related to inappropriate use of water resources and the inability of government organizations to rectify the situation. The Aral Sea crisis, which is one of the most significant ecological disasters in the world, was the result of inappropriate agricultural policy during the Soviet era in Central Asia and it has been exacerbated by the breakdown of the Soviet Union. About 35 million people, including a considerable part of the Uzbekistan's population, is now impacted.

Major efforts have been made by Central Asian governments and international organizations to develop instruments and ways to improve the water use in the region. Despite these efforts, adverse agricultural policies and wasteful irrigation and drainage management practices in some of the counties in Central Asia are at the principal cause of the region's water problems and continue to result in substantial losses in agricultural productivity (Central Asia Human Development Report, 2005).

If all countries in the region were to pursue better national water management policies in a coordinated and mutually reinforcing manner, it would lead to major economic and ecological benefits. A more efficient use of water resources would help reduce regional tensions about how to share limited natural resources. Ultimately, the countries in the region will need to develop a regional framework for management of these resources.

REFERENCES

Central Asia Human Development Report. 2005. Bringing down barriers: Regional cooperation for human development and human security.

Chub, V.E. and T.A. Ososkova. 1999. Climatic changes and surface water resources of the Aral Sea Basin. Bulletin #3, Information on fulfillment of commitments to the UNFCCC by Uzbekistan., Tashkent, Glavgidromet RUz.

Co-operative program on water and climate, 2004. Coping with climate change: the Syr Darya basin in Central Asia. Working paper. Govt. of Uzbekistan, TashkentCo-operative program on water and climate.

Djalalov, S. and B.S. Chandra. 2006. Policy Reforms and Agriculture Development in Central Asia.

Muminov F.A. and H.M. Abdullaev. 1997. Agricultural-and-climatic resources of the Republic of Uzbekistan. Tashkent: Glavgidromet RUz.

Ryan, J., P. Vlek. P., and R. Paroda. 2004. Agriculture in Central Asia: Research for Development.

State Statistics Committee of Uzbekistan 2004. Govt. of Uzbekistan, Tasshkent, Uzbekistan.

UNFCCC. 1999. Initial Communication under the UNFCCC is approved by the National Commission of the Republic of Uzbekistan on Climate Change. Main Administration of Hydrometeorology at the Cabinet of Ministers of the Republic of Uzbekistan.

UNU. 1998. Central Eurasian water crisis: Caspian, Aral, and Dead Seas. I. Kobori and M.H. Glantz (eds.), The United Nations University.

CHAPTER 6

Problems and management of the efficient use of soil-water resources in Central Asia with specific reference to Uzbekistan

Rasulmant Khusanov & Muhammad Kosimov
Research Institute of Market Reforms, Tashkent, Uzbekistan

1 INTRODUCTION

Cultivated land in Central Asia is concentrated in the region of the Aral Sea. Extensive plains and the adjacent foothills are the regions suitable for agricultural production. Moreover, the majority of the land is suitable for the production of a wide range of valuable agricultural crops, including corn, cotton, rice, maize, soybean, vegetables, lucerne, fruits and others. A limited amount of annual precipitation and the severe drought during the warmest period (approximately 2 months in the north corn belt region, and 3–4 months in the southern regions) make irrigation essential to crop production. Because of the limited precipitation, cultivated agricultural land in Central Asia is mainly located along the flood plains of large and small rivers, in foothills, and in mountainous and intermountain valleys (i.e., in regions with sufficient water supply for irrigation). However, the area of irrigated lands is increasing due to broadening of river flows and the construction of artificial water reservoirs and canals.

In the limited parts of the mountainous regions, including marginal rainfed areas with annual precipitation of 280 to 350 mm (450–750 mm at sea level) and regions with annual precipitation of 350 to 400 mm or more (above 600 m at sea level), the land is suitable for the cultivation of corn (*Zea mays*). The corn crop has time to utilize winter-spring moisture during its development stages and can be harvested during the dry season. The absence of the precipitation during the warmest period of the year excludes the possibility of continuous cropping because of severe drought. Therefore, irrigation is the basis of a successful agriculture.

Irrigated land in Uzbekistan occupies approximately 9% of its total land area, yet produces more than 95% of the total production. Agricultural lands comprise 26.7 million hectares (Mha), most of which is pastures located in the semi-desert and desert zones and characterized by a low productivity.

Therefore, an efficient management of soil-water resource is crucial to the development of agriculture. Meanwhile, decreasing productivity of the irrigated lands and decline in quality of irrigation water are major threats to agricultural sustainability.

2 FACTORS CONTRIBUTING TO DECLINE IN FERTILITY

2.1 *Salinization*

The irrigated cropland area in Uzbekistan was 3.83 Mha in 1991 and 3.73 Mha in 2001. The reduction in irrigated cultivated land area is due to new constructions, specifically for infra-structure such as roads, electric/power lines etc., in accord with increase in the population. Thus, each ha of irrigated cultivated land must produce more food, raw materials for industry and other products.

Research data show that the productivity of irrigated cultivated land is progressively declining. Over the 10 years between 1991 and 2001, soil grade (ball-classification) in Uzbekistan decreased by 3 points due to increase in soil salinity. The land area affected by salinity increased to 16.3% of the total area comprising slightly salinized area of 8.4%, moderately salinized area of 2.1%, and severely salinized area of 5.8%. The soil area effected in 2002 by salinization or salt affected soils included 35.4% of slightly salinized, 17.9% of moderately salinized, and 11.2% of severely salinized soils.

Salinization (in the Republic of Karakalpakstan) is more severe in the vicinity of Aral Sea than in other regions of Uzbekistan. Between 1997 and 2002, the salinized area in the Republic of Karakalpakstan increased to 6.0% for the moderate category, 3.3% for the severe category and 17.4% for the slight category.

The situation in the Republic of Karakalpakstan is the same as that in Uzbekistan with the additional problem of decline in the quality of irrigation water. Since shirkats, farmers, and dehkans reside in the higher part of Amudarya River and discharge the brackish collector waste water into the rivers, salinity of the river water has increased to 1.2–1.5 mg/l in vicinity of the Aral Sea.

The decline in investments in restorative projects is partly attributed to increase in land area prone to salnization in Uzbekistan. Individual farmers rarely invest in improving the irrigation system. Nonetheless, 30–35% of the collector drainage network must be cleaned on an annual basis.

2.2 *Inappropriate irrigation technology*

One of the problems of the present irrigation technology is the excessive water use, and the difficulty of regulating the water supply which depends on the crop water requirement and on the maintenance of an optimum level of soil moisture in the root zone. Research data shows that for the last 40 years water consumption of 13,700 m^3 ha^{-1} or 1.37 m ha^{-1}. More specifically, the problem of excessive water use is in the Horezm region and in the Republic of Karakalpakstan, where water consumption ranges from 20,000–24,000 m^3 ha^{-1}(2.0–2.4 m/crop).

Consumptive water use for 1 centner (1 quintal or 100 kg) of raw cotton ranges from 350 to 900 m^3. This level of water use amounts to 8,000–9,000 m^3, for 25–27 centners ha^{-1}. With a furrow irrigation system, 1 centner of raw cotton requires about 150–450 m^3 of water. Thus, the consumptive water use is 2 to 3 times the actual need.

It is the excessive irrigation which leads to over-consumption. The average volume of water used ranges 54 to 58 km^3 per annum, overflow_in the sources of the irrigation is about 1 km^3, use other than in irrigated lands amounts to 7 km^3, and total overflow amounts to 18 km^3 per annum. Consequently, actual use for irrigation does not exceed 40 km^3. Consequently, the water table is rising, and is causing poor drainage and anaerobiosis. Excessive use of water also leads to leaching or runoff of plant nutrients, reduction in soil organic matter content, and depletion of soil fertility.

The research conducted at the Uzbek Institute of Cotton show that 50 to 60 t ha^{-1} of soil is washed off during a single growing season due to furrow irrigation. The eroded soil contains as much as 1 t of N, 140 to 150 kg of P, 150–160 kg of K ha^{-1}. In addition, there is also a severe loss of micronutrients which are not being replenished. In severely eroded soils, agricultural production is 30–80% lower than in uneroded soil, even with the application of chemical fertilizers. Therefore, introduction of intensive plant-production technology without soil protection is not sustainable. Without adoption of conservation – effective measures, intensive agricultural, technologies involving in discriminate use of chemicals, are serious threat to environmental pollution. Such adverse consequences of intensive technology in agriculture are connected not only to accelerated soil erosion but also to an inappropriate technical infrastructure. In other words, intensive technologies are adopted without the provision of necessary supporting technical facilities.

Therefore, inappropriate irrigation technologies lead to a reduction in soil productivity and to an inefficient consumption of irrigation water. It also increases financial and labor resource costs for the delivery of irrigation water. It is thus necessary to construct multiple water reservoirs along with improving water delivery system during the vegetative periods to enhance soil fertility and to improve soil organic matter (SOM) content. The snowmelt runoff originating from mountains erodes a range of microelements and also preferentially transports SOM. However, most of the sediments are deposited in water reservoirs and do not reach the agricultural fields. Therefore,

Table 1. The contents of humus, phosphorus, nitrogen and potassium in soil of a cooperative farm "Dustlik" in the Tashkent region (the springtime before sowing of the cotton in 2002) (Annual Scientific Report, 2002).

Soil and zone of the selection	Depth (cm)	Humus (%)	Availabe nutrients (mg kg^{-1})		
			NO_3–N	P_2O_5	K_2O
R.3 Irrigated meadow soil	0–25	1.71	76.5	25.73	175
	26–54	1.22	67.7	6.79	100
	54–100	0.89	60.0	16.40	100
R.4 Irrigated meadow soil	0–30	1.89	63.3	34.79	200
	30–50	1.28	72.4	23.36	170
	50–79	0.73	70.5	21.60	100
	79–100	0.62	–	–	–

carbon in SOM is settled in the water reservoirs rather than in the fields. Thus, SOM and carbon pool in agricultural fields is progressively declining. Such a process repeated annually and systematically causes a drastic reduction in soil productivity.

2.3 *Negative balance of nutrients and enriching soil with nourishing elements*

Soil management research has been conducted under USDA funded project "Problems of effective use of land and water resources in conditions of market economy (coordinator of the Project was Dr. R. Khusanov from 2001–2003)" on the cooperative farm "Dustlik" located in the Tashkent region. The soil of the cooperative farm is meadow, and humus contents in the upper horizon during the springtime before sowing ranges from 1.71 to 1.89% with a gradual decrease in the sub-soil horizons (Table 1). Plant available concentrations of nitrogen, phosphorus, potassium are evenly distributed in the soil profile (Table 1), with the exclusion of P_2O_5, which declines to 6.79 mg kg^{-1} in the 26–54 cm depth in the R3 soil. This decline in P_2O_5 concentration is attributed to an uneven application of phosphatic fertilizers with the autumn plowing. Research data shows that soil is depleted of its nitrate and phosphate contents, but the concentration of soluble potassium ranges from 100 to 210 mg/kg. The concentration of NO_3–N is somewhat higher in spring than in autumn (Table 2), probably because of strong uptake of nitrogen by cotton in the early phase (initial growth and bearing stages) of development. The data in Table 2 also shows a drastic reduction in the concentration of available forms of NPK at the end of the vegetative period, particularly that of NO_3–N, which is drastically reduced.

Nutrient concentration of the cultivated soil is compared with that of the irrigated meadow soil. The humus contents in upper horizon range from 1.50 to 1.89%, with reduction to 1.32% at the end of the vegetative period. There is a progressive decline in the concentration of the nitrogen, phosphorus and potassium in the soil during the crop cycle (Figure 1).

The analyses of available nutrients show that uptake of nutrients by crop exceeded the rate of application of NPK fertilizers for the last three years. Additionally, the rate of application of potassium fertilizer declined over the last two years (Table 3).

Decline in humus and nutrient concentration in soil is attributed to the lack of adoption of recommended agricultural practices (RAPs). Continuous and longterm cultivation of cotton, as well as cotton-grain crop rotation, leads to depletion of soil fertility, reduction in humus pools, and decline in physical properties (e.g., increase in soil compaction). Lack of manure application, and of RAPs also lead to reduction in the humus contents. Thus, a strong imbalance between nutrient uptake and input leads to decline in soil fertility. Consequently, there has been a strong decline in production of raw cotton, and the supply of cotton has severely lagged behind the demand. In conclusion, continuous cultivation and the rate of input of fertilizers have not kept pace with the

Table 2. The contents of humus, phosphorus, nitrogen and potassium in soil of a cooperative farm "Dustlik" in the Tashkent region during the vegetation period of 2002 (Annual Scientific Report, 2002).

Soil and zone of the selection	Depth (cm)	Humus (%)	Availabe nutrients ($mg\,kg^{-1}$)		
			NO_3–N	P_2O_5	K_2O
Before sowing of cotton					
P.3. Irrigated meadow soil	0–25	1.71	76.5	25.73	175
	26–54	1.22	67.7	6.79	100
	54–100	0.89	60.0	16.40	100
P.4. Irrigated meadow soil	0–30	1.89	63.3	34.79	200
	30–50	1.28	72.4	23.36	170
	50–79	0.73	70.5	21.60	100
	70–100	0.62	–	–	–
The flowering stage					
P.3. Irrigated meadow soil	0–25	1.50	45.1	12.1	170
	26–54	1.32	38.5	11.7	100
	54–100	0.72	32.0	8.9	120
P.4. Irrigated meadow soil	0–30	1.61	31.5	10.7	210
	30–50	1.30	13.0	13.0	151
	50–79	0.70	17.1	7.8	100
	79–100	0.60	–	–	–
At the end of the vegetative period					
P.3. Irrigated meadow soil	0–25	1.32	0.085	22.7	150
	26–54	1.08	0.068	17.3	100
	54–100	0.79	0.061	7.4	100
P.4. Irrigated meadow soil	0–30	1.45	Trace	22.9	210
	30–50	1.19		18.1	150
	50–79	0.77		9.5	100
	79–100	0.71		7.3	–

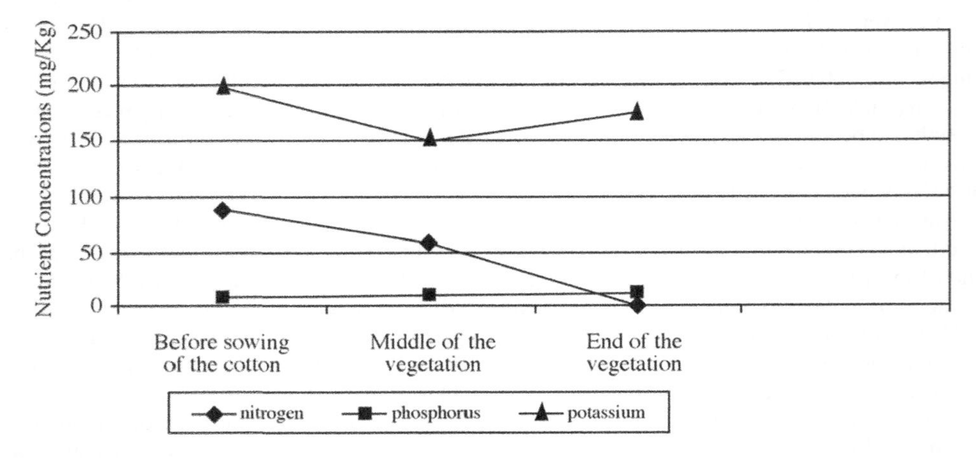

Figure 1. Temporal changes in nutrient concentrations.

nutrient uptake by cotton-based rotations. The imbalance has caused a decline in soil fertility and a reduction in the cotton production (Figure 2).

A similar experiment was conducted in the cooperative farm "Pakhtakor" in the Sirdarya region. The objective of this study was to assess the effects of different periods and rates of the application

Table 3. Indicators of the actual application and uptake of NPK per hectare of cotton production in cooperative farm "Dustlik" Annual Scientific Report, 2002.

	2000			2001			2002		
Indicators	N	P	K	N	P	K	N	P	K
Actual application (kg)	46.4	20.5	43.2	120.0	50.1	–	104.4	46.0	–
Actual uptake (kg)	62.8	20.7	49.6	130.0	53.0	–	117.0	55.0	–
Uptake as % of application	135.3	100.9	114.8	108.0	105.8	–	112.0	119.5	–
Productivity (centares haha^{-1})		15.7			26.2			23.4	

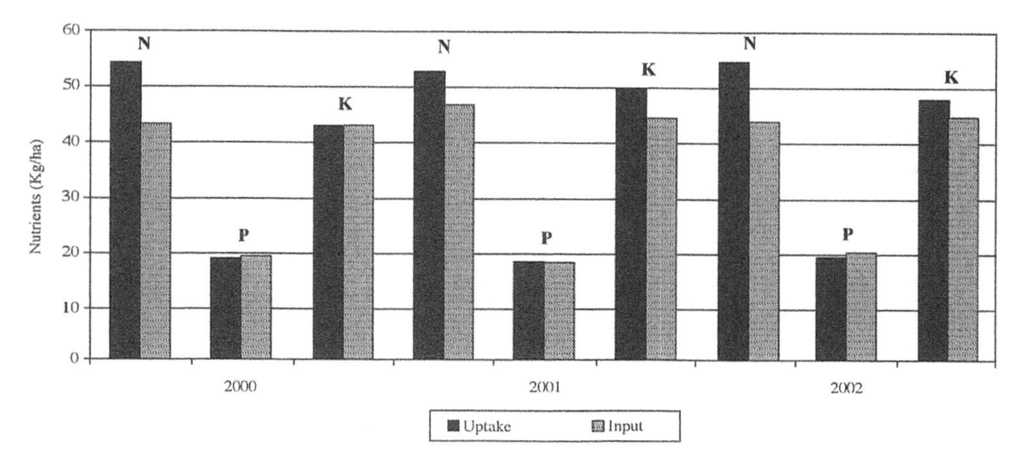

Figure 2. Factors of the actual application and uptake of NPK for cotton cultion in cooperative farm "Pakhtakor." (Annual Scientific Report, 2002).

of the mineral fertilizers on irrigated meadow soil under cotton production. The data show that the highest concentration of nutrient elements existed in spring before sowing and prior to bearing and flowering growth stages. Even during these phases, concentration of NO_3, P_2O_5 and K_2O in the upper (0–25 cm.) soil layer were not very high (Table 4). Furthermore, nutrient concentrations declined strongly during the vegetative and flowering phases due to rapid uptake of nutrients by plants. The data show that concentration of available (soluble) nutrients drastically declined at the end of the growing period.

The apparent increase in potassium concentration in the upper soil horizon is attributed to the local application of potassium fertilizers. Comparing the specific elemental concentration during the spring with that after the vegetative period shows that concentrations of humus and total nitrogen changed towards the end of the vegetative growth. Such changes in the concentrations of the available/soluble forms of these elements may be attributed to the application of mineral fertilizer during the vegetation period of the cotton (Figure 3).

Observations on uptake and application of nutrient elements were made over three years. The data on uptake of nutrient elements show that crop removal of NPK for the last three years systematically exceeded the amount applied as mineral fertilizers. Only in 2002 was the rate of application of phosphatic fertilizers equal to the uptake (Table 5).

Table 4. The contents of humus, nitrogen, phosphorus, and potassium in the soil of cooperative farm "Pakhtakor" in 2002 (Annual Scientific Report, 2002).

	Depth (cm)	Humus (%)	Amount (%)			Available nutrients (mg kg^{-1})		
			N	P	K	NO_3–N	P_2O_5	K_2O
Before sowing cotton								
P.1. Irrigated meadow soil (K-39)	0–25	1.72	0.126			89.1	9.47	200
	25–45	1.58	0.112			70.9	18.93	150
	45–75	0.97	0.088			55.9	5.33	75
	75–115	0.75	0.061			–	6.10	50
P.2. Irrigated meadow soil (K-12)	0–26	2.01	0.130			81.9	16.27	175
	26–45	1.65	0.101			71.3	12.13	125
	45–78	1.01	0.097			59.2	4.00	50
	78–110	0.51	0.048			–	6.80	100
At flowering stage								
P.1. (K-39)	0–25	1.55				57.5	10.6	150
	25–45	1.22				49.0	10.9	160
	45–80	0.87				45.7	6.0	70
P.2. (K-12)	0–26	1.72				56.4	12.2	140
	26–45	1.49				48.7	7.4	150
	45–75	0.75				44.5	6.0	100
At the end of vegetative period 110								
P.1. (K-39)	0–25	1.41	0.105			Trace	12.2	175
	25–45	1.24	0.093				7.4	110
	45–75	0.53	0.039				6.0	75
P.2. (K-12)	0–26	1.47	0.096			Trace	11.4	160
	26–45	1.26	0.078				6.0	100
	45–75	0.57	0.042				3.0	55

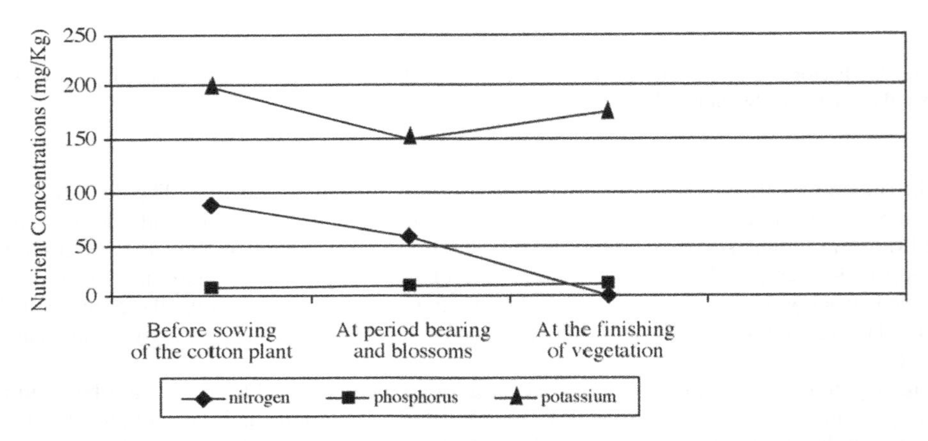

Figure 3. Temporal changes in nutrient concentrations of cooperative farm "Pakhtakor" (Section 1, contour 39, depth 0–25 cm) (Annual Scientific Report, 2002).

Therefore, the data showed that irrigated meadow soil is depleted of its humus pool, with humus concentrations ranging between 1.41 and 2.01%. Further, gross nitrogen concentration at the end of vegetation decreased throughout the soil profile. A slight increase in phosphate concentration in cultivated field may be due to immobilization of applied fertilizer and the reduced uptake

Table 5. Factors of the actual application and uptake of NPK for production of raw cotton in cooperative farm "Pakhtakor" (Annual Scientific Report, 2002).

	2000			2001			2002		
Indicators	N	P	K	N	P	K	N	P	K
Actual application (kg ha^{-1})	43.4	19.5	43.2	46.8	18.2	44.2	43.6	19.8	44.2
Actual uptake (kg ha^{-1})	54.6	19.1	49.5	52.9	18.4	49.6	54.6	19.1	47.5
Uptake (% of application)	125.8	102.1	114.6	113.0	101.0	112.2	125.2	96.5	107.5
Productivity (c ha^{-1})*		12.4			11.6			12.9	

*10 c = 1 metric ton.

by cotton. Overall, however, cultivation of cotton caused depletion of N, P, K and soil humus concentration.

The data in Table 5 show that excess of uptake over the application of nutrients is the principal cause of decline in soil fertility under cotton production. Moreover, the application of the mineral fertilizers alone (without manure or biosolids) reduces the use efficiency of fertilizers applied. It is important to note that with lack of adoption of RAPs, humus pool and nutrient concentrations are progressively diminished leading to an overall decline in soil fertility.

2.4 *Decline in soil fertility in the Aral Sea Basin (Republic of Karakalpakstan)*

Land mass of the Republic of Karakalpakstan is located downstream of the Amudarya River. Important factors which influence soil quality of the region include the following:

Construction of water reserviors in the Amudarya River limit deposition of carbon in the fields, because it is settled in the reservoirs:

1. Increase in salt contribution from the dried out Aral Sea bed;
2. Leaching of plant nutrients into the sub-soil horizons due to improper technology and excessive irrigation (24–25 thousand m^3 ha^{-1});
3. Nutrient imbalance caused by more crop uptake than input through application of chemical fertilizers;
4. Continuous cultivation of nutrient depleting crops (e.g., cotton, wheat, and rice) leading to decline in soil fertility and degradation of soil structure; and
5. Reduction in implmentation of soil restorative measures leading to rise in water table and increase in soil salinization.

The problem of soil degradation is exacerbated by the frequent passes of heavy machinery which also adversely affects soil structure and fertility. These processes have led to decline in soil fertility in the Republic of Karakalpakstan to grade 3 and to reduction in rating by 41 points during the last 10 years. (Table 6.).

In contrast, soil fertility in Nukus and Chimbay districts is not declining, and instead is improving, because of two factors. One, the farmers of these districts specialize in growing vegetables, follow crop rotations, and manage soil fertility by growing leguminous and oil seed crops. Two, these farmers have a high personal interest in their lands. They look after soil quality, maintain the proper NPK balance, and apply oranic matter and manures whenever necessary.

However, a sharp decrease in the rate of soil grade in the Muynak district is attributed to the drying of the Aral Sea and the subsequent movement (blowing) of salt and sands throughout region.

Based on the data of the State Nature Protection Committee of the Republic of Karakalpakstan, the humus concentration, which is the main indicator of soil fertility, decreased by 30–50% over the 15–20 year period. The data on soil classification/grade show that there is a reduction in highly productive soils in the region (Table 7).

In 1990 the soil of group 2, class III comprised of only 0.7% of the total area. In 1999, for comparison, the area under this class increased up to 5.4%, and area under soil in Class VIII, on

Table 6. Dynamic change of soil classification/grade in districts of Karakapakstan Republic for 1990–2002 (Annual Reports of the Land Resources Committee of Uzbekistan, 1990–2002).

Number	District	1990	2000	Change
1	Amudarya	51	46	−5
2	Beruniy	50	43	−7
3	Bozatau	43	41	−2
4	Karauzyak	42	41	−1
5	Kegeyli	43	39	−4
6	Kumgrad	40	37	−3
7	Kanlikul	40	38	−2
8	Muynak	40	27	−13
9	Nukus	40	43	+3
10	Hujayli	43	42	−1
11	Shumanay	45	39	−6
12	Chimbay	41	42	+1
13	Tahtakupir	44	40	−4
14	Turkul	51	50	−1
15	Ellikkala	49	42	−7
	Total	44	41	3

Table 7. Change in soil class/grade dynamics and estimation of irrigated land area in the Republic of Karakalpakstan (Annual Reports of the Land Resources Committee of Uzbekistan, 1990–2002).

			Irrigation areas (10^3 ha)			
Groups	Classes	Rates	1990	(%)	1999	(%)
1	I	0–10	–		–	
	II	11–20	–		1.9	0.4
2	III	21–30	3.1	0.7	25.1	5.4
	IV	31–40	239.9	50.8	242.2	52.3
3	V	41–50	136.9	29.0	70.7	15.2
	VI	51–60	41.3	8.8	48.0	10.4
4	VII	61–70	16.8	3,5	24.5	5.4
	VIII	71–80	16.1	3,4	6.8	1.4
5	IX	81–90	0.45	0.1	–	
	X	91–100	–			
6	Not estimated lands		17.5	3.7		9.5
	Total		472.3	100		100
	Average classification rate		44		41	

the other hand, decreased from 3.4% to 1.4% over the same period. Area under classes IV and V comprised 31–50% in 1999 and 67.5% in 1999 (Table 7).

Considering the fact that the soil of Karakalpakstan is sub-divided into northern and southern zones, it is also possible to assess soil fertility in these two distinct regions. In the northern zone, the majority of the soil (61%) is designated as Class-4, 40% of which is in the category of average to good soil. In southern zone, the good soil comprises 61% of the area, and average soil covers 39% of the total area. In other words, soil quality is improved in the southern regions of the Karakalpakstan Republic.

The ground water level changed considerably over the 6 years between 1996 and 2002,which has greatly influenced the overall quality of the land (Table 8). In comparison with 1996, the land area with ground water level of 0–1 m decreased by 31.1% in 2002. All of these changes are linked

Table 8. Temporal change in the ground water level in the Republic of Karaklpakstan between 1996 and 2002 (Annual Reports of the Land Resources Committee of Uzbekistan, 1990–2002).

	Land area (10^3 ha)						
Depth (m)	1996	1998	1999	2000	2001	2002	Area 2002 as % 1996
0–1.0	23.05	24.39	19.17	0.77	0.91	15.9	68.9
1.0–1.5	75.31	109.24	148.91	17.23	14.10	66.06	87.7
1.5–2.0	332.12	163.53	237.63	101.80	61.17	169.19	50.9
2.0–3.0	58.56	89.05	78.60	185.20	111.11	96.39	164.6
3.0–5.0	6.67	13.19	13.96	193.26	221.12	100.60	1508.2
>5.0 m	–	–	0.22	1.59	90.14	50.45	–

Table 9. Dynamic of change in salinity of irrigated lands in The Republic of Karakalpakstan from 1997 to 2002 (Annual Reports of the Land Resources Committee of Uzbekistan, 1990–2002).

		Salinization of soils							
	Irrigated land	Unsalinized		Weakly salinized		Moderately salinized		Extremely and strongly salinized	
Years	10^3 ha	10^3 ha	%	10^3 ha	%	10^3 ha	%	10^3 ha	%
1997	500.9	33.3	6.7	257.2	51.3	162.3	32.4	48.1	9.6
1998	500.0	28.4	5.7	254.0	50.8	168.7	33.7	48.9	9.8
1999	500.8	50.3	10.0	255.4	51.0	152.4	30.4	42.7	8.6
2000	500.1	47.6	9.5	246.1	49.2	158.3	31.7	48.1	9.6
2001	500.2	50.0	9.9	215.8	43.1	172.0	34.4	62.3	12.6
2002	500.2	73.8	14.8	169.7	33.9	192.2	38.4	64.5	12.9
2002 area as %1997	99.8		–	65.9	–	118.4	–	134.1	–

with recurring low water levels during the previous years. Thus, in 2001, the aggregated water supply was only 40.9% of the annual average supply. In low water years, the area sown to cotton, wheat and rice sharply declined.

As has been mentioned before, there is a sharp decline in the number of land ameliorative projects, which has raised the ground water level and increased secondary soil salinity. Reversing these trends would necessitate increasing the number of ameliorative projects as well as improvements in crop production technology.

In The Republic of Karakalpakstan, about 90% of the irrigated land is subject to some degree of salinization. Analysis of the data shows that the area under severely salinized soil is increasing (Table 9). Land area affected by moderate soil salinity increased from 32.4% in 1996 to 38.4% in 2002, and that affected by severe salinization doubled over the same period.

With severely salinized soils, it is impossible to achieve the production targets. As the Republic of Karakalpakstan is situated in the northern zone of the cotton belt, late harvest does not permit adoption of RAPs. Continuing some farming operations in late fall and early winter is not recommended. Further, leaching caused by excessive irrigation (3.5–4.0 thousand m^3 ha^{-1}) can transport soluble plant nutrients into the sub-soil horizon.

Drastic change in ground water levels also leads to destabilization of agriculture, and adversely impacts agricultural economy. Of the 500.8 thousand ha of irrigated lands in 2000, only 389.2 thousand ha were irrigable. Further, the amount of water available during 2001 was only 40.9% of that available in 2000. The irrigable land area decreased from 503 thousand ha in 1990 to 460.3 thousand ha in 2002, a reduction of 8.5% over the 12 year period (Table 9).

Adoption of inappropriate soil management technology also negatively influences soil fertility. The repeated passes of heavy equipment during soil preparation for sowing and during the growing cycle strongly compacts the soil and degrades its structure. Finally, these processes reduce productivity of irrigated land and lead to decline in revenue. Specific curative measures are routinely taken to minimize the adverse effects of these practices. Yet, adoption of routine practices does not produce the desired results. In the meanwhile, problems of soil degradation and productivity decline are being exacerbated.

3 MEASURES TO INCREASE THE USE EFFICIENCY OF WATER AND SOIL RESOURCES

3.1 *Periodic water shortages downstream of Amudarya*

Water shortage has occurred repeatedly for decades in Uzbekistan which, according to hydrologists, may reoccur in the future every 5–6 years. During years of water stress, water supply levels for agricultural land are merely 30–40% of the crop requirements. In these situations, a large area of irrigated land either remains unsown or crops fail. Soil fertility in agro-ecoregions declines as does the microbial activity, which adversely affects productivity. This is a serious problem in downstream regions of Amudarya (the Republic of Karakalpakstan and Khorezm region), as well as in regions with predominately steep terrains. Research data show that decline in fertility of irrigated soil is caused by salinization, drought stress, depletion in humus content, nutrient/elemental imbalance, and soil contamination by industrial pollutants.

Thus, there is a strong need for development of a research project to identify sustainable agricultural practices in the region. Under the conditions of a recurring water shortage, several research projects were conducted in Uzbekistan to assess agronomic productivity and quality of the produce. However, research aimed at assessing the productivity of different crop species (native and introduced) grown at a range of water supply (40%, 50%, 60% and 70% of optimum) is urgently needed.

3.2 *Measures to enhance soil microbiological activities*

Decline in humus concentration in soil is also accompanied by reduction in agronomic and agrophysical quality of soil. There is also a decline in nutrient concentration, both macro (NPK) and micro (Cu, Mn, Zn), in irrigated soils. Deficit or an absence of any one of these elements can adversely impact the metabolic processes and retard the growth and development of plants, reduction in crop yield and, deterioration in the quality of agricultural produce.

Therefore, research is being conducted on the influence of electrical conductance of soil on the agricultural, chemical characteristics of soil, transport of chemical elements, and microbiological activity of soil in the soil-plant system under a range of water supply. These experiments have been conducted in three-phases involving cultivation of cotton, soybean, corn, sunflower, and millet. These experiments have been conducted in principal soil types of the region. Soils of the experimental site, which is a meadow, is generally of loamy texture, with moderate salinity, and with a mean ground water depth of 1.5 to 2.5 m at the end of the growing season.

With regards to the mineral nutrition of crops, importance of N, which is mostly concentrated in the humus fraction, cannot be overemphasized. Soils of the experimental site contain low humus concentration of less than 1%. Some cultivated soils may contain as much as 1.4% of humus concentration. Thus, soil fertility of one region differs from that of another because of differences in humus concentration.

Depending on textural composition of the plow layer, humus concentration ranges from 0.5–0.8%, and as high as 1.1–1.5%. Humus concentration decreases with depth to as low as 0.2 to 0.9%. In irrigated meadow soils, humus content can increase to 50–60 cm depth. Soils of the

Amudarya delta are of low inherent fertility, because of low humus concentration and the attendant low concentration of essential plant nutrients.

Adverse hydrologic and management conditions of the region are attributed to the ground water flow. These adverse conditions are caused by a gentle slope gradient of the delta, coarse (sandy) texture of soils, and construction of large embankments with complete absence of impermeable layers. Predominately high summer temperatures, low relative humidity of the air, and high wind velocity exacerbate soil evaporation which mostly exceeds the precipitation. Furthermore, high soil salinity increases plant water requirements. Soil analyses of the experimental plots show moderate to average chlorides and sulphate concentration in these soils. Thus, predominant type of soil salinity is chloride-sulphate with some areas affected by sulphate-chloride salts. Experiments have been conducted on the effect of macronutrients and micronutrients in relation to the dynamics of soil microflora under different crops as influenced by electroprocessing under varying levels of water supply (40%, 50%, 60% and 70% of the optimal level). The data show that during moisture deficiency, electroprocessing has a positive influence on agronomic productivity, along with improvements in chemical and microbiological properties of soils.

Soil analysis has shown that concentration of soluble forms of nitrogen, phosphorus and potassium with 70% of the water supply in plow layers of cultivated soils planted to cotton is more under RAPs than under control. For example, at the beginning of summer and with 70% of the water supply, concentrations of NO_3–N in the plow layer ranged from 71.4 to 95.2 mg/kg in control compared with 108.0 to 148.0 mg kg^{-1} with electroprocessing. Concentrations of P_2O_5 under cotton also increased under electroprocessing, and ranged from 40.0 to 60.0 mg kg^{-1} soil under control compared with 50.0 to 70.0 mg kg^{-1} with electroprocessing. In the beginning of autumn, phosphorus concentration decreased to 27.0 mg kg^{-1} with electroprocessing compared with an increase to 31.8 mg kg^{-1} in control. The best results of electroprocessing in cotton are obtained for the soluble potassium concentration for all levels of water supply. Concentration of mobile form of K_2O may ranges from 219.0 to 398.8 mg kg^{-1} in the humus-rich layer, and 94.0 to 350.8 mg kg^{-1} soil in the sub-soil horizon. It is only during the mid-spring when the amount of potassium becomes less with the electroprocessing treatment (232.0 mg kg^{-1} of soil) than without electroprocessing (398.0 mg kg^{-1} soil). Therefore, under conditions of water shortage, high yield of raw cotton can be obtained in salt-affected soils of the Aral Sea with 60 and 70% of water supply but with electroprocessing of seeds and vegetative components of the cotton at all phases of vegetative growth.

Some experiments have also been conducted with regard to the nutrient uptake by sunflower under the impact of electroprocessing. The data shows an increase in the uptake of NO_3–N and K_2O by electroprocessing, regardless of the water supply, which is also in contrast with the control. The quantity of mobile forms of N under the impact of electroprocessings increased from 119.0 to 166.0 mg kg^{-1}, compared with 38.1 to123.8 mg kg^{-1} without electroprocessing with water supply level of 60%. Higher concentrations of mobile potassium were formed under 60% of moisture availability with electorprocessing (from 150 to 220 mg kg^{-1}) than without electroprocessing (from 125 to 200 mg kg^{-1}). However, concentration of P_2O_5 was an exception, because increase in its concentration was less at 40% of the moisture availability with electroprocessing (from 25.0 to 33.8 mg kg^{-1}) compared with only 11.4 to 25.0 mg kg^{-1} in the control. Micronutrients in the soil reserve pool can be used to enhance soil fertility.

The dynamics of micro-nutrients in soils of the Aral Sea Basin have not been studied in Uzbekistan. The concentrations of the mobile forms of copper, zinc, and manganese in meadow-alluvial soil formed on alluvial deposits have been studied to some extent. Distribution of these micronutrients in the soil profile is influenced by chemical and mechanical processes, including the degree of salinity and influence of anthropogenic factors.

Research experiments show that meadow soil in the desert zone differs in concentrations of mobile micronutrients from those in meadow soils of sierozem oasis. Soil analyses show that concentrations do not vary significantly from that of the control even under the influence of electroprocessing. Concentrations of Cu in soils of the experimental plots range from 0.30 to 0.50 mg kg^{-1}, but sometimes can be as high as 0.70 to 1.27 mg kg^{-1}. There may be some increase in control and with electroprocessing under cotton in 3 and 4 stages of vegetative growth. The concentrations of copper in the experimental soils were lower on the whole compared with those in the hyohomorphic soils containing hydrogen sulfide in the Tashkent region. In these soils concentration of available copper in the humus-rich horizon may range from 0.9 to 0.22 mg kg^{-1}. This low concentration of copper in the upper horizon reduces any additional impact of salinization on its concentration. The amount of Cu in these soils exists in form unavailable to plants.

Experimental data on zinc show high concentrations in the upper humus-rich horizon with a gradual decrease with increase in soil depth. However, concentration of micronutrients was studied only in the plow layer of arable lands. The concentration of available zinc in all experimental treatments ranged from 3.30 to 5.30 mg kg^{-1}. Such concentrations in meadow soil of the desert zone are similar to those in meadow – alluvial soils of the hydrogen sulfide oasis. The normal cotton growth is affected when zinc concentrations are in the 1.5–2.5 mg kg^{-1} range.

Manganese concentrations in experimental soils were insufficient in the upper horizon in all treatments in the first phase and ranged from 29.0 to 31.0 mg kg^{-1}. In the second phase, (the butonization stage) concentrations of manganese ranged between 24.4 mg kg^{-1} and 31.3 mg kg^{-1}. Analytical data showed that in the reproductive phase, concentrations of manganese decreased significantly both in the control treatment and with electroprocessing, and ranged from 17.4 to 32.2 mg kg^{-1}. These trends suggest that the experimental soil contains only trace amounts of manganese. Thus, using manganese as microfertilizers is recommended. Soils derived from alluvial material differ substantially with regard to the concentrations of mobile Zn, Mn, Cu from those of the meadow soil formed from hydrogen sulfide oasis depending on climate, humus content, level of micronutrient composition, and the nature of their distribution within the soil profile under the influence of electroprocessing.

Increase in availability of micronutrients to crops may be due to the applied fertilizers, but also due to increase in microbial activity. Soluble forms of micronutrients available for plant growth are strongly influenced by activity of specific micro-organisms in the rhizoshere.

Several experiments were conducted to assess the impact of electroprocessing on soil, seed and the microorganisms within the rhyzosphere of different crops under a range of water availability (40, 50, 60 and 70%). The principal groups of microorganisms which affect the cycling of nitrogen, carbon and phosphorus during the growth cycle of cotton, corn, sunflower, millet and sesame, were identified. In addition, ammonicator, oligonitrophyl bacterias, micromicet (the fungus) and actinomyctes in the rhyzosphere of different crops, as well as dynamics of development of soil microflora with and without electroprocessing in cotton were also identified.

The microbial count in each physiological group was assessed taking into account as much as is possible because microbial community is actively decomposing the biomass and soil humus. The data obtained supports the conclusion that soil moisture in the rhyzosphere of plants grown in saline soils greatly influences the nutrient concentration and microbial biomass. Even with moisture deficit, electroprocessing increases the concentration of beneficial microorganisms in soil and reduces the amount of harmful microorganisms. It also improves phytosanitary conditions of soil by reducing splash fungus and increasing the amount of actinomyces. The data from these studies show that at crop maturity the impact of electroprocessing successfully alters the dynamics of gross microorganisms under various levels of water availability. For example, during a vegetative period of cotton with $<40\%$ moisture availability, there are as many as 1872 thousand microorganisms per g of soil with electroprocessing compared with 1311 thousand per g soil without electroprocessing. The difference in microbial content with and without electroprocessing is not significant under 50% of moisture availability in the cotton rhizosphere. The gross amount of microorganism is 507 k/g soil without and 5829 k/g soil with electroprocessing.

The number of actinomys is significantly higher in soil under cotton than under corn. Prevalence of actinomys under these crops includes Cinereous and Roseus. The amount of fungal species is low in contrast with other microorganism in all soils. The higher concentrations of fungal species are observed in soil under sunflower and sesame. Dominant fungal species include Penicilium, Aspergillus.

Oligonitrophyls microorganisms are commonly observed in experimental soils. The ability of oligonitrophyls to grow under very low NO_3–N level in sub-soil enables them to develop under conditions unsuitable for other microorganisms. A large number of Bac. micoides is observed in these soils.

The total amount of ammonifying microbes is comparatively low in these soils. The largest amount of the ammonifying microbes occurs in the humus fraction. Higher concentrations are observed in soil under cotton and sesame crops.

The electroprocessing influences microbiological activity of soil differently at different vegetative stages of cultivated crops. Therefore, study of growth of microorganisms in soil depends on soil type and organic material (humus) concentration, degree of salinity, and the dominant vegetation. Soil properties greatly influence the number of microorganisms and their compositions, but not necessarily the presence of one or more physiological groups. The general microbiological activity depends on the concentration of organic matter and the degree of soil salinity.

The development of microbial community is higher in soils with high moisture content and humus concentrations than in soils with low moisture content. Thus, it is necessary to grow those crops which are tolerant to drought and high salt concentrations in the soil.

3.3 *Improved cropping systems – an important factor in efficient use of soil-water resources*

More than 85% of irrigated cultivated land in Uzbekistan is sown under cotton and wheat. Several experiments have been conducted to improve cotton-based cropping systems and to avoid a monoculture. However, rather than cotton monoculture, cotton-grain monoculture is now being widely practiced, and both of these cropping systems are the primary reasons for the decline in soil fertility. Between 1990–2004, area under lucerne declined from 467.8 thousand ha to 143.4 thousand ha, or by a factor of 3.2. Indeed, the area under legume cultivation has decreased drastically. Intensive cropping and decline in area under restorative plants (e.g., lucerne) hinder restoration of soil fertility.

Thus, it is crucial to improve the cropping system. On the basis of long-term experiments, Uzbek scientists recommend the following cropping systems: 40% cotton, 30% wheat, 20% lucerne and other leguminous species, and 10% vegetables, potatoes and melon (Avliakulov et al., 1998). In addition, it is important to restore the traditional practice of growing intermediate crops (or catch crops) with winter vegetation (rape, forages). Such a land use and cropping system enhances restoration of soil fertility.

3.4 *Using compound fertilizers is important factor to sustaining agronomic productivity*

Throughout Central Asia including Uzbekistan, N:P:K are used separately to achieve the desired ratio of 1:0.7:0.5. In reality, however, it is difficult to obtain such a ratio of N:P:K. There are numerous adverse effects when the fertilizers are not applied in the desired ratio (section 2). Thus, it is essential to manufacture compound fertilizers so that industry can supply N:P:K in the ratio desired, and the compound fertilizers also include Zn, Mn, and Cu which are generally not supplied. This strategy will increase the use efficiency of mineral fertilizers and also supply N at the desired rate.

3.5 *Using natural factors to lower the ground water level*

A large part of the soils of Uzbekistan and most of those of Central Asia are either saline or prone to salinization. High concentrations of salt exist in the sub-soil, which move upward into the plow layer upon dissolution by the irrigation water and rise in the ground water level. Therefore, preventing the rise of the ground water is important to reducing salinization. Several options to lowering to the ground water table are: (i) extending the collector-drainage network; (ii) improving methods of irrigation by regulating and reducing the rate of irrigation; and (iii) improving the cropping system by combining crops of high and low water requirements.

The first two strategies require higher capital investment. Although administrators prefer these strategies, they are not effective measures because of insufficient infra-structure and low rate of fertilizers. In the meanwhile, the 3rd strategy is not sufficiently used. Grouping crops according to their stages of vegetative growth can decrease the irrigation requirement. Increase in the area sown to wheat and cotton increases water use throughout the year, especially when these two crops occupy more than 85% of the irrigated cropland area. Yet, the process can be regulated by improving the cropping system (i.e., irrigation in August, September and October to decrease water requirement and achieving the natural decline in the ground water level). Moreover, this strategy does not require large investment, because it is based on the regulation of irrigation water flow (or a natural factor). Ignoring this important factor is one of the causes of the increase in salinization from 48.2% in 1990 to 64.4% in 2002.

4 CONCLUSIONS

Agriculture in Central Asia is based on irrigation, and soils are prone to severe salinization. Intensive cropping since 1950s has depleted soil carbon, humus, and microelements reserves with the attendant decline in soil fertility.

Traditional cropping systems neither restore nor conserve soil fertility. Inefficient restorative measures used and lack of funding support are hindrance to improving the collector-drainage network which is in disarray. Inappropriate irrigation soil management practices have adversely impacted soil fertility.

Research data show that decline in fertility of irrigated soils is caused by salinization, drought stress, decline in humus and nutrient reserves and soil pollution. Thus, there is a strong need to conduct research on restoration of agricultural soils in the arid regions.

Field studies support the conclusion that soil moisture plays an important role in microbial biomass, activity and processes. Despite the drought stress, electroprocessing increases the concentrations of useful microorganisms and decreases those of harmful microorganism. Thus, electroprocessing is a beneficial practice. It influences microbiological activity of soil differently at different vegetatinoal/growth stages of crops. Increase in microbial activity in a specific soil type depends on humus concentration, degree of salinization and the vegetation type.

More than 85% of irrigated cropland in Uzbekistan is sown to cotton and wheat. Several experiment have been conducted to improve cropping system and to avoid cotton monoculture. Rather than cotton monoculture, cotton-grain monoculture is now widely practiced in Uzbekistan. These cropping systems are primary causes of the decline in soil fertility. Between 1990–2004, the area sown to lucerne decreased from 467.8 thousand ha to 143.4 thousand ha, by a factor of 3.2. The area planted to other leguminous species also decreased. Such an intensive land use and absence of soil restorative crops in the rotation cycle do not promote restoration of soil fertility.

Thus, it is crucial to improve the cropping system. Recommended cropping system for Uzbekistan include the following: 40% cotton, 30% wheat, 20% lucerne and other leguminous species, and 10% vegetables, potatoes and melon. Moreover, it is also necessary to restore the traditional practice of sowing the winter catch crops (rape, forages). These cropping systems restore soil fertility and sustain agronomic productivity.

REFERENCES

Annual Reports of the Land Resources Committee of Uzbekistan. 1990–2002. Tashkent, Uzbekistan.

Annual Scientific Report. 2002. Problems of efficient use of land and water resources in conditions of market economy. Tashkent, Uzbekistan.

Annual Scientific Report. 2003. Problems of stabilization development of agriculture in droughty zones and low of Amu Darya in less water condition. Tashkent, Uzbekistan.

Avliakulov A., A. Tsamutali, R. Khusanov and G. Bezborodov. 1998. System of husbandry in the conditions of radical change of structure of agricultural production. Agrosanoat akhboroti Tashkent, Uzbekistan.

CHAPTER 7

Underground and surface water resources of Central Asia, and impact of irrigation on their balance and quality

R.K. Ikramov
Central Asian Research Institute of Irrigation, Tashkent, Uzbekistan

1 GEOGRAPHIC SITUATION

The Aral Sea Basin is a good representation of the Central Asia region. It consists of the basins of Amudarya and Syrdarya rivers, and is shared among five former Soviet Union republics: Kazakhstan, Kyrgyzstan, Turkmenistan, Tajikistan and Uzbekistan. These countries cover 87% of the total area of the Aral Sea basin, and the remainder 13% lies in Afghanistan and Iran. The entire basin spans between longitudes 56° and 78° East and latitudes 33° and 53° North. Total area of the basin is 1.55 million km^2, of which 0.59 million km^2 (or 38%) is arable land (Table 1) (UNESCO, 2000).

Land-locked situation of Central Asia in the Eurasian continent is responsible for the sharply continental climate and small amount of unevenly distributed precipitation. The potential evaporation (evaporation from water surface) varies between 1000–12000 mm in the foothills and 1500–1600 mm per annum in the desert zone: Precipitation-evaporation ratio in the region varies between 0.06 and 0.2.

Average temperature in July in the low elevations, in the valleys and deserts varies from 26°C in the north to 30°C in the south, with the maximum of 45–50°C. Precipitation in the lowlands and valleys is 80–200 mm per annum, and most of the rainfall is received during winter and spring. At the same time precipitation is 300–400 mm on foothills and 600–800 mm on the southern and south-western faces of mountain ranges.

Climate of the region is influenced by the geographic and geomorphological conditions. These are also the factors which influence the difference in irrigation water demand.

There are two major geomorphological-landscape zones based on the climatic characteristics and geomorphological features: desert lowlands and foothill plains.

The desert lowland regions include:

- fluvial terrace (medium and lower), poorly drained under natural conditions;
- alluvial flats, closed basins without groundwater outflow;

Table 1. Land resources of the Aral Sea basin, FAO, 1997 (Chub, 2000).

Country	Area (ha)	Arable land (ha)	Cultivated area (ha)	Irrigated area (ha)
Kazakhstan*	34,440,000	23,872,400	1,658,800	786,200
Kyrgyzstan*	12,490,000	1,257,400	595,000	422,000
Tajikistan	14,310,000	1,571,000	769,900	719,000
Turkmenistan	48,810,000	7,013,000	1,805,300	1,735,000
Uzbekistan	44,884,000	2,544,770	5,207,800	4,233,400
Aral sea basin	154,934,000	59,161,500	10,036,800	7,895,600

* Includes only provinces in the Aral Sea Basin.

- drainless coastal deltas;
- dry deltas of shallow and medium rivers;

The foothill plains include:

- fluvial terraces of upper and medium reaches, foothill plains well drained under natural conditions;
- undulating foothill plains, drained under natural conditions;
  detrital cone with well-drained upper part, and no groundwater outflow in the lower part.

Geographic-geomorphological structures of mountain plans, foothill plains, undulating foothill plains, upper and medium reaches river terraces and upper and medium parts of valleys consist of automorphic soils with deep groundwater and good natural drainage. These landscape-geographic-geomorphologic areas do not require complex measures of ecologic-reclamation processes under any possible land development and irrigation. Development of irrigation in these areas most often causes erosion processes, soil washing out and high soil losses. In other areas with poor natural drainage or with no groundwater runoff, irrigation leads to complex ecologic-reclamation processes: waterlogging, saline seepage, depletion of surface and ground waters, etc. Thus, there is a need for adoption of more complex water conservation/protection measures, and irrigation and drainage technologies in managing ecologic-reclamation processes in these areas (UNESCO, 2000).

2 SURFACE WATER RESOURCES

Large rivers of Central Asia (i.e., Amudarya and Syrdarya) belong to the Aral Sea basin. These are the main sources of surface water and flow directly into the sea, hydrographically connected to the basin and situated within the Aral basin. There are also small rivers: Zarafshan, Kashkadarya, Tedgen and Mougab. The latter two are linked to the basin of Amudarya and Karakum channel. Rivers Chu, Talas, and Assa flow into the Issyk Kul lake and many small rivers flow down from the mountainsides of Pamirs-Alay and Tien Shan but do not reach the main river systems.

Data on the surface water resources of the Aral Sea basin based on many years of observations are given in the Table 2. Total perennial water flow is 116 km^3.

The maximum flow in the upstream of these rivers occurs in the late spring and summer, and the minimum in winter. Yet in the lower reach of Amudarya as well as in Syrdarya, flow declines significantly as the most of it is used for irrigation. That is why the late summer flow of these rivers in their lower reaches is rather small. Estimates of surface water resources of the countries in the basin are given in the Table 3.

Due to discharge of saline wastewaters into the river main stream,irrigation water salinization in the middle reach of the rivers has increased to 1–1.1 $g\ l^{-1}$. The salt content during some periods in the lower reaches can be as high as 2 $g\ l^{-1}$ and more in comparison with the initial concentration of 0.2–0.3 $g\ l^{-1}$ (Table 4).

3 GROUND WATER RESOURCES

Ground water resources of the Aral Sea basin, including those in the territory of Uzbekistan, are formed under the influence of sediment infiltration, filtration from water reservoirs, river beds, channels, lakes, as well as from the irrigated lands. The latter is a significant factor affecting the ground water. Hydrological estimates show that total annual ground water resources potential in the Aral Sea basin is about 31.5 km^3. Distribution of the available ground water resources among the countries and the current usage of mined underground sources is shown in the Table 5.

The data in Table 5 show that ground water plays a significant role as a source of irrigation water only in Tajikistan and Uzbekistan. The total volume of groundwater comprising around 4 km^3 per annum used for irrigation in the Aral Sea basin appears to be quite insignificant compared to

Table 2. Water resources of the Aral Sea Basin in km^3. Adapted from data of SANIGMI (Central Asia Institute of Hydrometeorology); Haskoning et al, 2003; and Yakubov et al., 2001).

River	Flow ($km^3\,y^{-1}$)
Amudarya basin	
Pyange	36.0
Vakhsh	20.8
Kafirnigan	5.9
Sirljamdarua amd Sjerabad	4.0
Kashkadarya	1.6
Zerafshan	5.3
TOTAL	73.6
Syrdarya basin	
Naryn	13.8
Ferghana valley rivers	12.8
Akhangaran river basin	1.2
Chirchik river basin	7.8
Arys river basin	2.0
Other	1.2
TOTAL	38.8
Rivers of Turkmenistan (Tedgen, Murgab, Atrek, etc.)	3.2
TOTAL of the Aral Sea Basin	115.6

Table 3. Surface water resources of the Aral Sea basin as measured by the average annual flow ($km\,yr^{-1}$). Adapted from Sarsebekov et al., 2004.

	River basin		Total for the Aral Sea Basin	
Country	Syrdarya	Amudarya	$km^3\,yr^{-1}$	%
Kazakhstan	2.5	–	2.5	2.2
Kyrgyzstan	27.5	1.7	29.2	25.2
Tajikistan	1.0	58.7	59.7	51.5
Turkmenistan	–	1.4	1.4	1.2
Uzbekistan	5.6	6.8	12.4	10.6
Afghanistan and Iran	–	10.8	10.8	9.3
Total for Aral Sea Basin	36.6	79.4	116.0	100

116 km^3 of water taken annually from the surface water sources. The salt concentration of ground water is usually higher than that of the surface waters and its utilization requires a lot of input for pumping, pump equipment and well maintenance. It is apparent, however, that there is a limited potential for more intense usage of ground water for irrigation needs, and probably it won't become a significant part of the total water consumption.

4 RETURN WATERS

Return waters are a combination of collected-drainage and runoff water flow from the irrigated areas along with industrial, agricultural and residential waste waters.

In the present situation, conditions return water flow in the Aral Sea basin is about of 30–32 km^3, of which 16–20 km^3 is from the Amudarya river region (Table 5).

Table 4. Current levels of mineralization of river water across the Aral Sea basin (Haskoning et al., 2003).

Country	Location	Mineralization levels (g l^{-1}) in the period 1991–2000	
		Average	Maximum
Amudarya basin			
Pyange	Lower reach	0.45	0.73
Vakhsh	Lower reach (Kurgan-Tubeh)	0.78	0.93
Kafirnigan	Lower reach (Tartki)	0.36	0.46
Amudarya	Termez	0.63	0.98
	Atamurad (former Kerki)	0.74	2.4
	Ilchik	0.87	1.40
	Tuyamuyun	0.82	1.32
	Samanbay	1.05	2.23
Syrdarya basin			
Naryn	Lower reach (Uchkurgan)	0.31	0.60
Karadarya	Lower reach (Uchtepe)	0.50	0.85
Syrdarya	Low part of Kayrakum water reservoir	1.10	1.22
	Low part of Chardarin water reservoir	1.04	1.18
	Kyzlorda	1.12	2.00
	Kazalinsk	1.14	2.80

Table 5. Characteristic of the return waters flow of the Aral Sea basin for 1992–1997 (Yakubov et al., 2001).

Year	Flow (km^3)	Salinity (g l^{-1})	Return flow/water intake rate* (%)
Amudaraya river basin			
1992	17.67	2.40	44
1993	16.17	2.72	47
1994	16.47	2.58	45
1995	16.72	2.17	41
1996	18.46	2.75	38
1997	19.14	2.84	39
Range of variation for 6 years	16.17–19.14	2.58–2.40	38–44
Syrdarya river basin			
1992	14.4	2.42	70
1993	13.9	2.40	67
1994	13.5	2.34	64
1995	14.6	2.21	60
1996	15.8	2.17	62
1997	14.0	2.21	57
Range of variation for 6 years	13.9–15.5	2.21–2.4	54–70

* Salinity and return flow/water intake rate are given for Uzbekistan, for the flow quantity representing 65–75% of the total quantity.

5 FLOW REGULATION THROUGH WATER RESERVOIRS

There are 60 water reservoirs in the Aral Sea Basin which were constructed and are now in use. The useful capacity of each reservior is about being 10 million m^3. The gross total capacity of the

Table 6. Toktogul water reservoir operation regime in the past (UNESCO, 2000).

	Inflow	Losses	Discharge: Non-vegatation	Discharge: Vegetation	Total	Average annual balance of the reserve
	km^3					
As per design (early 1970s)	11.8	0.3	2.8	8.5	11.3	0.2
Average annual: 1975–1991 (16 years)	11.3	0.3	2.7	8.1	10.8	0.2
Average annual: 1991–2001	13.0	0.3	7.2	6.1	13.3	−0.6
2000–2001	12.8	0.3	8.4	5.9	14.3	−1.8

water reserves is 64.8 km^3, of which useful capacity is 46.8 km^3, including 20.2 km^3 in Amudarya basin and 26.6 km^3 in Syrdarya basin.

There are 45 hydroelectric power stations constructed in the Aral Sea basin with the total plant output of 34.5 Gigawatt, each plant's output varying between 50 and 2,700 Megawatt. The largest power plants are at Nurek (in Tajikistan, river Vakhsh), with output of 2,700 Megawatt, and Toktogul (in Kyrgyzstan, River Naryn) with output of 1,200 Megawatt. Hydroelectric power comprises 27.3% of the total power consumption in the Aral Sea Basin.

Flow regulation degree of water reservoirs (guaranteed output) is 0.94 for Syrdarya (i.e, natural flow is almost totally regulated), and 0.78 for Amudarya (i.e., regulation can be extended further). Presently water reservoirs are at various degree of sedimentation (on average at 30 %) which causes reduction of the design conservation zone (Dukhoviniy, 2001).

Water reservoirs were constructed during the Soviet era with a purpose to serve the entire basin. With independence, however, several conflicts emerged between water requirements for irrigated agriculture of the countries in the lower and medium reaches of river basins and electric power production in the upstream. This problem is especially critical in operating Naryn-Syrdarya cascade of Toktogul water reservoir. The data in Table 6 show the operation practices of Toktogul water reserve during the past and in the present.

Starting from 1992, around 27 km^3 (on average 3 km^3 per annum) of water had to be discharged into Arnasay saddle during winters due to limited conveyance capacity of water flow below Chardary. Water discharged into the saddle were not only lost for use as water resources in the future but also contributed to the destruction of infrastructure and inundation of lands.

In the following years almost every year negotiations were undertaken to conclude agreements on the concord and compensation related to summer and winter discharges of water that would take into account mutual interests of the countries of the lower reach and upstream of Syrdarya.

As experience shows, managing water resources on the regional level in each specific case shall be determined by a separate cooperation agreement. A basin-based water resource management will be integrated at the national level through establishment of water consumer associations at the lower level of hierarchy.

6 USAGE OF WATER AND LAND RESOURCES

Use of water resources for residential purposes and irrigation has been practiced in Central Asia since time immemorial. Expansion of irrigation agriculture started during the late 19th century, and development of new lands expanded rapidly between 1956–1990 years stipulated by a quick increase of population, development of industry and irrigation for cotton production (Table 7).

However, excessive use of water for irrigation led to the gradual drying of Aral Sea and desertification of its deltas (Table 8).

Shrinkage of the Aral Sea led to the drying of deltas and river beds which not only had high value for aquatic life, but also provided means of subsistence for the local population. Restoration

Table 7. Main indicators of water and land resources usage in the Aral Sea Basin (Dukhovniy, 2001; UNESCO, 2000).

Indicator	Measurement unit	1960	1970	1980	1990	1999
Population	millions	14.1	20.1	26.8	33.6	55
Irrigated lands area	thousand ha	4510	5150	6920	7600	7890
Total water intake	km^3 per annum	60.61	94.56	120.69	116.27	103.86
Of that, for irrigation	km^3 per annum	56.15	86.84	106.79	106.4	93.56
Specific water intake per irrigated hecatar	m^3per ha	12450	16860	15430	14000	11858
Specific water intake per capita	m^3 per capita per year	4270	4730	4500	3460	1888
Gross domestic product	US$ (billions)	16.1	32.4	48.1	74.0	54.0

Table 8. Major parameters of the Aral Sea.

Years	Water depth (m)	Volume (m^3)	Area (1000 km^3)	Salinity ($g\,l^{-1}$)	Supply of water into Aral Sea (km^3)		
					Syrdarya	Amudarya	Total
1960–1970	52.5	1047	63	10	11.2	33.6	44.8
1971–1980	48/7	875	55	13	3.5	14.1	17.6
1981–1990	41.8	415	44	25	2.3	5.6	7.9
1991–2000	35.8	239	32	51	5.6	13	18.6
2001	32.6	161	20.8	60	2.7	0.38	3.08

of the sea in the current conditions is not possible, thus, development of a strategy for sustainable management of water reservoirs in the delta and the dry bed has been recognized as the most important priority for nature conservation and vital to supporting livelihood of the population of the Aral region.

Irrigation systems developed before 1955–1960 were based on sound engineering design, yet all components were based on the earthen channels characterized by huge losses of water during its delivery, often without consideration of specifics, geochemical flows and natural salt deposit zones in the vicinity where reservoirs were built. Low efficiency coefficient of irrigation systems (0.4–0.5) created in this period led to disturbance of the area's natural water-salt balance and raised the ground water level and secondary salinization of soils in several regions of central Asia.

New development occurred at an especially high rate between 1956 and 1990. During these years, the area of irrigated lands increased to 7.4 million hectares (Mha), and by January 1, 1999 irrigated area within the Aral Sea Basin was estimated at 7.95 Mha. At the same time, 70–75% of lands newly converted to agricultural production had either salinized or were prone to salinization. That is why all irrigated lands developed during the last 40–50 years were provided with irrigation-drainage systems engineering services based on the modern land reclamation requirements. At the same time, many areas irrigated using the old technology beginning with the mid-1950s were reclaimed by wide-scale adoption of measures on reconstruction of irrigation-drainage systems and reclamation of saline lands.

Well developed engineering irrigation-drainage system in Central Asia includes a large gravity irrigation network covering an area of over 6 Mha, and some canals with discharge of 700 $m^3 sec^{-1}$ and length of 1400 km. Similarly, pumping irrigation systems cover an area of over 2 Mha with unique cascades of channels with delivery lift of 350 m and flow reaching 350 $m^3 sec^{-1}$. One of the integral parts of this system is high-capacity collector-drainage network constructed on about 4.5 Mha of land area.

Table 9. Area of medium and heavily saline lands on the irrigated territories for 1990–1999 (Haskoning et al., 2003).

Zone of planning	Irrigated area as of 1990 (thousand ha)	Areas with high and medium Salinity levels (thousand ha)		% of increase 1990–1999
		1990	1999	
Syrdarya basin				
Kyrgyzstan (total)	410	9.1	8.4	−8
Uzbekistan (total)	1860	199	330	66
Tajikistan	250	15	54	250
Kazakhstan (southern)	780	119	215	80
Total for Syrdarya basin	3300	342	608	78
Amudarya basin				
Tajikistan (total)	690	18	18	0
Uzbekistan (total)	2400	505	638	26
Total for Tajikistan and Uzbekistan	3090	524	656	26
Turkmenistan (total)	1310	636	1166	83
Total for Amudarya basin	4400	1160	1822	57
Total for the Aral Sea basin	7700	1502	2430	62

Table 10. Irrigated lands with groundwater level close to the surface (Haskoning et al., 2003).

Zone of planning	Irrigated area as of 1990 (thousand ha)	Area with groundwater level <2 m (thousand ha)		% of increase in 1999 compared to 1999
		1990	1999	
Syrdarya basin				
Kyrgyzstan (total)	410	11	14	27
Uzbekistan (total)		413	566	37
Tajikistan	250	26	31	19
Kazakhstan	780	98	294	200
Total for Syrdarya basin	3300	548	905	65
Amudarya basin				
Tajikistan	690	92	111	21
Uzbekistan		670	801	20
Turkmenistan	1310	528	654	24
Total for Amudarya basin	4400	1290	1566	21
Total for the Aral Sea basin		1838	2471	35

However, agricultural productivity of irrigated lands is declining. The irrigation network is on the decline because of the age and wear, and areas prone to salinization and ground water close to the surface are increasing (Tables 9 and 10).

Soil moisture in the irrigated lands is often maintained outside of the optimal range. It varies between 45% and 95% of soil field capacity (SFC) and in most cases between 45–60% SFC (optimum is 70–75 or 97%) in periods between irrigation.

Not maintaining the soil water in the root zone of agricultural crops at optimal level of SFC salinity and decline of irrigation water quality causes significant yield reductions. Soil salinity also causes extra expenditure on water, materials and equipment and labor cost. Aggregate losses to the national economies across the Aral Sea Basin caused by high groundwater level and secondary salinization are estimated at USD 1750 millions annually, approximately equal to 32% of the market value of potential agricultural crop produce.

Table 11. Extension of irrigation channels broken down by Republic (Yakubov and Ikramov, 2003).

Republic	Unlined channels	Lined channels	Total extension	%
	thousand km			
Kazakhstan	4.0	0.6	4.6	10
Kyrgyzstan	1.6	1.1	2.7	6
Tajikistan	3.3	2.0	5.3	11
Turkmenistan	7.8	0.5	8.3	16
Uzbekistan	18.7	9.3	28.0	57
Total	35.4	13.5	48.9	100

Table 12. Actual efficiency rates of irrigation broken down by indicative zones.

Zone of planning	Efficiency rate* of irrigation systems (%)	Efficiency rate** of watering techniques (%)	General efficiency (%)
Upstream			
Namangan/Uzbekistan	63	62	39
Gorno-Badakhshan/ Tajikistan	62	64	40
Old/middle reach-plain zones			
Chakir/Kazakhstan	63	70	44
Ferghana/Uzbekistan	55	73	40
Mary/Turkmenistan	58	70	41
Osh/Krygyzstan	59	70	41
Recent/middle reach-plain zones			
Golodnaya Steppe/Kazakhstan	63	70	44
Syrdarya/Uzbekistan	73	71	52
Old lower reaches			
Khorezm/Uzbekistan	52	65	34
Dashoguz/Turkmenistan	53	70	37
Recent lower reaches			
Karakalpakstan/Uzbekistan	48	70	34

* Efficiency rate of irrigation systems = quantitative rate of the water supplied to the field against water drawn from the river; ** Efficiency rate of watering techniques = quantitative rate of water used for watering of plants against water supplied to the field.

The reasons for not maintaining soil water at the optimum range in the root zone for agricultural crops is scarcity of water resources on the one hand and not following the recommended practices of irrigated crops on the other (low frequency irrigation with large volumes of water), insufficient readiness of irrigated lands for watering and on average low technical level of irrigation systems (Table 11).

The efficiency rate of irrigation systems throughout the region varies between 0.48 and 0.73, and that of the watering techniques between 0.62 and 0.70 (Table 12).

Thus, major field planning has been undertaken since the mid 1990s at only a few places where difference of levels and irregularities are $\pm$20–30 cm compared with normal range of $\pm 3 \ldots \pm 10$ cm. Furrow and strip irrigation are practiced, leading to uneven soil moisture, and inadequate planning of the fields causes significant losses. Only 30–35% of water drawn from the source is used for irrigation of agricultural crops root layer. Partially these losses return to the main stream as return flow and are used downstream. However, they affect quality of irrigation water.

Reasons of salinization of irrigated lands. An important cause of salinization is the existence of saline ground water close to the soil surface due to insufficient drainage of irrigated lands (caused by poor technical installation and inadequate collector-drainage systems). In addition, elimination of flushing water regime necessitated by the limited water usage in response to a limited or reduced water supply from year to year but especially during the inter-vegetation period. The flushing water regime coefficient (water supply to the field plus precipitation must exceed aggregate evaporation and transpiration) currently equals

$$\frac{\sum B+O}{\sum U+T}=0.64+1.0.$$

Moreover, the process is aggravated by the inferior technical level in irrigation systems and increase in salinity of irrigation water (Ikramov, 2001).

Technical level of collector-drainage systems. Out of the total irrigated area of 7.9 Mha, over 72.5% is in need of construction of drainage system (5.72 Mha). The collector-drainage network has been constructed on the 5.35 Mha, including horizontal subsurface drainage (790,000 ha) and vertical drainage (834,600 ha). Drainage system needs to be installed in the area of over 376,700 ha.

Construction of the drainage network involved installation of over 200,548 km of drainage network, of which 44,997 km are main and inter-farm drains, 155,551 km are on-farm drainage networks (including 48,600 km underground horizontal drainage), and 7,762 km of vertical drainage.

Full-scale examination of technical condition of underground drainage revealed that nearly 75% of drains are not functional. Physical and chemical sealing of filters reduced specific yield of vertical drainage wells by 37–79% from the initial level. The efficiency rate of the systems decline to less than 27–34% on average per year, against 60–65% recommended by SANIIRI due to the outage of a significant part of wells and pumps and deterioration of materials and technical resources supply in the recent years. Restoring and maintaining irrigated lands according to the norms requires cleaning and repair of inter-farm and intra-farm collector-drainage network. In recent years, however, due to problems in economical and financial area these works are performed at the level of 50–70% of the norm. Repair of underground horizontal drainage remains one of especially difficult challenges to be addressed.

Ways to improve. Virtually in all regions where artificial drainage was installed, desalination of soils, increase in crop harvests, and economic benefits have occurred through its normal use and for flushing watering. Soil desalination necessitates reviving the existing collector-drainage systems and simultaneously undertaking installation of advanced drainage types (vertical, underground horizontal and combined), to create an automorphic regime for maintaining ground water at 3 m depth. Using flushing regime of watering in combination with effective drainage reclaims salinized soils. Desalinized soils produce higher yields of agricultural crops and save water resources used for flushing salts from soils (Ikramov, 2001).

The main reason of increasing salinity of irrigation water and decline in its quality is the recycling of collector-drainage waters (which inevitably occur in irrigated agriculture) into irrigation sources – the rivers. The volume of return collector-drainage water for the Aral Sea Basin is 36–38 km^3, of which 92–95% comes from the irrigated lands. Salt concentration of the collector-drainage waters from the irrigated lands in the upstream area is 1.5–3 $g\ l^{-1}$, and that in the middle and lower reach is varies within limits of 3.5–6 to 5–7 $g\ l^{-1}$. Collector-drainage water from the irrigated lands flow into rivers and is used for irrigation of the lands situated downstream. This repeated recycling occurs several times across the main stream, thus increasing salt concentration in the river, contaminating it with pesticides, and severely damaging the river ecosystems.

In addition, compulsory use of very saline water for irrigation in soils prone to water deficit causes salinization of soils and decline in their quality.

Water and salt balances studies in the recent years (especially in the low-water years) show positive balance with accumulation of salt of 0.6 to 10 $t\ ha^{-1}$ in the aeration zone on the irrigated fields of middle and lower reaches of river basins. Significant worsening was observed in the lower reaches of Amudarya – in Karakalpakstan, due to significant (up to 2 times) reduction of specific water supply, which caused positive balance with accumulation of up to 8 t ha^{-1} of salt even during the high-water years of 1988 and 1989. In the Syrdarya River basin, a similar situation occurred in its middle reaches, in Golodnaya Steppe where accumulation of salt rate was 5.3 t ha^{-1}.

In general, across the middle and lower reaches of the rivers, increase in salt concentration by 0.1 g l^{-1} causes losses through decline in fertility of irrigated lands, and reduction of gross product between 71 to 158 rubles ha^{-1} (for the 1993 prices) (Vakubov et al., 2001).

Productivity of irrigation water in the current conditions is about USD 0.03–0.10 m^{-3} in comparison with USD 0.18–0.25 in 1990. In countries with advanced water conservation, water productivity is about USD 0.50–0.55 m^{-3}.

Ways to improve. Creation of highly effective, reliable and advanced drainage system, introduction of water conservation and irrigation technologies, reducing seepage losses from delivery of channels, and improving water usage would decrease the drainage flow. Maximum usage of collector-drainage waters for irrigation in light textured soils for cultivation of agricultural crops, and afforestation of barren lands would reduce the rate of return into rivers and salinization of the river water.

REFERENCES

Chub, V.E. 2000. Climate fluctuation and its effect on the natural resource potential of the Republic of Uzbekistan. Central Asian Scientific Research Institute of hydrometeorology named after V.A. Bugayov.(Rus).

Dukhovniy, V.A. 2001. Diagnostic report, Sustainable and effective usage of water resources in Central Asia/ Special Economic Program for Central Asia, UN; Tashkent – Bishkek, June 2001. (Rus).

Haskoning, B.V. et al. 2003. Regional report No. 3 Component A-1, Water Resources and Environment Management Project, GEF.

Ikramov, R.K. 2001. Principles of management of water/salt and fertilizing regime of irrigated lands of Central Asia in the conditions of water resource deficiency. Tashkent, GIDROINGEO. 192 p. (Rus).

Sarsebekov, T.T., A.N. Nurushev, A.E. Kozhakov, and M.O. Ospanov. 2004. Usage and protection of transboundary rivers in the countries of Central Asia. Atamura publ., Almaty. 271 p. (Rus).

UNESCO. 2000. Water-related vision for the Aral Sea Basin for the year 2025. Division of water sciences with the cooperation of the scientific advisory board for the Aral Sea Basin, Paris, France.

Yakubov, H.I., A. Usmanov, and M.O. Yakubov. 2001. Prospects of usage of return waters of collector-drainage flow in irrigated agriculture and desert development. p. 40–77. Proc. of SANIIRI Modern issues of land reclamation and water industry and the ways to solve them. Tome 3, Tashkent. (Rus).

Yakubov, H.I. and R.K. Ikramov. 2003. Modern issues of land reclamation. Seminar of the Training Centre of Intergovernmental Coordinating Commission on Water Industry (MKVK) on Integrated Water Resource Management. (Rus).

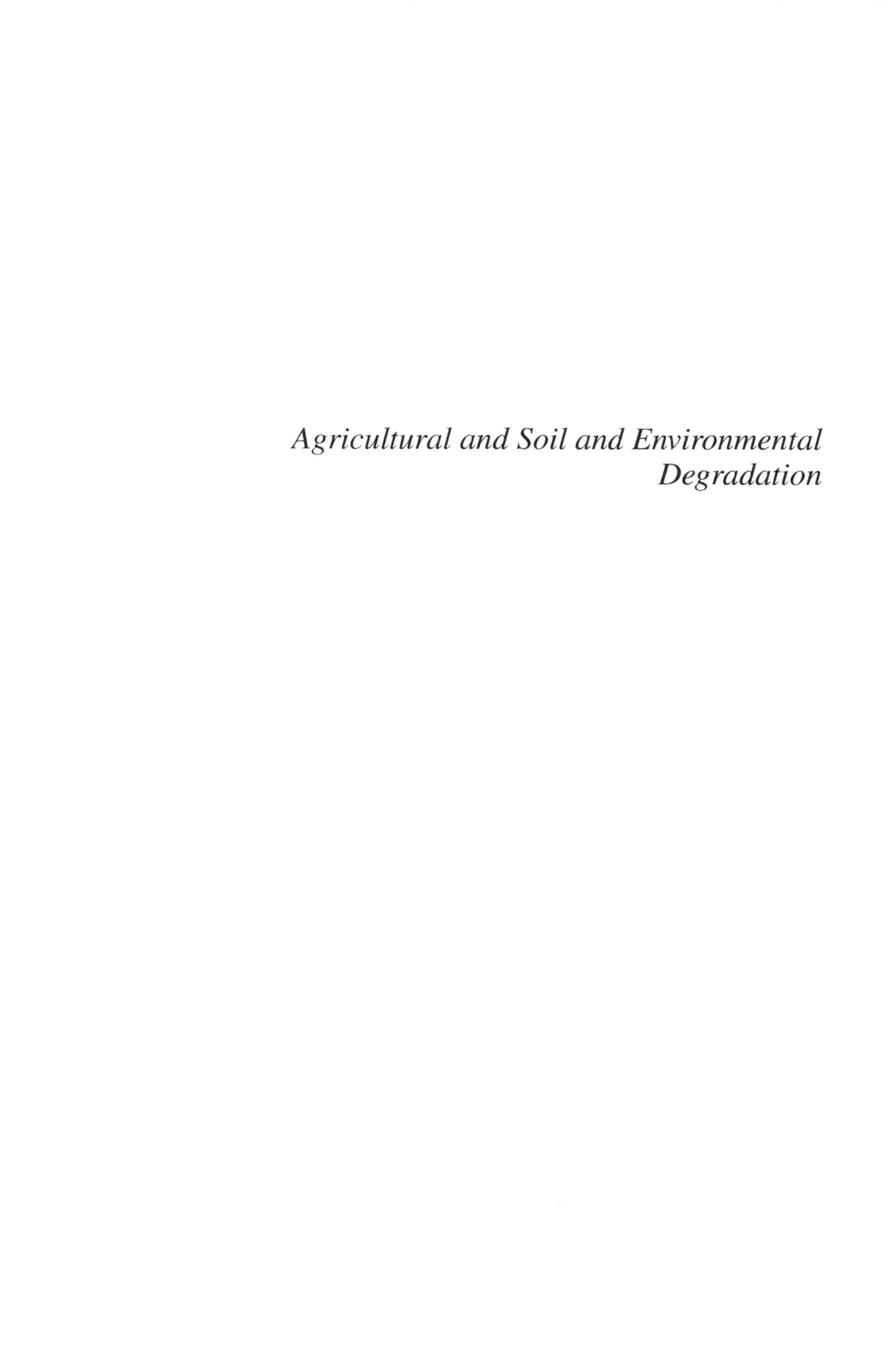

Agricultural and Soil and Environmental Degradation

CHAPTER 8

Addressing the challenges for sustainable agriculture in Central Asia

Raj Paroda

Program for Central Asia and Caucasus, International Center for Agricultural Research in Dryland Areas, Tashkent, Uzbekistan

1 INTRODUCTION

The region of Central Asia (CA), encompassing Kazakhstan, Kyrgyzstan, Tajikistan, Turkmenistan and Uzbekistan (Figure 1), has undergone tremendous economic and social perturbations following the dissolution of the former Soviet Union. The economies shrunk thus causing fall-off of incomes, increase of poverty and a major concern for food security.

All these countries became independent from the former Soviet Union in 1991. Gross National Income (GNI) per capita declined by an average of almost 50% between 1991 and 2000, and despite some signs of recovery, shown in the recent five years, the GNI per capita for three countries of the region is still less than in many of the low income countries(World Bank, 2005, 2006) (Table 1).

Figure 1. Map of the Central Asian Region.

Table 1. Gross national income per capita in 2005.

Countries	Gross national income per capita (US dollars)
Kazakhstan	2,930
Kyrgyzstan	440
Tajikistan	330
Turkmenistan	1,340
Uzbekistan	510
Low income (global average)	580
Middle income (global average)	2,640

Source: World Bank, 2005 and 2006.

Table 2. Population below the poverty line.

Countries	Population below poverty line (%)
Kazakhstan	27.9 (2002)
Kyrgyzstan	40.8 (2003)
Tajikistan	64.0 (2006)
Turkmenistan*	58.0 (2004))
Uzbekistan	26.0 (2003)

Sources: ABD, 2006 and Basic Statistics and IMF Reports, 2006 (*Since the latest available data for the World Bank and ADB for Turkmenistan are dated 1998, the estimates provided by CIA Factbook and Wilkipedia have been used).

Even in those countries, where GNI per capita seems to have improved up to the lower middle-income level, almost 26–52% of the population, most of whom agricultural workers, still live below poverty line because of growing income inequalities (Table 2). Poverty remains widespread throughout the region, where some 20% of the population is now living on less than US$ 2 a day, compared with around 3% prior to independence. Absolute poverty is as high as 68% in Tajikistan and 50% in Kyrgyzstan (ADB, 2006; The Economist Intelligence Unit, 2005; Library of Congress, 2005; East Agri., 2006).

Prior to independence, the Central Asian republics were economically interdependent within the centrally managed Soviet economy. The agriculture contributed between 15–45% of GDP and employed between 20–50% of the national labor force. Each republic was specialized according to its agro-climatic conditions, natural resources, with production distributed throughout the Soviet trading system. Following the dissolution of the USSR and collapse of existing trading arrangements, exporters lost their markets and importers were left with lack of cash reserves and unstable food supply. The newly independent countries were confronted with the task of developing their own economic systems while facing the challenge of globalization of agriculture.

This transition has had severe consequences. The agricultural sector contracted significantly; agricultural output fell by almost 50% between 1991 and 1999. This trend is still persisting in some of the Central Asian countries, though there have been signs lately of some economic revival and increased food production.

However, the income inequality between the agricultural and non-agricultural population is growing in most of the countries due to low economic efficiency of agricultural production, little capital investments in agriculture, including agricultural research, low level of institutional and infrastructural development beside much needed policy commitment for ARD.

With the opening up of national economies to global agricultural trade, the livelihoods of agricultural populations could be further threatened unless urgent measures are taken to disseminate

Table 3. Indicators of country profiles in 2005.

Country	Total area (M ha)	Arable land (M ha)	Population (million)	GNI per capita ($US) – Atlas method	Agriculture (% of GDP)
Kazakhstan	271.3	24.0	15.1	2,930.00	6.5
Kyrgyzstan	19.9	1.4	5.1	440.00	34.1
Tajikistan	14.3	0.9	6.9	330.00	24.2
Turkmenistan	48.8	1.8	6.5	1,340.00	26.0
Uzbekistan	44.7	4.9	26.3	510.00	28.1

Sources: The World Development Indicators, 2005 and 2006; Economic Intelligence Unit, 2005; Federal Research Division of the Library of Congress of USA, 2005; Asian Development Bank, 2006.

Table 4. Production dynamics of main crops in Central Asian countries in 1992 and 2005.

Major crops		Kazakhstan		Kyrgyzstan		Tajikistan		Turkmenistan		Uzbekistan	
		1992	1995	1992	1995	1992	1995	1992	1995	1992	1995
Cotton	Area, mln. ha	0.11	0.19	0.02	0.05	0.29	0.29	0.57	0.60	1.67	1.39
	Prod., mln. t	0.23	0.35	0.04	0.15	0.53	0.49	1.18	0.66	3.86	3.79
	Yield, t ha^{-1}	2.10	1.80	1.90	3.20	1.80	1.70	2.10	1.10	2.30	2.70
Wheat	Area, mln. ha	13.72	11.50	0.25	0.42	0.18	0.32	0.20	0.90	0.63	1.40
	Prod., mln. t	18.29	11.10	0.68	0.95	0.17	0.63	0.38	2.83	0.96	5.75
	Yield, t ha^{-1}	1.30	1.00	2.70	2.20	0.90	2.00	1.90	3.10	1.50	4.10
Rice (paddy)	Area, mln. ha	0.12	0.08	0.00	0.01	0.01	0.01	0.03	0.05	0.18	0.05
	Prod., mln. t	0.47	0.31	0.00	0.02	0.02	0.05	0.06	0.12	0.54	0.15
	Yield, t ha^{-1}	4.00	4.10	1.50	2.90	2.00	5.20	2.30	2.40	3.00	3.20
Barley	Area, mln. ha	5.63	1.50	0.26	0.10	0.06	0.05	0.06	0.07	0.30	0.11
	Prod., mln. t	8.51	1.55	0.62	0.21	0.04	0.06	0.13	0.07	0.29	0.10
	Yield, t ha^{-1}	1.50	1.00	2.40	2.10	0.80	1.30	2.10	1.00	0.90	0.90

Source: FAOSTAT, 2006.

economically more efficient farming practices, which would enable the farmers to save on resources and increase their income.

The model of agricultural growth followed in the past in all the countries of the region was extensive, i.e., producing more by using higher levels of inputs, which is no more sustainable in the present context. Moreover, negative effects of this approach are already evident in growing environmental and ecological problems such as degradation of soil and water resources with increased level of salinity, rise in soil water tables and decline in soil fertility. Therefore, there is an urgent need to have a paradigm shift from extensive to resource-efficient agriculture that is based on sustainable agricultural practices.

This issue is key for the overall development of the countries in the region because agriculture is an important component of the GDP and of national labor employment. Its percentage contribution to the GDP remains important in many countries of the region since they gained independence (Table 3) (ADB, 2006). About 70% of the total area of 418 million hectares (M ha) is classified as agricultural land. Of this, only 15% is arable, of which 24% is irrigated. Wheat, cotton and livestock are the most important agricultural commodities (Table 4). Some 85% of the agricultural land is considered as permanent rangeland and used largely for livestock production. The agriculture in the region is diverse and has a great potential to revitalize the withered economies of the CAC countries by means of improved productivity and higher production through agricultural research for development (FAO Stat., 2006).

2 EMERGING CHALLENGES

As already stated, the economies of the CA countries are primarily based on agriculture. The food and agriculture sectors represent both the promise and the challenge for the future. There is a huge potential to improve the productivity of agriculture in order to meet national demands and generate additional income for farmers (Ryan et al., 2004). The breakup of the Soviet Union brought many problems and challenges for the countries of CA. Some of them are discussed in the following sections.

2.1 *Food security and poverty*

Since the breakup of the Soviet Union, the countries in Central Asia have experienced major socio-economic shocks that have resulted in increased food insecurity, malnutrition, and poverty. Each country has undergone economic reforms, in varying degrees, in order to transform their centrally planned economies into market economies. However, economic instability, incomplete land reforms, inadequate institutional transformations, collapse of regional trade, and depleted foreign reserves have resulted in food insecurity and inefficient use of natural resources for food production.

Rightly too, the countries, therefore, focused on food self-sufficiency, particularly through increased production of wheat. The area sown to cereals increased by 24% since 1992; 40% of the arable area of CAC is now sown to cereals, of which 80% is under wheat. Conversely, in Kazakhstan, which was by far the largest producer and a major exporter of cereals (around 4 to 5 mt annually) in the region, the area under wheat decreased from 16 M ha to 12 M ha over the last decade, while in the rest of the region, grain production increased by almost 124%, due primarily to area expansion.

Projections from IFPRI's global food model IMPACT (International Model for Policy Analysis of Commodities and Trade) indicate that demand for cereals will continue to increase in the region. Though wheat is the major cereal, demand for rice is projected to grow the fastest, on the contrary, the area under rice has declined. Increasing domestic production to meet these demands from limited land under irrigation will require investment in productivity enhancing and resource saving technologies. However, it is also to be scientifically understood as to whether the current shift towards cereals is the most efficient way. Research is also needed to identify means to achieve food security according to each country's comparative advantage and resource endowments.

2.2 *Soil fertility and salinity*

Recent estimates show that more than 50% of irrigated soils in Central Asia are salt-affected and/or waterlogged. Since 1970s, the levels of salts in both major rivers (Sirdarya and Amudarya) have increased steadily as a result of the discharge of drainage waters from irrigated schemes back into the river systems. In the past, an extensive drainage network was developed which covered an area of 5.7 M ha. However, currently, the actual coverage and effectiveness of the system have been reduced to half of its capacity due to lack of investment in operation and maintenance of existing drainage network.

2.3 *Deteriorated irrigation system*

The most glaring example of an international environmental conflict is the Aral Sea basin, where five nations compete for water for irrigation, industrial, and urban use. During the Soviet era, a system of river basin structures was constructed primarily for the purpose of irrigation in downstream regions. Over the last two decades, excessive use of water upstream by each country and consequent mismanagement has led to the shrinking of the Aral Sea, cited as one of the worst ecological disasters on the planet. Since the breakup of the Soviet Union, the five Central Asian countries have attempted to cooperatively allocate water and manage the system without effective agreements in place. The region also needs adequate legal and regulatory policies on water pricing and water quality standards. At the local level, there are no organizations to effectively operate available irrigation systems or to provide incentives to the farmers to conserve water (FAO, 2000).

2.4 *Degraded rangelands*

A large share of the rural inhabitants of Central Asia relies on livestock production for their livelihood. Reversing the trend of rangeland degradation and land erosion require policies that encourage sustainable use of rangelands, provide livelihoods to the halt rural households, and reduce their current migration to urban areas.

Covering roughly 260 M ha, the vast rangelands of Central Asia, form the world's largest contiguous area of grazed land. Central Asian rangelands thus have the potential to sequester considerable amounts of carbon, especially if overgrazed and desertified areas can be improved through effective management and reclamation efforts (Gintzburger et al., 2003).

Central Asia's rangelands have long provided the forage to support large numbers of livestock, a mainstay of the region's economy. During the Soviet era, livestock production – like agriculture – was geared towards maximization of output, rather than ecological sustainability. The region is also home to a large diversity of livestock breeds, especially small ruminants, horses, camel and donkeys. Based on the data from on-going activities, it is apparent that this diversity is under threat as a result of large reductions in livestock populations, especially in Kazakhstan and Kyrgyzstan, and the collapse of animal breeding schemes after the dissolution of the Soviet Union. Also tremendous diversity for sheep germplasm exists, which needs to be characterized and conserved.

2.5 *Disintegration of markets*

Under the Soviet agricultural production system, large-scale collective and state farms controlled some 95% of agricultural land and produced the bulk of the commercially marketed output. Product markets and input supply channels were also largely controlled by State organizations. Commercial production from State enterprises was supplemented by household plots that relied on part-time family labor and produced mainly for subsistence or local farmers' markets. These household plots achieved relatively high levels of productivity, producing 20% of gross agricultural output from 2% of the land.

Achieving food security in the long run will depend not only on improving the productivity of agriculture, but also on re-establishing regional trade and identifying policy options that increase producer incentives to intensify and diversify production. The Former Soviet Union (FSU) was characterized with isolated economy in which export-import strategies and practices were limited to bulk imports through state channels. The word marketing did not exist in Russian vocabulary, and the producers were never worried about markets as any produce would be procured by the government irrespective of its quality.

The land reform process has not been adequately accompanied by market reforms. The centralized product markets and input supply channels have largely disappeared and the local markets are the only channels available to smallholders. In the beginning of the transition period, the centrally controlled marketing system for agricultural products was replaced by the barter system, mainly through exchange of inputs. This collapse of the marketing system has added to the disincentives in agricultural production. Assistance is needed in building market institutions, in reforming the markets and establishing trade links with a view to reducing transaction costs, mitigating risks, building social capital and redressing missing markets. This would help in diversifying farm production to generate income and meet domestic food demands. Deterioration of storage, processing and distribution facilities add to the problem. Given the lack of markets, there is no incentive for investment (public or private) in the processing sector.

2.6 *Land tenure and rights*

A key factor for agricultural development and food security is the existence of adequate institutional arrangements to determine rights of access to rural resources, whether land, water, minerals, trees or wildlife. The initial period of transition in Central Asia from central planning was characterized by reallocation of property rights through state-led actions aimed at de-collectivization, privatization and restitution to the original owners, often with substantial assistance from the international community. Although approaches varied among new Central Asian Republics, the reform programs, now being accelerated, have in many cases resulted in vast numbers of small, fragmented land holdings that are not economically viable. The fragmented layout has complicated farming and impeded the provision of rural infrastructure.

Transition countries have had to redirect the focus of land record systems from controlling land use to providing security of tenure to citizens. This will require changes in policy and legislation, and the development of new institutions and services to protect property rights and to stimulate private ownership of land by farmers in order to achieve increased investments in future.

2.7 *Inappropriate policies*

As stated earlier, the countries of the Central Asian region have undergone a series of transitions from centrally planned economies to market-orientated systems. Despite great efforts by these countries and external advice of international and bilateral agencies, policy reform has been rather slow, mainly due to high priority on addressing food insecurity and malnutrition. Unless efforts are made to jumpstart their economies, the Central Asian countries may face increased poverty, food insecurity, and malnutrition, the social and political costs of which could be enormous. One major constraint to deeper economic reform has been the lack of information on the impact of alternative policy options. Reforms continue to be externally designed without adequate policy analysis and without local ownership.

Generating momentum to reorient the nature, approach, and sequence of policy reform packages will require revisiting the entire policy reform process. Involving the local policy research community in identifying critical issues and challenges, setting priorities among them for food, agriculture, and natural resource policy research, and implementing joint research studies is the best way to build local capacity and to increase the ownership of policy packages for quick implementation.

2.8 *Human resource development*

The collapse of the Soviet Union had considerable impact on the research systems. While the newly independent republics have inherited a wealth of national research institutions from the Soviet System, which provides a foundation on which to build collaborative research, they are ill-equipped to respond to changes in the structure of their agricultural sectors and emerging market economies. In particular, they lack the institutional and human capacity for policy formulation and implementation. Moreover, links between research and producers have been broken with the breakdown of the central planning system. Researchers now have no formal channels through which to disseminate information, knowledge and new technologies.

The national agricultural research institutes in CAC are staffed by qualified scientists, but have suffered from isolation and lack of resources. They need immediate assistance in short-term specific training in new and advanced research methods and techniques, training of technical support staff and exposure to international seminars, workshops, symposia, etc. Training of young scientists is thus a priority as they embody the future of the region's research capacity, especially in view of the fact that average age of staff in most of the institutions is more than 60 years and young generation is not attracted towards agriculture due to lack of incentives and better infrastructure for research.

3 OPPORTUNITIES

The CAC region offers tremendous opportunities for agricultural development, since the two important cradles of agriculture exist in this region i.e. institutional infrastructure and human resource.

Unlike Sub Saharan Africa, Central Asia has tremendous potential for faster agricultural growth because of well developed agricultural production system, infrastructure (irrigation, electricity, roads, machinery, research institutions etc.) and the trained human resource – all critical for faster uptake of agricultural innovations.

Analyses of the first ten years of transition in Eastern Europe and the former Soviet Union have highlighted that market-oriented policy reforms play a critical role in promoting subsequent economic growth (World Bank, 2002). The development of market-oriented agricultural production systems will require improving the efficiency and productivity of large farms, which still produce the bulk of marketable output, and at the same time orienting towards small holdings for producing a marketed surplus, all within a unified policy and institutional environment. This fact is being recognized by most of the Governments and steps have been accelerated, with varying degrees, to achieve this goal.

The CAC region is the center of origin and genetic diversity of a number of economically important crop species and neglected and under-utilized species of regional and global economic potential and possesses a wealth of plant and animal genetic diversity. There exists one of the world's best collections of fruits and nuts, as well as wild relatives and progenitors of regionally and globally important food crops including cereals and grain legumes. With changes in farming systems, intensification of cereal cropping, abandonment of orchards and tree crop production, and increased pressure on pasture areas, the region is at risk of losing its landraces and wild relatives of crop plants. The germplasm base in the global collections from these Republics is limited and there is growing concern for the viability status of the germplasm held in local collections. As members of the FAO Commission on Genetic Resources for Food and Agriculture, some countries have already been engaged in discussions dealing with multilateral exchange of plant material and thus represent a potential in-region source of expertise with regard to plant genetic resources issues.

The region is also home to a large diversity of livestock breeds. However, as it was mentioned earlier, this diversity is under threat as a result of large reductions in livestock populations and the collapse of breeding schemes after the dissolution of the Soviet Union.

A Consortium of CG Centers and other International Institutions is operating in the region for over a decade to provide links with the international scientific community and support NARS of the eight CAC countries for sustainable agricultural research for development. Opportunity, therefore, exists for strengthening partnerships both at the regional and global levels.

The current global attention that the CAC region has attracted offers opportunities for forging new partnerships that link research with development and also address the problems of economies in transition as well as democracies in making. In some countries, Competitive Grant Projects or Farm Advisory Services have been established through assistance of donor organization. Development grants are being provided by the World Bank, Asian Development Bank, European Union, IFAD etc. An ambitious program called Central Asian Countries Initiative on Land Management (CACILM) has recently been launched jointly by donor organizations mainly led by ADB and UNEP (GEF), which offers great opportunity to address all Research and Development issues for sustainable agricultural development in Central Asia.

Most of the CAC countries are now on their way towards integration into the international community. More opportunities for the development of stable and credible market and direct foreign investments are emerging for those CAC countries which have joined the World Trade Organization (Armenia, Georgia and the Kyrgyz Republic). Other countries (Azerbaijan, Kazakhstan, Tajikistan and Uzbekistan) are presently negotiating their membership in the WTO, and, therefore, need assistance in re-orienting their economies to fit into liberalized international trade.

Ratification of several international treaties by the countries of the region, especially those relating to environmental protection and genetic resource conservation, also offers good prospects for addressing the issues of global importance, such as climate change, desertification and loss of biodiversity. All countries in the region have already ratified the UN Conventions on Biodiversity (CBD), Combating Desertification (CCD) and the Framework Convention on Climate Change (FCCC), but need assistance in meeting their requirements. Opportunities exist for linking with the implementing agencies of these conventions on regional research on global issues.

In conclusion, the opportunity exists to create a synergistic partnership of diverse research groups in applying science to address the problems of food security, poverty reduction, conservation of natural resources and stabilization of national economies in the CAC region.

4 CGIAR PROGRAM FOR CAC

At the request of CAC countries, CGIAR decided to extend its geographic mandate in the Lucerne meeting of members and senior policy makers. Consequently, a consultation meeting between the CGIAR Centers and Central Asian and Caucasian NARS was organized by ICARDA during September 1996 in Tashkent. The consultation reinforced the need for collaboration in agricultural research. The Task Force recommended to the CGIAR to extend support for agricultural research and human resource development by expanding its geographic mandate to the CAC region. The CGIAR accordingly approved this recommendation in AGM, 1996 and encouraged the CG Centers to develop partnerships with the NARS in CAC in their respective mandated areas of agricultural research. Since 1998 nine CG Centers are participating in the Program: CIMMYT, CIP, ICARDA, ICRISAT, IFPRI, ILRI, IPGRI, IRRI, and IWMI. Besides, three other members, AVRDC-World Vegetable Center, International Center for Bio-saline Agriculture (ICBA) and Michigan State University (MSU), have recently joined the Consortium (CSAA, 2004; ICARDA, 2002; 2006; PFU, 2001).

The main goal and objectives of the program are:

GOAL: The proposed Program will contribute to achieving the overall goal of food security, economic growth, environmental sustainability and poverty alleviation in the countries of Central Asia and the Caucasus.

OBJECTIVE: The immediate objective of the Program is to assist the CAC countries in achieving sustainable increases in the productivity of crop and livestock systems through the adoption and transfer of production technologies, natural resource management and conservation strategies, by strengthening agricultural research and fostering cooperation among the CAC countries and international agricultural research centers.

The main objective of the Consortium is to develop a strategy for inter-center partnership for collaborative research with greater involvement of the national programs in the region. The Consortium eventually developed a CGIAR Collaborative Research Program for CAC in May, 1998 which was launched in September, 1998 with its approval by the Steering Committee in a meeting held in Tashkent, Uzbekistan. Also a Program Facilitation Unit (PFU) was established in the CACRP Office of ICARDA in Tashkent. PFU has since then been providing facilitation function for the inter-center and inter-NARS initiatives in the region.

5 REGIONAL RESEARCH PRIORITIES

Since the establishment of Consortium, there have been extensive consultations between CG Centers and the NARS to identify and prioritize thematic research areas (for instance, germplasm improvement, soil and water management, livestock development, policy orientation, and NARS institutional strengthening).

In line with the bottom-up priority setting approach adopted by the CGIAR under "Plank 4" of the CGIAR's vision and strategy, the CAC Regional Forum and ICARDA organized a brainstorming meeting with NARS leaders and representatives of concerned CG Centers and donors on 20 September, 2001 in Tashkent and again in Aleppo on 8–10 May, 2002 in order to revisit earlier priority-setting efforts. According to the broad consensus, the following research priorities were identified:

- Productivity enhancement of crops and thrust on seed development, with major emphasis on cereals, legumes, oilseeds, fruits and vegetable crops. Breeding of high yielding varieties with resistance to diseases, pests, drought and salinity. Search for genotypes that can withstand better low input conditions.

Table 5. Research prioritites for Central Asia and Caucasus.

	Germplasm Management	Natural resource management	Socio-economics	Cross-cutting issues
Priority 1	Germplasm improvement and biotechnology; genetic resource conservation	Water Soils Rangeland	Marketing, commerce and trade; post-harvest technologies	Human resource development; capacity building and communication technology
Priority 2	Seed production: diversification	Biodiversity	Quality and added value; institutional policies	Intellectual property rights; crisis and risk management
Priority 3	Integrated pest management	Integrated crop management	Impact assessment	Biosafety and quarantine indigenous knowledge

- Soil and water management for sustainable agriculture, including thrust on nutrient- and water-use efficiency, on-farm water management, marginal water utilization, salinity, drainage, conservation tillage etc.
- Conservation of genetic resources (crops and livestock).
- Livestock improvement and management with emphasis on market oriented breed improvement, health, feed management and rangeland resources.
- Crop diversification, with greater emphasis on incorporation of legumes in cropping systems for long-term sustainability, better income generation and household nutrition security.
- Post-harvest management, storage and value addition of crops, livestock, fruit and vegetable products that can help small and marginal farmers in rural areas.
- Socio-economic and policy research for infrastructure development, economic feasibility of technologies, resource evaluation, marketing, finance, and policy interventions for required adjustments in market economy.
- Strengthening of NARS and human resource development, with greater emphasis on training, including language training.

Later on, these priorities were reassessed more globally when researchers and research administrators from throughout Central and West Asia and North Africa (CWANA) met at ICARDA headquarters from 8–10 May, 2002 to integrate the regional agricultural research priorities into the CGIAR agenda. Therefore, based on the revised analysis of constraints and opportunities for agricultural research and development in the CAC region (CSSA, 2004), the research priorities for CAC region have been identified as shown in Table 5.

More recently, a Round Table Discussion on ICARDA Strategic Plan for Dry Areas in CWANA Region was held on 12 February, 2006, in Tashkent, Uzbekistan, in which the agricultural research priorities were revisited in the light of new CGIAR Science Council research priorities that are fully compatible with the Millennium Development Goals (MDGs). The participants recognized several common needs of the national agricultural research systems in Central Asia and the Caucasus and identified ten out of twenty CGIAR priorities that are being addressed by the CAC Program, including (i) Conservation of plant genetic resources for food and agriculture; (ii) promoting conservation/ characterization of underutilized plant genetic resources (UPGR) for income; (iii) maintaining and enhancing yield of staples; (iv) tolerance of abiotic stresses; (v) income increases from fruit and vegetables; (vi) income increases from livestock; (vii) improving water productivity; (viii) dynamics of rural poverty; (ix) science and technology policy and institutions; and (x) rural institutions and their governance.

It was reaffirmed by the NARS leaders and other participants that priorities identified earlier for CAC hold good and that this region has its own peculiarities that are different from West Asia and

North Africa (WANA) region, such as: (i) this region has potential to be self-sufficient in wheat production and even export it, (ii) salinity is emerging as a major problem, (iii) it is not water availability but efficient use of water that will be a major issue in future, (iv) land degradation and management of rangelands for livestock development is the key area for research thrust, (v) genetic resource management is high priority in view of rich genetic diversity, and (vi) human resource development will remain a major need for the future.

Annual meetings have been held regularly among all participating CG Centers and NARS at which work plans are reviewed and adjusted according to re-assessed priorities.

6 MAJOR ACHIEVEMENTS

The CGIAR Consortium for CAC has made some significant progress during the last eight years of its existence and the major achievements are presented below.

6.1 *Germplasm enhancement*

Major constraints to improving crop production are the lack of improved high yielding varieties and poor access to germplasm. Since independence, practically very few new varieties have been released for want of improved germplasm. Research activities on germplasm enhancement have focused on testing different crop varieties to identify promising breeding material with resistance to both biotic and abiotic stresses (ICARDA, 2006; Morgunov et al., 2005). In this context, ICARDA, CIMMYT, ICRISAT, CIP, ICBA and World Vegetable Center have provided very timely support to all the CA countries, leading to very encouraging results, such as:

(i) Varietal improvement:

- Annually, about 5000 entries from 80 different nurseries of cereals, legumes, groundnut and potato supplied by ICARDA, CIMMYT, ICRISAT, CIP and ICBA are tested in CAC region. Practically in all the eight CAC countries, the new promising breeding material has been identified, which is being used now either for improvement of the local germplasm or for direct multiplication and introduction on farmers' fields.
- Research activities on germplasm enhancement have focused on testing and identifying most promising breeding materials with resistance to both biotic and abiotic stresses. Under this collaborative program, twenty one new promising varieties consisting of winter wheat (10), triticale (2) spring barley (1), chickpea (4), lentil (1), lathyrus (1) and groundnut (2) have recently been released in the region. These varieties have recorded consistently higher yield with superior quality and disease resistance over the local checks. Some of them are now covering large areas and getting popular with the farmers.
- In addition, more than 70 promising entries of various crops including wheat (33), barley (10), triticale (4), chickpea (11), lentil (7) grass pea (3), and groundnut (2) are presently being tested by the State Variety Testing Commissions (SVTC), and are awaiting decision for their release and wide scale adoption.
- In countries of Central Asia and the Caucasus, the important role of legumes is being increasingly felt since testing of chickpea, lentil, vetch, groundnut, etc. started under the program. The collaboration during the last 3–4 years has resulted in identification of some promising breeding material, which has been taken for seed multiplication and for on-farm trials and demonstrations.

(ii) Seed production

Production of quality seeds of high-yielding varieties is critical for faster varietal dissemination. Hence, special emphasis has been laid on the seed development activities in the region. On-farm trials and demonstration plots turned out to be the most important activity for increased agricultural production. The scientists and farmers are now keen to test new varieties. To have an impact on farmers' fields and for wide spread of promising varieties, efforts have been directed towards

Table 6. The details of seed availability and approximate area covered under some newly released varieties.

No.	Country	Name of variety	Available seed (t) 2005	Available seed (t) 2006	Released year	Area covered (ha) 2005	Area covered (ha) 2006
Wheat							
1	Armenia	Armcim	10	24	2006	50	90
2	Azerbaijan	Azametli 95	1,000	1,800	2004	25,000	170,000
3	Azerbaijan	Nurlu 99	400	680	2004	30,000	50,000
4	Azerbaijan	Qobustan	25	50	2006	125	500
5	Georgia	Mtsheta-1	30	100	2002	30	450
6	Kyrgyzstan	Djamin	27	50	2004	300	420
7	Kyrgyzstan	Zubkov	12	30	2004	100	200
8	Kyrgyzstan	Azibrosh	20	60	2004	300	500
9	Kyrgyzstan	Almira	2	5	2005	15	25
10	Kyrgyzstan	Dostlik	7,500	15,000	2002	12,000	20,000
11	Turkmenistan	Bitarap	6,320	15,772	2004	300	9,038
Barley							
1	Armenia	Mamluk	1,000	1,100	2002	5,000	6,500
2	Kyrgyzstan	Adel	6	8	2006	20	30
Triticale							
1	Kyrgyzstan	Alesha	1.0	3.0	2005	10	100
2	Kyrgyzstan	MISCIM	1.2	3.5	2006	10	100
Chickpea							
1	Azerbaijan	Narmin	3.0	20.0	2005	20	43.0
2	Georgia	Elixsir	8.0	16.0	2001	10	25.0
3	Kazakhstan	ICARDA-1	2.5	6.0	2005	4	15.0
4	Kyrgyzstan	Rafat	0.3	0.2	2005	1	2.0
Lentil							
1	Georgia	Pablo	2.6	3.2	2001	8	12
Grass pea							
1	Kazakhstan	Ali Bar	1.8	5.8	2005	5	18
Groundnut							
1	Uzbekistan	Mumtaz	100 kg	1.5	2005	1.5	7
2	Uzbekistan	Salomat	80 kg	0.72	2005	0.75	4

seed multiplication in collaboration with NARS partners. The details of the seeds of new varieties multiplied in the region are provided in Table 6.

(iii) Integrated pest management

Integrated disease and pest management is an important part of germplasm improvement program. Scientists from ICARDA have studied the overall situation for controlling yellow rust – the most important wheat disease in the region.

Identification of physiological races of yellow rust was undertaken in Kyrgyzstan and Uzbekistan. Data for mapping the distribution frequency of new races and the effective resistance genes to yellow rust have been undertaken. Recommendations for replacement of varieties susceptible to yellow rust have been made in view of release of new high yielding disease resistant winter wheat varieties in different countries under the existing research collaboration.

A Wheat-Cereal Leaf Beetle Nursery (WCLBK03) has been established at Kyrgyz Research Institute of Agriculture and the Galla-Aral Branch of Andijan Research Institute of Grain, Uzbekistan, where 144 selected wheat lines were tested for their resistance.

For IPM, a new project supported by USAID under IPM-CRSP, has been initiated with partnership of Michigan State University (MSU), ICARDA and the University of California (Davis). It would mainly focus on biological control of Sunn Pest in wheat and other key pests of selected vegetable crops.

6.2 *Crop diversification*

The CA region is known for monocropping of wheat and cotton. Traditionally, other crops are not grown, whereas crop diversification is a key for increased cropping intensity, sustainability, as well as increased income for farmers. Therefore, under ADB project on "Improving Rural Livelihoods through Efficient On-farm Water and Soil Fertility Management in Central Asia", ICARDA took major initiative to test and demonstrate the potential of crop diversification in the region. Some significant results are:

In spring wheat based cropping systems in northern Kazakhstan, field pea, chickpea, lentil and buckwheat are the best options for inclusion into existing rotations. Also, oat has been found to be high yielding than barley. During field days, farmers showed interest in grain legumes, but desired to have better marketing options before large-scale adoption, as well as exposure to integrated crop production technologies.

In rainfed winter wheat based cropping systems, there are many opportunities to diversify crop production. Out of spring cereals, oat is found to be the most productive and with the highest water use efficiency in southeastern Kazakhstan. Alfalfa is also very suitable for sustainable farming in semi-arid conditions of southern Kazakhstan. Under rainfed conditions, the most successful crop appeared to be safflower; area under which has increased significantly (up to 100,000 ha). This crop is becoming popular in Uzbekistan and Kyrgyzstan.

In winter wheat based irrigated cropping systems, a number of alternatives have been identified for more economical and sustainable farming. The most profitable are food legumes. Successful results were obtained in southeastern Kazakhstan with soybean, in Kyrgyzstan with field pea, common bean and soybean. Safflower can also be grown under supplemental irrigation. In Kyrgyzstan and southeastern Kazakhstan, sugar beet and maize are also good alternatives for crop diversification. In southeastern Kazakhstan, most successful crop for diversification is soybean. Its area has increased from 3,000 ha in 2002 to around 40,000 ha in 2005. The major reason for such a large increase in area is the establishment of organized market in view of new soybean processing plants.

In Fergana Valley of Uzbekistan the crops for diversification after wheat are: maize (both for feed and forage), mungbean, melons and carrots. Rice is also used for double cropping using low saline drain water. In Termez area, southern Uzbekistan, maize and mungbean are widely accepted by farmers for double cropping, covering around 7,000 ha and 5,000, ha respectively. Other alternative crops grown by the farmers are sesame, melons, groundnut and vegetables but rather on smaller scales. In Tajikistan, double cropping is widely adopted by small farmers, where maize and mungbean are widespread followed by common bean, soybean, vegetables, buckwheat, millet, tobacco, groundnut, and sesame. Rice is also grown where water availability is good.

6.3 *Soil and water management*

For the new countries of Central Asia and the Caucasus, to attain long-term sustainable growth and to increase agricultural productivity, it is critical that available natural resources are managed rather effectively using an integrated approach. Integrated natural resource management is thus the key issue for sustainable agriculture. Two CGIAR centers, ICARDA and IWMI, are addressing the issues of natural resource management at on-farm and basin levels, respectively. Some of the major findings are:

(i) Conservation Agriculture

In irrigated cotton-wheat system in Tashkent and Termez provinces, Uzbekistan, compared to deep ploughing, the broadcasting of wheat seeds followed by shallow cultivation was found to be economical with no significant difference in grain yield. Wheat planting into standing cotton using minimum tillage has become a generally accepted practice in almost 40% of irrigated wheat areas in Tajikistan, Turkmenistan and Uzbekistan, thus enabling increased cropping intensity. Recently, sowing of winter wheat in standing cotton using a no-till drill, imported from India, has shown promise in not only obtaining high yields but also in reducing seed rate by 50 kg/ha, due to placement of seed at proper depth and spacing. No-till planting has shown encouraging results for double cropping of mungbean after the harvest of wheat.

A newly designed equipment for planting winter wheat on cotton stubble was tested during recent years. The equipment was found to be good for conservation tillage as compared to local practice of deep ploughing. In Turkmenistan, in cotton-wheat rotation, reduced tillage was found appropriate for wheat sowing after cotton, whereas deep tillage proved useful for planting of cotton.

In rainfed spring wheat based system in northern Kazakhstan, zero tillage proved to be profitable and energy saving, provided nitrogen fertilizer was also applied. Direct seeding of spring wheat using modern no-tillage equipment has lately been adopted on large areas. In Kyrgyzstan, conservation tillage is quite economical in rainfed agriculture and must be promoted for sustainable agriculture. Under rainfed farming in Galla-Aral, Uzbekistan, direct seeding provided yield increase on summer fallow over the last three years due to better moisture accumulation. However, for wider adoption by the farmers, conservation tillage machinery need to be made available in local markets.

Raised bed planters, both local and those imported from Turkey and India, were tested in Azerbaijan, Kazakhstan, Kyrgyzstan and Uzbekistan. This practice helped in reducing the seed rate by almost half, saving water by about 30% and provided higher wheat yields than the traditional practice of wheat planting.

(ii) On-Farm Water Management

On sloping areas, soil erosion is a big problem. In Tajikistan, it was found that terracing and mulching reduce soil erosion and enable production on slopping lands. Further, research activities carried out in Tajikistan revealed considerable potential of micro-furrow (zig-zag furrows) irrigation technology to combat soil erosion on sloping lands. On areas with slope gradients of 0.1 and 0.06, soil erosion got reduced as much as by 86 and 84%, respectively. As compared with conventional furrow irrigation, use of micro-furrow irrigation also increased water use efficiency (WUE) and cotton yield by 29 and 21%, respectively (Karajeh et al., 2002).

In southern Kazakhstan, cutback alternate furrow irrigation increased yield of raw cotton by about 17% and WUE by about 37% as compared to control. Alternate furrow irrigation and cutback alternate furrow irrigation technologies for growing cotton are being disseminated on a large scale every year for the last six years. In south-eastern Kazakhstan, raised bed planting technology, which saves 50% of seed and 15–30% of irrigation water, was tested on large areas on farmers' fields and this technology is being picked up by the farmers. Raised bed planting and cutback irrigation provided higher moisture uniformity, WUE and wheat yield, and lowered surface runoff than conventional planting and strip irrigation.

Portable plastic chutes designed for sloping areas and used in Uzbekistan in winter wheat, potato and onion, increased WUE as high as 40% and increased yield by almost 10%. Under lowland conditions, portable chutes used in wheat and cotton in Uzbekistan, provided better uniformity of water distribution, helped reducing water losses by about 30% and contributed to yield increase, on an average, by 12%. Micro-furrow (zigzag furrow) irrigation in cotton not only reduced surface runoff losses and soil erosion losses but also increased WUE and yield as compared with conventional furrow irrigation.

The technology of drain water reuse was adopted in small farms in Uzbekistan. A special device was installed to lift water from the drain to adjacent fields. Farmers used the system for irrigation of potato, onion, tomato, maize, alfalfa and poplar trees. Therefore, drain water as well as blended water (drain water + fresh water) can effectively be used for irrigation purposes, which allows growing some crops on those areas which otherwise could remain fallow due to lack of irrigation water.

Technology of alternate furrow irrigation in cotton using polyethylene film on 50 and 75% of soil surface increases WUE greatly as compared to alternate furrow irrigation with no mulching. This technology helps in saving water by 30–40%. This technology was disseminated in Djizak Province, Uzbekistan, resulting in 14–24% increase in cotton yield.

In Azerbaijan, micro-sprinkler system, used to grow soybean, reduced irrigation water requirement by 50% as compared to furrow irrigation. Further, combined with raised bed planting and application of fertilizers, micro-sprinkler irrigation also provided the highest yield of soybean.

Water users are not only forming into Water Users Associations for better water distribution among themselves, but also taking part in the governance of the main supply canal through their elected representatives.

6.4 *Feed and livestock management*

An Integrated Feed and Livestock Management project funded by IFAD was implemented by ICARDA in collaboration with NARS of the Central Asia, USAID-GL-CRSP and ILRI during 1999–2002. In 2003, a few selected activities were continued on a no-cost extension basis.

A Regional planning workshop for the second phase of the project was organized by ICARDA-CAC office in Tashkent, Uzbekistan on 5–6 September, 2006, gathering 19 participants from ICARDA, NARS from Kazakhstan, Kyrgyzstan, and Tajikistan, an NGO representative from Tajikistan, and faculty members of Kazakh, Kyrgyz, and US universities. The workshop helped in developing working plans for the research activities. The project would consolidate research on promising options in Kazakhstan and Kyrgyzstan in Central Asia and expand activities to Tajikistan, initiate a new program of research in Pakistan, develop linkages with other countries in South Asia, strengthen national research institutions, and link with key development projects for more rapid achievement of goals. The project goal is to improve the livelihoods of rural communities in Central and South Asia. The project would focus on emerging small farm enterprises in countries of Central Asia, and resource-poor livestock (small ruminant) producers in rainfed areas of Pakistan.

Results of the three major components of the project are briefly summarized below:

(i) Socio-economic Studies

- Big scale farms (various types of cooperatives) are mainly involved in the production of strategic crops often with no livestock production. Emerging individual farms also do not play important role in livestock production as crops provide them with more benefits. Rural households are, therefore, major livestock producers supplying most livestock products to the markets.
- The main problem for all types of livestock producers is the lack of forage, especially during the winter period. Forage deficit was caused by elimination of subsidies and reduction of the forage cropland aimed to increase more profitable grain production.
- Small-scale farms and households have relatively poor or no access to remote rangelands, which is constraining effective livestock production, causing social conflicts among different farming groups, and leading to further degradation of rangelands around settlements.
- Local markets are the main source of income for producers compared with selling products to processing companies, as before, the latter offering less attractive price. In countries with more liberal policies like Kazakhstan, private companies are increasing their share in processing of livestock products. Among all types of products, live animals generate the largest share of producers' income. Fresh milk is being sold in cities and towns mostly by milk women who also supply various dairy products directly to customers at their doorstep.

(ii) Range Management

- Appropriate options for a rational utilization of rangelands have been developed through organized mobile flocks moved for grazing to the remote ranges in Kazakhstan.
- A rotational grazing system was introduced by dividing the village rangeland into two equal parts and using alternative grazing practice on each of these. As a result, every year one part could be in the "resting" stage either in spring or in autumn.

- The utilization of drain water in Turkmenistan confirmed the possibility for production of halophytes in saline soils.

(iii) Livestock Management
To improve the milk yield of local sheep for increasing farmers' income at Boykozon Farm, Uzbekistan, ewes of non-indigenous semi-fine wool breed were inseminated during October–November, 2002, using the semen of dairy East Friesian and Lacaune breeds, received from the University of Wisconsin, USA. Out of 557 ewes inseminated using a laparoscopic method, 306 ewes got pregnant (55% success) and by mid-April, 2003 376 lambs were born. All the lambs are being monitored and records on their growth and development are being kept on monthly basis. Average live body weight of new crossbred sheep at the age of one year was about 35 kg (30–42), which is higher by almost 8 kg than those of the control group.

6.5 *Genetic resource conservation*

Central Asia is the center of origin of a number of economically important crop species and the region possesses rich genetic diversity. Due to lack of continued links with the Vavilov Institute in St. Petersburg and due to financial constraints, the plant genetic resources programs remained rather weak. Hence, IPGRI, ICARDA and CIMMYT started working with the national systems to provide need-based support to collect, conserve, and document local and exotic genetic resources. Some of the important achievements are:

- A CAC network on PGR has been established with the support of IPGRI and ICARDA, consisting of plant genetic resources documentation units established in each country. All units are engaged in making inventory and documentation. Much needed equipment and training on documentation have been provided.
- So far, a total of 18 collection missions to all five CA countries have been organized. Around 2500 accessions of cereals, food legumes and their wild relatives and forage and range species have been collected. The collected germplasm is being kept by the host country and stored "in-trust" in ICARDA's gene bank, providing safety duplication.
- The web site of the CGIAR Collaborative Program for CAC (www.icarda.cgiar.org/cac/index.htm) is available on the Internet since September, 2001. In this web site, PGR webpage has also been established: www.cac-biodiversity.org. This webpage describes some of the plant genetic resources initiatives in the region, and highlights an ICARDA-led initiative to develop genetic resources units in each of the five CA countries.
- Also, some exciting developments have taken place with regard to establishment of Gene Banks in almost all five CA countries. These are:
  Kazakhstan: The Government of Kazakhstan has taken a bold decision to establish a new Institute of Genetic Resources in Almaty to conserve valuable plant and animal genetic resources. IPGRI and ICARDA are providing technical backstopping with regard to design and plans for the new building, construction of which is likely to start during 2007.
  Kyrgyzstan: The Kyrgyz Plant Genetic Resources Center has been provided with 5,000 seed containers, an electronic balance, a cooling system, an oven, and a dehumidifier, shelves for seed samples, necessary furniture and two computers.
  Tajikistan: The Plant Genetic Resources Center, established in September, 2002, is now fully functional. Since then, the facility has been upgraded with 10,000 seed containers, an electronic weighing scale, shelves, a cooling system, a dehumidifier, a standby generator and an oven. So far, around 2400 accessions have been rejuvenated and stored in the Medium Term storage facility.
  Turkmenistan: In March, 2005, the Turkmen National Gene Bank (TNGB) was opened. It is located in the newly constructed National Museum of White Wheat. ICARDA provided office furniture, laboratory equipment, including a computer and printer, and seed containers to make the Gene Bank operational. Recently, under GCDT Project, a dehumidifier, two deep freezers,

aluminum foil packets, an oven for moisture testing, etc. have been provided. To-date, around 398 accessions of wheat have been stored in the deep-feezers.

Uzbekistan: Through the joint efforts of the Ministry of Agriculture and Water Management of Uzbekistan, USDA, ICARDA and IPGRI, Uzbek Gene Bank was renovated and made fully functional. ICARDA-PFU provided technical support for its renovation, designing, and purchasing different equipment. Beside PCs for documentation, 11,000 plastic containers, shelves in the storage room, a generator for cooling system, a dehumidifier have been provided. Under GCTD project, the Gene Bank has recently received additional equipment such as a dehumidifier, a moisture tester, additional 15,000 plastic bottles and support for internet connection, in addition to PCs. The renovated Gene Bank is now storing around 48 000 samples of different crops.

6.6 *Strengthening of NARS*

(i) Human resource development

All centers have made considerable efforts in the area of human resource development. It includes training, study tours, participation in the international, regional and national scientific meetings and workshops, supply of computers and other research equipment. From September 1998 to October 2006, the CAC Program has arranged 112 short and long term training courses with participation of 1327 scientists, 96 study visits with participation of 220 scientists, 79 regional and national workshops with participation of 2241 scientists and farmers, 34 International Conferences with participation of 479 scientists, and 78 consultation planning meetings with participation of 1280 scientists. A total of about 4600 scientists and farmers from the Central Asian countries have so far benefited from different training and human resource development activities, of which 400 have received English training as well.

(ii) Information technology

All Central Asian countries are in due need to have access to global systems of ARD related information as most of them remained isolated in the past due to geographic, political and language barrier. Information exchange has thus been improved through participation in the international and regional workshops, establishing e-mail links and providing Internet access to the collaborating institutions and through CAC Newsletter and various publications, reports etc. The web site of the CGIAR Collaborative Program for CAC is available for consortium partners on the Internet since September, 2001. It contains all relevant information on CG Centers and NARS partners involved in the Program as well as all major achievements made so far, both in English and Russian. The site can be accessed at: www.icarda.org/cac. It is also linked to the CGIAR website: www.cgiar.org.

7 FUTURE OUTLOOK

As it is evident from the above results, the CGIAR Program for CAC has made good progress during the last nine years. It has catalyzed NARS for reorientation of their research agenda and established better interface among CG Centers as well as scientific institutions in the CAC region. It has also built much needed human resource and helped considerably in capacity building. Future strategy would demand the following:

(i) Strengthening Partnership:

Establishment of CAC Association of Agricultural Research Institutions (CACAARI), as a neutral forum, is a true reflection of strengthened partnership for R&D in the region. CACAARI is now building linkages with The Global Forum on Agricultural Research (GFAR) and other regional fora like the Association of Agricultural Research, institutions in the Near East and North Africa (AARINENA) and Asia Pacific Association of Agricultural Research Institutions (APAARI). Involvement of other stakeholders such as farmers, NGOs, and the private sector is critical for future success. Therefore, strengthening of partnership will have to be ensured as we move forward.

Also sustainability of these institutions will be a challenge, especially in an environment of declining support for agricultural R&D in the region.

(ii) Catalyzing the Policy Makers:
In order to bring the agricultural research for development (ARD) agenda upfront, catalyzing the policy makers will be an important objective to be accomplished. Policy support will be critical both for strengthening NARS and for effective technology transfer. Earlier effort through this program to bring together policy makers in a Ministerial Level Meeting held in June 2001 in Issyk-Kul proved useful in catalyzing the Ministries for enhanced support to implement Issyk-Kul Declaration. In future, similar efforts will be needed for increased funding support by the host Governments and donors as well as greater role of the International Agricultural Research Centers (IARCs), Agricultural Research Institutions (ARIs), Non-Governmental Organizations (NGOs), the private sector and farmers.

(iii) Knowledge Sharing:
The CA region is not only land locked but also knowledge locked in view of existing language barrier and the lack of ICT networking. The existing "digital divide" will have to be addressed on priority basis. Farmers demand access to information on reliable technologies, weather conditions, input availability, prevailing market prices as well as possible marketing opportunities. The CGIAR Program for CAC is addressing this concern to a certain extent through various publications, newsletters, leaflets and program website (www.icarda.cgiar.org/cac/index.htm) providing details on new technological advancements as well as strengthened research partnerships.

(iv) Technology Transfer:
During Soviet era, technology transfer (TT) followed the "top down" approach and had a unified system of command for implementation. Hence, once a decision taken by the policy makers, adoption and dissemination of technologies on large scale was never a problem since technology related knowledge and required inputs were simultaneously made available and the instructions were followed by farm workers having practically no role in decision making. In the present context of transition from centralized to a decentralized system towards greater flexibility and independence of farmers, a paradigm shift in the technology transfer mechanism to make it both "top-down" as well as "bottom-up" is indeed a challenge before the governments, farmers associations, non-governmental organizations (NGOs), research institutions and the private sector. Linkage mechanisms among key stakeholders for technology transfer are non-existent at the present juncture. In view of the lack of extension services, role of NARS to link "lab to land" or "scientists to farmers" through on-farm demonstrations, field visits, traveling workshops, participatory research etc. becomes most relevant in addressing this concern. Also linkages of research projects with those of development oriented programs will accelerate the TT process, involving key stakeholders, especially NGOs, in the present context. Scientific institutions can provide much needed technical backstopping to the NGO community for taking new technologies to the end users. For availability of inputs and marketing, private sector initiatives will be important, including support by the Government for needed incentives to the resource poor farmers, especially in the context of crop diversification, natural resource conservation and enhanced benefits to both producers and consumers

8 EPILOGUE

The CGIAR Consortium on Eco-Regional Program for Sustainable Agriculture in CAC has made a good beginning. Strong partnership for ARD has been built among all the stakeholders. All this was achieved through research collaboration between the national scientists and those of different CG centers involved. The Program also helped in strengthening the National Agricultural Research Systems (NARS), especially in human resource development. The research priorities for the CAC region have been identified on a bottom-up basis and initiatives have been taken to address them in a collaborative mode, with greater emphasis on a systemic vision involving an interdisciplinary and inter-institutional approach. In future, the existing partnerships will be extended to ensure active involvement of other stakeholders, especially non-governmental organizations, farmers' associations, the private sector and the advanced research institutions (ARI's), various research institutions, including those in Russia, with whom the links got broken since the Independence of CA countries in 1991. The results achieved so far have also encouraged the Consortium members to expand their activities to meet specific challenges of food security, poverty, and the conservation of natural resources in the CAC region.

The need to strengthen Program activities has also been discussed in a number of donor meetings as well as those of policy makers and the Directors General and senior staff of CGIAR Centers participating in the Consortium. During a recent Luncheon Meeting held in Morocco on 5 December, 2005, the progress achieved so far in such a short time was commended by all the donors concerned and they appreciated the need for continued support to the Program. It was specifically emphasized that the impact of the Program is already visible and it can serve as a model for the systemwide/ eco-regional initiatives elsewhere. It was also emphasized that these results need to be linked with the various development projects through active involvement of policy makers at the national level, including their commitment for a greater support to agricultural research and development in the region.

For required continuity and growth of this important and the only regional agricultural research program, it is envisaged that the NARS, CG centers and other stakeholders, including donor organizations, would continue strengthening their support for the on-going efforts under this initiative. Efforts so far have helped in catalyzing the process. Hence, continued support by all concerned would accelerate the process for achieving the main goal of sustainable agriculture in Central Asia.

REFERENCES

ADB, 2006. Country Key Indicators: Kazakhstan, Kyrgyzstan, Tajikistan, Turkmenistan and Uzbekistan, www.adb.org.

Babu. S., Tashmatov (eds.). 2000. Food Policy Reforms in Central Asia: Setting the Research Priorities. IFPRI, Washington, D.C.

CIA, The World Factbook, 2006, https://www.cia.gov/cia/publications/factbook/index.html

CSSA, 2004. Challenges and Strategies of Dryland Agriculture, Special Publication Number 32.

EastAgri, Country profiles: Kazakhstan, Kyrgyzstan, Tajikistan, Turkmenistan and Uzbekistan, 2006, http://eastagri.org/countries.asp.

Faostat, 2006, http://faostat.fao.org/default.aspx.

FAO Statistical Databases (2000), http://www.apps.fao.org.default.html.

Federal Research Division of the Library of Congress of USA, 2005. Country studies: Kazakhstan, Kyrgyzstan, Tajikistan, Turkmenistan and Uzbekistan, , http://lcweb2.loc.gov/frd/cs/

Gintzburger, G., K. Toderich, B. Mardonov, and B. Mahmudov. 2003. Rangelands of The Arid And Semi-arid Zones In Uzbekistan, CIRAD and ICARDA

ICARDA CAC. 2006. The Status of Varietal Development and Seed Sector Activities in Uzbekistan. FAO/TCP/UZB/3002 (A).

ICARDA. 2002. ICARDA in Central Asia and the Caucasus. Ties that Bind, No. 12 (revised). ICARDA, Aleppo, Syria, 36 pp. En.

Karajeh, F., J. Ryan, and C. Studer. 2002. On-farm Soil And Water Management In Central Asia, ICARDA

Morgunov, A., A. McNab, K.G. Campbell, and R. Paroda. 2005. Increasing Wheat Production In: Central Asia through Science and International Cooperation: Proceedings of the First Central Asian Wheat Conference, CYMMYT, MOA Rep. of Kazakhstan.

PFU. 2001. CGIAR Collaborative Research Program for Sustainable Agricultural Development in Central Asia and the Caucasus. Program Facilitation Unit (PFU), P.O. Box 4564, Tashkent 700 000, Uzbekistan. 24 pp.

Ryan, J., P. Vlek, and R. Paroda. 2004. Agriculture In Central Asia: Research For Development, ICARDA, ZEF

The Economist Intelligence Unit, Country profiles: Kazakhstan, Kyrgyzstan, Tajikistan, Turkmenistan and Uzbekistan, 2005, http://www.eiu.com

The World Bank, World Bank Development Indicators Database (2000), http://www. worldbank.org/data/countrydata/countrydata.html.

The World Bank, World Development Indicators, 2005 and 2006, www.worldbank.org

CHAPTER 9

Soil and environmental degradation in Central Asia

R. Lal

Carbon Management and Sequestration Center, The Ohio State University, Columbus, OH, USA

1 INTRODUCTION

Five countries in Central Asia have a total land area of about 400 million hectare (Mha), population of about 60 million, arable land area of 32.6 Mha, and average per capita arable land area of 0.55 ha (Table 1). The population growth rate ranges from 1.2% yr^{-1} to 3.0% yr^{-1} of the total land area of 400 Mha, about 148 Mha or 37.0% is desert (Table 2). Desert area, as a percent of total land area, is 27.5% in Kazakhstan, 35.4% in Kyrgyzstan 17.5% in Tajikistan, 79.3% in Turkmenistan and 55.7% in Uzbekistan.

Despite the low population, Central Asia is an important region historically, strategically and environmentally. Historically, the region comprises ancient civilization which strongly influenced South Asia and Southwestern Europe. Being on a silk route, the region was the trade link between China and Middle East and Europe. As newly independent states, all five countries are strategically important in global economy, especially for trading carbon (C) credits. Environmentally, the region can be a major terrestrial sink for sequestering atmospheric CO_2 in trees and soil thereby off-setting

Table 1. Land resources and population of Central Asia (World Bank, 2000; FAO, 2004).

Country	Total area (Mha)	Arable land (Mha)	Total population (10^6)	Per capita arable land (ha)
Kazakhstan	272.5	24.0	16.8	1.43
Kyrgyzstan	20.0	1.4	5.2	0.27
Tajikistan	14.3	0.8	6.2	0.13
Turkmenistan	48.8	1.4	5.2	0.27
Uzbekistan	44.7	5.0	26.1	0.19
Total	400.3	32.6	59.5	0.55

Table 2. Land area covered by deserts in Central Asia (Adapted from Jumashov, 1999).

Country	Desert area (Mha)	% of total area
Kazakhstan	74.7	27.5
Kyrgyzstan	7.0	35.4
Tajikistan	2.5	17.5
Turkmenistan	38.7	79.3
Uzbekistan	25.0	55.7
Total	147.9	37.0

Table 3. Land use in Central Asia (FAO, 2004).

Country	Agricultural land	Cropland	Rangeland	Forest	Other land	Irrigated land
	Mha					
Kazakhstan	206.6	21.5	185.1	9.6	35.7	2.35
Kyrgyzstan	10.7	1.3	9.4	0.7	3.0	1.10
Tajikistan	4.1	0.4	3.2	0.5	9.1	0.72
Turkmenistan	32.6	1.9	30.7	4.0	11.5	1.80
Uzbekistan	26.7	4.5	22.2	1.3	14.8	4.28
Total	280.7	30.1	250.6	16.1	79.1	10.35

Table 4. Land area under irrigation (Adapted from Babaev and Zonn, 1999; Macklin, 2000; Glantz, 2000).

Country	1950	1960	1965	1970	1975	1980	1985	1986	1995
	Mha								
Kazakhstan	–	–	–	–	–	–	–	–	0.76
Kyrgyzstan	–	–	–	–	–	–	–	–	0.46
Tajikistan	2.28	2.57	2.64	2.75	3.00	3.53	3.91	4.17	4.28
Turkmenistan	0.36	0.43	0.44	0.52	0.57	0.63	0.66	0.70	0.72
Uzbekistan	0.45	0.50	0.51	0.67	0.86	0.96	1.16	1.35	1.74
Total	3.09	3.50	3.59	3.94	4.43	5.12	5.73	6.22	7.94

the industrial emission. In addition, degradation of the Aral Sea and its Basin has been a major concern which must be addressed.

Human–induced soil and environmental degradation are widespread problems in the region. These problems are exacerbated by the predominantly arid climate of the region. The climatic aridity, characterized by low precipitation and high evaporative demand, reduces ecological resilience and renders soils and ecosystems highly prone to degradation. Thus, the objective of this chapter is to discuss predominant land uses, and the impact of land use and soil/water/vegetation management practices on the extent and severity of soil and environmental degradation in Central Asia.

2 LAND USE

Agricultural lands, comprising cropland and rangeland, are the predominant land use covering 281 Mha or 70% of the total land area (Table 3). Being an arid climate, rangelands occupy 63% of the land area compared with only 7% under cropland. Of the 30.1 Mha of cropland, 10.3 Mha or 34.0% is irrigated. Forested area is only 16.1 Mha or 4.0% of the total land area.

Land area under irrigation expanded drastically between 1950 and 1990. The increase in irrigated land area between 1950 and 1986 occurred from 2.3 Mha to 4.2 Mha (83% increase) in Uzbekistan, 0.36 Mha to 0.70 Mha (94% increase) in Tajikistan, and from 0.45 Mha to 1.35 Mha (300% increase) in Turkmenistan (Table 4). This rapid increase in irrigation strongly affected the water balance of the region, especially that of the Aral Sea and the two rivers feeding it: Amu Darya and Syr Darya.

3 THE ARAL SEA

Prior to the implementation of large scale irrigation schemes by the Soviet authorities during the 1950's and 1960's, the Aral Sea was the World's fourth largest lake with a surface area of 6.7 Mha,

Table 5. Temporal changes in the Aral Sea over a 40 year period from 1960 to 2000 (Adapted from Macklin and Williams, 1996; Micklin, 2000).

Year	Level (m)	Area (Mha)	Volume (Km^3)	Salinity $g\ l^{-1}$
1960	53.4	6.70	1090	10
1991	51.1	6.02	925	11
1976	48.3	5.57	763	14
1989				
(i) Large lake	39.4	3.69	341	30
(ii) Small lake	40.2	0.28	23	30
1994				
(i) Large lake	36.8	2.89	298	35
(ii) Small lake	40.8	0.31	273	25
2000				
(i) Large lake	33.4	2.18	186	60
(ii) Small lake	41.6	0.34	26	20
2010				
(i) Large lake	30.9	1.84	128	>60
(ii) Small lake	42.6	0.35	29	15–20

Table 6. Water use for agricultural production in Amu Darya and Syr Darya Basins during the late 1980s (Adapted from Tsutsu, 1991).

	Water use (%)	
Crop	Amu Darya ($44.8\ km^3$)	Syr Darya ($33.6\ km^3$)
Cotton	51.5	34.2
Fodder	18.5	28.7
Rice	12.1	18.6
Orchard	4.3	6.0
Vineyard	2.8	1.9
Maize	2.2	4.0
Wheat	3.0	2.5
Potatoes	0.8	0.9
Vegetables	2.4	2.3
Melons	1.6	0.8
Other cereals	0.8	1.0

water level of 53.4 m, total volume of 1090 $Km^3$from the Amu Darya and 37–40 Km^3 from the Syr Darya (Center for International Projects, 1991; Micklin, 1994; Tanton and Heaven, 1999). Developments of hydropower generation and irrigation schemes in early 1960's, increased the irrigation area in the Basin from 3.5 Mha in 1960 to 6.2 Mha in 1986 (Table 4) and to 7.5 Mha by 1994 (Tanton and Heaven, 1999; TACIS, 1995). Between 1950 and 1986, the irrigated land area increased from 2.28 Mha to 4.17 Mha in Uzbekistan, 0.36 Mha to 0.70 Mha in Tajikistan, and 0.45 Mha to 1.35 Mha in Turkmenistan. Consequently, the Aral Sea progressively shrunk (Table 5) with strong adverse environmental impacts. The large irrigation canal built under the project was the Karakum Canal: a 1,100 km long canal completed in 1965. A large fraction of the total flow of Amu Darya and Sry Darya was used for the production of cotton, fodder and rice (Table 6). With increase in irrigation, land area under cotton production increased by 122% in Uzbekistan, 196% in Tajikistan and 330% in Turkmenistan (Table 7). The unlined irrigation canal caused severe seepage losses through highly permeable desert sand, with the overall water use efficiency of about 30%.

Table 7. Temporal changes in land area under cotton production in Central Asian countries (Adapted from Critchlow, 1991).

Country	1940	1971–75	1976–80	1981–85	1985	1986	Increase from 1940–86 (%)
	Mha						
Tajikistan	0.11	0.26	0.30	0.31	0.31	0.31	182
Turkmenistan	0.15	0.44	0.50	0.53	0.56	0.65	330
Uzbekistan	0.92	1.72	1.82	1.93	1.99	2.05	122
Total	1.18	2.42	2.62	2.77	2.86	3.01	155

Table 8. The extent of salinization of irrigated land in Central Asia (Adapted from Pankova and Solovjev, 1995).

Country	Irrigated land (Mha)	Saline soils (Mha)
Kyrgyzstan	1.0	0.1
Tadijkstan	0.7	0.1
Turkmenistan	1.2	1.1
Uzbekistan	4.1	2.1
Total	7.0	3.4

Table 9. Salt affected soils in Central Asia (Adapted from Essenov and Redgephaev, 1999 and Funakawa et al., 2000).

Country	Salt affected soils (Mha)
Kazakhstan	1.40
Kyrgyzstan	0.01
Tajikistan	0.12
Turkmenistan	1.09
Uzbekistan	2.13
Total	4.75

Because of the diversion of flow, the annual inflow into the Aral Sea decreased to 7 km^3 yr^{-1}. By 1996, the Aral Sea had become saline and devoid of fish. The shrinkage and drying of Aral Sea led to the drying of extensive marginal swamp ponds and flood plains of the deltas, decline in domestic water supply of towns in the region, and the loss of agronomic productivity due to salinization of croplands, and a drastic decline in biodiversity (Frederick, 1991; Glantz, 2000).

4 SOIL SALINIZATION

With increase in area under irrigation, there has also been a problem of secondary salinization of irrigated cropland. Of the total irrigated land use of 7.0 Mha, almost half (3.4 Mha) had been salinized by mid 1990's (Table 8). The problem of salinization is especially severe in Turkmenistan and Uzbekistan where 92% and 51% of the irrigated land, respectively, had been salinized by 1995 (Pankova and Salovjev, 1995, Table 8). Along with 1.4 Mha of salt affected soils in Kazakhstan, total area of salt affected soil in Central Asia is 4.75 Mha (Table 9) or 68% of the total irrigated land area of 7.0 Mha. Salinization of the irrigated land in the Aral Sea Basin is rather high both

Table 10. The extent and severity at desertification in the Aral Sea Basin (Modified from Babaev and Muradov, 1999).

Degradation process	Total degraded area (Mha)	Degraded area as % total area
Vegetation degradation	108	77
Sand deflation	2	2
Water erosion	8	6
Salinization of irrigated land	13	9
Salinization due to shrinkage of sea	5	3
Desertification	3	2
Water logging	1	1
Total	140	100

Table 11. Estimates of desertification in Central Asia (Modified from Dregne and Chou, 1992).

Severity	Irrigated land	Rainfed cropland	Rangeland
	Mha		
Slight	10	29	130
Moderate	9	21	160
Severe	1	4	135
Very severe	0.2	–	5
Total area	20	54	430
% Desertified	51	45	70

due to secondary salinization and shrinkage of the Sea which has been source of salt through wind transport to the adjacent land (Table 10).

5 SOIL DEGRADATION AND DESERTIFICATION

Water mismanagement, excessive irrigation and monocropping caused severe problems of soil and environmental degradation, especially, in the Aral Sea Basin. Water logging and soil salinization affected 2.8 Mha of land area in the Aral Sea Basin. In addition, vast areas were affected by development of salt pans from irrigation seepage water especially in shallow depressions. Salt/dust storms arising from the exposed sea bed adversely affected soil and vegetation of the adjacent areas, leading to destruction/desertification of the ecosystems. The problem was exacerbated by degradation of the deltas of Amu Darya and Syr Darya by severe diminution of their flow (Frederice, 1991). The data in Table 10 show the extent and severity of soil and vegetation degradation.

Desertification of soil and vegetation are severe problems throughout Central Asia (Table 11). Dregne and Chou (1992) estimated that land area prone to desertification in Central Asia comprises 51% of irrigated land, 45% of the rainfed cropland and 70% of the range land. Soil and environmental degradation were caused by poor water management, and collapsing irrigation systems especially in the Karakum Canal. The causative factors are complex and driven by the interactive effects of biophysical, socio-economic and political drivers. Uncontrolled and excessive grazing were responsible for degradation of range lands. Vladychensky et al. (1995) observed that soils of the grazed pasture in Kyrgyzstan were severely degraded.

6 SOIL ORGANIC CARBON POOL

In general, soils of the arid and semi-arid regions have low soil organic carbon (SOC) pool. The SOC pool is depleted by land misuse and soil mismanagement, including practices such as excessive grazing, plowing, and residue removal. In Kyrgyzstan, Vladychensky et al. (1995) assessed the impact of grazing on the SOC pool in 0–15 cm depth. In the protected/ungrazed area, the SOC pool was 36.4, 49.7 and 43.9 Mg ha^{-1}. In contrast, adoption of soil restorative measures can enhance SOC concentration and restore the depleted pool.

Afforestation of degraded soils can enhance the SOC pool (Lalymenko and Shadzhikov, 1996; Kuliev, 1996; Faituri, 2002; Zayed, 2000). Similarly, restoration of degraded pastures and range lands enhances SOC pool. Establishment of grasses in pastures can improve soil fertility (Mirzaev, 1984), and increase addition of biomass-carbon to the soil. In Western Kazakhstan, Teryukov (1996) observed that improved pastures comprised a mixture of shrub species and perennial grasses. Restoring degraded soils and ecosystems in the Aral Sea Basin can also enhance the SOC pool (Baitulin, 2000; Dimeyova, 2000; Karibayeva, 2000; Meirman et al., 2000; Muradov, 2000). Martius et al. (2004) outlined several strategies for developing sustainable land and water management options for the Aral Sea Basin through an inter-disciplinary approach. These strategies include mapping of ground water and soil salinity, afforestration and assessing performance of tree growth, enhancing irrigation efficiency through improved methods, and soil fertility management.

Conversion of Steppe and natural ecosystems into cropland can deplete the SOC pool. In this regard, recycling of drainage water can be a useful strategy (Koloden and Robachev, 1999). The rate of SOC depletion can be severe with extractive farming practices (e.g., no fertilizer use, crop residue removal) and monoculture. Use of manure can maintain and even enhance the SOC pool (Table 12). Improved systems of soil and crop management in arable land can also enhance the SOC pool. Use of crop residue mulch is one option of enhancing the SOC pool. The data in Table 13

Table 12. Cropping systems and fertility effects on soil organic carbon pool (Recalculated from Nasyrov et al., 2004).

Treatment	Soil organic carbon pool[1] (Mg ha^{-1}) 1930	1976	Rate of change (Mg ha^{-1} yr^{-1})
Cotton monoculture, no fertilizers	30.8	20.3	−0.23
Cotton monoculture, with fertilizers	32.3	25.1	−0.16
Cotton monoculture, with manure	37.9	37.9	0
Crop rotation[2] with fertilizers	28.9	15.8	−0.28

[1] Assuming 30 cm depth and bulk density of 1.25 Mg m^{-3}.
[2] Rotation involved 3 years of alfalfa and 7 years of cotton.

Table 13. Mulching effects on soil organic carbon pool in Sierozems (Calculated from Sanginov et al., 2004).

	Soil organic carbon pool (Mg ha^{-1})				Rate of change (Mg ha^{-1} yr^{-1})	
	Fallow		Mulching			
Depth (cm)	1986	2001	1986	2001	Fallow	Mulching
0–30	27.4	19.4	27.9	63.2	−0.56	2.69
30–50	10.4	10.6	14.0	32.9	0.01	1.26
0–50	37.8	30.5	41.9	101.1	−0.48	3.95

Values based on assuming 58% carbon in soil organic matter.

show that long-term application of crop residue mulch enhanced the SOC pool in a Sierozem soil at the rate of about 4.0 Mg C $ha^{-1}yr^{-1}$. Judicious use of irrigation, which increased biomass production with a high water use efficiency (Kg $ha^{-1}mm^{-1}$ of water) can also increase the SOC pool. In this regard, recycling of drainage water can be a useful strategy (Koloden and Robachev, 1999). The data in Table 14 show that irrigation enhanced the SOC pool of a Typical Sierozem and Light Sierozem but not that of a Meadow.

Reducing intensity and frequency of plowing in conjunction with the use of crop residue mulch through conservation farming enhances the SOC pool (Lal, 2004). Application of conservation farming, however, is rather site specific (Barajev and Suleimenov, 1974; Suleimenov and Lysenko, 1997; Hemmat and Oki, 2001). In addition, elimination of summer fallowing can be extremely useful to reducing risks of erosion by wind and water and enhancing the SOC pool (Suleimenov et al., 1994; 1997; 2003; Barajev and Suleimenov, 1979). Assessment of the rate of SOC sequestration, however, must be based on long-term experimentation. The SOC pools of those soils are highly variable (Yanai et al., 2005). The temporal variability in SOC pool can be marked by high spatial variability, which must be taken into consideration through geostatistics and other techniques based on sampling for variable soils.

7 OFF-SETTING INDUSTRIAL EMISSIONS OF CO_2 THROUGH SOIL CARBON SEQUESTRATION

Temperal changes in CO_2 emission for 5 Central Asian countries are shown in Table 15. It is apparent that the collapse of former Soviet Union led to disruption in the industrial production

Table 14. Irrigation effects on soil organic carbon pool in some soils of Uzbekistan (Calculated from Nasyrov et al., 2004).

	Depth (cm)	Soil bulk density ($Mg\,m^{-3}$)	Soil organic carbon pool ($Mg\,ha^{-1}$)		
			Virgin soil	Rainfed	Irrigated
Typical Sierozem	0–30	1.25	39.2–54.4	17.4–23.9	23.9–37.0
	30–60	1.30	–	11.3–15.8	13.6–20.4
Meadow	0–30	1.20	52.2–73.1	37.6–73.1	31.3–52.2
	30–60	1.25	–	26.1–73.1	2.2–39.2
Light Sierozem	0–30	1.25	32.6	13.1–17.4	19.6–36.8
	30–60	1.30	–	9.0–13.6	13.6–22.6

Values based on assuming 58% carbon in soil organic matter.

Table 15. Temporal changes in CO^2 emissions in Central Asia (Adapted from Marland et al., 2001).

	Kazakhstan	Kyrgyzstan	Tajikistan	Turkmenistan	Uzbekistan	Total
Year	$Tg\,C\,yr^{-1}$					
1992	69.0	3.0	5.6	7.7	30.9	116.2
1993	58.4	2.3	3.7	7.5	32.4	106.3
1994	53.7	1.8	1.4	9.1	30.5	96.5
1995	45.2	1.4	1.4	9.2	27.0	86.2
1996	38.3	1.7	1.6	8.3	27.7	77.6
1997	37.1	1.8	1.4	8.0	27.7	76.0
1998	33.5	1.8	1.4	7.6	29.8	76.1
1999	30.7	1.3	1.4	9.3	32.0	74.7
2000	33.1	1.3	1.1	9.4	32.4	77.3

with an attendant decline in CO_2 emission. The annual $CO_2 - C$ emission of 116 Tg C yr^{-1} in 1992 declined to 77 Tg C yr^{-1} in 1996, and has stabilized over the last decade ending in 2005.

The industrial emission of CO_2 can be off-set by terrestrial C sequestration in soil and biota. The potential of soil C sequestration in Central Asia is 10 to 22 Tg C yr^{-1} (16 ± 6 Tg C yr^{-1}) (Lal, 2004) which is about 20% of the total industrial emission. In addition, there is also a potential of C sequestration in the biomass through afforestation. While off-setting industrial emissions, terrestrial C sequestration is an essential strategy of restoring degraded soils and ecosystems.

8 CONCLUSION

Adoption of extensive farming systems, monoculture of irrigated cotton and expansion of livestock industry, caused the severe problem of soil and environmental degradation. Principal soil degradation processes are desertification, salinization, waterlogging, defoliation and degradation of vegetation. Degradation processes have been exacerbated by shrinkage of the Aral Sea and drying up of the deltas of Amu Darya and Syr Darya.

Restoring degraded/desertified ecosystems is important to enhancing productivity and improving the environment. Improving soils and environment of the Aral Sea Basin, increasing inflow into the Sea, afforestation of degraded soils and improvement of rangeland through establishment of better species and controlled grazing are important options.

Enhancing SOC pool in cropland soils is necessary to the sustainable management of soil and water resources. The SOC pool can be enhanced by mulch farming techniques, use of conservation farming, elimination of summer fallow, and maintenance of soil fertility. The rate of SOC sequestration can be high with mulch farming and manuring techniques. Conservation farming can be site specific. The potential of SOC sequestration, 22 Tg C yr^{-1} or 20% of the annual industrial emission, provides another income source for the farmer through trading of carbon credits regionally and internationally.

Soil quality can be enhanced by adoption of conservation farming with crop residue mulching, elimination of summer fallowing, manuring and judicious use of irrigation water. Afforestation and restoration of range lands also enhance the soil carbon pool. The potential of carbon C sequestration in soils of Central Asia is 10 to 22 Tg C yr^{-1}. In addition, there is also a potential of C sequestration in biomass through afforestation. Realizing the potential of soil C sequestration can offset 20% of the annual industrial emission of 77 Tg C yr^{-1} in 2002.

REFERENCES

Babaev, A.G. and I.S. Zonn. 1999. Desertification: its consequences and central strategies. In: A.G. Babaev (ed.), Desert Problems and Desertification in Central Asia. Springer Verlag, Berlin: 257–265.

Babaev, A.G. and C.O. Muradov. 1999. The problems of Aral Sea and Caspian Sea. In: A.G. Babaev (ed.), Desert Problems and Desertification in Central Asia. Springer Verlag, Berlin: 231–245.

Baitulin I.O. 2000. National strategy and action plan to combat desertification in Kazakhstan. In: *Sustainable Land Use in Deserts*, Breckle S-W, Veste M, Wucherer W (eds.). Springer Verlag: Berlin, 441–448.

Barajev A.I. and M.K. Suleimenov. 1979. The role of sub-surface soil tillage in controlling wind erosion and drought in the conditions of northern Kazakhstan. *Proceeding of the 9th ISTRO Conference*, Stuttgart, Germany.

Center for International Projects. 1991. The modern condition of nature, population and economy of the Aral Basin: Dianostic study. Center for International Projects, Moscow, Soviet Union.

Critchlow, J. 1991. Nationalism in Uzbekistan. A Soviet Republic's Road to Sovereignty. Westview Press, Boulder, Colorado.

Dimeyova, L.A. 2000. Methods of conservation and restoration of vegetation cover on the Aral Sea coast. p. 69–73. In: *Sustainable Land Use in Deserts,* Breckle S-W, Veste M, Wucherer W (eds.). Springer Verlage: Berlin.

Dregne, H.E. and N-T. Chou. 1992. Global desertification dimensions and costs. p. 249–281. In: H.E. Dregne (ed.) Degration and Restoration of Arid Lands, Texas Tech Univ. Lubbock, TX.

Essenov, P.E. and K. R. Redjepbaev. 1999. The reclamation of saline soils. p. 167–177. In: R. Lal, J. M. Kimble, H. Eswaran and B.A. Stewart (eds.), Climate Change and Peodogenic Carbonates. Lewis Publishers, Boca Raton, FL 167–177.

Faituri, M.Y. 2002. Soil organic matter in Mediterranean and Scandinavian forest ecosystems – dynamics of organic matter, nutrients and monomeric phenolic compounds. *Acta Universitais Agriculturae Sueciae Silvestria* 236: 39.

FAO. 2004. Production yearbook. FAO, Rome, Italy.

Frederick, K.K. 1991. The disappearing Aral Sea resources. Resources for the Future, Winter Issue, Washington, DC. p. 11–14.

Glantz, M. (ed.). 2000. Creeping Environmental Problems and Sustainable Development in the Aral Sea Basin. Cambridge Univ. Press, Cambridge, U.K.

Glantz, M.H. 2000. Sustainable Development and Creeping Environmental problems in the Aral Sea Region. p. 1–25. In: M.H. Glantz (ed.), Creeping Environmental Problems and Sustainable Development in the Aral Sea Basin. Cambridge Univ. Press, Cambridge, U.K.

Hemmat A. and O. Oki. 2001. Grain yield of irrigated winter wheat as affected by stubble tillage management and seeding rate in central Iran. *Soil & Tillage Research* 63: 57–64.

Karibayeva, K.N. 2000. Environmental problems of the southern region of Kazakhstan. p. 427–440. In: S-W Breckle, M. Veste and W. Wucherer (eds.), Sustainable Land Use in Deserts. Springer-Verlag, Berlin.

Koloden, M.V. and G.I. Rabochev. 1999. The use of drainage water. p. 201–209. In: A.G. Baaev (ed.), Desert Problems and Desertification in Central Asia. Springer, Berlin.

Kuliev, A. 1996. Forests – an important factor in combating desertification. *Problems of Desert Development* 4: 29–33.

Lal, R. 2004. Carbon sequestration in soils of Central Asia. Land Degredation of Development 15: 563–572.

Lalymenko, N.K. and K.K. Shadzhikov. 1996. Forest improvement on the foothill plain of the Bol'shoi Balkhan range. *Problems of Desert Development* 6: 48–73.

Marland, G., T. Boden and R. Andres 2001. National CO_2 emissions from fossil fuel burning, cement manufacture and gas flaring: 1751–1998. CO_2 Information Analysis Center, Oak Ridge National Laboratory, Oakridge TN, USA.

Martiuis, C., J. Lameras, P. Wehrheim, A. Schoeller-Schletter, R. Eshchanov, A. Tupitsa, A. Khamzina, A. Akramkhanov and P.L.G. Vlek. 2004. Developing sustainable land and water management for the Aral Sea Basin through an inter-disciplinary approach. p. 45–60. In: Proc. Int. Conf. "Research on Water in Agricultural Production in Asia for the 21st Century". ACIAR, Canberia, Australia.

Meirman, G.T, L. Dimeyeva, K. Dzhamantykov, W. Wucherer and S.W. Breckle. 2000. Seeding experiments on the dry Aral Sea floor for phytomelioration. p. 318–322. In: S-W Breckle, M. Veste, and W. Wucherer (eds.), Sustainable Land Use in Deserts,Springer-Verlag, Berlin.

Micklin, P.P. 1994. The Aral Sea Problem. Proc. Instn. Div. Engrg., London 102(8): 114–121.

Micklin, P. and W.D. Williams (eds). 1996. The Aral Sea Basin. NATO ASI Series, The Environment, Vol. 12. Springer Verlag, Berlin.

Micklin, P. 2000. The Aral Sea Crisis: Anatomy of a Water Management Disaster. Western Michigan University, Kalamazoo, MI (unpublished).

Mirzaev, M.M. 1984. The role of grasses in increasing soil fertility in the orchards of Uzbekistan. *Vestnik-Sel'Skokhozyaistvennoi-Nauki* 1: 87–93.

Muradov, C. 2000. Activity of the Consultation Center to combat desertification in Turkmenistan. p. 415–417. In: S-W Breckle, M. Veste, and M. Wucherer (eds.), Sustainable Land Use in Desert. Springer-Verlag: Berlin; 415–417.

Nasyrov, M., N. Ibragimov, B. Halikov and J. Ryan. 2004. Soil organic carbon of Central Asia's agro ecosystems. p. 120–139. In: J. Ryan, P. Vlek and R. Paroda (eds.), Agricultural in Central Asia: Research for Development, ICARDA/ZEF, Aleppo, Syria.

Pankova, E.I. and D.S. Solovjev. 1995. Remote sensing methods of dynamics of irrigated land salinity for construction of geotechnical system. Probliemy Osvoieniia Pustyn 1995, No. 3.

Sanginov, S., S. Nurmatov, J. Sattorov and E. Jumabekov. 2004. Harvesting the soil resources of Central Asia for development. p. 116–125. In: J. Ryan, P. Vlek, and R. Paroda (eds.), Agriculture in Central Asia: Research for Development. ICARDA/ZEF, Aleppa, Syria.

Suleimenov, M.K., S.P. Singh and C. Prasad. 1994. Development of soil conservation practices in the steppe regions of northern Kazakhstan. p. 213–217. Technologies for Minimizing Risk in Rainfed Agriculture.

Suleimenov, M.K. and N.E. Lysenko. 1997. Soil tillage systems in northern Kazakstan. p. 621–624. In: Proceedings of the 14th ISTRO Conference, Pulawy, Poland.

Suleimenov, M.K., H.J. Braun, F. Altay, W.E. Kronstad, S.P.S. Beniwal and A. McNab. 1997. Continuous spring wheat grown under semi-arid conditions of northern Kazakhstan. p. 481–484. In: Proceedings of the 5th International Wheat Conference, Ankara, Turkey, 10–14 June 1996.

Suleimenov, M.K., M. Pala, F. Karajeh, L. Garcia-Torres, J. Benites, A. Martinez-Vilela and A. Holgado-Cabrera. 2003. ICARDA's network on conservation agriculture in central Asia. p. 165–168. In: Conservation Agriculture: Environment, Farmers Experiences, Innovations, Socio-economy, Policy. Kluwer Academic Publishers: The Netherlands.

Tanton, T. W. and S. Heaven. 1999. Worsening of the Aral Sea Basin Crisis: can there be a solution? J. Water Resources planning and Management 125: 363–368.

Technical Assistance to the Commonwealth of Independent States (TACIS). 1995. Irrigation crop production systems. In: Water Resources Management and Agricultural Production in the Central Asian Republics. Vol. IV. Rep. of European Commission. DGI. WARMAP Project, Phase 1. European Commission, Brussels, Belgium.

Teryukov, A.G. 1996. Phytomelioration of degraded pastures in wester Kazakhstan. Problems of Desert Development 5: 64–69.

Tsutsui, H. 1991. Some remarks on the Aral Sea Basin irrigation management. Nara, Japan (mimeo).

Vladychensky, A.S., T.Y. ul. Yanova, S.A. Balandin and I.N. Kozlov. 1995. Effect of grazing on Juniper forest soils of southwestern Tya-Shan region. Euasian Soil Science 27: 1–9.

World Bank. 2000. World Bank Development Indicators Database. Washington, DC. (http://www.worldbank.org/data/countingdata/countingdata.html.).

Yanai, J., A. Mishima, S. Funkawa, K. Akshalov and T. Kasaki. 2005. Spatial variability of organic matter dynamics in the semi-arid croplands of northern Kazakhstan. Soil Sci. Plant Nutr. 51: 261–269.

Zayed, M.A. 2000. Spatial distribution of soil nutrients and ephemeral plants underneath and outside the canopy of Artemisia monosperma (Asteraceae) shrubs in the Egyptian deserts. Acta Botanica Hungarica 42: 347–353.

CHAPTER 10

Land degradation by agricultural activities in Central Asia

B. Qushimov
Tashkent State University of Economics, Tashkent Uzbekistan

I.M. Ganiev
Samarkand Agricultural Institute, Samarkand, Uzbekistan

I. Rustamova & B. Haitov
Tashkent State Agrarian University, Tashkent, Uzbekistan

K.R. Islam
The Ohio State University South Centers, Piketon, Ohio, USA

1 INTRODUCTION

Central Asia (CA) is situated in the heart of the Eurasian continent with the total area of 3.9×10^6 km^2 and the population of over 53 million. It encompasses five countries, namely Kazakhstan, Kyrgyzstan, Tajikistan, Turkmenistan, and Uzbekistan (Figure 1). The main natural features of the region include the highest ridges of the Pamir, the Tien Shan and the Altai, vast deserts and steppes, large rivers and such as the Amu-Darya, the Syr-Darya, the Irtysh and the Ily, and inland seas, such as the Caspian Sea, the Aral Sea and a range of lakes. The regional landscapes differ by their continental character and aridity. Over the years, irrigated agriculture and livestock production have formed the core of the socio-economic conditions within these vulnerable ecosystems and water deficit environment of the CA region.

Prior to independence from Soviet Union in 1991, the CA republics were socio-economically interdependent within the centrally managed Soviet economy. Agriculture contributed 10 to 45% of gross domestic product (GDP) and employed 20 to 50% of the labor force. Each republic was agriculturally specialized according to its agroclimatic and soil conditions, with production distributed through the Soviet trade system. Since independence, the CA republics have faced a tremendous task of developing their own market-based economies. The transition from the centrally planned Soviet economy to independent market-oriented economies has had severe consequences. The agricultural sector has contracted leading to a significant decline in agricultural output. Following independence, the large state farms inherited the problems manifested in the high-input, energy-intensive, traditional agricultural production methods adopted under the Soviet system. Intensive flood irrigation with poor drainage networks resulted in waterlogging and secondary salinization of soils. Furthermore, due to the lack of resources, the former state-operated, large-scale irrigation system collapsed. With the demise of supply and subsidy systems, there have been dramatic decline in the use of input for crop production such as fertilizers, chemicals, and farm machinery.

Agriculture still remains the most important sector of the economics in the region, accounting for more than 20% of GDP and employing about 40% of the labor force. A large part of the agricultural production is now generated by individual household, which accounts for a substantial income for households struggling to cope with the economic transition. However, agricultural production is constrained by decline in quality of soil and water resources due to poor management of former state-operated large-scale irrigation systems.

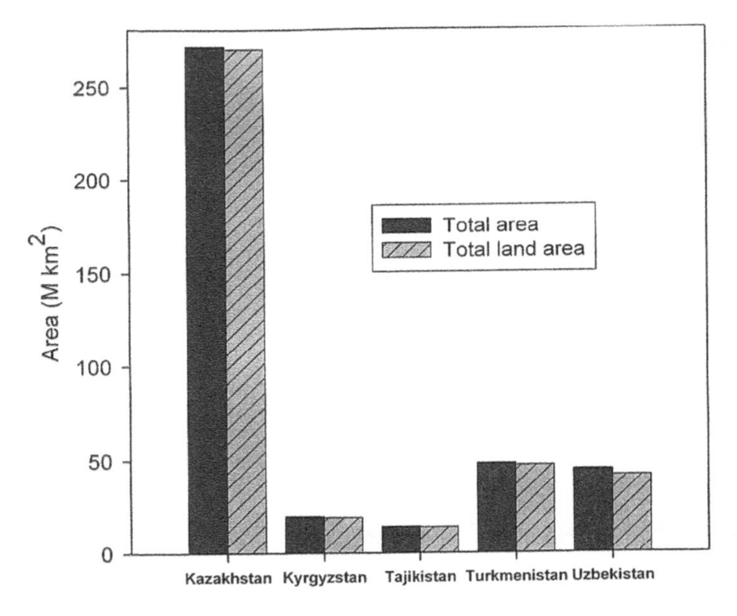

Figure 1. Land area distribution of Central Asian countries (adapted from www.http//faostat.fao.org).

More than 95% of the gross agricultural output is produced on irrigated lands. However, increasing water shortage is already having an adverse effect upon agricultural production. During low rainfall years, water levels in the deltas of the Amu-Darya and the Syr-Darya are significantly dropped. Since 1961, the water level of the Aral Sea has been declining progressively at the rate of 20 to 90 cm/yr. Accelerated salinization and desertification of land along with the severe degradation of water ecosystems are occurring in the Amu-Darya and Syr-Darya deltas. The former bed of the Aral Sea is now a source of windblown dust, pesticides, and salt. The decline in the area of the Aral Sea is also causing a climate change in its Basin. The water deficit is growing with time, especially in view of the population growth in CA, the increase in water use by Afghanistan and the aggravated desertification process and climate change. Accelerated soil erosion reduces soil fertility and accentuates degradation of vast land areas, which reduce crop production and worsen the environment. Therefore, the objectives of this chapter are to discuss temporal changes in population, land-use, agricultural production, water availability and consumption, and their effects on soil degradation in CA countries in between 1992 to 2002.

2 DATA SOURCE AND METHODOLOGY

Long-term data (1992 to 2002) on population, land-use practices, agricultural production, chemical use and production, water resources, and soil degradation processes were collected from a number of websites (http://faostat.fao.org; http://www.fao.org/landandwater/agll/glasod/glasodmaps.jsp; http://www.fao.org/ag/agl/agll/terrastat; http://www.isric.org; and http://enrin.grida.no/aral/aralsea). Data were processed, transformed and analyzed to explain agriculture and soil degradation processes and/or causes in CA countries.

3 RESULTS AND DISCUSSION

3.1 *Population and land-use*

The population in the CA region is unevenly distributed throughout the region (Table 1). Of the total population of 53 million, 40% lives in Uzbekistan and 32% in Kazakhstan. The population makes

Table 1. Temporal changes of total and agricultural population (%) in Central Asia.

	Total population					Agricultural population				
	1992	2002	Change	2010	Change	1992	2002	Change	2010	Change
Kazakhstan	32.5	27.0	−4.5	24.5	−2.5	21.9	18.8	−4.1	15.8	−3.0
Kyrgyzstan	8.6	8.9	0.3	9.1	0.2	30.7	24.5	−6.2	20.4	−4.1
Tajikistan	10.6	10.8	0.2	10.9	0.1	39.3	32.4	−6.9	27.4	−5.0
Turkmenistan	7.5	8.4	0.9	8.8	0.4	36.4	32.6	−3.8	29.6	−3.0
Uzbekistan	41.4	44.9	3.5	46.7	1.8	33.4	26.3	−7.1	21.4	−4.9
Average			<0.1		0.0	32.3	28.8	−5.6	22.9	−4.0

Adapted from FAOSTAT (2005).

Table 2. Temporal changes of total agricultural and irrigated lands in Central Asia.

	Agricultural land (% of total land)			Irrigated land (% of agricultural land)		
Country	1992	2002	Change	1992	2002	Change
Kazakhstan	82.0	77.0	−5.0	9.8	15.7	5.9
Kyrgyzstan	54.2	55.9	1.7	74.6	82.0	7.4
Tajikistan	32.1	30.4	−1.7	83.5	77.5	−6.0
Turkmenistan	68.6	69.4	0.8	100	97.3	−2.7
Uzbekistan	62.0	60.5	−1.5	94.7	95.5	−0.8
Average	59.8	58.6	−0.2	72.5	73.6	1.1

Adapted from FAOSTAT (2005).

up about 11% in Tajikistan, 9% in Kyrgyzstan and 7% in Turkmenistan. The population decreased by 5% between 1992 and 2002 in Kazakhstan and increased by 0.2 to 3.5% in other countries. The projected population in 2010 is expected to decrease in Kazakhstan but increase in Uzbekistan. More than 32% of the population was involved in agriculture in 1992, but only 27% in 2002, with a net decrease of 5% over one decade (Table 1). The decrease in agricultural population was more pronounced in Uzbekistan and Tajikistan followed by Kyrgyzstan. The agricultural population in 2010 is also expected to decrease strongly in Uzbekistan and Tajikistan and slightly in Kyrgyzstan, with an average decline of about 4%.

In 1992, 32 to 82% of the total lands in the CA countries were used for agricultural production (Table 2). Kazakhstan had the highest percentage (82%) of lands under agriculture. On average, 60% of the total land was used for agricultural production in 1992, and the area declined to 59% by 2002, a net change of 1% over a decade. Decline in 1992 to 2002 was 5% in Kazakhstan, 20% in Tajikistan and 2% in Uzbekistan. Of the total agricultural land, about 72% was irrigated in 1992, and 74% in 2002, a net increase in 2% in 10 years. The highest increase in irrigated land area occurred in Kyrgyzstan followed by Kazakhstan. Irrigated land area decreased over time in Tajikistan, Turkmenistan, and Uzbekistan.

Averaged across all the CA countries, total arable land area was decreased by 0.4% over a decade (Table 3). The largest decrease in arable land area was observed in Kazakhstan (4.9%). In contrast, however, an increase in arable land area was observed in Tajikistan (2.7%), and Turkmenistan (1.5%). Total land area under perennial crops did not change substantially over this time, but a change in pasture land was noted. An increase in pasture area was recorded in Kazakhstan and Kyrgyzstan with a considerable decrease in Tajikistan and Turkmenistan.

Decline in the agricultural population was most probably caused by migration. Kazakhstan has been losing its population as a result of migration for more than a quarter of a century. The majority of

Table 3. Temporal change of agricultural lands in Central Asia.

	Arable land (%)			Perennial crops (%)			Pasture (%)		
	1992	2002	Change	1992	2002	Change	1992	2002	Change
Kazakhstan	15.8	10.9	−4.9	0.1	0.1	0	84.1	89.0	4.9
Kyrgyzstan	13.0	12.2	−0.8	0.5	0.5	0	86.5	87.3	0.8
Tajikistan	19.2	21.9	2.7	2.8	3.0	0.2	78.0	75.1	−2.9
Turkmenistan	4.2	5.7	1.5	0.2	1.6	−0.04	95.6	94.1	−1.5
Uzbekistan	16.1	16.6	0.5	1.4	1.3	−0.01	82.5	82.1	−0.4
Average	13.9	13.5	−0.4	1.0	1.0	0.03	85.3	85.5	0.2

Adapted from FAOSTAT (2005).

the emigrants left for various destinations within the CA countries. The crisis in post-independence agricultural sector led to a decline in the rural population because of growing unemployment. As the CA countries were passing through the "transitional phase," from primarily collective agricultural to industrial economy, large-scale migrations of rural populations occurred to towns and cities. If this trend continues, Kazakhstan will be facing a tremendous labor shortage in agriculture.

3.2 *Agriculture, chemicals and water resources*

There was an increase in production of cereals, legumes, vegetables, melons and potatoes, but decrease in cotton, barley, and pastures between 1952 and 2002 (Table 3 and 4). On average, total area under cereal production increased by 38% to 47% except in Kazakhstan. The increase in arable land area was greater in Turkmenistan (24.5% in 1992 to 50.8% in 2002) followed by Tajikistan (30.9% in 1992 to 42.2% in 2002) and Uzbekistan (27.4% in 1992 to 37.2% in 2002). On the whole, there was a tremendous increase in arable land area for growing wheat at the expense of barley, pasture, perennials, and cotton. Under large-scale irrigation, salt-affected and marginal soils were brought under wheat in all CA countries. The highest shift to wheat was observed in Turkmenistan and Uzbekistan. Consequently, the barley production decreased in all CA countries, and the highest decline was recorded in Kyrgyzstan and Kazakhstan and the lowest in Tajikistan and Turkmenistan. Cotton production was decreased in Turkmenistan followed by Uzbekistan and Tajikistan, but increased slightly in Kazakhstan and Kyrgyzstan.

Among the crops, cereal production per ha had increased by more than 1/3rd especially wheat yield which doubled between 1992 and 2002 (Table 4). Cereal yield increased tremendously in Uzbekistan (1.78 Mg ha^{-1} in 1992 to 3.51 Mg ha^{-1}in 2002) followed by Tajikistan (1.04 Mg ha^{-1} vs. 2.19 Mg ha^{-1}) and Turkmenistan (2.21 Mg ha^{-1} vs. 2.96 Mg ha^{-1}) but decreased in Kazakhstan (1.34 Mg ha^{-1} vs. 0.89 Mg ha^{-1}). Among the cereals, wheat production was dominant crop in all CA countries. Between 1992 and 2002, the wheat yield increased from 1.54 to 3.66 Mg ha^{-1} in Uzbekistan, 1.91 Mg ha^{-1} to 3.25 Mg ha^{-1} in Turkmenistan and 0.93 Mg ha^{-1} to 1.96 Mg ha^{-1} in Tajikistan but decreased from 2.73 Mg ha^{-1} to 2.43 Mg ha^{-1} in Kazakhstan, and 1.33 Mg ha^{-1} to 1.09 Mg ha^{-1} in Kyrgyzstan. Cotton yield slightly decreased in all CA countries, except in Kyrgyzstan.

Change in agricultural production was related to a similar change in use of the fertilizers and pesticides over time (Table 5). Total fertilizer use decreased from 1,540,000 Mg y^{-1} in 1992 to 938,046 Mg y^{-1} in 2002, and the decrease by 86% was most pronounced in Kazakhstan followed by 78% in Tajikistan. The amount of fertilizer use did not change in Uzbekistan. Among fertilizers, N use increased tremendously (14.9% in 1992 to 45.4% in 2002), and the highest use was recorded in Kazakhstan and the lowest in Kyrgyzstan. Phosphorus use also decreased in all countries, except in Uzbekistan. The use of Potassium slightly increased in Kazakhstan but decreased in all other countries. However, total amount of fertilizer production decreased. Highest decrease of 80% in fertilizer production was recorded in Kazakhstan, and the lowest of 20% Turkmenistan.

Table 4. Temporal use of arable lands and changes in major agricultural crops in Central Asia.

Country	Major crops	Arable land (%)			Yield (Mg ha^{-1})		
		1992	2002	Change	1992	2002	Change
Kazakhstan							
	Barley	16.0	7.2	−8.8	1.51	0.85	−0.66
	Maize	0.34	0.44	0.1	3.14	4.58	1.44
	Rice	0.33	0.34	0.01	3.99	3.62	−0.37
	Wheat	39.1	52.0	12.9	1.33	1.09	−0.24
	All cereals	63.2	61.4	−1.8	1.34	0.89	−0.45
	Cotton	0.32	0.97	0.65	0.68	0.64	−0.04
Kyrgyzstan							
	Barley	19.5	7.8	−11.7	2.34	2.28	−0.06
	Maize	4.1	5.6	1.5	5.13	6.19	1.06
	Rice	0.15	0.46	0.31	1.47	3.01	1.54
	Wheat	18.4	31.3	12.9	2.73	2.43	−0.30
	All cereals	42.8	45.4	2.6	2.77	2.88	0.11
	Cotton	1.6	3.5	1.9	0.64	1.04	0.40
Tajikistan							
	Barley	6.4	5.3	−1.1	0.75	1.29	0.54
	Maize	1.3	0.97	−0.33	2.96	5.27*	2.31
	Rice	1.2	1.2	0	2.02	4.90	2.88
	Wheat	21.4	34.7	13.3	0.93	1.96	1.03
	All cereals	30.9	42.2	11.3	1.04	2.19	1.15
	Cotton	33.3	31.6	−1.4	0.61	0.59	−0.02
Turkmenistan							
	Barley	4.5	3.0	−1.5	2.10	1.09	−1.01
	Maize	3.2	1.4	−1.8	3.70	0.52	−3.18
	Rice	2.1	3.2	1.1	2.28	1.83	−0.45
	Wheat	14.6	43.2	28.6	1.91	3.25	1.34
	All cereals	24.5	50.8	26.3	2.21	2.96	0.75
	Cotton	42.0	28.4	−13.6	0.69	0.42	−0.27
Uzbekistan							
	Barley	6.8	1.7	−5.1	0.94	1.38	0.44
	Maize	2.2	0.78	−1.42	3.72	4.49	0.77
	Rice	4.1	1.5	−2.6	2.96	2.74	−0.22
	Wheat	14.0	32.8	18.8	1.54	3.66	2.12
	All cereals	27.4	37.2	9.8	1.78	3.51	1.73
	Cotton	37.3	32.5	−4.8	0.81	0.80	−0.01

Adapted from FAOSTAT (2005).

The N fertilizer production increased in all countries but P and K production decreased (Table 5). Total use of pesticides in CA region decreased over time (data not shown).

Water resources for Central Asia are strategic, vital natural resources and have interstate significance. Natural existence and national economy function, especially agriculture, completely depend on water supply and demand among the neighboring countries. Agricultural irrigation water is generally withdrawn from the two main rivers, the Amu Darya and Syr Darya (Table 6). Both discharge into the Aral Sea, an inland sea without an outlet. The water supply in the Amu Darya basin is about 79 km^3 compared to 37 km^3 in Syr Darya basin. However, the Syr Darya no longer reaches the Aral Sea because of excessive use enroute. All of its discharge is now being used for irrigation and other purposes. Only about 10% of the Aquaria's surface water supply reaches the Aral Sea. The average long-term water availability in the Aral Sea basin from Amu Darya ranges from more

Table 5. Temporal fertilizer consumption and production in Central Asia.

Country	Types of fertilizers	Consumption (%)			Production (%)		
		1992	2002	Change	1992	2002	Change
Kazakhstan							
	Nitrogen	31.6	77.0	45.4	13.3	17.4	4.1
	Phosphorus	66.3	18.5	−47.8	86.7	82.6	−4.1
	Potassium	2.1	4.5	2.4	–	–	–
	Total (Mg y^{-1})	475,000	64,900	−86.3	461,500	94,400	−79.5
Kyrgyzstan							
	Nitrogen	79.1	94.0	14.9	0	0	0
	Phosphorus	5.3	5.3	0	0	0	0
	Potassium	15.6	0.7	−14.9	0	0	0
	Total (Mg y^{-1})	32,000	29,146	−8.9	0	0	0
Tajikistan							
	Nitrogen	50.0	76.7	26.7	100	100	0
	Phosphorus	43.7	23.0	−20.7	0	0	0
	Potassium	6.3	0.3	−6.0	0	0	0
	Total (Mg y^{-1})	128,000	27,900	−78.2	20,000	12,200	−39.0
Turkmenistan							
	Nitrogen	57.1	82.0	24.9	66.5	100	33.5
	Phosphorus	37.1	4.7	−32.4	33.5	0	0
	Potassium	5.7	13.3	7.6	0	0	0
	Total (Mg y^{-1})	175,000	97,800	−44.1	104,600	84,000	−19.7
Uzbekistan							
	Nitrogen	56.2	77.8	21.6	75.0	81.1	6.1
	Phosphorus	35.6	17.6	18.0	25.0	18.9	−6.1
	Potassium	8.2	4.6	−3.6	0	0	0
	Total (Mg y^{-1})	730,000	718,300	−1.6	1,357,700	768,400	−43.4

Adapted from FAOSTAT (2005).

than 75% in Tajikistan to 6% in Uzbekistan (Table 6). On the other hand, 74% of total water from Syr Darya is available for Kyrgyzstan followed by 17% for Uzbekistan, and 7% for Kazakhstan. However, the present water use from the two rivers accounts for 52% of the total water use in Uzbekistan, 20% in Turkmenistan, 11% in Tajikistan, and 10% in Kazakhstan. About 94% of that water is used for irrigated agriculture compared to only 3% for municipal use.

More than 95% of the gross agricultural output is produced on irrigated lands. The sharp continental climate, high evaporation (up to 1,700 mm per year), small and very non-uniform seasonal precipitation (on average 150–200 mm), high summer temperature (up to 40°C) make agriculture impossible without irrigation. As a result, salt accumulation in the soil is a serious problem. Thus, water is used both for irrigating crops and washing out salts. Because most of the water is now being used, further expansion of irrigation is difficult. Water deficit is observed in the entire region except Kyrgyzstan and Tajikistan because of the arid climate and large water losses especially in the Soviet style irrigation systems. Principal reasons of water deficit are inefficient irrigation and water distribution systems, outdated equipment, and a lack of water saving technologies. Further, the disastrous situation caused by the drying up of the Aral Sea emerged as a result of a collective agricultural policy based on development of flood irrigation and a corresponding increase in water consumption. However, increasing water shortage has a strong adverse effect upon social and economic situation. For example, during the last several years, water supply downstream the Amu-Darya amounted to 50% of the agreed water draw off limit, which was lower than required. The shortage will grow with time, because of the population growth in CA, increase in water use by

Table 6. Average long term availability of water in the Aral Seas basin of Central Asia.

	River basin						
Tran boundary	Syr Darya ($km^3\ yr^{-1}$)	(%)	Amu Darya ($km^3\ yr^{-1}$)	(%)	Total water availability ($km^3\ yr^{-1}$)	(%)	Water use (%)
Countries							
Kazakhstan	2.4	6.5	0	0	2.4	2.1	10
Kyrgyz Republic	27.6	74.2	1.6	2.0	29.2	25.1	5
Tajikistan	1.0	2.7	59.9	75.5	60.9	52.3	11
Turkmenistan	0	0	1.6	2.0	1.6	1.4	20
Uzbekistan	6.2	16.7	4.7	5.9	10.9	9.4	52
Afghanistan & Iran	0	0	11.6	14.6	11.6	10.0	–
Total	37.2		79.3		116.5		

Adapted from Micklin (1994); and http://enrin.grida.no/aral/aralsea/english/water/water.htm.

Afghanistan, the aggravated desertification process, and the projected climate change. However, the development of irrigation is limited by water availability. Further expansion of irrigated areas is possible only with the introduction of new water-saving techniques, improved irrigation technology; radical improvement and amelioration of old irrigated lands. In years with low precipitation, water levels in the deltas of the Amu-Darya and the Syr-Darya are significantly decreased. From 1961, the level of the Aral Sea has been falling at the rate of 20 to 90 cm/yr. Intensive drying up and salinization of land along with the severe degradation of water ecosystems is occurring in the deltas of these rivers. The former bed of the Aral Sea is now a source of dust and salt. The decrease in the sea area is also causing climate change.

3.3 *Soil-related constraints in Central Asia*

A large portion of the agricultural lands was degraded to varying degrees in response to accelerated human-induced activities. Major soil-related constraints which presently affect land resources are: wetlands (waterlogging) from excessive irrigation and impeded drainage, low cation exchange capacity, secondary salinity and sodicity, and shallow and eroded soils (Table 7). Most of the soils are shallow (27%) and prone to salinity and sodicity (19%), and accelerated erosion (13%). Among the CA countries, soil degradation by secondary salinization is dominant in Kazakhstan (~50%) followed Uzbekistan (24%), and in Turkmenistan (19%). About 50% of soils in Kyrgyzstan and Tajikistan are shallow, and 26 to 30% are prone to soil erosion.

On average, about 10% of the land area is totally degraded in the entire CA region (Table 8). Kazakhstan has the highest area of degraded soils (17%) followed by that in Uzbekistan and Turkmenistan (13%). About 62% of the soils in Kyrgyzstan are prone to light and moderate levels of degradation, compared to 31% in Kazakhstan, and only 10% in Uzbekistan, Turkmenistan and Tajikistan. More than 22% of the arable land in Uzbekistan is prone to severe degradation, compared with 21% in Tajikistan, 19% in Turkmenistan and only 2 to 4% in Kazakhstan and Kyrgyzstan (Table 9). Overall, 14% of the soils in the CA region are totally degraded (Table 9).

Principal causes of soil degradation are those related to intensive and traditional agricultural practices, including over-grazing and deforestation (Table 8). While agriculture is the prominent cause of soil degradation in Uzbekistan, Tajikistan, Kyrgyzstan, and Turkmenistan, deforestation and over-grazing are prominent causes in Kyrgyzstan and Kazakhstan. The prominent soil degradation types in CA countries are chemical degradation, water and wind erosion, and physical degradation (Table 8). Overall, soil degradation is most prevalent in Kazakhstan (>30%) followed by that in 20% in Turkmenistan (20%), Uzbekistan, (16%) and in Kyrgyzstan (10%) (Figure 2).

Soil degradation processes from intensive and traditional agricultural practices are increasingly affecting more and more of the land area. In addition, agricultural lands are also affected by

Table 7. Major soil-related constraints in Central Asia.

	Kazakhstan	Kyrgyzstan	Tajikistan	Turkmenistan	Uzbekistan	Average
Soil constraints	Percent of Agricultural Land					
Wetlands	8.7	1.0	3.6	3.2	8.7	5.0
Low CEC	0	0	0	2.3	0	0.5
Salinity	8.0	0.5	5.0	15.5	14.1	8.6
Sodicity	39.7	0	0	3.6	10.3	10.7
Shallowness	14.3	55.8	48.6	7.4	8.7	27.0
Erosion	2.9	29.2	26.4	1.5	2.9	12.6
Total	73.6	86.5	83.6	33.5	44.7	64.4

Adapted from http://www.isric.org (2006); http://www.fao.org/landandwater/agll/glasod/glasodmaps.jsp (2006); and http://www.fao.org/ag/agl/agll/terrastat/wsr.asp.

Table 8. Severity of human-induced degradation of agricultural soils in Central Asia.

	Kazakhstan	Kyrgyzstan	Tajikistan	Turkmenistan	Uzbekistan	Average
Soil degradation category	Percent of Total Land Area					
None	52	36	83	75	75	64.2
Light	12	62	0	0	2	15.2
Moderate	19	0	10	12	9	10
Severe	15	2	7	9	13	9.2
Very severe	2	0	0	4	0	1.2
Total degradation	17	2	7	13	13	10.4
Degradation cause	D* and O	O and A	A	A	A	
Degradation types	N and W	C and W	C, N & W	C and P	C and P	

*D = Deforestation; C = Chemical deterioration; N = Wind erosion; O = Overgrazing; P = Physical deterioration; and W = Water erosion.

Adapted from http://www.isric.org (2006); http://www.fao.org/landandwater/agll/glasod/glasodmaps.jsp (2006); and http://www.fao.org/ag/agl/agll/terrastat/wsr.asp.

Table 9. Extent of soil degradation in Central Asia.

	Kazakhstan	Kyrgyzstan	Tajikistan	Turkmenistan	Uzbekistan	Average
Soil degradation category	Percent of Total Agricultural Land					
Severe	1.4	3.7	21.2	13.2	22.2	12.3
Very severe	0.2	0	0	6.4	0	1.32
Total	1.6	3.7	21.2	19.6	22.2	13.7

Adapted from http://www.isric.org (2006); http://www.fao.org/landandwater/agll/glasod/glasodmaps.jsp (2006); and http://www.fao.org/ag/agl/agll/terrastat/wsr.asp.

the high salinity of irrigation water. The poor availability of water resources and lack of water supply are also among the principal factors of soil degradation. Inappropriate irrigation methods are among the reasons for accelerated soil erosion and secondary salinization. The Soviet-style civil engineering standards for irrigating vast areas of arable lands for cotton production have caused secondary salinization and development of marshy areas within the irrigated territories. One of the main reasons of salinization and waterlogging is the rise in water table as a result of poor drainage

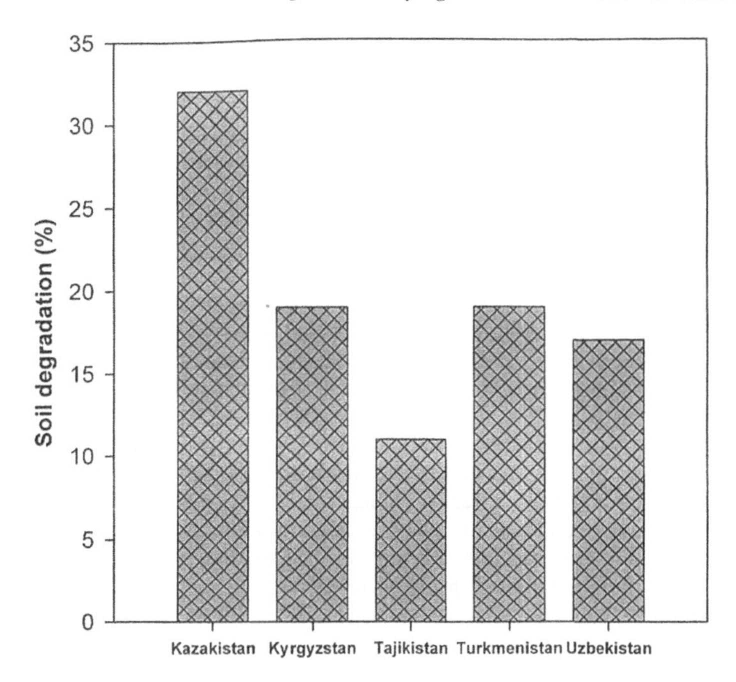

Figure 2. Land degradation severity in Central Asia. Adapted from www.fao.org/ag/agl/agll/terrastat/wsr.asp.

network. Erosion processes are one of the reasons for the degradation of environment in highlands, where pasture is the predominant land use. Excessive stocking rate and uncontrolled grazing has caused the decline in carrying capacity of pasture by a factor of 3 to 3.6 in comparison with 1992. Average pasture productivity decreased by 40% over the 25 years between 1980 and 2005.

The problem of soil degradation is confounded by the drying off of the Aral Sea, and susceptibility to wind erosion of the vast territories of the former seabed. Annually 43 million tons of salts are blown out of the Aral Sea basin and spread on 1.5–2.0 million km^2 causing considerable damage to neighboring agricultural regions. Humus content in soil decreased over the last decades by 30–40%. The humus content of plow layer had declined by 20 to 45% due to accelerated erosion. Soils with low or very low humus content have occupied about 40% of the total irrigated area. Since desertification is connected mainly with vegetation cover degradation, accelerated wind and water erosion, and secondary salinization, a vast area of arable land is prone to human-induced desertification, and is rapidly becoming unfit for agricultural production.

4 CONCLUSIONS

The CA region is a vast area of desert, steppe and mountains. Agriculture and livestock production form the core of the socio-economics setting within its vulnerable ecosystems and water limitations. All CA republics (e.g., Kazakhstan, Kyrgyzstan, Tajikistan, Turkmenistan, and Uzbekistan) had undergone tremendous socio-economical changes following their independence from the USSR in 1991. To address the post-independence changes in socio-economical aspects of agricultural environment, long-term data (1992 to 2002) on population, land-use practices, agricultural production, chemical use and production, water resources, and soil degradation were collated, processed, and analyzed to explain changes in agricultural practices and the attendant soil degradation processes. An analysis of data showed that population involved in agriculture decreased in all CA countries, and the decrease was especially pronounced in Tajikistan and Uzbekistan. While total agricultural land area decreased by more than 1%, irrigated land area increased by 2% between 1995 and 2005. Decrease in the agricultural land area was the highest in Kazakhstan (~5%). On average, cereal production increased by more than 1/3rd, especially the wheat grain yield which doubled in between 1992 to 2002. Over time, the cereal crop production increased tremendously in Uzbekistan, Tajikistan, and Turkmenistan, but decreased in Kazakhstan. There was an increasing trend of area under wheat at the expense of barley, pasture, perennials, and cotton production. Cotton production was decreased slightly in all CA countries except Kyrgyzstan. Fertilizer use decreased, and the decrease was especially pronounced in Kazakhstan and Tajikistan. Among the fertilizers, N use increased but that of phosphorus and potassium decreased. However, fertilizer production decreased, and the highest decrease (80%) was recorded in Kazakhstan. In general, N fertilizer production increased but that of P and K production decreased over time. Water withdrawal from Amu and Syr Darya was 52% in Uzbekistan, 20% in Turkmenistan, 11% in Tajikistan, 10% in Kazakhstan, and 94% of that water withdrawn was used for irrigated agriculture. Indiscriminate use of saline and low productive soils for crop production by flood irrigation under a continental climate exacerbated the problem of soil degradation by salinization, erosion, water-logging and desertification. The estimated cropland area affected by degradation processes is 22% in Uzbekistan, 21% in Tajikistan, 19% in Turkmenistan, and 2–4% in Kazakhstan and Kyrgyzstan. Averaged across the CA countries, 14% of the cropland soils are totally degraded. Soil degradation by secondary salinization and sodicity is a dominant problem in Kazakhstan (~50%), Uzbekistan (24%), and Turkmenistan (19%). While agricultural practices have been the prominent cause of soil degradation in Uzbekistan, Tajikistan, Kyrgyzstan, and Turkmenistan, deforestation and over-grazing have been the prominent causes of soil degradation in Kyrgyzstan and Kazakhstan. Soil degradation is a serious issue in Kazakhstan, Turkmenistan, Tajikistan, Uzbekistan, and Kyrgyzstan.

REFERENCES

FAO. 2006. Statistics Division. http://faostat.fao.org/default.aspx.

FAO. 2006. National Soil Degradation. http://www.fao.org/landandwater/agll/glasod/glasodmaps.jsp.

FAO. 2006. Terrastat – land resource potential and constraints statistics at country and regional level. Land and Water Development Division. http://www.fao.org/ag/agl/agll/terrastat/wsr.asp.

ISRIC. 2006. World Soil Information. http://www.isric.org.

Lal, R. 2004. Carbon sequestration in soils of Central Asia. Land Degradation and Development. 15: 563–572.

Micklin, P.P. 1994. The Aral Sea problem. Proc. Paper 10154. Instn. Civ. Engrs. Civ. Engng. 114: 114–121.

Mirzaev, S.Sh., R.M. Razakov, and V.G. Nasonov. 2000. The future of the Aral Sea basin in flourishing oasis or fruitless desert, Collection of scientific articles: Water resources, the Aral problem and environment. Tashkent University, Tashkent, Uzbekistan.

Muminov, F.A. and I.I. Inagatova. 1995. The changeability of the Central Asian climate. SANIGMI, Tashkent, Uzbekistan.

Razakov, R.M. 1991. Use and protection of water resources in Central Asia. Tashkent, p. 203.

Ryan, J., P. Vlek, and R. Paroda. 2004. Agriculture in Central Asia: Research for Development. International Center for Agricultural Research in the Dry Areas (ICARDA), Aleppo, Syria.

Sokolov, V.I. 2000. Definition of boundaries of water collection basins of trans-bordering, local and mixed types of surface water resources in the basin of the Aral Sea and quantitative assessment. Collection Scientific Works of SRC ICWC, Tashkent, Issue 2. p. 35–53.

UNDP. 1995. The Aral crisis. Tashkent, Publishing house of UNDP.

UNEP. 2005. Water resources. State of Environment of the Aral Sea Basin. http://enrin.grida.no/aral/aralsea/english/water/water.htm.

World Bank. 1997. Aral Sea Basin. Project 3.1.B. Agricultural water quality improvement. EC IFAS., World Bank, Tashkent, p. 51.

CHAPTER 11

Salinity effects on irrigated soil chemical and biological properties in the Aral Sea basin of Uzbekistan

D. Egamberdiyeva & I. Garfurova
Tashkent State University of Agriculture, Tashkent, Uzbekistan

K.R. Islam
Ohio State University South Centers, Piketon, Ohio, USA

1 INTRODUCTION

Accelerated secondary salinization of irrigated soil is recognized as the major agricultural problem in Uzbekistan. The history of irrigation development in Central Asia has more than 2500 years in the deltas of the Amu Darya and SyrDarya. The Amu Darya basin in the south covers about 86.5% and the Syr Darya basin in the north about 13.5% of Uzbekistan.

Prior to independence in 1991, Uzbekistan was socio-economically integrated within the centrally managed Soviet economy for agricultural production. Uzbekistan within Soviet Union was agriculturally specialized according to its agroclimatic and soil conditions for cotton (*Gossipium hirsutum*) production. A massive expansion of irrigated agriculture was initiated making Uzbekistan one of the largest cotton producing countries in the world. However, the expansion of irrigated agriculture was limited by available water resources. The success was achieved through massive construction of long networks of irrigation canals and the diversion of the waters from the Syr Darya and the Amu Darya away from the Aral Sea, the 4th largest terminal lake (without surface outflow) in the world. Simultaneously land development began for planting rice (*Aryza sativa*) in the Amu Darya basin in Karakalpakstan. Total volume of water flows in the Amu Darya and Syr Darya is about 116 km^3 yr^{-1}, and about 19% of that water is generated within Uzbekistan.

Irrigated lands produced over 96% of the gross agricultural output in Uzbekistan. Of that, about 44% of the total irrigated area is concentrated in the Syr Darya basin and 56% in the Amu Darya basin. Because of large scale water withdrawal from both rivers for irrigated agriculture, especially for high water consumptive cotton production, the water flow reaching the Aral Sea is limited (<15% in the driest years) over time. Since 1961, the Aral Sea has been declining progressively with increasing water salinity. The former bed of the Aral Sea is now a source of windblown dust, pesticides, and salt, adversely affecting the surrounding ecosystems. Extensive flood irrigation with poor quality water and inefficient drainage facilities have accelerated salinization of irrigated soils with severe degradation of water ecosystems in the Amu Darya and Syr Darya deltas. Out of total available water flow, Uzbekistan is now using about 42 km^3 yr^{-1} of the transboundary rivers flow, and >80% of that volume of water is from Amu Darya and Syr Darya. Since the demand for water is growing in the Central Asian region with time especially in view of the population growth and irrigated agriculture; a possible increase in transboundary water use by Afghanistan and other countries may accelerate soil salinization of irrigated soils, and subsequently affect agricultural production in Uzbekistan.

Soil biological properties are one of the most important components to evaluate functional stability of agroecosystems in response to environmental degradation. They are very sensitive to environmental changes and can be used as indicators to evaluate the effects of soil degradation on agroecosystems. Therefore, the objectives of this chapter are to discuss soils, temporal changes in irrigation and soil salinization, and their effects on selected chemical and biological properties of soils in the Sayhunobod district of the Syr Darya basin in Uzbekistan.

2 DATA SOURCE AND METHODOLOGY

This study was conducted on a silty clay loam soil (typical sierozems) at farmer's fields in the Sayhunobod district of Syr Darya province in the republic of Uzbekistan. The irrigated soils in this region were severely affected by pollution and secondary salinization from Syr Darya and Aral Sea degradation. The mean values of the selected soil properties were: pH 7.9; cation exchange capacity 1.98 meq $100\,g^{-1}$; sand $186\,g\,kg^{-1}$; silt $514\,g\,kg^{-1}$; clay $300\,g\,kg^{-1}$; particle density $2.65\,g\,cm^{-3}$; bulk density $1.41\,g\,cm^{-3}$; total porosity $0.47\,m^3\,m^{-3}$; and field moisture capacity $3.25\,mm\,cm^{-1}$(need reference).

The existing farming system includes a range of irrigated crops including cotton, corn (*Zea mays*), (*Triticum aestivum*), wheat, fodder, vegetables, and pulses. The average annual rainfall is 324 mm and more than 90% of the rainfall occurs during October to May. The mean annual minimum temperature is −2°C in January and the mean maximum temperature is 34°C in July. The average highest soil temperature is 35°C in July and the lowest is −2°C in January. The mean relative humidity is 66% with a maximum of 82% in December and January, and a minimum of 48% in June.

Soil cores (divided into 0–28, 28–49, 49–96, and 96–120 cm depths) were randomly collected from weak and saline irrigated fields in spring (April), summer (July), and autumn (September) in 2004. Soil cores were pooled and mixed to obtain a composite sample for each depth. Field-moist soils were then gently sieved through a 2-mm mesh, and a portion of the soil was incubated or analyzed to measure selected soil biological properties within 72-h of sampling and processing. Another portion of the field-moist soil was air-dried at room temperature, ground and analyzed for soil chemical and physical properties. Soil biological properties such as basal respiration (CO_2 evolution), and catalyse, invertase, and urease activities were measured using *in vitro* static incubation of unamended field-moist soil at room temperature. Soil chemical properties such as pH, total organic C, total N, P_2O_5, K_2O, Na_2O, CaO, MgO, chloride, sulfate, and various salts [$Ca(HCO_3)_2$, $CaSO_4$, $MaSO_4$, $MgCl_2$, Na_2SO_4 and NaCl] were measured. [How?] Soil physical properties such as particle size analysis (sand, silt and clay), particle density, bulk density and moisture content were determined. [How] All the analyses were done by following standard procedures. The mass of total organic C, total N, P_2O_5, K_2O, Na_2O, CaO, MgO, chloride, sulfate, and salts at different depths of soil was calculated by multiplying their concentration with each sampling depth and concurrently measured soil bulk density.

Regression and correlation were performed to predict Aral Sea degradation using data from Micklin (1994), and establish relationship between salt and total organic C content with biological properties for different soil depths.

3 RESULTS AND DISCUSSION

3.1 *Soils of Uzbekistan*

On the basis of surface territory, Uzbekistan is divided into two unequal parts. About 78.7% of the territory of Uzbekistan is desert plains which are predominantly covered with brown meadow and sierozems soils; the remaining 21.3% is high mountainous ridges and intermountain valleys which comprise of sands, takir, marsh and Solonchaks soils (Table 1). Typical sierozems are the dominant (~7%) group of soils in high belt and Grey brown soil covers about 25% of the desert zones.

3.2 *Major water resources and the Aral Sea*

Almost all of the water used for irrigating crops is withdrawn from the two main rivers in Uzbekistan, the Amu Darya and Syr Darya. Both discharge into the Aral Sea, an inland sea without an outlet.

Table 1. Soil types of Uzbekistan.

Soil types	Area (km^2)	Area (%)
Mountainous regions		
1. Light-brown, meadow steppe	540	1.2
2. Brown mountain forest	1660	3.7
3. Dark sierozems	1050	2.4
4. Typical sierozems	3050	6.8
5. Light sierozems	2590	5.8
6. Meadow sierozems	780	1.8
7. Meadow sierozems belt	670	1.5
8. Marsh-meadow sierozems belt	70	0.2
Total	10410	23.4
Desert zones		
9. Grey brown	11025	24.8
10. Desert sandy	1370	3.1
11. Takir soils and takirs	1780	4.0
12. Meadow takir and takir meadow	460	1.0
13. Meadow desert soils	1790	4.1
14. Marsh meadow desert zones	50	1.0
15. Solonchaks	1270	2.9
16. Sands	12100	27.2
Total	29845	67.2
Other land and rocks	4155	9.4
Total all soils	44410	100

Source: Umramov (1989).

The water supply is about 79 km^3 in the Amu Darya basin compared to 37 km^3 in Syr Darya basin. However, the Syr Darya no longer reaches the Aral Sea because of excessive use enroute. Most of its upstream discharge is now being used for irrigation. Only about 10% of the Aquaria's surface water reaches the Aral Sea. Because in arid regions, where irrigation is essential for agriculture, the highest volume of water is used for irrigated agriculture. As a result, the Aral Sea has lost much of its volume (from 1090 km^3 to 196 km^3) and area (from 67,000 km^2 to 23,000 km^2) between 1960 and 2000 (Figure 1). At the same time, the water level fell by more than 16 m (from 53 m to 37 m) and its salinity quadrupled (from 10 $g\,L^{-1}$ to 40 $g\,L^{-1}$) over time (Figure 1). Presently, the dry seabed exposed to weathering has increased desertification around the sea, and accelerated soil salinization in the Amu Darya and Syr Darya deltas. Salinization of irrigated lands resulting from over irrigation and poor drainage as well as the wind transport of salts from the exposed sea bed confounded the environmental problems within these rivers basin. Annually, $15–75 \times 10^6$ Mg of dust and salt are carried over long distances of up to 400 km from the Aral Sea dry bed. The total mass of sediment deposited in one of the major storms was estimated as 1.68×10^6 Mg in the surrounding areas. If everything goes as it is, a projection on the basis of available data suggests that by 2030, the Aral Sea will be completely dried out and will not exist on earth (Figure 1).

3.3 *Irrigation and secondary soil salinization*

Progressive degradation of Aral Sea has intensified, and will continue to intensify the secondary salinization of irrigated soils (~65%) in the Amu Darya and Syr Darya deltas (Table 2). Consequently, most of the arable lands are suffering from various degrees of soil salinity. In 1990, about 48% of the total irrigated lands were suffering from soil salinity, by 2000, salinity increased to 64% of the lands. A decrease in 80,000 ha of irrigated lands with 16% increase in soil salinity

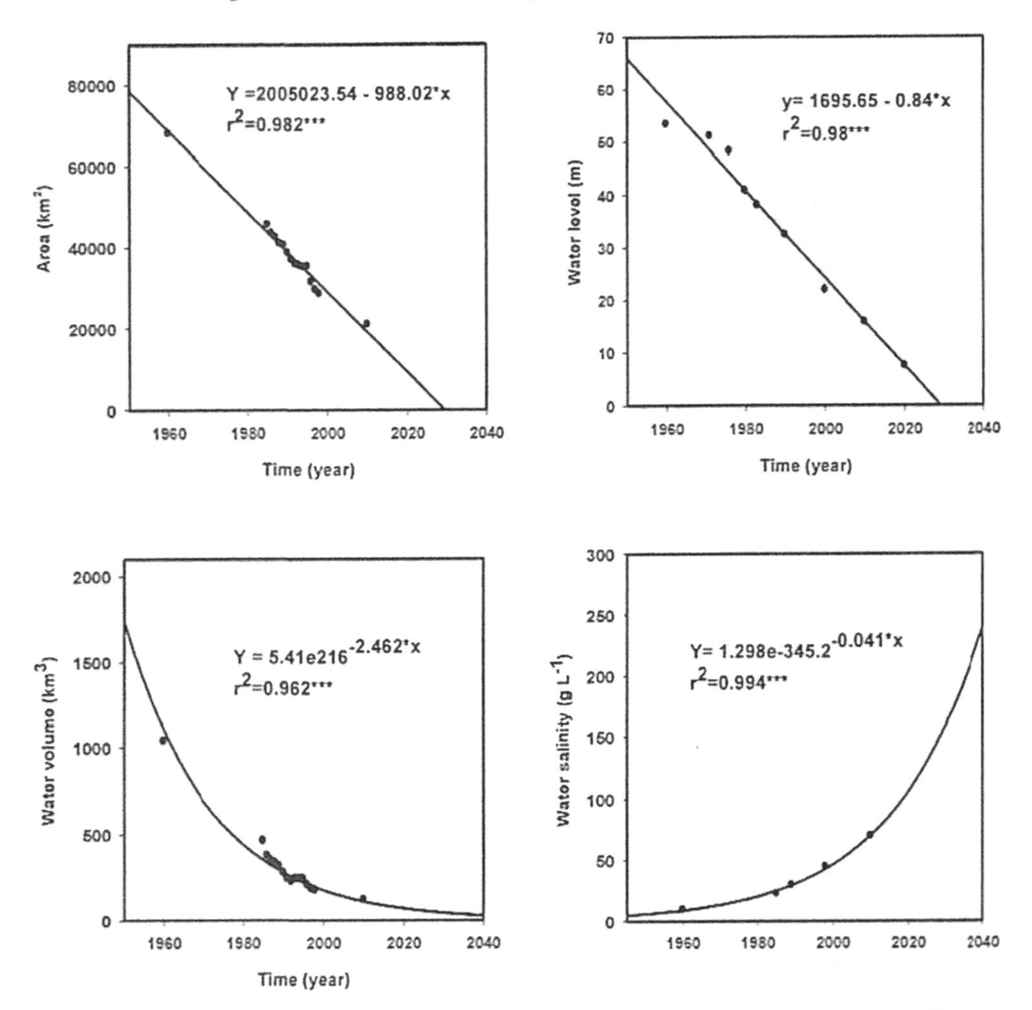

Figure 1. A projection of Aral Sea degradation over time (Adapted and projected from Micklin, 1994).

occurred within a decade (Table 2). Irrigated lands substantially decreased in Andijan, Namangan, Samarkand, Surkhandarya, Syr Darya, Tashkent and Fergana regions between 1990 and 2000. However, soil salinity in Andijan, Djizak, Namangan, Samarkand, Surkhan Darya, Syr Darya, Tashkent, Fergana, and Kashka Dariya regions increased.

Out of total irrigated lands affected by various degrees of soil salinity, 35% were weakly saline, 18% were moderately saline, and 11% were strongly saline in 2000 compared to 27% weakly saline, 16% moderately saline, and 5% were strongly saline in 1990. In other words, about 7% increase in strongly saline soil occurred between 1990 and 2000. In fact, 80–95% of the irrigated lands were saline in Karakalpakya, Bukhara and Syr Darya regions, and 60–70% of the lands were saline in Kashka Dariya and Khorezm regions. The degrees of irrigated soil salinization in the Syr Darya region were presented in Table 3. Out of 216,101 ha of salinized lands in the Syr Darya region, about 47% were weakly saline followed by 35% moderately saline and 18% strongly saline. A widespread salinization of irrigated soil occurred in Arnasoy, Forish, Gallaorol, and Zomin where more than 20–40% of the lands were strongly saline. The change from moderate to strong salinity severely affected soil fertility and agricultural productivity in the Syr Darya basin.

Table 2. Temporal change in irrigated lands and soil salinization in Uzbekistan (1990 to 2000).

Region	Years	Irrigation ($10x^3$ ha)	Degree of salinization: Weak ($10x^3$ ha)	(%)	Moderate ($10x^3$ ha)	(%)	Strong ($10x^3$ ha)	(%)	Total salinized lands ($10x^3$ ha)	(%)
Republic of Karakalpakstan	1990	457.2	167.3	36.6	183.7	40.2	74.6	16.3	425.6	93.1
	2000	462.1	110.4	23.9	151.7	32.8	142.9	30.9	405.0	87.6
Andijan	1990	245.1	42.3	17.3	16.5	6.7	4.8	2.0	63.6	25.9
	2000	227.4	51.8	22.8	20.3	8.9	4.9	2.2	77.0	33.9
Bukhara	1990	228.1	133.2	58.4	57.3	25.1	16.5	7.2	207.0	90.7
	2000	229.2	125.8	54.9	48.2	21.0	31.2	13.6	205.2	89.5
Djizak	1990	267.3	61.8	23.1	20.0	7.5	8.4	3.1	90.2	33.8
	2000	275.7	101.0	36.6	75.7	27.5	38.8	14.1	215.5	79.2
Navoi	1990	102.1	17.5	17.1	71.7	70.2	3.3	3.2	92.5	90.6
	2000	108.1	49.8	46.1	19.6	18.1	6.7	6.2	76.1	70.4
Namangan	1990	239.7	28.1	11.7	17.5	7.3	6.8	2.8	52.4	21.8
	2000	236.1	51.1	21.6	18.1	7.7	13.1	5.5	82.3	34.9
Samarkand	1990	356.5	39.1	11.0	5.6	1.6	0.1	0.0	44.8	12.6
	2000	309.5	104.3	33.7	19.9	6.4	4.6	1.5	128.8	41.8
Surkhan Darya	1990	287.0	65.2	22.7	44.7	15.6	7.2	2.5	117.1	40.8
	2000	279.3	108.4	38.8	70.0	17.0	48.9	8.1	178.5	63.9
Syr Darya	1990	283.0	129.8	45.9	59.3	21.0	38.5	13.6	227.6	80.4
	2000	273.8	115.7	42.3	47.6	25.6	22.5	17.8	234.6	85.7
Tashkent	1990	351.1	29.6	8.4	2.9	0.8	0.3	0.1	32.8	11.4
	2000	337.4	67.6	20.0	13.07	3.9	5.3	1.6	86.0	25.5
Fergana	1990	307.7	33.2	10.8	10.8	3.5	2.8	0.91	46.8	15.2
	2000	296.0	108.0	36.5	67.5	22.8	42.9	14.5	218.4	73.8
Khorezm	1990	234.3	119.0	50.8	35.7	15.2	14.8	6.3	169.5	72.8
	2000	240.1	106.8	44.5	50.6	21.1	23.2	9.7	180.6	75.2
Kashka Dariya	1990	452.5	163.3	36.1	76.6	16.9	28.4	6.3	268.3	59.3
	2000	452.2	216.9	48.0	63.3	14.0	31.5	7.0	311.7	68.9
Total in Republic	1990	3811.6	1029.4	27.0	602.3	15.8	206.5	5.4	1838.2	48.2
	2000	3726.9	1317.6	35.4	665.6	17.9	416.5	11.2	2399.7	64.4

Table 3. Various degrees of irrigated soil salinity in the Syr Darya region of Uzbekistan.

Region	Total salinized lands (ha)	Weakly saline (%)	Moderately saline (%)	Strongly saline (%)
Arnasoy	32911	24.9	32.5	42.6
Bahmal	334	91.0	9.0	–
Gallaorol	761	42.2	32.4	25.4
Djizak	17975	74.7	24.2	1.1
Dustlik	33986	39.1	44.7	16.2
Zomin	30952	43.9	35.0	21.1
Zarbdor	26367	50.7	32.2	17.1
Mirzachul	29073	22.3	59.9	17.8
Zafarobod	19394	88.4	11.6	–
Pahtakor	20210	67.9	24.5	7.6
Forish	3104	29.5	30.6	39.9
Yangiobod	1053	35.9	64.1	–
Total	216101	46.8	35.2	18.0

Source: Anonymous (1994).

Table 4. Chemical properties of farmer's field soil (typical sierozems) in the Sayhunobod district of Syr Darya province of Uzbekistan, 2004.

Soil depth (cm)	TC	TN	P_2O_5	K_2O	CaO	MgO	Na_2O	Cl^-	SO_4^{-2}
	g kg^{-1}								
0–28	6.17	0.96	4.8	27.8	9.0	63.1	18.4	3.29	71
28–49	2.4	0.92	4.8	26.0	7.8	50.2	24.5	25.5	145.8
49–96	2	0.87	5.2	29.7	10.1	44.5	24.5	16.1	82.1
96–120	1.64	0.79	0.42	27.8	6.7	24.3	22.1	20.3	71.8
Total (Mg ha^{-1})	49.8	15	69.7	478.1	151.2	755.2	390.7	266.4	1507.8

TC = Total organic C; TN = Total N; Cl = Chloride; and SO_4 = Sulfate.

A greater accumulation of salts in the irrigated soils in the Syr Darya basin is due to accelerated evaporation from inefficient flood irrigation and poor drainage systems. During hot summer spells (air and soil temperatures range from 35 to 45°C), soil water evaporates and salts move upward to the surface by capillary movement from the saline groundwater. The salts are leached into the groundwater in response to irrigation and rainfall, and are then transported upward, eventually increasing salt accumulation in both soil and groundwater.

3.4 *Soil chemical properties*

Progressive soil salinization in response to inefficient flood irrigation and poor drainage managements for cotton production under continental climates has affected degradation of soil fertility. Total organic C and N, available P, and K of the soil (typical Sierozems) in the Sayhunobod district of the Syr Darya province of Uzbekistan were severely affected by salinity (Table 4). A total mass of only 50, 15, and 69 Mg ha^{-1} of organic C, N, and P was present up to 120 cm depth. More than 475, 150, and 75 Mg ha^{-1} of K_2O, CaO, and MgO accumulated in the soil. However, Na, chloride and sulfate concentration were even higher. More than 260 Mg chloride ha^{-1} accumulated in the soil from irrigation over time. Total mount of Na was 400 Mg ha^{-1} within 120 cm soil depth. Abnormally high accumulation of sulfate (1508 Mg ha^{-1}) was also observed. However, higher concentrations of chloride and sulfate were at 28–49 cm soil depth. Soils in the plains of Uzbekistan are characteristically low in total organic C, total N and P concentrations due to lack of organic amendments and accelerated mineralization of organic matter under arid climate.

Among the salt types, Ca, Mg and Na associated sulfates, chlorides and bicarbonates were dominant within soil (Table 5). The concentrations of various salts were higher in surface soil layers. Among the salt types, greater amount of Na-sulfate (>40%) was accumulated within the soil profile (0–120 cm depth) followed by Ca-sulfate (~ 29%), NaCl (18%), and MgCl (10%). A total amount of 328 Mg ha^{-1} of salts was accumulated in the soil. Out of that total amount, 70% of the salts (Mg and Na sulfates and chlorides) were highly toxic to plant growth. In this region, the groundwater is saline and contains 5–24 gL^{-1} of salts. As a result, salt accumulation in irrigated soil differs seasonally. In spring, the salt accumulation at 0–100 cm depth is 180 Mg ha^{-1} followed by 140 Mg ha^{-1} at 1–2 m depth. In autumn, the salt content increased up to 200 Mg ha^{-1} in 0–1 m depth and 160 Mg ha^{-1} in 1–2 m depth. In arid or semiarid climates of Uzbekistan, salinity is usually combined with high soil pH, because of Na, Ca, and Mg associated carbonate, chloride and sulfate salts accumulation in the uppermost soil horizons.

3.5 *Soil biological properties*

Soil biological activities were influenced greatly by salinity and seasonal variations (Figures 2–5). Average across soil depth and season, base respiration (BR) (CO_2 evolution) rate was 5 mg kg^{-1} d^{-1}

Table 5. Toxic and non-toxic salt species in irrigated soil (typical sierozems) in the Sayhunobod district of Syr Darya province of Uzbekistan, 2004.

Soil depth (cm)	$Ca(HCO_3)_2$	$CaSO_4$	$MgSO_4$	$MgCl_2$	Na_2SO_4	NaCl	Total	Non-toxic	Toxic
					g kg^{-1}				
0–28	0.32	12.99	5.86	0.19	7.6	6.52	33.48	13.31	20.17
28–49	0.24	5.74	0.95	0.18	8.77	2.70	18.58	5.98	12.60
49–96	0.28	3	1.03	0.17	7.24	2.93	14.65	3.28	11.37
96–120	0.24	1.16	1.04	0.18	7.91	1.08	11.61	1.40	10.21
Total (Mg ha^{-1})	4.4	92.6	36.9	3.07	133	57.6	327.6	97	230.6

$Ca(HCO_3)_2$ = Calcium Na_2SO_4 bicarbonate; $CaSO_4$ = Calcium sulfate; $MgSO_4$ = Magnesium sulfate; $MgCl_2$ = Magnesium chloride; Na_2SO_4 = Sodium sulfate; and NaCl = Sodium chloride.

in moderately saline irrigated soil compared to 3.91 mg kg^{-1} d^{-1} in weakly saline irrigated soil (Figure 2). In other words, about 22% increase in BR rate was measured in response to increasing soil salinity. However, the BR rates differ among over seasons. The highest BR rate (12.5 and 9.55 mg kg^{-1} d^{-1}) was recorded in spring (April), intermediate (5.2 and 4.5 mg kg^{-1} d^{-1}) in summer (July), and lowest (2 and 1.97 mg kg^{-1} d^{-1}) in autumn (September) in both moderately and weakly saline irrigated soils. Higher BR rate was recorded in surface layer (0–28 cm) of both soils.

Similarly, the catalase activity was affected by soil salinity, and also different among seasons (Figure 3). The average catalase activity was higher in moderately saline irrigated soil (1.46 mL O_2 kg^{-1} min^{-1}) compared to weakly saline irrigated soil (1.24 mL O_2 kg^{-1} min^{-1}). On average, there was a 15% increase in catalase activity from weakly to moderately saline of soil. Averaged across soil depth, catalase activity was recorded highest (1.97 and 1.68 mL O_2 kg^{-1} min^{-1}) in spring, intermediate (1.63 and 1.47 mL O_2 kg^{-1} min^{-1}) in summer, and lowest (0.78 and 0.56 mL O_2 kg^{-1} min^{-1}) in autumn for both moderately and weakly saline soils. The decrease in catalase activity was more pronounced in autumn which is about 50% less than that in spring. In both soils, the catalase activity decreased with increase in soil depth in all seasons.

Averaged across seasons and soil depths, the invertase activity was higher (1.31 mg glucose kg^{-1} d^{-1}) in moderately saline soil than in weakly saline (0.91 mg glucose kg^{-1} d^{-1}) irrigated soil (Figure 4). On average, the moderately saline soil had ~30% more invertase activity than weakly saline soil. Highest invertase activity was recorded at surface depth of both weakly and moderately saline soils. The activity was more for all depths in moderately saline soil than weakly saline irrigated soil. The invertase activity was higher (1.56 and 1.06 mg glucose kg^{-1} d^{-1}) in summer, intermediate (1.36 and 0.9 mg glucose kg^{-1} d^{-1}) in autumn, and lower (1 and 0.77 mg glucose kg^{-1} d^{-1}) in spring.

Likewise, the urease activity was higher (0.85 mg NH_4 kg^{-1} d^{-1}) in moderately saline soil than in weakly (0.59 mg NH_4 kg^{-1} d^{-1}) saline soil (Figure 5). In other words, more than 30% increase in urease activity occurred due to increase in soil salinity. Averaged across soil depth, urease activity was higher in summer (0.96 and 0.68 mg NH_4 kg^{-1} d^{-1}) followed by that in autumn (0.83 and 0.56 mg NH_4 kg^{-1} d^{-1}), and spring (0.77 and 0.54 mg NH_4 kg^{-1} d^{-1}). Greater urease activity was measured in surface depth of both soils.

Biological activities are potential indicators of the extent to which soil disturbance by a given activity may affect the immediate environment (Figures 6–8). Adverse effects of soil salinity on soil biological properties have long been recognized. Higher BR rates and enzyme activities in moderately saline irrigated soil during summer and autumn than in weakly saline irrigated soil in spring may be attributed to several factors. During hot summer months (with air and soil temperature ranging from 35 to 45°C) under continental climate, salt moves to the surface by upward capillary from saline ground water in response to high evaporation, and results in greater accumulation of salts in summer and autumn than in spring.

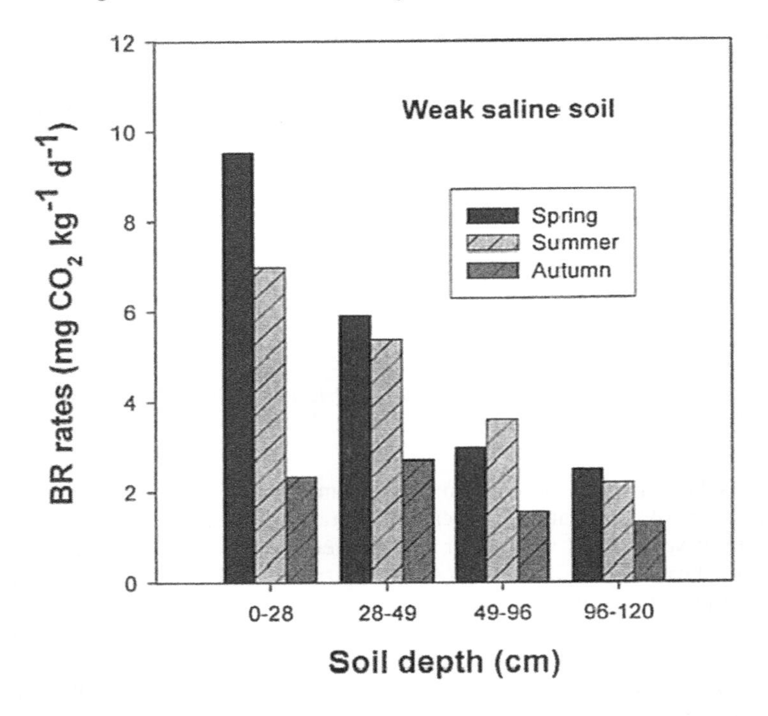

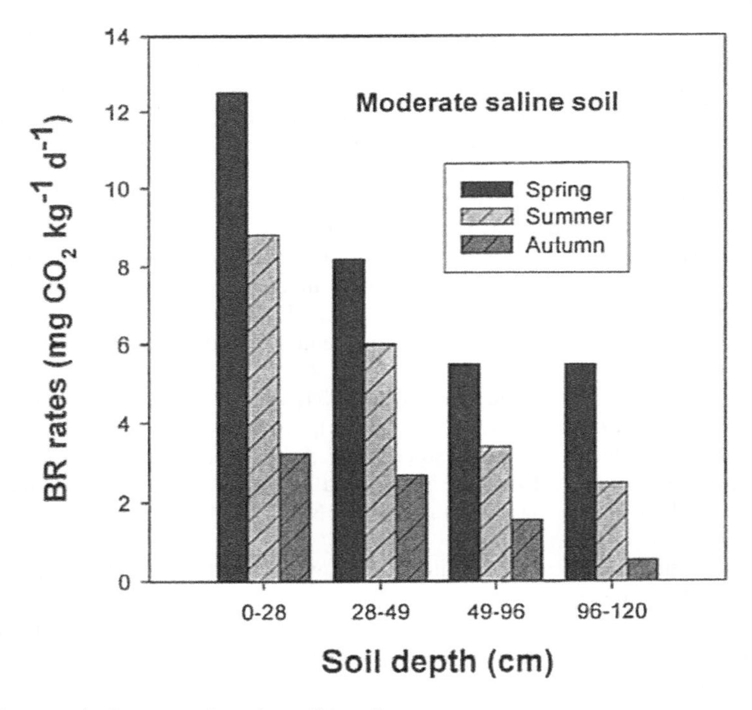

Figure 2. Base respiration rates of varying saline soils.

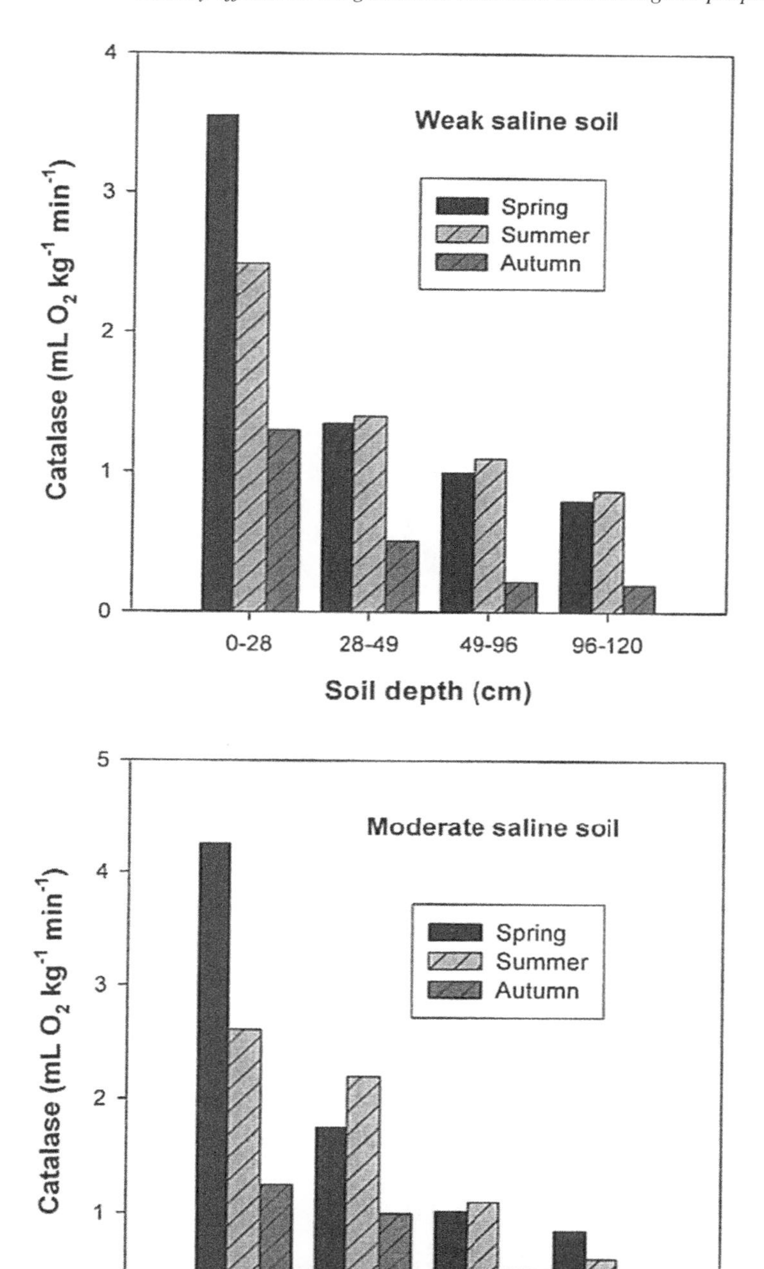

Figure 3. Catalase activity rates of varying saline soils.

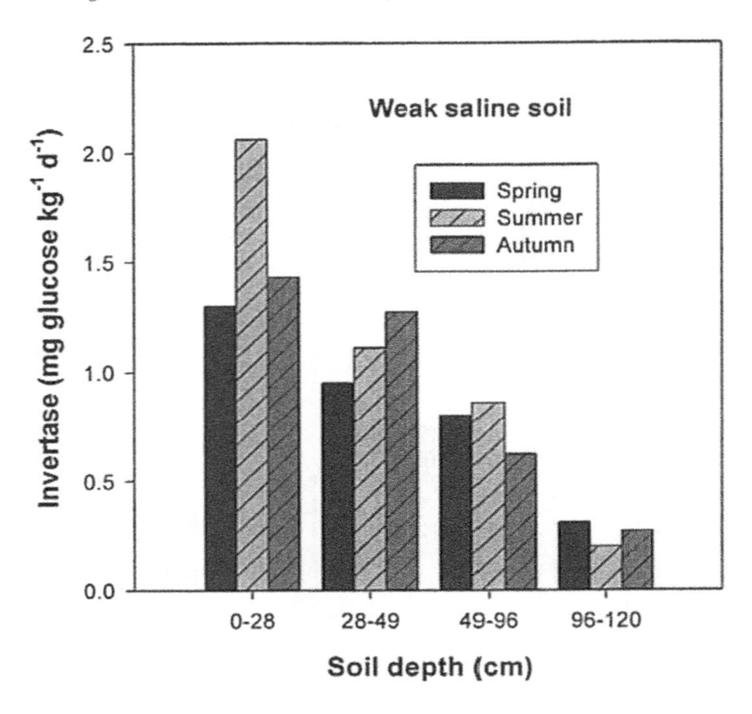

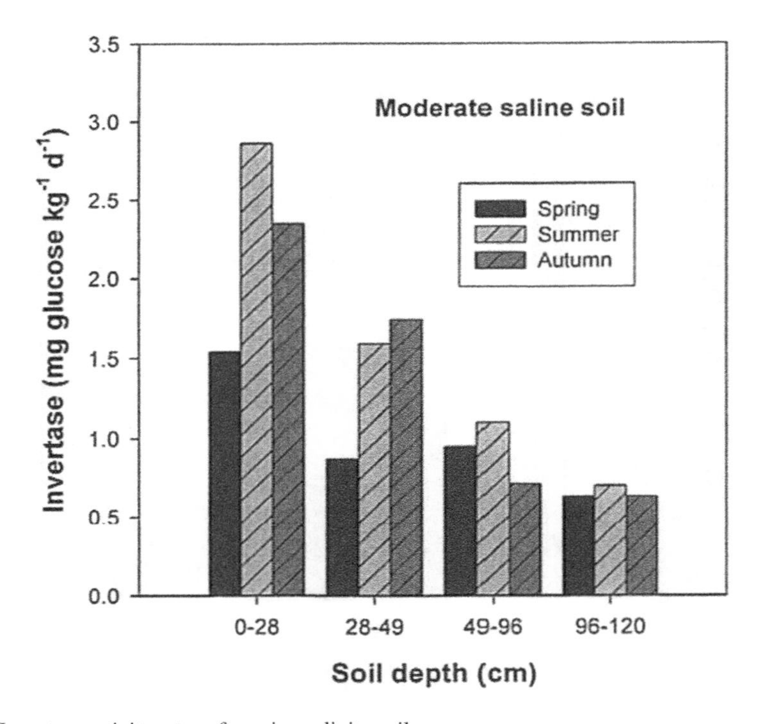

Figure 4. Invertase activity rates of varying salinity soils.

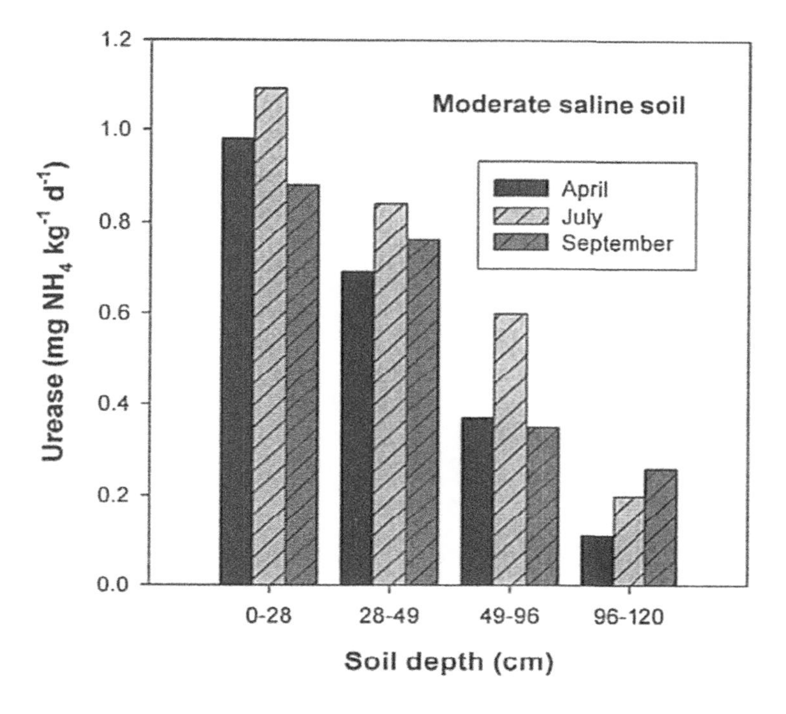

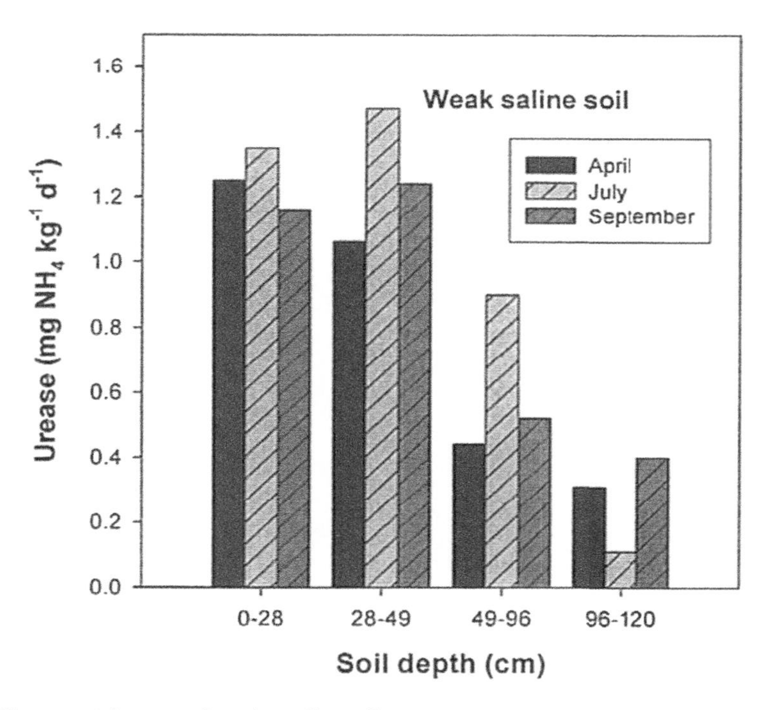

Figure 5. Urease activity rates of varying saline soils.

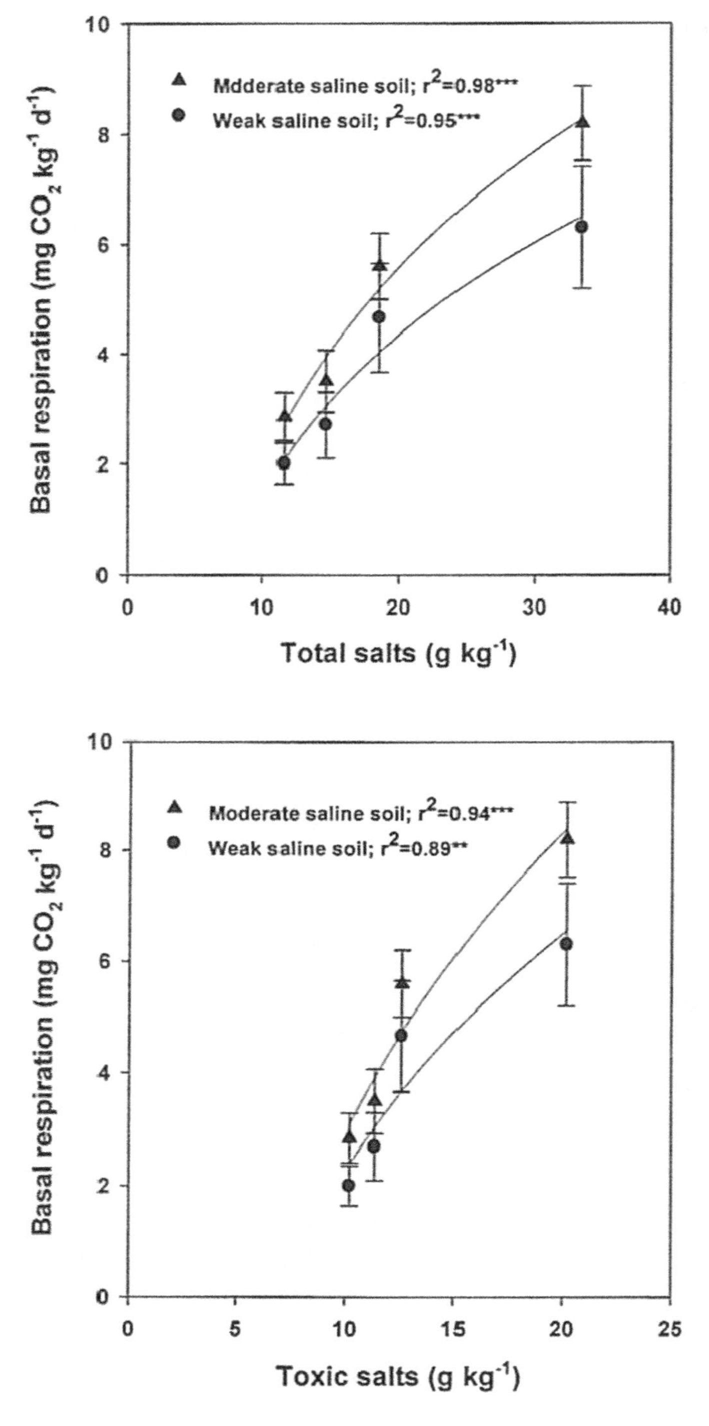

Figure 6. Basal respiration rates for toxic and total salts in varying saline soils.

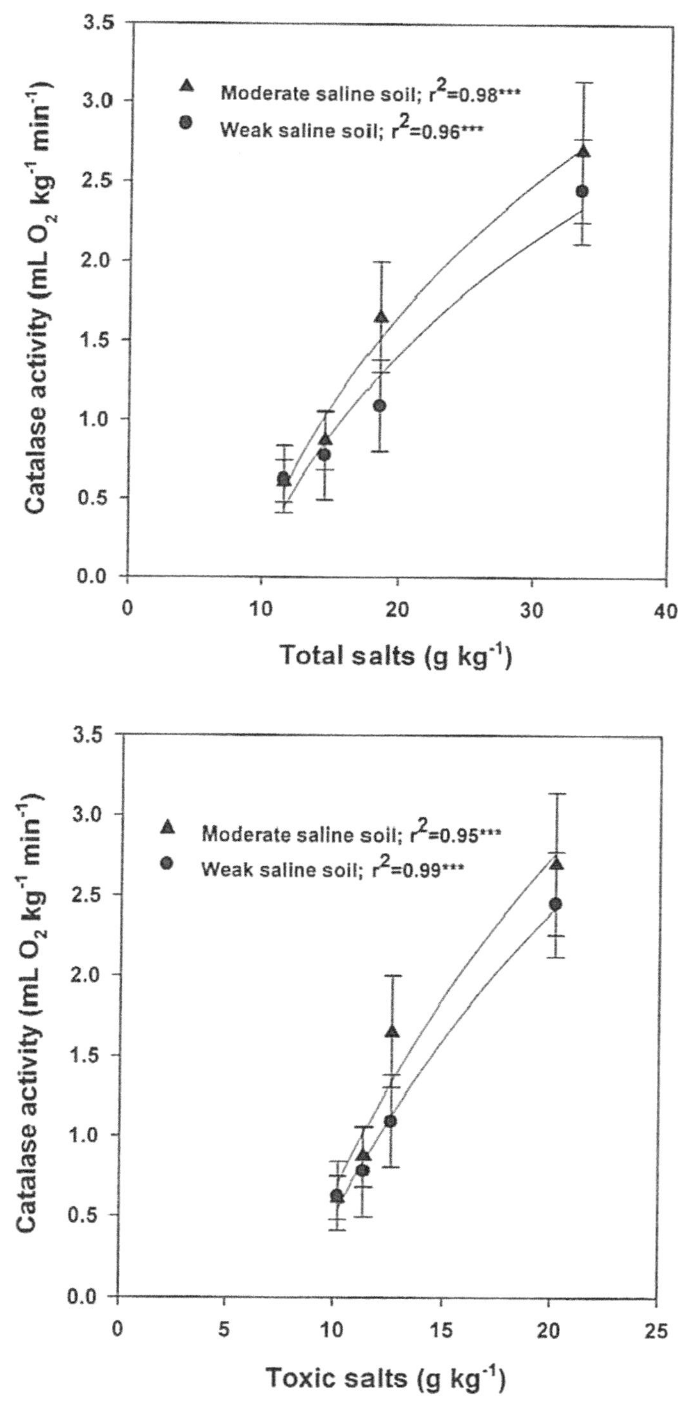

Figure 7. Catalase activity rates for toxic and total salts in varying salinity soils.

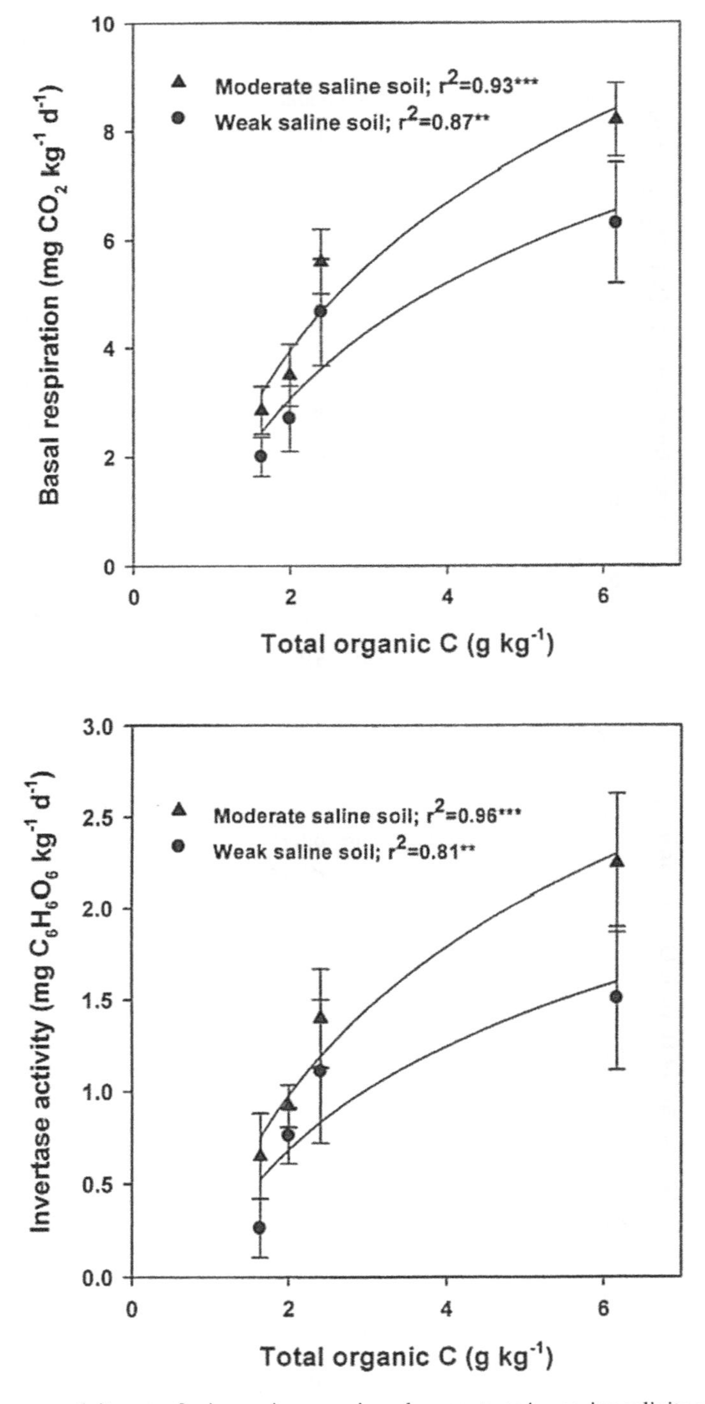

Figure 8. Invertase activity rates for increasing organic carbon contents in varying salinity soils.

Since salt content of soil is considered to be the limiting factor for the survival of microorganisms, higher BR rates suggest that moderately saline soil environment is unfavorable for microorganisms due to the osmotic effects (Figure 6). In high stress environments (e.g., moderately saline soil); microbial cells may be forced to maximize their catabolism for survival than anabolism (growth) in response to increasing salt content (Figures 6 and 7). The catabolic effect is more pronounced in response to toxic contents of salt (e.g., $MaSO_4$, $MgCl_2$, Na_2SO_4 and NaCl) than total salt contents. A significant relationship between biological activities and salt contents (total and toxic) supports the conclusion that increasing salt content increases microbial catabolism for survival under saline environment (Figures 6 and 7).

Significant relationships between BR rates and invertase activity with increasing C substrate availability suggest a greater microbial catabolism to use organic matter for sustaining biological activity in saline soil (Figure 8). Greater invertase and urease activities during the summer are most probably related to greater transformations of organic matter by microorganisms for their increasing energy and nutrients demand for survival than assimilation in response to high temperature and salinity stresses.

4 CONCLUSIONS

Irrigation is essential to high agricultural production in the Aral Sea basin. About 44% of the land is concentrated in the Syr Darya and 56% in the Amu Darya river basin. In response to large scale withdrawal of water for irrigated agriculture especially for cotton production, the river water flows to the Aral Sea severely decreased over the years. As a result, the Aral Sea, once the fourth largest inland body of water in the deserts of Central Asia, has lost much of its volume (82%) and area (~66%) and quadrupled in salinity. Indiscriminate flood irrigation with poor quality of water and inefficient drainage systems accelerated secondary salinization of irrigated soils with severe degradation of ecosystems. Aral Sea degradation has caused varying degrees of salinity of irrigated soil in the Amu Darya and Syr Darya basins. Increase in soil salinity (16%) spread to Andijan, Djizak, Namangan, Samarkand, Surkhan Darya, Syr Darya, Tashkent, Fergana, and Kashka Dariya regions between 1990 and 2000. In Syr Darya region alone, about 47% of the irrigated lands are weakly saline followed by 35% moderately saline, and 18% strongly saline. Among the salts, Ca, Mg, and Na associated sulfates, chlorides, and bicarbonates are dominant in the soil profile. Highest amount of Na-sulfate (>40%) followed by Ca-sulfate (~29%), NaCl (18%), and MgCl (10%) have accumulated within 0–100 cm depth of typical sierozems soils. More than 70% of the salts are highly toxic to plant growth and soil biological activities. As a result, the BR and enzyme activities are influenced strongly by soil salinity in different seasons. There were 22, 15, and 30% increase in BR rates, catalase, invertase, and urease activities in moderately saline soils compared to weakly saline irrigated soils. The BR rates and catalase activities were higher in spring than in summer in both weakly and moderately saline irrigated soils. However, the invertase and urease activities are higher in summer than in spring. Abnormally high biological activities in moderate saline irrigated soil during summer and autumn than in weakly saline irrigated soil during spring may be due to osmotic effects on microorganisms. Under high salinity stress, microbial cells use more energy and nutrients (catabolism) for survival than growth (assimilation). The catabolic effects on microbial cells were more pronounced in moderately saline soil and in response to toxic salt concentration ($MaSO_4$, $MgCl_2$, Na_2SO_4 and NaCl) than total salt concentration in both weakly and moderately saline soils.

REFERENCES

Anonymous. 1994. Annual reports on land reclamation and water use. 1975 to 1994. Ministry of Water Resources of the Republic of Uzbekistan. Tashkent. Uzbekistan. 220 p. (in Uzbek).

Anonymous. 1995. Environment conditions and natural resources use in Uzbekistan. National Report. Tashkent, 80 p. (in Russian).

Anonymous. 1996. Suggestions for a national water management strategy of the Republic of Uzbekistan. Design Institute 'Vodproekt' of the Ministry of Water Resources of Uzbekistan Report. Tashkent, 233 p. (in Russian).

Dadenko, E.V. 2006. Some Aspects of Soil Enzyme Activity Application. Rostov State University, B. Sadovaya str., 105, Rostov-on-Don, 344006, Russia. 18th World Soil Science Congress, Philadelphia, PA.

FAO. 2006. National Soil Degradation. http://www.fao.org/landandwater/agll/glasod/glasodmaps.jsp.

FAO. 2006. Terrastat – land resource potential and constraints statistics at country and regional level. Land and Water Development Division. http://www.fao.org/ag/agl/agll/terrastat/wsr.asp

Green, D.M. and M. Oleksyzyn. 2002. Enzyme activities and carbon dioxide flux in a Sonoran Urban Ecosystem. Soil Sci. Soc. Am J. 66: 2002–2008.

Killham, K. 1985. Assessment of stress to microbial biomass. In: A.P. Rowland (ed.), Chemical Analysis of Environmental Research. ITE symposium no. 18, pp. 79-83. Marlewood Research Station, Grange-over-Sands, Cambria, UK.

Micklin, P.P. 1994. The Aral Sea problem. Proc. Paper 10154. Instn. Civ. Engrs. Civ. Engng. 114: 114–121.

Minshina, N.G. 1996. Soil environmental changes and soil reclamation problems in the Aral Sea basin. Eurasian Soil Sci. 28: 184–195.

Mirzaev S.Sh., R.M. Razakov, and V.G. Nasonov. 2000. The future of the Aral Sea basin in flourishing oasis or fruitless desert, Collection of scientific articles: Water resources, the Aral problem and environment. Tashkent University, Tashkent, Uzbekistan.

Muminov, F.A. and I.I. Inagatova. 1995. The changeability of the Central Asian climate. SANIGMI, Tashkent, Uzbekistan.

Razakov, R.M. 1991. Use and protection of water resources in Central Asia. Tashkent, p. 203.

Rietz, D.N and R.J. Haynes. 2003. Effects of irrigation-induced salinity and sodicity on soil microbial activity. Soil Biology and Biochem. 35: 845–854.

Ryan, J., P. Vlek, and R. Paroda. 2004. Agriculture in Central Asia: Research for Development. International Center for Agricultural Research in the Dry Areas (ICARDA), Aleppo, Syria.

Sokolov, V.I. 2000. Definition of boundaries of water collection basins of trans-bordering, local and mixed types of surface water resources in the basin of the Aral Sea and quantitative assessment. Collection Scientific Works of SRC ICWC, Tashkent, Issue 2. p. 35–53.

Tabatabai, M.A. 1994. Soil Enzymes. p. 778-834. In: J.M. Bigham (ed.), Methods of Soil Analysis; Part 2: Microbiological and Biochemical Properties. Soil Science Society of America, Madison, WI.

Tripathi, S., S. Kumari, A. Chakraborti, A. Gupta, K, Chakraborti, B. Kumar, and A. Gupta. 2006. Microbial biomass and its activities in salt-affected coastal soils. Biol. Fert. Soils. 42: 273–277.

Umramov, M. 1989. Soils of Uzbekistan. Tashkent, Uzbekistan.

UNDP. 1995. The Aral crisis. Tashkent, Publishing house of UNDP.

UNEP. 2005. Water resources. State of Environment of the Aral Sea Basin. http://enrin.grida.no/aral/aralsea/english/water/water.htm

Uzbekistan Academy of Sciences. 1979–1981. Irrigation of Uzbekistan. Tashkent. Vol.1: 382 p., 1979; Vol.2: 368 p., 1979; Vol. 3: 359 p., 1979; Vol. 4: 448 p., 1981. (in Russian).

World Bank. 1997. Aral Sea Basin. Project 3.1.B. Agricultural water quality improvement. EC IFAS., World Bank, Tashkent, p. 51.

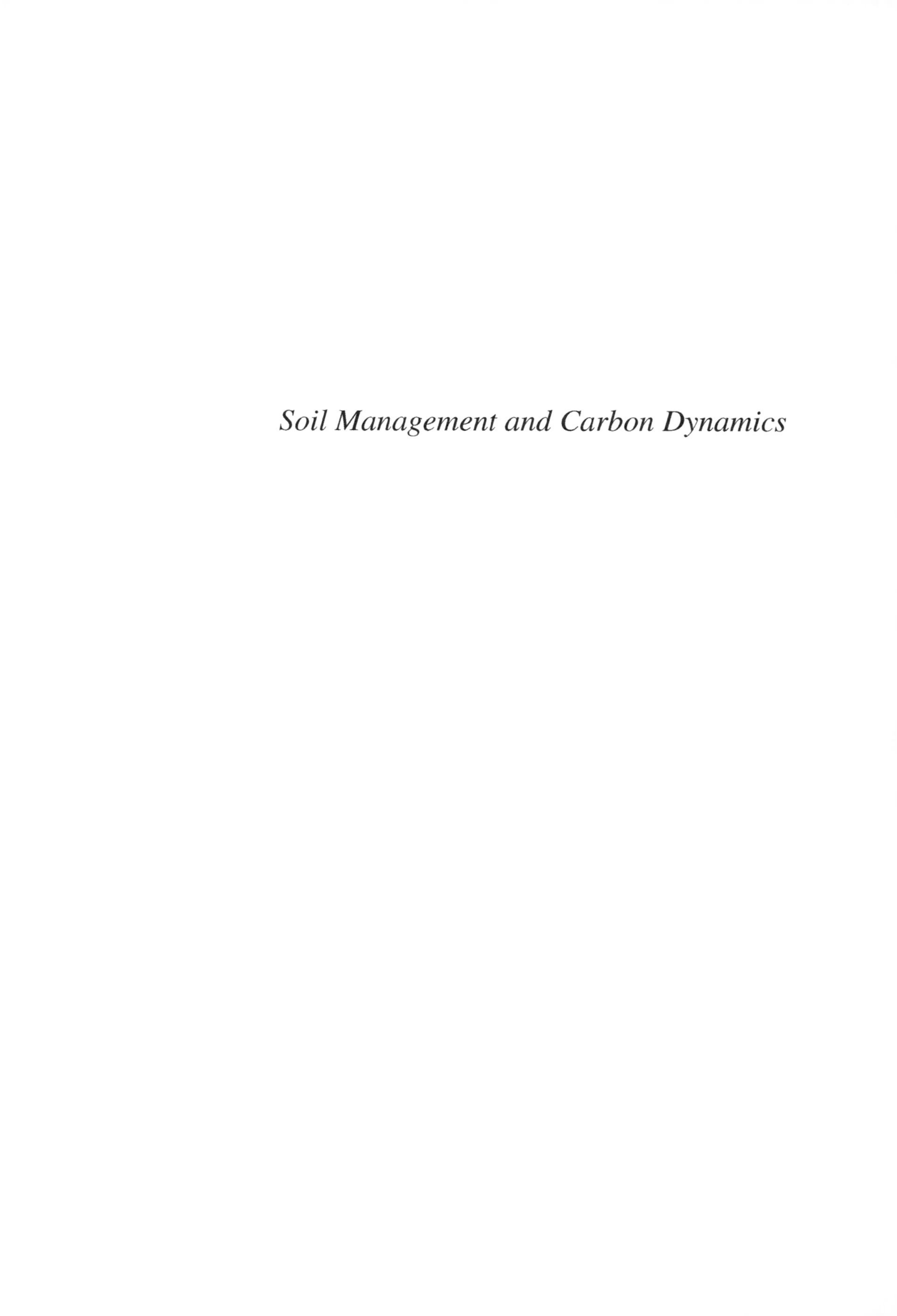

Soil Management and Carbon Dynamics

CHAPTER 12

Central Asia: Ecosystems and carbon sequestration challenges

Mekhlis Suleimenov
International Center for Agriculture Research in Dryland Areas, Tashkent, Uzbekistan

Richard J. Thomas
International Center for Agriculture Research in Dryland Areas, Aleppo, Syria

1 INTRODUCTION

Countries of the Former Soviet Union (FSU) including Russia, the republics of Central Asia and the Caucasus represent vast areas that can have a substantial role in carbon sequestration. Agriculture in the FSU was developed somewhat independently from the rest of the world, was production oriented but was characterized by relatively low crop and livestock productivity. Moreover it was remarkably inefficient as compared to many developing and industrial countries. During the transition to market oriented economies, agricultural production in all countries of the FSU was subject to enormous pressure associated with the disintegration of the Soviet Union and the transformation of government controlled farm sectors into more liberal based private and individual farming businesses. During this transition crop and livestock production declined because of lower input use and less support from the governments. Some areas have been abandoned because of lack of resources and low returns. Poor management of both crop and livestock sub-sectors resulted not only in lower productivity but also in further land degradation reflected in more severe soil erosion and soil salinization leading to large losses of soil carbon on both cropland and rangeland. In this paper, the issues of land management for better carbon sequestration and challenges in addressing these issues are discussed.

1.1 *Eco-systems*

The regions and eco-systems considered here include Central Asia, the Caucausus and adjacent areas of Russia. They can be subdivided as follows:

1. The steppe belt in northern Kazakhstan (NK) and adjacent areas of Russia including southern part of western Siberia (WS);
2. Southern Ural (SU) and the Volga area (VA), located between 52–56° NL;
3. The dry steppe belt in central Kazakhstan and in southeast of Russia including the south Volga area just above the Caspian Sea located between 48–52° NL;
4. The northern Deserts of Betpakdala, Moyinkum and other deserts in Kazakhtan located be desert between 44–48° NL;
5. The Kyzylkum desert, mostly in Uzbekistan located mainly between 40–44° NL;
6. The Southern Garagum desert, mostly in Turkmenistan, located between 36–40° NL;
7. The mountains of Central Asia and the Caucasus.

1. The NK is a territory between 50° and 54° N latitude and between 60° and 78° E longitude. It covers an area of 57 M (million) ha, and is comprised of 4 provinces: Akmola, Kostanay, Pavlodar and North-Kazakhstan. Soils are fertile, but there are many solonetzic soils and almost all soils are subjected to erosion. The northern area consisting of semiarid southern forest-steppe is situated on gray forest soils and leached black soils occupying 8% of cropland; semiarid steppe on ordinary black soils occupies 22% of cropland. These are the most fertile soils of the region, medium and

heavy textured, with soil organic matter (SOM) content of 5–7%. The semiarid steppe zone on southern calcareous chernozem is located in the south and occupies 28% of the cropland. This soil has SOM content of 3–4%, with soil texture of heavy clay loam.

The southern part of WS area is located adjacent to and north of the NK. The dryland area is located between 50–55° N latitudes and 62–86° E longitudes. There are 9 M ha of cropland; 70% of these are on chernozem soils (Khramtsov et al., 2000). The area produces over 10% of Russia's grain. In the steppe zone, cropland occupies 80% of the area. The major crop is spring wheat produced in the semiarid steppe and southern forest-steppe with average grain yields of 1.0–1.5 t ha^{-1}.

2. The Steppe belt of the SU is located west of WS and northwest of NK between 51–55° latitudes and 52–62° longitudes. Orenburg is the largest province in SU with an average area planted to grains of around 4 M ha with average grain yields of 1 t ha^{-1}. The western part of the area has a climate similar to the Volga area while the eastern part is similar to NK and WS. Cropland is located on black and dark chestnut soils. The Volga area (VA) is located west of the SU. The dryland agriculture belt is located between 48–52° N latitudes and 44–52° longitudes. The total agricultural land of the area comprises about 30 M ha in the Soviet era, including 20 M ha of cropland. The area produced 16% of grain, and 24% of sunflower of Russian production. The share of irrigated cropland was about 3%.

3. The Dry Steppe belt has chestnut heavy textured soils occupying 40% of the cropland in Kazakhstan. It extends into the lower VA (Volgograd region). The soils have SOM content of 2–3.5%. The main crop is spring wheat producing rather low grain yields but of high baking quality (protein content 14% up) that is widely used for blending with low protein wheat. Significant areas of these lands are marginal and have been abandoned during the transition period. The southern part of Dry Steppe belt is semi-desert represented by rangelands in central Kazakhstan and above the Caspian Sea, the so called Black lands in Kalmykiya. The semi-desert, light chestnut soils cover 2% of the cropland; these soils have less than 2% SOM in the arable layer, with sandy loam and loamy sand textures. The Black lands have been severely degraded because of intensive overgrazing and use of some of these marginal lands for grain production.

4–6. The deserts of Central Asia extend from the Kopetdag and Paromiza mountains in the south up to 48° N and from the Caspian Sea in the west to the foothills of Jungar Alatau mountains in the southeast and the deserts of China in the east. There are many deserts types in this vast area; sandy deserts of Garagum (360,000 km^2), Kzylkum, Moyunkum, Saryyesik-Atyrau, Near Aral Garagum, Great and Small Barsuk, Naryn sands in Volga-Ural area; and smaller size deserts, stony Betpakdala, stone chippings Usturt, clayey Golodnaya steppe; saline desserts Kelkor, Mertviy Kultuk. There is also a new desert, Aralkum, emerging as a result of the drying of the Aral Sea. If the Sea disappears then the Garagum and Kyzylkum deserts will merge. The area covered with desert occupies 79% of the total area of Turkmenistan while in Uzbekistan, Krgyzstan, Kazakhstan and Tajikistan deserts occupy 56%, 35%, 28% and 17% respectively (Bbayev, 1995).

7. The Central Asian region is divided orographically into two parts; an eastern part occupied by the mighty mountain ranges of Tien Shan and Pamir-Alay, while the western part is occupied by lowland plains. The Caucasus stretches for about 1000 km from the Black Sea to the Caspian Sea, separating Europe from Asia. North of the main Caucasus ridge there are mountain ridges that extend to the Russian steppes. The fragile highlands of Central Asia and the Caucasus are threatened by overgrazing, cultivation of steep slopes, non-sustainable fuel wood and timber harvesting that result in severe soil erosion.

1.2 *Climate*

Moving from the east of NK-WS westwards and further into middle VA the amount of rainfall gradually increases from 320 mm in Omsk to 400 mm in Saratov. The pattern of precipitation distribution is also gradually changing from more pronounced rainfall in July in the east to more even rainfall distribution in the VA region. The amount of rainfall and its distribution in the Volga area is similar to the Canadian prairies of Saskatchewan. Rainfall in July amounts to 12% of the total annual rain in Orenburg, 19% in Chelyabinsk and 21% in Omsk. This explains why spring wheat in Siberia and Kazakhstan is sown much later than in the VA.

The annual precipitation in NK and WS decreases southwards; in the northern semiarid steppe zone, precipitation is between 320 and 350 mm, whereas in the southern semiarid steppe zone it is between 280 and 320 mm. In the dry steppe, annual precipitation varies between 220 and 280 mm, the water deficit during summer averages about 250 mm. The distribution of precipitation throughout the year in all regions is similar. About 30% of precipitation falls evenly as snow during November–March. In VA annual precipitation decreases from about 350–400 mm in semi-arid steppes (Saratov) to 220–300 mm in the more southerly dry steppes (Volgograd). Below the dry steppe belt, stretches a semi-desert belt with 150–220 mm annual precipitation with gradual transition into desert with annual precipitation below 150 mm.

Deserts of Central Asia are characterized by a continental climate of dry and hot summers across the deserts, cool winter in the south and cold winter in the north (average monthly temperature in January is −15°). Rainfall amounts are small (100–150 mm a year) and are evenly distributed throughout the year in the north, with prevailing rainfall in winter in the south. Evaporation is high, from 1300 mm in the north up to 2850 mm in the south. Sandy deserts comprise 60% of all the arid territory of Central Asia (Babayev, 1995). Grey and brown-grey soils with various degree of salinity are widespread in Central Asia.

Temperatures gradually increase moving from the NK and WS areas westwards to the SU and WA. The NK and WS areas are characterized by a continental climate with hot summers and very cold winters. The average annual air temperature is 1.6°C (at Astana). Winter begins in November; the coldest months are December, January and February with average monthly temperatures of 16–18°C below zero. In May, the average is 12.6°C but by the end of June, average temperatures go up to 20°C. July is the hottest time of the year and the only month with no frost. The average temperature is about 20–21°C, while daily temperatures exceed 30°C. In August, temperature falls to an average to 16–18°C with a high probability of night frost that can damage grain at the end of the month. WS has the same temperature regime as the Northern Kazakhstan. Omsk, situated in the forest-steppe zone and 600 km north of Astana, has an average annual temperature of 0.4°C, due to slightly lower temperatures throughout the year compared to Astana. In the SU and VA the average annual temperature is higher than in Siberia. Temperatures in the semi-desert and northern desert belt in winter are just as cold as in the dry steppe but summers are drier and hotter. Winters are milder only in the southeast with average annual temperatures of around 6–8°C (Almaty, Bishkek) and are much warmer in southern Kazakhstan and in Uzbekistan (14°C in Tashkent and Dushanbe), going up to 18°C in southern deserts (Ashgabat, Turkmenistan).

2 AGRICULTURAL POLICIES AND PRODUCTION IN CENTRAL ASIA

The development of agriculture in the FSU differs a great deal from country to country. Essentially, the FSU countries may be subdivided in two groups; (1), those that followed a relatively radical approach to reforms (Russia, Kazakhstan, Kyrgyzstan, Armenia, Azerbaijan and Georgia), and (2), those that followed a relatively gradual approach to reforms, maintaining governmental control of agricultural production (Tajikistan, Turkmenistan and Uzbekistan). Different policies strongly affect not only agricultural production but also the sustainability of agriculture and the viability of livelihoods.

In the semi-arid steppes of NK and WS with black and dark chestnut soils almost all the agricultural land was plowed up for grain production about fifty years ago. This is predominantly a spring wheat zone grown continuously for four-five years with summer fallow occupying from 15–20% of cropland. In SU and VA, agriculture has a longer history with spring wheat predominating although winter wheat has been grown on summer fallow. All other crops occupy small areas. Recent developments associated with the liberalization of production, markets and prices have resulted in major changes in production systems. In both regions, especially in NK, bankruptcy of the restructured collective farms in the mid-nineties resulted in the transfer of these enterprises to large grain trading companies that decided to invest in agricultural production. Nowadays these companies own grain processing and storage facilities and control the agricultural production on at least 100,000 ha. Individual farmers with farm sizes of several hundred hectares do not play an important role in the overall production scenario although some of them are involved in integrating grain production with processing and marketing.

In the dry steppe across Kazakhstan and the Volga area, rangelands occupy large areas while cropland is less suitable for grain production. Nevertheless, during the Soviet era, most of the drylands including marginal lands were used for grain production. Since independence and market oriented developments, large areas of marginal lands especially in Kazakhstan (over 10 M ha) were abandoned and left as weedy fallow in the mid-nineties. One of the reasons for abandoning marginal lands was the poor financial status of farms. Some of these earlier abandoned lands (10–15%) have been brought back to production after agricultural business in Kazakhstan became profitable over the last three-four years and investments in agriculture increased.

In semi-desert and desert areas agricultural lands are being used for livestock grazing except for pockets of irrigated agriculture and limited areas of rainfed agriculture. A general trend in livestock production is more reliance on rangeland grazing rather than integration with forage production. In all countries under radical and gradual reforms, forage production almost came to a standstill because large and small farms decided to get rid of livestock and concentrate on grain production. Finally, in all FSU countries livestock was mainly restricted to households that owned cow and calf operations and small flocks of sheep and goats. Radical reforms resulted in dramatic cutbacks of livestock populations first of all in small ruminants. The distribution of large flocks of animals to small household operations resulted in an enormous concentration of livestock around villages leaving remote large rangelands unused especially in Kazakhstan and Kyrgyzstan. In the past some neighboring countries shared rangelands, for example Georgia used winter rangelands on the territory of Azerbaijan. This practice was discontinued. In countries with gradual reforms in agriculture, livestock numbers were not cutback so dramatically and in Turkmenistan population numbers increased remarkably. FAO statistics indicate a stable figure of some 6 M heads of small ruminants during the last five years whereas Turkmenistan national statistics indicate a small ruminant number of 15 M heads (Turkmen Mallary, 2005). Even if the national statistics are rather optimistic the livestock population in this country increased substantially and their influence on range status and finally on carbon sequestration can not be overemphasized.

Irrigated agriculture in southern (Garagum) and the Kyzylkum desert areas have changed a great deal during the transition. First, dryland rainfed agriculture decreased more markedly in countries with liberal economies as it became unprofitable. In irrigated agriculture, dramatic changes occurred in all countries driven by different forces. In countries with liberal economies the market has been a major factor effecting cropping structure. In Kazakhstan, cotton-alfalfa rotations have been replaced by continuous cotton as the most profitable cash crop. In Azerbaijan, on the contrary, policy liberalization caused dramatic cutback of cotton growing areas because small farmers couldn't take advantage of good world market prices for cotton and cotton was replaced by wheat. In Kyrgyzstan, the area dedicated to forages was reduced and was replaced by wheat. In addition, sugar beet areas increased driven by good local prices for this product. Some new crops such as common bean were introduced as a result of a favorable market. In all countries with government controlled agriculture the priority crop for expansion was wheat that replaced alfalfa and other forage crops. All these changes in countries with different policies invariably resulted in severe cutbacks of the areas under forage crops including perennial forages like alfalfa. This change has negatively affected soil fertility in general and SOM content in particular.

3 CROPLAND AND RANGELAND MANAGEMENT FOR CARBON SEQUESTRATION

3.1 *Cropland management*

It is well known that major losses of SOM are associated with soil erosion. In the steppe belt of the FSU conservation tillage was introduced in the 1960s in the NK. Later it spread to WS and across this region conservation tillage became a generally adopted practice because it not only protected soil from wind erosion but also increased crop yields thanks to better soil moisture accumulation. In SU and VA conservation tillage did not improve crop yields significantly and its adoption was not widespread. However even with conservation tillage, if it is done several times during the summer fallow period, no residue is left on the soil surface to protect soil and soil erosion occurs by wind and water. In spite of this finding the rotation of small grains with summer fallow became a practice for moisture conservation and weed control across the steppe belt regions even though the practice appears to be unsustainable.

In the middle of the VA, SOM content amounts to 200–600 t ha^{-1} in natural grasslands versus 150–200 t ha^{-1}in croplands (Medvedev et al., 2003). On average, annual losses of SOM from cropland on black soils amount to 1–1.5 t ha^{-1} resulting mainly from soil erosion. The cropland area of black soils affected by varying extents by water erosion is about 55% with summer fallow contributing greatly to soil erosion. Global climate change, increases in areas under summer fallow, the introduction of row crops, reduced application of fertilizers and removal of crop residues combined with soil erosion by water, have all contributed to increased land degradation on black soils. Retrospective analysis of soil quality monitoring data indicated that during agricultural use black soils lost 40–50% of SOM. In the process, deep fertile chernozems with SOM contents of 12–14% have been transferred into medium humus common cherenozems with SOM contents of 6–8%. In turn, common chernozems have been transformed into dark chestnut soils. The active fraction of the SOM of cherenozem soils has been reduced from 60–65% to 35–45%.

Alongside soil erosion, SOM losses have been caused by the decomposition of organic matter especially in fields left to summer fallow (Medvedev et al., 2003). Annually, on black soils depending on soil texture, between 1.5–2.5 t ha^{-1} of SOM has been decomposed during the summer fallow period. As a result of periodic leaching, a significant amount of nutrients accumulated during summer fallow is leached beyond the crop root zone in spring during snow thawing into ground waters. In eroded soils the availability of easily available nitrogen was reduced by 38.8%, moderately available nitrogen by 15.9% and stable nitrogen by 9.8% when compared with non-eroded soils.

The annual loss of SOM from intensively used cropland in rotations of grains with row crops and fallow in VA was 1.1 t ha^{-1} with no fertilizer application while the loss of SOM was reduced to 0.52–0.64 t ha^{-1} when fertilizers were applied at a rate of 60 kg N, 60 kg P and 40 kg K/ha. The application of 40 t ha^{-1} of manure once in six years reduced the SOM losses to 0.40–0.32 t ha^{-1}. In rotations of grains with perennial forages there was an accumulation of SOM even with no fertilizer applied (Medevedev et al., 2003). The authors however failed to point out that cropping systems including clean summer fallow are not sustainable because of the loss of SOM.

In the Omsk region, WS the original SOM contents in black soils in southern forest-steppe and steppe zones were 6.78% and 5.48% respectively (Khramtsov and Kholmov, 2003). During 16 years of cultivation, losses of SOM in 0–20 cm layer were 1.51 t ha^{-1} in a rotation of wheat with fallow (once in 3 years), 0.98 t ha^{-1} in rotations with 25% under summer fallow and 0.76 t ha^{-1} with fallow on 16% of cropland. Interestingly, in a four year rotation when summer fallow was replaced by annual forage crop there was almost no loss of SOM. In spite of these findings crop rotations with 20–25% summer fallow are recommended.

In the Kulunda steppe of Altay, WS, theadoption of soil conservation practices on sandy loams has helped to control wind erosion but crop rotations including summer fallow have been causing increased losses of SOM (Gnatovskiy, 2003). Over 15 years, the annual loss of SOM in the 0–20 cm soil layer was 0.68 t ha^{-1} compared with no significant losses under continuous wheat or perennial grasses. Although recognizing that summer fallow results in the loss of SOM and is not particularly efficient in conserving moisture, the author indicates that summer fallow is a means to control weeds.

There is a general belief that in any steppe region summer fallowing is essential. In Kostanay, one of the largest grain growing regions of NK with grain sown on 3 M ha, there is about 800,000 ha of summer fallow which is planned to increase up to 1 M ha in order to control weeds, insect pests and plant diseases (Tuleubayev, 2003). There is evidence however that soil erosion on bare summer fallow fields is 6 times higher than on fallow with oats as a cover crop (Gilevich, 2003). Some authors advocating summer fallowing agree that it should not be bare but covered. Shramko (2003) came to conclusion that on dark chestnut soils, the black fallow should be replaced with fallow with application of manure or green manure.

Loss of SOM during soil erosion should be considered as a loss of crop yields (Zhunusov, 2003). On sloping lands with weakly washed out soils crop yields have been decreased by 15–25%, on medium washed soils where SOM has decreased by 25–45%, crop yields have been reduced by 35–45%, on strongly washed out soils with SOM losses of 50%, crop yields decreased by 45–75%.

The data obtained in a long-term study in Shortandy (Suleimenov and Akshalov, 2004) provides evidence to test the efficiency of cropping systems based on summer fallow practice. The summer fallow is believed to be very important in soil moisture conservation providing greater water storage in soil before planting spring wheat in mid-May. The long-term observations show that water storage in 1 m soil layer before planting of spring wheat depends on inputs applied for different snow management technologies and amounted to the following; on fallow – from 60 to 170 mm, on stubble land – from 40 to 150 mm. This indicates that using similar approaches to management practices on fallow and on stubble one can expect only 20 mm of additional available water under summer fallow. This does not justify keeping land without a crop for one year.

Weed control is the most frequently emphasized factor to advocate summer fallow. However there is no evidence that weeds could not be controlled without summer fallow. Moreover research on weed control by tillage and chemicals conducted in Shortandy has concluded that the frequency of herbicides application needed to control weeds was the same with summer fallow occupying in crop rotations from 16 to 33% of cropland (Shashkov, 2003).

Summer fallow is reported to improve the nutrient regime of soils for growing grains. In terms of nitrogen, nitrate availability is significantly higher under summer fallow compared with stubble land. But one has to recognize that it is at the expense of SOM and in the long run frequent summer fallow will lead to greater losses of organic matter. Rennie et al. (1976) estimated that the SOM content of soils of Saskatchewan, Canada had declined to 35–44% of their original values. They estimated that crop removal could account for only 30–33% of the N lost from SOM, and that the missing N was equivalent to approximately 36 M Mg to 1974. They further estimated that about 10 M Mg of this lost N (600 kg N/ha) was leached below rooting depth, mainly during fallow periods.

The most important integral element of the assessment of cropping practices is crop yield. On average for sixteen years (1988–2003) the advantage of wheat sown first year after summer fallow was 15% higher than on stubble land. The advantage of summer fallow in increasing wheat yield was the highest under low input technologies in dry years. However it is more important to calculate grain production from the total cropland area including fallow which does not produce a crop. In our experiment, the control was the rotation of "fallow-wheat-wheat-barley-wheat". In continuous grain rotation fallow was replaced by oat for grain. Comparative data for 13 years demonstrated that grain production from the total area always was higher when fallow fields were replaced by oats. Under low, medium and high input technologies yields increased by 6, 18 and 28%, respectively.

The first clear and well publicized call to Saskatchewan farmers to re-examine their summer fallowing frequency was issued by Dean D.A. Rennie to a University of Saskatchewan Farm and Home Week audience in 1973 (Austenson, 1998). In the FSU, after the first publication on the possibility of no-fallow farming practices (Suleimenov, 1988), it was thought that there was insufficient evidence to draw firm conclusions. Since then further data is available that incorporates more weather data. This data indicates that the yield advantage of wheat sown after fallow as compared to continuous wheat increased with increased drought. On average, there was no yield decline in continuous wheat on stubble after twenty years provided adequate cultural practices were applied. This new finding contrasts with the belief that grain yield declines gradually with continuous wheat cropping in northern Kazakhstan (Shiyaty, 1986). In spite of these findings many agricultural specialists recommend an increase summer fallow area which may result in increased soil erosion, loss of SOM and serious land degradation.

Thus, over fifteen years there was no publication that questions the efficiency of the cropping system based on summer fallow, moreover some scientists and specialists are keen to increase the share of summer fallow in cropland up to 33%. However, a publication in Russia (Shnider, 2004) relates the findings of a very successful innovative large farmer in Western Siberia on an area adjacent to north Kazakhstan. The farm is well equipped with modern tractors, combines and equipment including no-till planters. Strikingly, he states that he has never used summer fallowing on some 28,000 ha over a ten year period although it was recommendation that farmers should keep on average about 15% of cropland under fallow. During an eight year period he was growing spring wheat continuously and harvesting on average 2.11 t ha^{-1} against 1.59 t ha^{-1} in the neighbouring district. During recent years he has diversified his crops, planting 3,000 ha of dry pea, 4,200 ha of sunflower and 1,200 ha of maize in 2004. Based on economical efficiency of crop diversification in 2005 he planted 9,000 ha of sunflower, 2,500 ha of both maize for grain and canola.

In 2004–2005, two traveling workshops were organized by ICARDA to NKWS to see soil tillage and crop diversification practices on research stations and farms. On the research farms crop rotation was practiced with one third of the cropland under summer fallow while on commercial farms the share of summer fallow was around fifteen percent. Mr. Vadim Shnider was the only farmer out of tens of thousands in the whole region of NKWS who did not use summer fallow. Certainly his farming practices can be considered as very intensive technologies as he has been practicing application of recommended full rates of fertilizers and controlling weeds successfully by applying up-to-date chemicals with modern sprayers. Thus, the farmer Shnider proved on a large scale that under adequate grain growing technologies there is no need to fallow up to 33% of cropland every year (Suleimenov, 2004).

Of course continuous grain cropping is not the best alternative. Actually, rotation of small grains with fallow one year out of every 3–5 is a form of monoculture. Recently experiments have been laid out testing crop rotations with no fallow. Alternative crops which could possibly replace summer fallow are food legumes such as field pea, chickpea, lentil (Paroda et al., 2004). In western Kazakhstan chickpea was found to be one of the best preceding crops for grains. When planted after winter wheat chickpea is the most drought resistant high yielding crop among spring crops (Vyurkov, 2003).

Long-term data indicated that annual losses of SOM amounted to 1.4 t ha^{-1}: in typical fertile black soils of northeastern region of Kazakhstan, 1.2 t ha^{-1} in common black soils, and 0.9 t ha^{-1} on southern black soils, and 0.5–0.8 t ha^{-1} on rainfed sierozems and light chestnut soils in the south (Kenenbayev et al., 2004). To maintain soil fertility on rainfed sierozem soils in rotations of grains with fallow once in four years it is recommended to leave all residues of grain crops during harvest and to apply manure (20–40 t ha^{-1}) or green manure (4–5 t ha^{-1}). Certainly, leaving crop residues during grain harvest is a feasible practice in Kazakhstan but applications of manure or green manure are unlikely to be adopted.

SOM can be maintained using well known agronomic practices. On rainfed sieroziems the SOM content decreased from 1.2% to 1.09% after five years under the generally adopted practice of straw removal. Where straw was spread over the soil surface, SOM increased up to 1.4% while the application of 30 t ha^{-1} of manure once in five years resulted in an increase of SOM up to 1.6% (Omarov, Kozhakhmetiov, 2003).

In irrigated light chestnut soils in southern Kazakhstan, rotation of grains with sugar beet with no fertilization decreased SOM content by 0.2–0.3% but it did not change with the application of adequate fertilizer rates. In crop rotations of grains with alfalfa there was no reduction of SOM with or without fertilizers. On sloping croplands, annual soil losses by runoff amounted to 47 M t with a reduction in grain yield of 16–80%. Alfalfa should take an important share in crop rotations (30–35%) for soil fertility management in irrigated agriculture (Kenenbayev et al., 2004).

In southeastern Kazakhstan, during 40 years of irrigated agriculture SOM content in 0–20 cm and 20–40 cm layers reduced from 2.60–2.50% to 2.23–2.16% respectively in crop rotations including wheat, alfalfa and sugar beet and with no fertilization (Ramazanova et al., 2004). With commercial fertilization rates SOM decreased slightly to 2.50–2.35%. For complete soil fertility management with no losses of SOM it was necessary to apply both commercial fertilizers and manure at a rate of 40 t ha^{-1} once in an eight year rotation.

Soil erosion is a major factor for SOM management in the dry sub-tropics of Uzbekistan. Soil losses from water erosion may reach up to 500 t ha^{-1} on slopes with a gradient of 5° or more in a year including annual losses of SOM of 500–800 kg ha^{-1}, 100–120 kg of N ha^{-1}, 75–100 kg of P ha^{-1} (Gafurova et al., 2004). On areas subject to soil erosion by water, losses of nitrogen are equivalent to 70% of that applied with fertilizers and losses of phosphorus exceed the rates of fertilizer applied.

SOM may be lost by widespread practices of burning stubble. On typical sierozem soils a loss of biomass in terms of above ground and root residue as a result of burning of crop residue amounted to 35–38% while on dark sierozems the loss of biomass amounted to 57–59% (Riskiyeva et al., 2004). With the grain crop residue burnt, losses of nutrients amounted to 100–160 kg of N ha^{-1}, 900–1400 kg of C ha^{-1}.

The main causes of land degradation in mountain areas are soil erosion by wind and water, soil salinization and water-logging (Shokirov, 2001). Intensive development of soil erosion with gulley formation has been observed in all natural zones of Tajikistan. The main reason for soil salinization is unsatisfactory maintenance of collector and drainage systems with as much as 30% of horizontal and vertical drains non-functional. The area covered with forest decreased from 392,100 ha in 1988 to 171,500 ha in 2000 (Shokirov, 2001) as a result of uncontrolled wood cutting. This status is a characteristic feature of all mountain countries in Central Asia and the Caucasus. Poor ecological status of soil cover is a result of inadequate soil management by farmers who have only short-term outputs in mind and who use monocultures (Jumabekov, 2001). Thus 70% of soil area is subjected to erosion including 80% of the cropland in Kyrgyzstan. Intensive use of rainfed foothill grey-brown soils in Azerbaijan over 50–100 years under 400–500 mm annual rainfall reduced SOM contents in the 0-25 cm layer from 2.5–3.8% to 1.5–2.0% (Gasanov et al., 2001).

Land management using sowings of perennial forages is an efficient practice to maintain soil OM. Studies on sloping lands in Tajikistan with alfalfa in row spacings between almond rows and irrigated by sprinklers, indicated an increase of SOM in 0–50 cm layer from 0.27–1.62% (Kabilov and Sanginov, 2001). Also, annual mulching with straw increased SOM significantly. Studies in Tajikistan demonstrated that application of green manure for maize (mixture of clover with rapeseed) resulted in a reduction of nitrogen application rates by 50 kg ha^{-1} (Rashidov et al., 2001).

There is a varying attitude to the role of shelter belts in dryland agriculture in the different areas of the steppe belt. In the NK, the shelter belts were not recommended because they were believed to contribute to uneven snow distribution in between windbreaks (Barayev, 1960). In WS, some scientists favored shelter belts but it was not a widely adopted practice (Kashtanov, 2003). In VA, shelter belts have been recognized as important part of farming systems in dryland agriculture (Arkhangelskaya, 2005).

3.2 *Range management*

Although arid and semi-arid ecosystems are known to have substantially lower productivity than forests, we hypothesize that the vast landscapes of Central Asia, dominated by rangeland ecosystems, could be an important contributor to the "missing carbon sink". Thus we studied the daily magnitudes and growing season dynamics of net ecosystem CO_2 exchange (NEE) in representative rangelands.

The magnitude of daily NEE at the peak of growing season was highest in the steppe site at Shortandy as compared to semi-desert and desert sites. Results from 1998–2000 measurements of daily and growing season NEE indicated that rangelands of Central Asia are potential sinks for atmospheric CO_2. Among the three study sites, differences in magnitude of 20-min and daily NEE at peak season were consistent with variations in CO_2 sink capacity for the whole growing season. Long-term precipitation data from nearby weather stations indicated that mean annual precipitation at Shortandy was much higher than in arid steppe in Uzbekistan and desert in Turkmenistan where parallel studies were conducted. This trend of annual precipitation agreed with the pattern of total NEE for the growing season. This suggests that annual precipitation is a key environmental factor influencing the capacity of Central Asian rangelands to act as sinks for atmospheric CO_2(Saliendra et al., 2004).

Kazakhstan has a huge rangeland area of 180 M hha, which could be a carbon sink of global importance under proper management. During the transition period from centrally commanded to market oriented economical policies, livestock and rangeland management were greatly affected. During the first period of reforms from 1992–1998 small ruminant numbers were reduced dramatically from 34 to 10 M heads as a result of fragmentation of large farms and distributing livestock to small holders. Small ruminant populations during 1928–1934 decreased from 19.2 M to 2.3 M (Asanov et al., 2003). In six years dedicated to collectivization, Kazakhstan lost about 17 M heads of sheep and goats, while during six years of de-collectivization the country lost 24 M heads of sheep and goats.

The striking problem of new production systems emerging after the break up of centrally organized farming system was the failure to adequately and timely use the rangelands – the cheapest feed source although reduced livestock herds were left on the vast rangeland area. Large flocks of small ruminants used to be managed by different seasonal and rotational grazing strategies, some involving the movement of flocks through several thousand of km in the year to graze seasonal ranges allocated by government for each large state farm. Nowadays the expansive movement of flocks by small producers is virtually impossible, as this is not economical. As a consequence animals roam nearby ranges around the villages causing intense degradation of rangelands nearby villages and leaving remote rangelands ungrazed.

The consequence of overgrazing is the decreased productivity around villages even in spring when fodder availability is greatest. Fodder availability is 1.5 times greater at 6 to 12 km away from homesteads and 2.5 times in remote ranges (Iniguez et al., 2004). Such conditions apply to all countries, even those where the reforms were not applied extensively, and are particularly critical during fall and winter. The consequence of this is direct land degradation and the rise of conflicts near farms due to increased competition for resources.

To benefit from adequate feed resources access to the remote ranges should be considered urgently. Community action is required in order to overcome the financial limitations for individual farmers. This can be achieved by the promotion of associations for collective range grazing and consolidation of flocks.

The other area which may play important role in carbon sequestration is marginal lands used during the Soviet period for grain production and nowadays abandoned. During 1992–1998 over 10 M ha were left as weedy fallows. The major reason for this strategy was the low productivity of these lands. But the other reason was low resource capacity of agricultural producers who were not able to cultivate these lands. Over the last five-six years the grain industry is recovering and is becoming a profitable business. Hence, some profitable companies would like to put back these marginal lands to grain production when grain prices will be favorable. Therefore, it is a great challenge to introduce government policies to prevent the use of the abandoned marginal lands for grain production.

Overstocking of mountain rangelands and intensive haymaking reduced plant cover by 36–48% and reduced rangeland productivity by 30%, followed by reduced percentage of edible plants and increased share of poisonous and non-edible plants. The process has been exacerbated in recent years as a result of increased use of shrubs for fuel (Aknazarov, 2001).

In the rainfed area of the Gissar valley, 93% of rangelands were found to be affected by erosion, including 45% severely eroded and 38% moderately eroded (Sanginov and Aliyev, 2001). In southern Tajikistan, a survey conducted in 1969–1990 indicated that 60% of rangeland soils were subjected to erosion, including 12.7% very severely eroded with very poor productivity (Boboyev, 2001).

4 CHALLENGES

In different ecosystems the potentials for carbon sequestration and the challenges can be summarized as follows:

- In semiarid steppe belt: a reduction of summer fallow and its elimination in the future, reducing tillage to minimum with the final goal of zero tillage, introduction of shelter belts in accordance with recommendations of forestry science.
- In dry steppe belt: reducing summer fallow, replacing clean summer fallow with chemical, covered and partly occupied types of fallow, reducing tillage to minimum with the final goal of zero tillage, introduction of shelter belts in accordance with recommendations of forestry science.
- In semi-desert and desert rangeland areas: introducing better grazing management including limiting stock numbers to prevent rangeland degradation, establishing shelter belts in accordance with recommendations of forestry science, introducing policies preventing using of shrubs and trees for fuel as well as using marginal lands for crop production.
- In irrigated agriculture: addressing issues of soil salinization through better management of irrigation and drainage systems, increasing crop diversification by adopting more food and forage legumes into crop rotations, introducing conservation agriculture practices preventing soil erosion and maintaining soil fertility.
- In mountainous areas: introducing policies preventing intensive forest cutting for timber and for fuel, introducing integrated feed and livestock production to prevent overgrazing of rangelands, introducing conservation agriculture approaches including no-tillage, strip cropping, mulching and terracing to prevent soil erosion.

REFERENCES

Arkhangelskaya, G.P. and O.I. Zhukova. 2005. Ecological potential of forestations of the Volgograd province. p. 48–54. In: Issues of the Volgograd State University. Proc. of the Round Table, (Rus).

Aknazarov O. 2001. Desertification of anthropogenic landscapes of Pamir and measures to soften its negative consequences. p. 143–145. In: Proceedings of Soil Science Congress of Tajikistan, , Dushanbe, Tajikistan (Rus).

Asanov, K., B. Shakh, I. Alimayev, and S. Prianishnikov. 2003. Pasture Farming in Kazakhstan (Textbook). Gylym publishers, Almaty, Kazakhstan translated into English.

Austenson, H.M. 1988. A continuing reexamination of summerfallowing in Saskatchewan. p. 814–816. In: Challenges in dryland agriculture. A global perspective. Proc. of Intl. Conf. on Dryland Farming, Amarillo, Bushland, Texas.

Babayev, A. 1995. Problems of desert development. Ylym Publishers, Ashgabat, Turkmenistan (Rus).

Barayev, A. 1960. Cropping structure and crop rotations in grain farming of northern Kazakhstan. J. Zemledeliye, #7 (Rus).

Boboyev, R. 2001. Ecological aspects of soil status of rangeland territory of Southern Tajikistan. P. 288–290. In: Proc. of Soil Science Congress of Tajikistan. Dushanbe, Tajikistan (Rus).

Gafurova L., K. Mirzazhanov, K. Makhsudiv, and O. Khakberdiyev. 2004. Influence of mineral fertilizers on winter wheat on eroded sierozem soils. p. 183–187. In: Proc. of First National Conference on Wheat Improvement and Agronomy Technologies in Uzbekistan. Tashkent State University, Uzbekistan (Rus).

Gasanov, B., N. Novruzova, R. Aslanova, and C. Galandarov. 2001. Low-productive rainfed soils of foothills of Azerbaijan and restoration of their fertility. p. 14–150. In: Proc. of Soil Science Congress of Tajikistan. Dushanbe, Tajikistan (Rus).

Gnatovskiy V. 2003. Some peculiarities of adaptive landscape farming systems in Kulunda zone of Altay area. p. 86–93. In: Proc. of Intl. Conf. on Development of soil conservation practices in new socioeconomic conditions. Shortandy, Kazakhstan (Rus).

Gilevich S. 2003. Crop diversification in crop rotations of Northern Kazakhstan. p. 239–250. In: Proc. Intl. Conf. on Development of soil conservation practices in new socioeconomic conditions. Shortandy, Kazakhstan. (Rus).

Iniguez, L., M. Suleimenov, S. Yusupov, A. Ajibekov, M. Kineyev, S. Kherremov, A. Abdusattarov, and D. Thomas. 2004. Livestock production in Central Asia: Constraints and Research Opportunities. p. 278–302. In: J. Ryan and J. Vlek, and R. Paroda (eds.). Proc. of a Symposium on Agriculture in Central Asia: Research for development. ASA Annual Meeting, Indianapolis, Indiana. ICARDA, Aleppo, Syria.
Kabiliov, R., and S. Sanginov. 2001. Changes in organic matter and nitrogen content in sloping lands under long-term cropping irrigated with sprinklers. p. 312–313. In: Proc. of Soil Science Congress of Tajikistan. Dushanbe, Tajikistan (Rus).
Kashtanov, A.N. 2003. A.I. Barayev and modern agriculture. p. 3–7. In: Proc. of Intl. Conf. on Development of soil conservation practices in new socioeconomic conditions. Shortandy, Kzakhstan (Rus).
Khramtsov, I., and V. Kholmov (eds.), 2003. Farming practices on lowland landscapes and agro-technologies in Western Siberia. Novosibirsk, Russia, 412 p. (Rus).
Kenenbayev, S., A. Iordanskiy, and K. Balgabekov. 2004. Main outcomes of research on soil fertility management. p. 66–80. In: Proc. of Conf. "70 years to Scientific Production Center of Soil and Crop Management." Nurly Alem Publishers, Almaty, Kazakhstan (Rus).
Medvedev, I., N. Levitskaya, and I. Ryabova. 2003. Resources availability and agroecological sustainability of various agro-landscapes in black soil zone of Volga area. p. 104–116. In: Proc. Intl. Conf. on Development of soil conservation practices in new socioeconomic conditions. Shortandy, Kazakhstan (Rus).
Omarov, B., and M. Kozhakhmetov. 2003. Soil biologization – to conservation agriculture. p. 297–305. In: Proc. Intl Conf. on Development of soil conservation practices in new socioeconomic conditions. Shortandy, Kazakhstan (Rus).
Paroda, R., M. Suleimenov, H. Yusupov, A. Kireyev, R. Medeubayev, and L. Martynova. 2004. Crop diversification for dryland agriculture in Central Asia. p. 139–151. In: S. Rao and J. Ryan (eds.). Challenges and strategies of dryland agriculture. CSSA Special Publication 32, Madison, WI.
Ramazanova, S., G. Baymaganova, E. Suleimenov. 2004. Agrochemistry research in Kazakh Research Institute of Soil Management. p. 80–89. In: Proceedings of Conference "70 years to Scientific Production Center of Soil and Crop Management." Nurly Alem Publishers, Almaty. (Rus).
Rashidov, Kh., D. Gaziyev and S. Gaforov. 2001. Influence of green manure on soil fertility and grain yields of maize on stony soils of northern Tajikistan. p. 328–330. In: Proc. of Soil Science Congress of Tajikistan. Dushanbe, Tajikistan (Rus).
Rennie, D.A., G.J. Racz, and D.K. Mcbeath. 1976. Nitrogen losses. p. 325–353. In: Proc. of the western Canada nitrogen symposium. Alberta Agriculture, Edmonton, Alberta, Canada.
Riskiyeva, Kh., A. Bairov, R. Riskiyev, and M. Mirsadykov. 2004. Change of fertility and ecologic status of soil as affected by stubble burning. p. 325–353. In: Proc. of First National Conf. on Wheat Improvement and Agronomy Technologies in Uzbekistan. Tashkent State University, Uzbekistan (Rus).
Saliendra, N.Z., D.A. Johnson, M. Nasyrov, K. Akshalov, M. Durikov, B. Mardonov, T. Mukimov, T. Gilmanov, and E. Laka. 2004. Daily and growing season fluxes of carbon dioxide in rangelands of Central Asia. p. 140–154. In: J. Ryan, P. Vlek, and R. Paroda (eds.). Proc. of a Symposium on Agriculture in Central Asia: Research for Development. ASA Annual Meeting, Indianapolis, IN, 2002. ICARDA, Aleppo, Syria.
Sanginov, S. and I. Aliyev. 2001. Ecologic and soil quality conditions and factors of desertification in Gissar valley. p. 152–160. In: Proceedings of Soil Science Congress of Tajikistan. Dushanbe, Tajikistan (Rus).
Shashkov, V. 2003. Integrated weed management in grain crops. p. 274–279. In: Proc. of Intl. Conf. on Development of ideas of conservation agriculture in new socioeconomic conditions. Shortandy, Kazakhstan, (Rus).
Shnider, V. 2004. Large Grain Farm in Siberia. D. Vermel (ed.). Moscow. 27 p.
Shramko, N. 2003. Role of crop rotations in addressing issues of biologization and ecologization of conservation agriculture. p. 145–153. In: Proc. of Intl. Conf. on Development of soil conservation practices in new socioeconomic conditions. Shortandy, Kazakhstan. (Rus).
Shokirov, U. 2001. Status of land resources of republic Tajikistan, its conservation and rational use. p. 22–26. In: Proc. of Soil Science Congress of Tajikistan. Dushanbe, Tajikistan (Rus).
Shiyaty, E. 1996. Three years grains-fallow rotations in steppe of Kazakhstan. J. Zemledeliye, Number 6 (Rus).
Suleimenov, M.K. 1988. About theory and practice of crop rotations in Northern Kazakhstan. J. Zemledeliye, Number 9 (Rus).
Suleimenov, M. 2004. Farmer's folly. In newspaper "Kazakhstanskaya Pravda." 3 September, 2004, Astana, Kazakhstan (Rus).
Suleimenov, M. and K. Akshalov. 2004. Issues of transition from grains-fallow rotations to no-fallow crop rotations. p. 33–39. In: V. Loshakov (ed.). Proc. Intl. Conf. on Crop rotation in modern farming practices. Moscow Agricultural Academy, Moscow. (Rus).

Tuleubayev, T. 2003. Main factors of sustainable development of agriculture in Kostanay region in new socioeconomic conditions. p. 127–140. In: Proc. Intl. Conf. on Development of soil conservation practices in new socioeconomic conditions. Shortandy, Kazakhstan (Rus).

Viurkov, V. 2003. Crop rotations of conservation agriculture of Ural area. p. 116–127. In: Proc. of Intl. Conf. on Development of soil conservation practices in new socioeconomic conditions. Shortandy. (Rus).

Zhunusov, K. 2003. Influence of slope exposition on runoff and soil losses. p. 83–86. In: Proc. Intl. Conf. on Development of soil conservation practices in new socioeconomic conditions. Shortandy. (Rus).

Zhumabekov, E. 2001. Soil fertility management in Kyrgyzstan. p. 43–44. In: Proc. Soil Science Congress of Tajikistan. Dushanbe, Tajikistan (Rus).

CHAPTER 13

Dynamics of soil carbon and recommendations on effective sequestration of carbon in the steppe zone of Kazakhstan

A. Saparov
Soil Science Institute, Academgorodok, Almaty, Kazakhstan

K. Pachikin, O. Erokhina & R. Nasyrov
Soil Research Institute, Academgorodok, Almaty, Kazakhstan

1 INTRODUCTION

Maintenance of the carbon (C) balance in the atmosphere is one of the most urgent global ecological problems of the 21st century. Increasing concentrations of carbonic acid in the lower layers of the atmosphere are caused not only by industrial emissions but also by the irrational use of natural resources. This leads to the destruction of natural biogeochemical cycles of ecosystems, with the destructive processes of biomass transformation by mineralization occurring more frequently than the productive processes of photosynthesis and humification. Uncontrolled economic activity causes soil alienated through degradation from the biological cycle, and inefficient agricultural systems deplete the soil fertility, which is determined by humus content and soil C levels, among other variables.

The steppe zone is one of the regions in The Republic of Kazakhstan (RK), where the problem of soil degradation is especially serious. Factors responsible for intensive development of agriculture in the region include favorable natural-climatic conditions and sufficient availability of water resources.

During the first half of the 20th century, the local population had traditionally used these regions as highly productive pastures. However, these ecosystems were converted to arable lands following either natural ecosystems or pastoral land use. The area of arable lands in the steppe zone of the RK comprised 29.3 million ha (Mha) or 75% of the total arable land of the RK by the early 1990s. Despite the lack of profitable grain production, these lands intensively farmed until the disintegration of the former Soviet Union. The long-term exploitation of lands under minimal field management and the intensive application of mineral fertilizers and herbicides caused a drastic decline in soil C levels, and acceleration of erosion and deflation processes. Consequently, the plowed layers of cultivated soils in the steppe zone of RK lost up to 30% of their antecedent C pool in comparison with the antecedent level in 1954.

The area under arable land sharply declined in the 1990s, primarily due to the economic difficulties during the transition period, and because of a considerable decline in soil fertility caused by soil degradation. Large land areas used for pastures are now set aside as lay lands overgrown to varying degrees by natural fallow vegetation. Although these lands are covered by weeds and unpalatable species not suitable or grazing, they are still categorized as arable lands.

Biological diversity of the unique steppe landscapes has strongly declined. The specific structure of vegetation has changed because of the overgrazing of pastures. Rich virgin steppes have been transformed into secondary weed communities, which have lead to an appreciable decline in soil C pool.

The problems of land resources management for a steady development of the regions, control of soil cover under diverse ecological conditions, conservation and enhancement of soil fertility, and prognosis of soil anthropogenic transformations are of special importance at the dawn of the 21st century. There is a pressing need to develop scientifically proven recommendations on the stabilization and increase of soil C content, as a parameter, defining soil bio-productivity, and finally, social and economic well-being of a society. Therefore, the objective of this chapter is to describe the impact of land use conversion on soil C pool, and identify soil management options to restore degraded ecosystems and enhance soil C pool in different ecoregions of RK.

2 PRINCIPAL ECOREGIONS AND FACTORS DETERMINING THE DYNAMICS OF C POOL IN SOILS OF THE STEPPE ZONE

The steppe zone covers the northern and central part of RK and occupies an area of 126.2 Mha or 42.6% of the total land area. The large land area encompasses a variety of landscapes and diverse soils.

The northern part of the steppe zone, located in the West-Siberian lowland, is characterized by the lowest absolute heights and flat terrain. Oval uplands and plains (Obshy Syrt, Poduralskoye plateau, Zauralskoye plateau, Turgai tableland) prevail in the western part of the zone. These are alternated by denudation plains or plateaus (Kazakh hummocky topography) with the massifs of insular low mountains (Ulutau, Kokchetau, Bayan-Aul, Karkaralinsk and other) in the central and eastern regions.

Zonal sequence of soils, namely progression of dry conditions from the north to the south, is the principal factor that determines the heterogeneity of soil cover in the steppe zone of RK and, the resultant C pool in different soils. Furthermore, provincial peculiarities, connected with lythology-geomorphologic conditions of soil cover formation influence the spatial position of zonal boundaries. The vertical zonality is limited, and is primarily located on the insula mountains of the Kazakh hummocky topography and mountainous Mugodzhary.

From the soil-geographic point of view, the steppe zone of RK comprises five distinct sub-zones: regular chernozems of moderately arid steppes, southern chernozems of arid steppes, dark-chestnut soils of moderately-dry steppes, chestnut soils of dry steppes and light-chestnut soils of desert steppes (Figure 1).

The climatic conditions of soil cover formation, the mean value of principal parameters shown in Table 1, appreciably differ within the specified sub-zones, but have several common features.

Average annual air temperature is positive for the whole steppe zone with natural increase to the south with the corresponding decrease in annual precipitation. Average annual amplitudes of temperature vary widely. The warmest month is July (20–22°C), and the coldest is January (−17–20°C).

The depth of soil freezing in winter is the greatest in the central part of the zone (up to 2 m); it is connected with a small amount of autumn precipitation, the lowest snow cover and increased heat conductivity of the soils. Average parameter for the zone is 120–150 cm.

Precipitation is more frequent during the warm period of the year, and up to 70–80% of the annual amount is received during the summer. Thus, it leads to a significant loss of water by evaporation. Snow cover in winter is low throughout the region, and a rapid increase of temperatures in the period of snow melt after a significant freezing of soils in winter is accompanied by a redistribution and drainage of thawed snow into the low lying valleys of the landscape. These climatic peculiarities create diverse soil covers and structures along with a wide range of soil combinations (complexes, combinations, mottles), connected with a different meso- and microrelief soil occurrence, increasing with the increase in climatic aridity.

The moderately arid and arid steppe zone of chernozems is characterized by a flat terrain, and it occupies about 25.3 Mha of land area. The vegetation cover is represented by grass-Stipa vegetation (decreasing in species and percentage ratio further south), and birch-aspen kolki are observed in the northern part of the zone in closed depressions of the relief. Regular and southern chernozems

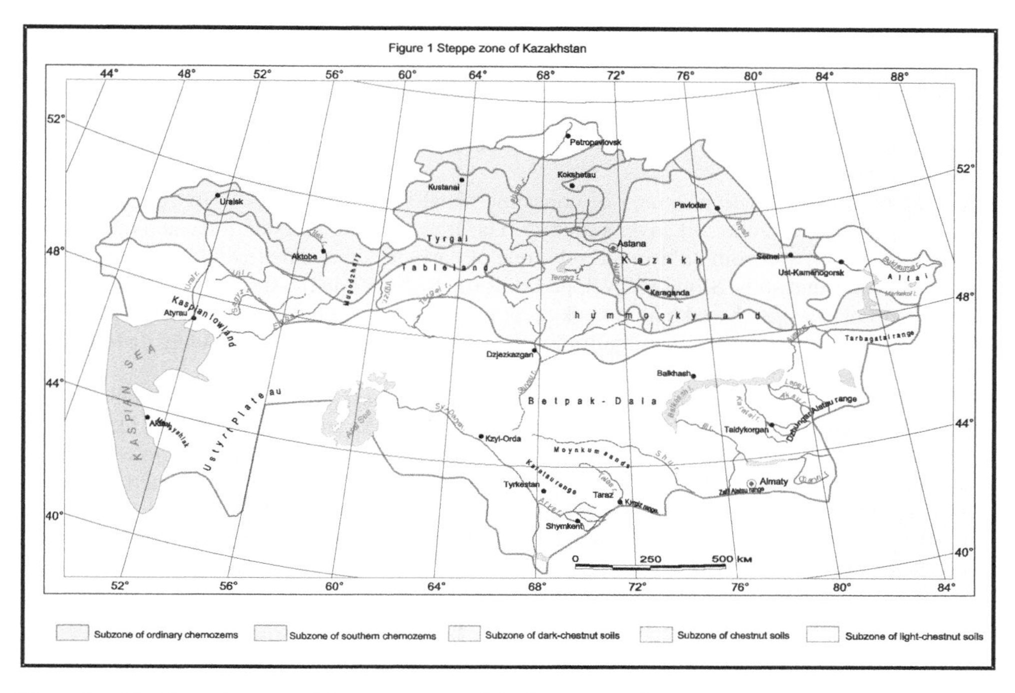

Figure 1. The five sub-zones of the steppe zone of the Republic of Kazakhstan (RK).

Table 1. Mean climate indices of soil sub-zones in the steppe region of Kazakhstan.

Sub-zones	Avg. T(°C)	Sum of T > 10°C	Frost free period (days)	Average annual amplitude of T (°C)	Depth snow cover (cm)	Annual Pct. (mm)	Pct. when T > 10°C (mm)	Hydrothermic coefficient for yearly warm period
Regular chernozems of moderately arid steppe	0.6–0.8	2070	114–118	37	23–26	320–350	220–230	0.9–1.0
Southern chernozems of arid steppe	1–1.5	2160	107–119	38	21–26	300–330	160–180	0.8–0.9
Dark-chestnut soils of moderate dry steppes	1.5–2	2310	128–131	39	20–23	270–300	150–170	0.7–0.8
Chestnut soils of dry steppes	2–4	2380	131–147	40	18–23	220–270	130–150	0.6–0.7
Light-chestnut soils of desert steppes	4–7	2490	131–158	40	16–21	180–210	90–100	0.4–0.5

(normal, calcareous and alkaline) are the dominating soil subtypes. These soils cover 60–65% of the whole territory, 20–22% are comprised of complexes of zonal soils with steppe alkali, and about 10% are hydromorphic and semihydromorphic soils (meadow-cherno-zemic, meadow, meadow-boggy). Alkaline soils (steppe, meadow-steppe and meadow) and solonchaks occupy only a small area of about 3–4%.

Moderately-dry, dry and desert-steppe zone of chestnut soils, located further south, occupy more than 90 Mha and are characterized by more complex surface structure. Xerophytic turf-like cereals (*Stipa, Festuca*) with small presence of grasses (*Linosyris, Potentila, Piretrum*) and different spices of *Artemisia* prevail in vegetation cover of zonal soils in the northern part of the zone. Presence of *Artemisia* L. in the vegetation cover increases further south with decline in vegetation cover from 60–80% to 40–50%.

The proportional reduction in area of nonsaline zonal soils from 40–50% to 10–20% further south is typical for the zone soil cover structure. The areas of complexes of zonal soils with alkali characteristics increase from 30% in the sub-zone of dark-chestnut soils to 45 to 50% in the sub-zone of light chestnut soils. Slightly developed soils (up to 7–15% in the structure of soil cover) are widely distributed within Kazakh hummocky topography. The area under solonchaks is about 3%. The soils of hydromorphic and semihydromorphic moisture regimes are not widely distributed in this zone.

The soil C pool changes according to the heterogeneity of soil cover structure, as the balance of organic matter characterizes each taxonomic soil unit down to the soil series level, as is determined by the complex soil-forming factors. The soil C pool is a parameter which reflects an equilibrium condition, and is and the result of the interactive effect of two diverse yet interconnected processes: humification and mineralization of plant residues.

The research data show that the ratio of annual average temperatures and the sum of precipitation, appropriate to moderately arid steppe sub zone of the regular chernozems, is the best for accumulation of carbon in soil. The soil C pool decreases because of low humidity and high temperatures especially when the hydrothermal coefficient is less than 1.

Vegetation cover is one of the most important factors that determine the amount and composition of organic residues input into a soil. These residues are the main source of C accumulation in soil. The above ground and root residues serve as a source of biomass for the humification and these may be considered as one of the parameters affecting potential soil fertility.

Reduction in the general stocks of vegetative biomass and corresponding C level in soil is observed with the changes in climate from moderately arid steppes to desert steppes. The research data show that the maximum amount of biomass (above ground and below ground) among the natural biomes of the steppe zone in RK accumulates under the grass-Stipa, Festuca-Stipa vegetation (580–600 c ha^{-1}), followed by that under Stipa-Festuca (470–490 c ha^{-1}), Artemisia-Festuca (350–360 c ha^{-1}) associations, and progressively declines with decrease in precipitation (Gromyko et al., 1964). The data on deccomposition show that only about 10% of the vegetation residues returned to the soil is humified and the remaining 90% of organic matter is mineralized into CO_2 (Titlyanova et al., 1998).

The root biomass is the principal factor determining C accumulation in soil. The principal component of vegetation biomass is represented by roots in the plant associations of the steppe zone. Due to the increase in climatic aridity, the ratio of above-ground to below-ground biomass is 10.3 for regular chernozems, 13.4 for southern chernozems, 16.0 for dark-chestnut soils, and 19.22 for light chestnut soils (Kolkhodzhayev, 1974).

The highest amount of C is stored in the upper horizons of the soil. Soil C stock in 0–20 cm layer contributes 40–42% of the total C stock to 1-m depth, and that in the 0–50 cm layer contributes 75% of the total C stock to 1-m depth (Vishnevskaya, 1974).

Soil C stock is not constant during a year, and varies among seasons. It decreases at the beginning of summer because conditions are favorable for mineralization of organic matter. The maximum intensity of decomposition is observed at soil moisture content of about 60–80% of the total moisture capacity, and temperature reaches about 30°C (Vishnevskaya, 1974). The loss of C in the upper root layer of 0–20 cm at those conditions can be as much as 15–18 t ha^{-1}, taking into account an intensive biomass increase in vegetation which uses nutrients or mineral nutrition for the vegetation growth (Ponomaryova et al., 1980).

The level of C in soil is stabilized at the end of summer, when the plant growth practically stops and the microbiological activity of soils decreases. At this time, new humic substances are not involved in the biological processes but they are bound with the mineral fraction of soil, forming strong organo-mineral compounds. Both root and the above-ground biomass debris and root secretions (organic acids, amino acids, lactose, ferments), the proportion of which makes up more than 10% of the vegetation biomass, are the sources of the increase in soil C stock by the end of summer (Ponomaryova et al., 1980).

The amplitude of seasonal fluctuations of soil C stock decreases with the increase in soil depth, and no change is observed below 50 cm depth. Together with common bioclimatic conditions, lythology-geomorphologic conditions of soil occurrence and the moisture regime strongly influence the soil C stock. C stock in zonal soils of one sub zone differs naturally among soils of different texture. It is high in soils of heavy texture (clayey, heavy loam) and low in those of light texture. Decrease of C stock in light-textured soils is caused by an intensive decomposition of organic matter and weak fixation of newly formed substances by the mineral component of the soil.

The variability in the C stock is caused by soil characteristics as influenced by the forms of topography in conditions of a dissected terrain. It is natural that the C stock is lower on the slopes due to the accelerated erosion, and the soils that occur on the concave slope accumulate more organic matter transferred from the adjacent upslope soils.

The average values of the C stock in principal soils of the steppe zone of RK are given in the Table 2. The data in Table 2 show that the C stock in the layer 0–50 cm in soils of the steppe zone in RK ranges within considerable limits, decreasing from the value of 136–170 t ha^{-1} for zonal soils of moderately arid steppe subzone up to 49–52 t ha^{-1} in the desert-steppe subzone soils according to the zonal changes of bioclimatic conditions of soil cover formation. Alkaline soils have the lowest level of inherent fertility and C stock among the zonal soils. It is due to poor hydrological and physical properties, and unfavorable conditions for vegetation growth and development.

Calcareous soils don't appreciably differ from the normal zonal soils in the humus content, but the C stock is somewhat higher either in the 0–50 cm layer or in the 0–20 cm layer. This trend in C stock is attributed to the heavy texture of calcareous soils. The lowest C stock values of 25–35% less than in old soils are typical for undeveloped soils which have a shallow humus horizon, scarce vegetation cover, and consist of a large skeletal fraction.

More favorable conditions for vegetation development exist where there is additional ground cover and surface wetness. These conditions promote the accumulation of organic matter in soils. As a result, the hydromorphic and semihydromorphic soils are rich in humus content and C stock in comparison with the automorphic zonal soils. The difference in C stock is higher under arid than humid conditions.

Salinization and extremely unfavorable hydrological-physical properties are the limiting factors to the development of vegetation in alkaline soils. Thus, C stock in these soils is low, and is in the range of 30–75 t ha^{-1}, which is 40–50% less than that of the normal zonal soils.

3 DYNAMICS OF C STOCK IN SOILS UNDER AGRICULTURAL USE

The main factor which changes the soil C balance is the conversion of steppe zone soils to agriculture. The most cardinal changes are brought about by the agricultural land use and related activities. The steppe zone of chernozems is characterized as non-irrigated agricultural zone for grain crop production. The subzone of regular chernozems is relatively stable, but the subzone of southern chernozems is not stable. The frequency of droughts in the last subzone is 25%, or one in four years. Regular and southern chernozems are the best arable lands among the zonal soils of the steppe zone in RK, and they are almost entirely cultivated for grain crop production.

The subzones of dark-chestnut and chestnut soils, or the so-called zone of "risky agriculture", are characterized by climatic instability. Frequency of droughts ranges from 35 to 50% or once in two or three years. Thus, agronomic yield decreases by 3–4 times in case of the extended period of droughts. Sometimes, there is a total crop failure. A wide spread problem of alkalization in those soil covers is a limiting factor to agricultural development.

Table 2. Carbon stock in soils of the Kazakhstan steppe zone.

	Carbon stock				
	0–50 cm layer			0–20 cm layer	
Soils	Type of land	($t\,ha^{-1}$)	% of virgin soil	($t\,ha^{-1}$)	% of virgin soil
Regular normal chernozems	Virgin	165		98	
	Arable	149	90	83	85
Regular calcareous chernozems	Virgin	170		85	
Regular alkaline chernozems	Virgin	136		78	
	Arable	120	88	68	87
Regular slightly developed chernozems	Virgin	107		73	
Southern normal chernozems	Virgin	115		75	
	Arable	103	90	63	83
Southern calcareous chernozems	Virgin	132		56	
	Arable	113	86	48	85
Southern alkali chernozems	Virgin	101		60	
	Arable	90	89	52	86
Southern slightly developed chernozems	Virgin	93		51	
Meadow-chernozemic typical	Virgin	192		134	
Dark-chestnut normal	Virgin	102		52	
	Arable	91	89	43	82
Dark-chestnut calcareous	Virgin	117		60	
	Arable	98	85	50	83
Dark-chestnut alkali	Virgin	93		47	
	Arable	86	92	43	90
Dark-chestnut slightly developed	Virgin	95		54	
Chestnut normal	Virgin	75		40	
	Arable	68	90	32	80
Chestnut calcareous	Virgin	73		36	
	Arable	62	85	30	83
Chestnut alkaline	Virgin	62		33	
Chestnut slightly developed	Virgin	51		33	
Light-chestnut normal	Virgin	53		28	
	Arable	49	92	22	78
Light-chestnut calcareous	Virgin	5245		30	
	Arable		86	24	80
Light-chestnut alkaline	Virgin	49		25	
Light-chestnut slightly developed	Virgin	46		31	
Meadow-chestnut typical	Virgin	131		75	
	Arable	114	87	70	93
Meadow typical	Virgin	257		145	
Alkali steppe	Virgin	74		46	
Alkali desert-steppe	Virgin	34		21	

Table 3. Distribution of arable lands in the sub-zones (Adapted from Dyusenbekov, 1998).

	Area (Mha)		Arable land (% of total area)	
Sub-zones	Of the whole sub-zone	Arable lands in 1977	Of sub-zones	Of all arable Kazakhstan land
Regular cheronozems of moderately arid steppe	11.7	7.3	59.8	24.3
Southern chernozems of arid steppe	13.6	7.2	52/9	24.0
Dark-chestnut soils of moderately dry steppes	27.7	10.0	36.1	30.3
Chestnut soils of dry steppes	24.3	3.6	14.8	12.0
Light-chestnut soils of the desert steppes	38.4	1.5	3.9	5.0

The lands of the desert-steppe zone are mainly used as pastures due to extreme aridity of the climate. Frequency of droughts is up to 75%, or in 3 out of 4 years. Normal zonal soils (without alkalinity and salinization) are not widely distributed and are attributed to the parent material of light textured mechanical composition. The areas of arable lands in the steppe zone of RK, percent of total area and distribution in the subzones, are given in Table 3.

The anthropogenic changes in cultivated soils are mainly related to the decrease in humus content, caused by mixing of the surface horizons, erosion and deflation of humus with the fine earth, humus mineralization and removal of nutrient elements by harvested crops. It is also caused by mechanical disturbance of surface horizon (A and partially B1) structure and strengthening or compaction of subsurface horizon. The research data show that losses of C occur in arable non-irrigated soils and in the sub-soil layer, up to 1-m depth. These losses essentially vary depending on the soil morphological properties, relief, crop rotation, tillage system, application of fertilizers and can be as much as 30% in the plow layer (Rubinstein and Tazabekov, 1985). On average, chernozems have lost 19–22% and chestnut soils 28–30% of humus over 50 years of cultivation. Average annual loss of C is about 4.6–5.8 t ha^{-1} (Akhanov et al., 1998).

Soil C stock in natural ecosystems, amount of organic matter formed by biomass and release of C into the atmosphere by mineralization of plant residues, is at a steady state, and the processes of organic matter accumulation are unremitting as natural biomes consist of plants of different phenological rhythms.

The ecological regime of soils sharply changes under arable land use. The rhythm of agroecosystem production process is set artificially, and the interaction of plants with soil is limited in time. A considerable part of organic matter of agroecosystems is removed irreversibly, and the biochemical equilibrium is disturbed, leading to the depletion of soil C stock. Up to 50–60% of organic matter is removed with harvests and economically useful production in agroecosystems (Titlyanova et al., 1998).

An accelerated decomposition of organic matter occurs in arable lands. It is caused by the change in water, air and temperature regimes, strengthening of microbiological activity, and weakening or breakdown of soil aggregates. The major part of plant residues of the former virgin or climax vegetation is decomposed in 8 years of cultivation (Gromyko et al., 1964). This process occurs more intensively on fallow fields. Inactive or recalcitrant humus fraction is involved in the biological cycle; nutrient elements are transformed into an available form, absorbed by plants, and removed with crop harvests.

The research data shows that the roots of the crops in the 0–50 cm layer of cultivated soils decompose 1.4–1.6 times faster than roots of vegetation in virgin soils (Vishnevskaya, 1974). The labile component of C stock increases, because the biomass input of the cultivated vegetation

(120–140 c/ha) is lower than that of the virgin soil (350–360 c/ha for the dark-chestnut soils). The vegetation biomass of a fallow field, represented only by the roots of the pervious crop, is about 70–90 c/ha (Gromyko et al., 1964), and the losses of C in this system are especially high and may be as much as 20–25% (Titlynova et al., 1988; Aderikhin, 1964; Akhanov, 1996). For example, it was observed that the upper 20-cm layer of regular chernozem lost 30.2 t ha^{-1} of C after 12 year of fallowing in comparison with the virgin soil (Ponomarkova, 1980). The sharp fluctuations of hydrothermal conditions (alterations between dry hot and warm humid periods) accelerate the mineralization processes. It's typical for the climate of the steppe zone.

The rate of loss of C in agroecosystems eventually decreases and it is related to the decrease in humus level. At the same time, the humus, left after the long-term cultivation, changes by its group and fractional composition and is characterized by a high proportion of recalcitrant fraction. The recalcitrant form of humus is attributed to strong chemical bonding with the mineral fraction of soil. The new steady state equilibrium level is attained in about 25 years (Ponomrayova et al., 1980).

The data in Table 2 on the C stock in virgin soils permit estimation of the C losses caused by cultivation in the soils of the steppe zone in RK in the 0–50 cm and 0–20 cm layers. The losses of C in the 0–50 cm layer for normal zonal soils under cultivation are small and consist about 9–10% of the total stock, but increase in alkali (11–12%) and calcareous (14–15%) soils. The loss of C in the surface plow layer (0–20 cm) more strongly reflects its interrelation with inherent properties of soils, caused by type and genera differences.

The losses of C in normal arable soils are likely to accelerate with increase in climatic aridity – from 15% for regular chernozems up to 23% for light-chestnut soils. Slightly lower values of C losses (15–20%) in the calcareous compared with the normal soils are caused by decrease in humus solubility in calcareous soils. Further, a sharp decrease of humus with depth is not so typical of alkaline soils as is the case for normal and calcareous soils because of a high-absorbing capacity of the alkaline horizons. It slightly decreases the negative effects of mixing in the 0–20 cm layer caused by tillage operations. The loss of C in the 0–20 cm layer in this type of soil exceeds the losses in the 0–50 cm layer, and can be as much a 10–13% of the total C stock.

A high loss of C in the 0–50 cm layer (up to 23–25%) and in the 0–20 layer cm (about 5%) is typical for intrazonal arable soils of semihydromorphic and hydromorphic types. It is related to the trasport of the soluble humus components down the profile.

The climatic factors of the steppe zone (e.g., shallow depth of snow cover and deep freezing of soils in winter, large fluctuations of seasonal and diurnal air temperatures, frequent occurrence of soil and atmospheric drought, high wind velocity, and low relative humidity) exacerbate soil's susceptibility to wind erosion. Ancient alluvial plains (Priirtysh plain, North-turgai, Podural plateau) and sand massifs are the areas of soil deflation by wind erosion in the steppe zone of RK. The decrease of C stock in cultivated soils of the steppe zone is caused by deflation of fine particles (less than 0.01 mm in diameter) enriched in organic matter. Thus, cultivated surface horizon becomes progressively sandy because these soils are not resistant to erosion. The C concentration decreases sharply, and often as much as 15–30% (Akhanaov et al., 1998).

In general, soils prone to severe erosion contain more than 50–60% of the fine fraction (<1 mm) in the surface layer (Barayev, 1964). The soils of light texture and calcareous soils (among the soils of heavy texture) are more prone to wind erosion than others. Wind erosion is especially severe in autumn, winter and early spring, when there is no protective vegetation cover on the fields.

Deflation of the cultivated soils has been almost prevented as a result of "soil protective system development" (Barayev, 1964). But water erosion still is also an important cause of soil degradation. Principal factors which increase surface runoff in soils of the steppe zone are intense but short duration of summer precipitation and heavy snow melt runoff on deeply frozen soils. In such conditions, water erosion occurs even on gentle slopes as low as 0.5 degrees. The arable land use on these soils covers 4.2 Mha. The area of the eroded arable lands has increased by 40% over the 25 years of measurements (Akhanov et al., 1988). Water erosion is not severe in soils of Podural plateau, Kokchetau upland, Obshy Syrt, residual denudation plains of the RK.

Excessive overgrazing of pastures is also an important factor that depletes the soil C stock. Overgrazing leads to disturbance of the natural vegetation cover, formation of the secondary weeds ecosystems, and complete destruction of vegetation cover in some places. This decrease in biomass and vegetal cover is inevitably accompanied by loss of soil C stock.

Vast areas of steppes are ploughed within the chernozem zone and subzone of dark-chestnut soils, and pastures are established on land unsuitable for agriculture, such as those with more dissected topography or widespread occurrence of salt-affected soils. Overgrazing causes pasture degradation, and exacerbates water erosion, especially on the edges of the river valleys and lake basins.

The area of arable lands is not large in the chestnut and light-chestnut subzones, but scarce vegetation cover, high wind activity, insufficient soil moisture and overgrazing exacerbate the degradation processes. Even a single run of cattle disturbs the surface cover of soil; and the fine dust thus created is easily removed by wind (4–5 m/sec) (Mukhametkarimov and Smailov, 2001). The intensive overgrazing causes C loss in the 0.10 m layer by as much as 15–18%.

4 ACTIONS ON STABILIZATION AND INCREASE OF C LEVEL IN SOILS

Soil fertility restoration and increase in soil C stock in the steppe zone of RK is a complex challenge. The first step is to use the soil cover inventory and identify the most degraded soils with initially low level of inherent fertility. These soils must be taken out of agricultural production farming.

It is important to maintain a sufficient C balance in cultivated soils by adopting the balanced crop rotations through replacement of fallows and fallow-intertilled crops by grasslands along with the application of organic matter including ciderates. Due to a high percentage of eroded and severely erosion-prone soils in the structure of arable lands in the steppe zone of RK, it is also important to establish erosion-prevention measures, adapted to zonal and regional nature-climatic and lythology-geomorphologic conditions.

During the middle of the 20th century, the processes of soil deflation caused the loss of soil C stock by cultivation of vast areas of virgin lands in the steppe zone of RK. In order to protect the soils from degradation, the All-Union Research Institute of grain management has developed a soil protective system of agriculture, the main method of which is a shallow (12–16 cm) surface soil tillage (Barayev, 1964). The speed of wind is reduced near the surface due to roughness of the surface and the residue cover. It protects fine particles rich in organic matter from deflation, enhances snow retention, and increases available soil moisture reserves. Elimination of moldboard plowing controls wind erosion in the steppe zone of RK. However, this method is not suitable for all the subzones of the steppe zone, especially if the problem is associated with the conservation of soil C.

More than 50% of soil C is conserved in chernozems after shallow surface cultivation in comparison with the moldboard plowing because the surface soil is loosened without being subjected to mixing. The 0–15 cm depth is the most active soil layer from the biological point of view (Kolkhodzhayev, 1974; Vishnevskaya, 1974). However, field experiments conducted on dark-chestnut and chestnut soils show that in arid conditions deep plowing (up to 30–35 cm) and turning over followed by non-moldboard tillage lead to the formation of a fertile buried horizon. The buried horizon conserves organic matter, and increases its C stock under favorable soil moisture conditions (Matyshyuk, 1964). It also promotes a prolific and deep root system development. The B horizon, brought to the surface, is low in humus, compact, less subjected to erosional processes than the surface horizons. Maximum benefits of deep plowing are normally observed in droughty years because more favorable conditions for the development of the root systems are created.

The system of agriculture, based on combination of different methods of moldboard and non-moldboard cultivation, periodic alternation of these operations, chosen with due consideration to the genetic peculiarities of the ploughed soils and climatic conditions of their formation is an optimum system of farming for the steppe zone soils of RK.

As has already been indicated, the surface erosion causes severe losses of C in the cultivated soils of the steppe zone. Therefore, adoption of some special actions depending on the degree of relief dissection and steepness of slopes, is necessary for conservation and enhancement of soil C stock. Periodic deep plowing (without turning over of low fertility horizons) gives good results on the fields with slopes of less than 4–5 degrees. It provides the destruction of furrow bottom. In addition, the following measures prevent soil erosion: deep plowing along with sowing of coulisses, non-moldboard cultivation to maintain crop residues cover, change of bare fallow to weed fallow, bedding and hole-digging, and introduction of perennial herbs in the rotations with herbaceous cover of up to 50%. Adoption of these practices increases soil moisture reserve by 15–25 mm in spring (Dzhanpeisov et al., 1978; Akhanov et al., 1998; Rubinshtein et al., 1964).

Contour plowing, bunding, chiseling, leveling of ephemeral gullies and installation of permanent buffer zones are necessary to reduce risks of soil erosion on slopes about 5–6 degrees. The optimum width of zone spacing is 400–500 m for chernozems, 300–400 m for dark-chestnut soils, and 200–300 m for chestnut soils (Dzhanpeison et al., 1978; Akhanov et al., 1998; Rubinshtein et al., 1964). Complete grassing of slopes and transformation of arable or lay-lands into hay lands, and afforestation of gullies and steep slopes are necessary for soils with slopes of >6 degrees.

Establishing forest shelter belts is one of the most important measures for conservation and restoration of fertility of cultivated soils in arid climate and high wind velocity in the steppes zone. Shelter belts can reduce the wind speed at 3 meters above the surface by 44–65%. Shelter belts also capture snow drifts which can be 30–70 cm deep along the tree line. Increase of snow cover thickness considerably reduces soil freezing in winter. Most soils in the open field are generally frozen to 120 cm depth. In contrast, the soil freezing occurs to 60–90 cm depth at the distance of 100 m from the forest belts and up to 15–24 cm depth in close proximity of the belts. The soil moisture reserve increases in the top 1 m by 114 mm after snow melt in the open steppe, and by 175, 225 and 252 mm at the distance of 100, 50 and 25 m from the forest belts, respectively.

Deposition of fine soil, blown by wind and enriched in organic matter, increases the thickness of the humus horizon. Crop yields increase by 20–40% within the belt zone of 250–300 m. There are also notable changes in soils within the belts where the C stock in the 0–40 cm depth is increased by 20–25% and may appreciably exceed the analogous parameters for zonal virgin soils after 8–10 year of planting trees and shrubs (Dzhanpesov et al., 1978; Rubinshtein et al., 1985; Sobolev et al., 1964).

A large development of irrigated agriculture is needed in the southern part of the sub-zone where the probability of droughts is very high. The high expenses for installation of the irrigation network and wells in the regions prone to water deficit are fully justified by the increase in agricultural production. Development of irrigation would raise the production efficiency of each hectare of arable land by 3 to 5 times in the dry steppe zone on dark-chestnut soils. Local supplemental irrigation enhances, through indirectly, the ecosystem C stock in the steppe ecosystems because it reduces the total cultivated land area and converts the surplus land to natural vegetation cover.

The strategy of restoring lay land, taken out of cultivation, is very important to enhancing the humus content. Self-restoration of technologically-disturbed soils is a slow process because of the low soil moisture reserve. Weed communities represent the dominant vegetation of the lay lands whose NPP is less than those of natural ecosystems during a long period of up to 10 or more years depending on the subzonal type. The root systems of weedy vegetation do not form a dense turf, and thus do not protect against the processes of water erosion.

Planting of the lay lands to perennial grasses is an effective method of enhancing the C stock. Application of organic matter or biosolids is also desirable. The residues, harvested from the virgin sites after the seeds mature, applied to the lay lands are very effective in the restoration of natural ecosystems. In this case, the seeds serve as a propagating material and the above ground parts of the plants serve as an organic amendment.

The pastures of the steppe zone also require radical improvement because these are also severely degraded. Grazing management is the first step, and overgrazing must be avoided. Pastures must be subdivided to facilitate rotational grazing in different seasons. Lowland steppes, sub-mountainous and piedmont regions must be grazed during summer. In contrast, sandy soils and the flood plains of large rivers must be grazed during winter. Rotational grazing and outlet restoration wedge must be used to minimize erosion, deflation and decline of ecosystem bioproductivity (Akhanov et al.,

1998). Grasses must be established on severely degraded pastures, and appropriate measures of soil deflation and erosion control must be implemented on the eroded pastures.

5 CONCLUSION

The steppe zone of RK covers a large area and comprises diverse bioclimatic conditions and complex of the surface structure which result in heterogeneous soil cover and a wide range of soil genetic properties. These diverse ecoregions lead to a wide range of soil C stock because the organic matter balance depends on soil type, and differs among down to the soil series level.

Total soil C stock decreases progressively from north to south. Changes in the C stock within the limits of subzones are caused by heterogeneity of soil cover as determined by lythology-geomorphological conditions and the soil moisture regime.

Dynamics of soil C stock is mainly determined by anthropogenic factors, in conjunction with the agricultural land use. Extensive system of agriculture and inappropriate expansion of arable lands, indiscrininate agrotechnical methods of soil tillage, high doses of chemical fertilizers and low rates of application of organic amendments, have degraded soil fertility and depleted C stock.

The steppe zones are a special type of natural environment. Specific processes of soil organic matter dynamics increase vulnerability of soil cover to anthropogenic perturbations. The risk of continuing C losses in soils is especially high because the self-restoration of the disturbed ecosystems in the arid climate of the steppe zone is extremely slow.

A revolutionary reorganization of the land resources management in the steppe zone of RK is necessary based on the principle of landscape-ecological agriculture and adoption of the principles which enhance the soil organic matter budget. It is necessary to consider the soil genetic properties while choosing an optimum system of agriculture. It is equally important to adopt soil protective technologies of vegetative cover, and to implement moisture conservation and erosion control measures.

REFERENCES

Aderikhin, P.G. 1964. Researches of chernozems fertility in the Central chernozemic zone under cultivation. Fertility and amelioration of soils in the USSR. Doklady for VIII International Congress of Soil Science. Nauka. p. 31–38.

Akhanov, Z.U., R.E. Yeleshev, T. Dzhalankuzov, M.I. Rubinshtein, and A.I. Iorgansky. 1998. Problems of soil fertility restoration RK. Condition and rational use of soils in the Republic Kazakhstan. Almaty. Tethys. p. 8–14.

Akhanov, Z.U., K.M. Pachikin, and A.A. Sokolov. 1998. Modern condition of soils and soil cover of Kazakhstan, their rational use, protection and objects of further study. Condition and rational use of soils in the Republic Kazakhstan. Almaty, Tethys. p. 4–8.

Akhanov, Z.U. 1996. Theoretical bases of soil fertility restoration in Kazakhstan/Izvestiya MS-AS RK. Ser. Biological. No. 3. p. 13–24.

Barayev A.I. 1964. Protection of soils from wind erosion. Fertility and amelioration of soils in the USSR. Doklady for VIII International Congress of Soil Science. Nauka. p. 223–230.

Dyusenbekov, Z.D. 1998. Land resources of RK, problems of their rational use and protection in the conditions of market economy. Condition and rational use of soils in the Republic Kazakhstan. Almaty, Tethys. p. 18–25.

Dzhanpeisov, R.D., A.K. Alimbayev, and K.B. Balgabekov. 1978. Development of soil deflation and erosion, scientific bases and practice in soil protection amelioration in Kazakhstan. Almaty. Nauka. p. 80–97.

Gromyko I.D., E.V. Kulakov, A.P. Mershin, and N.P. Panov. Biological cycle and fertility of chernozems and chestnut soils of Tseliny krai. Fertility and amelioration of soils of the USSR. Doklady for VIII International Congress of Soil Science. Nauka. p. 38–47.

Kolkhodzhayev M.K. 1974. Soils of Kalby and adjacent territories. Alma-Ata. Nauka. 254 p.

Matyshyuk I.V., 1964. Problems of cultivated soils formation in arid agriculture of Kazakhstan // Soil researches in Kazakhstan. Alma-Ata: AS Kaz SSR. p. 118–143.

Mukhametkarimov, K.M. and K. Smailov. 2001. Change of physoco-chemical properties of soils under different regimes of grazing on natural pastures // Scientific bases of fertility restoration, protection and rational use of soils in Kazakhstan. Almaty. p. 228–231.

Ponomaryova, V.V. and T.A. Plotnikova. 1980. Humus and soil formation. L.: Nauka. 1980. 221 p.

Rubinshtein, M.I. and T.T. Tazabekov. 1985. Anthropogenic changes of humus in arable soils of Kazakhstan / Archievements of Dokuchayev's soil science in Kazakhstan. Alma-Ata, p. 33–43.

Rubinshtein, M.I., B.K. Borangaziyev, and A. Bektemirov. 1985. Influence of forests field protective zones on the fertility of dark-chestnut soils in the west Kazakhstan // Fertility of soils in Kazakhstan. Issue 1. Alma-Ata: Nauka. p. 5–12.

Sobolev, S.S., I.D. Braude, and A.M. Byaly. 1964. Zonal systems of erosion control measures in the USSR in conditions of agriculture intensification. Fertility and amelioration of soils in the USSR. Doklady for the VIII International Congress of soil Science. M.: Nauka. p. 206–215.

Titlyanova, A.A., N.A. Tikhomirova, and N.G. Shatokhina. 1988. Productional process in agrocoenoses. Novosibirsk: Nauka. 185 p.

Uteshev, A. S. 1959. Climate of Kazakhstan L.: Gydrometeoizdat. 367 p.

Vishnevskaya, B.Y. 1974. Content and composition of organic matter. Southern chernozems of the north Kazakhstan. Alma-Ata: Nauka. p. 148–190.

CHAPTER 14

Carbon dynamics in Saskatchewan soils: Implications for the global carbon cycle

A. Landi
University of Ahwaz, Faculty of Agriculture, Department of Soil Science, Ahwaz, Iran

A.R. Memut
University of Saskatchewan, Department of Soil Science, Saskatoon, Saskatchewan, Canada

1 INTRODUCTION

The pedosphere has a significant influence on the gaseous composition of the atmosphere. Soil organic C (SOC) and inorganic C (SIC) are the two major pools in the pedosphere. The total amount of global SOC has been estimated at 1500 Pg, which is three times more than the biotic pool (living plants), and two times more than the atmospheric pool. Estimation on the total amount of SIC varies between 700–1700 Pg. This is at least partly due to the difficulty of separation between pedogenic and lithogenic carbonates (Lal et al., 1998). The lack of reliable information about the size and dynamics of the SIC pool is emphasized in the recent literature.

Lal et al. (2000a; b) expressed the importance of knowledge on the SIC pool, its cycling, and relationship to SOC within the global C cycle. The SIC sequestration occurs over tens to hundreds of thousands of years and plays a very significant role in the global C cycle (Mermut et al., 2000). Answers to the questions are quite important for complete understanding of the C cycle. Carbon sequestration can be highly cost effective and environmentally friendly mitigation technique (Mermut and Eswaran, 2001).

The soils of Saskatchewan, with a well-documented deglaciation history and a regular zonal ecological gradient, provide an ideal opportunity to estimate the amount of C flux in soils of the dry, cool boreal region. The arid climate of the region is similar to that of the Central Asian region. Thus, the objective of this chapter is to determine the amount and accumulation rate of pedogenic carbonates and SOC using stable isotope geochemistry (^{13}C:^{12}C ratio of SOC and SIC in a series of zonal boreal soils of Saskatchewan, Canada.

2 MATERIALS AND METHODS

A southwest to northeast transect of about 500 km in length across five different soil-climatic zones (Dry Brown, Brown, Dark Brown, Black, and Gray) in Saskatchewan, between 49° 13′ and 53° 63′ latitude and 104° 92′ and 107° 66′ W longitude was selected for the study and total of 18 sites, with three replicates, representative of the major soil association on glacial till, and three from highly calcareous Rego Black Chernozem soils were used in this study (Figure 1). At each site, one soil profile was excavated to >120 cm depth, and described according to the Canadian System for Soil Classification (Soil Classification Working Group, 1998).

In the vicinity of each profile two more pedons were sampled by auger and the data for soil chemical properties represent the mean of three measurements. A Global Positioning System (GPS) was used to record the elevation, longitude, and latitude of each profile. Soils were sampled

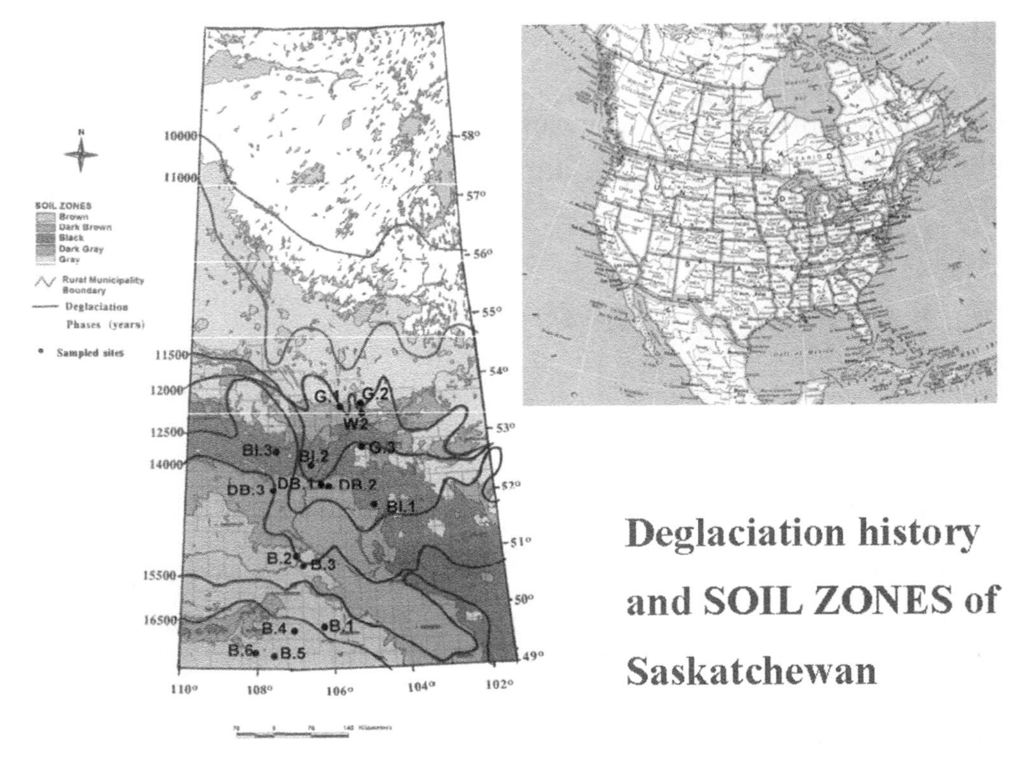

Figure 1. Ice margin positions during the deglaciaion of Saskatchewan, the study region and soil sampling sites (produced from the map of the Saskatchewan Land Resource Center (1999) and deglaciation maps (Christiansen, 1978)).

on well-drained, level to slightly undulating upper slopes with a moderate and dominantly vertical downward flow of soil moisture, in order to minimize the influence of landscape. All the soils were developed on glacial till parent materials, and were never cultivated.

2.1 *Physical and chemical properties*

Air-dried soil samples, passed through 2 mm sieve and saturated with deionized water, were used to prepare a saturated paste, and an extract from this paste was used to determine pH, electrical conductivity (EC), and concentration of soluble ions.

Two to four core samples were taken from each horizon for bulk density measurement. Total initial and oven-dried (105–110°C) weights were recorded. Bulk density was calculated by dividing the dry weight by the sample volume, and the mean bulk density for each horizon was calculated. Particle size analysis of the samples <2 mm soil was performed using the pipette method after removal of carbonates and organic matter (Sheldrick and Wang, 1993).

The cation exchange capacity (CEC) and exchangeable cations were determined using 0.25 M $BaCl_2$-triethanolamine, buffered at pH 8 as a saturation solution. Barium was replaced from the exchange site using 0.1 M NH_4-acetate buffered at pH 7. Calcium (Ca), Mg, and Ba were determined by using atomic absorption spectrometry while K and Na were determined by atomic emission spectrometry. Exchangeable cations were corrected for soluble cations (Hendershot and Lalande 1993). The samples for SOC measurement were ground to pass a 60-mesh Sieve (<250 μm) and C contents were measured at 840°C in a Leco CR-12 Carbon analyzer (Wang and Anderson, 1998a).

2.2 *Sample preparation and measurement of $\delta^{13}C$ of organic carbon*

The δ^{13}C value of SOC was measured for all the horizons with >0.2% organic C. Horizons containing both SOC and SIC were treated with 3 M HCl to remove carbonates, then washed with deionized water using 0.22 μm Millipore filters to remove excess HCl, until a negative test was obtained for chloride in the filtrate using $AgNO_3$. Samples were dried and then ground to a fine powder by using a ball mill.

The isotopic composition (δ^{13}C value) of soil organic matter was completed using Europa Scientific Instruments consisting of "Automated Nitrogen, Carbon Analyzer Gas/Solid/Liquid" (ANCA g/s/l) elemental analyzer coupled to a 20/20 mass spectrometer. These instruments work with "continuous flow" technology. The reproducibility of this method for C isotope ratio is ±0.2‰.

The results of the isotope analyses are expressed in term of δ value (‰):

$$\delta^{13}C(\%) + (\frac{R_s}{R_{st}} - 1) \times 1000$$

Where $R_s = {}^{13}C/{}^{12}C$ in sample, and R_{st} is the corresponding stable isotope ratio in the reference standard. The δ values for δ^{13}C are reported relative to Vienna Pee Dee Belemnite (VPDB).

2.3 *Rego Black Chernozem soils*

In the modern soil surveys, most "Wooded Calcareous" soils are classified as Rego Dark Gray, Calcareous Dark Gray, or Rego Black Chernozems (Soil Classification Working Group, 1998). Wooded Calcareous soils occur mainly on extensive lowlands such as the Debden, Shellbrook, and the Saskatchewan lowlands. A typical area is the Weirdale Plain, just to the north and east of Prince Albert (Mitchell et al., 1950).

Rego Black Chernozem soils may occur adjacent to Orthic Black soils under grassland, but they are typically under poplar forests (Fuller et al., 1999). These soils, first described by Mitchell et al. (1950), are formed on calcareous deposits under a wooded to peat vegetation. The Rego Black Chernozem soils have only limited indication of typical Chernozomic soil development. In general they are somewhat lower in organic matter than comparable Black soils; they are only slightly leached as compared to Degraded Black soils (Dark Gray Chernozomic). Their profile usually consists of Ahk, AC, Csca, and Ck horizons.

3 RESULTS AND DISCUSSION

3.1 *Soil development*

The thickness of Ah horizons, depth of solum, organic matter, and pedogenic carbonate contents and rates of accumulation increases from the Dry Brown zone to the Black zone, despite the fact that the actual time for soil formation decreases. In the studied landscape, soils from knoll (shoulder complex) to footslope become deeper, the thickness of Ah and B horizons, and depth to secondary carbonate layer increase.

3.2 *Organic carbon*

The amount of SOC to 1.2 m depth increases with an average of 9.1, 11.7, 14.9, 9.6, and 21.0 kg m^{-2} (90.8, 117.2, 148.8, 96.3, and 210.2 Mg C ha^{-1}) for Brown, Dark Brown, Black, Gray, and Rego Black Chernozem soils (LFH is included for Gray and Rego Black Chernozem soils), respectively (Figure 2). The amount of SOC pool for Gray soils is 99.2 Mg C ha^{-1}, which is close to the value reported by Huang and Schoenau (1996) for Gray Luvisols (Alfisols) under aspen forest in Saskatchewan.

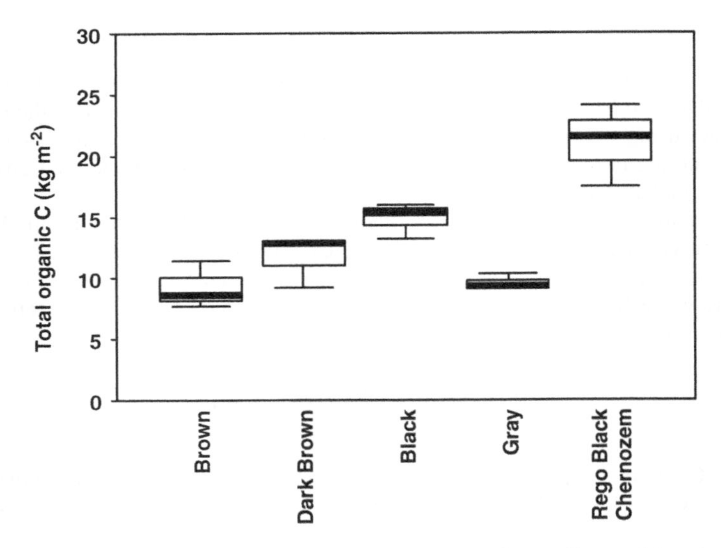

Figure 2. Total organic carbon storage in a sequence of Brown to Gray soils to a 1.2 m depth. The boxes show the central 50% of the values with the median. Whiskers show 25th and 75th centiles. There were three or more replicates of each soil group.

At regional scale, these values are similar to the data reported by McGill et al. (1988) for the same soil zones in Alberta. However, the current values are lower than those reported for Brown soils, and higher than Dark Brown and Black soils by Anderson (1995) for Saskatchewan soils. The current values agree with estimates for soil orders in Canada reported by Tarnocai (1995) as 9.3 kg m^{-2} for Luvisols and 12.4 kg m^{-2} for Chernozems. Those values for SOC pool storage are 9.6 kg m^{-2} and 11.9 kg m^{-2}, respectively. In comparison to the global data (12.5, 14.8, and 13.4 kg C m^{-2} for Alfisols, Inceptisols, and Mollisols, respectively) reported by Eswaran et al. (2000), present values are lower for Luvisols (Alfisols), but similar for the Black Chernozem and Dark Brown soils (Mollisols).

Annual SOC accumulation rates in the soils studied are 0.57, 0.90, 1.18, 0.84, and 1.83 g m^{-2} yr^{-1} for Brown, Dark Brown, Black, Gray, and Rego Black Chernozems, respectively. At local scale the average rate of SOC accumulation in the landscape was 1.25 g m^{-2} yr^{-1}, which is close to the average of soils in Black zone. The rate of SOC accumulation for Gray soils is higher, and for Black and Dark Brown soils are lower than those calculated by Harden et al. (1992) for the glaciated area of North America. Their calculation shows that Mollisols (similar to Black and Dark Brown Chernozems) store 2 g m^{-2} yr^{-1}, Alfisols, similar to Gray, sequester 0.27 g m^{-2} yr^{-1}.

Many researchers have suggested that prairie soils have lost about 30% of their antecedent SOC pool under cultivation. The loss is estimated to be about 2 kg m^{-2} by Anderson (1995), and 1.5 kg m^{-2} by Mann (1986). Considering these amounts of losses for 80 years of agricultural practices, the rate of loss is about 19 to 25 g C m^{-2} yr^{-1}. This rate is higher at the early stage of C losses and then reaches to the steady state, which is less than the calculated value. This is 33–4, 21–28, 16–21, 23–30 times more than the accumulation rate for Brown, Dark Brown, Black, and Gray soils. Therefore, it may take only a few hundred years to lose majority of the SOC pool.

At regional scales, the mean $\delta^{13}C$ values of SOC range from −22.9‰ for the Dry Brown, −24.3‰ for Brown, −24.8‰ for Dark Brown, −25.3‰ for Black, and −26.8‰ for Gray soil zones. These gradual changes are considered to be a consequence of increasing input to organic matter from C4 plants in drier and warmer regions, with an estimated 30% C4 plants characteristic of Dry Brown soils, and none in the Gray soils. The Dry Brown and Brown soils are more depleted in ^{13}C at depth, suggesting a greater proportion of C3 plants in the past.

3.3 *Pedogenic carbonates as a store for C*

X-ray diffraction analysis indicated that randomly selected carbonate pebbles from parent material were entirely dolomite. Calcite with a d-spacing of 3.01 Å was the dominant mineral in pendants from carbonate accumulation zones (Cca horizons). A shift in the calcite peak from 3.03 Å to 3.01 Å is an indication of the presence of Mg-bearing calcite (data not shown).

The internal part of the pendant was much more crystalline than the surface layers as shown by SEM. The outer layers are more recently deposited (Amundson et al., 1989), soft and porous, and consist of calcite with minor amount of quartz and dolomite that are likely inclusion. The inner laminae were denser and more crystalline than the outer layer. Different morphological forms of pedogenic carbonate (PC, calcite) are typical of Saskatchewan soils (Mermut and St. Arnaud, 1981a; b).

The amount of PCs ranges between 100.8 and 161.8 $kg\,m^{-2}$ with an average value of 133.9 $kg\,m^{-2}$ for Brown soils. For Dark Brown soils (Mollisols), the range is between 102.1 $kg\,m^{-2}$ and 177.2 $kg\,m^{-2}$ with an average of 146.3 $kg\,m^{-2}$. For the Black soil zone (Mollisols) the minimum, maximum, and average values are 117.5, 156.7, and 138.2 $kg\,m^{-2}$, respectively. In Gray soil zone (Alfisols) the values are 151.0, 181.5, and 164.5 $kg\,m^{-2}$. For Rego Black Chernozem soils the amount of PCs range between 98.7 and 120.9 $kg\,m^{-2}$ with an average of 110.3 $kg\,m^{-2}$.

Within a typical, hummocky landscape there is a marked variation in PC stores in the upper 1 m. Some soils (generally in concave, locally moist positions) have no carbonate, either lithogenic carbonate (LC) or PC, indicating complete removal during soil formation. Other soils, often immediately adjacent to soils with substantial losses, contain considerably more carbonate than the beginning of soil formation, with marked gains of PC.

The rate of PC accumulation increases from southwest to northeast between 8.3, 11.4, 11.2, 14.3, 9.6 $g\,m^{-2}\,yr^{-1}$ for Brown, Dark Brown, Black, Gray, and Rego Black Chernozem soils. Despite the fact the time for soil formation decreases from 17,000 yr in the southwest to 11,500 yr in northeast, there is a gradual increase in the rate of $CaCO_3$ formation from SW to NE. This is primarily due to changes in climate. In the landscape scale, the rate of accumulation was about 10.4 $g\,m^{-2}\,yr^{-1}$ which is close to estimates for Black and Dark Brown soils. No regional data for the rate of PC accumulation have been documented so far for Canadian Prairies. The present data for dry to sub-humid boreal soils are much higher than the estimates for arid and semi arid soils reported in the literature (Eghbal and Southard, 1993; Mcfadden, 1982; Bachman and Machette, 1977; Buol and Yesilsoy, 1964). It is important to note that all these estimates were made without the use of the stable isotope technique.

3.4 *Association of organic and inorganic carbon*

Using the net primary production (NPP) data (Saskatchewan Agriculture and Food 2001) and rate of C accumulation as SOC, it is estimated that about 0.46% of NPP accumulate as SOC. This value appears to be less than the world average. Considering the long-term average rate for other soils such as Spodosols and peat soils, which are between 10 to more than 50 $g\,m^{-2}\,yr^{-1}$, present estimate seems to be quite reasonable for boreal soils. It must be noted that the annual C accumulation rates calculated are an average for soils that may have been more or less at equilibrium for thousands of years.

It is estimated that about 0.46% of NPP can be accumulated as SIC. On average the rate is about 1.4 times higher than that of calculated value for SOC (0.46%) (Table 1). This rate shows that in long term PC plays an important role in carbon cycle, and can sequester more C than SOC. Adding SOC and SIC together, less than 1% NPP C can be accumulated in the soil. This is consistent with Chadwick et al. (1994) findings, suggesting that in temperate grassland more than 98% of NPP eventually returns to atmosphere through soil respiration.

The $\delta^{13}C$ values of PC generally show a similar trend with the $\delta^{13}C$ values of SOC. The $\delta^{13}C$ value of PC becomes slightly more negative to 50 cm depth, below which the values are almost constant.

Table 1. Relationship between NPP and organic and inorganic carbon accumulation.

Soil zone	Aboveground NPP ($g\,m^{-2}\,yr^{-1}$)	Belowground[a] NPP ($g\,m^{-2}\,yr^{-1}$)	NPP accumulation as organic C (%)	NPP accumulation as inorganic C (%)	Other losses[b] (%)
Brown	323.8	136.0	0.47	0.28	91.25
Dark brown	464.65	195.2	0.46	0.30	91.24
Black	490.12	205.9	0.43	0.38	91.19

[a]Calculated based on 42.6% belowground reported by Slobodian (2001).
[b]Return to atmosphere, losses through the soil system, etc.

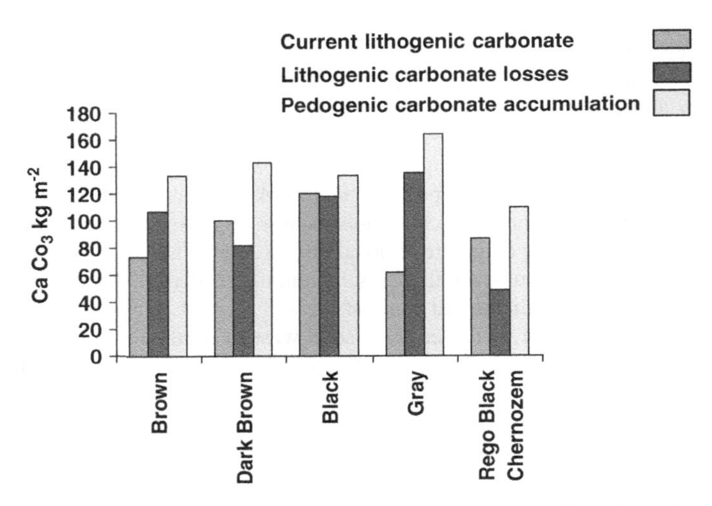

Figure 3. Pedogenic carbonates in comparison to the amount of lithogenic carbonate lost by weathering for Brown to Gray soils in Saskatchewan.

There is a high correlation between C and O compositions of PCs. Based on the data we can differentiate three different groups: (1) lithogenic group, which has more positive values both for $\delta^{18}O$ and $\delta^{13}C$, (2) soils of the grassland, which have more negative value of $\delta^{18}O$ and $\delta^{13}C$ than first group, and (3) forest soils which has more negative value than Prairie soils. The $\delta^{18}O$ value of PC decreases with decreasing $\delta^{13}C$ value. Soil water mainly controls the oxygen isotopic composition of PC (Cerling, 1984; Amundson and Lund, 1987; Cerling and Quade, 1993).

The present data shows that there is more PC formation than the loss of LC in all soil zones (Figure 3). The amount of LC for each soil zone calculated on the bases of the amount of original carbonate in the parent material (equal to the present Ck horizon). The LC lost is calculated by subtracting the total carbonate from the current LC levels in the soil. The rate of losses is obtained by dividing the amount of lost to the age of the soil. The rate of lithogenic loss is lower than the rate of formation of PC, suggesting that Ca was also provided from other sources such as weathering of non-carbonatic Ca-bearing minerals and Ca adding through precipitation, to fix biogenic CO_2 (Figure 4).

The decrease in temperature and increase in precipitation from southwest to northeast direction encourage C3 plants becoming more dominant, that result in depletion of ^{13}C. Meteoric waters become more depleted in ^{18}O in moving toward higher latitude, because of cooler temperature (Cerling and Wang, 1996). These findings are consistent with other worker (Schlesinger et al., 1989; Cerling and Quade, 1993; Khademi and Mermut, 1999). Koehler (2002), and McMonagle (1987) reported the annual average value of $\delta^{18}O$ as $-15‰$. The average value for the O isotope of PC for this area was $-10‰$ which is about 5‰ enriched in comparison to the precipitation value,

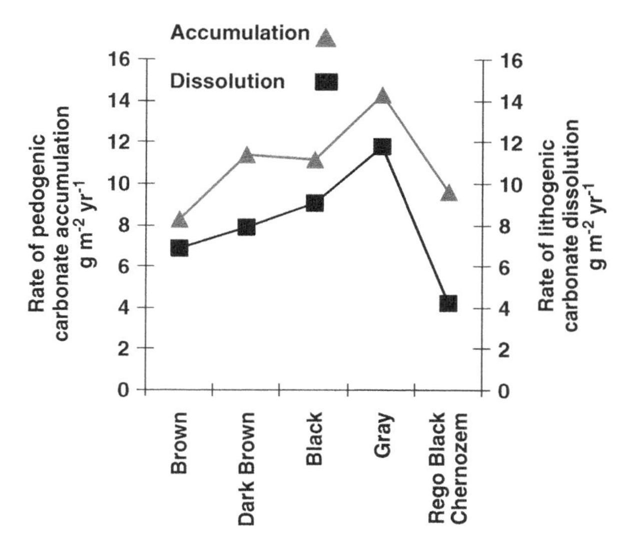

Figure 4. Relationship between rate of dissolution and precipitation of carbonate in different soil zones.

consistent with findings that enrichment of ^{18}O by evaporation takes place in the soil solution prior to carbonate precipitation (Quade et al., 1989; and Schlesinger et al., 1989).

The rate of PC formation increases from the Brown to Gray soils, consistent with increases in effective precipitation and the degree of soil formation as measured by thickness of solum. The data presented show that the availability of Ca related to the weathering of Ca-bearing minerals in the A and B horizons is the rate-limiting step in this group of dry boreal soils. Using the simple linear regression ($Y = 17.46X + 201.83$, $R^2 = 0.89$), where Y is the annual precipitation (mm), and X is the rate of PC accumulation ($g\,m^{-2}\,yr^{-1}$) from the plotted data, it is possible to use the mean annual precipitation in semiarid and sub-humid regions (200–700 mm) to estimate the rate of PC accumulation.

4 CONCLUSIONS

Soil organic matter and carbonates are the two major C pools in the pedosphere, which are interconnected to the biosphere and atmosphere. The objective of this study was to determine the amount and accumulation rate of pedogenic carbonate (PC) in soils of the boreal grassland and forest regions of Saskatchewan, Canada.

A southwest to northeast transect of about 500 km length across five different soil-climatic zones (Dry Brown, Brown, Dark Brown, Black, and Gray) in Saskatchewan, between 49° 13′ and 53° 63′ latitude and 104° 92′ and 107° 66′ W longitude, was selected for the study. Three replicates from each soil zone, and three profiles from the Rego Black Chernozem soils (Mollisols), totally 18 profiles, were described in detail and sampledto 120-cm depth to ensure that calculations are based on a constant depth. Bulk density was measured on replicate samples from each horizon. A Global Positioning System (GPS) was used to record the elevation, longitude, and latitude of the soils.

The mass of organic C was $9.1\,kC\,m^{-2}$ for Brown, $11.7\,kg\,C\,m^{-2}$ for Dark Brown, and $14.9\,kg\,C\,m^{-2}$ for Black soils, decreasing to $9.6\,kg\,C\,m^{-2}$ for Gray soils. A Rego Black Chernozem soil (carbonated phase) contained $21.0\,kg\,C\,m^{-2}$. The $\delta^{13}C$ values of organic C was −22.9‰ for Dry Brown soils, −24.3‰ for Brown soils, −24.8‰ for Dark Brown soils, −25.3‰ for Black soils, and −26.8‰ for Gray soils.

The storage of pedogenic carbonate increases from $134\,kg\,m^{-2}$ in semi-arid grassland (Brown soils) in the southwest to $165\,kg\,m^{-2}$ in the northeast, under forest (Gray soils), within the time decreasing from 17,000 yr in the southwest to 11,500 yr in the northeast. The rate of pedogenic carbonate accumulation likewise increases from 8.3 to 14.3 $g\,m^{-2}\,yr^{-1}$ in the same direction. The results show that the soils of the prairies and forests have sequestered 1.4 times more C in the form of pedogenic carbonates than as organic matter.

Stable carbon isotope values of pedogenic carbonate decreases from southwest to northeast. This is consistent with decreasing representation of C4 plants in the vegetation in the same direction. The rate of pedogenic carbonate accumulation increases with increasing annual precipitation. This suggests that the rate-limiting factor to precipitate with CO_2 is Ca, in the Boreal region of Canada. Silicate weathering is more significant in Luvisols (Alfisols), suggesting that they may be most effective in truly sequestering additional amount of C in the soil. Soils of grasslands and forests of the boreal regions have considerable potential to store C in organic form in the short to medium term, and a long term potential as pedogenic carbonate.

REFERENCES

Amundson, R.G., O.A. Chadwick, J.M. Sowers, and H.E. Doner. 1989. The stable carbon isotope chemistry of pedogenic carbonate at Kyle Canyon. Soil Sci. Soc. Am. J. 53:201–210.

Amundson, R.G., and L.J. Lund. 1987. The stable isotope chemistry of a native and irrigated Typic Natrargid in the San Joaquin Valley of California. Soil Sci. Soc. Am. J. 51:761–767.

Anderson, D.W. 1995. Decomposition of organic matter and carbon emission from soils. P. 165–175. In: R. Lal, John Kimble, Elissa Levine, and B.A. Stewart (eds.). Soils and global change. Advances in Soil Science. CRC Press. U.S.

Bachman, G.O., and M.N. Machette. 1977. Calcic soils and calcretes in the southern United States. U.S. Geol. Survey, Open File Report 77–794.

Buol, S.W., and M.S. Yesilsoy. 1964. A genesis study of a Mohave sandy loam profile. Soil Sci. Soc. Am. Proc. 28:254–56.

Cerling, T.E., 1984. The stable isotopic composition of modern soil carbonate and its relationship to climate. Earth Planet. Sci. Lett. 71, 229–240.

Cerling, T.E., and J. Quade. 1993. Stable carbon and oxygen isotopes in soil carbonates.p. 217–231. In: P. Swart, J.A. McKenzie and K.C. Lohmann (eds.), Climate change in continental isotopic records. American Geophysical Union, Washington D.C.

Cerling, T.E., and Y. Wang. 1996. Stable carbon and oxygen isotopes in soil CO_2 and soil carbonate: Theory, practice, and application to some prairie soils of upper midwestern North America. p. 113–131. In: T.W. Boutton, T.W. and S. Yamasaki (eds.), Mass spectrometry of soils. Marcel Dekker, Inc. U.S.A.

Chadwick, O.A., E.F. Kelly, D.M. Merritts, and R.G. Amundson. 1994. Carbon dioxide consumption during soil development. Biogeochemistry 24:115–127.

Christiansen, E. A. 1978. The Wisconsinan deglatiation of southern Saskatchewan and adjacent areas. Can. J. Earth Sci. 16:1300–1314.

Eghbal, M.K., and R.J. Southard. 1993. Stratigraphy and genesis of Durorthids and Haplargids on dissected alluvial fans, western Mojave Desert, California. Geoderma 59:151–174.

Eswaran, H., P.F. Reich, J.M. Kimble, F.H. Beinroth, and E. Padmanabhan. 2000. Global carbon stocks. p. 16–25. In: R. Lal, J.M. Kimble, H. Eswaran, and B.A. Stewart (eds.) Global climate change and pedogenic carbonates. CRC Press. USA.

Fuller, L.G., D. Wang, and D.W. Anderson. 1999. Evidence for solum recarbonation following forest invasion of a grassland soil. Can. J. Soil Sci. 79:443–448.

Harden, J.W., E.T. Sundquist, R.F. Stallard, and R.K. Mark. 1992. Dynamics of soil carbon during deglaciation of the Laurentide Ice Sheet. Science 258:1921–1924.

Huang, W.Z., and J.J. Schoenau. 1996. Forms, amounts, and distribution of carbon, nitrogen, phosphorus and sulfur in a boreal aspen forest soil. Can. J. Soil Sci. 76:373–385.

Hendershot, W.H., and H. Lalande. 1993. Ion exchange and exchangeable cations. p. 167–176. In: Martin R. Carter (ed.) Soil sampling and methods of analysis. Lewis Publisher USA.

Khademi, H., and A.R. Mermut. 1999. Submicroscopy and stable isotope geochemistry of carbonates and associated palygorskite in Iranian Aridisols. European J. Soil Sci. 50, 207–216.

Koehler, G. 2002. Stable isotope mass spectrometry data. National Water Research Institute, National Hydrology Research Centre. Saskatoon, Saskatchewan. Canada.

Lal, R., Kimble, J.M., and R. Follett. 1998. Pedospheric processes and the carbon cycle. p. 1–8. In: R. Lal, J.M. Kimble, H. Eswaran, R. Follett, and B.A. Stewart (eds.). Soil processes and the carbon cycle. CRC Press. USA.

Lal, R., and J.M. Kimble. 2000a. Pedogenic carbonate and the global carbon cycle. p. 1–14. In: R. Lal, J.M. Kimble, H. Eswaran, and B.A. Stewart (eds.) Global climate change and pedogenic carbonates. CRC Press. USA.

Lal, R., and J.M Kimble. 2000b. Inorganic carbon and the global carbon cycle: Research and development priorities. p. 291–302. In: R. Lal, J.M. Kimble, H. Eswaran, and B.A. Stewart (eds), Global climate change and pedogenic carbonates. CRC Press. USA.

Mann, L.K. 1986. Changes in soil carbon storage after cultivation. Soil Sci. 142:279–288.

McFadden, L.D. 1982. The impacts of temporal and spatial climatic changes on alluvial soil genesis in Southern California. Ph. D. Dissertation, University of Arizona, Tucson, 430 p.

McGill, W.B., J.F. Dormaar, and E. Reinl-Dwyer. 1988. New perspective on soil organic matter quality, quantity and dynamics on the Canadian prairies. p. 30–48. In: Proc. 34th Meeting, Can. Soc. Soil Sci., Calgary, AB. Land degradation and conservation tillage.

McMonagle, A.L. 1987. Stable isotope and chemical compositions of surface and subsurface waters in Saskatchewan. M.Sc. Thesis. University of Saskatchewan. Saskatoon. Canada.

Mermut, A.R., and R.J. St. Arnaud, 1981a. A study of microcrystalline pedogenic carbonates using submicroscopic techniques. Can. J. Soil Sci. 61:261–272.

Mermut, A.R., and R.J. St. Arnaud. 1981b. A micromorphological study of calcareous soil horizons in Saskatchewan soils. Can. J. Soil Sci. 61:243–260.

Mermut, A.R., R. Amundson, and T.E. Cerling. 2000. The use of stable isotopes in studying carbonates dynamics in soils. p. 65–85. In: R. Lal, J.M. Kimble, H. Eswaran, and B.A. Stewart (eds.). Global climate change and pedogenic carbonates. CRC Press. USA.

Mermut, A.R., and H. Eswaran. 2001. Some major developments in soil science since the mid-1960s. Geoderma. 100:403–426.

Mitchell, J., H.C. Moss, and J.S. Clayton. 1950. Soil survey of southern Saskatchewan. Soil survey Report No. 13. University of Saskatchewan, Saskatoon, Saskatchewan.

Quade, J., T.E. Cerling, and J.R Bowman. 1989. Systematic variations in the carbon and oxygen isotopic composition of pedogenic carbonate along elevation transects in the southern Great Basin, United States. Geol. Soc. Am. Bul. 101:464–475.

Sakatchewan Land Resource Center. 1999. Soil zones of Saskatchewan. Department of Soil Science. University of Saskatchewan, Canada.

Saskatchewan Agriculture and Food. 2001. 2001 Saskatchewan forage crop production guide. Regina, Canada.

Sheldrick, B.H., and C. Wang. 1993. Particle size distribution. p. 499–512. In: M.R. Carter (ed.). Soil sampling and methods of analysis. Lewis Publisher USA.

Schlesinger, W.H., G.M. Marion, and P.J. Fonteyn. 1989. Stable isotope ratios and the dynamics of caliche in desert soils. p. 309–317. In: P.W. Rundel, J.R. Ehleringer, and K. Nagy (eds.). Stable isotopes in ecological research. Springer Verlag.

Slobodian, N. 2001. The landscape-scale distribution of belowground plant biomass in a fescue prairie and wheat field. M.Sc. Thesis. University of Saskatchewan, Saskatoon Canada.

Soil Classification Working Group. 1998. The Canadian System of Soil Classification, third edition. NRC Research Press.

Tarnocai, D. 1995. The amount of organic carbon in various soil orders and ecological provinces in Canada. p. 81–92. In: R. Lal; John Kimble; Elissa Levine; and B.A. Stewart (eds.). Soils and global change. Advances in Soil Science. CRC Press. U.S.A.

Wang, D., and D.W. Anderson. 1998. Direct measurement of organic carbon content in soils by the Leco CR-12 carbon analyzer. Commun. Soil Sci. Plant Anal. 29:15–21.

CHAPTER 15

Conservation agriculture: Environmental benefits of reduced tillage and soil carbon management in water-limited areas of Central Asia

D.C. Reicosky
North Central Soil Conservation Research Laboratory, USDA Agricultural Research Service, Morris, Minnesota, USA

1 INTRODUCTION

Conservation agriculture (CA) aims to conserve, improve and make more efficient use of natural resources through integrated management of available soil, water and biological resources combined with external inputs. The CA contributes to global environmental conservation as well as to enhanced and sustained agricultural production and can play a central role in global agricultural policy. Food security and sustainability are important for all citizens. Agriculture, the major industry for food and fiber production, is known to cause emission and storage of greenhouse gases (GHGs). Intensification of agricultural production has been an important factor influencing GHG emission and affecting the water balance. Agricultural activities contribute to carbon dioxide (CO_2) emissions released to the atmosphere through the combustion of fossil fuel, soil organic matter (SOM) decomposition and biomass burning. Improved CA practices, especially in water-limited areas, have great potential to increase soil organic carbon (SOC) sequestration, available water storage, and decrease net emissions of CO_2 and other GHGs.

World soils, an important pool of active C, play a major role in the global C cycle and contribute to changes in the concentration of GHGs in the atmosphere (Lal et al., 1998). Intensive agriculture is believed to cause some environmental problems, especially related to water use, water contamination, soil erosion and greenhouse effect (Houghton et al., 1999; Schlesinger, 1985; Davidson and Ackerman, 1993). The soil contains two to three times as much C as the atmosphere. In the last 120 years, intensive agriculture has caused a C loss between 30 and 50%. Minimizing the increase in ambient CO_2concentration through soil C management, reduces the production of GHGs and minimizes potential for climate change. In fact, agricultural practices have the potential to store more C in the soil than agriculture releases through land use change and fossil fuel combustion (Lal et al., 1998).

There is a vast potential for C sequestration in dryland ecosystems of Central Asia (Lal, 2004). With limited rainfall, much of the land area is irrigated resulting in secondary salinization and desertification suggesting that present land use and management systems are not sustainable. An important strategy of soil C sequestration is to reverse soil degradation. The CA and phytoremediation of degraded soils and ecosystems are important options in enhancing crop production and C sequestration. The CA with residue mulch can increase the available water storage in the root zone by increasing infiltration and decreasing soil temperature, reducing evaporation losses and improving water use efficiency (WUE). Lal's (2004) estimates of the potential of soil C sequestration in the Central Asian countries indicate a range of 10 to 23 $Tg\,C\,yr^{-1}$ over 50 years. To achieve this amount of C sequestration, improved management practices will require less fallowing. Karbozova-Saljnikov et al. (2004) concluded that a frequent fallow system cultivated four to five times for weed control during the season depletes SOM via accelerated mineralization.

The potentially mineralizable C was inversely proportional to the frequency of the fallow period and was highest in continuous wheat system. They found the nitrogen (N) dynamics were closely related to the recent influence of plant residues while C dynamics was more related to the long-term residue addition. These results agree with those by Campbell and Zentner (1993) in similar climatic areas of the Canadian prairies.

The SOC is a major determinant of soil quality and is the fundamental foundation of environmental quality in water-limited areas. Soil quality is largely governed by SOM content, which is dynamic and responds effectively to changes in soil management, primarily tillage and C input. This review will primarily address soil C and water conservation as they relate to environmental benefits. The close coupling of the water, N and C cycles is discussed along with the duel role of crop residues for C input and soil water evaporation reduction. Throughout the following discussion, the terms "SOC" and "SOM content" are used synonymously. (See other recent reviews on the role of C sequestration in conservation agriculture presented by Reicosky et al. (1995), Robert (2001), Uri (1999), Tebrugge and Guring (1999), Lal et al. (1998) and Lal (2000)).

2 KEY ROLE OF SOIL CARBON (SOIL ORGANIC MATTER)

The SOM generated from crop residues, is the main determinant of biological activity because it is the primary energy source. The primary chemical in SOM is C that represents a key indicator for soil quality, both for agricultural functions (production and economy) and for environmental functions (C sequestration and air quality). The amount, diversity and activity of soil fauna and microorganisms are directly related to SOM content and quality. Both SOM and the biological activity that it generates, have a major influence on the physical and chemical properties of the soil. Soil aggregation and soil structure stability increases with increase in SOM and surface crop residues. These factors in turn increase the infiltration rate and available water-holding capacity (WHC) of the soil as well as resistance against erosion by wind and water. The SOM also improves the dynamics and bio-availability of main plant nutrient elements.

3 WATER-C-N LINKAGES IN CONSERVATION AGRICULTURE

In water-limited areas (<400 mm of annual rainfall), like many of the Central Asian countries, water and C management are interdependent. All farming systems reflect in some way the fact that photosynthetic productivity involves simultaneous water loss through transpiration by plants. In areas where water supply is limiting, several basic strategies are relied on to bring crops to maturity within the available supply. One strategy is to insure that a large portion of the available water goes to transpiration, a second is to achieve high level of production per unit of transpiration, and the third involves achieving a balance between seasonal water use and seasonal supply. As part of this balance, farming practices must also contend with the losses of water through runoff, percolation, and evaporation from the soil surface. It is in managing this water that conservation agriculture has the greatest opportunity for enhancing crop biomass production and soil C storage.

The exchange of the two main gases of interest, CO_2 and water vapor, are so critically important and related to solar-energy-dependent processes for C fixation and soil storage. The critical link between water and C starts with the process of photosynthesis represented by the chemical formula given below.

$$6CO_2 + 6H_2O \xrightleftharpoons{\text{Light energy}} C_6H_{12}O_6 + 6O_2$$

CO_2 comes through the stomata into the leaf via photosynthesis and water vapor goes out through the stomates to the atmosphere via transpiration. The combination of CO_2 plus water in the presence of light with chlorophyll in the plant leaves yields a sugar or carbohydrate plus oxygen. This is the start of the C cycle in agricultural ecosystems as a part of the global C cycle. These carbohydrates are transformed into many different chemical compounds and structures that make photosynthesis so important to the existence of life on earth. Water availability is a major factor in plant productivity because of the linkage of CO_2 and H_2O flux through the plant stomatal openings. Water transports nutrients from the soil into the plant and acts as a solvent for transport of various chemical compounds within the plant.

The reverse of photosynthesis is respiration where the plant material is oxidized and CO_2 is released to the atmosphere. Water is essential, along with optimum temperatures, for respiration which is the oxidation of the plant material and SOM. Efficient agricultural water use is critical in an era of increasing competition for limited water resources that in turn reflects the biomass productivity as a precursor for SOC input. So it is clear that water is an important factor critical for C fixation in the plant material and subsequent input into the soil C pools. This is where tillage and residue incorporation have a negative impact on SOM and has resulted in 30–60% soil C loss in the last 150 years of intensive agriculture in the US. Returning the respired CO_2 to the atmosphere completes the final part of the global C cycle. The C and water cycles are intimately coupled with strong interdependence on each other and in dry and warm climates, soil C sequestration continues to be a major scientific challenge.

In general, dry matter production and water use are more closely correlated than grain yield and water use (Ritchie, 1983). The information on yield and water use relationships enables predictions of crop yield that then can be used to estimate C yield for soil C input. There is little question that soil water supply for transpiration causes major variation in crop yields and biomass. The relationship between transpiration and total dry matter production of plants under field conditions is generally the same for plants grown in containers provided the dry matter production of plants in the fields is limited by the availability of water. Thus by knowing the seasonal transpiration, it is possible to estimate the amount of biomass that may be generated by specific crops and with that information it is possible to calculate potential C input to the soil system. The strong linear relationship supports the concept that biomass production is primarily water limited, which also implies that C input to the soil is water limited. Power et al. (1961) observed that biomass yields of wheat were linearly related to evapotranspiration and were affected by phosphorus (P) fertilization. Biomass yields, hence C input, can be easily calculated from grain yield and straw:grain ratio often referred to as the harvest index (HI). The evapotranspiration efficiency (Power et al., 1961), that is the slope of the line of biomass versus water use, was about 1.89 and 2.13 kg biomass m^{-3} water as a result of the P fertilization treatment. Jensen and Sletten (1965) showed similar effects of nitrogen (N) application. Effective nutrition is a part of C fixation through its impact on photosynthesis and biomass accumulation. Knowledge of these relations is essential to estimate the C input based on seasonal rainfall data within different geographic regions.

The devastating effects of water stress on plant productivity have been recognized since the beginning of agriculture. The central question with respect to C sequestration is the nature of the ratio of biomass productivity to crop water use and its inherent variability. Dry matter production rates under water-stress conditions are a function of source and sink activity (size and strength). As water stress develops, reductions occur in both leaf area and photosynthetic rate, reducing the total assimilating capacity of the plant (Hsiao, 1973). In essentially every case, whole-plant C assimilation capacity declines as water stress develops. The resultant effect on biomass yield depends on the relative contribution of newly assimilated C, compared with the utilization of previous accumulated dry matter. Much remains to be accomplished in terms of increasing photosynthesis and crop productivity to meet future C input demands. By understanding the order of limitations in crop growth and development in determining the degree of genetic variability in existing crop species, it is possible to develop higher-yielding, more stress-tolerant varieties which can enhance soil C accumulation.

Soil water deficits may affect photosynthate utilization by altering either the efficiency with which the photosynthate converted to new growth or the rate of photosynthesis used in maintenance of existing dry matter. The production of photosynthetically active leaf area by field crops is one of the most important factors affecting crop activity (Hsiao, 1973). The reduced rate of leaf area accumulation usually associated with growth in dry land environments may be associated with smaller sized leaves or with the production of fewer leaves. In addition, plant-water deficits may alter light interception through effects on the display and duration of the green leaf area.

The close linkage of the water and C cycles is clear, but the linkage of these two cycles extends to the N cycle. All three cycles are closely coupled in agricultural production systems. Intensive tillage can disrupt all three cycles because it buries the crop residue that when left on the surface can serve many other functions. Sometimes additional N may be required for high residue crops to assist in residue control by providing the optimum C:N ratio for crop residue decomposition. The belowground root biomass and exudates of legumes play a significant role in providing energy for the soil fauna and for modifying the subsurface soil physical properties (Wilts et al., 2004). However, there are limited management opportunities for these forms of C and N.

Water deficit can simultaneously reduce N acquisition and advance the onset of foliar senescence. The efficient use of water for crop production can also be enhanced in the semi-arid areas by adding fertilizer to deficient soils. This technique usually increases the total aboveground biomass without increasing total water use. Added fertilizer shifts the balance between soil and plant evaporation more toward plant evaporation. There is a strong relationship between water availability and N-fertilizer responses and that changing one of these factors can greatly affect responses to the other. Smika et al. (1965) showed how native grass production responded to N fertilization as available water supplies vary when the water supply was very limited; grass production was low and little response to N fertilizer occurred at any N rate. As available water supply increased, yield from unfertilized grass increased. However, as the rate of annual N fertilizer increased, the slopes of the regression lines relating yield to available water also increased. Thus, while N rate had very little effect on a biomass production when water supply was limited, responses of several thousand kg ha^{-1} of biomass were recorded when adequate water was available. Similar responses to increased water were very modest with no N fertilizer, but increased greatly as the rate of N fertilizer increased. Thus, water has both direct and indirect effects on biomass production and requires close coordination with N fertilizer rates for optimal biomass and C production.

4 A CASE FOR CONSERVATION AGRICULTURE AND ZERO TILLAGE

Tillage or soil preparation has been an integral part of traditional agricultural production. Tillage is also a principal agent resulting in soil perturbation and modification of the soil structure with soil degradation. Intensive tillage loosens soil, enhances the release of soil nutrients for crop growth, kills the weeds that compete with crop plants for water and nutrients and modifies the circulation of water and air within the soil. Intensive tillage can adversely affect soil structure and cause excessive breakdown of aggregates making it vulnerable to erosion. Intensive tillage causes soil degradation through C loss and tillage-induced GHG emissions that impact productive capacity and environmental quality. Intensive tillage also causes a substantial short-term increase in soil evaporation to rapidly deplete the surface layer.

Recent studies involving a dynamic chamber, various tillage methods and associated incorporation of residue in the field indicated major C losses immediately following intensive tillage (Reicosky and Lindstrom, 1993, 1995). The moldboard plow had the roughest soil surface, the highest initial CO_2 flux and maintained the highest flux throughout the 19-day study. High initial CO_2 fluxes were more closely related to the depth of soil disturbance that resulted in a rougher surface and larger voids than to residue incorporation. Lower CO_2 and water fluxes were caused by tillage associated with low soil disturbance and small voids with no till having the least amount of CO_2 and water loss during 19 days. The large gaseous losses of soil C following moldboard plowing compared to relatively small losses with direct seeding (no till) showed why crop production systems using moldboard plowing decreased SOM and why no-till or direct-seeding crop production systems are stopping or reversing that trend. The short-term cumulative CO_2 loss was related to the soil volume disturbed by the tillage tools. This concept was explored when Reicosky (1998) determined the impact of strip tillage methods on CO_2 and water loss after five different strip tillage tools and no till. The highest CO_2fluxes were from the moldboard plow and subsoil shank tillage. Fluxes from both slowly declined as the soil dried. The least CO_2 flux was measured from the no-till treatment. The other forms of strip tillage were intermediate with only a small amount of CO_2 detected immediately after tillage. These results suggested that the CO_2 fluxes appeared to be directly and linearly related to the volume of soil disturbed. Intensive tillage fractured a larger depth and volume of soil and increased aggregate surface area available for gas exchange that contributed to the vertical gas flux. The narrower and shallower soil disturbance caused less CO_2 and water loss suggests that the volume of soil disturbed must be minimized to reduce C loss and impact on soil and air quality. The results suggest environmental benefits and water and C storage of strip tillage over broad area tillage that needs to be considered in soil management decisions.

Plowing or excessive tillage is another factor that exacerbates the problem of soil degradation and reduces the SOM pool. Plowing depletes the SOM pool increasing mineralization and the risks of soil erosion. Reicosky (1997) reported that average short-term C loss from four conservation tillage tools was 31% of the CO_2 loss from the moldboard plow. The moldboard plow lost 13.8 times more CO_2 as the soil not tilled while conservation tillage systems averaged about 4.3 times more CO_2 loss. The smaller CO_2 loss from conservation tillage systems was significant and suggests progress in equipment development for enhanced soil C management. Conservation tillage reduces the extent, frequency and magnitude of mechanical disturbance caused by the moldboard plow and reduces the large air-filled soil pores to slow the rate of gas exchange and C oxidation. With tillage depths of 30 to 45 cm and adequate soil water, the long-term differences in evaporation were negligible.

The C loss associated with intensive tillage is also associated with soil erosion and degradation that can lead to increased soil variability and decreased yield. Tillage erosion or tillage-induced translocation, the net movement of soil downslope through the action of mechanical implements and gravity forces acting on the loosened soil, has been observed for many years. Papendick et al. (1983) reported original topsoil on most hilltops had been removed by tillage erosion in the Paulouse region of the Pacific Northwest of the USA. The moldboard plow was identified as the primary cause, but all tillage implements contribute to this problem (Grovers et al., 1994; Lobb and Kachanoski, 1999). Soil translocation from moldboard plow tillage can be greater than soil loss tolerance levels (Lindstrom et al., 1992; Grovers et al., 1994; Lobb, Kachanoski and Miller, 1995; Poesen et al., 1997). Soil is not directly lost from the fields by tillage translocation, rather it is moved away from the convex slopes and deposited on concave slope positions. Lindstrom et al. (1992) reported that soil movement on a convex slope in southwestern Minnesota, USA could result in a sustained soil loss level of approximately 30 t ha^{-1} yr^{-1} from annual moldboard plowing. Lobb et al. (1995) estimated soil loss in southwestern Ontario, Canada from a shoulder position to be 54 t ha^{-1} $yr^{-1}$from a tillage sequence of moldboard plowing, tandem disk and a C-tine cultivator. In this case, tillage erosion, as estimated through resident ^{137}Cs, accounted for at least 70% of the total soil loss. The net effect of soil translocation from the combined effects of tillage and water erosion is an increase in spatial variability of crop yield and a likely decline in soil SOC related to lower soil productivity (Schumacher et al., 1999).

5 CROP RESIDUES AND EVAPORATION

Water management is an integral part of any agricultural production system. Changes in total crop water use may occur through modification of crop residue management. To use the limited water resources more efficiently for crop production, it is important to develop new technology for effectively capturing and retaining rainfall using available resources. Agricultural residues serve dual purposes, one minimizing soil evaporation with improved residue management and second, residues are the primary form of soil C input. Crop residues improve soil and water conservation when retained on the surface (Unger, 1994). However, the residues are usually destroyed or incorporated with the soil through the use of intensive tillage methods.

Reduced tillage and crop residue management systems were initially developed to protect the surface from wind and water erosion, but they also increased soil water storage under a wide range of climates and cropping systems. Unger (1978) showed that high wheat residue levels resulted in increased storage of fallow season precipitation, which subsequently produced higher sorghum grain yields in the field studies in the Southern Great Plains of the USA. High residue levels of 8 to 12 Mg ha^{-1} resulted in about 80 to 90 mm more stored soil water at planting and about 2.0 Mg ha^{-1} more of sorghum grain yield than a no residue treatment. Similarly, Smika (1976) showed pronounced tillage effects on soil water profiles following 34 days of drying in field experiments where no tillage treatment that maintained surface residue cover resulted in more water storage in the soil profile below a depth of 5 cm. Smika and Unger (1986) and Unger et al. (1988) provide excellent reviews of the effects of reduced tillage and increased residues on water conservation. Thus, with improved crop residue management and less tillage, soil physical conditions improve with time after initiation of conservation tillage (Peterson et al., 1998). Emphasis on improved residue management and less intensive tillage systems in conservation agriculture combines the beneficial effects of water conservation and soil C enhancement important in water-limited areas.

A major advantage of maintaining crop residue on the soil surface, especially in subhumid and semiarid regions, is improved soil water conservation (Steiner, 1994). This is a result of reduced runoff of surface water and improved soil surface condition that allows more time for and permits greater water infiltration. Crop residues also reduce evaporation, which reduces the loss of stored water. Improved water conservation with less intensive tillage is also an important in humid regions where the short-term drought can severely limit crop yields on soils that have low water-holding capacities. As SOM levels increase, soil physical factors such as aggregate stability, bulk density and porosity that affect water infiltration and flow are positively influenced.

The manner in which the residue is left on the surface can play a big role in controlling water loss through evaporation. Residue that lays flat, provides continuous surface coverage and reduces evaporation much more than upright residue of the same mass. Residue lying in strips, like that found in strip till or following planting operations with aggressive residue managers, is less effective in reducing evaporation in comparison with complete residue coverage. The way the residue is left on the surface can have a big role in infiltration and soil water recharge. Stubble left upright over winter is important to snow catchments and water conservation in much of the Great Plains of the USA. Greb et al. (1967) reported that snow-melt moisture is more than 66% effective in moisture storage compared with 0 to 15% effectiveness of moisture from a July rainstorm. In addition, crop residue can reduce soil freezing, provide infiltration channels into the soil, and reduce the evaporation rate from the wetted soil as the snow melts. Mannering and Meyer (1961) demonstrated the value of the form of the crop residues on runoff and soil losses. They evaluated corn stalks as left by the corn picker (check treatment), stalks that were shredded, and stocks that were shredded and disked once. Runoff with the check and shredded corn stalk treatment was intermediate, but shredding the corn stalks decreased soil losses to about one half of those of the check. Although runoff was not affected, the study showed that increasing the surface coverage by shredding was an effective soil-loss control practice.

Crop residue on the soil surface reduces evaporation. Most of the evaporation occurs when the soil is wet, within a few days after rain or irrigation. The residue insulates the wet soil from solar energy and reduces evaporation. In instances where the soil is wetted more frequently, as in the case of sprinkler irrigation, evaporation increases and crop residue can control evaporation. Todd et al. (1991) measured mean daily evaporation from soil under a corn canopy during the growing season at North Platte, Nebraska, USA to document the effects of residue on soil evaporation in irrigated corn. The wheat straw residue (13.4 Mg ha^{-1}) left lying flat on the surface, produced complete cover and reduced soil evaporation by 50 to 64 mm during the growing season. While these results may have limited practical use, they do identify potential water savings and C input from biomass production accomplished with improved residue management in rainfall-limited areas.

Unger and Parker (1968) studied the effectiveness of stubble-mulch farming on water conservation during the fallow period. They found that mulches conserved water during long, dry periods. Evaporation from soil over a 16-week period was reduced 57% by straw applied and mixed with the soil surface, and 19% by straw buried 30 mm deep. Other findings showed a significant increase in fallow moisture efficiency with 1.68 and 6.72 Mg ha^{-1}of surface straw mulch, compared with bare fallow.

Removal of crop residues likewise affects soil nutrient availability and water relations. Barnhart et al. (1978) showed that continued removal of corn silage from an Iowa soil resulted in decreased SOM and total N content, when compared with plots where grain only was removed. Similarly, Reicosky et al. (2002) found that 30 years of fall moldboard plowing reduced the SOC whether the aboveground corn biomass was removed for silage or whether the stover was returned and plowed into the soil. Their results suggest that no form of residue management will increase SOC content as long as the soil is moldboard plowed. Hooker et al. (2005) also found that within a tillage treatment, residue management had little effect on SOC in the surface soil layer (0–5 cm). Tillage tended to decrease the SOC content, although only no till combined with stover return to the soil resulted in an increase in SOC in the surface layer compared with moldboard plowed treatments.

Removal of crop residues exposes the soil to water and wind erosion. Larson et al. (1978) calculated that removal of crop residues from Minnesota corn land would result in the removal of substantial amounts of N directly in the crop residues in addition to the N loss in accelerated soil erosion. Such losses of N over a period of years would eventually reduce fertility levels and reduce WUE. Understanding the SOC dynamics in bioenergy crops is important since C sequestration can influence biomass production, ecosystem sustainability, soil fertility, and soil structure. Continued crop residue removal for biofuels also raises concern about the long-term sustainability of this management system (Wilhelm et al., 2004).

Increasing SOM content has been considered effective for increasing its available water-holding capacity. Smaller additions of crop residue can result in smaller increases in water retention, but is often more practical to grow the materials in place. Returning most or all the crop residues from well-managed, high-residue crops to the soil should maintain or gradually increase SOM; thus, increasing water-storage capacity over a long period. Generally, this requires reduced tillage intensity or no tillage at all and increased cropping intensity (Peterson et al., 1998).

6 ENVIRONMENTAL BENEFITS OF SOIL C

The main benefit of CA or direct seeding is the immediate impact on SOM and soil water interactions. The SOM is so valuable for what it does in soil, it can be referred to as "black gold" because of its vital role in physical, chemical and biological properties and processes within the soil system. Agricultural policies are needed to encourage farmers to improve soil quality by storing C which also leads to enhanced air quality, water quality and increased productivity as well as to help mitigate the greenhouse effect. The SOC is one of the most valuable resources and may serve as a "second crop" if global C trading systems become a reality. While technical discussions related to C trading are continuing, there are several other secondary benefits of SOC impacting environmental quality that should be considered to maintain a balance between economic and environmental factors.

The importance of SOC can be compared to the central hub of a wagon wheel. The wheel represents a circle, which is a symbol of strength, unity and progress. The "spokes" of this wagon wheel represent incremental links to SOC that lead to the environmental improvement that supports total soil resource sustainability. Many spokes make a stronger wheel. Each secondary benefit resulting from SOC conservation management contributes to an overall improvement of environmental quality. Soane (1990) discussed several practical aspects of soil C important in soil management. Some of the "spokes" of the environmental sustainability wheel are described in following paragraphs.

The primary role of SOM in reducing soil erodibility is by stabilizing the surface aggregates through reduced crust formation and surface sealing, which increases infiltration (Le Bissonnais, 1990). Under these situations, the crop residue acts as tiny dams that slow down the water runoff from the field allowing the water more time to soak into the soil (Jones et al., 1994). Worm channels, macropores and plant root holes left intact increase infiltration (Edwards et al., 1988). Water infiltration is two to ten times faster in soils with earthworms than in soils without earthworms (Lee, 1985). The SOM contributes to soil particle aggregation that makes it easier for the water to move through the soil and enables the plants to use less energy to establish to the root systems

(Chaney and Swift, 1984). Enhanced soil water-holding capacity is a result of increased SOM that more readily absorbs water and releases it slowly over the season to minimize the impacts of short-term drought. In fact, certain types of SOM can hold up to 20 times its weight in water. Hudson (1994) showed that for each one percent increase in SOM, the available water-holding capacity in the soil increased by 3.7% of the soil volume. The extra SOM prevents drying and improves water retention properties of sandy soils. In all texture groups, as SOM content increased from 0.5 to 3%, available water capacity of the soil more than doubled. Increased water-holding capacity plus the increased infiltration with higher SOM and decreased evaporation with crop residues on the soil surface all contribute to improve crop water-use efficiency.

Ion adsorption or exchange is one of the most significant nutrient cycling functions of soils. Cation exchange capacity (CEC) is the amount of exchange sites that can absorb and release nutrient cations. SOM can increase CEC of the soil from 20 to 70% over that of clay minerals and metal oxides present. In fact, Crovetto (1996) showed that the contribution of SOM to CEC exceeded that of the kaolinite clay mineral in the surface 5 cm. Robert (1996; 2001) reported a strong linear relationship between organic C and CEC of his experimental soil. The CEC increased four-fold with an SOC increase in SOC from 1 to 4%. The toxicity of other elements can be inhibited by SOM, which has the ability to adsorb soluble chemicals. The adsorption by clay minerals and SOM is an important means by which plant nutrients are retained in crop rooting zones.

Soil erosion leads to degraded surface and ground water quality. Another secondary benefit of higher SOM is decrease in water and wind erosion (Uri, 1999). Crop residues on the surface help hold soil particles in place and keep associated nutrients and pesticides on the field. The surface layer of organic matter minimizes herbicide runoff, and with conservation tillage, herbicide leaching can be reduced as much as half (Braverman et al., 1990). The enhancements of surface and ground water quality are accrued through the use of conservation tillage and by increasing SOM. Increasing SOM and maintaining crop residues on the surface reduces wind erosion (Skidmore et al., 1979). Depending on the amount of crop residues left on the soil surface, soil erosion can be reduced to nearly nothing as compared to the unprotected, intensively tilled field.

The SOM can decrease soil compaction (Angers and Simard, 1986; Avnimelech and Cohen, 1988). Soane (1990) presented different mechanisms where soil "compactibility" can be decreased by increased in SOM content: (1) improved internal and external binding of soil aggregates; (2) increased soil elasticity and rebounding capabilities; (3) diluted effect of reduced bulk density due to mixing organic residues with the soil matrix; (4) created temporary or permanent root networks; (5) localized change electrical charge of soil particles surfaces, and (6) changed soil internal friction. While most soil compaction occurs during the first vehicle trip over the tilled field, reduced weight and horsepower requirements associated with forms of conservation tillage also minimize compaction. Additional field traffic required by intensive tillage compounds the problem by breaking down soil structure. The combined physical and biological benefits of SOM can minimize the effect of traffic compaction and improve soil tilth.

Maintenance of SOM contributes to the formation and stabilization of soil structure. Another spoke in the wagon wheel of environmental quality is improved soil tilth, structure and aggregate stability that enhance the gas exchange properties and aeration required for nutrient cycling (Chaney and Swift, 1984). Critical management of soil airflow with improved soil tilth and structure is required for optimum plant function and nutrient cycling. It is the combination of many little factors rather than one single factor that results in comprehensive environmental benefits from SOM management. The many attributes suggest new concepts on how to manage the soil for the long-term aggregate stability and sustainability.

7 SUMMARY

Agricultural carbon (C) sequestration may be one of the most cost-effective ways to slow processes of global warming and enhance plant-available water in water-limited areas of Central Asia.

Numerous environmental benefits and enhanced water-use efficiency result from agricultural activities that sequester soil C and contribute to crop production and environmental security. Increase in surface residues and soil C increases infiltration, decreases runoff, increases water-holding capacity, and decreases evaporation. As part of no-regret strategies, practices that sequester soil C also help reduce soil erosion and improve water quality and are consistent with more sustainable and less chemically-dependent agriculture. While we learn more about residue management and soil C storage and their central role in direct environmental benefits, we must understand the secondary environmental benefits and what they mean to production agriculture. Increasing soil C storage in water-limited areas can increase fertility and nutrient cycling, decrease wind and water erosion, minimize compaction, enhance water quality, decrease C emissions, impede pesticide movement and generally enhance environmental quality. The sum of each individual benefit adds to a total package with major significance on a regional scale. Incorporating C storage in conservation planning in areas of limited water resources demonstrates concern for our global resources and presents a positive role for soil C that will have a major impact on our future quality of life.

REFERENCES

Angers, D.A. and R.R. Simard. 1986. Relationships between organic matter content and soil bulk density. Can. J. Soil Sci. 66:743–746.

Avnimelech, Y. and A. Cohen. 1988. On the use of organic manures for amendment of compacted clay soils: Effects of aerobic and anaerobic conditions. Biol. Wastes 29:331–339.

Barnhart, S.L., W.D. Schrader and J.R. Webb. 1978. Comparison of soil properties under continuous corn grain and silage cropping systems. Agron. J. 70:835–837.

Braverman, M.P., J.A. Dusky, S.J. Locascio and A.G. Hornsby. 1990. Sorption and degradation of thiobencarb in three Florida soils. Weed Sci. 38(6):583–588.

Campbell, C.A. and R.P. Zentner. 1993. Soil organic matter as influenced by crop rotations and fertilization. Soil Sci. Soc. Am. J. 57:1034–1040.

Chaney, K. and R.S. Swift. 1984. The influence of organic matter on aggregate stability in some British soils. J. Soil Sci. 35:223–230.

Crovetto Lamarca, C. 1996. Stubble over the soil: The vital role of plant residue in soil management to improve soil quality. ASA, Madison, WI. 245 pp.

Davidson, E.A. and I.L. Ackerman. 1993. Changes in soil carbon inventories following cultivation of previously untilled soils. Biogeochemistry 20:161–193.

Edwards, W.M., M.J. Shipitalo and L.D. Norton. 1988. Contribution of macroporosity to infiltration into a continuous corn no-tilled watershed: Implications for contaminant movement. J. Contam. Hydrol. 3:193–205.

Greb, B.W., D.E. Smika and A.L. Black. 1967. Effect of straw mulch rates on soil water storage during summer fallow period in the Great Plains. In: SSSA Proc. 31:556–559.

Grovers, G., K. Vandaele, P.J.J. Desmet, J. Poesen and K. Bunte. 1994. The role of tillage in soil redistribution on hillslopes. Eur. J. Soil Sci. 45:469–478.

Hooker, B.A., T.F. Morris, R. Peters and Z.G. Cardon. 2005. Long-term effects of tillage and cornstalk return on soil carbon dynamics. Soil Sci. Soc. Am. J. 69:188–196.

Houghton, R.A., J.L. Hackler and K.T. Lawrence. 1999. The U.S. carbon budget: Contributions from land-use change. Science 285:574–577.

Hsiao, T.C. 1973. Plant responses to water stress. Ann. Rev. Plant Physiol. 24:519–570.

Hudson, B.D. 1994. Soil organic matter and available water capacity. J. Soil Water Conserv. 49(2):189–194.

Jensen, M.E. and W.H. Sletten. 1965. Evapotranspiration and soil moisture-fertilizer interrelations with irrigated grain sorghum in the southern High Plains. USDA/TAES Conservation Res. Report no. 5.

Jones, O.R., V.L. Hauser and T.W. Popham. 1994. No-tillage effects on infiltration, runoff, and water conservation on dryland. Trans. ASAE 37:473–479.

Karbozova-Saljnikov, E., S. Funakawa, K. Akhmetov and T. Kosaki. 2004. Soil organic matter status of Chernozem soil in North Kazakhstan: Effects of summer fallow. Soil Biol. Biochem. 36:1373–1381.

Lal, R. 2000. A modest proposal for the year 2001: We can control greenhouse gases in the world with proper soil management. J. Soil Water Conserv. 55(4):429–433.

Lal, R., J.M. Kimble, R.F. Follett and V. Cole. 1998. Potential of U.S. cropland for carbon sequestration and greenhouse effect mitigation. USDA-NRCS, Washington, D.C. Ann Arbor Press, Chelsea, MI.

Lal, R. 2004. Carbon sequestration in soils of Central Asia. Land Degrad. Develop. 15:563–572.
Larson, W.E., R.F. Holt and C.W. Carlson. 1978. Residues for soil conservation. In W.R. Oschwald (ed.). "Crop Residue Management Systems," ASA Special Publication 31. ASA, Madison, WI: 1–15.
Le Bissonnais, Y. 1990. Experimental study and modelling of soil surface crusting processes. In: B. Bryan (ed) "Soil Erosion: Experiments and Models," Catena Verlag: Cremlingen-Destedt: 13–28.
Lee, K.E. 1985. Earthworms: Their ecology and relationship with soils and land use. Academic Press, New York. 411 pp.
Lindstrom, M.J., W.W. Nelson and T.E. Schumacher. 1992. Quantifying tillage erosion rates due to moldboard plowing. Soil Tillage Res. 24:243–255.
Lobb, D.A. and R.G. Kachanoski. 1999. Modelling tillage translocation using steppe, near plateau, and exponential functions. Soil Tillage Res. 51:261–277.
Lobb, D.A., R.G. Kachanoski and M.H. Miller. 1995. Tillage translocation and tillage erosion on shoulder slope landscape positions measured using 137Cesium as a tracer. Can. J. Soil Sci. 75:211–218.
Mannering, J.V. and L.D. Meyer. 1961. The effects of different methods of corn stock residue management on runoff and erosion as evaluated by simulated rainfall. Soil Sci. Soc. Am. Proc. 25:506–510.
Papendick, R.I., D.K. McCool and H.A. Krauss. 1983. Soil conservation: Pacific Northwest. In H.E. Dregne and W.O. Willis (eds) "Dryland Agriculture," Agronomy 23. ASA, Madison, WI.
Peterson, G.A., A.D. Halvorson, J.L. Havlin, O.R. Jones, D.J. Lyon and D.L. Tanaka. 1998. Reduced tillage and increased cropping intensity in the Great Plains conserves soil C. Soil Tillage Res. 47:207–218.
Poesen, J., B. Wesenael, G. Govers, J. Martinez-Fernadez, B. Desmet, K. Vandaele, T. Quine and G. Degraer. 1997. Patterns of rock fragment cover generated by tillage erosion. Geomorphology. 18:193–197.
Power, J.F., D.L. Grunes and G.A. Reichmann. 1961. The influence of phosphorus fertilization and moisture on growth and nutrient absorption by spring wheat: I. Plant growth, N uptake, and moisture use. Soil Sci. Soc. Am. Proc. 25:207–210.
Reicosky, D.C. 1998. Strip tillage methods: Impact on soil and air quality. In P. Mulvey (ed) "Environmental Benefits of Soil Management," Proc. ASSSI National Soils Conf., Brisbane, Australia:56–60.
Reicosky, D.C. 1997. Tillage-induced CO_2 emission from soil. Nutrient Cycling Agroecosystems. 49:273–285.
Reicosky, D.C., S.D. Evans, C.A. Cambardella, R.R. Allmaras, A.R. Wilts and D.R. Huggins. 2002. Continuous corn with moldboard tillage: Residue and fertility effects on soil carbon. J. Soil Water Conserv. 57(5):277–284.
Reicosky, D.C., W.D. Kemper, G.W. Langdale, C.W. Douglas, Jr. and P.E. Rasmussen. 1995. Soil organic matter changes resulting from tillage and biomass production. J. Soil Water Conserv. 50:253–261.
Reicosky, D.C. and M.J. Lindstrom. 1993. Fall tillage method: Effect on short-term carbon dioxide flux from soil. Agron. J. 85:1237–1243.
Reicosky, D.C. and M.J. Lindstrom. 1995. Impact of fall tillage and short-term carbon dioxide flux. In R. Lal (ed) "Soil and Global Change," Lewis Publishers, Chelsea, MI: 177–187.
Ritchie, J.T. 1983. Efficient water use in crop production: Discussion on the generality of relations between biomass production and evapotranspiration. In H.M. Taylor, W.R. Jordan and T. R. Sinclair (eds), "Limitations to Efficient Water Use in Crop Production," ASA-CSSA-SSSA, Madison, WI: 29–43.
Robert, M. 1996. Aluminum toxicity a major stress for microbes in the environment. In P.M. Huang et al (eds), "Environmental Impacts, Vol. 2, Soil component interactions," CRC Press, Boca Raton, FL: 227–242.
Robert, M. 2001. Carbon sequestration in soils: Proposals for land management. AGLL, FAO, United Nations, Rome. Report No. 96. 69 pp.
Schlesinger, W.H. 1985. Changes in soil carbon storage and associated properties with disturbance and recovery. In J.R. Trabalha and D.E. Reichle (ed), "The changing carbon cycle: A global analysis," Springer Verlag, New York: 194–220.
Schumacher, T.E., M.J. Lindstrom, J.A. Schumacher and G.D. Lemme. 1999. Modelling spatial variation and productivity due to tillage and water erosion. Soil Tillage Res. 51:331–339.
Skidmore, E.L., M. Kumar and W.E. Larson. 1979. Crop residue management for wind erosion control in the Great Plains. J. Soil Water Conserv. 34:90–94.
Smika, D.E. 1976. Seeds zone soil water conditions with reduced tillage in semiarid Central Great Plains. In Proc. 7th Conf Intl. Soil Tillage Res. Organ., Uppsala, Sweden 37:1–9.
Smika, D.E., H.J. Haas and J.F. Power. 1965. Effects of moisture and nitrogen fertilizer on growth and water use by native grass. Agron. J. 57:483–486.
Smika, D.E. and P.W. Unger. 1986. Effect of surface residues on soil water storage. In B.A. Stewart (ed), "Advances in Soil Science," Vol. 5. Springer-Verlag New York: 111–138.
Soane B D. 1990. The role of organic matter in the soil compactibility: A review of some practical aspects. Soil Tillage Res. 16:179–202.

Steiner, J.L. 1994. Crop residue effects on water conservation. In P. Unger (ed), "Managing Agricultural Residues," CRC Press/Lewis Publishers, Boca Raton, FL: 41–76.

Tebrugge, F. and R.A. Guring. 1999. Reducing tillage intensity – a review of results from a long-term study in Germany. Soil Tillage Res. 53:15–28.

Todd, R.W., N.L. Klocke, G.W. Hergert and A.M. Parkhurst. 1991. Evaporation from soil influenced by crop shading, crop residue, and wetting regime. Trans. ASAE. 34(2):461–466.

Unger, P.W. 1978. Straw-mulch rate effect on soil water storage and sorghum yield. Soil Sci. Soc. Am. J. 42:486–491.

Unger, P.W. and J.J. Parker, Jr. 1968. Residue placement on decomposition, evaporation, and soil moisture distribution. Agron. J. 60:469–427.

Unger, P.W. 1994. Managing agricultural residues. CRC Press/Lewis Publishers, Boca Raton, FL.

Unger, P.W., G.W. Langdale, and R.I. Papendick. 1988. Role of crop residues-improving water conservation and use. In W.L. Hargrove (ed). "Cropping Strategies for Efficient Use of Water and Nitrogen," Special publication No. 51. ASA, Madison WI: 69–79.

Uri, N.D. 1999. Conservation tillage in U.S. agriculture. In "Environmental, Economic, and Policy Issues," The Haworth Press, Inc., Binghamton, NY: 130.

Wilhelm, W.W., J.M-F. Johnson, J.L. Hatfield, W.B. Voorhees and D.R. Linden. 2004. Crop and soil productivity response to corn residue removal: A literature review. Agron. J. 96:1–17.

Wilts, A.R., D.C. Reicosky, R.R. Allmaras and C.E. Clapp. 2004. Long-term corn residue effects: Harvest alternatives, soil carbon turnover, and root-derived carbon. Soil Sci. Soc. Am. J. 68:1342–1351.

CHAPTER 16

Conservation agriculture for irrigated agriculture in Asia

Ken Sayre
International Maize and Wheat Improvement Center (CIMMYT), El Batan, Texcoco, Mexico

1 INTRODUCTION

It is self-evident that the current level of food security that has existed in many countries in Asia, particularly in south Asia (India, Pakistan and Bangladesh) and in China since the mid 1970s, and to some extent in Central Asia since the mid 1990s following the collapse of the Soviet Union (Uzbekistan for example) has been the result of increased production of staple food crops like wheat, rice and maize from irrigated lands. The remarkable increase in food production from irrigated land when combined with well-managed and well-fertilized, high-yielding modern cultivars has provided a stark contrast to the persistent threat of famine that was common in many developing countries before the early 1970s.

Irrigated production systems in Asia can be broadly differentiated into two main categories: flood and basin irrigation, and furrow irrigation.

1.1 *Irrigated production systems where flood and basin irrigation is used*

Flood and basin irrigation is the main gravity-base irrigation category used in Asia (Figure 1). One of the main cropping systems in this category is the puddled, transplanted rice-wheat system, which covers 32% of the total rice and 42% of the total wheat area across the Indo-Gangetic Plain (IGP), which encompasses parts of Pakistan, India, Bangladesh and Nepal. This system occupies 10 million hectares (Mha) in India, 2.2 Mha in Pakistan, 0.8 Mha in Bangladesh, 0.5 Mha in Nepal, and is also extremely important in many areas in China covering more than 7 Mha as well as in substantial areas in Central Asia (Hobbs and Morris, 1996).

Other crops such as winter maize, potatoes, winter oil seeds and pulses may replace wheat in the rotation with rice but are of less importance and, if water is available, some farmers may include a third crop like mungbean in the spring before the monsoon rice crop or they may also plant sugar cane, harvesting many ratoon crops over several years, before reverting back to rice production. This production system can also include more than one rice crop per year in rotation with other crops or continuous rice culture, especially in the warmer areas in the eastern Gangetic Plains in India and Bangladesh and in southern China.

Flood and basin irrigation is also used in some regions of south Asia and China where rice is not a component in the cropping system. Crop rotations like wheat-maize (Yellow River basin of China), wheat-soybean (central India and parts of northern China) and wheat-cotton in China, India and Pakistan can be flood irrigated. In some cases there may be some "hilling-up" of cotton and other row crops for weed control that can lead to a semblance of furrow irrigation later in the crop cycle.

A universal, key characteristic of flood and basin irrigated, rice-based production systems is that all crops are normally planted on the flat in variable sized, bordered basins whose dimensions largely depend on field size and shape and the degree of leveling. As a consequence of the presence

Figure 1. Flood and basin irrigation in northwest Mexico.

of rice, typically extensive dry and wet tillage (puddling) is used for rice, which leads to important soil management problems for other crops grown sequentially after rice.

1.2 *Irrigated production systems where furrow irrigation is used*

Furrow irrigation (Figure 2) is the other main gravity-based irrigation system used in Asia. Normally all crops in the various production systems are consistently irrigated with furrows. Interestingly, the use of furrow irrigation in Asia has traditionally been linked to many areas with cotton production, especially in west Asia (Turkey, Armenia, Azerbaijan and Iran), in Central Asia (Uzbekistan, Turkmenistan, Tajikistan, Kyrgyzstan and southwest Kazakhstan as well as in parts of western China. The cotton-wheat rotation is perhaps the most significant irrigated system in Asia using furrow irrigation but furrow irrigation is also used for other crops like maize, alfalfa, fodder sorghum, sunflower and/or soybean which may also be included in this production system.

The historical rationale to use flood and basin irrigation versus furrow irrigation in the different, geographical areas in Asia is somewhat difficult to comprehend except that for the rice-based production systems, shifting from flood and basin irrigation required for puddled, transplanted rice to furrow irrigation for succeeding crops would entail considerable effort. There are interesting and promising attempts, however, in India, Bangladesh and Pakistan to grow rice as well as all other crops in rotation on beds with furrow irrigation.

The fact that furrow irrigation has traditionally been more commonly used where cotton was the main crop possibly indicates that cotton probably performs better on the raised bed or ridge between the irrigation furrows as opposed to being flooded with each irrigation or that mechanical weed control was more straightforward.

Figure 2. Furrow irrigation in northwest Mexico.

It is also likely that field size has played a role. Flood irrigation is far more prevalent in south Asia and China where individual irrigated fields (even those not used for rice) are smaller and therefore may not be conducive for furrow irrigation whereas in west and central Asia, where furrow irrigation is more common, field size tends to be larger.

There are other irrigation systems, including sprinkle irrigation and drip irrigation systems that are observed throughout Asia, although on a very limited scale and usually for high value crops like vegetables. This chapter will not include a discussion of the application of conservation agriculture (CA) technologies for these irrigation systems. The chapter discusses how the application of CA to flood and basin irrigation systems versus furrow irrigation systems requires a somewhat different slant for the development and use of CA technologies. Examples are also presented of efforts that have been made to develop appropriate CA technologies for irrigated production systems in Asia.

2 APPLICATION OF CA FOR IRRIGATED PRODUCTION SYSTEMS IN ASIA

Conservation Agriculture implies the adoption of:

- marked reductions in tillage (with zero till/controlled till seeding as an optimal goal);
- rational retention of sufficient levels of crop residues on the soil surface to provide soil protection and contribute to the enhancement of soil quality (with elimination of burning as a residue management option);
- diversified, biologically viable crop rotations, which are economically feasible for farmers to employ.

Farmers who have already adopted sound CA technologies can generally be characterized as follows:

- The vast majority follow rainfed cropping systems;
- Most are normally larger-scale farmers with ready access to appropriate zero till seeders and other implements required for the successful adoption of CA technologies;
- Most are located in developed countries like Canada, the USA and Australia or in developing countries like Brazil and Argentina that have more advanced agricultural systems and/or well organized and motivated farmer organizations that have stimulated efforts to develop and extend appropriate CA technologies.

However, in nearly all developing countries, with limited exceptions like Brazil, Argentina, Paraguay, Bolivia and Ghana, the adoption of CA technologies, even by larger-scale farmers following rainfed production systems has been very restricted and this lack of adoption is even more dramatic for small and medium scale rainfed farmers in nearly all countries.

It is troubling that the extension and adoption of CA technologies for rainfed conditions in most developing countries has not progressed more rapidly given that CA technologies and appropriate implement designs already exist for rainfed production systems. This chronic, low level of adoption by rainfed producers in most developing countries must be carefully considered when assessing how best to develop and extend appropriate CA technologies for farmers following irrigated production systems since there has been negligible adoption of CA technologies by farmers (large, medium and small scale) who follow irrigated production systems in both developed as well as developing countries. This is particularly the case where surface delivery of irrigation water, either by flood and basins or by furrows, are the common practices in west, central and south Asia and China.

At present virtually all farmers that utilize gravity based irrigated production systems, continue to employ extensive tillage combined with either the incorporation of the minimal, left over crop residues that remain after partial or full removal for fodder, fuel or other purposes. In addition, lamentably, burning of crop residues is becoming an increasingly common residue management practice in many irrigated areas. It is quite apparent that all of these widely used crop residue management strategies based on residue removal/burning result in insignificant levels of soil surface cover needed to reduce wind and water erosion, to reduce evaporation and to enhance soil sustainability as well as bestowing dramatically diminished opportunities to even maintain let alone increase soil organic matter (SOM) levels. Therefore, it seems clear that a dichotomy exists concerning the current available knowledge and technologies to afford opportunities to extend CA to farmers using irrigation in developing countries, especially those that follow irrigated wheat-based production systems.

First, less than 5% of total wheat production for developed countries (and some advanced developing countries like Brazil and Argentina) comes from irrigated production systems whereas there has been extensive development and adoption of CA technologies for rainfed conditions signifying that there should be considerable potential for adoption. Improved technology is already on the shelf for introduction, modification and extension to developing country, especially to rainfed farmers producing wheat.

In contrast, well over 50% of total developing country wheat production comes from irrigated production systems, and this proportion reaches far higher levels for the major Asian wheat producers like China, India, and Pakistan and for several Central Asian and Caucasus Republics. Yet there has been extremely limited development and adoption of CA by irrigated farmers anywhere, including developed countries, leaving the technology shelf rather bare.

The numerous well documented, dramatic examples of wind and water driven soil erosion have occurred in many rainfed production areas (See Figure 3) associated with extensive tillage, crop residue removal and mono-cropping which have resulted in dramatic reductions in crop productivity from the associated decline in SOM and the subsequent degraded soil chemical, physical and biological parameters. These occurrences have been the main motivation for the development of sound CA technologies for rainfed systems in many locations.

Figure 3. Example of soil erosion due to wind from a tilled and fallowed, rainfed wheat field near Haymana, Turkey (Photo by Ken Sayre).

3 CONSERVATION AGRICULTURE TECHNOLOGIES FOR IRRIGATED PRODUCTION SYSTEMS

As indicated above, throughout Asia there are two main categories for delivery of irrigation water at the field-level – flood and basin irrigation and furrow irrigation. Examples of the adoption of CA technologies in Asia for these two categories are given below.

3.1 *CA Technologies for flood and basin irrigation – rice-wheat systems in India and Pakistan*

As indicated above, the rice wheat system is likely the most widely use irrigated production system in Asia. It is a system characterized by many small-scale farmers who largely have only marginal access to mechanization. It is of immense interest, therefore, to note that currently, by far the largest concentration of farmer adoption of CA technologies for irrigated production conditions anywhere in the world involves zero till wheat in the rice-wheat system in the IGP (mainly in the states of Haryana, Punjab and UP in northwest India and in the Punjab state in Pakistan). This adoption has occurred at an explosive rate for the time period from the 1998–99 to the 2003–04 crop cycles as illustrated in Figure 4, and was estimated to be nearly 2 Mha for the 2004–05 cycle.

In contrast to essentially all other examples throughout the world where extensive farmer adoption of CA has occurred, the adoption of zero tilled wheat in the IGP has predominantly involved medium and small scale farmers who have benefited from a strong farmer participatory effort supported by the Rice Wheat Consortium (RWC), which includes the National Research Organizations in India, Pakistan, Bangladesh and Nepal together with the associated international research centers including CIMMYT and IRRI, to develop sound CA technologies and the corresponding

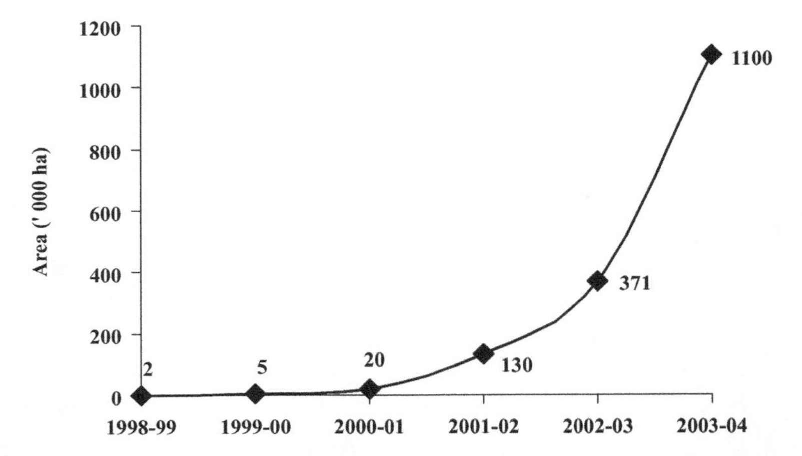

Figure 4. Rate of adoption for zero till wheat in the rice-wheat cropping system in the IGP (Gupta et al., 2003).

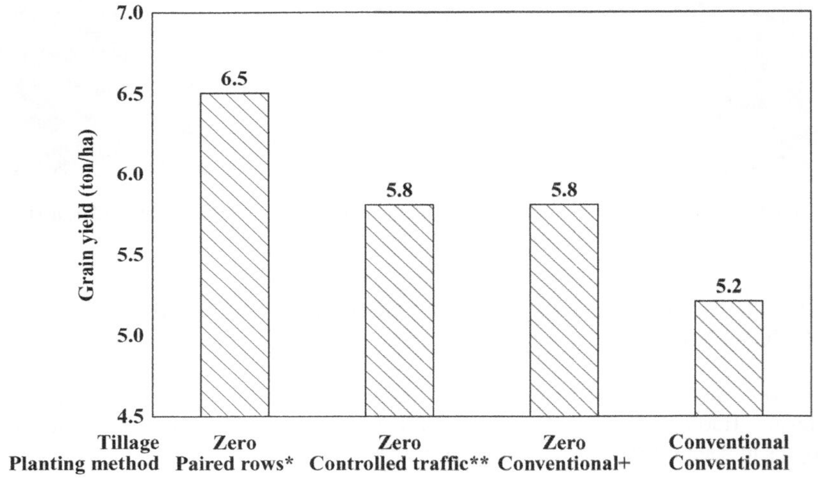

Figure 5. Effect of planting method and tillage on wheat yield after rice in Ghazabad, India during the 2000/01 crop season (Rice-Wheat Consortium, 2003).

appropriately-scaled zero till seeders needed by these farmers. The stimulation of wide-scale production and sales of zero till seeders, mainly small, privately owned machinery manufacturers has been paramount to the rate of farmer adoption shown in Figure 4.

The rapid adoption of zero till wheat by farmers in the IGP has been stimulated by the yield advantage from zero till seeding as compared to conventional tillage as illustrated in Figure 5 (personal communication from Dr. Raj Gupta, facilitator of the RWC) which also compares different planting methods for zero tilled wheat. Several factors have been attributed for the higher yield level for zero till seeded wheat including reduced grass weed problems and more efficient fertilizer (especially N) and water use but the opportunity to plant wheat on time (harvest rice today and plant wheat tomorrow) by eliminating the normal 7+ days required for conventional tilled wheat probably is of most importance (Gupta et al., 2003).

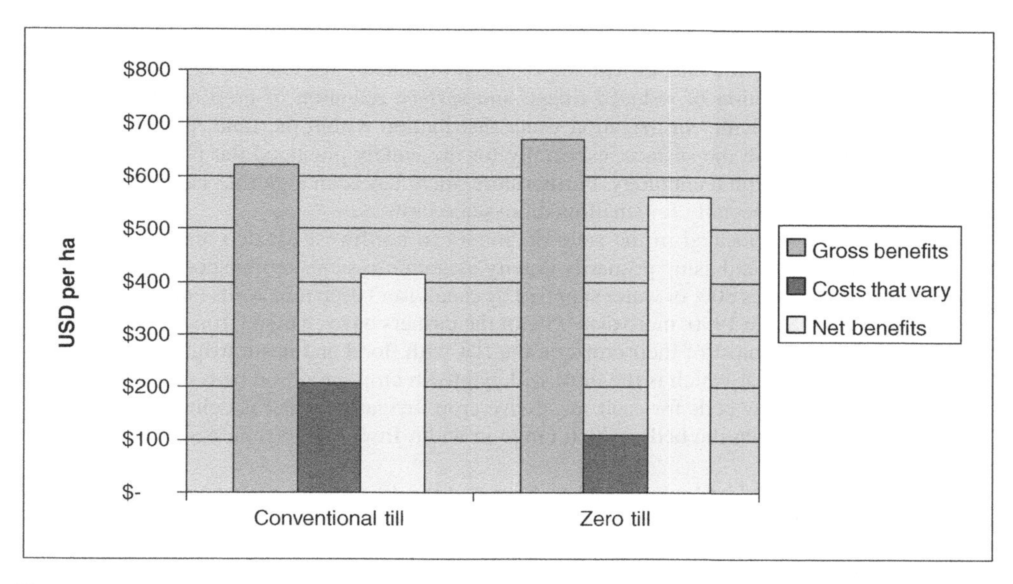

Figure 6. Economics of zero till wheat versus conventional till wheat in India (Gupta et al., 2003).

The significantly higher wheat yields that farmers are able to achieve with zero till seeding combined with the dramatic reductions in tillage costs have provided farmers with remarkable increases in economic net benefits which has been the main stimulus for farmer adoption as illustrated by Gupta et al. (2003) in Figure 6.

However, the application of CA technologies for the rice-wheat system in the IGP provides an excellent example how farmer adoption of CA technologies is many times a step-by-step adoption process, not an immediate, all-inclusive adoption of the CA tenets across each component of the overall cropping system. So far, the adoption of CA technologies in India and Pakistan has only impinged upon the wheat phase of the irrigated rice-wheat system (although efforts to apply CA to rice by developing zero till, direct seeding technologies for rice are underway). The extensive, heavy tillage used prior to planting the puddled, transplanted rice crop is still practiced by nearly all farmers who have adopted zero till wheat seeding which likely constrains or perhaps even reverses potential improvements in sustainability issues related to soil physical, biological and chemical parameters associated with the zero till wheat establishment. This situation exists, however, simply because the technologies (including appropriate rice varieties and crop management practices for zero till rice are still under development.

In addition, considerable partial burning of loose rice straw where rice is combine harvested is still common, neglecting another CA tenet mainly due to the lack of zero till seeders that can readily plant into high levels of rice straw retained on the surface without problems associated with raking/dragging the loose straw. However, rapid progress in seeder development to handle this is underway in the region.

Similarly, it is unfortunate that, as wheat has been brought under zero till planting, there has been only marginal emphasis towards the much needed crop diversification of the rice-wheat rotation. However, government crop and irrigation water pricing and subsidy policies combined with the lack of appropriate CA technologies for other crops have largely stymied diversification of the rice-wheat system. The RWC, however, is now more fully addressing these issues.

3.2 *Permanent raised bed planting systems for furrow irrigated systems*

The development of permanent raised bed planting systems for use in existing furrow irrigated systems and perhaps as a superior CA technology to replace flood and basin irrigation is a new innovation in some regions. This has been a goal for CIMMYT agronomists working in Mexico and in collaborative efforts with various national research organizations. One of the main constraints that has limited application of reduced tillage and surface retention of crop residues has been associated with the difficulty in irrigation water distribution within the field, particularly when loose residues are left on the surface, especially for the widely practiced flat planted, flood and basin irrigation production technology. Furthermore, there has been a glaring lack of appropriate seeding implements, especially for small/medium scale farmers.

The Yaqui Valley is located in the state of Sonora in northwest Mexico and includes about 255,000 ha of irrigated land using primarily gravity irrigation systems to transport irrigation water to fields from canals (over 80% of water supplied by canals) and deep tube wells (around 20% water supply from wells). Since 1990, more than 95% of the farmers have changed from the conventional technology of planting most of their crops on the flat with flood and basin irrigation, to planting all crops, including wheat which is the most widely grown crop, on raised beds albeit with tillage and the formation of new beds for each successive crop. Irrigation water is delivered through the fields by furrows between the beds, which range in width from 70–100 cm, center bed to center bed (Aquino, 1998).

A single row is planted on top of each bed for row crops like maize, soybean, cotton, sorghum, safflower and dry bean with 1–2 rows per bed planted for crops like chickpea and canola but 2–3 rows, spaced 15–30 cm apart depending on bed width, are used for wheat. Wheat yields for the Yaqui Valley have averaged over six tons per ha for the past several years. Farmers growing wheat on beds obtain about 8% higher yields with nearly 25% less operational costs as compared to those still planting conventionally on the flat, using border/basin flood irrigation (Aquino, 1998).

Most farmers practice conventional tillage by which the beds are destroyed after the harvest of each crop by several tillage operations before new beds are formed for planting the succeeding crop. This tillage is often accompanied by widespread burning of the crop residues although some maize and wheat straw is baled-off for fodder and, when turn-around-time permits, some crop residues are incorporated during tillage (Meisiner et al., 1992).

There has been intense farmer interest in the development of new production technologies that will allow marked tillage reductions combined with retention of crop residues which may lead to potential reductions in production costs, improved input-use efficiency, more rapid turn-around-time between crops and more sustainable soil management while allowing continued use of the more inexpensive gravity irrigation system. Therefore, a long term experiment was initiated in 1992 in the Yaqui Valley to compare the common, farmer practice based on extensive tillage to destroy the existing beds with the formation of new beds for each succeeding crop versus the permanent raised bed system where new beds are initially formed after a final tillage cycle and are then reused as permanent beds with only superficial reshaping as needed before planting of each succeeding crop (Sayre and Moreno Ramos, 1997).

It is common that conversion from conventional tillage to a reduced or zero till seeding system with residue retention may require several crop cycles before potential advantages/disadvantages begin to become apparent (Blevins et al., 1984) and the results from this long-term trial confirm this observation.

Figure 5 presents the yield trends over the twelve wheat crop cycles when 300 kg N/ha was applied at the first node stage. Only small yet significant yield differences between the tillage/residue managements occurred from 1993 to 1997. However, beginning with the 1998 wheat crop, large and significant differences between the tillage/residue management options occurred even though year-to-year yield effects were large (Figure 5). The permanent bed treatment with continuous crop residue retention demonstrated the highest average yield followed by conventional tillage with residue incorporation, then the permanent bed treatment with partial removal for fodder and lastly, with markedly reduced yields after 1997, the permanent beds with residue burning.

Following the 1997 wheat crop, the lowest yields have been continually observed for permanent beds with residue burning and in particular for a poor yielding crop cycle like 2004, the wheat yield for permanent beds with residue burning suffered larger, relative yield reduction as compared with

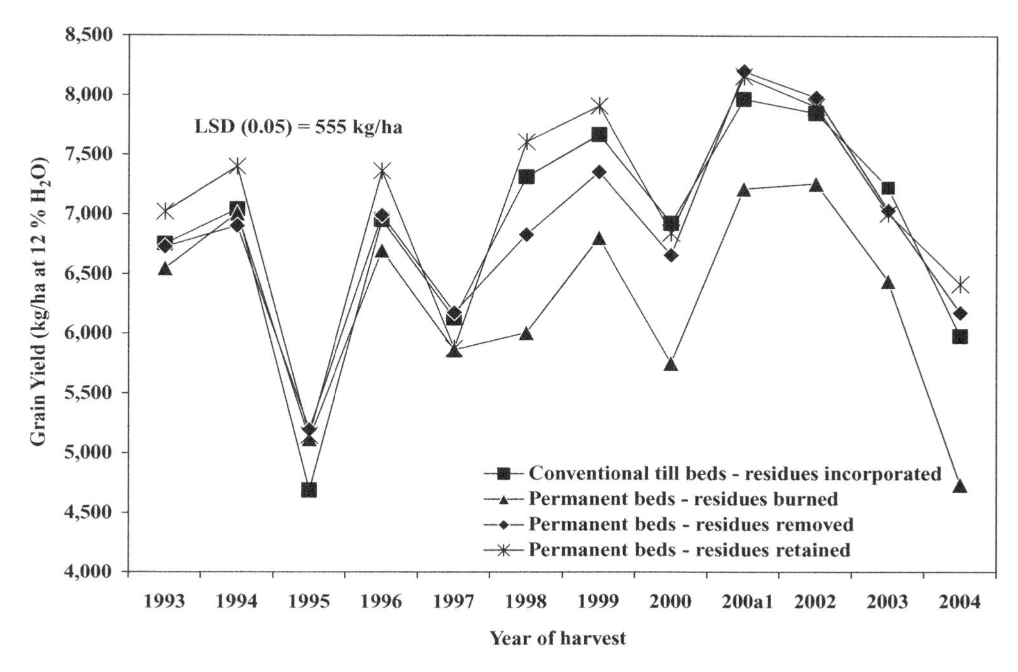

Figure 7. Effect of tillage and residue management over years on wheat yield at CIANO/Cd. Obregon.

the other management practices (Figure 7). It is clear that residue burning is not compatible with the permanent bed technology under these production conditions.

Permanent beds, with partial residue removal for fodder, have resulted in much smaller or no yield reductions in most years as has occurred with residue burning, especially when compared to permanent beds with full residue retention. Since at least 30–40% of the loose residues and/or standing stubbles are not removed for fodder, they appear to provide adequate ground cover to benefit soil quality-related properties. In addition, the economic value of the residue removed for fodder will likely override the associated, small grain yield reductions, at least in the short term. The yield differences between permanent beds with residue retention and conventional till beds with residue incorporation, were consistently small for most years. More importantly, however, the permanent beds provided a 20–30% savings in production costs. There was only a small interaction for year by tillage/residue management for wheat yield.

A number of soil chemical, physical and biological parameters have been regularly monitored throughout the experimental period for this long-term trial. Table 1 presents a brief summary of some of these parameters that were measured during the 2001/02 wheat-growing season.

The soil chemical parameters indicate that while pH was not different for the tillage/residue management treatments, Na was significantly less for permanent beds where part or all of the crop residues had been left on the soil surface. Higher levels of Na occurred with the conventional tilled beds with residue incorporation but highest Na levels occurred for permanent beds with residue burning. This is an exceedingly important result because it indicates that for soils which may tend towards the development of salinity problems, the use of permanent beds with retention of crop residues on the soil surface may help ameliorate this trend towards saline conditions by reducing Na accumulation in the beds through minimizing evaporation from the soil surface due to the mulching effect of the retained residues and by the improvement of soil physical properties that may enhance Na removal by enhanced leaching action.

The SOM levels were lowest for conventional tillage with residue incorporation and higher for permanent beds, especially with full of crop residue retention (Table 1). P levels were similar across the tillage/residue management treatments.

Table 1. Effect of tillage and crop residue management on soil properties for a long-term bed planted trial initiated at CIANO, Cd. Obregon, Sonora in 1993. (Results reported here are from 0–7 cm soil samples taken during the Y2001/2002 crop cycle).

Tillage/residue management	PH (H_2O) 1:2	% OM	% Total N	% Olsen	Na ppm	MWD* mm	SMB** mg kg^{-1} soil	SMB** mg kg^{-1} soil
Conventional till/ incorporate residue	8.13	1.23	0.069	10.6	564	1.26	464	4.88
Permanent beds/ burn residues	8.10	1.32	0.071	9.9	600	1.12	465	4.46
Permanent beds/ remove residue for fodder	8.12	1.31	0.074	12.1	474	1.41	588	6.92
Permanent beds/ retain residue	8.06	1.43	0.079	10.3	448	1.96	600	9.06
Mean	8.10	1.32	0.073	10.7	513	1.44	529	6.33
LSD (P = 0.05)	0.13	0.15	0.004	5.4	53	0.33	133	1.60

*Mean weight diameter; **Soil microbial biomass.

Table 1 also presents the values for the wet aggregate stability. Permanent beds with residue burning showed the poorest aggregate stability followed by tilled beds. Permanent beds with partial or full residue retention (especially the latter) had the best, wet soil aggregate stability.

Soil microbial biomass determinations of C and N in the biomass clearly indicated the obvious superiority of permanent beds with some or all residue retention compared to either permanent beds with residue burning or conventional tilled beds with residue incorporation (Table 1). This measure of potential soil health favors the permanent beds with residue and correlates with the observations that have been made on root disease scores and soil, pathogenic nematode levels which have been consistently higher for the permanent bed treatment with residue burning (data not shown). It seems clearly evident that the inferior grain yield performance of the permanent bed treatment with crop residue burning as shown in Figure 7 is strongly linked with the unremitting degradation associated with residue burning for most of the soil parameters that have been monitored and which are considered to be associated with the sustainability soil productivity.

The permanent raised bed planting system offers tremendous potential as a way to apply CA technologies to irrigated production systems. Irrigation water savings have consistently been substantial as compared to flood and basin irrigation. CIMMYT, through collaborative efforts with national agricultural development agencies and institutions in several countries, has begun to develop and extent permanent raised bed systems to farmers, including the development of appropriate permanent bed planting implements. It has been successfully applied to a number of crops besides wheat including maize, soybean, chickpea, lentils mungbean, cotton, many vegetables and even including rice, among other crops. It offers a more integrated approach to enhancing diversification of crop rotations. Figure 8 presents an example of the yields for rice produced on raised beds with intermittent furrow irrigation in comparison with zero till rice seeded on the flat and transplanted rice with conventional tillage with puddling, both with flood basin irrigation. As can observed in Figure 8, rice direct seeded on raised beds did comparatively poorly as did zero tilled, direct seeded rice on the flat. Lack of availability of appropriate rice varieties for these planting systems combined with weed control problems explain these relative poor performances. The yield of transplanted rice on beds was best as compared to all other systems indicating its potential.

The cotton-wheat rotation in Uzbekistan provides a good example of how permanent raised beds offers potential improvements for this important, furrow-irrigated system. Table 2 presents some preliminary results from trials in farmer fields that are being conducted within a project involving

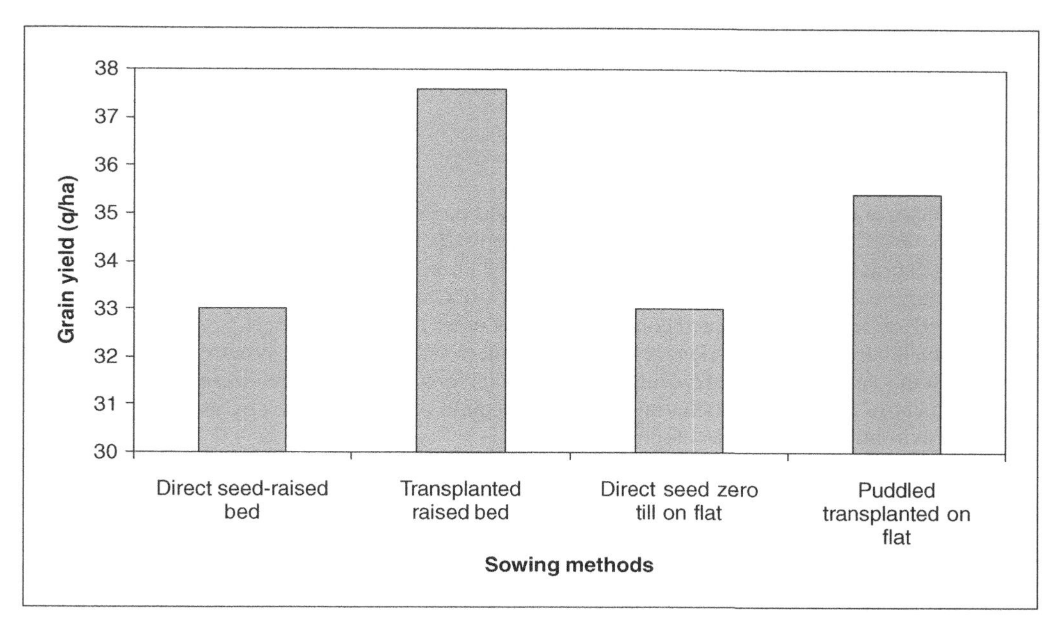

Figure 8. Rice yields form different tillage and planting methods in India during the 2003 monsoon season (Data provided by Dr. Raj Gupta, RWC facilitator) (one q = 100 kg).

Table 2. Cotton and wheat yield results from different tillage/planting systems in Khorezm, Uzbekistan (Data provided by Mehriddin Tursunov and Oybek Egamberdiev – graduate students associated with the project).

	2004 Cotton Lint Yield kg ha^{-1}		2005 Winter Wheat Yield kg ha^{-1}	
Tillage system	+Residues	−Residues	+Residues	−Residues
Conventional farmer practice	2500	2800	4200	5500
Intermediate bed system*	2700	2400	5500	5300
Permanent beds	2100	2050	6000	5600
Zero till on the flat	1200	2100	5400	4800
Mean	2125	2338	5275	4300

*Conventional tillage for cotton with formation of bed before planting; beds are reshaped and reused for other crops in the rotation until the next cotton crop when tillage is again performed to repeat the cycle.

the Center for Development Research, University of Bonn, Germany, the Tashkent Irrigation and Engineering Institute (TIIM) and CIMMYT.

The bed planting system for wheat and other crops has been widely adopted by farmers in northwest Mexico where they depend on the widespread, gravity irrigation systems, however, with continued use of intensive, conventional tillage combined with considerable burning of crop residues. The research reported here indicates that the extensive tillage with its associated high costs and long turn-around- time can be dramatically reduced by the use of permanent raised beds with furrow irrigation. The results, however, also indicate that surface retention of crop residues with permanent beds may be essential to enhance chemical, physical and biological soil parameters that are crucial to insure that long-term production sustainability can be achieved. Farmers must clearly realize this before they attempt to adopt permanent bed planting systems for the primary objective to simply lower production costs inherent with the reductions in tillage while continuing to remove all crop residues for fodder or other purposes or by burning.

4 CONCLUSIONS

The application of sound CA technologies to irrigated production systems everywhere, including Asia, has been extremely miniscule. However, progress is being made as outlined here and it is highly encouraging that so far, the most progress for farmer adoption has been made with small and medium scale farmers in the IGP. This provides a fine and rather unique example that CA can be used effectively under irrigated conditions and by other than large farmers.

The small but growing experiences with permanent raised bed planting systems clearly demonstrate that this approach offers new, innovative methods for water and surface-retained crop residue management in addition to facilitating the diversification of crop rotations by providing a better growth environment for root development.

Without a doubt, through the collaboration of farmers, researchers, extension agents and policy makers, CA will play an ever more important role in enhancing the sustainability of irrigated production systems Asia.

REFERENCES

Aquino, P. 1998. The adoption of bed planting of wheat in the Yaqui Valley, Sonora, Mexico.Wheat Special Report No. 17a. Mexico, D.F.: CIMMYT.

Blevins, R.L., Smith, M.S., Thomas, G.W. 1984. Changes in soil properties under no – tillage. In: R.E. Phillips and S.H. Phillips (Eds.) No – Tillage Agriculture – Principles and Practices. pp. 190–230. Van Nostrand Rheinhold Company, New York.

Gupta R.,C. Meisner, K. Sayre, J.K. Ladha, J.F. Rickman, and L. Harrington. 2003. Annual Progress Report, 2003, Rice Wheat Consortium for the Indo-Gangetic Plains CIMMYT-India, Pusa, New Delhi.

Hobbs, P.R. and M.L. Morris. 1996. Meeting South Asia's future food requirements from rice-wheat cropping systems: priority issues facing researchers in the post green revolution era. In NRG Paper 96-01, pp 1-45. CIMMYT, Mexico, D.F., Mexico.

Meisner, C.A., E. Acevedo, D. Flores, K. Sayre, I. Ortiz-Monasterio, and D. Byerlee. 1992. Wheat production and grower practices in the Yaqui Valley, Sonora, Mexico. Wheat Special Report No. 6. Mexico, D. F.: CIMMYT.

Sayre, K.D. and O.H. Moreno Ramos. 1997. Applications of raised-bed planting systems to wheat. Wheat Special Report No. 31. Mexico, D. F.: CIMMYT.

CHAPTER 17

Syria's long-term rotation and tillage trials: Potential relevance to carbon sequestration in Central Asia

John Ryan & Mustafa Pala
International Center for Agricultural Research in the Dry Areas, Aleppo, Syria

1 INTRODUCTION

Historically, soil and its fertility have been associated with providing enough food to support man's survival on earth. After many millennia since the beginning of settled agriculture, that perception of the primary function of soils has shifted at least for a large swathe of the globe. Although hunger is still stalking many of the world's poor (Borlaug, 2003), soil resources have taken on a broader dimension. With the world's population currently at unprecedented levels, the major global concern of the 21st century are food security, soil degradation by land misuse and soil management, and anthropogenic increases in atmospheric greenhouse gases (Lal, 2001). As available land area per capita decreases, a major challenge to society is to sustainably use the world's soil and water resources (Lal, 2000). The extent to which any ecosystem can withstand land-use pressure or its resilience or potential to recover from a degraded state is a function of the ecosystem in question, with semiarid and arid regions being particularly vulnerable (Stewart and Robinson, 1997). The extent to which any land can be sustainably farmed depends on soil quality from the physical, chemical and biological perspectives (Karlen et al., 2001).

Land use is a complex, dynamic issue; the driving force is how man uses, or abuses the soil within the context of constraints imposed by climate, notably moisture (rainfall) and temperature. While concepts of sustainability may vary according to the users' perspective (Zimdahl, 2006), a key factor that determines soil degradation (Lal, 2001) or cropping productivity (Haynes, 2005; Karlen et al., 1994; Rasmussen et al., 1998) is the relatively small but disproportionately influential component, soil organic matter (SOM) or soil carbon (C) fraction. At the global scale, the equilibrium level of SOM varies with agro-climatic zones, being a net function of the amount of organic matter inputs to the soil, i.e., roots and residues of crops, which are related to crop yields, and the extent to which climatic factors, such as high temperatures and soil moisture promote soil C loss through mineralization (Parton et al., 1987).

While temperate climates favour SOM accumulation, even though tillage can greatly influence the changes in SOM with time, areas of the world with a Mediterranean climate are characterized by low levels of SOM, thus imposing constraints with respect to soil vulnerability and productivity. Consequently, the focus of this chapter is SOM in soils of the Mediterranean region, more specifically the broader area of North Africa and West Asia (WANA). The latter region was the core mandate area of the International Center for Agricultural Research in the Dry Areas (ICARDA), based in the rainfed, cereal-producing belt in northern Syria. Because of the common agro-ecological similarities and constraints, the newly independent countries of Central Asia have come under ICARDA's research mandate in the past decades (Figure 1). As a prelude to a consideration of SOM in soils of the WANA region, brief mention is made of the general climatic and agricultural characteristics of the region, with specific reference to Syria where most of the information on soil C was generated.

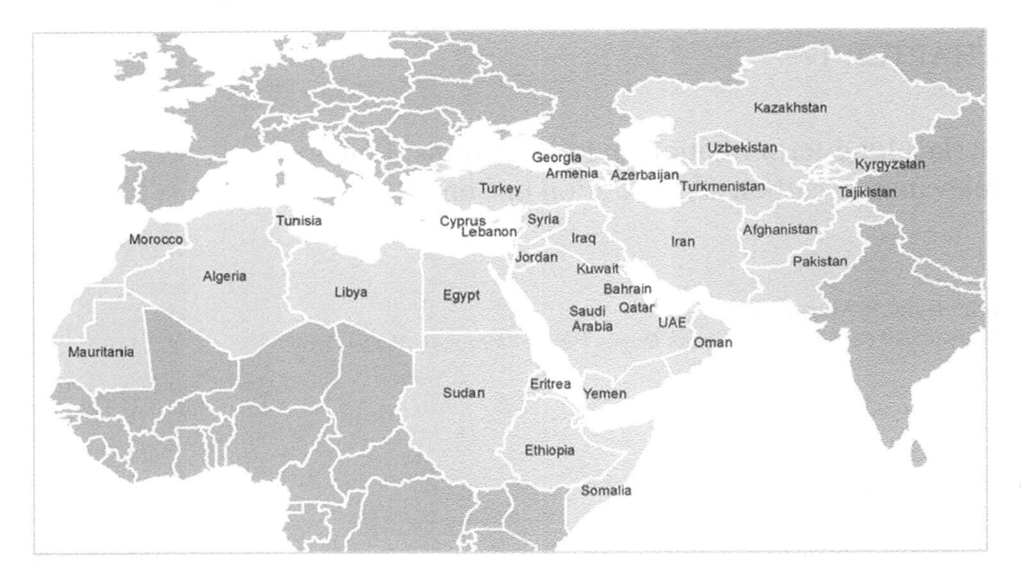

Figure 1. Schematic map of the North Africa-West Asia and Central Asia region.

2 THE MEDITERRANEAN REGION

The lands bordering the Mediterranean Sea have a long history of human settlement and cultivation. The Fertile Crescent of West Asia, comprising Lebanon, Syria, Turkey, and Iraq, is credited with being the center of origin of some of the world's most important crops, notably cereals, legumes, and nuts (Damania et al., 1997). The region is characterized by a typical Mediterranean climate in lowland coastal areas (Kassam, 1981), merging to a continental one of greater climatic extremes inland and in high-elevation areas. Typically, the region has a "wet" cool season from late fall (October-November) to late spring-early summer (April-May) during which rainfed farming is practised (Figure 2) (Cooper et al., 1997; Gibbon, 1981). The crops are mainly cereals, wheat, either bread wheat (*Triticum aestivum*) or durum wheat (*T. turgidum* var *durum*) as well in rotation with a range of crops, i.e., food legumes such as chickpea (*Cicer arietinum*) and lentil (*Lens culinaris*), forage legumes such as vetch (*Vicia sativa*) and medic (*Medicago* spp) and oil-seed crops such as safflower (*Cartamus tinctorus*) and sunflower (*Helianthus annus*). During the long, dry, hot summer-fall period, cropping is only possible with irrigation or in some cases where residual moisture is adequate to grow short-season crops such as watermelon (*Citrullus vulgaris*).

Traditionally, most cropping was rainfed. However, in the past few decades, there has been increasing intensification (Ryan, 2002). The major trends in most countries in WANA have been increasing supplemental irrigation to stabilize wheat yields in rainfed areas (Oweis et al., 1998); this has given rise to concerns about the sustainability of groundwater supply for irrigation. Other trends include increasing mechanization, increased use of chemicals, especially nitrogen (N) and phosphorus (P) fertilizers, and the introduction of new and high- value horticultural crops.

The changes that have been witnessed in the WANA region as a whole have been reflected in Syria, where significant changes have occurred in the relative importance of the cropping systems within the normal rainfall range of less than 200 mm yr^{-1} in the driest area to over 600 mm yr^{-1} in the most favorable zones (Harris, 1995). Thus, as one progresses up the rainfall ladder (Figure 3), deserts and scrubland and irrigation, when water sources are available (<100 mm), native pasture and steppe land (<200 mm), barley/livestock (sheep) from 200–350 mm, wheat-based cropping (350–500 mm), and horticultural crops (olives, pistachio) above 500 mm. With the increased irrigation, fertilizer use has shown a 20-fold increase since the 1970s; the increase for P was about 15-fold, but there was little use of potassium fertilizers (Ryan et al., 1997a).

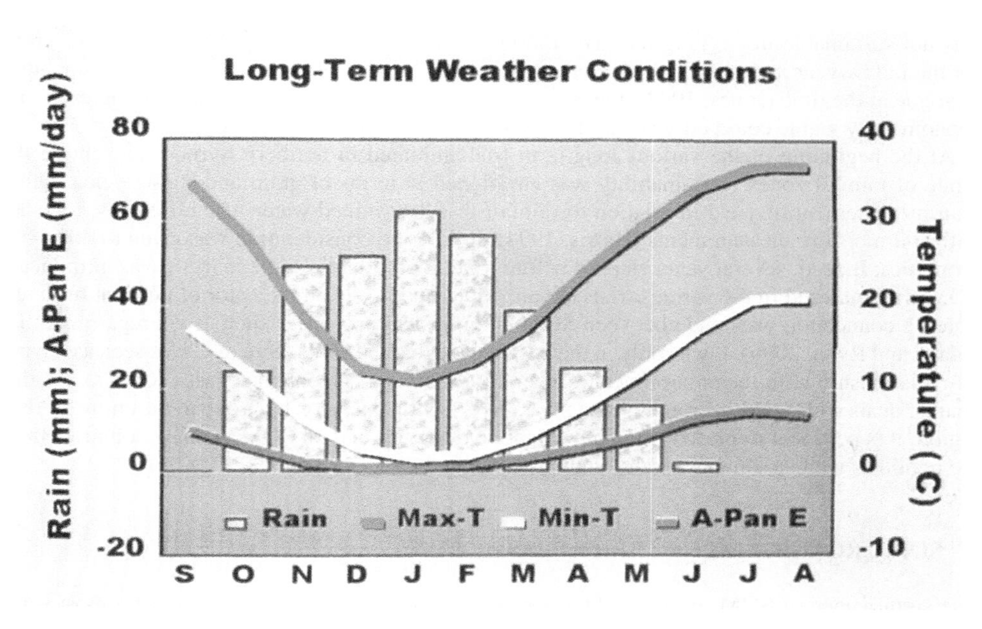

Figure 2. Typical seasonal weather parameters in the Mediterranean-type climate: rainfall, maximum and minimum temperatures, and class A pan evaporation.

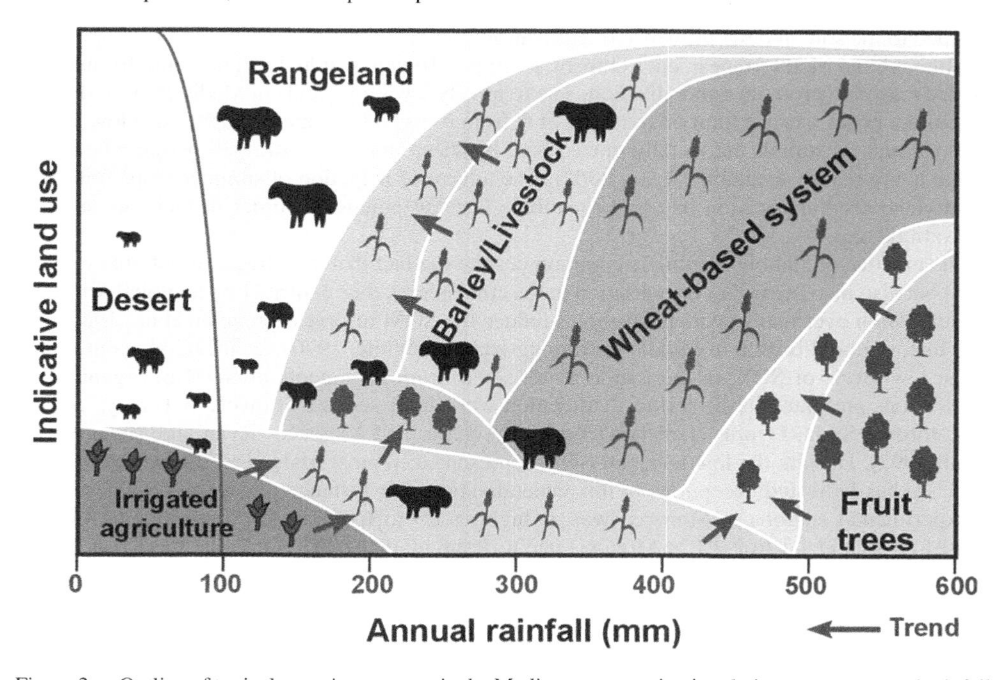

Figure 3. Outline of typical cropping systems in the Mediterranean region in relation to mean annual rainfall.

The rapid pace of change in agriculture in the 1980s led to the establishment of long-term agronomic trials mainly involving crop rotations either with wheat or barley (Ryan and Abdel Monem, 1998). With increasing land use pressure, fallow, which had been traditionally used as a hedge against drought, began to disappear and be replaced by continuous cereal cropping, which was not sustainable in the long run. The rationale for the trials was to provide alternative crops for the fallow year instead of cereal after cereal (Harris, 1995). With the concept of sustainability in vogue at the time (Jones, 1993), the extent to which any cropping system was biologically and economically viable could only be validly assessed over many years.

At the beginning of the various long-term trials initiated in northern Syria, representing the range of rainfall zones, sustainability was envisioned in terms of grain and straw yields. Other than measurement of yield in relation to rainfall that determined water- use efficiency, together with soil moisture measurements (Harris, 1994), little or no consideration was given to other soil properties. Indeed, several years elapsed before routine pre-planting soil analysis was introduced. SOM was analyzed in the various trials (Ryan, 1998) and it was an indicator of nutrient reserves. Later, a connection was made between SOM and physical properties such as aggregate stability (Masri and Ryan, 2006). It was only in the past decade that the soil's store of C was seen as having any relationship with the environment, especially in terms of greenhouse gases (GHGs). As this chapter deals with SOM in cropping systems in Syria, with extension by proxy to the entire WANA region, it is pertinent to present some representative background on SOM in a few countries from the region as well as some generalities about SOM in the region's soils.

3 SOIL ORGANIC MATTER: BRIEF PERSPECTIVE

The significance of SOM in the Mediterranean area hinges around the low levels in most soils, i.e., frequently less than 1%, and generally between 1 to 2%. The question is why is SOM "low" and what does "low" mean? The dominating factors on SOM are environmental, primarily high temperature and sufficient moisture to promote mineralization of SOM, the amount of residues returned to the soil, and the influence of cultivation.

Although the Mediterranean environment, there is sufficient residual soil moisture for much of the dry season to promote mineralization. The relatively low crop yields in Mediterranean dryland agriculture poses a major limit on the input of C from roots and residues. Despite such low inputs, residues such as stubble are usually grazed bare by sheep during the summer period when fresh forage is not readily available (Ryan, 2002). The advent of irrigation of summer crops following the dryland cropping season has lead to stubble burning practice in order to facilitate summer cultivation.

The practice of stubble burning is common despite the fact that it is illegal; it not only reduces SOM but also increases CO_2concentration in the atmosphere, thus contributing to global warming. As cultivation over many years is known to reduce the SOM reserves (Freyman et al., 1982), the long history of cultivation in Middle Eastern agriculture (White, 1970) has inevitably contributed to the low levels of SOM in the region as reflected in measurements of SOM or organic C in publications presented in the soil test cultivation workshops series that involved Turkey, Yemen, Iran, Jordan, Syria, Cyprus, Lebanon, Tunisia, Algeria, and Morocco (Ryan and Matar, 1992; Ryan, 1997). Despite the low levels of SOM represented in most trials, e.g., around 1% or less, some studies indicated exceptions to this generalization. For instance, the mean organic C level at experimental stations in Morocco was as high as 2.3 to 6.0% at Ain Zagh, 1.9 to 4.0% at Sidi El-Aydi, and 1.1 to 2.2% at Monim Lucein (Ryan et al., 1990). Similarly, there was a relationship of organic C with soil type, i.e., 1.9% in a Rendoll, 1.2% in a Calcixeroll, and 1.0% in a Chromoxerert (Abdel Monem et al., 1990).

In contrast, the range of organic C in soils of Lebanon (Ryan et al., 1990) reflected the norm for Mediterranean soils. Similarly, SOM values were relatively low across a rainfall transect in northern Syria with little obvious trends (Ryan et al., 1996, 2002a). However, a survey of agricultural experiment stations and field sites in northern Syria (Ryan et al., 1997b) showed an increase from the drier zone (<200 mm) to the more favorable zone, i.e., from 0.5% SOM at Ghrerife to 1.1% at Jindiress. Despite these differences in SOM in the topsoil, sampling with depth to 200 cm showed the characteristics decrease in values more or less following the same differences as seen in the top soil. A survey of an area irrigated with wastewater, mainly from untreated municipal sewage,

Table 1. Overall mean effects of rotations, nitrogen, and grazing on soil organic matter.

Rotations	Fallow, 1.07%; Melon, 1.07%; Continuous wheat, 1.12%; Lentil, 1.13%; Chickpea, 1.17%; Vetch, 1.2%; Medic, 1.32%
Nitrogen	Control, 1.12%; 30 kg, 1.13%; 60 kg, 1.19%; 90 kg, 1.20%
Stubble grazing	None, 1.20%; Medium, 1.15%; High, 1.14%

indicated higher average values for SOM compared with the norm for rainfed agriculture in the surrounding province (Ryan et al., 2006).

Notwithstanding reference to SOM in most studies from the WANA region (Ryan and Matar, 1992; Ryan, 1999), few if any have elaborated on the significance of this parameter, despite the well-known fact that it serves as a reservoir for soil nutrients, especially N, and is a source substrate for micro-organisms, in addition to influencing physical properties such as soil structure, aggregation and bulk density and related water behavior (infiltration, permeability), and soil erosion (Haynes, 2005). Only two studies dealt with the actual role of SOM soils of the region. Selective removal of the organic component in soils from Lebanon led to the conclusion that SOM is an aggregating agent in soils, while iron oxides also have a role in the process (Arshad et al., 1980). A later study (Habib et al., 1994) showed that SOM mineralization can lead to an increase in available soil P and exhibits a seasonal pattern of availability.

These generalizations are by no means intended to comprehensively reflect the literature on SOM status and all studies of this important constituent in soils from the Mediterranean area. However, it is intended to indicate the content of SOM in Mediterranean soils in general and provide an example of its behavior in relation to soil properties.

4 SOIL ORGANIC MATTER IN LONG-TERM ROTATION TRIALS

Of the 19 long-term rotation trials established at ICARDA in 1980's (Ryan and Abdel Monem, 1998), none was designed with any consideration of SOM; their main focus was *sustainability* of crop yields. Consequently, the parameters recorded were grain and straw yields in the case of cereals, grain in the case of lentil and chickpea, cut forage for vetch, and animal weight gain or offtake in the case of grazed forages. Nevertheless, when a systematic program of soil analysis was instituted for the trials, a new dimension for sustainability in terms of soil quality became apparent. The purpose of this section is to highlight the extent to which the independent variables in ICARDA's long-trials influenced the SOM status.

4.1 *Cropping system productivity trial*

The Cropping Systems Productivity trial was the institution's "flagship" cereal-based rotation trial (Harris, 1995). Started in 1983/84 and terminated in 1997/98, it involved rotations of durum wheat grown in sequence with fallow, watermelon as a summer crop, continuous wheat, lentil, chickpea, vetch, and medic. From 1986/87 onwards, subsidiary treatments were imposed. These treatments involved applying N fertilizer to the cereal phase at 0, 40, 80, and 120 kg ha^{-1}, and three grazing regimes of the cereal stubble following harvest: no grazing or stubble retention, medium grazing, and intensive grazing. Both cereal and non-cereal phases were present each year.

During the period in which SOM was routinely measured (1989–97) there was a gradual overall increase in the SOM in the rotations, which varied with the crop in question. In the last 2–3 years of the trial, an equilibrium seems to have been reached. Despite the absence of initial baseline measurements of SOM at the start of the trial, clear differences had emerged with respect to SOM in the rotations as well as the subsidiary treatments, i.e., N and stubble grazing (Table 1).

The fallow and the pseudo-fallow (melon) had the lowest SOM levels, continuous wheat, lentil, and chickpea were intermediate, while the highest was from medic and second-highest with vetch, both forage crops. Nitrogen fertilizer application produced a small but consistent increase in overall SOM levels across all treatments, while the opposite effect was experienced with grazing intensity. Differences due to grazing were manifested in the last 6 years of the trial, whereas the rotation and N effects were evident from the first sampling period in 1989. While limited depth wise sampling was done, it was clear that the enrichment of SOM in the medic plots extended down to at least 60 cm in comparison with the other rotations; probably reflecting the deep root system of medic.

An interesting feature of the trial was that the SOM in the rotations was directly related to an effect on physical properties (Masri and Ryan, 2006). Thus, the extent of dispersion was highest in the continuous wheat plots followed by the fallow and melon, then with chickpea, lentil and vetch being similar, and medic least. In other words, medic with the highest SOM had the highest aggregation. Infiltration and hydraulic conductivity followed an increasing trend with SOM and the degree of aggregation.

In a sub study, conducted in a few representative plots over a period of 3 years, measurements of biomass C and labile C were taken at various times throughout the cropping seasons. While there were variations in labile C as the season progressed, and values declined during the post-cropping summer period, the values for medic and vetch were relatively similar and both were considerably higher than continuous wheat and fallow. However, biomass C, though a much smaller fraction of total SOM, was more variable than labile with fallow being lowest, but all values decreased to almost zero in the summer period.

The SOM is a complex entity derived from the remains of plants and animals. It have has varying degrees of biodegradability or solubility phases from biomass to labile C forms to recalcitrant fractions that are resistant to decomposition. As it represents C that has been incorporated into the soil throughout its existence, it also varies in age. The only known study on SOM age was that of Jenkinson et al. (1999) who, using C as a radioactive isotope, estimated the age of organic matter in some plots from this trial to be 300–400 years.

4.2 *Grazing management trial*

The focus of this trial was pasture grazing, although it did have some features in common with the "Cropping Systems" trial. This trial involved an initial phase with wheat (1987–92) in rotation with medic pasture grazed with 4, 7 and 10 sheep ha^{-1} or low, medium and high grazing intensity; common vetch; lentil, clean fallow, and watermelon. The trial was modified in 1993 to replace wheat with barley, which is more adapted to this marginal rainfall zone and more appropriate to rotations with forage legumes. As watermelon was essentially the same as fallow, this rotation was dropped. Similarly, lentil was omitted as it was included in other rotations. The main modifications, in addition to barley, were treatments involving grazing of vetch by sheep, taking the crop as mown hay, and leaving the vetch to maturity and harvesting the seed. In addition, as N fertilization has now become a common practice, an N treatment of 60 kg ha^{-1} was introduced as well as the control.

While the first phase with wheat produced differences in grain and straw yields, the order being fallow, medic, vetch and lentil, as well as in N concentration in the grain and straw (White et al., 1994), the aspect relevant to this paper was the extent to which the rotations influenced SOM, as measured in the top 0–20 cm (Table 2). As was also shown in the "Cropping Systems" trial, the legume forages, vetch and medic, increased the SOM content relative to the fallow, melon (essentially fallow as it is a summer crop) and lentil, which is a short growing-season crop with a shallow root system, and therefore limit return of organic biomass to the soil. During the wheat phase of this trial, there was a consistent increase with time in SOM in the vetch and medic rotations, but not in the others. As increasing stocking rate for the medic pasture removes more above-ground vegetative biomass, it was not surprising that SOM values decreased as grazing intensity increased. On the contrary, as N fertilizer application rate increased cereal yields, the SOM values were consistently higher where N was applied compared to the unfertilized plots. The increase of SOM in the medic rotation was probably due to both soil incorporation of medic shoots trampled in the grazing process by sheep and the turnover, or mineralization of the medic roots (White et al., 1994). The same explanation probably applied to vetch.

Table 2. Soil organic matter concentration in the topsoil (0–20 cm).

	+N (%)		−N (%)	
Rotation	1987	1991	1987	1991
Medic*				
Low	0.99	1.28	0.96	1.24
Medium	1.02	1.20	1.03	1.18
High	1.01	1.11	1.02	1.08
Vetch	1.04	1.24	1.03	1.22
Lentil	1.06	1.00	0.95	1.04
Fallow	1.10	1.08	1.10	0.94
Melon	0.97	0.92	0.99	0.91

*Indicates grazing intensities for medic.

Table 3. Organic matter in the topsoil in relation to rotation and N fertilizer application.

Rotation	+60 kg N ha^{-1} (%)	Zero N (%)
Continuous barley	1.01	0.86
Fallow	1.18	1.02
Medic	1.25	1.23
Vetch	1.26	1.18

In the modified barley phase (1993–2005) grazing systems trial, the effect of rotation on crop yields and soil properties were also expressed (Ryan et al., 2002b). Again, fallow yielded highest (but only once in every two years), followed by vetch and medic, with continuous barley being lowest. As expected N fertilizer also increased cereal yields. Following trends in the initial wheat phase of the trial analysis of SOM up to both rotation and N fertilizer (Table 3). Highest values were again observed in the medic and vetch rotations and least with continuous barley rotations, but had little effect in the forage legumes, which contributed their own N through symbiotic fixation.

As with the cropping systems wheat-based trial, there was considerable annual variation in SOM values, possibly due to changes in the more labile C fractions in addition to sampling variation, but differences between rotations were relatively consistent. With respect to the effect of N, added N consistently increased the labile C compared to the control, except in the case of vetch, where the effect was similar.

4.3 *Conservation tillage/compost management trial*

This is the most recent of the institution's long-term trials, which was started in 1996, and the only one surviving. What is unique about this trial is the incorporation of two aspects that are increasing in importance, *conservation tillage* and *residue management*. While no-till or minimum tillage research is well established throughout the world, other than the work of Mrabet and colleagues in Morocco (Bessam and Mrabet, 2003; Mrabet, 2000), conservation tillage is still only a concept that has yet to be implemented in the Mediterranean area. Similarly, the notion of managing cereal straw and stubble other than *in situ* grazing bare by sheep is gaining momentum as land use intensifies, especially with summer irrigation which requires the removal of stubble as soon as possible to facilitate cultivation.

The trial was designed to run for 12 years or several cycles of a two-case rotation (barley, vetch hay) or a four-course rotation (barley, vetch hay, wheat). It involved barley in rotation with vetch,

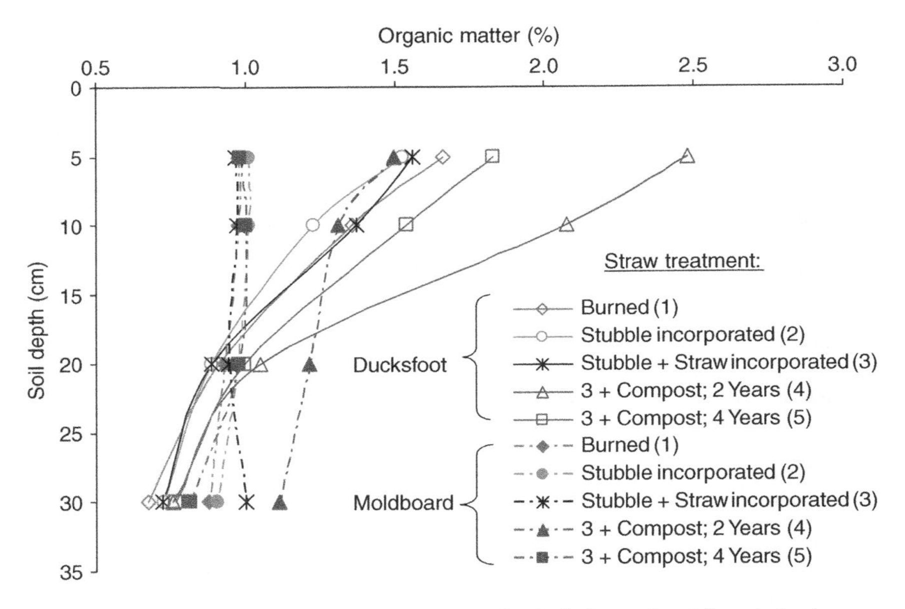

Figure 4. Organic matter in relation to straw treatment under ducksfoot and moldboard plowing.

conventional and tillage and shallow (12 cm) or conservation tillage, and incorporation of a straw-based compost at 10 t ha^{-1} every 2 and 4 years. Other treatments involved straw removed and stubble incorporated, incorporating all straw and stubble, and burning all straw and stubble. While initial observations were made on the trial (Ryan et al., 2003), the trends observed in relation to SOM are of particular interest. Indeed, the treatment effects have become accentuated after 7 years composed to the initial observation after 4 years (Ryan et al., 2003). Although final assessments in terms of crop yields awaits completion of the complete rotation cycles, the soil observations are independent of cycles and reflect the cumulative effect of these factors.

Tillage had a major impact on SOM concentration and its distribution within the soil to 30 cm depth; regardless of the straw or compost treatments, all values with the shallow, conservation-type tillage were significantly higher than these with the conventional moldboard plough (Fig. 4). It was clear that the main effect on SOM was with compost addition, particularly when applied once every 2 years compared to once every 4 years with the moldboard plow, all treatments were similar except the compost every 2 years. While treatments differences were accentuated under the conservation tillage, there was no difference between the straw being incorporated or burned.

Labile C generally followed the same pattern as total SOM, but the reverse was true for the biomass C values; these treatments were higher for the stubble incorporated and stubble compost treatments under conventional compared to conservation tillage.

4.4 *Other rotation trials*

Measurement of SOM was a minor part of most other trials; indeed, in some, it was only an item at the level of soil pH in describing the soil properties. One trial described as the "New Rotation", established in 1992/93, in ICARDA's listing of long-term trials (Ryan and Abdel Monem, 1998) indicated SOM of barley and fallow at two sites (Ryan, 1998), Breda, a dry site (280 mm), and Tel Hadya, a more favourable one (340 mm), in relation to fertilization regime. At the Breda site, SOM after barley was higher than after fallow; there was a slight increase with fertilization (N at 40 kg ha^{-1} and P at 60 kg ha^{-1}) with continuous barley, but no difference with fallow at the

Tel Hadya site, continuous barley again had higher SOM values than fallow, with slight increases in each rotation with N and P fertilization.

In a variant of this rotation "continuous barley", established in 1986/87, the increase in SOM with continuous cropping, compared to fallow, extended down to 50–100 cm in the soil profile. As in other trials, these differences are attributed to mineralization of SOM in the fallow year and continuous root biomass input in the continuous cropping rotation. As in other trials, fertilization with P and N had a positive influence on SOM (Masri et al., 1998); the increases in SOM were accompanied by similar increases in aggregation, infiltration and hydraulic conductivity.

5 DISCUSSION AND CONCLUSION

This brief overview of soil organic C, as reflected in SOM determinations, in the Mediterranean's WANA region revealed a number of important generalizations. Clearly, most soils have relatively low SOM levels, mainly 1–2%, with some notable exceptions at a local level. It can be assumed that such equilibrium levels of SOM are a reflection of the low amount of crop residues returned to the soil and, more importantly, the environmental conditions that favor SOM mineralization (Parton et al., 1987). While it is unclear to what extent to which SOM levels were drawn down over the millennia of soil cultivation in the Mediterranean area, especially the Fertile Crescent area of west Asia, it is reasonable to conjecture, based on cultivation data from North America (Freyman et al., 1982), that current SOM levels are probably no more than half of what they were originally before the pressure of land use in the Middle East today (Lal, 2001). The quest for sustainable cropping systems (Harris, 1995), which focuses mainly on crop yields and animal off-take, has expanded to consider sustainability in terms of maintenance of soil quality.

One consistent fact emerged from a consideration of the various long-term trials in a typical rainfed cereal belt in northern Syria, an area representative of much of the WANA region (Kassam, 1981), and that is that legume-based rotations are not only biologically and economically viable, but are also compatible with improved soil properties as a result of enhanced SOM levels (Masri and Ryan, 2006). An interesting feature of the trials was that fertilization, which is now a feature of all but the most environmentally harsh areas of the Mediterranean region (Ryan, 2002), consistently increased SOM and related properties. However, the studies indicated that this gain in soil quality could be offset by the age-old practice of stubble grazing with sheep flocks.

Another interesting development was the extent to which conservation or shallow tillage enhanced SOM accumulation, largely through reducing conditions that promote SOM mineralization. That compost addition combined with reduced tillage could greatly increase SOM levels clearly showed that disposal of straw and organic farm waste when necessary could have beneficial effects on the soil without any negative influence of crop yields. Notwithstanding the fact that introduction of conservation tillage (minimum, reduced, no-till) is still in the experimental stage in the Middle East region (Mrabet, 2000; Bessam and Mrabet, 2003) the likelihood is for considerable adoption of this approach which is now widely used in North and South America. Even if conservation tillage does not always increase yield, it is less costly in terms of energy consumption (Pala et al., 2000).

Indeed, developments in Central Asia, under relatively similar agro-ecological conditions, but with colder winters and limited winter cropping (De Pauw et al., 2004), suggest that conservation tillage is far more advanced than in North Africa or West Asia (Suleimenov et al., 2004). With current efforts to promote rotational cropping systems to replace summer fallow, there is likelihood for greater adoption of cropping systems that are economically attractive and also improve SOM. The overview of SOM in soils of Central Asia by Nasyrov et al. (2004) showed familiar trends: significant differences between cultivated and virgin soils in similar agro-ecological zones, and a sharp decline in SOM with cropping intensification, particularly with cotton (*Gossypium hirsutum*). Only where organic manures were added was SOM maintained. Indeed, after 7 years continuous cotton, 3 years of alfalfa was not sufficient to reverse the SOM decline (Nasyrov et al., 2004). This clearly indicated that for rotations to be effective for SOM maintenance, both the type of rotation and the duration of the rotation have to be considered.

In conclusion, the SOM data that have been presented here, mainly from long-term cereal-based rotation trials in northern Syria have relevance to Central Asia, an adjacent area of the world with many agro-climatic features in common with the WANA region (De Pauw et al., 2004). The findings related to legume-based rotation (or the absence of fallow) and the impacts of fertilization and stubble grazing (i.e., reducing crop biomass return to the soil) all have messages to Central Asian agriculture. That any improvement in SOM is likely to evoke better soil physical quality (Masri and Ryan, 2006), with obvious implications for sustainable crop yields is an added bonus. While there is much debate about the global significance of SOM in relation to C sequestration (Lal, 2002), the potential of cropland to sequester within the context of normal crop production is of particular relevance in Central Asia in view of the large area of land involved, in addition to rangeland which, if properly managed, can sequester large amounts of carbon dioxide from the atmosphere (Saliendra et al., 2004). The findings on SOM from the WANA region make a compelling argument to include SOM measurements, including inputs/outputs, and various SOM stability fractions (e.g., labile, soil microbial biomass) in cropping systems assessment, as well as in rangelands.

Organic matter has a mediating influence on the physical, chemical and biological properties of soils, particularly as it relates to crop production and the environment. As levels of soil organic matter (SOM) are dictated by climatic factors, mainly rainfall and temperature, as well as man's influence through tillage, soils of the Mediterranean climatic zone are inherently low in this component. This brief and selective overview of SOM in the West Asia-North Africa region, which is characterized by a Mediterranean-type climate, provides some background information of SOM in some soils from the region, as well as studies of SOM in relation to soil properties and behavior. The main focus of the overview is on the relationships of SOM in long-term cereal/legume rotation trials with a focus on cropping systems productivity (including stubble management and nitrogen fertilization), grazing management of pastures, and tillage systems, both traditional and conservation, including the use of compost. These trials showed that not only do legumes in rotation with cereals provide an alternative to fallowing or continuous cropping in a biologically and economically sustainable manner, but also improve SOM status along with its attendant benefits for soil structure and nutrient availability. As the area of northern Syria where these trials were conducted shares several features, climatically and agriculturally, with the Central Asia region, the findings from this study are of relevance to that area of the world. In view of the huge expanse of land in Central Asia, any cropping system that has even a modest effect on soil organic carbon would have significant effects on carbon sequestration, and thus help to mitigate the effect of greenhouse gases, in addition to promoting soil nitrogen accumulation and sustainability of crop yields.

ACKNOWLEDGEMENT

I wish to acknowledge the invaluable assistance that I have had in this and all my work at ICARDA from the late Mr. Samir Masri who passed away suddenly at work on February 15, 2006.

REFERENCES

Abdel Monem, M., J. Ryan, and M. El Gharous. 1990. Preliminary assessment of the soil fertility status of the mapped area of Chaouia. Al-Awamia 72: 85–107.

Arshad, M., J. Ryan, and R.C. Paeth. 1980. Influence of organic matter, free iron, and calcium carbonate on soil particle size distribution. Agrochimica 24: 470–477.

Besssam, F., and R. Mrabet. 2003. Long-term changes in soil organic matter under conventional tillage are no-tillage systems in semi-arid Morocco. Soil Use and Manage. 19: 193–143.

Borlaug, N. E. 2003. Feeding a world of 10 billion people: The TVA?IFDC legacy. Travis P. Hignett Memorial Lecture. March 14, Muscle Shoals, AL, USA.

Cooper, P. J. M., P.J. Gregory, D.Tully, and H.H. Harris. 1987. Improving water-use efficiency of annual crops in the rain-fed farming system of West Asia and North Africa. Expl. Agric. 23: 113–158.

Damania, A.B., J. Valkoun, G. Wilcox, and C.O. Qualset. 1998. The origins of agriculture and crop domestication. The Harlan Symposium. International Center for Agricultural research in the Dry Areas, Aleppo, Syria.

De Pauw, E., F. Pertziger, and L. Lebed. 2004. Agro-climatic mapping as a tool for crop diversification in Central Asia and the Caucasus. p. 21–43. In: J. Ryan, P. Vlek, and R. Paroda (eds.) Agriculture in Central Asia: research for development. International Center for Agricultural Research in the Dry Areas, Aleppo, Syria, and Center for Development Research, Bonn, Germany.

Freyman, S., C.J. Palmer, E.H. Hobbs, J.F. Dormaar, G.B. Schalje, and J.R. Moyer. 1982. Yield trends on ling-term dryland wheat rotations at Lethbridge. Can. J. Plant Sci. 62: 609–619.

Gibbon, D. 1981. Rainfed systems in the Mediterranean region. Plant and Soil 58: 59–80.

Habib, L., S. Hayfa. and J. Ryan. 1994. Temporal change in organically amended soil: implications for phosphorus solubility and adsorption. Commun. Soil Sci. Plant Anal. 25 (19 and 20): 3281–3290.

Harris, H.C. 1994. Water use efficiency of rotations in a Mediterranean environment. Aspects Appl. Biol. 38: 165–172.

Harris, H. 1995. Long-term trials on soil and crop management at ICARDA. Adv. Soil Sci. 19: 447–469.

Haynes, R.J. 2005. Liveable organic matter fractions as central components of the quality of agricultural soils: An overview. Adv. Agron. 85: 221–268.

Jenkinson, D.S., H.C. Harris, J. Ryan, A. M. McNeill, C. J. Pilbeam, and K. Coleman. 1999. Organic matter turnover in a calcareous clay soil form Syria under a two-course cereal rotation. Soil Biol. Biochem. 31: 687–693.

Jones, M. 1993. Sustainable agriculture: an explanation of a concept. p. 30–47. *In*: Crop Protection and Sustainable Agriculture. Wiley, Chichester, UK.

Karlen, D.L., G.E. Varvel, D.G. Bullock, and R.M. Cruse. 1994. Crop rotations for the 21th century. Adv. Agron. 53: 1–45.

Karlen, D.L., S.S. Andrews, and J.W. Doran. 2001. Soil quality: Current concepts and applications. Adv. Agron. 74: 1–40.

Kassam, A. 1981. Climate, soil and land resources in North Africa and West Aisa. Plant and Soil. 58: 1–29.

Lal, R. 2000. Soil management in developing countries. Soil Sci. 165: 57–72.

Lal, R. 2001. Managing world soil for food security and environmental quality. Adv. Agron. 74: 155–192.

Lal, R. 2002. Carbon sequestration in dryland ecosystems of West Asia and North Africa Land Degradation and Development 13: 45–59.

Masri, Z., S. Masri., and J. Ryan. 1998. Changes in soil physical properties with continuous cereal cropping in northern Syria. p. 463–648. Symposium on Arid-Region Soils. Izmir, Turkey, Sept. 21–25.

Masri, Z., and J. Ryan. 2006. Soil organic matter and related physical properties in a Mediterranean wheat-based rotation trial. Soil and Tillage Res. 87: 146–154.

Mrabet, R. 2000. Differential responces of wheat to tillage management systems in a semi-arid area of morocco. Field Crops Res. 66: 165–174.

Nasyrov, M., N. Ibragimov, B. Habikov, and J. Ryan. 2004.. Soil organic carbon of Central Asia's agro-ecosystems. p. 126–139. In: J. Ryan, P. Vlek, and R. Paroda (eds.) Agriculture in Central Asia: research for development. International Center for Agricultural Research in the Dry Areas, Aleppo, Syria, and Center for Development Research, Bonn, Germany.

Oweis, T., M. Pala, and J. Ryan. 1998. Stabilizing rainfed wheat yields with supplemental irrigation in the Mediterranean region. Agron. J. 90: 672–681.

Pala, M., H.C. Harris, J. Ryan, R. Makboul, and S. Dozom. 2000. Tillage systems and stubble management in a Mediterranean-type environment in relation to crop yield and soil moisture. Expl. Agric. 36: 223–242.

Parton, W.J., D.S. Scimel, C.V. Cole, and D.S. Ojima. 1987. Analysis of factors controlling soil organic matter levels in Great Plains grassland. Soil. Sci. Soc. Am. J. 51: 1173–1179.

Rasmussen, P.E., K.W.T. Goulding, J.R. Brown, P.H. Grace, H.H. Janzen, and M. Korchens. 1998. Long-term agro-ecosystem experiments: assessing agricultural sustainability and global change. Science 282: 893–896.

Ryan, J. 1990. Soil and fertilizer phosphorus studies in Lebanon. p. 6–28. In: J. Ryan and A. Matar (ed.) Proceedings, Third Regional Soil Test Calibration Workshop. Amman, Jordan. 2–9 Sept. 1988, ICARDA, Aleppo, Syria.

Ryan, J. (ed.) 1997. Accomplishments and future challenges in dryland soil fertility research in the Mediterranean area. Proceedings, International Soil Fertility Workshop, Nov. 19–23, 1995, ICARDA, Aleppo, Syria.

Ryan, J. 1998. Changes in organic carbon in long-term rotation and tillage trials in northern Syria. p. 285–296. In: R. Lal, J.M. Kimble, R.F. Follett, and B.A. Stewart (eds.) Management of carbon sequestration in soil. Adv. Soil Sci. CRC Press, New York, N.Y., USA.

Ryan, J. 2002. Available soil nutrients and fertilizer use in relation to crop production in the Mediterranean area. p. 213–246. In: K.R. Krishna (ed.) Soil fertility and crop production. Science Publ., Inc., Enfield, New Hampshire, USA.

Ryan, J., and A. Matar. 1992. Fertilizer use efficiency under rainfed agriculture in West Asia and North Africa. Proceedings, Fourth Regional Workshop. May 5–10, 1991. Agadir, Morocco. ICARDA, Aleppo, Syria.

Ryan, J., and M. Abdel Monem. 1998. Soil fertility for sustained production in West Asia-North Africa region: Need for long-term research. p. 55–74. In: R. Lal (ed.) Soil Quality and Agricultural Sustainability. Ann Arbor Press, Chelsea, Michigan, USA

Ryan, J., M. Abdel Monem, and M. El-Gharous. 1990. Soil fertility assessment at agricultural experiment stations in Chaouia, Abda, and Doukkala. Al-Awamia. 72: 1–47.

Ryan, J., S. Masri, and S. Garabet. 1996. Geographical distribution of soil test values in Syria and their relationship with crop response. Commun. Soil Sci. Plant Anal. 27: 1579–1593.

Ryan, J., S. Masri, and Z. Masri. 1997a. Potassium in Syrian soils: implications for crop growth and fertilizer needs. p. 134–145. In: A.E. Johnston (ed.) Food security in the WANA region: The essential need for balanced fertilization. Bornova, Turkey, May 26–30, 1997. International Potash Inst. Basel, Switzerland.

Ryan, J., S. Masri, S. Garabet, J. Diekmann, and H. Habib. 1997b. Soils of ICARDA's agricultural experiment stations and sites: Climate, classification, physical-chemical Properties, and land use. ICARDA, Aleppo, Syria.

Ryan, J., S. Masri, and E. De Pauw. 2002a. Assessment of soil carbon in Syria and potential for sequestration through crop management. Intern. Soil Science Congress. Bangkok, Thailand. Symposium 45: Abstracts 1304

Ryan, J., S. Masri, M. Pala, and M. Bounejmate. 2002b. Barley-based rotations in a typical Mediterranean agro-ecosystem: crop production trends and soil quality. Options Mediterraneenees Series A, No. 50: 287–296.

Ryan, J., S. Masri, J. Diekmann, and M. Pala. 2003. Organic matter and nutrient distribution following conservation tillage and compost application under Middle Eastern cropping conditions. ASA Meetings, Program p. 95. Denver, CO.

Ryan, J., S. Masri, and M. Qadir. 2006. Nutrient monitoring in sewage water for irrigation: impacts for soil quality and crop production. Commun. Soil Sci. and Plant Anal. 37: 2185–2198.

Saliendra, N.Z., D.A. Johnson, M. Nasyrov, K. Akshalov, M. Durikov, B. Mardonov, T. Mukimov, T. Gilmanov, and E. Laca. 2004. Daily and growing season fluxes of carbon dioxide in rangelands of Central Asia. p. 140–153. In: J. Ryan, P. Vlek, and R. Paroda (eds.) Agriculture in Central Asia: research for development. International Center for Agricultural Research in the Dry Areas, Aleppo, Syria, and Center for Development Research , Bonn, Germany.

Suleimenov, M.K., K.A. Akhmetov, J.A. Kasbarbayov, F. Khasanova, A. Krieyev, L.I. Martinova, and M. Pala. 2004. p. 188-211. In: J. Ryan, P. Vlek, and R. Paroda (eds.) Agriculture in Central Asia: research for development. International Center for Agricultural Research in the Dry Areas, Aleppo, Syria, and Center for Development Research, Bonn, Germany.

Stewart, B.A., and C. A. Robinson. 1997. Are agro-ecosystems sustainable in semi-arid regions? Adv. Agron. 60: 191–228.

White, H. 1970. Fallowing, crop rotations and crop yields in Roman times. Agric. Hist. 44: 281–290.

White, P.F., N.K. Nersoyan, and S. Christiansen. 1994. Nitrogen cycling in a semi-arid Mediterranean region: Changes in soil N and organic matter under several crop/livestock production system. Aust. J. Agric. Res. 45: 1293–1307.

Zimdahl, R.L., 2006. Agriculture's ethical horizon. Academic Press/Elsevier, Amsterdam, The Netherlands.

CHAPTER 18

Potential for carbon sequestration in the soils of Afghanistan and Pakistan

Anwar U.H. Khan
University of Agriculture, Faisalabad, Pakistan

R. Lal
Carbon Management and Sequestration Center, Ohio State University, Columbus, OH, USA

1 INTRODUCTION

Human activities have led to global climatic change and increase in the global temperature by $0.6 \pm 0.2°C$ at an average rate of increase of 0.17°C per decade since 1950 (IPCC, 2001). This climatic change is attributed to increase in concentration of greenhouse gases (GHGs)(e.g., CO_2, CH_4, N_2O) by fossil fuel combustion, land use change and deforestation, and soil degradation because of inappropriate land use and soil management practices (FAO, 1994; Lal, 2004). The GHG emissions, particularly that of CO_2 from terrestrial ecosystems, can be decreased by: (1) increasing C sinks in soil organic matter (SOM) and the above-ground biomass, (2) avoiding carbon (C) emissions from farm operations by reducing direct and indirect energy use, and (3) increasing renewable-energy production from biomass that either substitutes for consumption of fossil fuels or replaces inefficient burning of fuel wood or crop residue (Pretty et al., 2002). The Bio-sequestration of C, both by soil and biota, is also important to improving quality of soil and the environment. At present, the soil organic carbon (SOC) concentration in developing countries is low because of high soil temperatures during summer accentuating the rate of mineralization of SOM; low rainfall and lack of availability of good quality irrigation water exacerbating the drought problem; little or no crop residue and/or manure returned to the soil because of numerous alternate uses as fodder or fuel; and excessive tillage, extractive cropping systems, and improper crop rotations which accentuate the problem of soil degradation. Thus, the objective of this manuscript is to review the soil and water resources of Afghanistan and Pakistan, and discuss the importance of restoration of degraded/desertified soils, and use of recommended soil and crop management techniques in soil C sequestration and improving soil quality.

2 PREDOMINANT CLIMATE AND ECO-REGIONS

Afghanistan is a landlocked country sited in northwest of Pakistan between 29 and 37°N, and 61 and 71°E. The climate of Afghanistan is arid and semi-arid with cold winters and hot summer and annual rainfall varies between 0.05 m in southwest to about 1.00 m in the northeast mountains. The rainy season in Afghanistan usually occurs between Octobers through April. Windy season lasts up to 120 days (between June and September). Strong winds create dust storms and it is not generally possible to grow any thing because winds beat down everything.

Pakistan is located between 24 and 37°N and 61 and 75°E and the climate is arid to semi-arid with low rainfall as well as humidity and high solar radiation over most parts of the country. The rainfall varies from 0.01 m in some parts to more than 0.5 m in other parts. Most areas receive less than 0.2 m annual rainfall except for the districts of Sialkot, Chakwal and Rawalpindi and the

Table 1. Population and land use in Afghanistan and Pakistan (Adapted from FAO, 2004; GOP, 2004, 2005; PAD, 2004; NCA, 1988; PARC, 1980).

Parameter	Afghanistan	Pakistan
Population (millions)	29.93	162.42
Land use (Mha)		
Total area	65.21	83.42
Land area	64.56	79.61
Arable and permanent crops	8.05	22.74
Arable land	7.91	21.45
Permanent crops	0.14	0.67
Permanent pasture	Not available	5.00
Forest and woodlands	1.90	3.78
Irrigated land	3.21	18.23*
Rainfed cropland	4.52	4.81
Rangeland/grassland	29.18	45.20**

*3.37 Mha irrigated by tube wells and 0.21 Mha by wells also included; ** Permanent pasture of 5.0 Mha also included.

higher altitude northern mountains which receive more than 0.5 m annually. However, only 9% of the country receives more than 0.5 m annual rainfall (Qureshi and Barrett-Lennard, 1998). About 60% of total rainfall occurs in Punjab and Sindh provinces during the monsoon season, i.e., from July to mid September. The northern areas and Baluchistan province are located out of the monsoon rain range and receive maximum rainfall during October to March (FAO, 1987).

3 LAND USE AND SOIL TYPES

Total land area of Afghanistan is 65.2 million hectares (Mha) and had a total population of 29.0 million in July, 2005 (Table 1). Total area of Pakistan is 83.4 Mha with a population of 162.4 million. The total land area of Pakistan of 79.6 Mha comprises 21.5 Mha of arable land, 5 Mha of permanent pasture, 3.6 Mha of forest and woodlands, 0.7 Mha of permanent crops and 40 Mha of rangeland. Of the total land area of 64.6 Mha of Afghanistan, 7.9 Mha is arable, 29.2 Mha is rangeland, 1.9 Mha is forest and woodlands, 0.14 Mha is permanent crops and 25.3 Mha comprises other lands. Based on reconnaissance soil survey data and information covering about 62 Mha of Pakistan, about 26 Mha of soils are Aridisols, 18 Mha are Entisols and 3 Mha are Inceptisols. Vertisols, Alfisols and Mollisols are also present but their areal extent is rather small (Rashid and Memon, 1996). The ochric epipedon is the most extensive surface diagnostic horizon present because soil organic matter concentrations are usually less than 5 g kg^{-1} in most soils. The large arable (30 Mha) and rangeland (70 Mha) areas in both countries have a potential to enhance productivity and to sequester C while improving environment quality. In comparison, the total arable land in 5 Central Asian countries is only 32.6 Mha (WBS, 2004).

4 SOIL AND WATER RESOURCES

Afghanistan is known for its mountainous terrain and deserts, and only limited area comprising plains. The mountain range divides Afghanistan into three geographic regions: the central highlands, the northern plains and the southwestern plateau. The altitude, climate and soil conditions vary widely depending on the specific eco region. Two major rivers of the country are the Kabul and the Amu Darya (formerly known as Oxus River) which have created the country's fertile alluvial soils along their flood plains and are principal sources of water for irrigation and drinking. The plains and rivers have large impact on country's agriculture. Total irrigated area was 3.2 Mha prior to 1982 and before 23 years of civil conflict, of which 1.55 Mha was extensively irrigated and 1.65 Mha

Table 2. Principal agro-ecological regions of Pakistan.

Agro-ecological regions	Area (Mha)	% of Total
Northern Dry Mountains	7.43	8.9
Wet Mountains	2.20	2.6
Western Dry Mountains	15.38	18.4
Rainfed Lands	5.12	6.1
Sulaiman Piedmont	3.74	4.5
Dry Western Plateau	21.03	25.2
Southern Irrigated Plains	6.93	8.3
Indus Delta	1.98	2.4
Northern Irrigated Plains	9.89	11.9
Sandy Deserts	9.72	11.7
Total	83.43	100

Source: PARC (1980).

intermittently (i.e., irrigation done on irregular basis). Irrigation rehabilitation/improvement is envisaged to be completed on 1 Mha by year 2010 and 2.2 Mha by 2015.

Pakistan is divided into three distinct geographical regions, i.e. the northern highlands, the flat Indus river plains in east and the Baluchistan plateau in the west. These regions are subdivided into 10 agro-ecological regions (Table 2). The soils in irrigated Punjab, North West Frontier Province (NWFP) and Sindh range from clay loam to silt loam and are generally developed from the alluvial materials. In most rainfed areas of Pakistan, soils are derived from loess, old alluvial deposits, mountain out-wash and recent stream valley deposits. Soils of huge Thal desert, Cholistan and Tharparker belt deserts (9.7 Mha) are alluvial with coarse-textured sand dunes covering 50 to 60% of the area (Khan, 1993). Major sources of fresh surface water in Pakistan are five rivers (e.g., Indus, Jehlum, Chenab, Ravi and Sutlej). Pakistan has one of the largest single contiguous gravity irrigation systems in the world. The Indus Basin irrigation system comprises three major reservoirs, 16 barrages, 12 inter river link canals and 44 canal systems. The system also utilizes 6.8 Mha-m of ground water pumped through more than 0.7 M tube-wells to supplement the canal supplies (Anonymous, 2002; OFWM, 2005). The seepage from link canals and the irrigation network has led to the development of fresh water layer over the relatively saline ground water layer in the entire Indus basin (Kahlown and Kemper, 2004). Thickness of fresh ground water zone ranges from less than 60 m along the margin of *Doabs* (land between two rivers) to 30 m or more in the central part of *Doabs* (Sufi and Javed, 1988).

5 SOIL DEGRADATION AND DESERTIFICATION

Soil degradation is the temporary or permanent loss of the productive capacity of agricultural land (UNEP, 1992). The prevalent management system is not suitable and causes a severe problem of degradation (Table 3) and desertification (FAO, 1994). Different types of soil degradation include water erosion, salinization, water-logging, soil fertility decline, depletion of SOM (i.e. less than 5 g kg^{-1}), deforestation and forest degradation, and crusting, compaction and rangeland degradation. A serious problem of ground water depletion also occurs in the Punjab province of Pakistan. Water erosion is a major problem that affects 11.2 Mha in Afghanistan and 7.2 Mha in Pakistan. The total area affected by wind erosion is 2.1 Mha in Afghanistan and 10.7 Mha in Pakistan. Land area affected by desertification in Pakistan is estimated at 6.1 Mha of irrigated cropland, 3.36 Mha of rain fed cropland and most of 45 Mha of rangeland. Therefore, majority of land area is affected by some degree of desertification (Lal, 2002). Secondary salinization of irrigated land is also a severe problem. Soil crusting is a serious problem on 4.7 Mha of clayey soils under irrigation and moderate to minor problem on 2.3 Mha of rainfed cropland (Rafiq, 1990). Soil crusting reduces seedling emergence and rate of water infiltration (Shafiq et al., 1994; Hassan, 2000).

Table 3. Area (Mha) affected by land degradation in Afghanistan and Pakistan.

		Light	Moderately	Strongly	Total
Type	Country	Mha			
Water erosion	Afghanistan	8.6	2.6	0.0	11.2
	Pakistan	6.1	1.1	0.0	7.2
Wind erosion	Afghanistan	1.9	0.0	0.2	2.1
	Pakistan	4.0	6.7	0.0	10.7
Soil fertility	Afghanistan	NA	NA	NA	–
decline	Pakistan	5.2	0.0	0.0	5.2
Salinization	Afghanistan	1.3	0.0	0.0	1.3
	Pakistan	1.9	1.2	3.0	6.1
Water-	Afghanistan	0.0	0.0	0.0	0.0
logging	Pakistan	0.8	0.4	0.8	2.0
Lowering of	Afghanistan	0.0	0.0	0.0	0.0
water table	Pakistan	0.1	0.1	0.0	0.2

Source: Khan, 1993; FAO (1994); WAPDA (2005).

6 CAUSES OF SOIL DEGRADATION

Principal causes of soil degradation are inappropriate irrigation, steep slopes of the mountain, low soil permeability or poor drainage, predominantly coarse-textured soils, removal of crop residue, excessive plowing and excessive tillage, deforestation and over pumping of ground water (FAO, 1994; Lal, 2004). The causes of low SOC pool of most soils include extractive farming practices, conventional tillage, removal of crop residue for fodder and fuel, use of dung as household fuel, and climatic aridity. Plowing has depleted the SOC pool through rapid mineralization and accelerated soil erosion. There is a strong correlation between low SOC concentration and soil physical degradation (Lal, 2004). Due to low SOC and/or high exchangeable Na^+, soil structure is weak, resulting in crusting, compaction and erosion. Soil crusting also occurs due to flood irrigation (Hassan, 2000). Sandy soils of Thal, Cholistan and Thar regions have low resistance to wind erosion. Loss of rainwater through surface runoff and soil erosion is a serious problem in rainfed agriculture in Pakistan (Shafiq et al., 1994). Susceptibility to water erosion is high in northeastern Pakistan due to steep slopes, high silt and/or fine sand contents and low SOC concentrations (Ellis, 1994). Rates of soil erosion are estimated at 150 Mg ha^{-1} y^{-1} on recently deforested land in high altitude, and 50–70 Mg ha^{-1} y^{-1} in overgrazed rangelands in lower altitudes. Low rainfall, little crop cover and high temperatures during summer cause severe water loss from soil surface through evaporation, which in turn increases salinity in the root zone. In Pakistan, continuing large negative balance has been reported for K and moderate deficit for P (FAO, 1986), Zn and S (FAO, 1992). In several arid and semi-arid regions of the world, pumping exceeds recharge, resulting in lowering of the ground water at alarming rates (Pimentel et al., 1999). Decline in the water table in Punjab in Pakistan and India is of particular concern (Seckler et al., 1999). A drastic depletion of water resources has an adverse effect on the sustainability of agriculture.

7 WATER QUANTITY AND QUALITY

Soil and water are two basic resources for crop production. However, lack of adequate management of these resources decreases the potential productivity of soils. Lack of water is the limiting factor for plant growth in both countries. In Afghanistan, there are limited fresh water resources and availability of water is unreliable. The existing irrigation systems are poorly maintained and operating at low efficiency of about 25%. Only 30% of total arable and permanent croplands were irrigated in 1999. Fresh water resources are of 5 types: modern surface system (10%), traditional

Table 4. Change in water storage capacity of three major reservoirs in Pakistan.

Tarbela		Mangla		Chashma	
	Effective storage		Effective storage		Effective storage*
Year	Mha-m	Year	Mha-m	Year	Mha-m
Designed	1.192	Designed	0.657	Designed	0.088
2000	0.897	1985	0.609	1973	0.078
2001	0.889	1995	0.576	1983	0.061
2002	0.881	1999	0.570	1988	0.054
2003	0.877	2002	0.569		
2004	0.875	2004	0.564		

Source: WAPDA (2005). *Effective storage = gross content − dead storage.

surface systems (80%), springs (5%), and Krazes and wells (5%). The reduction of land cover on the sloping watersheds has caused severe and destructive floods and reduced water retention in the aquifers. There is a widespread use of unsustainable water resource management practices by local commanders who do not respect water rights or the authority of the *mirabs* (water manager). *Mirab* is a person known for honesty and hard work and is appointed to oversee the management of water and its distribution, and canal maintenance.

Out of 18.23 Mha irrigated area in Pakistan, 7.06 Mha is irrigated by canals, 7.17 Mha by canals and tube wells, 3.42 Mha by tube wells, 0.38 Mha by wells and canals and 0.20 Mha by other sources. The storage capacity of three major reservoirs is decreasing due to severe silted up of dams (Table 4). The total inflow of the western rivers fell to 11.22 Mha-m in 2001–02 from 19.57 Mha-m in 1992–1993 and the water shortages registered during last few years were as high as 40–50%. There is a significant decrease in per capita water availability which was 5650 m^3 in year 1951, 4500 m^3 in 1967, 3300 m^3 in 1974, 1700 m^3 in 1992, 1400 m^3 in 2002 and 1000 m^3estimated for year 2012. The pumped water in the shallow fresh ground water has become more saline with time and many of deep tube wells are being shut down (Hafeez et al., 1986). This limits the exploitation of ground water for agricultural purposes. The only option for supplementing canal irrigation systems of the irrigated area is shallow wells which can extract water up to the safe limit. The shortfall of 0.43 Mha-m during the Rabi (winter) season in 1990s could increase to 1.6 Mha-m by the year 2017 (Qutab and Nasiruddin, 1994). About 70–75% of the pumped ground water in Punjab province is unfit for irrigation because of high electrical conductivity (EC), sodium adsorption ratio (SAR) and/or residual sodium carbonates (RSC), which adversely affects crop yields (Ghafoor et al., 2001a and 2001b). In arid and semi-arid regions (e.g., Pakistan and Afghanistan) with scarcity of fresh water, farmers are compelled to use brackish water for irrigation which exacerbates soil salinization.

8 FOREST AND RANGELANDS

Forests in Afghanistan have been adversely affected during the period of civil strife for more than 20 years, and there is little reliable information on the extent and status of forests. In 1880, the forest area in Afghanistan was about 2.2 Mha (3.4% of the land area). There is widespread concern about extensive harvesting of old growth cedar forest (*Cedrus deodara*) for lucrative export market and depletion of the pistachio forests (*Pistacia vera*) by neglect and for firewood. The UNEP satellite data showed that 1.3 Mha of southeast forest has been reduced by an average of 50% since 1978 and pistachio forest from 0.3 Mha to only 70,000 ha. At this rate, there will be no forest remaining in 25 years between 2005 and 2030. Rangelands cover about 45% of the total land area and produce about half of the yearly 20 Mg of food for livestock in Afghanistan. The rangeland suffered by severe overgrazing especially during 4 years of drought between1998 and 2002 and

Table 5. Regional distribution of rangeland in Pakistan.

	Total area	Rangeland area	Percentage of Provincial area
Province		Mha	
Balochistan	34.7	27.4	79 (5.0%*)
Sindh	14.1	7.8	55 (23.0%)
Punjab	20.6	8.2	40 (55.6%)
NWFP	10.2	6.1	60 (13.4%)
Northern Areas	8.3	2.7	33 (2.4%)
Total	88.0	45.2	51

Source: NCA (1988). *Percent of the country population in the Province.

lack of management. Pastureland is the principal source of conflict among settled and nomadic land users, and ethnic and territorial concerns. The loss of forests and herbaceous vegetation, and excessive grazing expose soil to accelerated erosion, desertification and fertility depletion.

In Pakistan, rangelands cover 56.8% of land area or 51% of total area (Table 5). Increases in human and livestock population have lead to cutting of shrubs and trees for domestic fuel consumption and progressive elimination of palatable grasses, legumes, herbs, shrubs and trees (Umrani et al., 1995). Thus, misuse and centuries of over-grazing, have adversely affected productivity of rangelands in Pakistan. At present, rangelands are producing only at 10–15% of their potential capacity (Muhammad, 2002). Rangeland productivity can be increased by adopting recommended management practices (RMPs) such as periodic closures, re-seeding of degraded rangeland (Khan, 1999) and improved grazing management. During winter, *Pawindas* (Nomadic tribe) arrive from Afghanistan due to scarcity of fodder and extremely low temperature in their native land. Thus, they leave their country in search of forages, stay for 4 months until spring. Their animals eat leaves of trees, stubble, roughage from fruits and vegetables and depend partly on grazing in rangelands.

9 SOIL CARBON SEQUESTRATION

There is a little information available regarding soil C-sequestration in Afghanistan and Pakistan. Whatever little information exists, the data have some analytical problems. In some cases, the particulate organic matter (POM) was included during SOC determination, which leads to erroneous data. A five-year field study was conducted near Lahore, Pakistan, to evaluate the effectiveness of growing salt-tolerant *Leptochloa fusca* (Kallar grass) in improving the physical and chemical characteristics of a saline-sodic soil irrigated with brackish ground water (Akhter et al., 2003, 2004). The data indicated a linear increase in SOC concentration and significant improvement in properties of saline-sodic soil (except bulk density) with the increase in duration since establishment of Kallar grass. However, the SOC sequestration rate was exceptionally high, i.e. 9.33 Mg $ha^{-1}y^{-1}$ (Table 6), probably due to the incorporation of POM in the soil sample. It is also possible that the baseline data in the control treatment was erroneous or there was a high spatial variability. Recently we assessed SOC sequestration rate at three different sites planted to Kallar grass. Site A is a state land and it has been prepared after clearing the acacia (*Acacia nilotica*). The soil is saline and water-logged due to saline water intrusion. Now the site is under vegetation of salt tolerant plants for more than two years. Relatively large plots are irrigated with brackish ground water without any schedule. The SOC sequestration rate in these plots was often negative (Table 7).

The Site B is close to a village and belongs to a farmer. The grass has been grown for more than five years without any management and is irrigated without any schedule with canal water. Some time rainwater run-on enters the site during monsoon inundation. The grass is not harvested routinely but goats graze occasionally. The SOC sequestration rate determined at this site is

Table 6. Soil organic carbon sequestration rate through restoration of degraded soil by growing Kallar grass for five years.

Growth duration	0–20 cm		40–60 cm		80–100 cm		TSOCP	Rate of sequestration
year	BD*	SOC**	BD	SOC	BD	SOC	$Mg\,ha^{-1}$	$Mg\,ha^{-1}y^{-1}$
0	1.62	1.91	1.73	1.10	1.68	1.04	21.98	
1	1.61	1.86	1.72	5.16	1.60	1.62	45.98	24.00
2	1.58	3.19	1.65	6.79	1.59	1.97	61.26	19.64
3	1.55	4.23	1.59	6.21	1.56	1.51	58.71	12.24
4	1.54	3.65	1.53	6.90	1.55	1.68	58.95	9.24
5	1.53	4.29	1.53	7.71	1.54	2.20	68.63	9.33

Soil organic carbon sequestration rate is 8.67 $Mg\,ha^{-1}y^{-1}$. *Bulk density in $Mg\,m^{-3}$; ** Soil organic carbon in $g\,kg^{-1}$. (Recalculated from Akhter et al., 2003, 2004).

Table 7. Soil organic carbon sequestration through growing Kallar grass (*Leptochloa fusca*) in saline and water-logged soil, and normal soils in Punjab, Pakistan.

Location	SOC Sequestration rate ($Mg\,ha^{-1}\,y^{-1}$)
Site A (Irrigated with brackish ground water, salt-affected soil), more than 2 years of Kallar grass	−0.56
Site B (Irrigated with canal water, not regular basis, grazed by animals), more than 5 years of Kallar grass	2.17
Site C (Farmers's field, cut for fodder, irrigated with canal water), more than 2 years of Kallar grass, irrigated	0.98

2.17 $Mg\,ha^{-1}\,y^{-1}$(Table 7). This rate is also high because grass has been growing for more than 5 years and organic C is also added from dung addition during grazing as well as village waste water run-on during rainy seasons.

The Site C is also farmer's field which is well managed and is also irrigated with canal water. The grass is harvested as hay. The rate of SOC sequestration is 0.98 $Mg\,ha^{-1}$ per year (Table 7). Thus, for a normal soil irrigated with canal water, the rate of SOC sequestration is much higher than that of Site A.

National Fertilizer Development Centre (NFDC) and FAO conducted integrated plant nutrient system (IPNS) studies between 1991 and 1996 at the Barani Agriculture Research Institute, Chakwal, under rainfed conditions with wheat (*Triticum aestivum*)-fallow rotation. Manure treatments applied each year were: farm manure (FM) @ 20 $Mg\,ha^{-1}$ and crop residues (i.e., dry wheat straw and old mung bean plants (*Phaseolus mungo*)) @ 2.8 $Mg\,ha^{-1}$. Superimposed on these was the fertilizer treatment of different rates of N (0, 60 and 90 $kg\,ha^{-1}$) and P (0, 40 and 60 $kg\,ha^{-1}$). Extremely low rainfall was received in two out of five wheat seasons. The data in Table 8 show higher rate of SOC sequestration of 0.61 $Mg\,ha^{-1}\,y^{-1}$ with FM and NP application than with either crop residue alone or with fertilizer. This was expected as large quantity of FM was added and also it contained much more nutrients than crop residues.

In a long-term experiment, Iqbal (2006) applied dairy manure at 0, 10 and 20 $Mg\,ha^{-1}$ to corn (*Zea mays L*) crop every year following corn-wheat rotation along with recommended dose of chemical fertilizer. The rate of SOC sequestration in 0.3 m depth was 0.89 and 1.04 $Mg\,ha^{-1}\,y^{-1}$ for 10 and 20 $Mg\,ha^{-1}$ manure treatments, respectively.

Table 8. Effect of organic matter additions on soil organic carbon sequestration in 0.2 m depth after five wheat crops in rainfed area of Chakwal, Punjab.

	Farm manure		Crop residue		Mung bean	
Fertilizer nutrient	SOC* (g kg^{-1})	SOCSR** (kg ha^{-1}y^{-1}	SOC (g kg^{-1})	SOCSR (kg ha^{-1}y^{-1})	SOC (g kg^{-1})	SOCSR (kg ha^{-1}y^{-1})
0–0–0	3.074	522.0	2.378	110.4	2.378	104.4
60–40–0	3.364	609.0	2.436	52.2	2.494	87.0
90–60–0	3.422	609.0	2.494	0.522	2.552	87.0

Data modified from NFDC (1998). Farm manure was added at 20 Mg ha^{-1}y^{-1} and crop residue at 2.8 Mg ha^{-1} y^{-1} dry wheat straw and old Mung bean plants at 2.8 Mg ha^{-1} y^{-1}. *Soil organic carbon in g kg^{-1}; **Soil organic carbon sequestration rate in kg ha^{-1} y^{-1}.

Table 9. Effect of different treatments on soil organic carbon sequestration after growing rice and wheat crops in a salt-affected soil.

Treatment	Soil organic carbon sequestration rate (Mg ha^{-1} y^{-1})
T_1 Tube well water alone	2.8
T_2 Gypsum at 25% gypsum requirement	2.4
T_3 Farm manure at 10 Mg ha^{-1}	2.8
T_4 Combination of $T_2 + T_3$	2.5

Data modified from Ali (2004); Ghafoor et al. 2005.

Another long-term field experiment was conducted in the Fourth Drainage Project Area (FDPA), Faisalabad, Punjab to determine the effectiveness of brackish water in reclamation of saline-sodic soils and assess the rate of SOC sequestration in rice (*Oryza sativa L*)-wheat crop system. Four treatments used were viz. T_1) Tube-well brackish water only, T_2) gypsum application @ 25% SGR to each of first two crops, T_3) FM application @ 10 Mg ha^{-1} annually before transplanting rice, and T_4) Combination of $T_2 + T_3$ applied to the rice-wheat rotation (Ghafoor, 2004; Ali, 2004). There was a substantial increase in yield of rice and wheat, and the POM was also not separately analyzed in SOC determination. The rate of SOC sequestration ranged between 2.4 and 2.8 Mg ha^{-1} y^{-1} (Table 9). In addition to POM, high rates of SOC sequestration may also be because soils are severely depleted of their SOC pools and thus have a high sink capacity.

Another study was conducted in the FDPA to evaluate the treatment effectiveness for reclamation of saline-sodic soils and assess the rate of SOC sequestration using brackish ground water. The three treatments used for rice-wheat rotation were: T_1) Tube well water alone, T_2) Green manuring every year, and T_3) Gypsum @ Water gypsum requirement + green manure every year (Ghafoor, 2004; Bilal, 2004). The mean rate of SOC sequestration was high (4.1 Mg ha^{-1}y^{-1}), probably because POM was included in SOC determination (Table 10). There was also a tremendous increase in yield of rice and wheat as crops were grown on highly saline-sodic and barren soils. The data supported the conclusion that restoration of barren saline-sodic soils is agronomically productive and environmentally friendly.

10 TECHNOLOGICAL OPTIONS FOR SOIL RESTORATION AND SOIL CARBON SEQUESTRATION

Agriculture in Pakistan and Afghanistan is primarily dependent on irrigation which encompasses 85% and 41%, of arable land, respectively. Soil degradation has severe adverse effects on the

Table 10. Soil organic carbon sequestration during reclamation of a saline-sodic soil after 3 years of rice-wheat rotation.

Treatment	Soil organic carbon sequestration rate (Mg ha^{-1} y^{-1})
T_1 Tube well water alone	4.4
T_2 Green manure (Janter) every year before rice	3.7
T_3* Gypsum at WGR* on RSC basis + green manure	4.1

*Water gypsum requirement on residual sodium carbonate basis. Data modified from Bilal (2004).

SOC pool because of reduction in biomass production and the low amount of crop residues and plant roots returned to the soil. Adopting RMPs and restorative measures can increase biomass production as well as the amount of biomass returned to the soil, thereby decreasing SOC losses by erosion, mineralization and leaching. Emission of CO_2 from agricultural activities must be reduced by decreasing deforestation, controlling biomass burning, restoring soil structure and decreasing energy based inputs, e.g. tillage, pumping irrigation water, N-fertilizer, pesticides, etc. through enhancing their use efficiency and decreasing losses. Reducing intensity and frequency of plowing and conversion of marginal cropland to perennial grasses (pasture) or trees (forest) are very important in SOC sequestration in croplands.

Combination of the ethnic/political conflict (23 years), civil disorders, massive population movements and severe drought (1998–2002) have degraded soils and the environment of Afghanistan. Achieving political stability and the external financial assistance since the fall of Taliban regime in 2001 has helped in establishment of Afghan conservation corps to implement labor-based tree plantations and pistachio forest regeneration. The new government has given a high priority to improving forest and rangeland and to sustainable management of natural resources. Agriculture sector also recovered during 2003 with the end of 4 years of drought, but drought recurred in the southern half of the country during 2004. Desertification control is also very important to SOC-sequestration. In Pakistan, dieback of Shisham (*Dalbergia sissoo*) in irrigated soils has adversely affected tree plantations and other perennial vegetation (Gill et al., 2001).

Lack of water is the major limiting factor in crop production in both countries. Restoring wastelands by supplying irrigation water or using RMPs for water conservation and recycling can increase crop production and enhance SOC pool. Using sprinkler and drip irrigation systems on 9.62 Mha of sandy soils and lining irrigation water conveyance system can increase plant growth and enhance SOC concentration in Pakistan. Reuse of saline and/or sodic agricultural drainage water via cyclic or sequential strategies of crop production, where possible or practical (Qadir et al., 2003), can also increase the SOC pool. Reclamation of salt-affected (Ali, 2004; Bilal, 2004; Ghafoor, 2005) and water-logged soils, effective erosion control (Lal, 2004) and establishing vegetation on agriculturally marginal soils can also increase the SOC pool and enhance soil quality.

There is a strong need to develop genotypes of increased tolerances to salinity and hypoxia for improving productivity on saline soil (Qadir, 2000), enhancing SOC pool, improving aggregation and increasing leaching of salts out of the root zone. Phytoremediation is an efficient, inexpensive and environmentally acceptable strategy of soil restoration (Qadir et al., 2001). Growing salt-tolerant species such as Sesbania (*Sesbania bispinosa*), Kallar grass (*Leptochloa fusca*) and Bermuda grass (*Cynodon dactylon*) are effective in soil amelioration through phytoremediation (Qadir and Oster, 2002; Akhter, 2004). Calcareous sodic soils with adequate drainage can be effectively restored through phytoremediation (Qadir, 1996; Oster et al., 1999). In addition to being economical, this strategy promotes soil aggregation, improves plant nutrient availability, increases crop growth and sequesters soil C.

Pakistan Atomic Energy Commission (PAEC) and Higher Education Commission (HEC) have launched a pilot project in all the four provinces to reclaim 4.5 Mha of wastelands by the Bio-Saline Agriculture Technology (BSAT) wherever other techniques can not be used in the country. This technique will improve the environment and increase farm income. The BSAT establishes salt-tolerant crops, grasses, bushes and trees on saline soils (Qureshi and Barrett-Lennard, 1998). *Dalbergia sissoo* plantation rehabilitates degraded sodic soils and also enhances the SOC pool (Mishra et al., 2002; 2003, Tripathi and Singh, 2005).

Nutrient disorders and elemental imbalances are also important factors limiting crop production, and are constraints second only to the drought stress (Rashid and Ryan, 2004). Fertilizer use in Pakistan is predominantly nitrogenous (mainly as urea), whereas use of P is relatively less (Twyford, 1994) and little K or micronutrients are ever applied, except by the progressive farmers. Increasing crop yields through adequate and balanced crop nutrition also increases addition of SOM into the system through addition of roots and crop residues. The IPNS studies based on combined use of organic and inorganic nutrients must be practiced for long-term sustainability. Viable alternatives to farm manure and crop residues as domestic fuel must be identified (NFDC, 1998). About 50% of all farm manure collected is used as cooking fuel (HESS, 1993). Growing green manure crop of Dhaincha (*Sesbania aculeata*) increases SOC concentration and can fit into cropping systems without loss of time (NFDC, 1998). Productivity level of rainfed and irrigated farming are low in Afghanistan compared to regional average, indicating a considerable potential of productivity improvement.

Reseeding of degraded rangeland with grasses such as Gorkha (*Lasiurus sindicus*) and Buffel grass (*Cenchrus ciliaris*) and controlled grazing produces 2–10 times more forage than the native range even with 0.04 m or less of total rainfall (Khan, 1999). Agro-forestry, urban forestry, afforestation of marginal lands, and judicious use of inputs (i.e., irrigation, fertilizer, pesticides) can increase the SOC pool. Household and corporate lawns, sports arenas and urban parks must be managed well and return of their clippings to soil can increase the SOC pool. Bio-gas production has a profound effect on crop nutritional resources and on denuding of woodland and forestland (NFDC, 1998). In this system, all organic refuse collected is placed in a tank buried in soil, anaerobic fermentation produces methane gas which can be used for cooking, lighting and heating. In contrast, burning dung is a waste because the sludge after the bio-gas production can be used as manure. Replacing inefficient burning of fuel wood or crop residue with use of improved cooking stoves and biogas digester can reduce emission of CO_2 and other obnoxious gases (Pretty et al., 2002).

Late planting of wheat (due to late maturing of rice and cotton varieties, excess or lack of soil water and lack of appropriate mechanization) is a major factor responsible for low yields. Grain yield of wheat declines linearly at the rate of 1% to 5% day^{-1} with delayed sowing in November (Ortiz-Monasterio et al., 1994). Use of the no-till technology in the rice-wheat system increases grain yields by early sowing of wheat, reduces cost of production, increases return of crop residues to the soil, conserves the residual moisture available after rice harvest for wheat establishment and increases SOC concentration in the surface soil (Sheikh et al., 2000; Khan and Hashmi, 2004; Gill and Ahmad, 2004). There is a significant increase in wheat grain yield by no-till compared to conventional tillage system (Aslam et al., 1993). It results in saving of 21–63% (Mean 38.5%) of irrigation water during the first irrigation after wheat sowing and 15–20% in the subsequent irrigations (Gill et al., 2000). However, no-till system is not good for corn-wheat rotation because corn is sensitive to poor soil aeration which often occurs in a no-till system (Iqbal, 2006).

Cotton (*Gossipium hirsatum*)-wheat system is the most important rotation in Pakistan and is followed on 3.2 Mha of irrigated land (PAD, 2005). Cotton grows best under conditions of low relative humidity due to less vegetative growth and minimal insect infestation. Sowing on raised beds rather than on normal flat seedbed improves wheat and cotton yields on soils prone to crusting, suggesting that furrow irrigation is better than flood irrigation (Hassan, 2000). On-Farm Water Management Program has recommended bed and furrow irrigation system for cotton-wheat system and sprinkler/drip irrigation system for undulating rainfed areas and sandy soils where significant wastage of water is caused by seepage (OFWM, 2005). The alternate furrow irrigation method is the best for cotton because it saves more than 40% irrigation water (Mumtaz Manais, personal communication, 2005). However, this system can only be practiced by progressive farmers who can afford the machinery required.

Relay cropping (i.e., planting of wheat into standing cotton or rice) without tillage is best suited to increase wheat yield and increase the SOC pool (Pathic and Shrestha, 2004; Paroda, 2005). Of the various tillage options, permanent raised beds are promising as they improve productivity of all crops in the rotation cycle. Raised beds also reduce water use (50–60% of that used in flood irrigation) and improve use efficiency of fertilizer N (Duxbury et al., 2004).

11 SOIL CARBON POOL AND POTENTIAL OF C SEQUESTRATION

The total soil C pool comprises of SOC and soil inorganic carbon (SIC). The SOC pool, comprising of highly active humus and relatively less active charcoal carbon, is major source or sink for atmospheric CO_2 and a key determinant of soil quality (Lal, 2004). The SIC pool includes elemental C and carbonate minerals such as calcite and dolomite and is an important constituent of soils of the arid and semi-arid regions (Nettelton et al., 1983; Lal and Kimble, 2000) such as those in Pakistan (Qadir et al., 2001). Inner portion of calcite nodules found in soils of Punjab, Pakistan are presumed to have formed in equilibrium with soil CO_2, reflecting a large influence of atmospheric CO_2 and a very low soil respiration rate (Pendall and Amundson, 1990). Inner nodules are secondary or pedogenic carbonate based on morphological evidence. Formation of pedogenic or secondary carbonates plays a significant role in C-sequestration through formation of $CaCO_3$, $MgCO_3$ or $(Ca,Mg)CO_3$ or decalcification (i.e., leaching of $Ca(HCO_3)_2$, especially under irrigated conditions). Leaching of bicarbonates into ground water and transport of alkalinity into rivers and eventually to the oceans over periods of thousands of years, is an important mechanism of SIC sequestration. In arid and semi-arid regions, pedogenic carbonates accumulate because of high evapo-transpiration (ET), and precipitation prohibits significant leaching but most of SIC is lost in leachate to ground water in humid environment as rainfall exceeds ET (Nordt et al., 2000). Under irrigated conditions in arid and semi-arid climates, carbonates are subject to dissolution and leaching, thus enhance atmospheric CO_2 sequestration, at least temporarily. The SOC and SIC sequestration rates, used to estimate SOC and SIC sequestration potential presented in Table 11 through 13, are based on the data by Lal (1999; 2001; 2002; 2004a and b); Lal and Kimble (2000) and Mrabet et al. (2001a and b).

Total potential of soil C sequestration is estimated at 3 to 6 Tg y^{-1} in Afghanistan and 8 to 15 Tg y^{-1} in Pakistan (Table 14). This sequestered soil C can be traded in the international market under the Kyoto Protocol or the World Bank. At EU price of \$20 Mg^{-1} of CO_2, the soil C sequestration potential has a total monetary value of \$220–440 million y^{-1} for Afghanistan and \$580–1100 million y^{-1} for Pakistan. While off-setting fossil fuel emission, it will decrease the rate of atmospheric CO_2 enrichment, enhance soil quality and improve productivity.

Table 11. Soil organic carbon sequestration through restoration of degraded soils.

Degradation Process	Area (Mha)		SOC sequestration rate ($kg\,ha^{-1}\,y^{-1}$)		Total SOC sequestration potential (Tg C)	
	Afghanistan	Pakistan	Afghanistan	Pakistan	Afghanistan	Pakistan
Water erosion	11.2	80–120	80–120	80–120	0.90–1.34	0.58–0.86
Wind erosion	2.1	40–60	40–60	40–60	0.08–0.13	0.43–0.64
Water-logging	NA	–	100–200	100–200	–	0.20–0.40
Salinization	1.3	120–150	120–200	120–200	0.16–0.20	0.73–1.22
Soil fertility decline	NA	–	120–150	120–150	–	0.62–0.78
Lowering of water table	NA	–	40–60	40–60	–	0.01–0.01
Total					1.14–1.67	2.57–3.91

Table 12. Potential for SOC sequestration by adopting improved management practices in Afghanistan and Pakistan.

	Area (Mha)		SOC sequestration rate (kg ha^{-1} y^{-1})		Total SOC sequestration potential (Tg C y^{-1})	
Land use	Afghanistan	Pakistan	Afghanistan	Pakistan	Afghanistan	Pakistan
Forests and woodland	1.9	3.8	100–200	100–200	0.19–0.38	0.38–0.76
Permanent pasture	–	5.0	–	50–100	–	0.25–0.50
Rangelands	14.6	20.0	50–100	50–100	0.73–1.46	1.00–2.00
Rice-wheat system	NA	2.6	–	100–200	–	0.26–0.52
*Irrigated farmland	3.21	10.42*	100–200	100–200	0.32–0.64	1.04–2.08
Total					1.24–2.48	2.93–5.86

*Irrigated farmland after subtracting area under rice-wheat and affected by soil fertility decline.

Table 13. Potential for SIC sequestration by adoption of recommended management practices (RMPs).

	Area (Mha)		SOC sequestration rate (kg ha^{-1} y^{-1})		Total SOC sequestration potential (Tg C y^{-1})	
Land use	Afghanistan	Pakistan	Afghanistan	Pakistan	Afghanistan	Pakistan
Irrigated farmland	3.21	18.22	100–200	100–200	0.32–0.64	1.82–3.64
Forests and woodland	1.9	3.6	20–50	20–50	0.04–0.10	0.07–0.18
Rangeland*	14.6	22.5	20–50	20–50	0.29–0.73	0.45–1.13
Total					0.65–1.47	2.34–4.95

*Half of total rangeland is used in estimation.

Table 14. The total potential of soil carbon sequestration (Tg C y^{-1}).

Land use/management	Afghanistan	Pakistan
Restoration of degraded soils	1.14–1.67	2.57–3.91
Adoption of RMPs (SOC)	1.24–2.48	2.93–5.86
Adoption of RMPs (SIC)	0.65–1.47	2.34–4.95
Total	2.93–5.62	7.84–14.72

12 CONCLUSIONS

Lack of fresh water and low soil organic C are the principal constraints in agriculture sustainability. The prevalent management system for agriculture in both countries is not suitable as there is severe problem of soil degradation and desertification. Various forms of soil degradation include accelerated erosion, salinization, water-logging, soil fertility decline, depletion of soil organic carbon, deforestation, soil crusting and compaction, lowering of water table and rangeland degradation. The soil organic carbon (SOC) pool of most soils has been depleted due to conventional tillage practice, removal of crop residue for fodder and fuel, use of manure as fuel, subsistence agriculture based on low external inputs, aridity and mining of soil fertility. The SOC sequestration can be enhanced by restoring degraded soils, controlling desertification, growing improved grass species on rangeland and salt-tolerant crops, grasses, bushes and tree species on saline soils, providing alternate sources for fuel and fodder, adopting no-till technique in rice-wheat system, using integrated plant nutrition systems and adopting recommended residues and water management technologies. The SIC sequestration is also important in irrigated areas. Estimated total potential of soil C sequestration is 3–6 Tg C y^{-1} in Afghanistan and 8–15 Tg C y^{-1} in Pakistan which can be traded in the international market. At EU price of \$20 Mg^{-1} of CO_2, the soil C sequestration potential has a total monetary value of \$220–440 million y^{-1} in Afghanistan and \$580–1100 million y^{-1} in Pakistan.

ACKNOWLEDGEMENTS

We gratefully acknowledge the literature provided by Engineer Ijaz Ahmad Khan, Managing Partner, NDC, 62-M, Gulberg III, Lahore and Dr. Nisar Ahmad, Chief, National Fertilizer Development Centre (NFDC), Islamabad.

REFERENCES

Ahmad, B. 2004. Effect of rice-wheat crop rotation receiving gypsum on CO_2 assimilation and accumulation of organic matter during reclamation of saline-sodic soils. M.Sc. Thesis, Institute of Soil and Environmental Sciences, University of Agriculture, Faisalabad, Pakistan.

Akhter, J., K. Mahmood, K.A. Malik, S. Ahmed, and R. Murray. 2003. Amelioration of a saline- sodic soil through cultivation of a salt-tolerant grass *Leptochloa fusca*. Environmental Conservation 30(2): 168–174.

Akhter, J., R. Murray, K. Mahmood, K.A. Malik, and S. Ahmed. 2004. Improvement of degraded physical properties of a saline-sodic soil by reclamation with Kallar grass (*Leptochloa fusca*). Plant and Soil 258: 207–216.

Ali, M.K. 2004. Sequestration of CO_2 during the reclamation of salt-affected soils: A helps head-off global warming. M.Sc. Thesis, Institute of Soil and Environmental Sciences, University of Agriculture, Faisalabad, Pakistan.

Anonymous, 2002. Pakistan Statistics Year Book. Federal Bureau of Statistics Division, Govt. Pakistan, Islamabad.

Asian Development Bank. 1992. Pakistan: Forestry sector master plan. Upland degraded watershed component. Asian Development Bank, Manila, Philippines.

Aslam, M., A. Majid, N.I. Hashmi, and P.R. Hobbs. 1993. Improving wheat yield with rice-wheat cropping system of the Punjab through no-tillage. Pakistan J. Agri. Res. 14:8–11.

Bilal, M. 2004. Effect of rice-wheat crop rotation receiving gypsum on CO_2 assimilation and soil C sequestration. M.Sc. Thesis, Institute of Soil and Environmental Sciences, University of Agriculture, Faisalabad, Pakistan.

Duxbury, J.M., J.G. Lauren, M.H. Devare, A.S.M.H.M. Talukder, M.A. Sufian, A. Shaheed, M.I. Hossain, K.R. Dahal, J. Tripathi, G.S. Giri, and C.A. Meisner. 2004. Opportunities and constraints for reduced tillage practices in the rice-wheat cropping system. p. 121–131. In: R. Lal, P.R. Hobbs, N. Uphoff, and D.O. Hansen (eds.). Sustainable Agriculture and the International Rice-Wheat System. Marcel Dekker, Inc., New York.

Ellis, S., D.M. Taylor, and K.R. Masood. 1994. Soil formation and erosion in the Muree hills, Northeastern Pakistan. Catena 22(1): 69–78.

FAO. 1986. Status report in plant nutrition in fertilizer programs of counties in Asia and Pacific region. FAO, Rome.

FAO. 1992. The state of food and agriculture 1991. FAO, Rome.

FAO. 1994. Land Degradation in South Asia: Its Severity, Causes and Effects upon the People. World Soil Resources Reports 77. FAO, Rome, Italy.

FAO. 2004. Yearbook of Agriculture. FAO, Rome, Italy.

Ghafoor, A., M.A. Gill, A. Hassan, G. Murtaza and M. Qadir. 2001. Gypsum: An economical amendment for amelioration of saline-sodic waters and soil and for improving crop yields. Int. J. Agri. Biol. 3: 266–275.

Ghafoor, A. 2005. Farmer participation in technology development and transfer for using agricultural drainage water for growing grain crops during reclamation of saline-sodic soils. Final Technical Report. June 2001 to May 2004. ISES, University of Agriculture, Faisalabad.

Gill, M.A., M.A. Kahlown, M.A. Choudhary, and P.R. Hobbs. 2000. Evaluation of resource conservation techniques in rice-wheat system of Pakistan. Water and Power Development Authority Report. Lahore, Pakistan.

Gill, M.A., I. Ahmad, A.U. Khan, M. Aslam, S. Ali, M.M. Khan, R.M. Rafique, and T. Mehmood. 2001. Phytophthora cinnamomi – A cause of Shisham decline in the Punjab, Pakistan. Proc. of National Seminar on Shisham diebeck. Oct. 27, 2001, Faisalabad. p. 21.

Gill, M.A. and M. Ahmad. 2004. The role of the South Asian Conservation Agriculture Network (SACAN) in no-till farming in Pakistan. p. 479–494. In: R. Lal, P.R. Hobbs, N. Uphoff, and D.O. Hansen (eds.). Sustainable Agriculture and the International Rice-Wheat System. Marcel Dekker, Inc., New York.

Hassan, A. 2000. Application of soil physics for sustainable productivity of degraded lands: A review. Int. J. Agri. Biol. 2(3): 263–268.

Hobbs, P.R. 2001. Tillage and crop establishment in South Asian rice-wheat system: Present practices and future options. J. Crop Production 4: 1–22.

HESS. 1993. Household Energy Strategy Survey project, Word Bank, Islamabad. 1991–1993 Reports. Dung Cake. Islamabad (Mimeo).

IPCC. 2001. Climate change 2001: The scientific basis. Inter-government panel on climate change. Cambridge University Press, Cambridge, UK.

Iqbal, M. 2006. Soil physical properties and growth of maize and wheat as affected by tillage, farm manure and mulch. Ph.D. Dissertation. Institute of Soil & Environmental Sciences, University of Agriculture, Faisalabad.

Kahlown, M.A. and W.D. Kemper. 2004. Seepage losses as affected by condition and composition of channel banks. Agric. Water Management 65 (2): 145–153.

Khan, G.S. 1993. Characterization and genesis of saline-sodic soils in Indus Plains of Pakistan. Ph. D. Dissertation. Department of Soil Science. University of Agriculture, Faisalabad.

Khan, K.F., D.M. Anderson, M.I. Nutkani, and N.M. Butt. 1999. Preliminary results from reseeding degraded Dera Ghazi Khan Rangeland to improve small ruminant production in Pakistan. Small Ruminant Research 32(1): 43–49.

Khan, M.A. and N.I. Hashmi. 2004. Impact of no-tillage farming on wheat production and resource conservation in the rice-wheat zone of Punjab, Pakistan. p. 219–228. In: R. Lal, P.R. Hobbs, N. Uphoff, and D.O. Hansen (eds.). Sustainable Agriculture and the International Rice-Wheat System. Marcel Dekker, Inc., New York.

Lal, R. 1999. Soil management and restoration for C sequestration to mitigate the greenhouse effect. Progress in Environmental Science 1: 307–326.

Lal, R. and J.M. Kimble. 2000. Inorganic carbon and the global carbon cycle: Research and development priorities. p. 291–302. In: R. Lal, J.M. Kimble, H. Eswaran, and B.A. Stewart (eds.) Global climatic change and pedogenic carbonates. CRC/Lewis Publishers, Boca Raton, FL.

Lal, R. 2001. World cropland soils as a source or sink for atmospheric carbon. Advances in Agronomy 71: 145–191.

Lal, R. 2002. Carbon sequestration in dryland ecosystems of West Asia and North Africa. Land Degrad. Develop. 13: 45–59.

Lal, R. 2004a. Carbon sequestration in soils of Central Asia. Land Degrad. & Develop. 15: 563–572.

Lal, R. 2004b. Soil carbon sequestration in India. Climatic Change 65: 277–296.

Mishra, A., S.D. Sharma, and M.K. Gupta. 2003. Soil rehabilitation through afforestation: Evaluation of the performance of *Prosopis juliflora, Dilbergia Sissoo* and *Eucalyptus tereticornis* plantations in a sodic environment. Arid Land Research and Management 17(3): 257–269.

Mishra, A., S.D. Sharma, and G.H. Khan. 2002. Rehabilitation of degraded sodic lands during a decade of *Dalbergia sissoo* plantation in Sultanpur district of Uttar Pradesh, India. Land Degradation Development 13(5): 375–386.

Mrabet, R., K. Ibno-Namr, F. Bessam, and N. Saber. 2001a. Soil chemical quality changes and implications for fertilizer management after 11 years of no-tillage wheat production systems in semi-arid Morocco. Land Degradation Development 12: 505–517.

Mrabet, R., N. Saber, A. El-Brahli, S. Lahlou, and F. Bessam. 2001b. Total, particulate organic matter and structural stability of a Calcixeroll soil under different wheat rotations and tillage systems in semi-arid area of Morocco. Soil and Tillage Research 57: 225–236.

Muhammad, D. 2002. Grassland and Pasture crops. Country pasture/forage resources profile. Pakistan. FAO, Rome, Italy.

NFDC. 1998. Integrated Plant Nutrition Systems (IPNS): Combined use of organics and inorganics. Technical Report 3/98. National Fertilizer Development Centre. Islamabad, Pakistan.

Nordt, L.C., L.P. Wilding, and L.R. Drees. 2000. Pedogenic carbonates transformation in leaching soil systems: Implications for the global C cycle. p. 43–64. In: R. Lal, J.M. Kimble, H. Eswaran, and B.A. Stewart (eds.) Global climatic change and pedogenic carbonates. CRC/Lewis Publishers, Boca Raton, FL.

Ortiz-Monasterio, J.I., S.S. Dillon, and R.A. Fischer. 1994. Date of sowing effects on grain yield and yield components of irrigated spring wheat cultivars and relationships with radiations and temperature in Ludhiana, India. Field Crops Res. 37: 169–184.

PAD. 2005. Provincial Agriculture Departments, Pakistan.

PARC. 1980. Pakistan Agricultural Research Council, Islamabad.

Paroda, R. 2005. The challenges of agriculture in Central Asia. The University Distinguished Lecture delivered at Carbon Sequestration in Central Asia Conference, at OSU, OH. Nov. 02, 2005.

Pathic, D.S. and R.K. Shrestha. 2004. No-till in rice-wheat system: An experience from Nepal. p. 209–218. In: R. Lal, P.R. Hobbs, N. Uphoff, and D.O. Hansen (eds.). Sustainable Agriculture and the International Rice-Wheat System. Marcel Dekker, Inc., New York.

Pendall, E. and R. Amundson. 1990. The stable isotope chemistry of pedogenic carbonate in an alluvial soil from the Punjab, Pakistan. Soil Sci. 149(4): 199–221.

Pimentel, D., O. Bailey, P. Kim, E. Mullaney, J. Calabrese, L. Walman, F. Nelson, and X. Yao. 1999. Will limits of the earth's resources control human numbers? Environ. Sustainability Dev. 1: 19–39.

Pretty, J.N., A.S. Ball, X.Y. Li, and N.H. Rivindranath. 2002. The role of sustainable agriculture and renewable resource management in reducing greenhouse-gas emissions and increasing sinks in China and India. Philos. T. Roy. Soc. A. Math. Phys. Eng. Sci. 360(1797): 1741–1761.

Qadir, M., R.H. Qureshi, N. Ahmad, and M. Ilyas. 1996. Salt-tolerant forage cultivation on a saline-sodic field for biomass production and soil reclamation. Land Degradation Development 7: 11–18.

Qadir, M., A. Ghafoor, and G. Murtaza. 2000. Amelioration strategies for saline soils: A review. Land Degradation Development 11: 501–521.

Qadir, M., S. Schubert, A. Ghafoor, and G. Murtaza. 2001. Amelioration strategies for sodic soils: A review. Land Degradation Development 12: 357–386.

Qadir, M. and J.D. Oster. 2002. Vegetative bioremediation of calcareous sodic soils: History, mechanisms and evaluation. Irrigation Sci. 21: 91–101.

Qureshi, R.H. and E.G. Barrett-Lennard. 1998. Saline agriculture for irrigated land in Pakistan: A handbook. Australian Center for International Agric. Res., Canberra.

Qutub, S.A. and Nasiruddin. 1994. Cost effectiveness of improved water management practices. p. 43–61. In: C. Inayat-Ullah (ed.). Water and community: An assessment of on-farm water management program. SPDI, Islamabad, Pakistan.

Rafiq, M. 1990. Soil resources and soil related problems in Pakistan. p. 16–23. In: Proc. Int. Symp. on Applied Soil Physics in Stress Environment. January 22–26, 1989, Islamabad.

Rashid, A. and K.S. Memon. 1996. Soil Science. National Book Foundation. Islamabad, Pakistan.

Rashid, A. and J. Ryan. 2004. Micronutrient constraints to crop production in soils with Mediterranean-type characteristics: A review. J. Plant Nutrition 27(6): 959–975.

Seckler, D.W., R. Barker, and A. Singhe. 1999. Water scarcity in the twenty-first century. Int. J. Water Resour. Dev. 15: 29–43.

Shafiq, M., A. Hassan, S. Ahmad, and M.S. Akhter. 1994. Water intake as influenced by induced compaction and tillage in rain fed areas of Punjab (Pakistan). J. Soil Water Conservation 49(3): 302–305.

Sufi, A.B. and S.M. Javed. 1988. Review of existing research on skimming wells. International Water-logging and Salinity Research Institute. Wapda, Lahore, Pakistan.

Tripathi, K.P. and B. Singh. 2005. The role of revegetation for rehabilitation of sodic soils in semi-arid subtropical forest, India. Restoration Ecology 13(1): 29–38.

Twyford, I. 1994. Fertilizer use and crop yields. Proceedings of 4th National Congress of the Soil Science Society of Pakistan, Islamabad, 1992.

Umrani, A.P., P.R. English, and D. Younis. 1995. Rangeland in Pakistan. Asian Livestock. Bangkok, Thailand XX(3): 30–36.

UNEP. 1992. United Nations Environment Program (UNEP) report.

WAPDA. 2005. Water and Power Development Authority reports, Pakistan.

WBS. 2005. World Bank Statistics.

CHAPTER 19

Improvement of soil physical and chemical conditions to promote sustainable crop production in agricultural areas of Kazakhstan

Warren Busscher & Jeff Novak
Coastal Plains Soil, Water and Plant Research Center, USDA-ARS, Florence, SC, USA

Flarida Kozybaeva, Temirbulat Jalankuzov & Beibut Suleymenov
Akademgorodok, Institute of Soil Science, Almaty, Kazakhstan

1 INTRODUCTION

Soils are a source of food, clothing, and energy for Kazakhstan, the USA, and the world, so their deteriorated condition is a cause for concern. Soils are not only losing the organic carbon (Lal, 2002) that help make them fertile, but they are also losing their productivity at an estimated worldwide rate of 15 Mha per year (Buringh, 1981). They are degraded by water and wind erosion, salinity, desertification, land development, and pollution, to mention a few. Though this paper is limited to the physical sciences, soils should be studied in their entirety (physical, biological, and economic/social sciences) to develop new, rational, realistic management systems that can save them from degradation or loss while improving their beneficial use to society. To solve these problems, researchers can build on the works of their predecessors from both the east (Kononova, 1951; Kostychev 1951; Dokuchayev, 1952; Tyurin, 1965) and the west (Van Schifgaarde et al., 1956; Black, 1957; Klepper et al., 1973) who helped shape the knowledge of present-day soil science. Though the problem is worldwide and solutions have global implications, soil issues described here focus on parts of Kazakhstan.

2 GEOGRAPHY

Kazakhstan has a population of about 17 million people. It borders Russia on the north, China on the east, Kyrgyzstan and Uzbekistan on the south, and Turkmenistan and the Caspian Sea on the west (Figure 1). It covers 2.72 million square kilometers, ranging about 1600 km from north to south and 3000 km from east to west, making it the 9th largest country in the world.

Kazakhstan is a country of vast socio-economic diversity ranging from rustic camp to cosmopolitan city. It has an industrial base and important natural resources of oil, coal, iron ore, manganese, chromite, lead, zinc, copper, titanium, bauxite, phosphate, sulfur, gold, and silver. Because of years of neglect under former administrations, its industrial base and its agricultural sector are undergoing much-needed repair and updating, which are fueled by a healthy economy. Recently, Kazakhstan's economy had double-digit or near double-digit growth as a result of its energy sector aided by economic reforms, good harvests, and foreign investments.

Kazakhstan's agriculture is diverse, ranging from small farms to large cooperatives raising crops and livestock. The country's agricultural base is a historically important national resource containing fertile soils and extensive irrigation. Its renovation is fueled by a need to change from crop management systems dictated by the former centralized government to current-day market-driven forces.

Figure 1. Satellite photo of Kazakhstan showing its oblasts (states) and neighbors. Object from http://www.fao.org/countryprofiles/Maps/KAZ/19/im/index.html.

3 TERRAIN AND CLIMATE

Because the country is so large, it has a wide variety of climate, terrain, vegetation, and soil parent material and type. Though the southeast and east are ringed by the picturesque, snow-capped Tjan Shan and Altai Mountains, most of the country is flat lowlands, less than 500 m in elevation. Other areas of the country include plateaus and lesser mountain ranges that rise 200 to 500 m above the surrounding terrain to elevations of 1,000 m above sea level.

Most of Kazakhstan has a continental climate with cold winters and hot summers. Except for the mountains of the southeast and east, most of the country is also dry. Annual precipitation for northernmost Kazakhstan is 315 mm. In central Kazakhstan, annual precipitation is about 150 mm. In the foothills of the mountains, precipitation increases to 880 mm on the forested moutainsides. Mountainside precipitation not only waters the soils locally, but excess amounts of rainfall and snowmelt flow onto the lowlands where they are used for irrigation (Brown, 2006).

The Transili Alatau piedmont in southeastern Kazakhstan is near the former capital of Almaty and is one of Kazakhstan's areas of dense agricultural production. The piedmont is characterized by large daily and annual temperature fluctuations, cold winters, and hot summers. Its minimum winter temperatures are −14°C in January with relative humidity 66–67%, average snow thickness of 26–30 cm, and soils freezing to depths of 15–55 cm. As the temperature rises in spring and summer to highs of 37–43°C, the piedmont's humidity falls during the region's 140–170 frost-free-day growing season.

4 SOILS

About 8% of Kazakhstan's soils are arable, 12% of which are irrigated, and another 70% of Kazakhstan's soils are pasture (Wikipedia, 2006). Soils are distributed throughout the country generally in zones changing from north to south that correspond roughly to the vegetation (Figure 2). In the northern forest-steppe, soils are deep, dark chernozems. These northern chrenozems, along with a mixture of meadow soils and small areas of saline soils, cover about 1 million ha in an area where precipitation and evaporation are approximately equal.

South of this are extensive grassland steppes where soils are moderately-droughty chernozems, less fertile chestnut soils, and brown soils. These soils are interspersed with saline soils because

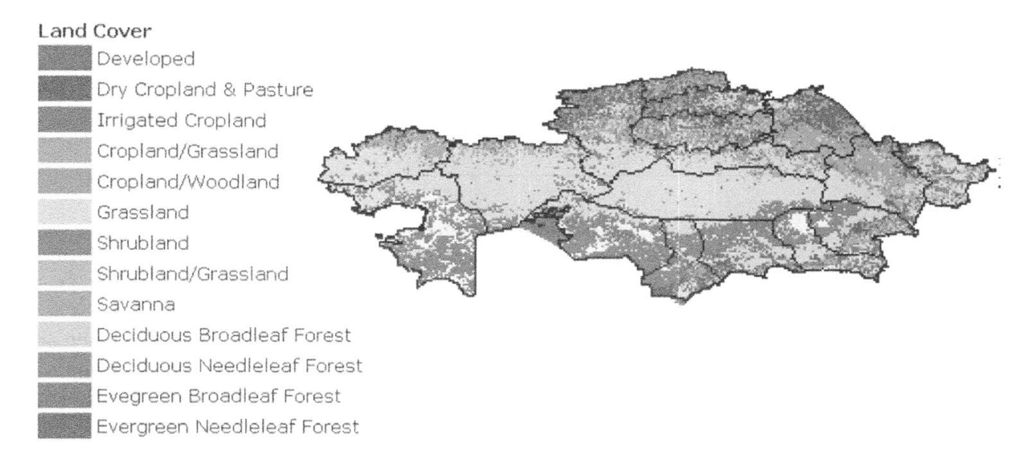

Figure 2. Zones of land cover (http:www.fao.org/countryprofiles/).

evaporation exceeds rainfall. Steppe soils can be compacted, low in organic carbon, and shallow, at times shallow enough to be unsuited for cultivation, though suitable for pasture.

Further south is the desert steppe with brown soils that typically develop in semiarid climates. These soils are rich in nutrients and low in organic carbon. Still further south is the desert where sandy and gray-brown soils develop under dry conditions but can be productive with irrigation. In the southeast and east, the foothills of the Tjan Shan Mountains have sierozem soils with low organic carbon and sparse vegetation. These soils developed in areas of warm, wet springs; hot, dry summers; and moderately cold winters.

In the Transili Alatau piedmont, soils are dark-chestnut. These soils are distributed in a 447,000 ha east-to-west band across the piedmont plains. They can be divided into leached, calcareous, and eroded. They differ from the plain chestnut soils by being thinner, not alkaline, and higher in surface-horizon humus. These soils are arable enough to grow cereals, forage, and high-value vegetables. Sloping soils are used for hay and pasture. An 80 to100 ha research farm in the plateau has been used for the past 50 to 70 years to compare virgin soils to pasture, vegetable production, and various rotations.

At the experimental site, tillage has reduced surface-horizon aggregates of size $>250\,\mu m$ from 42%, as seen in the virgin soil, to 22–25%. It has also reduced micro-aggregates ($>250\,\mu m$) to 67% of that seen in virgin soil. As expected, reduced aggregation also reduces infiltration (Sokolova et al., 2001) and increases bulk density, which in turn reduces yield. The addition of organic carbon through manure has been shown to improve soil properties.

Humus plays an important role in the improvement of soil physical, biological, and chemical characteristics, which in turn improve soil productivity. The humus content of virgin dark-chestnut soils is 3.6%, decreasing gradually with depth. Humus contents of production soils are 1.7 to 2.5%, also decreasing with depth to a value of 0.24% at 140- to 150-cm depths. The reduced humus contents are large when compared to other studies where humus contents decreased by 20–30%. Reductions in humus are related to decreases in cation exchange capacities (CEC). CECs are 21.62 c $mol_c\,kg^{-1}$ for virgin soils, while they are 15 c $mol_c\,kg^{-1}$ for the tilled soils. This is significantly lower not only because of the reduced organic carbon contents but also because of lower clay contents and increased erosion, though more research needs to be performed to verify these results.

5 DEGRADATION AND REMEDIATION

Because soil is a significant resource of Kazakhstan, it needs to be developed or remediated to provide food, clothing, and exports. Yet soil also has to be maintained as an environmentally sustained

Table 1. The land area and population of Kazakhstan (Kharin and Tateishi, 1996).

Land area	Population ($\times 10^3$)					
($\times 10^3$ km^2)	1959	1970	1979	1989	1999	2005
2,713	9.3	13	14.7	16.5	16.9	15.2

resource for future use. New management systems that take into account inherent soil properties and limitations can improve yield. Limitations include salinity (Suleimenov, 2002), reduced fertility (Suleimenov, 2000; Suleimenov and Rubinshtein, 2001), and desertification (Dregne, 1986). These problems can be ameliorated with the addition of soil organic carbon which can buffer salinity, increase cation exchange capacity, and increase water holding capacities (El-Hage Scialabba and Hattam, 2002). Management systems such as reduced tillage or green manure crops that improve soil organic carbon can help ameliorate physical, chemical, and associated biological problems while improving soil productivity (Nabiyev, 1996; Condron et al., 2000).

Another reason for improving management and increasing organic carbon is the increased demand on the land from Kazakhstan's population growth (Suleimenov, 2000), which took place in the late 1900's (Table 1). To support the increased population and ensure its security, marginal lands were brought into production to increase the food supplies and develop textile exports. Bringing marginal land into production led to land degradation, including both desertification caused by overgrazing and unsustainable cultivation caused by increased compaction, wind and water erosion, and salinity. Salinity in Kazakhstan is so severe that 60 to 70 percent of the irrigated land is affected (Suzuki, 2003). Overgrazing was the result of the previous administration's policy of centralization of the animal sector onto small farms. This increased degradation of soil physical properties (Suzuki, 2003).

Other forces degrading the land were unsustainable production practices, including deep plowing of fragile soils, cultivation of erosion-prone crops such as maize, and the use of heavy machinery that destroys soil structure through compaction. These practices led to the loss of soil organic carbon and soil fertility (German Advisory Council on Global Change, 1994; European Environment Agency, 1999). Many of these degradation processes have a direct impact on the global carbon cycle, particularly through the decrease of organic carbon and the subsequent release of greenhouse gasses into the atmosphere. Declines in organic carbon contents are attributed to tillage and agricultural management practices that fail to maintain soil organic carbon. Tillage techniques that invert the soil play a major role in this process. Lal (2002) estimated that agricultural activities in the five Central Asian countries (Kazakhstan, Kyrgyzstan, Tadjikistan, Turkmenistan, and Uzbekistan) have caused the combined loss of between 1 to 2 Pg of soil organic carbon.

From the late 18th to the early 20th centuries, country-wide changes, especially the influx of immigrants and development of settlements, disrupted the Kazak nomadic lifestyle. The most dramatic changes in land use were the establishment of centralized animal production and cultivation of the fragile rangelands. Concentration of animal production facilities near water sources caused grazing pressure on nearby lands because it eliminated the movement of large animal herds across the vast rangeland. Because animals were kept immobile, local land vegetation was unable to recover as it had in the past when the herd moved on. With limited rainfall and overgrazing, land productivity for grazing declined dramatically. According to Schillhron van Veen et al. (2005), this and other problems reduced productive rangeland to half of its original 186 Mha.

As part of the Soviet "Virgin Lands Project", the amount of marginal land brought under cultivation increased dramatically between 1950 and 1960; cultivated land went from about 8 to 28 Mha. During this period, transformation of the steppes into arable land was coined "Conquest of the Deserts" (FAO, 1995). Land was planted with cotton and wheat to provide grain and raw product for textiles to meet the demands of the growing population (Table 1). Cotton production in Kazakhstan rose to a level of almost 315 million t in 1992 (FAO, 1995). Because the region has low rainfall (100 to 200 mm y^{-1}, FAO, 1995), maintaining these yields required intensified irrigation. Large-scale irrigation consumed vast amounts of water with some areas using up to 12,800 m^3 ha^{-1}

per year (FAO, 1995). Water was taken from the Aral Sea, now known more for its dry bed than its water, and from other river and ground water sources. Because water evaporates faster than it is replenished by precipitation in this arid to semi-arid environment, irrigation with salt-laden water promoted soil salinity that reduced crop yields. The reductions could be substantial. Gardner (1997) reported that between the late 1970s and the late 1980s in the Central Asian republics, salinity reduced cotton yields from 280 to 230 t km^{-2}. Other sources list that Kazakhstan's wheat harvest declined from about 13 Mt in 1980 to 8 Mt in 2000 – an economic loss of $900 million per year (http://www.earth-policy.org/Books/Eco/EEch1_ss2.htm, accessed January 23, 2006). Loss of income from declining yields placed economic hardships on producers. This decline plus fluctuations in the country's gross national product eventually caused vast areas of cropland to be abandoned. According to the Land Resource Management Agency (FAO, 1995), 12.8 Mha that once grew wheat were no longer used in 2000.

Over-utilization of irrigation water in Kazakhstan also had a major environmental impact on erosion. From 1961 to 1988, irrigation water removed from the Aral Sea caused an average 15-m drop in water depth (FAO, 1995). This drop exposed 25,000 km^2 of the former seabed, which in turn increased the amount of soil lost to dust storms and sand deflation (sand-size material movement). Soil losses in the area were also heightened by erosion from the abandoned cotton and wheat fields. Soil loss to wind erosion was so severe that certain areas lost 2 to 7 cm of topsoil per year in dust storms that blew former topsoil as far away as Poland and Hungary (FAO, 1995). Over-utilization of irrigation water, salinity, and other ill-advised management practices contributed to desertification of an estimated 60% of the rangeland in Kazakhstan (UNECE, 2000). Overexploitation of marginal lands never suited for large scale cultivation was one of several agricultural management practices of arid and semi-arid zones that had a detrimental effect on the Kazakhstan landscape.

Desertification in Kazakhstan can have a significant influence on the global carbon cycle, particularly through the decline in organic carbon levels and release of carbon dioxide to the atmosphere. Because of extensive above- and below-ground vegetation in the steppe regions of Kazakhstan, as much as 1.27 t ha^{-1} of organic carbon can be sequestered, mainly during the period of May to October (USAID-CRSP, 2002). Indeed, the five Central Asia countries have the potential to sequester between 1 to 2 Pg of organic carbon over a 50 yr period (Lal, 2002). Assuming sequestration across the entire Kazakhstan rangeland of 186 Mha, the amount of carbon sequestered annually can be as high as 0.24 Pg. Unfortunately, this sequestered carbon is dynamic and can easily be diminished by grass fires, overgrazing, conversion to cropland, or urban development (Schillhorn van Veen et al., 2005). Nonetheless, because of the sheer size of Kazakhstan's steppes, carbon sequestration is substantial on a global scale.

6 CONCLUSIONS

Kazakhstan is a large country with diversity of terrain and climate. Though most of the country has a dry climate, it contains a wealth of natural resources including soils, which can be managed in an environmentally safe and sustainable manner to help provide food and clothing for the country's growing population and to help provide exports to sustain the country's economy.

Because of various politically motivated programs from previous administrations, soils have undergone renovation much like the rest of the infrastructure of the country. Renovation includes increased soil carbon content. Low levels of organic carbon put Kazakhstan at risk for future production. Nevertheless, the country is in a position to sequester significant amounts of carbon, which may help reduce the greenhouse effect even if the amount sequestered fluctuates annually. Improved organic carbon through research on innovative management practices can help increase soil productivity.

REFERENCES

Black, C. 1957. Soil-Plant Relationships. Wiley and Sons, New York, NY.

Brown, L.R. 2006. Chapter 4: Rising Temperatures and Rising Seas. In: Plan B 2.0: Rescuing a Planet Under Stress and a Civilization in Trouble. W.W. Norton & Co., NY. Accessed online 19 Jan. 2006. <http://www.earth-policy.org/Books/PB2/PB2ch4_ss4.htm>

Buringh, P. 1981. An assessment of losses and degradation of productive agricultural lands in the world. Agric. Univ., Wageningen, The Netherlands.

Condron L.M., K.C. Cameron, H.J. Di, T.J. Clough, E.A. Forbes, R.G. McLaren, and R.G. Silva. 2000. A comparison of soil and environmental quality under organic and conventional farming systems in New Zealand. New Zealand Journal Agricultural Research 43(4): 443–466.

Dregne, H.E. 1986. Desertification of arid lands. In: F. El-Baz and M.H.A. Hassan (eds.) Physics of desertification. Dordrecht, The Netherlands.

Dokuchaev, V.V. 1952. Russian chernozems. Selkhos. 636 p.

European Environment Agency. 1999. Environment in the Eurpoean Union at the turn of the century. Environmental Assessment Report No. 2. Office of Official Publications of the European Communities, Luxembourg.

El-Hage Scialabba, N, and C. Hattam. 2002. Organic agriculture, environment and food security. FAO Yearbook of Fishery Statistics. 258 p.

FAO. 1995. Desertification and drylands development in CIS countries. pp. 1–203. In: Dryland development and combating desertification. Environment and Energy Papers-14. (Available on the internet at: http://www.fao.org/documents/t441eoh.html).

Gardner, G. 1997. Preserving global cropland. In: Brown, L. (ed.) State of the world 1997. W.W. Norton, New York.

German Advisory Council on Global Change. 1994. World in transition: The threat to soils. 1994 Annual Report. Economica, Bonn.

Kharin, N.R, and R. Tateishi. 1996. Degradation of the drylands of Asia. Chiba University, Japan.

Klepper, B., H. Taylor, and M. Huck. 1973. Water relations and growth of cotton in drying soil. Agron. J. 65:307–310.

Kononova, M.M. 1951. Soil humus problems and tasks of studying it. AS SSSR. 389 p.

Kostychev, P.A. 1951. Tillage and fertilization of chernozems. Doklady AS SSSR. 668 p.

Lal, R. 2002. Conservation agriculture and soil carbon sequestration in limited water resources areas of Central Asia. In Proceedings of the International Workshop on Conservation Agriculture for Sustainable Wheat Production in Rotation with Cotton in Limited Water Resources Areas. Tashkent, Uzbekistan. 8 p.

Nabiyev, M. 1996. Contamination of environment by mineral fertilizers is not permissible. Agriculture of Uzbekistan, Tashkent.

Schillhorn van Veen, T.W., I.I. Alimaev, and B. Utkelov. 2005. Kazakhstan: Rangelands in Transition, The Resources, the Users, and Sustainable Use. World Bank Technical Report No. 31384. Available at: http://www.worldbank.org/eca/Kazakhstan/pr/rangelandeng.pdf, accessed January 2006.

Sokolova T.M., T.R. Ryspekov, T.M. Sharypova, D.S. Makhmutova, and G.A. Saparov. 2001. Changes in water-physical properties of dark-chestnut irrigated soil under the influence of fertilizers. In: Scientific bases of fertility rehabilitation, protection and rational use of soils in Kazakhstan. Almaty, Kazakhstan, p. 125–128.

Suleimenov B.U. 2000. Transformation of sierozems as related to their cultivation. Nauka and Obrazovaniye of the south Kazakhstan, ser.ecol. Shymkent N1. p. 2–5.

Suleimenov, B.U. 2002. Pecularities in formation of sierozems. Izvestiya NAS RK.

Suleimenov, B.U. and M.I. Rubinshtein 2001. Genetic characteristics of chernozems in the south Kazakhstan. Collection of articles. Scientific bases of fertility rehabilitation, conservation and rational use of soils in Kazakhstan. Almaty p. 34–41.

Suzuki, K. 2003. Sustainable and environmentally sound land use in rural areas with special attention to land degradation. pp. 1–14. In: Asia-Pacific Forum for Environment and Development Meeting. 23 January 2003. Gulian, People's Republic of China. (Available on the internet at: http://www/iges/or.jp/en/ltp/pdf/APFED3_EM_doc4.pdf).

Tyurin, I.V. 1965. Soil organic matter and its role in fertility. Nauka, 320 p.

UNECE. 2000. Environmental performance review of Kazakhstan. United Nations Economic Commission for Europe (UNECE). Geneva.

USAID-CRSP. 2002. Review of Global Livestock CRSP. University of CA, Davis. Available on the internet at http://glcrsp.ucdavis.edu/project_subpages/CAoverview.html).

Van Schilfgaarde, J., D. Kirkham, and R.K. Frevert. 1956. Physical and mathematical theories of tile and ditch drainage and their usefulness in design. Agricultural Experiment Station, Iowa State University.

Wikipedia. 2006. Kazakhstan. accessed online 21 Jan. 2006. http://en.wikipedia.org/wiki/kazakhstan

CHAPTER 20

Technological options to enhance humus content and conserve water in soils of the Zarafshan valley, Uzbekistan

Sh. T. Holikulov & T.K. Ortikov
Samarkand Agricultural Institute, Samarkand, Uzbekistan

1 INTRODUCTION

Organic carbon (C) in soil is primarily found in the humus, which also partially determines soil fertility. Thus, the concentration of organic C in soils is a major contributor to increased soil fertility of soils which is strongly influenced by the balance of humus. Carbon becomes organic when the content of humus is increased which results in a reduction of carbon dioxide (CO_2) gas in the air. This in turn positively influences the ecological condition of the atmosphere. On the other hand, measures adopted to increase the humus have positive effects on water use efficiencies in desert areas. Increasing the humus content promotes the accumulation of C in the form of organic matter and conservation of water in moisture deficit conditions. It is especially important in Uzbekistan, where the soil and air temperatures are high and moisture is limited.

The process of soil deulmification has been prevalent in Uzbekistan as elsewhere (Berestetskiy, 1986; Dobrovolskiy et al., 1985; Zonn, 1989; Kovda, 1981; Riskiyeva and Toshkenboyev, 2002; Rozanov et al., 1989; Sattarov, 1990), and as a result organic C is oxidized, CO_2 gas is released into the atmosphere, and the natural balance of C is broken in the agricultural landscapes. Consequently, reduction in humus adversely impacts the water balance of soils and all agricultural practices become less effective (Efimov and Osipov, 1991; Elyubayeva and Tursunkulova, 1987 etc.).

Soils with high humus concentration are rapidly decreasing in Uzbekistan. This reduction in humus concentration in soils is mostly observed in fertile lands, which is why the mineralization of humus is observed in the meadow soils rather than in Sierozem. Because meadow soils do not have more humus than Sierozem soils, the content of organic C has become equal in both soils. Identical measurements of change in soils conducted in different types and subtypes of soil suggest that anthropogenic factors are of prime significance in explaining changes. When impacted by anthropogenic factors, different types and subtypes of soil become equal in their chemical composition. Reductions of humus during the last 30–50 years have resulted from agricultural production practices and can be observed in the context of Samarkand province. Comparison of the results of agrochemical maps of soils were made in 1971, 1991 and 2001 and shows that the humus content in the soils of the province declined rapidly. For instance, in 2001 the organic content of soils on the "Saidbekmuridov" farm, Pastdargom district, was greatly reduced in comparison the organic content of the same soils in 1971. In 1971 the content of humus was about 0.81–2.0% (1.21–2.0% in 99% of soil, 0.81–2.0% in 1% of soil). Crop cultivation has resulted in the reduction of humus concentration and the carbon pool over this period. In 2001, the average humus concentration was 0.81–1.2% in 43% of arable lands, and only 6% of soils had more than 1.2% of humus. Even over such a short period of time, many suffered reduced humus concentration. Soils with average humus concentration increased from 1% to 43%. Soils with high humus concentration decreased from 44.5% to 4%, and with the highest concentration from 30.9% to 2%. By 1991 soils with high humus concentration were almost eliminated.

Intensive cropping of lands reduces humus content and results in the emission of CO_2 gas and reductions in soil fertility levels. Increased emission of CO_2 gas and depletion of humus concentration can be observed, for example, on the "Saidbekmuridov" farm. In 1971 the average humus concentration in every 100 hectares (ha) was:

$$\frac{(99 \times 1.605\%) + (1 \times 1.005\%)}{100} = 1.599\%$$

In 1991 it was:

$$\frac{(4 \times 0.20\%) + (48 \times 0.605\%) + (43 \times 1.005\%) + (4 \times 1.405\%) + (2 \times 1.805\%)}{100} =$$

$$\frac{0.8 + 29.04 + 43.215 + 5.62 + 3.61}{100} = 0.82285\%$$

The humus contents of farm soils decreased by 0.78% over a period of 20 years. This suggests that if a meter layer of soil contained 120 t ha^{-1} of humus reserve in 1971, it would contain an average of 62 t ha^{-1} today, with a reduction in the humus reserve of 58 t ha^{-1}. The loss of the 58 t ha^{-1} humus is associated with the emission of 123.35 t ha^{-1} of CO_2 gas.

The results of our research show that mineral and organic fertilizers are of great importance to the humus concentration, mineralization and ulmification, tillage of soil, irrigation, sowing and other factors. Types of crops affect the humus content and reserve as is shown in Table 1. For example, the humus concentration associated with the continuous tillage of crops (cotton and tobacco plants) was less than 1% and rapidly reduced along the soil profile. Low humus content was also observed in soils that were used for continuous wheat growing. Lucerne and perennial plants enhanced the content of organic matters. These fluctuations were also observed among the entire profile of soils, bound up with the annual plant fall off, with the deep penetration of lucerne (alfalfa) roots and fruit crops, and with different soil cultivation techniques. The root system of lucerne is well developed and protects upper soil layers from erosion. It also creates conditions conductive to humus formation, by noticeably reducing the process of mineralization of organic matter. Nitrogen is conserved within the soil in large quantities because lucerne uses it in trace amounts. This allows the organic nitrogen to mineralize slowly.

Interrow cultivation of cotton is frequently practiced with the excessive application of nitric fertilizers which promotes humus mineralization. Wheat production yields limit plant residues for ulmification. Continuous cotton-plant, tobacco-plant and maize growing reduce the content of carbon, but the cultivation of lucerne and fruit trees enhances the carbon content. Likewise, cereal crops do not enhance soil organic matter content.

Agricultural farming systems may be divided into three groups according to their influence on humus concentration:

1. Intertilled crops (cotton-plant, tobacco-plant, maize);
2. Narrow-row sown crops, interrow crops which are not tilled, namely cereal crops such as wheat barley, oats, rye; and
3. Perennial grasses such as lucerne.

These crops differ from each other by their botanical and biological peculiarities, methods of cultivation, and their effect on humus concentration.

Usually after crop tillage, organic C content in soils is very low. It is mostly due to the fact that the interrows are tilled several times which exacerbates aeration and oxidizing-restoration processes. It results in the intensification of humus decomposition and a reduction in humus formation from different organic residues, which are oxidized fully in good aeration and transferred into mineral matters. Humus formation is reduced under these conditions due to heavy application of nitric fertilizers and insufficient use of organic fertilizers while cultivating the intertilled crops. Cultivating cereal crops like wheat and barley reduces the need for frequent soil tillage and mineral fertilizers. It also reduces humus mineralization to some extent and increases the efficiency of

Table 1. Humus content of different genetic horizon of sierozems in the farm "Zarafshan", Urgut district, Samarkand province.

Number of section	Type of plant	Precursor	Genetuc horizont	Depth, cm	Mechanical content	Humus content, %
1.	Wheat	1. Tobacco – 6 years	A_a (arable)	0–27	Average loamy	1.12
			$A_{u/a}$ (underar)	27–44	Heavy loamy	0.83
		2. Lucerne – 3 years	B_1	44–83	Average loamy	0.67
			B_2	83–137	Light loamy	0.61
			B_3	137–217	Light loamy	0.36
2.	Autumn plough land	1. Tobacco – 6 years	A_a	0–40	Heavy loamy	0.86
			$A_{u/a}$	40–56	Heavy loamy	0.84
		2. Wheat – 1 year	B_1	56–106	Average loamy	0.64
			B_2	106–156	Light loamy	0.35
			B_3	156–196	Heavy loamy	0.54
			B_4	196–263	Average loamy	0.46
3.	Lucrne 2 years	1. Tabacco – 4 years	A_a	0–37	Heavy loamy	1.08
			$A_{u/a}$	37–58	Light clay	1.07
		2. Lucerne – 1 year	B_1	58–99	Light loamy	0.76
			B_2	99–126	Heavy loamy	0.72
			B_3	126–148	Sandy	0.35
			B_4	148–203	Average loamy	0.46
4.	Apple tree	1. Apple tree – 30 years	A_a	0–26	Light clay	1.34
			$A_{u/a}$	26–39	Light clay	1.10
			B_1	39–52	Light clay	0.83
			B_2	52–72	Heavy loamy	0.81
			B_3	72–94	Light clay	0.87
			B_4	94–144	Light clay	0.88
5.	Tobacco	1. Tobacco – 4 years	A_a	0–33	Average loamy	1.27
			$A_{u/a}$	33–44	Loamy sand	0.76
			B_1	44–89	Loamy sand	0.55
			B_2	89–124	Sandy	0.37
			B_3	124–191	Average loamy	0.43

organic residue ulmification. That is why the reduction of humus concentration occurs very slowly when cultivating cereal crops only, as compared to intertilled crops. Lucerne positively affects the quality and quantity of humus because of its powerful root system and technology used to cultivate it. Lucerne-sown areas are not cultivated for many years and nitric fertilizers are not used. This reduces humus decomposition, increases the efficiency of plant residue ulmification and creates a positive humus balance. It explains why the humus concentration increased in soils of the Zarafshan valley after the production of lucerne.

Mineral and organic fertilizers affect the concentration and reserve of organic fertilizers differently. Overdose applications of mineral and nitric fertilizers cause reductions in the content of organic C. Thus, humus content after 7 years of experimentation without fertilizers (controlled variant) is 1.08%; in variant No250P175K125 it is 1.02%; and in variant No. 350P263K175 it is 0.95%. This shows that heavy applications of mineral fertilizers results in substantial decreases in humus concentration. The quantity of one time nitric fertilizer applications highly affects the content and mineralization of humus. If the one time application is reduced in quantity, it reduced to some extent the negative affect of nitric fertilizers and increases the efficiency of its use. In order to reduce the negative impact of heavy nitric fertilizer applications, a balanced concentration of phosphorus and potash fertilizers needs to be mixed with it. Results of our experiments show that applying inhibitors of nitrification (i.e., slowing down the process of transferring ammonium C into nitric C) prevented reduction of humus content in Sierozems.

Table 2. Coefficient of ulmification of plant mass in the light sierozems in cultivatng cotton-plant.

		Ulmification coefficient				
Variants	Depth, cm	Hullo acid	Humino acid	Total	Average	Comparison to control
PK – Control	0–15	10.50	6.75	17.25	16.32	–
	15–30	12.30	6.70	19.00		
	30–45	8.37	4.35	12.72		
Control + N220	0–15	6.23	14.73	20.96	14.19	−2.13
	15–30	7.07	3.74	10.82		
	30–45	6.68	4.10	10.78		
Control + N165	0–15	6.90	23.70	30.60	19.29	18.20
	15–30	12.58	4.68	17.26		
	30–45	3.93	6.09	10.02		
Control + N165 + KMP	0–15	12.90	25.67	38.57	24.39	49.45
	15–30	11.41	6.19	17.60		
	30–45	8.15	8.84	16.99		
Control + N165 + CG	0–15	17.77	39.56	57.33	28.48	74.51
	15–30	6.55	5.93	12.48		
	30–45	9.84	5.80	15.64		
Control + N165 + ATG	0–15	8.38	15.87	24.25	20.34	24.63
	15–30	7.83	8.61	16.44		
	30–45	7.82	12.51	20.33		

The concentration and reserve of humus depends on the two opposing processes of ulmification and mineralization. Studies of the ulmification process of lucerne residues on the base of isotope ^{14}C indicated that the nitric fertilizer applications of 165 kg h^{-1} improved the efficiency of ulmification of lucerne mass. Increasing the dose of nitrogen (fertilizer) to 220 kg h^{-1} reduced the transferof lucerne mass into humus as compared to the control group. As shown in Table 2, applying inhibitors of nitrification KMP (karbomoil metil pirazol), CG (cyanogen guanidin), and ATG (aminotriasol) increased the efficiency of plant mass ulmification. It may be concluded, therefore, that an average application of nitric fertilizers and inhibitors of nitrification positively affect the process of ulmification. On the other hand, heavy applications of nitric fertilizers slow the humus formation process. Reductions of organic residue ulmification efficiency and the intensification of humus mineralization when applying large amounts of nitric fertilizers is related to the reduced covariation of carbon with nitrogen (C:N), which enhances humus usage by microorganisms. Their relationship is is reduced because microorganisms use more organic matter of soil if nitrates are reduced, which in turn causes the decomposition of humus. Consequently, the humus concentration is reduced when to much nitric fertilizer is applied because of the contraction of C:N and increased nitrate content in the soil. In order to prevent the negative effect of over application of nitric fertilizers on humus, it should not all be applied at the same time. Rather the fertilizer should be applied several times in small portions together with inhibitors of nitrification and organic fertilizers.

Results of our research show that the introduction of organic fertilizers and composts causes a rapid increase in humus concentration. The humus concentration was 1.12% when 30 t/ha of manure was applied (1.08% for the control field). Applying a reduced level of mineral fertilizer with manure (N165P125K85 + 30 t ha^{-1} manure) prevented a reduction in humus. In this particular instance the humus concentration was 1.09%. Similar results were obtained when applying composts prepared from city and industrial solid wastes. Composts prepared from wastes can restore carbon in soil in the context of organic soil matter.

The annual introduction of 20 t/ha of compost prepared from both city and industrial solid wastes to the old irrigated light Sierozems greatly impacts the humus concentration in the 0–50 cm layer.

Humus concentration increases in the 1–30 cm layer, by 16–30% and in 30–50 cm layer by 7–9% compared to the control. This indicator is higher in compost with 50% manure, especially when it is mixed with N150P100. Using composts in old typical Sierozems also caused the humus concentration change in soil. In spite of the fact that old irrigated typical Sierozem differs from old irrigated light Sierozem with its high humus content, absolute indicators of humus are reduced. They are approximately 9–12% in the arable layer and 6–8% underneath the arable layer, which creates 2.4–4.1 and 0.4–2.7 t/ha of humus. Under these conditions, 5.104–8.721 and 0.851–5.743 t ha^{-1} of carbonic acid transfer into organic matter.

Organic fertilizer at the rate of 17–18 t ha^{-1} needs to be applied in order to balance the humus concentration in the case of Uzbekistan. Thus, 340–360 t of organic fertilizers need to be applied over 20 h of land in order to keep this balance, which equates to keeping 2–3 herds of beef cattle or 3–4 thousand hens on each ha of irrigated land. However the number of cattle per ha of land is currently less than what is required. This deficit of organic fertilizer in Uzbekistan agriculture facilitates a reduction in humus concentration. The storage of organic fertilizers can be partly recovered at the expense of different wastes used to prepare composts. This is confirmed by our research. Straw from cereal crops and stems of cotton plants negatively affect the balance of humus and cause an increase in the carbonic acid content in the atmosphere. Additionally, they reduce the ulmification process and intensify the greenhouse effect in the atmosphere.

One of the main sources for preparing composts is the silt of water reservoirs. There are 53 water reservoirs in Uzbekistan and 2.1 billion tons of silt stored in them. Using silt as compost enhances humus content in soil, which is necessary to increase the area under lucerne and other fodder crop production. This leads to more effective crop rotation in agriculture. Lucerne not only raises the fertility of soil; it also decreases the need for organic inputs to soils by 25–30%. Lucerne provides many of the inputs to soils provided by organic fertilizers. Including lucerne in crop rotation improves the outcome of fodder that promotes the development of stock farming. It increases the amount of organic fertilizer. All of these enhance humus concentration in soil and shift the balance towards ulmification (formation of organic matter). It improves the ecological conditions of both the soil and temperature.

The process of formation and decomposition of humus is carried out by microorganisms. It is not impossible to estimate ongoing processes in soil without studying the microbiological processes related to the formation and conservation of the humus concentration. Research results show that in cultivating virgin lands, the quantity of microorganisms increases in relation to the humus unit. This may lead to humus decomposition. This substantiated by the appearance of greater quantities of microorganisms to one unit of organic carbon when soils are converted from more humus meadow soil to less humus light Sierozem.

The introduction of mineral nitric fertilizers brings about a rapid increase of the amount of nitrifications and denitrificators, which negatively affects the humus balance in the soil. It is known that an increase of nitrification activity, especially denitrificators, intensifies the mineralization of organic matters. It is one determinant of humus concentration reduction in applied overdosed nitric fertilizers. The formation of "extra" nitrogen lends additional support to this interpretation since "extra" nitrogen is made up of extra decomposed organic matter.

Consequently, an increase in the amount and activity of microorganisms, especially nitrificators and denitrificators at the expense of mineral matters without applying organic fertilizers, causes an increase of humus mineralization. As stated before, nitric fertilizers initially raise humus mineralization efficiency levels and then lower them. This causes an increase of nitrogen within the soil. Some scientists, such as Beresteskiy et al. (1983) and Tuev (1989), claim that ammonium nitrogen increases ulmification efficiency while nitrate decreases it.

While applying organic fertilizers, the number of taxonomic and physiological groups of organisms increases together with the separate groups of microorganisms. These applications have no effect on other groups of microorganisms that have no effect on the natural humus balance in soil. Applying organic fertilizers (manure, compost) increases not only the number of microorganisms, but also the amount of dehumidified organic matters, which prevent extra humus decomposition.

Activity of microorganisms is also determined by the amount of carbonic acid in soil. An increase of CO_2 gas in soil air is correlated with the decomposition of humus matters. Therefore studying the soil air content gives an estimate of the effect of various factors to humus concentration in soil. In regard to CO_2 gas concentration in soils used to grow cotton, an applied overdose of nitric fertilizer causes an increase in CO_2 gas content only at the first phase of cotton plant growth, after which its content decreases. In meadow Sierozem soils, 165 kg ha^{-1} of nitric fertilizer increased the CO_2content in contrast to the PK variant. Increasing the nitrogen dose to 220 kg ha^{-1} raised the CO_2 content in soil air at the beginning stage of vegetation. Comparisons that focused on the CO_2 effects of different experimental variants on two types of soils indicate that meadow Sierozem soils are more buffered than less humidified, light Sierozem soils.

The influence of mulching (polyethylene film, compost) on the soil air regime in irrigated meadow soils and light Sierozems has been investigated by some. Irrigated meadow land is characterized by high CO_2 gas content, which is correlated with the amount of organic matter and water in soil. The content of carbonic acid in the 0–30 cm layer fluctuates between 0.1–0.2% and in the 30–50 cm layer 0.3–1.4% of the water volume. Carbonic acid content varies between 0.1–0.2% in the 0–30 cm layer and 0.2–0.6% in the 30–50 cm layer during the vegetation period in the old irrigated Sierozems (Table 3).

Mulching the old irrigated light Sierozem soils used to grow cotton with manure affects the oxygen and carbonic gas in the soil air less than transparent polyethylene film. Under these conditions, high carbonic acid content is explained by the rapid rise in soil temperature, which stirs up microorganisms and covering of soil surface during the sowing and germination periods of the cotton plant.

Mulching has both positive and negative effects. One positive effect is the enrichment of soil air with CO_2 gas in the cotton plant root system position and, most likely, in the layer of atmosphere closest to earth. It can be assumed that the mobility of some nutrient elements is increased and their condition is improved as is reflected by photosynthesis intensity and plastic matter accumulation. Oxygen content in the soil air decreases and it may negatively affect the undesirable carbonic transformation in soil. It suggests that denitrifying factors, molecular nitrogen formation and deterioration of nitrogen nutrition of the cotton plant are activated.

Several periods of intensification and reduction of the carbonate acid concentration during vegetation exist. High intensity is observed in early spring (April–May) and summer (July), while reduction is observed in June and August. This most likely depends upon the amount of organic waste, watering, changing levels of subsoil waters as well as temperature variations and microbiological changes in soil.

Mulching soil with polyethylene film increases the carbonate gas content in soil air. The reason for this surge of carbonate gas is explained not only by the increase of carbonate gas formation in the mulching process, but also by the retention of CO_2 under the polyethylene film and the deceleration of aeration. Acceleration of humus decomposition is not the only explanation for carbonate concentration increases in soil. With the rise in soil and air temperature, the humus concentration in soils is decreased. This process can be seen in soils in arid areas – where the temperature is high and water shortages can be frequently observed.

Applying composts prepared from city and industrial solid wastes causes an increase in CO_3 acid. It is mostly connected with the ulmification of fresh organic matters. When this occurs, CO_3 acids are also emitted. However, CO_2 gas concentration increases are not the only reason for decomposition of humus matters. Soil moisture also plays an important role in humus formation and the transformation of the CO_2 into the structure of organic matter. It is especially significant in arid areas. Research findings suggest that the soil water regime depends on mulching, fertilizer application, soil cultivation and also on the agro-physical and agro-chemical peculiarities of soils. Mulching soils with polyethylene film is critical to moisture storage. Increasing the organic C concentration in soils promotes storage and efficient use of moisture. Normal doses of mineral and organic fertilizers promote moisture use by plants in soil.

The areas in which research was carried out differ by level of subsoil water position. That is why mulching materials as well as subsoil water were shown to have important effects on the formation

Table 3. CO_2 and O_2 content in soil air of irrigated meadow soil and old irrigated light seroziem, %.

		20.05		30.05		10.06		20.06		30.06		10.07		20.07		30.07		10.08	
Types of experimennt	Depth, cm	CO_2	O_2	CO_2	O_2	CO_2	O_2	CO_2	O_2	CO_2	O_2	CO_2	O_2	CO_2	O_2	CO_2	O_2	CO_2	O_2
Irrigated meadow lands																			
Control (without mulching)	0–30	0.6	20.4	0.6	20.4	0.4	20.6	0.2	20.8	0.3	20.7	0.6	20.4	0.5	20.5	0.3	20.7	0.2	20.8
	30–50	1.4	19.6	0.8	20.2	0.8	20.2	0.7	20.3	0.6	20.4	1.2	19.8	0.9	20.1	0.6	20.4	0.6	20.4
Mulching	0–30	2.8	18.2	1.0	20.0	1.7	19.3	0.8	20.2	0.6	20.4	2.0	19.1	1.6	19.4	0.8	20.2	1.0	20.0
	30–50	5.6	15.4	1.8	19.8	2.4	18.6	1.4	19.6	1.0	20.0	1.0	20.0	2.2	19.0	0.9	20.1	1.2	19.8
Old irrigated light seroziem																			
Control (without mulching)	0–30	0.2	20.8	0.2	20.8	0.1	20.9	0.2	20.8	0.1	20.9	0.2	20.8	0.2	20.8	0.1	20.9	–	–
	30–50	0.6	20.4	0.4	20.6	0.4	20.6	0.3	20.7	0.2	20.8	0.4	20.6	0.5	20.5	0.4	20.6	–	–
Mulching (film)	0–30	1.8	19.2	0.6	20.4	0.4	20.6	0.5	20.5	0.4	20.6	0.6	20.4	0.8	20.2	0.4	20.6	–	–
	30–50	4.8	16.2	1.0	20.0	0.6	20.4	0.6	20.4	0.6	20.4	1.4	19.6	1.2	19.8	0.6	20.4	–	–
Mulching (manure)	0–30	0.4	20.6	0.4	20.6	0.2	20.8	0.3	20.7	0.2	20.8	0.2	20.8	0.3	20.7	0.1	20.9	–	–
	30–50	0.6	20.4	0.8	20.2	0.3	20.7	0.4	20.6	0.3	20.7	0.4	20.6	0.6	20.4	0.4	20.6	–	–

of water regimes. In irrigated meadow lands, the level of subsoil water is 1.5–2.0 m deep. After irrigation it is closer to the surface of the soil; but in irrigated light Sierozem it is 10 m deep. Because atmospheric precipitation mainly occurs in the winter and spring, it can be concluded that it does not affect the formation of water regimes during vegetation period of plants. It can also be concluded that moisture primarily depends on the type of mulching materials used and the period in wich they are applied. The impact of mulching materials is especially evident in spring, after sowing, before the first watering and between water applications. Soil moisture related to mulching is spent on transpiration, and in the open soil – on physical evaporation and transpiration. Research on water regimes of irrigated meadow soils shows that soil moisture in 50–100 cm layer is sufficient and changes less according to the period of vegetation than at the 30–50 cm and 0–30 cm layers. It shows that lower layers of soil become in direct contact with subsoil waters.

Research on levels of irrigated meadow soil moistures indicated that the reserves of total moisture in 0–5 cm layer was higher compared with those of the control sample at the beginning period of cotton plant vegetation. Soil moisture was distributed evenly under the polyethylene film in comparison with that in the open soil. The upper 0–30 cm layer of mulched soil remained moisturized when compared with unmulched soil.

Dynamic moisture changes occur in irrigated meadow soils. During the vegetation of cotton plants the use of transparent polyethylene film as a mulching material causes an increase in soil moisture. This is at the expense of preventing physical evaporation during the early spring – autumn period. The level of subsoil water in mulched areas gradually decreases from May to September, which is not observed in the unmulched area. Mulching the soil with transparent polyethylene film creates a favorable condition for cotton plants to use the moisture that is formed at the expense of the water coming out of the lower layer.

Results of field experiments conducted on irrigated meadow soil show that transparent polyethylene film mulching accumulates more water, thus preventing physical evaporation and resulting in optimum soil moisture during the germination of the cotton plant. Results of research conducted in the old irrigated light Sierozems show that the moisture reserves are higher in 0–30 cm layer in the soil mulched with transparent polyethylene film from the beginning of May to mid-June. During the second half of June, moisture reserves are less, which can be explained by quick growth and fruit bearing process of cotton plants in mulched areas. As for meadow soil, moisture reserves increase under the polyethylene film at the end of vegetation.

Mulching with manure generated an increase in the moisture reserve in old irrigated light Sierozems at the beginning stage of cotton plant vegetation compared with control. Subsequently, indices of both variants became equal.

As a whole, mulching the soil with transparent polyethylene film and manure allowed the upper arable layer to remain looser compared with a control soil that was subjected to optimum air, water-thermic regimes. The transparent polyethylene film has a greater effect on water regime than manure. Comparison of the effects of film as a mulch in irrigated meadow soil and old irrigated Sierozems suggest that it is more effective in irrigated meadow soil.

Soil water regimes depend greatly on the agrophysical and agrochemical characteristics of the soil. Moisture amount, water permeability and water retaining capabilities of the soil improve with the enhancement of humus. Incorporation of lucerne into sowing rotations positively affects the water regime because it improves the agrophysical and agrochemical characteristics of the soil.

Organic fertilizers have a great effect on agrophysical and agrochemical indicators of soil. They lead to a more effective use of moisture and prevent its loss. Among the mineral fertilizers, potassium has a more positive effect on the water regime of soil and plants. Mineral fertilizers promote the development and growth of plants, economical moisture use, and the reduction of transpiration efficiency.

Consequently, mulching practices, normal dose applications of mineral fertilizers, the sowing of Lucerne, and other measures that improve the agrophysical and agrochemical characteristics of soil, promote water regime improvements, more economical moisture use and reduced moisture loss.

2 SUMMARY

In summary, processes of humus content and humus quality reduction can be observed in present-day Uzbekistan as well as in many other countries. These processes increase the amount of carbonic acid in the air and deteriorate the water-air regime of soil. A shift in the equilibrium towards ulmification of organic fertilizers, requires the introduction of composts, including lucerne, into the sowing rotation, as well as nitric fertilizers in several normal dosage amounts. All techniques that increase humus content also improve agrophysical and agrochemical characteristics of soils and promote moisture conservation in them.

REFERENCES

Berestetskiy O.A. Actuality and practical significance of a microbiological research in the decision of problems of soil fertility increase. Proceedings of VNII Agricultural microbiology. 1986. P.56

Berestetskiy O.A., Voznyakovskaya I.M., Kruglov Y.V. et.al., Biological basis of soil fertility. Moskow, Kolos, 1984. P.287

Dobrovolskiy G.V., Grishina L.A., Rozanov V.G., Targulyan, V.O. Influence of man on soil as a component of biosphere. Soil Science. 1985. No.12. P.56–65

Elyubaev S.N., Tursunkulova P.H. Optimal content of humus in the irrigated typical serozems. Proceedings of Soil Science and Agrochemistry Institute. Academy of Science UzSSR. 1987, No.32. P.44–48

Efimov V.N., Osipov A.I. Humus and nitrogen in the crop farming of a non-chernozem zone. Soil Science., 1991. No.1. P.67–77

Zonn S.V. Significance of Vilyams's V.P. ideals in the basing soil protection role of forest and reproduction of soil fertility in the grass- tillage system of crop farming. Soil Science, 1989. No.9. P.25–31

Kovda V.A. Soil cover, its improvement, usage and protection. Moscow, Nauka, 1981. P.5–15

Riskieva H.T., Toshkenboev O. Change of humus condition of a meadow soils of desert zone of Zarafshan Valley under the influence of irrigation. Proceedings of influence of agrarian Science. Tashkent, 2002, No.1 (7). P.69–72

Rozanov V.G., Targulyan V.O., Orlov D.S. Global changes of soil and soil cover. Soil Science, 1989. No.5. P.5–18

Sattorov D.S. Main problems of improvement of a soil cover condition of Uzbekistan. Reports of I-delegate's meeting of soil Scientists of Uzbekistan. 14–17 November. 1990. Tashkent, 1990. P.3–10

Tuev N.A. Microbiological processes of humus formation. Moskow. Agropromizdat, 1989. P.239

CHAPTER 21

Eliminating summer fallow on black soils of Northern Kazakhstan

M. Suleimenov
International Center for Agricultural Research in the Dry Areas – Central Asia and Caucasus, Tashkent, Uzbekistan

K. Akshalov
Scientific Production Center of Grain Farming, Shortandy, Kazakhstan

1 INTRODUCTION

Large scale grain production in northern Kazakhstan began in mid-1950s, and it involved conversion of grasslands into cropland ecosystems. From the very beginning, the recommendation was to grow spring small grains, primarily spring wheat, continuously with summer fallow once every 5 years. This recommendation was based mostly on the Canadian practice of alternate summer fallow-wheat in dryland agriculture. Long-term data obtained on black soils have shown that frequent summer fallowing reduces total grain production from cropland and leads to rapid losses of soil organic matter (SOM). Long-term comparative data obtained for three crop management practices (simple, common and best) in Kazakhstan supported the conclusion that summer fallow is not justified for any of these technologies. Summer fallow, however, leads to a better supply of NO_3-N for plant growth and also to a better grain baking quality. Thus, the objective of this chapter is to describe the long-term effects of summer fallow on yield of spring grain crops, total productivity, and environment quality in black soils of northern Kazakhstan. Spring wheat is sown in the second half of May and harvested between mid-August and mid-September in northern Kazakhstan. Thus, the summer fallow period lasts 20–21 months. For a long period since development of new lands in mid-1950s, summer fallow was deemed necessary for the success of dryland farming. The only debatable point was the frequency of incorporating fallow in the rotation cycle. Canadian practice of alternate fallow-wheat was found to be unacceptable under conditions of Kazakhstan from the very beginning. In the 1960s, summer fallow was recommended to occupy 25% of the crop rotation cycle. In the 1970s, the duration of fallow was reduced to only 20%, and later on in 1980s the duration was further reduced to 16.7%. In other words, summer fallow was recommended once in four, five or six years of the rotation cycle during 1960s, 1970s and 1980s, respectively.

The first publication in the former Soviet Union casting doubts about the necessity of summer fallow on black soils in northern Kazakhstan was written by Suleimenov (1988). The report was published for discussion among agronomists working in dryland agriculture. All of them unanimously concluded that something was wrong either with the research methodology used, or with the interpretation of the data which lead to erroneous conclusions. Thus, additional data were obtained which also supported the earlier conclusions on the preference of annual cropping for obtaining economic and environmental benefits and eliminating the summer fallow (Suleimenov and Akshalov, 1996; Suleimenov et al., 2001).

It is true that the methodology used in that experiment was very different from the one used in previous studies. First, crop rotations without summer fallow were never studied because it was believed that dryland farming is impossible without using the summer fallow. Thus, growing spring wheat with different frequency of summer fallowing was compared with cultivation of continuous wheat. Most striking drawback of this methodology was, in fact, that continuous wheat was grown with use of neither fertilizers nor chemicals, thus resulting in very low grain yields. Based on these observations, a new long-term experiment was established with the purpose of comparing continuous cultivation of small grains with and without summer fallow using a range of technologies. This chapter describes the results obtained from this long-term study.

2 MATERIALS AND METHODS

A field selected in 1983 for establishment of the experiment was used for continuous cultivation of small grains since 1979 on heavy clay loam black soil. A 5-year rotation used as control was: summer fallow-wheat-wheat-barley-wheat. In the second rotation, summer fallow was replaced with oats sown in 1984 producing crop rotation of annually grown small grains as: oats-wheat-wheat-barley-wheat. Besides, continuous wheat was used as another treatment.

The second factor in the study was the agronomy or crop management practices. Three crop management practices used were: simple, common and best. A Simple Crop Management (SCM) included use of no fertilizers and chemicals, and no snow management. Summer fallowing included three tillage operations to control weeds. The Common Crop Management (CCM) included widely used practice of applying phosphorus fertilizers (80 kg ha^{-1} of P_2O_5) applied once in five years during the fallow period, and application of 2,4-D herbicide to control broadleaf weeds. In addition, snow ridging was done once in winter using special snow ridge plows to trap snow drifts. Nitrogen fertilizers were not applied in the CCM treatment because it is not a common practice till now, although there is a need for N application. Summer fallow in this treatment was tilled four times during the summer to control weeds. The Best Crop Management (BCM) included application of both phosphorus and nitrogenous fertilizers. Ammonium nitrate was applied annually before sowing at the rate of 30–40 kg N ha^{-1}. The BCM practice also included application of herbicides 2,4-D to control broadleaf weeds and Avadex BW to control wild oats, whenever necessary. These herbicides were applied continuously during the first four to five years, and were later applied only occasionally. Snow ridging, as 40–50 cm snow windrows, was done twice during winter to assure adequate snow accumulation. Also in BCM, summer fallow tillage was done as many times as necessary to control weeds: from 4–6 times in the first rotation and 2–3 times in the recent rotations because weeds were adequately controlled during the previous period and there was no need of numerous tillage operations. Besides, short barriers of mustard were established in early July to trap snow drift on summer fallow land during the second winter.

It is important to emphasize that the same type of management practices was used for all crops in all fields continuously since 1983. In the BCM practice, this type of cultivation technology was used for over twenty years continuously with adequate snow management, application of fertilizers and weed management in comparison with the application of SCM technologies with no chemicals and fertilizers during the entire period. The application of herbicides in the BCM technology included only the use of needed chemicals to keep crops free of weeds with reduced frequency of herbicides application over time. The widespread crop management during the transition period involves combination of SCM and CCM technologies.

The average annual precipitation in Shortandy is 320 mm, distributed throughout the year with maximum rainfall during summer and peak rainfall in July. Winter precipitation, as snow, is an important source of moisture and constitutes about one third of the total annual amount. Out of the 20 years, 4 years or 20% were extremely dry (1991, 1993, 1994 and 1995), 5 years or 25% were dry (1984, 1985, 1988, 1997 and 1998), 9 years or 45% were moderately dry (1986, 1987, 1989, 1990, 1992, 1996, 1999, 2000 and 2003), and 2 years or 10% were favorable in terms of the total amount of precipitation received. Crop yield data were included for discussion beginning with 1988 because it is the only data obtained after completing the full cycle of the five year crop rotation.

3 RESULTS AND DISCUSSION

Moisture conservation is supposedly one of the most important advantages of summer fallow. According to the long-term data, water storage in 0–100 cm soil layer prior to sowing spring wheat in mid-may depended on summer fallow or stubble and on intensity of snow management. During the summer fallow, moisture storage depended on snow management intensity as follows: 60 to 100 mm for simple, 120 to 140 mm for common, and 140 to 160 mm for intensive management. In the same years, water storage on stubble fields ranged from 40–70 mm for simple, 90–120 mm for common and 130–150 mm for intensive snow management. These data show no advantage of summer fallow in moisture storage for the same intensity of snow management. Fallowing increased only 10–30 mm of available water in one meter of soil layer. The data obtained also demonstrated large differences in favor of summer fallow with establishment of short barrier strips to trap snow against leaving stubble without any snow management. On the other hand, good water storage was obtained through double snow ridging in stubble in comparison with improperly managed summer fallow.

Weed control is one of widely recognized advantages of summer fallow, because numerous tillage operations during summer fallow period ensure weed-free crops for several years. The data presented showed that there was no need to practice fallow for weed control because weeds can be effectively controlled in both rotations with or without summer fallow including continuous wheat since 1979. Data obtained on weed infestation in 2002–2005 showed that continuous wheat under intensive weed control using chemicals was more effective than wheat sown after summer fallow with both simple and common technologies (Table 1).

For example, number of weeds in spring wheat sown on summer fallow was 53 m^{-2} in simple, 11 m^{-2} with common (control of broad leaves) and 4 m^{-2} with intensive (control of both broad leaves and grasses) cultural practices. In spring wheat sown after barley, weed population decreased from 89 m^{-2} to 28 m^{-2}. The data obtained on continuous wheat were similar. These data also showed that summer fallow had definite advantage in weed control in case of SCM when chemicals were not properly used. Number of weeds tripled in three years in wheat grown after summer fallow. This is exactly what many farmers and scientists have been emphasizing about the importance of weed control with summer fallow. Notably, weed infestation in continuous wheat under poor crop management was lower than on wheat sown after three years of summer fallow. The same trend was observed for CCM. The level of weed infestation reduced remarkably (3–5 times) under better crop management in all fields of the rotation, but it increased in barley field almost five times during three years after summer fallow. However, wheat crop in the fourth year after fallow was considerably cleaner than in the previous years. This is explained by the fact that preceding crop was spring barley sown normally one week later than wheat which provides good opportunity to control wild oats and other weeds. Under BCM practices, weeds were effectively controlled with no fallow. One can see that wheat sown after barley was even cleaner than wheat sown after a summer fallow.

Table 1. Weed infestation (weeds m^{-2}) as affected by crop rotation and management (average for 2002–2005 seasons).

	Crop management practice		
Year after summer fallow	Simple crop management	Common crop management	Best crop management
First (wheat)	53	11	4
Second (wheat)	112	36	4
Third (barley)	153	53	3
Fourth (wheat)	89	28	0
Continuous	99	39	0
LSD_{05}	23	15	3

Table 2. Spring wheat grain yield (t ha^{-1}) as affected by summer fallow and crop management practices in very dry years (average for four years).

	Crop management practices			Average yield	
Year after summer fallow	Simple crop management	Common crop management	Best crop management	t ha^{-1}	%
First	0.74	1.30	1.86	1.30	100
Second	0.40	0.95	1.48	0.94	72
Fourth	0.44	0.88	1.42	0.91	70
Continuous	0.32	0.98	1.44	0.91	70
LSD_{05}	0.20	0.23	0.20		

Table 3. Spring wheat grain yield (t ha^{-1}) as affected by summer fallow and crop management practices in dry years (average for three years).

	Crop management			Average yield	
Year after fallow	Simple crop management	Common crop management	Best crop management	t ha^{-1}	%
First	0.91	1.38	1.98	1.42	100
Second	0.73	1.01	1.49	1.08	76
Fourth	0.70	1.22	1.56	1.16	82
Continuous	0.59	1.10	1.59	1.09	77
LSD_{05}	0.11	0.31	0.41		

Analysis of soil sampled prior to sowing of crops showed that nutrient concentration depended on crop management, and there was a notable advantage of summer fallow. In spring 2004, prior to sowing spring wheat, NO_3-N concentration in 0–20 cm soil layer in summer fallow was 3.29 mg 100 g^{-1} with SCM, 4.76 mg 100 g^{-1} with CCM and 8.1 mg 100 g^{-1} with BCM. The NO_3-N concentration was lower on stubble land and ranged from 1.93 to 3.62 mg 100 g^{-1} of soil. These data show that nitrogen availability is substantially improved by summer fallowing. Before sowing in stubble land, nitrogen fertilizers were applied in BCM practices at the rate of 30–50 kg N ha^{-1}, which improved nitrogen availability up to the medium supply level.

Most important integrating assessment of pre-cropping management is its influence on crop yield. It is often emphasized that summer fallow drastically increases grain yields in dry areas. This is the reason for analyzing the grain yield data in groups of years with different moisture regimes. The driest year was 1991, in which grain yield for continuous spring wheat under SCM treatment decreased to 0.16 t ha^{-1}. On average, in extremely dry years, spring wheat yield for SCM practices was only 0.32 t ha^{-1}. The spring wheat grain yield on stubble land in rotation with fallow was a little better but still very low (Table 2).

Summer fallow demonstrated a significant yield advantage under SCM by more than doubling the grain yield. However, by improving crop management practices including better water, nutrients and weed management, advantages of summer fallow against stubble land decreased remarkably to merely 23–25%. The data presented show that, on average, grain yield of spring wheat on summer fallow as compared to that on stubble land increased by 28–30%. It was also obvious that advantage of growing wheat on stubbles in rotation with summer fallow was observed only in the SCM treatment. Under CCM and BCM agronomic practices, there was no difference in yield from the second year after summer fallow up to continuous cropping. The data obtained in dry years show that efficiency of summer fallow was lower than that in very dry years (Table 3).

Table 4. Spring wheat grain yield (t ha^{-1}) as affected by summer fallow and crop management practices in moderately dry years (average for seven years).

	Crop management			Average yield	
Year after fallow	Simple crop management	Common crop management	Best crop management	t ha^{-1}	%
First	1.11	1.92	2.57	1.87	100
Second	0.96	1.76	2.40	1.71	91
Fourth	1.03	1.78	2.33	1.71	91
Continuous	0.96	1.70	2.39	1.68	90
LSD_{05}	0.18	0.27	0.31		

Also, in dry years, summer fallow was quite efficient in the case of SCM treatment. The grain yield of the first wheat after fallow was 54% higher compared to that of the continuous wheat since 1979. The grain yield on stubble land in rotation with summer fallow was also 19–24% higher as compared to that of the continuous wheat. These data show that if water is not stored on the stubble land and weeds are not controlled with chemicals, fallowing is one of the techniques to minimize drought and improve nitrogen availability. When the crop management was improved by introducing CCM practices, the advantage of summer fallow over continuous wheat reduced from 54% to only 25%.The highest grain yield on stubble land was obtained when wheat was grown after barley, which indicates that barley was a useful preceding crop for wheat due to better weed control in delayed sowing than oats. Under the BCM, the crop yield increased dramatically (more than doubled in all crop fields) and was very high during the dry years. The advantage of summer fallow against continuous cropping on grain yield was as high as 24%. On average in dry years, increase in grain yield of wheat sown after summer fallow against stubble land was 24%. In these years, there was no grain yield decline on stubble land grown continuously while barley was a good crop preceding wheat. In moderately dry years, which are the prevalent climatic conditions, the advantage of summer fallow over continuous wheat was rather small (Table 4).

On average for seven moderately dry years, and under SCM, advantage of summer fallow against stubble land was as high as 15% with no difference in wheat yields on stubble land over the long duration. Notably, grain yield on stubble under continuous cultivation of spring wheat was quite high under CCM (1.70 t ha^{-1}) and very high under BCM (2.39 t ha^{-1}) which is much higher than that from summer fallow in SCM (1.11 t ha^{-1}). On average, wheat yield on summer fallow was higher than that on stubble land by about 10%.

Last five years of the experiment had favorable rainfall, which was also above average. In most favorable years (2001–2002), average grain yield under summer fallow practice increased from 1.29 t ha^{-1} to 2.55 t ha^{-1} due to better crop management. In continuous wheat, increase in yield ranged from 1.32 to 2.62 t ha^{-1}. On average, grain yield in three treatments of wheat sown on summer fallow was higher than that from stubble land by 4–7%, and didn't change with duration from the fallow year: it was 2.30 t ha^{-1} in the second year, 2.34 t ha^{-1} in the fourth year, and 2.37 t ha^{-1} with continuous cropping. Considering the average yield for 16 years (1988–2003), wheat sown in summer fallow treatment produced 15% more grain yield compared with that form continuous wheat grown during 24 years on one field (Table 5).

The long-term data indicated that the advantage of summer fallow was 15% with continuous wheat, comprising 21% with SCM and 13% in both CCM and BCM. These data support the conclusion that significant yield advantage in wheat grown in summer fallow over continuous wheat occurred with low input technologies, and especially in dry years. Also, there was no decrease in wheat grain yield during continuous wheat cultivation when the same cultural practices were used.

Impact of cultivating feed grains (e.g., barley and oats) in the rotation cycle was also studied. Both feed grains produced more grain yield than the spring wheat, and their yields depended on crop management practices (Table 6). Although the feed grain crops were higher yielding than

Table 5. Spring wheat grain yield (t ha^{-1}) as affected by summer fallow and crop management (average for 1988–2003).

	Crop management			Average yield	
Year after fallow	Simple crop management	Common crop management	Best crop management	t ha^{-1}	%
First	1.05	1.72	2.38	1.72	100
Second	0.85	1.51	2.06	1.47	85
Fourth	0.92	1.52	2.03	1.49	87
Continuous	0.83	1.49	2.07	1.46	85
LSD_{05}	0.19	0.26	0.34		

Table 6. Grain yield (t ha^{-1}) of barley and oats as affected by rotation and crop management (average for 1988–2003).

	Crop management practice		
Crops and their place in crop rotation	Simple crop management	Common crop management	Best crop management
Barley in F-W-W-B-W rotation	1.40	2.00	2.69
Barley in O-W-W-B-W rotation	1.29	1.90	2.63
Oats in O-W-W-B-W rotation	1.30	2.04	2.83
LSD_{05}	0.14	0.20	0.33

wheat, their yield also depended on weather and crop management. On average, grain yields of barley and oats were higher than that of spring wheat by 33–41%. Oats were higher yielding, and a better preceding crop for spring wheat than barley. The grain yield of wheat sown after oats was 8% higher than that of continuous wheat, and the yield was not affected when sown after barley.

The grain yield of barley was slightly lower in continuous grain crop rotation as compared to including barley in rotation with summer fallow and wheat. Notably, the difference was less in BCM practices and was about 8% in SCM, 5% in CCM and only 2% in BCM. Herein lies the evidence of improved grain yields in continuous cropping of dryland farming with BCM practices.

During extremely dry years, yield of feed grains were low in both rotations with or without fallow when poor cropping practices were adopted. However, grain yield increased dramatically with adoption of BCM practices. For example, in the driest 1991 year, grain yield of barley under SCM was just 0.25 t ha^{-1} in fallow-based rotation and zero in continuous grain cultivation. With improved cultural practices under BCM, the barley grain yield increased to 2.09 t ha^{-1} and 2.20 t ha^{-1} in both rotations with and without fallow, respectively. The highest grain yield of barley (4.39 t ha^{-1} and 4.77 t ha^{-1}) was obtained in 2001 in control and in continuous cultivation, respectively. These data also indicated that under BCM, grain yield of barley was even more when summer fallow was excluded from the rotation. The highest oats yield of 5.08 t ha^{-1} was obtained in 1990. On average, oats were higher yielding than barley by 1% under SCM, 7% under CCM and 8% under BCM practices.

Although grain yield is an important indicator for assessing the efficiency of crop rotations, total production from the entire area practicing summer fallow, which produces nothing, is also an important criterion. In control, a five year crop rotation included three crops of spring wheat and one of barley. In annual grain cropping, the fallow was replaced by oats. Thus, total production involved five fields in both rotations, which was divided by five in both cases. The comparative data of 13 years show an advantage of annual cropping in grain production for all crop management practices (Table 7).

Table 7. Grain yield (t ha^{-1}) from total cropland area as affected by crop rotation or continuous wheat under different crop management practices (on average for 1991–2003).

	Crop management practice			Average yield	
Cropping system	Simple crop management	Common crop management	Best crop management	t ha^{-1}	%
F-W-W-B-W	0.90	1.38	1.77	1.35	100
O-W-W-B-W	1.04	1.63	2.26	1.64	121
Continuous wheat	0.81	1.49	2.10	1.47	109
Average	0.92	1.50	2.04	1.49	
LSD_{05}	0.16	0.20	0.26		

Data on grain production from the total area show that, first and foremost, crop management is more important than crop rotation. Grain yield from total cropland area increased for CCM and BCM as compared to SCM by 1.63 and 2.17 times, respectively. In addition, crop management influenced the efficiency of crop rotation. Under poor crop management, average grain yield from total area decreased in continuous wheat by 10% because of dramatic yield reduction in continuous wheat during the dry years. However, even under SCM, the annual grain cropping including oats instead of summer fallow increased grain yield from the total cropland area by 15%.

Under the common crop management practice, even continuous wheat produced 8% more grains than the rotation based on summer fallow, whereas annual small grains cropping increased grain production by 18%. Under the best crop management, growing continuous wheat or annual small grains *vis-à-vis* the fallow based rotation increased grain production by 19% and 28%, respectively.

Economic assessment is the most important criterion for farmers to make decisions, although prices for inputs and outputs change over time. Approximate economic analyses were done for the five year data (1996–2000) at prices of US $100 t^{-1}$ for wheat and US$ 80 t^{-1} for feed grains. For SCM, CCM and BCM, net profit ha^{-1} was US $9–25–49 in fallow-based rotation, US $27–43–75 for annual rotation of small grains, and US $ 13–37–77 for continuous wheat, respectively. Hence, continuous wheat or annual small grains were economically more profitable under any crop management and not just for the best management. These conclusions contradict the common beliefs prior to conducting this study.

In some years, however, the best management practice can produce extremely high grain yield for which the main constraint is the lack of warm period during the grain filling stage. The first frost damage to grains usually occurs on the 5th of September, which risks grain quality under best crop management practices. The grain filling under BCM was delayed by one week compared to CCM and by two weeks compared to SCM. In addition, there was also a strong correlation between grain yield and its quality: the higher the grain yield the lower was the gluten content, which is a major indictor of the baking quality. Also, summer fallow with higher availability of NO_3-N can be an important factor in the gluten content of grains. Under SCM, grain quality of wheat sown on summer fallow was very high in most years (32.5%). It was also very high in continuous wheat in most years, but was low in wet years. Under CCM, the gluten content was a little lower on summer fallow, but was still high (31.1%). In continuous wheat, the gluten content decreased by 4%. In BCM, the gluten content in wheat grains dropped further in all crops including wheat sown on summer fallow.

An important consideration for sustainable agriculture is the maintenance of soil fertility, of which the main factor is the soil organic matter (SOM) content. In 2005, after 17 years of complete establishment of three cropping systems: F-W-W-B-W (control), O-W-W-B-W and continuous wheat, the SOM content in 0–30 cm soil layer under BCM slightly decreased in control or the fallow-based cropping system from 3.40% to 3.35% (Table 8).

In contrast, the SOM content increased to 3.80% in continuous grains and 4.03% in continuous wheat systems. Measurements of SOM contents were made in small plots which were not prone to

Table 8. Change of soil organic matter content under BCM in 0–30 cm soil layer during 17 years as affected by crop rotation or continuous wheat (1988–2005).

		SOM, %		
Cropping system	Soil layer	1988	2005	Difference
F-W-W-B-W	0–10	3.40	3.50	+0.10
	10–20	3.60	3.24	−0.36
	20–30	3.20	3.30	+0.10
	0–30	3.40	3.35	−0.05
O-W-W-B-W	0–10		3.90	+0.50
	10–20		4.00	+0.40
	20–30		3.50	+0.30
	0–30		3.80	+0.40
Continuous wheat	0–10		4.04	+0.64
	10–20		4.03	+0.43
	20–30		4.02	+0.82
	0–30		4.03	+0.63

soil erosion. But losses of SOM through erosion on summer fallow fields were much higher than those reported herein on small plots.

4 CONCLUSIONS

Ever since the first publication by Suleimenov (1988), it has been observed that research data on summer fallow were too scanty to make decisive conclusions. It was emphasized that crop response would be different in long-term studies including very dry years. This chapter summarizes the data of 20 years from these experiments. The data for 13–16 years were included in this chapter after completion of the five-year rotations. Data analyses were done separately for very dry, dry, moderately dry and favorable precipitation years.

The benefits of summer fallow were definitely more during dry years as compared to moderately dry and wet years. There was in increase in yield by summer fallow *vis-à-vis* stubble land during the dry years. On average, increase in yield of wheat by summer fallow in comparison with continuous wheat was as follows: 28–30% in very dry years, 18–24% in dry years, 8–10% in moderately dry years, and 4-7% in wet years.

Benefits of summer fallow also depended on the crop management. It increased in dry years under simple crop management without application of water, crop and weed management technologies. The average increase in spring wheat yield sown on summer fallow compared to that sown on stubble land was about 15%, and the grain yield was not affected by the duration of continuous wheat up to 25 years. This hypothesis was put forward in 1988, and was substantiated by the data from the long-term experiment.

Based on the data from many research stations and farms, it had been concluded that wheat yield decreased gradually and proportionally with continuous cultivation after summer fallow (Shiyatiy, 1985). The empirical relationship thus established was used to predict grain yields. Use of these equations would show that the grain yield would decrease to zero within seven to eight years of continuous cultivation. Based on this assumption, some scientists in Northern Kazakhstan and Western Siberia recommended transition to rotations of grains with summer fallow once in three years (Dvurechenskiy, 2003; and Shiyatiy, 1996).

It is important to understand why contradictory conclusions were made for similar soil conditions and even within the same organization. These contradictions may be attributed to differences in research methodology and data interpretation. When this long-term experiment on crop rotations was established in the Shortandy Grain Center in 1961, it was believed that summer fallow was a necessary component of successful dry land farming. Many scientists still believe in this paradigm. That being the case, it was important to determine how often summer fallow should be rotated with small grains. That is why most experiments studied 2, 3 and 4-year rotations of summer fallow with wheat. After a while, the 4-year rotation was found to be the best. Subsequently, a 5-year and 6-year rotations were included in field studies.

According to the methodology used in this experiment, phosphorus fertilizer was applied once in fallow field to be used by crops during the whole rotation. This recommendation was made for economical reasons, but it was beneficial to crops sown after the fallow. In the 1960s, the land was newly developed and there was an abundance of NO_3-N especially after the year of summer fallow. By 1980s, however, the availability of NO_3-N decreased. Yet it was adequate in soils under summer fallow but was deficient on stubble land (Lichtenberg, 1996). However, nitrogen fertilizer was not used in the long-term experiment. The lack of N also led to yield decline on stubble land in general and continuous wheat in particular. It became obvious that, if one would address issues of weed, water and nutrient management, the comparison might not be in favor of rotation with summer fallow. This was a justification of establishing the new long-term trial in 1983.

Economist B. Koshelev (2003), having said that grain production in steppes of Western Siberia is based on rotation of small grains with summer fallow, referred to the experience of a farmer named V. Shnider as follows: "Meanwhile alternative opinion about role of summer fallow are coming forward. For example, farmer from Odessa District, Omsk Province, Mr. V. Shnider farming 20,000 ha of grains in steppe zone has not practiced summer fallow in any form for several years. He prefers quality tillage, application of fertilizers and chemicals, growing new varieties and using imported modern tractors and equipment. This farmer obtains higher gain yields every year than his neighbors. This fact changes all existing theoretical foundations of grain production in dryland farming, and signifies its practical importance."

Mr. V. Shnider (2004) reported that he established a 1,700 ha farm in the beginning of the 1990s by collecting land shares of Germans leaving Russia and returning to Germany. During four years, farm area under his control increased up to 21,000 ha. Thanks to successful grain business, he became a John Deere dealer and purchased tractors, combines and other equipment. He improved the farming practices dramatically by successfully using this equipment. This farmer was using continuous grain cultivation until 2001, primarily spring wheat without using summer fallow. Without paying attention to theoretical considerations, he eliminated summer fallow because this practice interfered with usage of heavy tractors and large equipment. After eliminating fallow, he started growing wheat on very large fields with high efficiency. During 1994–2001, the average grain yield on his farm was 2.11 t ha^{-1} compared to 1.59 t ha^{-1} in the neighboring farms which on average maintained 14.3% of land under summer fallow. Thus, his grain yield was 0.52 t/ha higher and grain yield from total area was 0.74 t/ha or 35% higher than that in the District.

Recently, Mr. V. Shnider started crop diversification because of price fluctuations. In 2004, he increased farm size up to 28,000 ha by renting land from neighbors, of which 8,400 ha were allocated to sunflower, corn and dry peas. He farms without any summer fallow, although local scientists strongly recommended allocating 25% of cropland to summer fallow (Moshchenko and Bormotov, 2000). There are many advocates of 3-year rotation, with 33% of cropland under summer fallow (Dvurechenskiy, 2003). Shiyatiy (2000) suggested that "spring wheat should be grown in Asian parts of Russia on 8 M ha leaving 4 M ha under summer fallow."

In the dark brown soil zone of central Kazakhstan, Yushchenko (2006) reported that efficiency of summer fallow depended on landscape and moisture storage from snow distribution. Summer fallow was not efficient in landscape with good moisture storage from snow. In contrast, in sloping areas with poor snow accumulation and consequently low water storage, summer fallow was the only means to improve crop yields.

Many CIS scientists refer to the Canadian experience with regards to the necessity of increasing the area under summer fallow. Shiyatiy (2000), for example, noted without any supporting reference that "in recent years, the summer fallow area has exceeded the wheat sown area". This is not a correct statement. Even during the 1970s, area under fallow and wheat in western Prairies of Canada occupied only 40 and 39% of cropland, respectively (Larney et al., 2004). Recently, however, area under summer fallow has strongly decreased. In 1990s, the average areas under summer fallow and wheat were 20% and 35%, respectively, whereas on black soils these were 12% and 30%, respectively.

In contrast to the static systems, dynamic agricultural systems permit farmers in the Northern Plains to quickly adjust to major changes in international markets, weather, and the government policies (Elstein and Comis, 2005). "Dynamic agricultural systems" in Mandan, North Dakota refer to the new farming approach that allows farmers to choose from various management options such as diversified crop sequences and livestock to obtain the greatest economic return while minimizing expenses and the risks. A major component of the system is "a long-term strategy of planting crops every year while balancing economic and environmental factors."

Summer fallow supposedly improves water management, but the increased moisture storage under fallow occurred only in two years for one crop. In the present experiment, double snow ridging made water storage on stubble land similar to that under summer fallow. In Pavlodar Province, a technology of wheat grain stripping allowed enough snow to accumulate on stubble land by tall stubble and ensured the same water storage as on summer fallow (Mustafaev and Sharipov, 2006).

In southern Ural, sweet clover sown on summer fallow as green manure for 12 years in comparison to black fallow improved grain yield of durum wheat by 0.19 t ha^{-1} (Maksyuta et al., 2006). Application of green manure on fallow land maintained the SOM content. However, in black fallow, 2 t ha^{-1} of SOM was lost by decomposition and 3 t ha^{-1} by soil erosion. Another kind of summer fallow utilizing a cover crop included sowing of Sudan grass, which produced 16 t ha^{-1} of green forage and improved wheat grain yield by 0.23 t ha^{-1} compared to the black summer fallow.

Data on grain quality showed that best crop management under limited duration of crop growing season and shortage of nitrogen availability reduced the grain gluten content. Also, continuous cropping as compared to summer fallow reduced grain quality because of decline in nitrogen availability.

While complete elimination of summer fallow is not recommended immediately, cropping systems based on frequent summer fallow are unsustainable because summer fallow is rather destructive in any farming system due to severe soil erosion by water and wind and accelerated decomposition of SOM. Some soil scientists have reported high risks of soil erosion on summer fallow, and yet advocate this practice. According to Shiyatiy (1996à), soil losses by water erosion were 31 times more under summer fallow than those from a stubble land.

In another study on the same research site, Funakava et al. (2004) concluded that, given the possible adverse effects of the summer fallow on enhanced decomposition of SOM, snow management must be the main strategy for capturing water rather than the summer fallow practice.

Thus, agricultural research must address this problem by identifying a better replacement of summer fallow by either food legumes or oilseeds. As to farming practices, Shnider's experience indicates the right strategy is that the reduction of summer fallow under better crop management practices will improve benefits to farmers and make farming more sustainable. It is widely recognized that present crop management practices are at lower level of sustainability than those in the past. If present issues of nitrogen availability and weed control are addressed by increasing the area under summer fallow, it can have drastic consequences with regards to food production and environment quality.

REFERENCES

Dvurechenskiy, V. 2003. Water and soil conserving technologies of grain production in northern Kazakhstan. p. 84. In: Scientific bases of state program of agricultural production of republic of Kazakhstan for 2003–2005, Ministry of Agriculture, Astana. (Rus).

Elstein, D. and D. Comis. 2005. Moving away from Wheat-Fallow in the Great Plains. J. Agricultural Research, June, p. 14–17.

Funakava, S., I. Nakamara, K. Akshalov, and T. Kosaki. 2004. Water dynamics in soil-plant systems under grain farming in northern Kazakhstan. Soil Sci. Plant Nutr. 50 (8): 1219–1227.

Koshelev, B. 2003. Organization and economical bases of grain production in western Siberia. Omsk, Omsk Agricultural University. (Rus).

Lichtenberg A. 1996. Agrochemical research base of agricultural production. p. 105–115. In: A. Eskov and E. Shiyatiy (eds.). Soil conservation practice – challenges and future. Proceedings of KazNIIZKH, Shortandy, Kazakhstan. (Rus).

Larney F., H. Yanzen, E. Smith, and D. Anderson. 2004. Dryland agriculture on the Canadian Prairies: Current Issues and Future Challenges. p. 113–139. In: S. Rao and J. Ryan (eds.). Challenges and Strategies for Dryland Agriculture. CSSA Special Publication, Madison, Wisconsin, USA.

Maksyuta N., V. Zhdanov, V. Zakharov, and V. Laktionov. 2006. Soil and resources conserving methods and technologies in farming of Orenburg Province. p. 129–138. In: Z. Kaskarvayev, V. Skoblikov, and M. Karpenko (eds.). Modern challenges of conservation agriculture and ways to improve sustainability of grain production in steppe regions. Ministry of Agriculture, Kazakhstan, Astana-Shortandy. (Rus).

Moshchenko Y. and I. Bormotov. 2000. Development of principles of steppe farming system. p. 135–139. In: Energy and resources saving in farming systems of arid territories. RASKHN, Barnaul. (Rus).

Mustafayev B. B. Sharipov. Influence of rational utilization of natural resources for crop productivity and quality. p. 221–225. In: Z. Kaskarbayev, V. Skoblikov, and M. Karpenko (eds.). Modern challenges of conservation agriculture and ways to improve sustainability of grain production in steppe regions. Ministry of Agriculture, Kazakhstan, Astana-Shortandy. (Rus).

Suleimenov, M. 1988. About theory and practice of crop rotations in northern Kazakhstan. J. Zemledeliye, No. 9, Moscow. (Rus).

Suleimenov M. and K. Akshalov. 1996. Continuous spring wheat growing under semi-arid conditions of Northern Kazakhstan. p. 328–330. In: Abstracts of 5th International Wheat Conference. June 10–14, 1996, Ankara, Turkey.

Suleimenov M., K. Akshalov, K. Akhmetov, and B. Kanafin. 2001. Possibilities to reduce summer fallow for better soil conservation in northern Kazakstan. p. 205–209. In: L. Garcia-Torres, J. Benites, and A. Martinez-Vilela (eds.). Conservation agriculture, a world challenge. Ist World Congress on Conservation Agriculture, October 1–5, 2001, Vol. II.

Shiyatiy E. 1985. Yield model and stability of spring wheat grain production. J. Zemledeliye, No. 5: 15–18. Moscow. (Rus).

Shiyatiy E. 1996. Three year grain-fallow rotations in steppes of Kazakhstan. J. Zempledeliye, No. 6. Moscow. (Rus).

Shiyatiy, E. 1986. Improvement of conservation practices – base for progress in crop production. p. 20–33. In: A. Eskov and E. Shiyatiy (eds.). Soil conservation practice – challenges and future. Proceedings of KazNIIZKH. Shortandy, Kazakhstan.

Shiyatiy, E. 2000. Bases for optimizing technical, ecological and economic aspects of grain production in steppe regions. J. Agricultural science No. 6: 16–18. (Rus).

Shnider, V. 2004. Large grain farm in Siberia. In: D. Vermel (ed.). Moscow. (Rus).

Yushchenko D. and N. Yushchenko, 2006. Study of efficiency of zero tillage in central Kazakhstan. p. 246–257. In: Z. Kaskarbayev, V. Skoblikov, and M. Karpenko (eds.). Modern challenges of conservation agriculture and ways to improve sustainability of grain production in steppe regions. Ministry of Agriculture, Kazakhstan, Astana-Shortandy. (Rus).

CHAPTER 22

Dynamics of water and soil organic matter under grain farming in Northern Kazakhstan – Toward sustainable land use both from the agronomic and environmental viewpoints

Shinya Funakawa
Graduate School of Agriculture, Kyoto University, Kyoto, Japan

Junta Yanai
Graduate School of Agriculture, Kyoto Prefectural University, Kyoto, Japan

Yusuke Takata
Graduate School of Agriculture, Kyoto University, Kyoto, Japan

Elmira Karbozova-Saljnikov
Institute of Soil Science, Belgrade, Serbia and Montenegro

Kanat Akshalov
Barayev Kazakh Research and Production Center of Grain Farming, Shortandy, Kazakhstan

Takashi Kosaki
Graduate School of Global Environmental Studies, Kyoto University, Kyoto, Japan

1 INTRODUCTION

The arid and semi-arid vegetation zones comprise the main part of Central Asia. The nomadic land use system has been an integral part of the area for the past several thousands years. However, influence from settled societies of Russia and China began to impact the region during the 16th and 17th centuries. The most drastic changes in land use and land cover occurred in the 20th century after occupation of the region by former Soviet Union and China Republic, under which large-scale agricultural projects were developed by strong political initiatives (Chuluun and Ojima, 2002).

The Soviet part of Central Asia covers a total area of 400 million hectares (Mha) and an arable area of 31 Mha (FAO, 2005). The per capita arable land area ranges 0.17 (Tajikistan) to 1.40 ha (Kazakhstan). The ratio of irrigated land to arable land ranges from 10.8 (Kazakhstan) to 94% (Turkmenistan) and agricultural production is usually possible only by irrigation in southern part of the region.

Wheat is the major crop in Central Asia, especially in the rainfed agricultural zone (FAO, 1998). Kazakhstan counts for 70% of the total cereal harvested area, most of which is located in the northern region.

1.1 *Historical overview of agriculture development in northern Kazakhstan*

Kazakhstan is the most northerly country among the Central Asian nations. It accounts for approximately 90% of total area under wheat production in Central Asian countries (CIMMYT, 2000), and hence it is one of the most important regions for global food production. Wheat production in Kazakhstan is located in two broad agro-environments: northern spring wheat and southern winter wheat ecoregions (Longmire and Moldashev, 1999). While in the southern region (42°–48° latitude N), winter wheat is grown either under irrigated or rainfed condition, rainfed spring wheat is the major crop in the northern region (48° –55° latitude N). About 70% of national cereal production of Kazakhstan is produced in the northern region (Nobe, 1998).

Although northern Kazakhstan is characterized by soils of high fertility, climatic condition is generally severe because of low annual precipitation of about 300 mm. Hence drought stress is a major abiotic stress (Morgounov and Zuidema, 2001). Historically, wind erosion emerged as a serious problem. It is exacerbated by the light soil texture and use of heavy machinery for large scale dry farming operations (Gossen, 1998). In the late 1950's, a research station was developed in Shortandy (northern Kazakhstan) to conduct research to alleviate these problems. The research station was changed to the All-Union Research Institute for Cereal Production. The institute had tremendous impact on cereal production in the region through development of effective soil-conservation cropping systems (Morgounv et al., 2001). Principal methods developed by the Research Institute were: 1) snow management, a technology to decrease the snow drifting and to optimize soil temperature and water regimes by understanding the spatial redistribution of snow received and its thawing in spring (Shegebaev, 1998; Vladimir et al., 2001), 2) subsoil cutting, a technology to enhance infiltration of thawing-water as well as to prevent evaporation by cutting capillary rise from subsoils (Shegebaev, 1998), and 3) summer fallow, a technology to minimize the weed incidence, capture soil moisture, and accelerate mineralization of soil organic matter (SOM) for nutrient replenishment by not raising crops for one year (Shegebaev, 1998). The summer fallow was also common in rainfed agriculture in semi-arid steppes in northern America (e.g. Mathews and Army, 1960; Smika and Wicks, 1968; Black and Power, 1965; Farahani et al., 1998).

Though this was a soil conservation strategy, the improved technology was applied uniformly and intensively to achieving high production during the period of former Soviet Union. However, high production was achieved at the expense of sustainability of natural resources (Srivastava and Meyer, 1998). Numerous problems of sustainability were inevitable from the use of these practices including decline in soil fertility, soil erosion, and soil compaction (Srivastava and Meyer, 1998) with attendant changes in soil quality (Sorokina and Kogut, 1997). Both organic carbon and total nitrogen concentrations in the top 10-cm layer declined by 38–43% and 45–53%, respectively, in continuously cropped field of Chernozem soils in Russia (Mikhailova et al., 2000). Conversion of virgin soils to arable land reduced SOM content by 50% during the first few years of cultivation (Buyanovsky et al., 1987). In general, soil fertility declined by 50% in Chernozem soils in the former Soviet Union (Srivastava and Meyer, 1998).

However, in Kazakhstan, agriculture is presently not managed in a way that it was during the Soviet era because of financial problems caused by the collapse of the former Soviet Union in 1991. Thus, the rate of depletion of SOM is now lesser than it was during 1970s and 1980s. Due to severe budget cuts, extensive cropping systems became prevalent; e.g., application of fertilizers and/or herbicides decreased and the use of agricultural machinery became less frequent than during the Soviet era, along with an attendant decline in farm production (Longmire and Moldashev, 1999; Gossen, 1998; Morgounov et al., 2001). The highest crop yield (i.e., 1.00 Mg ha^{-1}) was achieved during 1986 to 1990, but it declined to 0.65 Mg ha^{-1} in 1994–1996 (Gossen, 1998). Since the agricultural sector is one of the principal elements of the country's economy, it is important to rectify this situation. Moreover, socioeconomic conditions are changing and farmers in former state or collective farms have the right to farm individual plots of lands (Meng and Morgounov, 2000). The support system of agricultural technology (e.g., research, education, and extension) in Kazakhstan entered a new era. Thus, an innovative approach of land management must be assessed in an objective, responsive and effective manner (Morgounov et al., 2001).

1.2 *Significance of C sequestration in Chernozem soils*

Conversion of grasslands into croplands and subsequent soil mismanagement has caused serious environmental problems including soil degradation. Several studies in the Central Asian Republics indicate that the grazing impact on vegetation has been progressively decreasing during the past several decades (Kharin et al., 1998; Lioubimtseva et al., 2005). The loss of SOM is exacerbated by deforestation, land-use conversion, and soil degradation and desertification, all of which are severe problems in Central Asia (Lal, 2004). Cultivation of virgin soils or conversion of natural to agricultural ecosystems leads to depletion of SOM pool accompanied with accelerated emission of greenhouse gases (GHGs) into the atmosphere. About 54% of soil humus was depleted during 67 years of cultivation of Chernozem in the Central Russian Uplands (Kokovina, 1979). In Kyrgyzstan, Glaser et al. (2000) reported that development of pastures led to a loss of about 30% of SOM compared with the native vegetation. The World Bank (2003) reported that 10–30% of soil humus was lost in the surface soil of croplands in Kazakhstan between 1960 and 1995, and that the most serious loss occurred in soils under rainfed agriculture.

The capacity of Chernozems to store a large amount of SOM has drawn considerable attention in terms of both the large source and sink of carbon dioxide (CO_2) in relation to the 'global warming'. Increase in concentration of GHGs in the atmosphere may already be changing climate, and may result in an increase in average global temperatures of 1 to 3.5°C by 2100 (Houghton et al., 1996). Since the global warming may adversely impact the agricultural production in Central Asia through increase in drought stress (Mizina et al., 1997; Yu et al., 2003), it is important to diminish CO_2 emission from these agricultural lands while sustaining the agricultural production.

FAO (2004) described several biophysical aspects of C sequestration in dry-land soils. Important options include crop residue management, conservation tillage (CT), and crop rotation.

Both the quantity and quality of plant residues are important factors determining the amount of C stored in soils. The quantity strongly depends on the environmental conditions and agricultural practices. Application and retention of crop residues on the soil surface strongly affects soil properties. Standing plant residues can trap the drifting snow and replenish soil moisture (Aase and Siddoway, 1980; Caprio et al., 1989). Crop residues, standing or as mulch, decrease the risk of wind erosion (Chepil et al., 1964; Poluektov, 1983). Mulch application reduces water loss and moderates soil temperature (Duiker and Lal, 2000), both of which are important factors for the success of dryland farming.

The importance of crop rotation in agricultural systems has long been recognized, and rotations constitute an integral component of many CT practices (FAO, 2004). Incorporation of N-fixing crops in a rotation increases soil N and reduces the need for high dose of N fertilizers. Suleimenov et al. (2005) compared continuous wheat cultivation based on summer fallow with a pea-based rotation in northern Kazakhstan. They observed that the pea-based rotation more strongly impacted the quantity of plant biomass produced than summer fallow-based rotation. Shkonde et al. (1982) reported similar results and concluded that replacing summer fallow by introducing sainfoin reduced SOM loss by 30% in a Ciscaucasian Chernozem. Drinkwater et al. (1998) estimated that application of the pea-based rotation in the maize/soybean-growing region of the United States would increase soil C sequestration by 0.01–0.03 Pg C yr^{-1}. Reducing fallow frequency or tillage intensity is often most effective in dry environment (Batjes and Sombroek, 1997).

The research on soil C sequestration in Central Asia is scanty (Lal, 2002). In order to realize the great potential of C sequestration in this region, additional research is needed to identify those soil-specific practices which can restore SOM in degraded soils. Moreover, the agronomic environment in northern Kazakhstan is similar to that in the Chernozem areas of western Siberia, Caucasus, and Mongolia. Thus, the results obtained in northern Kazakhstan can also be applied to other regions with similar soils and the environments.

1.3 *Problem setting*

N.K. Azarov reported that the geography has a strong correlation with snow depth, humus content, moisture content and cereal productivity, and indicated a strong possibility of growing cereals on the best agricultural landscapes (Gossen, 1998). Such a strategy can lead to change from soil conservation system to adaptive landscape system based on contour cultivation of the fields. Azarov called this strategy as 'an agro-landscape agricultural system'. The system is in accord with the concept of site-specific management. The latter involves dividing a field into small cells for more careful management, as was proposed by Dr. H.H. Cheng, a soil scientist at the University of

Minnesota (Shibusawa, 1999). These concepts also can be applied to an alternative agricultural system to address both the agroeconomic and environmental concerns

Therefore, the principal objectives of the present study were:

1) to re-evaluate the water-storage strategies such as summer fallow in the context of SOM conservation,
2) to quantify the SOM dynamics under the grain farming, and
3) to assess a possibility of 'site-specific management' as an alternative land use strategy to address both the economic and environmental issues.

The report describes the general characteristics of steppe soils in northern Kazakhstan and in southern mountain foothills, and explains field experiments conducted to study dynamics of water and SOM. These experiments were conducted at the experimental farm of Barayev Kazakh Research and Production Center of Grain Farming, Shortandy, northern Kazakhstan (51°35′N, 71°03′E). The SOM status under different fallow frequency in the experimental field is discussed in relation to management. The data from experiments on *in situ* SOM dynamics are extended to farmer fields near Shortandy taking into consideration the spatial variability of soils according to topographic factor, and to farmland in three different soil types in northern Kazakhstan.

2 CHARACTERISTICS OF STEPPE SOILS OF KAZAKHSTAN

Macrolandscape of Kazakhstan is characterized by an east-to-west extension of central desert zone along with Lake Balkhash, Aral Sea and Caspian Sea and the adjoining steppe zones in north and south. The southern steppe, located on northern foothill of Mt. Alatau, is characterized by relatively high temperatures and high rainfall, unlike the northern steppe which are characterized by severe drought after mid-summer and cold climate in winter (Figure 1). The general characteristics of the soils from both of these regions are outlined in Table 1.

Loess-derived soils are distributed east-to-west along the northern foothills of Mt. Alatau, between inner desert and mountain forest zones. Concentrations of soluble salts in the saturation extract, exchangeable Na, and gypsum are low throughout the profiles. The coarser fractions (e.g., fine sand and/or silt) are generally higher in these soils than in typical soils of northern steppe (Profile 3 in Table 1) In general, coarser soils are distributed near the desert rather than along the mountain-side, indicating that the risks of soil salinization are low. Major components of clay minerals are 1.4 nm-smectite, mica and kaolinite.

In the northern steppe of Kazakhstan, relatively fine-textured soils occur on quaternary deposits in the Astana regions of the capital, to Kokshetau from south-to-north. In contrast, coarse-textured

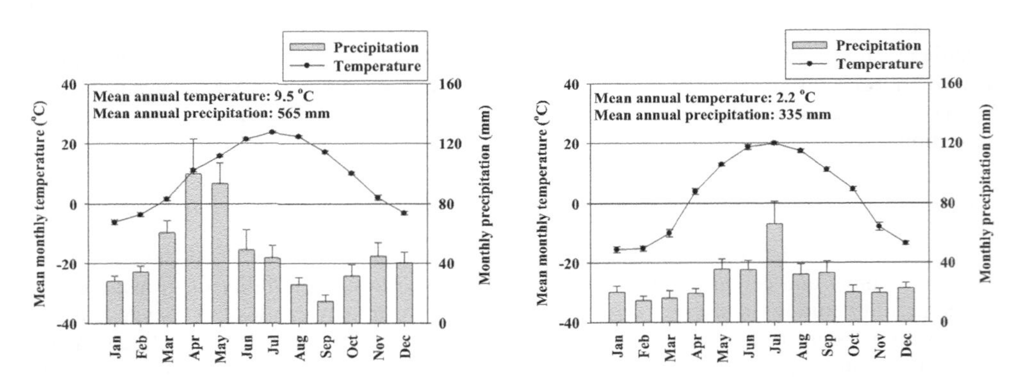

Figure 1. Fluctuation of monthly temperature and precipitation at Almaty (a) and Shortandy (b) during 1990–1999.

Table 1. Physicochemical properties of selected steppe soils in Kazakhstan.

			Particle size distribution									Saturation extract					
Horizon	Color	Depth (cm)	Coarse sand (%)	Fine sand (%)	Silt (%)	Clay (%)	Coarse clay (%)	Fine clay (%)	Organic C (g kg^{-1})	CO_3-C (g kg^{-1})	Gypsum (g kg^{-1})	pH	$Na^+ + K^+ + Mg^{2+} + Ca^{2+}$ ($cmol_c$ kg^{-1})	Sodium adsorption ratio (SAR) (mmol L^{-1})$^{1/2}$	Exch. Na ($cmol_c$ kg^{-1})	CEC ($cmol_c$ kg^{-1})	Exch. Na percentage (ESP) (%)
Profile 1 (Typic Calciudolls in souther steppe)																	
A1	10YR2/3	0–9	0.2	32.4	35.5	32.0	15.6	16.4	49.7	3.1	0.0	6.9	0.4	0.5	0.0	31.1	0.0
A2	10YR3/3	9–18	0.0	40.3	16.8	42.9	23.3	19.6	42.5	2.8	0.0	6.7	0.6	0.1	0.0	28.0	0.1
Bw1	10YR3/3	18–32	0.2	31.9	36.3	31.6	15.9	15.7	33.2	3.2	0.0	7.0	0.3	0.1	0.1	27.1	0.2
Bw2	10YR4/3	32–42	0.2	43.1	31.6	25.1	14.8	10.3	25.5	2.9	0.0	6.8	0.2	0.2	0.1	22.6	0.3
BC	10YR4/4	42–61	0.0	44.8	31.2	24.0	13.5	10.5	16.3	2.3	0.0	6.9	0.2	0.2	0.1	21.2	0.3
Ck	10YR6/3	61–90+	0.0	49.4	31.3	19.3	12.2	7.1	5.7	30.0	0.0	7.4	0.2	0.2	0.0	10.5	0.4
		200	0.0	46.2	38.7	15.2	12.8	2.3	2.1	25.1	0.0	7.8	0.2	0.3	0.1	8.6	0.7
		300	0.0	45.2	35.4	19.3	13.9	5.4	2.8	15.0	0.0	8.0	0.3	0.4	0.1	8.6	1.2
		400	0.0	45.0	38.8	16.2	13.1	3.1	3.2	16.2	0.0	8.1	0.3	0.8	0.1	9.3	1.4
		500	0.0	58.3	28.6	13.1	10.8	2.3	1.7	20.2	0.0	8.1	0.3	1.8	0.2	6.8	2.6
Profile 2 (Typic Calciustolls in southern steppe)																	
A1	10YR3/2	0–8	2.0	33.2	36.7	28.0	19.8	8.2	33.0	2.9	0.0	7.7	0.3	0.1	0.0	21.3	0.1
A2	10YR3/3	8–18	1.3	31.9	39.5	27.2	22.9	4.3	24.3	3.1	0.0	7.7	0.3	0.1	0.0	17.1	0.2
Bw	10YR3/4	18–40	1.3	36.8	40.6	21.3	19.0	2.3	17.8	12.3	0.0	7.7	0.3	0.1	0.0	14.1	0.2
BCk	10YR5/4	40–60	0.4	33.9	50.0	15.7	15.5	0.2	6.8	31.9	0.0	7.7	0.2	0.2	0.0	8.9	0.3
C	10YR6/4	60–90+	0.4	40.6	44.2	14.8	13.3	1.5	3.9	27.6	0.0	7.8	0.2	0.3	0.1	8.7	0.6
		200	0.0	44.4	42.2	13.3	11.2	2.1	1.5	19.1	0.0	8.1	0.3	1.2	0.1	6.1	2.1
		300	0.9	47.3	42.6	9.2	7.8	1.4	1.5	17.5	4.5	7.6	5.3	7.1	0.7	6.2	11.0
Profile 3 (Type Haplustolls, clayey, in northern steppe)																	
A1	10YR3/1	0–20	0.0	24.9	25.8	49.2	27.9	21.3	29.6	7.0	0.0	8.0	0.4	0.1	0.1	29.7	0.2
A2	7.5YR4/2	20–10	0.0	25.2	26.8	48.0	22.0	26.0	14.3	14.1	0.0	7.3	0.4	1.2	0.3	26.3	1.3
Bw	7.5YR4/2	40–50	0.0	27.1	28.9	44.0	23.1	20.9	14.1	14.4	0.0	7.8	0.4	4.5	1.6	24.7	6.4
BCk	7.5YR5/4	50–75	0.0	26.1	25.8	48.1	26.1	22.0	11.3	17.1	0.0	7.9	0.6	12.5	4.2	24.2	17.3
C1	7.5YR5/4	75–92	0.0	27.9	21.9	50.2	26.8	23.5	2.0	15.2	0.0	7.6	2.6	14.8	4.1	20.5	20.2
C2	7.5YR5/6	92–100+	0.0	29.0	22.9	48.0	27.7	20.4	1.5	12.7	121.8	7.3	4.0	10.0	2.1	16.6	12.8
		200	0.0	32.4	22.6	45.0	24.4	20.6	1.5	13.5	8.8	7.4	11.2	21.3	3.7	18.0	20.6
		300	0.0	30.4	25.2	44.4	24.2	20.2	1.3	11.0	0.0	7.5	10.2	26.3	2.5	24.7	10.0
Profile 6 (Typical Haplustolls, sandy, in northern steppe)																	
Ap1	2.5Y3/2	0–10	27.6	28.9	14.1	29.5	14.0	15.5	17.9	1.4	0.0	6.5	0.1	0.3	0.1	18.2	0.3
Ap2	2.5Y3/2	10–20	28.9	28.1	11.9	31.1	13.9	17.2	18.4	1.5	0.0	6.2	0.1	0.4	0.1	18.2	0.4
A	2.5Y3/2	20–30	30.4	26.1	12.7	30.8	13.2	17.5	14.8	1.4	0.0	6.7	0.1	0.6	0.1	20.6	0.6
AB	2.5Y3/2	30–40	28.8	23.5	13.8	34.0	16.1	17.9	9.8	1.8	0.0	6.8	0.3	1.1	0.2	20.4	1.1
Bk	2.5Y4/6	40–60	27.3	33.6	23.8	15.3	14.7	0.6	6.2	15.4	0.4	7.0	1.7	5.2	0.6	13.7	4.6
BC	2.5Y5/4	60–80	29.8	32.9	23.2	14.2	13.4	0.7	4.6	16.6	0.4	7.2	2.5	9.3	1.1	13.6	8.1
C	2.5Y5/6	80–100	43.9	25.1	14.3	16.8	16.2	0.6	2.4	12.6	0.0	7.5	2.4	14.7	1.6	10.6	15.0
		150	42.0	28.0	9.1	21.0	15.8	5.1	0.9	5.3	0.6	7.6	3.2	20.2	1.8	10.7	17.1
		200	16.6	22.4	21.4	39.6	30.8	8.8	1.6	11.1	2.5	7.5	5.1	23.9	4.0	19.6	20.4

soils are found along with R. Iritish (Pavlodar region) in the northeast and in the Kustanai region in the northwest. The southern limit of occurrence of Mollisols is Astana, in which climate is drier than in the upper north. In contrast, the soil color in the forest steppe zone north of Kokshetau is darker due to decreasing temperatures. Most of the soils of this region are classified as Ustolls according to Soil Taxonomy (Soil Survey Staff, 2003). Using the soil classification system of former Soviet Union, the soils of southern half of the region or those derived from sandy materials are classified as Southern Chernozems or Dark chestnut soils, while those of the northern region are classified as Ordinary Chernozems.

The soils derived from quaternary deposits in the northern steppe are characterized by predominately 1.4 nm-clay minerals with a high swell-shrink capacity. Such clay mineralogy reflects swelling and shrinkage of the soils by fluctuations in soil moisture. During the wet spring, soils have low permeability to snow melt water, resulting in excessive water loss through accelerated evaporation and/or surface runoff. This is one of the severe constraints to agricultural production. In addition, large amounts of soluble salts are often accumulated in deeper layers in the soils derived from the quaternary deposits in the north. It may be a cause of secondary soil salinization, if intensive irrigation agriculture were introduced in this area. Despite these problems, however, the SOM pool is 150 Mg C ha^{-1} (e.g., Profile 3 in Table 1), indicating high soil fertility for agricultural production in the region.

3 WATER DYNAMICS IN SOIL-PLANT SYSTEMS UNDER GRAIN FARMING IN NORTHERN KAZAKHSTAN

The area of northern Kazakhstan is strongly affected by the continental climate, being typically cold and dry. Because of such extremely dry conditions for wheat cultivation, the average grain yield is generally as low as 1.0 Mg ha^{-1} (during the period of 1986–90 for Kazakhstan) (Gossen, 1998), and water management is one of the major constraints to sustainable production. Principal water management practices include: 1) snow management, which is done mainly in February, to accumulate additional snowfall by making parallel snow-rows at specific intervals (Figure 2a), 2) summer fallow to store rainfall water for the next cropping cycle (Figure 2b), and 3) subsoil cutting (conservation tillage) in autumn to reduce the loss of water through evaporation by decreasing capillary rise (Figure 2c) (Shegebaev, 1998). Summer fallow is usually practiced in the rotation systems once in five years to store moisture in soils, decrease weed hazard, and accumulate N through mineralization of SOM. Surface soils under fallow are mechanically harrowed several times to maintain bare soil and to minimize ET during the cropping season. The sustainability of such management including summer fallow is, however, one of the most controversial topics because of the low water storage efficiency or increase in decomposition of SOM through repeated soil disturbance (Janzen, 1987; Mikhailova et al., 2000). The present study compared water budgets in fallow and cropped fields, and identified conditions under which a specific type of water management is effective.

The experiment was conducted over a two-year period from autumn 1998 to autumn 2000 at the experimental farm of Barayev Kazakh Research and Production Center of Grain Farming, Shortandy, northern Kazakhstan. The long-term meteorological data (i.e., 1936–2005) show that the mean annual temperature is 1.7°C and the mean annual precipitation is 324 mm (Table 2). The soils are classified as Typic Haplustolls according to Soil Taxonomy (Soil Survey Staff, 2003) and correspond to the Southern Chernozem soils according to the classification system of the former Soviet Union.

Five plots in 1998–1999 and seven plots in 1999–2000 were established at the experimental farm. These plots were laid out in experimental blocks in which long-term experiments were conducted since 1983 for improving farming technology. The size of each plot was 6 × 60 m and the plots were managed by different farming methods. Crop species and field management of the experimental plots are shown in Table 3. Field management included plots with crop rotation at different stages and mechanical management at different intensities (e.g., depth of primary tillage or degree of snow

Figure 2. Snow management (top) in mid-winter for accumulation of additional snowfall by making parallel snow-rows at certain intervals (February 1, 1998). Landscape of cropped field (middle) after harvest (left) and adjacent fallow field (right) (April 13, 2000). In the cropped field, plant residues were left standing in order to accumulate snowfall as much as possible. In the fallow field, on the contrary, almost no remaining plant residues were incorporated into soil. Attachment (bottom) for conservation tillage (subsoil cutting) (replica in the exhibition room of the Center).

capturing). The meteorological data recorded during the experiments are presented in Table 2 with long-term data collected at the Center. While the precipitation during the winter time (January to April) and cropping period (May to August) in 1999 and 2000 was almost similar to the long-term average data. However, the precipitation after the harvest season (September to December) in 1998 and 1999 was lower than the average.

In each of these 12 plots, gravimetric soil moisture content was determined by oven-drying and converting it to volumetric basis, using the bulk density for each layer. Soil bulk density ranged from 1.1 to 1.3 $Mg\,m^{-3}$ in the layers with SOM accumulation up to the 50-cm depth and from 1.3 to 1.5 $Mg\,m^{-3}$ below 50-cm depth. Daily rainfall was recorded at the Center. To calculate the water supply derived from thawing, maximum depths of snow coverage and snow density were measured on April 2, 1999 and March 17, 2000, except for Plots 11 and 12. In the representative plots, including both the fallow and cropped plots in the preceding year, fluctuations in soil temperature during springtime were recorded using dataloggers (CR–10X, Campbell Scientific, Inc.). During the cropping period, plant biomass was measured several times on a 1-m^2 subplot in triplicate. Grain yield was measured at harvest.

3.1 *Water dynamics in the pre-cropping seasons of 1998/1999 and 1999/2000 under different land use stages and types of field management*

The soil profile is characterized by a gypsiferous layer at about 1-m depth with a drastic increase in the amounts of soluble salts below 1 m, indicating that the water movement is at equilibrium at this depth (Profile 3 in Table 1). Assuming that the water budget is balanced at 90-cm depth, evapotranspiration was calculated as the difference between precipitation and soil moisture increment for a given period. Downward or upward movement of water beyond this depth may, therefore, result in a possible error by over- or under-estimation of the evapotranspiration.

Figure 3 shows the dynamics of soil moisture content and cumulative precipitation, including water derived from thawing and cumulative evapotranspiration estimated for the pre-cropping and cropping phases. Table 4 summarizes the water budget for the pre-cropping seasons of 1998/1999 and 1999/2000. The data in Table 4-1 show that total water contents up to 90-cm depth were 295 and 297 mm in Plot 3 and 4 on September 16, 1998, respectively. These plots w ere fallowed with several plowings during the preceding cropping season in 1998. Thus, these plots accumulated higher amounts of water than the cropped plots 1 (144 mm) and 5 (244 mm). Since Plot 2 was also fallowed but had been more extensively managed (Table 3), the soil did not store an appreciable amount of water. A similar trend was also observed for these plots in 1999/2000 (Table 4-2), in which the fallow plot in the preceding summer (Plot 8) stored higher amounts of soil water (232 mm) than the cropped plots (Plot 6: 174 mm, Plot 7: 175 mm, Plot 9: 171 mm, and Plot 10: 170 mm).

During the winter, 70.6 mm and 86.9 mm of snowfall were recorded in 1998/1999 and 1999/2000, respectively. From late January to early February, during which the snow depth reached 20 to 30 cm, snow management was carried out at different intensities (i.e., different heights of snow-rows) in order to accumulate the snow-cover by making parallel snow-rows at specific intervals (Fig. 2a). Total amount of snow-cover, which was expected to be added to the soils at the time of thawing in spring, ranged from 102 to 234 mm of water on April 2, 1999 and March 17, 2000 (*b* in Table 4).

After thawing, soils accumulated 206 to 299 mm of water in 1999 and 221 to 254 mm of water in 2000, respectively (*c* in Table 4). The increase in the soil water content since autumn of the preceding year was, however, quite variable, ranging from −40 mm (Plot 3) to 74 mm (Plot 1) in 1999 and −6 mm (Plot 8) to 84 mm (Plot 10) in 2000, respectively (*d* in Table 4). The difference between the highest and the lowest water catchment amounted to 114 mm in 1999 and 90 mm in 2000, respectively. Despite the snow management during winter, soil water storage decreased in some cases. Figure 4 shows that, during thawing, the magnitude of increase of soil moisture decreased (Figure 4a) and the loss of water increased (Fig. 4b), as the soil moisture storage in the preceding autumn increased.

Table 2. Climatic conditions in shortandy during the experiment for water budget.

	Monthly precipitation (mm)									Mean monthly temperature (°C)								
	1998	1999	2000	2001	2002	2003	2004	2005	1936–2005	1998	1999	2000	2001	2002	2003	2004	2005	1936–2005
Jan	4	21	21	36	23	15	6	12	17	−20.7	−13.2	−15.2	−14.5	−6.7	−14.1	−17.8	−15.8	−16.6
Feb	27	11	21	22	23	6	34	2	13	−14.6	−11.4	−11.4	−14.9	−6.5	−16	−11.5	−21.8	−16.4
Mar	1	15	11	2	15	13	5	26	13	−9.8	−17.0	−8.3	−3.4	−1.7	−10.5	−9.5	−5.3	−10.2
Apr	19	33	11	16	21	17	19	9	20	−3.3	3.5	7.7	6.3	3.5	4.2	3.9	4.2	3.3
May	41	32	64	2	44	34	13	42	32	12.2	14.1	10.6	17.2	11.5	15.5	15.7	13.9	12.5
Jun	25	72	61	34	88	41	24	73	39	21.0	14.0	19.4	17.7	16.8	17.3	19.1	19.7	18.3
Jul	86	42	35	97	57	39	32	63	57	23.0	20.5	20.2	17.9	18.8	17.7	20.4	20.2	20.0
Aug	6	5	22	46	40	30	32	58	40	20.9	19.3	18.4	18.0	21.6	20.1	17.4	17.1	17.4
Sep	4	23	31	19	40	19	11	3	25	10.5	14.3	10.5	10.2	13.4	12.7	13.4	11.5	11.3
Oct	15	8	27	78	19	31	27	0	28	4.5	6.6	0.0	3.2	5.1	4.7	4.9	5.5	2.9
Nov	12	30	22	44	35	13	24	17	21	−11.2	−9.8	−11.8	−2.4	−2.9	−8.9	−2.8	−3.4	−7.5
Dec	13	4	35	4	18	12	47	17	19	−10.8	−10.0	−11.1	−14.1	−18.7	−10.3	−14.7	−12.6	−14.1
Jan–Mar	33	46	53	61	61	34	45	40	43	−15.0	−13.9	−11.6	−10.9	−5.0	−13.5	−12.9	−14.3	−14.4
Apr–Jun	86	138	136	51	152	92	56	125	92	11.3	10.5	12.6	13.7	10.6	12.3	12.9	12.6	11.4
Jul–Sep	96	69	88	162	138	88	74	124	122	18.1	18.0	16.4	15.4	17.9	16.8	17.1	16.3	16.2
Oct–Dec	40	43	84	125	72	56	98	35	68	−5.8	−4.4	−7.6	−4.4	−5.5	−4.8	−4.2	−3.5	−6.3
Total/ average	**255**	**296**	**362**	**400**	**423**	**270**	**273**	**324**	**324**	**2.1**	**2.6**	**2.4**	**3.4**	**4.5**	**2.7**	**3.2**	**2.8**	**1.7**

Table 3. Description of study plots for determination of water budget.

Plot No.	Plot	Crop rotation system[1)2)]	Depth of main tillage in autumn (cm)	Maximum depth of snow cover (cm)	Remarks
1998–1999					
1	F0-C	**F**-W-W-B-W	20–25	30	Fallow
2	F1-S	F-**W**-W-B-W			After fallow
3	F1-C	F-**W**-W-B-W	20–25	30	After fallow
4	F1-I	F-**W**-W-B-W	25–27	45–50	After fallow
5	O1-I	O-**W**-W-B-W	25–27	45–50	
1999–2000					
6	F0-C	**F**-W-W-B-W	20–25	30	Fallow
7	O0-C	**O**-W-W-B-W	20–25	30	
8	F1-C	F-**W**-W-B-W	20–25	30	After fallow
9	O1-C	O-**W**-W-B-W	20–25	30	
10	F4-C	F-W-W-B-**W**	20–25	30	
11	CW-I	**W**-W-W-W-W	25–27	45–50	
12	P1-I	P-**W**-W-B-W	25–27	45–50	

1) F: fallow, W: wheat, B: barley, O: oats, P: chick pea
2) Bold letters denote cropping stage of each plot.

Fluctuations in soil temperature during thawing indicated the reasons why some soils did not store water (Figure 5). After subjecting soils to summer fallow (Plot 3 in 1998/1999 and Plot 8 in 1999/2000), which had accumulated larger amounts of water than the cropped soils, these soils were frozen over a longer period of time than others at the time of thawing. In 1999, soil temperature exceeded 0°C on April 13 and 14 at 15-cm depth in Plots 1 (after cropping) and 3 (after fallow). Soil temperature at 45-cm depth in Plot 3, however, remained below 0°C until April 28, whereas that in Plot 1 had already increased to above 0°C on April 17. Soil temperature fluctuations at 75-cm depth followed a trend similar to those at 45-cm depth, which is that the soil temperature in Plot 3 was below 0°C on April 30 while that in Plot 1 was increasing. This delay in thawing of soil frost was observed again in 1999/2000 for Plot 8, where the soil temperature remained below 0°C at 45-cm depth for about 10 days after the temperature of other plots (Plots 6, 7, 9, and 10) had increased to above 0°C around April 10. Higher moisture content in the frozen soil layer was considered to be the main cause of delayed thawing, resulting in a slower water percolation from the soil surface or overlying layers of the profile. The water from thawing, then, stagnated on the surface and was rapidly lost through evaporation and runoff. The loss of water amounted to 105 to 348 mm at the time of thawing. Thus the accumulation of larger amounts of soil water in autumn occasionally hindered water percolation at the next thawing stage and decreased the water-capturing efficiency of snow management (Table 4). In the autumn of 1999 (data not shown), soil temperature was below 0°C in early November at 15-cm depth, in late November at 45-cm depth, and in mid-December at 75-cm depth. Since the snow cover on the soil surface was limited during December due to the low precipitation and strong wind, soil frosted may have been very severe in these plots. Similar behavior of frost soils upon thawing was also observed by Johnsson and Lundin (1991) and Hardy et al. (2001) in northern USA. The loss of soil water can occur quite extensively in the northern steppe regions of inner continents.

3.2 *Water dynamics in the cropping seasons of 1999 and 2000 under different land use stages and types of field management*

Figure 3 shows the dynamics of the soil moisture content, cumulative precipitation including water derived from thawing, and estimates of cumulative evapotranspiration. After the end of thawing

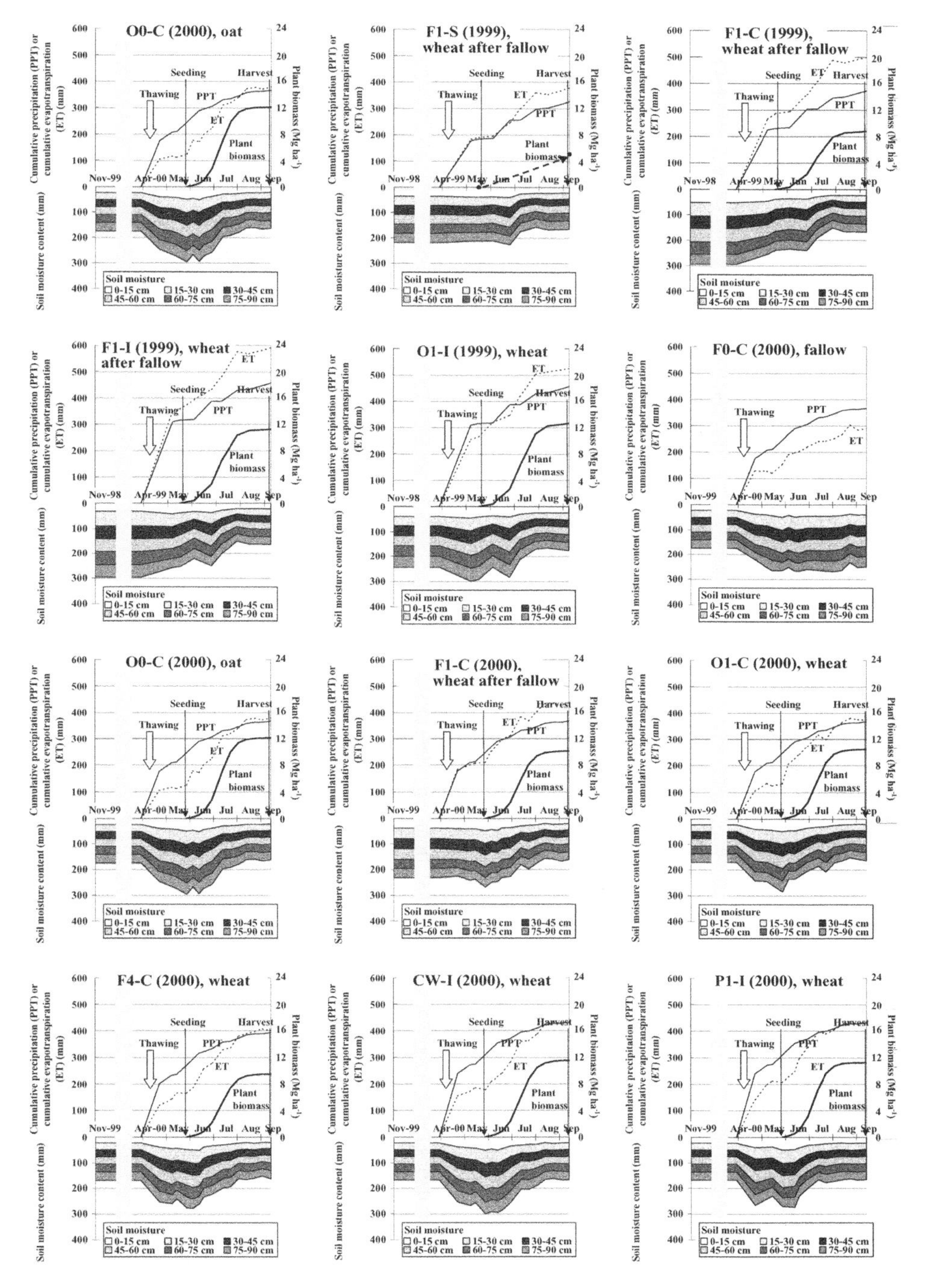

Figure 3. Dynamics of soil water and precipitation, including thawing and estimated cumulative ET (Plots 1 to 12 from left to right). For Plot 11, accumulated snow was estimated from the values in plots with same level of snow management (Plots 4 and 5 in 1999). For Plots 11 and 12, the water content in autumn, 1999 was estimated based on the average values determined in autumn, 1999 for cropped plots, i.e. Plots 2, 3, 4, and 5. For Plots 1 and 6, negligible amounts of weed biomass were detected.

Table 4-1. Water balance during the pre-cropping season of 1999 (from Sep., 1998 to Apr., 1999).

Plot No.	Plot	Soil moisture at 0–90 cm depth on Sep. 16 (mm) a	†Accumulation of snow on Apr. 2 (mm) b	Soil moisture at 0–90 cm depth on May 11 (mm) c	Increment of soil moisture during thawing (mm) d = c − a	Loss of water during thawing (mm) (b + †74.9) − d	Water-capturing efficiency (%) d/(b + †74.9)	Remarks
1	F0-C	144.1	116.7	217.8	73.7	117.9	38.5	Fallow
2	F1-S	218.2	102.7	206.2	−12.0	189.6	−6.8	After fallow
3	F1-C	294.6	150.7	254.2	−40.4	266.0	−17.9	After fallow
4	F1-I	296.5	234.2	257.6	−38.8	347.9	−12.6	After fallow
5	O1-I	244.3	234.2	299.2	54.9	254.2	17.8	

†During the period of Sep. 16 to May 10, 74.9 mm of rainfall and 70.6 mm of snowfall were recorded.
It is assumed that all the rainfall was directly supplied to the soil, whereas all the snowfall had been accumulated on Apr. 2 on the soil surface.

Table 4-2. Water balance during the pre-cropping season of 2000 (from Sep., 1999 to Mar., 2000).

Plot No.	Plot	Soil moisture at 0–90 cm depth on Nov. 11 (mm) a	†Accumulation of snow on Mar. 17 (mm) b	Soil moisture at 0–90 cm depth on Apr. 25 (mm) c	Increment of soil moisture during thawing (mm) d = c − a	Loss of water during thawing (mm) (b + †10.2) − d	Water-capturing efficiency (%) d/(b + †10.2)	Remarks
6	F0-C	174.3	165.5	221.1	46.8	128.9	26.6	Fallow
7	O0-C	175.4	165.5	246.2	70.9	104.8	40.3	
8	F1-C	232.1	166.3	225.9	−6.2	182.7	−3.5	After fallow
9	O1-C	170.7	166.3	241.9	71.2	105.3	40.3	
10	F4-C	170.2	191.6	253.9	83.7	118.2	41.5	

†During the period of Nov. 11 to Apr. 24, 10.2 mm of rainfall and 86.9 mm of snowfall were recorded.
It is assumed that all the rainfall was directly supplied to the soil, whereas all the snowfall had been accumulated on Apr. 2 on the soil surface.

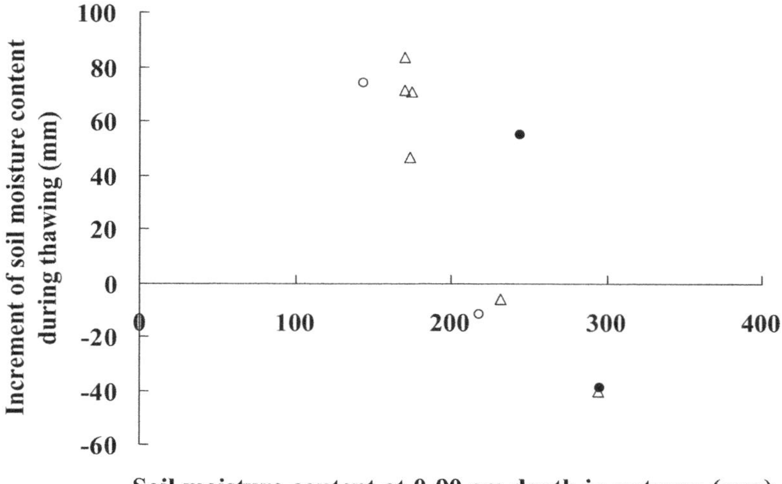

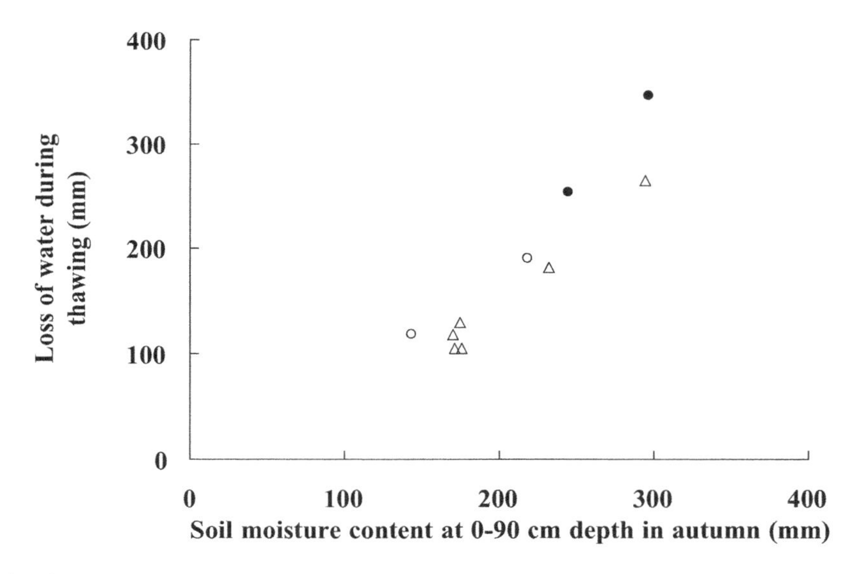

Figure 4. Relationships between soil moisture content in autumn and an increment (top) of soil moisture or loss (bottom) of water by evaporation and/or surface runoff during thawing. Accumulation of snow: ○ 100–150 mm, △ 150–200 mm, and ● >200 mm, respectively.

in April, the loss of soil moisture was not extensive during May, as was evidenced by the low evapotranspiration estimated for this period. Subsoil cutting was, therefore, effective in reducing evaporation loss during this period, though comparable data for non-treated soils were not available. Wheat or oats were seeded in late May in all plots except the fallow plots (Plots 1 and 6). Consequently, the evapotranspiration increased, and soil became drier during June and/or July. Total rainfall during the cropping periods was 142 mm in 1999 and 122 mm in 2000.

Table 5 summarizes the water budget during the cropping seasons of 1999 and 2000 and the data on crop yields. Compared to the soil moisture content at the time of seeding in May, it decreases drastically at harvest in September except for the fallow plots (Plots 1 and 6). The evapotranspiration

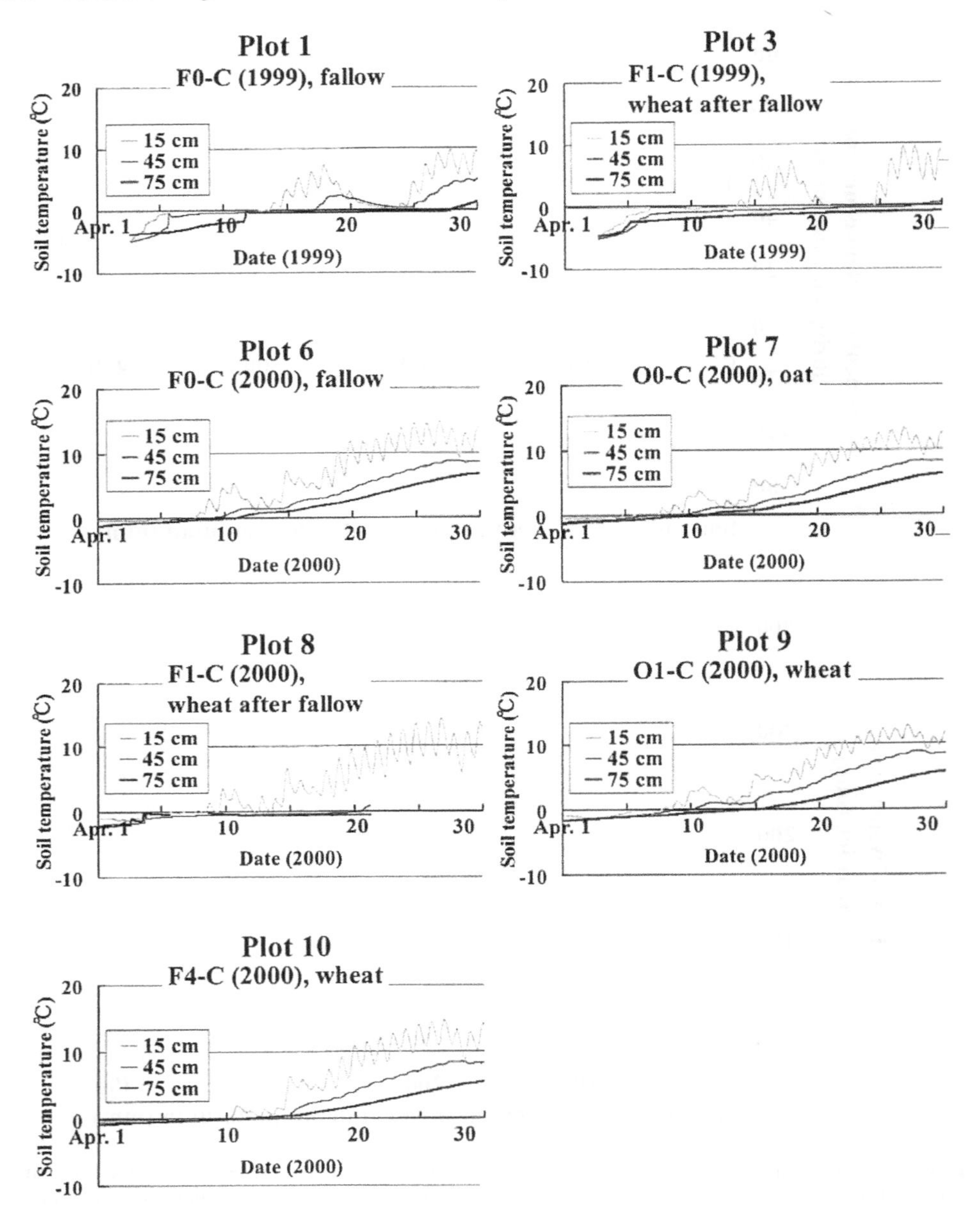

Figure 5. Fluctuations of soil temperature in April 1999 and 2000.

from the cropped plots was estimated between 194 and 259 mm (*d* in Table 5), equivalent to 1.67 and 2.31 mm/day, respectively. These values correspond to the upper limit of the amount of water that crops can use during the period. Although the crop yields were somewhat higher than those reported in farmers' fields for the whole Kazakhstan (i.e., 1.00 Mg ha^{-1} during 1986–90 and 0.65 Mg ha^{-1} during 1994–1996) (Gossen, 1998), presumably due to better management in the experimental field, a positive correlation between the evapotranspiration and the biomass or yield of wheat at the harvest time was obvious. Thus, crop yield was mainly determined by the amount of available water in the root zone. There was a 0.017 Mg ha^{-1} increase in yield for every 1 mm increase in water supply (i.e., estimated evapotranspiration) (Figure 6a), This increase in yield is similar to those reported for winter cereals ranging from 0.015 to 0.019 Mg ha^{-1} by Leggett (1959) and Cook and Veseth (1991). The relative contribution of the initial soil moisture to evapotranspiration ranged from 27 to 52%. Since the amount of precipitation in both cropping periods of 1999 and 2000 was not appreciably different, there was also a positive correlation between soil moisture content just before seeding and the crop yield or the biomass yield (Figure 6b).

The data in Figure 3 and Table 5 show the relative benefits of summer fallow to the accumulation of soil moisture, because the fallow plots (Plots 1 and 6) retained approximately 50 and 90 mm more water than the cropped plots at harvest time (*b* in Table 5), respectively. But even with fallow, 155 and 135 mm of water were already lost through evaporation in Plots 1 and 6, respectively, (*d* in Table 5), which exceeded the precipitation received during that period (142 and 122 mm). The difference in the evapotranspiration indicated that the fallow plots accumulated 39 to 104 mm more water in 1999 and 100 to 119 mm in 2000 than the cropped plots, respectively (*d* in Table 5). These values were comparable to the difference in moisture acquisition upon thawing under different conditions, that is, 114 mm in 1999 and 90 mm in 2000, which was affected by the soil moisture content in the preceding autumn.

The data shows that both summer fallow and snow management increase soil moisture storage by about 100 mm, but the benefit of snow management may be annulled by the effect of the summer fallow, since the moisture increase in autumn may decrease the water-capturing efficiency in the next spring through severe soil frost. Considering the possibly negative effect of the summer fallow on enhanced decomposition of SOM, it is recommended that snow management should be the main strategy of capturing water rather than by summer fallowing, at least from the viewpoint of water management.

4 SOM DYNAMICS UNDER GRAIN FARMING IN NORTHERN KAZAKHSTAN

Given the vast area of cereal production in northern Kazakhstan, it is important to maintain SOM for both environmental and agricultural reasons. A field study to provide information on SOM budget under cereal production was conducted in 2000 at the experimental farm of Barayev Kazakh Research and Production Center of Grain Farming, Shortandy, northern Kazakhstan. Dynamics of *in situ* soil respiration and microbial biomass, as well as soil environmental factors such as soil temperature and moisture, were analyzed for five plots (F0-C, O0-C, F1-C, O1-C, and F4-C) in the experimental farm in order to determine the SOM budget and the factors that affect the SOM decomposition rate. The crop species and land use practices for the experimental plots are summarized in Table 6. The F0-C plot was kept fallow in 2000. The O0-C and O1-C were included in the rotation system in which oat was substituted for summer fallow. Spring wheat was sown in the F1-C, O1-C and F4-C plots, whereas oats were sown in the O0-C plot. Since F1-C was the field just after fallow, it did not receive any residue input in the preceding year. The annual precipitation and average annual temperature in 2000 were 362 mm and 2.4°C, respectively (Table 2). Precipitation, recorded at the early stage of crop growth between May and June in 2000, was higher than the long-term average, which may account for the higher than average crop yield indicated in Table 6. In contrast, the temperature during summer (June, July, and August) in 2000 was similar to the long term average.

CO_2 emissions from the soil surface were measured in triplicate in these plots for 14 times at approximately two-week intervals between April and September, 2000. The confounding effect of temperature on fluctuations in CO_2 emissions within a day was minimized by using the alkali-trap method for estimating the daily respiration (Anderson, 1982). Simultaneous measurements were made for microbial biomass C and N in fresh soil samples from the surface 15-cm depth using the chloroform fumigation-extraction method (Brookes et al., 1985; Vance et al., 1987). Soil temperature at 5-cm depth and soil moisture for 0–30 cm depth were continuously monitored for each plot using a datalogger system (CR–10X, Campbell Scientific, Inc.). Plant biomass and grain yield were measured on a 1-m^2 subplot in triplicate at harvest.

Table 5-1. Water balance during the cropping season in 1999.

Plot No.	Plot	Soil moisture at 0–90 cm depth on May 21 (mm) a	Soil moisture at 0–90 cm depth on Sep. 14 (mm) b	Apparent decrease of soil moisture (mm) c = a–b	Evapo-transpiration (ET) (mm) d = †141.7 + a–b	Relative contribution of initial soil moisture to ET (%) c/d * 100	Yield (Mg ha^{-1}) e	Water use efficiency (kg ha^{-1} mm^{-1}) e/d * 1000	Remarks
1	F0-C	225.7	212.3	13.3	155.0	8.6	–	–	Fallow
2	F1-S	218.2	165.8	52.4	194.1	27.0	1.0	5.3	After fallow
3	F1-C	238.7	167.4	71.3	213.0	33.5	1.4	6.8	After fallow
4	F1-I	254.0	162.6	91.4	233.1	39.2	2.0	8.5	After fallow
5	O1-I	293.5	175.9	117.6	259.3	45.4	2.1	8.0	

†Rainfall during May 21–Sep. 13, 1999: 141.7 mm

Table 5-2. Water balance during the cropping season in 2000.

Plot No.	Plot	Soil moisture at 0–90 cm depth on May 31 (mm) a	Soil moisture at 0–90 cm depth on Sep. 18 (mm) b	Apparent decrease of soil moisture (mm) c = a–b	Evapo-transpiration (ET) (mm) d = †121.6 + a–b	Relative contribution of initial soil moisture to ET (%) c/d * 100	Yield (Mg ha^{-1}) e	Water use efficiency (kg ha^{-1} mm^{-1}) e/d * 1000	Remarks
6	F0-C	264.7	251.1	13.6	135.2	10.1	–	–	Fallow
7	O0-C	295.5	162.7	132.8	254.4	52.2	*3.2	12.6	
8	F1-C	268.6	162.8	105.9	227.5	46.5	1.4	6.0	After fallow
9	O1-C	284.5	164.5	120.0	241.6	49.7	1.9	7.7	
10	F4-C	277.0	163.4	113.6	235.2	48.3	2.0	8.6	
11	CW-I	300.4	167.9	132.5	254.1	52.2	2.3	9.0	
12	PI-I	271.1	165.7	105.4	227.0	46.4	1.7	7.5	

†Rainfall during May 31–Sep. 18, 2000: 121.6 mm
*Yield of oats

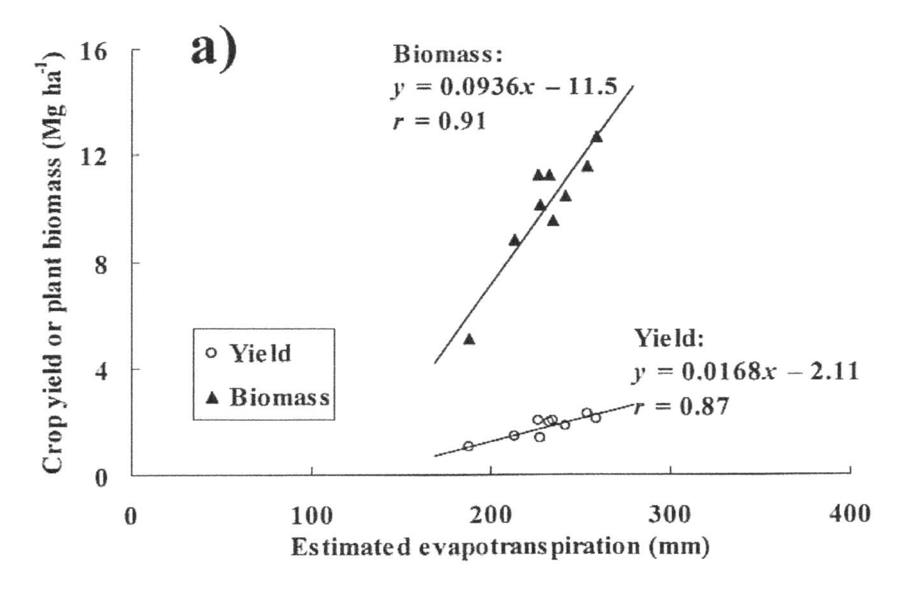

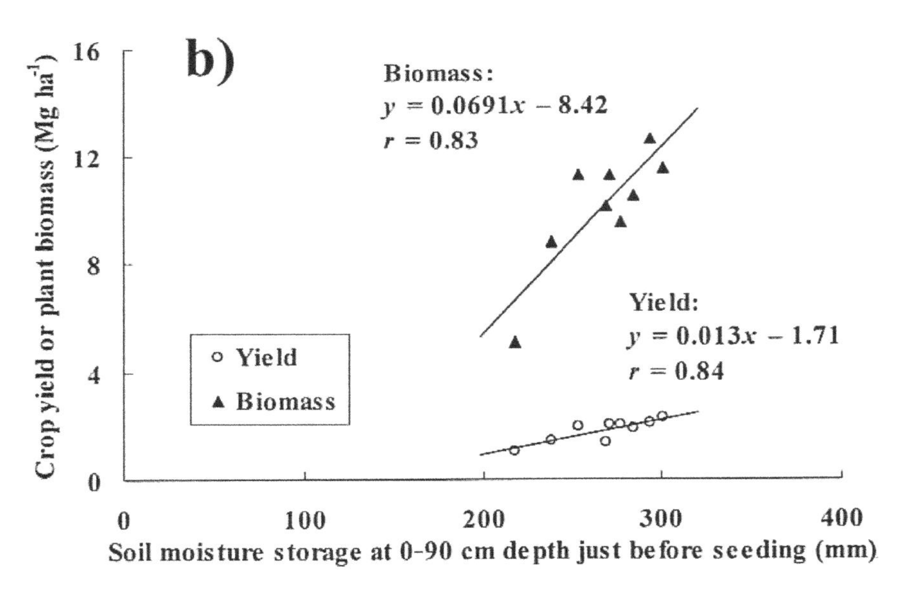

Figure 6. Relationships between crop yield or biomass and a) estimated evapotranspiration based on the water budget or b) soil moisture storage just before seeding in spring.

4.1 *Fluctuations of soil temperature, soil moisture content, soil respiration rate, and soil microbial biomass*

Figure 7 shows the fluctuations in soil temperature and soil moisture content during the experiment. Some data are missing due to mechanical treatment of the field at the time of seeding (May 25–June 7 in all the plots) and malfunction of the datalogger (July 4–19 in F4-C). Mean daily soil temperature increased to above 0°C in early April and remained >20°C from mid-June to mid-August. Then it sharply decreased to <5°C at the end of September. Most of the biological activities were

Table 6. Description of study plots for SOM dynamics.

Plot	Crop rotation system[1)2)]	Overall crop yield at the same stage of rotation: 2000 (Mg ha^{-1})	average in 1990–1999 (Mg ha^{-1})	Remarks
F0-C	**F**-W-W-B-W	0	0	Summer fallow
O0-C	**O**-W-W-B-W	3.01	1.82	Fallow in conventional rotation was substituted by oat cultivation
F1-C	F-**W**-W-B-W	1.52	1.63	1st year after fallow
O1-C	O-**W**-W-B-W	1.90	1.34	1st year after oats in modified rotation
F4-C	F-W-W-B-**W**	2.09	1.29	4th year after fallow

1) F: fallow, W: wheat, B: barley, O: oats.
2) Bold letter denote the cropping stage of each plot.

apparently limited between April and September. On the other hand, the soil moisture content in the surface layers remained high after thawing until mid-June. However, it continuously decreased in the cropped plots except during the rainfall events. In contrast, soil moisture content in the fallow plot (F0-C) remained at a certain level after July because of the lack of transpiration by plants. Soil moisture content of 0.18 L L^{-1} was equivalent to the permanent wilting point (-1.5 MPa) of the soils, based on data on the moisture retention curves (unpublished data). Hence, the cropped soils were subjected to extremely dry conditions during late summer (late June to August), along with a strong increase in soil temperature. These data were used for the calculation of annual CO_2 emissions.

Figure 8 shows *in situ* CO_2 emissions, (i.e., the soil respiration rates) during the 2000 cropping period. Maximum values of soil respiration were recorded on June 24 or July 4, during which soil was still moist despite the high temperature of >20°C, followed by decrease as soil progressively dried. The overall trend of CO_2 fluctuations, however, was similar to that of the soil temperature.

The data in Figure 9 show that concentrations of microbial C and N in soils were high during early summer followed by drastic decrease, indicating trends similar to those of the soil moisture content. Indeed, there was a highly positive correlation between soil moisture content and microbial biomass except for the fallow plot (F0-C), in which repeated plowing during the summer may have interfered with such a direct relationship (Figure 10).

The difference in the fluctuation patterns between soil respiration rate and the amount of soil microbial biomass indicated a unique dynamics in soil respiration rate / microbial biomass C content, which apparently increased during the late summer (Figure 11). These trends implied that, despite the decrease in the microbial biomass due to the very dry conditions, some microorganisms were still active and contributed to soil respiration, possibly by using dead microbial debris as additional substrates.

4.2 *Estimation of CO_2 emissions throughout the cropping period using the measured data of soil temperature, moisture content, and soil respiration rate*

Estimates of the total soil respiration rate throughout the cropping period were made by deriving an empirical equation to represent the relationship between the *in situ* daily soil respiration rate and climatic factors (e.g., soil temperature and moisture content) by multiple regression analysis. The daily soil respiration rate was calculated by substituting each parameter of the equation by using the monitored data, and summing up the daily soil respiration rates for a given period. In the first

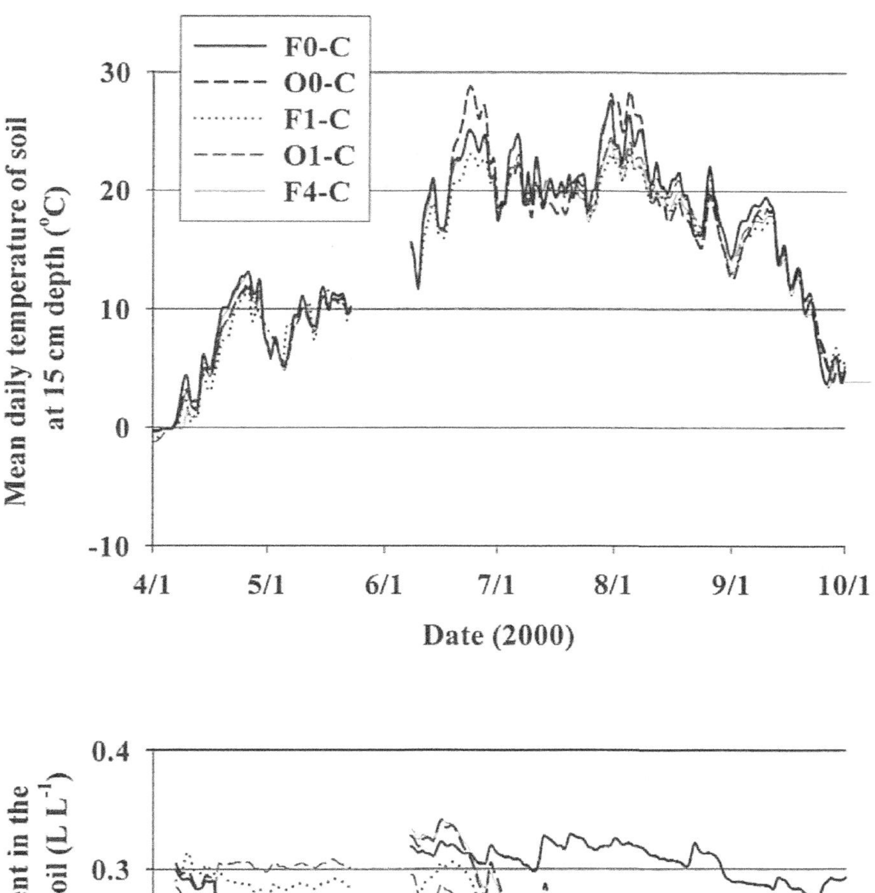

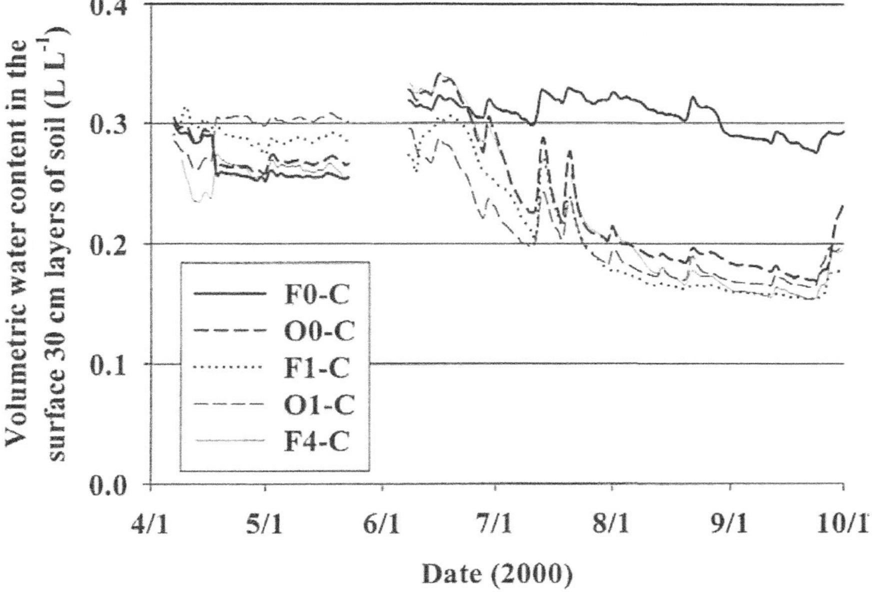

Figure 7. Fluctuations of soil temperature and soil moisture content during the experiment, measured by datalogger.

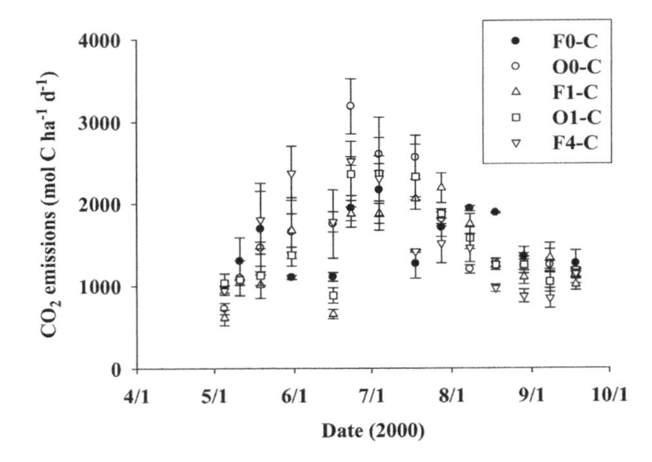

Figure 8. CO_2 emissions from the soil surface of the experimental plots.

step, we assumed that the Arrhenius relationship between the soil temperature and respiration rate was assumed valid as follows (Eq. 1):

$$C_{\text{em}} = aM^b e^{-E/RT} \tag{1}$$

where C_{em} is the daily soil respiration rate (mol C ha^{-1} d^{-1}), M is the volumetric soil moisture content (L L^{-1}), E is the activation energy (J mol^{-1}), R is the gas constant (8.31 J mol^{-1} K^{-1}), T is the absolute soil temperature (K), b is a coefficient related to the contribution of soil moisture, and a is a constant. The first order kinetic model, in which the direct proportion of the decomposition rate and the amount of substrates at a given time is assumed, has been most widely used to simulate laboratory data for SOM decomposition. However, we applied the 'zero-order' model in which a fixed amount of substrates is assumed throughout the process of decomposition, because of the possibility of the existence of an additional source of substrates caused by the fluctuations of the temperature and moisture content during the field experiment. The equation was then rewritten in the logarithm form (Eq 2):

$$\ln C_{\text{em}} = \ln a + b \ln M - E/RT \tag{2}$$

Then a series of coefficients (a, b, and E) were calculated by stepwise multiple regression analysis ($p = 0.25$) using the measured data, C_{em}, M, and T (SPSS Inc., 1998a).

The data in Table 7 show that a significant relationship at 1 or 5% level was obtained between the soil respiration rate and the activation energy, E, indicating a significant dependency of the soil respiration rate on the soil temperature. Based on the value of E, the Q_{10} values estimated from 10 to 20°C ranged between 1.3 and 2.0. In contrast, the contribution of moisture content was somewhat uncertain, except for the O0-C and F4-C plots, based on the fact that the moisture parameter was rejected even at the level of $p = 0.25$ in the stepwise regression. In some cases (especially O1-C), the r^2 value was unexpectedly low, presumably because the short-term effects of surface soil disturbance on seeding and harrowing, and occasional rainfall events during the dry summer may have been neglected.

The data showed a positive impact of soil temperature on the soil respiration rate. Although the reasons why soil moisture did not appreciably affect soil respiration rate are unknown, contribution of some soil microorganisms that were still active under severely dry conditions after summer may be an important factor (Figure 11). Such a dependency of soil respiration rate mainly on soil temperature was also reported for Chernozem soils in Russia by Kudeyarov and Kurganova (1998) and in Argentina by Alvarez et al. (1995). Using these regression equations and the data monitored by the dataloggers, the fluctuations in soil respiration rate during the cropping season were simulated (Figure 12), and cumulative soil respiration from April 10 to October 3 was calculated to be 2.9

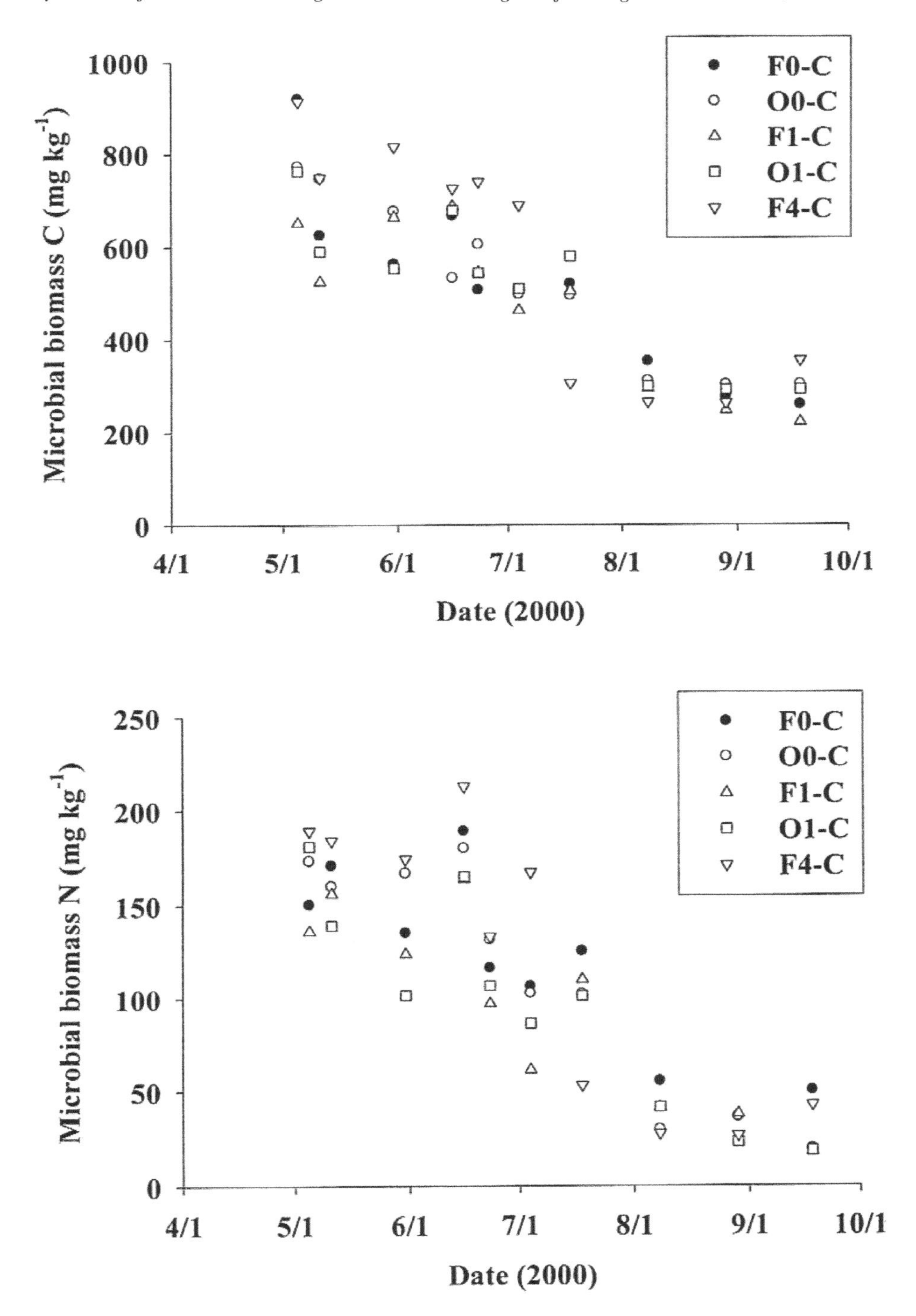

Figure 9. Fluctuations of microbial biomass C and N contents in the surface soils.

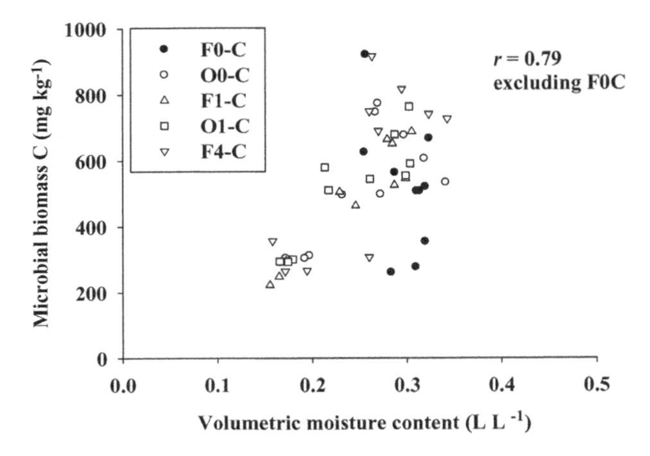

Figure 10. Relationship between the volumetric water content of soils and the microbial biomass C content.

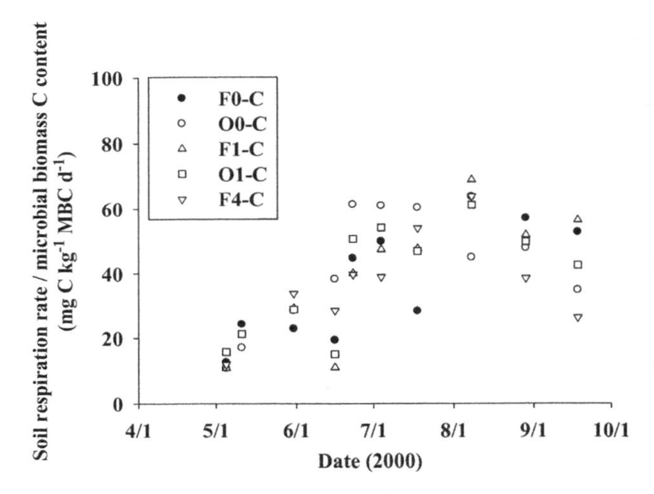

Figure 11. Fluctuations of the values of soil respiration rate / microbial biomass C content during the experiment.

(F0-C), 3.2 (O0-C), 2.5 (F1-C), 2.8 (O1-C), and 3.1 (F4-C) Mg C ha^{-1}. Since the monthly trend of air temperature during summer time in this year was similar to that of the long-term average (Table 2), the calculated values represent the conditions in normal years. The lower value of soil respiration rate estimated in the F1-C plot, just after summer fallow, than in others suggested the possible depletion of readily decomposable SOM due to the absence of crop residues in the preceding year. Excluding the fallow plot, F0-C, the average value of the remaining three cropped plots (O0-C, O1-C, and F4-C), which received crop residues at least in the preceding year, was 3.0 Mg C ha^{-1} during the cropping phase.

4.3 *General discussion on soil carbon budget in the experimental farm*

Although soil respiration was associated with both SOM decomposition by soil microorganisms and plant root respiration, the results obtained practically excluded a large part of root respiration due to the removal of nearby plant materials during the measurements. Table 8 summarizes the

Table 7. Coefficients determined by stepwise multiple regression analysis.

Site	Coefficients					$C_{em} = aM^b e^{-E/RT}$ (at $T = 298K$, $M = 0.2\,L\,L^{-1}$) ($mol\,C\,ha^{-1}\,d^{-1}$)	Cumulative CO_2 emission from Apr. 10 to Oct. 3 ($Mg\,C\,ha^{-1}$)
	ln$a$	b	E ($kJ\,mol^{-1}$)	r^2	n		
F0-C	17.5 ***	−2.01 *	30.6 ***	0.53 ***	13	4485	2.92
O0-C	22.0 ***	0.68 **	33.0 ***	0.59 ***	13	2014	3.19
F1-C	26.5 ***	–	46.6 ***	0.65 ***	12	2197	2.52
O1-C	19.0 ***	–	28.4 **	0.42 **	13	1927	2.76
F4-C	17.3 ***	0.90 ***	20.9 **	0.69 ***	10	1634	3.06

*, **, ***: Significant at 25%, 5%, and 1% levels, respectively.
$C_{em} = aM^b e^{E/RT}$
C_{em}: Rate of CO_2 emissions ($mol\,C\,ha^{-1}\,d^{-1}$)
T: Soil temperature (K)
M: Volumetric moisture content of soil ($L\,L^{-1}$)
a: Coefficient ($mol\,C\,ha^{-1}\,d^{-1}$)
E: Activation energy ($J\,mol^{-1}$)
R: Gas constant (8.31 $J\,mol^{-1}\,K^{-1}$)
This equation is converted to; $\ln C_{em} = \ln a + b \ln M - E/RT$

data on SOM budget in the experimental plots. Crop yields, which were primarily determined by the amounts of available water during the growing season, were 3.2 (O0-C, oat), 1.9 (F1-C), 1.4 (O1-C), and 2.0 (F4-C) $Mg\,ha^{-1}$. After harvest, 4.5 (O0-C), 2.3 (F1-C), 1.6 (O1-C), and 2.6 (F4-C) $Mg\,C\,ha^{-1}$, respectively, were returned to the soils as plant residues. Assuming that all of the soil respiration determined here was caused by the SOM decomposition, budget of the SOM pool was −2.9 (F0-C), 1.3 (O0-C), −0.2 (F1-C), −1.2 (O1-C), and −0.5 (F4-C) $Mg\,C\,ha^{-1}$. Except for the plot sown to oats (O0-C), in which an exceptionally higher residue biomass of oats than that of wheat contributed positively to the budget, SOM budget in the cropped plots was slightly negative for this year, that is, the soils lost their SOM pool. This trend may be more conspicuous during an average year since the crop yield in 2000 was considerably higher than the 10-year average (Table 6). In the fallow plot (F0-C), the SOM loss was much higher than that in the cropped plots because of the lack of residue input.

A significant relationship was observed between the amount of available water and wheat production in the same plots (section 3). Using the same data, the following relationship was established between the amount of evapotranspiration (*ET* in mm) and organic C content of the wheat residues (*CR* in $Mg\,C\,ha^{-1}$) (Eq. 3):

$$CR = 0.0201 \quad ET - 2.43 \qquad r^2 = 0.48,\ n = 9 \tag{3}$$

In order to obtain the wheat residues that could compensate for the CO_2 emissions, namely 3.0 $Mg\,C\,ha^{-1}$ in the corrected average of the present study or 2.0 $Mg\,C\,ha^{-1}$ under the assumption that one third of the CO_2 emissions in the present study was derived from root respiration, approximately 270 or 220 mm of water would be required for evapotranspiration, respectively (Eq. 3). Even in research farm where land and water management were optimal, the maximum value of evapotranspiration was 259 mm under intensive snow management in 1999/2000 (section 3). Thus, it would be difficult for farmers to secure 220 to 270 mm of available water for wheat production. In addition, burning of cereal husks in spring and/or cattle grazing, which were occasionally practiced in farmers' fields, would further reduce the input of crop residue into soils. Since the amount of potentially mineralizable C (PMC) of the surface 15-cm soils in these field, which was determined by application of the first order kinetic model for the dataset of the laboratory incubation experiment of fresh soils for 133 d under constant conditions (temperature and gravimetric soil moisture fixed to 30°C and 60%, respectively), was $5.44 \pm 0.14\,Mg\,C\,ha^{-1}$ soil ($n = 4$) and was significantly

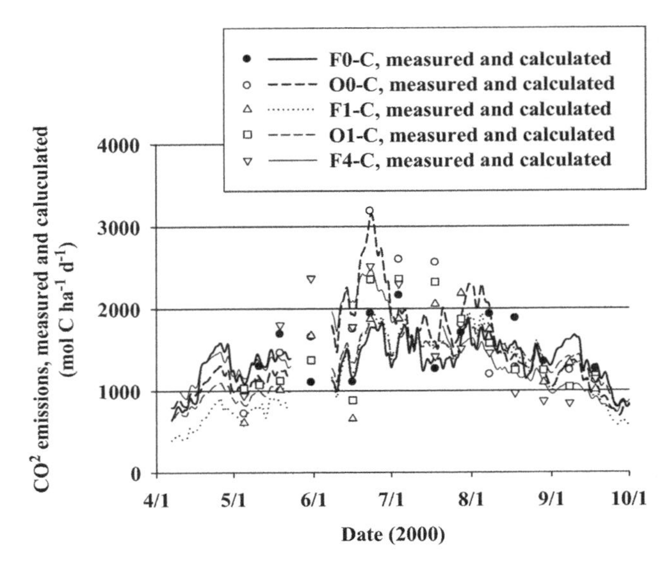

Figure 12. Estimation of CO_2 emissions throughout the cropping season using the regression equations obtained in Table 11.

Table 8. Budget of soil organic carbon during the period of April 10 to October 3, 2000, in the experimental field of Shortandy.

Site	Cumulative CO_2 emission from Apr. 10 to Oct. 3 (Mg C ha^{-1})	Crop yield (Mg ha^{-1})	Plant residues (Mg C ha^{-1})	Budget of SOM (Mg C ha^{-1})
F0-C	2.92	–	–	−2.92
O0-C	3.19	3.20 (0.19)	4.46 (0.27)	1.27
F1-C	2.52	1.86 (0.07)	2.28 (0.15)	−0.24
O1-C	2.76	1.36 (0.12)	1.60 (0.08)	−1.16
F4-C	3.06	2.03 (0.03)	2.61 (0.04)	−0.45

* Figures in parenthesis denote standard error.

higher than that in the nearest farm in Shortandy (2.72 ± 0.13 Mg C ha^{-1}, $n = 70$; Table 12 in section 6), both the C input and output as well as mineralizable pool of SOM in farmers' fields were expected to be lower than those in the present study. Although it is difficult to generalize the data on C budget among different years because of the large variations in crop growth due to the fluctuations in annual precipitation, the disadvantage of summer fallow is obvious from the data of SOM budget. The annual loss of SOM in the fallow plot (F0-C) of 2.9 Mg C ha^{-1} was approximately equivalent to 4% of the total SOM pool in the 30-cm plow layer (70 to 80 Mg C ha^{-1}).

In order to reduce the SOM loss, an extensive use of summer fallow must be reconsidered. Intensive snow management is an alternative approach to improve soil moisture conditions for some topographical locations. Since the results of the present study were obtained on a strictly managed experimental farm, it is still necessary to determine the actual relationship between the topographical characteristics and the possible water management or C dynamics. A general conclusion from this study, namely that soil respiration was mostly controlled by soil temperature while residue input was a function of moisture conditions, is useful to the development of an appropriate land use system in accordance with the topographical characteristics which would enable high yields while reducing the net C emission into the atmosphere.

5 SOM STATUS OF CHERNOZEMS IN NORTHERN KAZAKHSTAN: EFFECTS OF SUMMER FALLOW

Almost 50 years of monoculture of wheat has been practiced in northern Kazakhstan with summer fallowing practiced in crop rotation to store soil moisture, accumulate nutrients through mineralization, and control weed infestation. Fallowed fields are cultivated many times to keep the land bare during the whole cropping season. The adverse effects of following on SOM quality and quantity in relation to degradation of the fertility of Chernozem soils and subsequent agricultural sustainability must be addressed.

Ferguson and Gorby (1971), Clarke and Russell (1977), Dormar (1983), and others have demonstrated that fallowing significantly exacerbates the depletion of SOM. Janzen (1987) reported that the organic C and N contents of soil after 33 years of cropping decreased with increase in frequency of fallow in a rotation on Canadian soils. Rubinstein (1959) reported that the Southern Chernozems (Kazakh soil classification; Redkov, 1964) of northern Kazakhstan lost 11% of its antecedent SOM pool. Dzhalankuzov and Redkov (1993) reported 28–30% loss of the antecedent humus in the surface horizon of arable Chernozems of northern Kazakhstan.

The mineralization rate of SOM varies among crop rotations. K.A. Akhmetov (unpublished Ph.D. thesis, Kazakh Research Institute of Grain Production, 1999) reported that use of summer fallow in a rotation hastens decomposition of SOM and that differences in SOM among various crop rotations are mainly due to differences in the amount of plant residues returned to the soil.

The SOM is highly heterogeneous, consisting of fractions varying in turnover time from days to many centuries. Gregorich et al. (1994) reported that more than 75% of SOM exists as slowly decomposable compounds and the remainder is readily decomposable or 'mineralizable'. The amount of organic C contained in a soil is a function of the balance between the rate of deposition of plant residues in or on the soil and the rate of mineralization of the residue C by soil biota (Baldock and Nelson, 2000). Operationally defined fractions such as C and N mineralized under controlled conditions and 'light' fraction are good indicators of labile SOM because it affects nutrient dynamics within single growing seasons, organic matter content in soils under contrasting management regimes, and C sequestration over extended periods. Quality of SOM may also be characterized by estimates of kinetically-defined pools obtained by fitting simulation models to data on C and N mineralization (Elliott et al., 1996).

The SOM of less stable pools is decomposed with increase in cultivation intensity, as is indicated by decreasing portions of sand-sized SOM (2–0.05 mm) (Bird et al., 1996; Christensen, 1996; Amelung et al., 1998), or the so called light fraction C (Christensen, 1992; Trumbore et al., 1996). Organic compounds adsorbed to surfaces of clay particles may also become exposed to microbial attack after disruption of aggregates due to tillage.

Whereas there are numerous reports regarding the effects of wheat-fallow rotations on total SOM pool, the influence of summer fallow on mineralizable fractions has not been widely studied in semiarid regions of northern Kazakhstan. Thus, the objectives of the study were to examine the effects of summer fallow on the characteristics of SOM on a long-term (type of crop rotation with a range of frequencies of fallow) and a short-term bases (pre- and post-fallow phases) with special reference to readily decomposable fractions.

The crop rotation experiment including different frequencies of fallow was conducted at Barayev Kazakh Research and Production Center of Grain Farming, Shortandy. The experimental site was initially cultivated in 1933 and a variety of wheat-fallow crop rotation systems have been practiced since 1961. The local cultivars of spring wheat 'Tselinnaya 3C' were seeded at the rate of 125 kg ha^{-1}. Five representative cropping systems studied included spring wheat rotations with different frequencies of fallow; 6-y (6R), 4-y (4R), 2-y (2R) rotations, continuous wheat (CW) and continuous fallow (CF) systems. Soil samples were collected from pre- (2R-pre, 4R-pre and 6R-pre) and post-fallow (2R-post, 4R-post and 6R-post) phases in each rotation. One half of the composite sample was air-dried, and the remaining half was stored in field-moisture condition at 4°C for subsequent biological analysis. Grain yield of wheat was determined by using combine harvester 'SAMPO' (Finland), and each plot was harvested separately and the grain yield determined. A quantitative-weighing method was applied to assess the weed biomass (Lykov and Tulikov, 1976).

The air-dried soils were ground and analyzed for organic C concentration by the dichromate oxidation method (Nelson and Sommers, 1996), total N concentration (Shimadzu NC–800–13N), and mineral N (min-N) as NO_3^- and NH_4^+ ions (after extraction with 2 M KCl solution). The 'light fraction' organic matter (LF-OM) was determined after density separation using reagent-grade NaI solution adjusted to density of 1.8 g cm^{-3} (Spycher et al., 1981; Sollins et al., 1984; Janzen et al., 1992; Elliot et al. 1996). The fresh soils were assayed for labile SOM content using laboratory incubation techniques with a constant temperature of 30°C and moisture content at 50% of the water holding capacity (WHC) for 70 days. Potentially mineralizable C (PMC) was estimated from the rate of CO_2-C evolution during 70 days of incubation using nonlinear regression according to the following equation (SPSS Inc., 1998a) (Eq. 4):

$$C_{\min} = C_0(1 - e^{-kt}) \tag{4}$$

where, $C_{\min}$ is the quantity of mineralized C (mg kg^{-1} dry soil) at time t (d), C_0 is PMC (mg kg^{-1} dry soil), and k is a nonlinear mineralization rate constant, i.e., fraction mineralized d^{-1}. $N_{\min}$ was also determined after incubating soils for 14-, 28-, 42-, 56-and 70-d and analyzed for NO_3^-– and NH_4^+–N. Non-linear regression was used to describe N mineralization potential (PMN) according to the following equation (SPSS Inc., 1998b) (Eq. 5):

$$N_{min} = N_0(1 - e^{-k(t-c)}) \tag{5}$$

where, $N_{\min}$ is the quantity of mineralized N (mg kg^{-1} dry soil) at time t (d), N_0 is PMN (mg kg^{-1} dry soil), k is a nonlinear mineralization rate constant, (i.e. fraction mineralized d^{-1}), and c is an initial delay in mineralization (mg kg^{-1} dry soil). Because mineralization of N in the first 2 weeks was delayed for all the treatments, the initial delay factor c was introduced in the first order kinetic model for the best fit of the model.

5.1 *SOM status under different frequency of summer fallow*

Soil organic carbon (SOC) and total N (TN) concentrations: The SOC concentration was significantly affected by long-term fallow. The SOC concentration was the lowest in the CF system (21.9 kg Mg^{-1}), and the highest in 6R (31.0 kg Mg^{-1}) and CW (27.2 kg Mg^{-1}) treatments (Table 9). The SOC concentration was inversely proportional to the fallowing frequency, indicating the negative effect of fallow on long-term accumulation of SOM. The effect of the rotations on TN paralleled that of the SOC (Table 9). The highest TN concentrations were observed in the 6R (2.54 kg Mg^{-1}) and CW (2.38 kg Mg^{-1}) systems, and the lowest in the CF system (1.97 kg Mg^{-1}).

Potentially mineralizable C: Differences in PMC among the rotation systems ($p < 0.001$) were clearer than those for SOC concentration (Table 10). The PMC ranged from 3.6 (CF) to 5.8% (CW) of the SOC. The amount of PMC was impacted more by the long-term effect of fallowing than by the short-term effect and was inversely proportional to the frequency of following.

Soil mineral N: On a long-term basis, the CF system accumulated the highest amount of soil mineral nitrogen (min-N). However, the min-N was also strongly affected by summer fallowing on a short-term basis. Pre- and post-fallow phases had significant differences, with more min-N accumulating in post-fallow than in pre-fallow phases (Table 10). Post-fallow phases accumulated 3.0-, 1.9- and 1.9-fold amounts of min-N than the pre-fallow phases in 2R, 4R and 6R, respectively.

Potentially mineralizable nitrogen: The pattern of N mineralization showed a different trend between pre- and post-fallow phases in all rotations (Figure 13). Pre-fallow phases (Figure 13a, c and e) were characterized by a larger value of PMN (N_0), a smaller mineralization rate constant (k), and a shorter initial delay of mineralization (c) than in the post-fallow phases (Figure 13b, d, and f).

Table 9. Effects of fallow (F) frequency and rotation phase on soil organic C (SOC) and total N (TN) in surface soil of Southern Chernozem.

Rotation phase	Rotation phase, sampled[2]	SOC (kg Mg^{-1} soil)	TN (kg Mg^{-1} soil)	C-to-N ratio
CF	Cont. Fallow	21.9a[1]	1.97a	11
2R-pre	(F)-W	25.4b	2.26b	11
2R-post	F-(w)	25.1b	2.16b	12
4R-pre	(F)-W-W-W	26.1b	2.26b	12
4R-post	(F)-(W)-W-W	24.9b	2.19b	11
6R-pre	(F)-W-W-W-W-W	31.0c	2.57c	12
6R-post	F-(W)-W-W-W-W	30.6c	2.50c	12
CW	Cont. Wheat	27.2c	2.38c	13

[1] a-c: values within columns followed by the same letter are not significantly different ($p = 0.05$) as determined by LSD analysis.
[2] ( ) denotes rotation phase sampled.

Table 10. Effects of fallow frequency and rotation phase on labile fractions of SOM in surface soil of Southern Chernozem.

	Mineralizable C and N			Light fraction' OM		
Rotation phase	PMC (mg kg^{-1} soil)	min-N (mg kg^{-1} soil)	PMN (mg kg^{-1} soil)	LF-DM (mg kg^{-1} soil)	LF-C (mg kg^{-1} soil)	LF-N (mg kg^{-1} soil)
CF	794b[1]	46a	69a	0.9	240a	15a
2R-pre	1194ab	14b	166b	3.6	810b	51b
2R-post	1012b	42a	69a	2.5	660b	38bc
4R-pre	1224a	13b	86c	5.7	1330c	81d
4R-post	1215a	24c	67a	5.3	1250c	73d
6R-pre	1524c	16b	124b	6.4	1560d	74d
6R-post	1300ac	30c	82c	6	1500d	75d
CW	1581c	14b	93c	7.4	1730c	103c

[1] a-c: values within columns followed by the same letter are not significantly different ($p < 0.001$) as determined by LSD analysis.

Fallowing influenced accumulation of PMN on a short-term basis, that is, the pre-fallow phases (2R-pre, 4R-pre and 6R-pre) accumulated more PMN than post-fallow (2R-post, 4R-post and 6R-post) phases (Table 10). The lowest PMN was observed under the CF system (69 mg kg^{-1}) and the highest under 6R-pre (124 mg kg^{-1}). Pre-fallow phases accumulated 2.4-, 1.3- and 1.5-fold amounts of PMN compared to the post-fallow phases in 2R, 4R and 6R, respectively.

'Light fraction' organic matter: The amount of LF-OM was highly responsive to fallowing frequency, accounting for 1.1(CF)–6.3(CW) % of the SOC and 0.8(CF) –4.3(CW) % of the TN (Table 11). The LF-OM, expressed on the dry matter basis (LF-DM), C (LF-C) or N (LF-N), was inversely related to the fallowing frequency. For example, the LF-C content of the CW system was 7.2-fold higher than that in the CF system. These results are in accord with other studies (e.g., Janzen et al., 1992; Haynes, 2000), where LF content was the highest under continuous cropping and the lowest under a high frequency of summer fallowing. Additionally, LF-C was affected by the rotation phase, showing larger amounts in pre- than in post-fallow phases in 4R and 6R rotations.

Weed biomass and grain yields: Weed biomass was linearly proportional to the duration of a rotation (Table 11). Average yearly inputs of weed biomass were 132 in 2-y, 155 in 4-y, 162 in

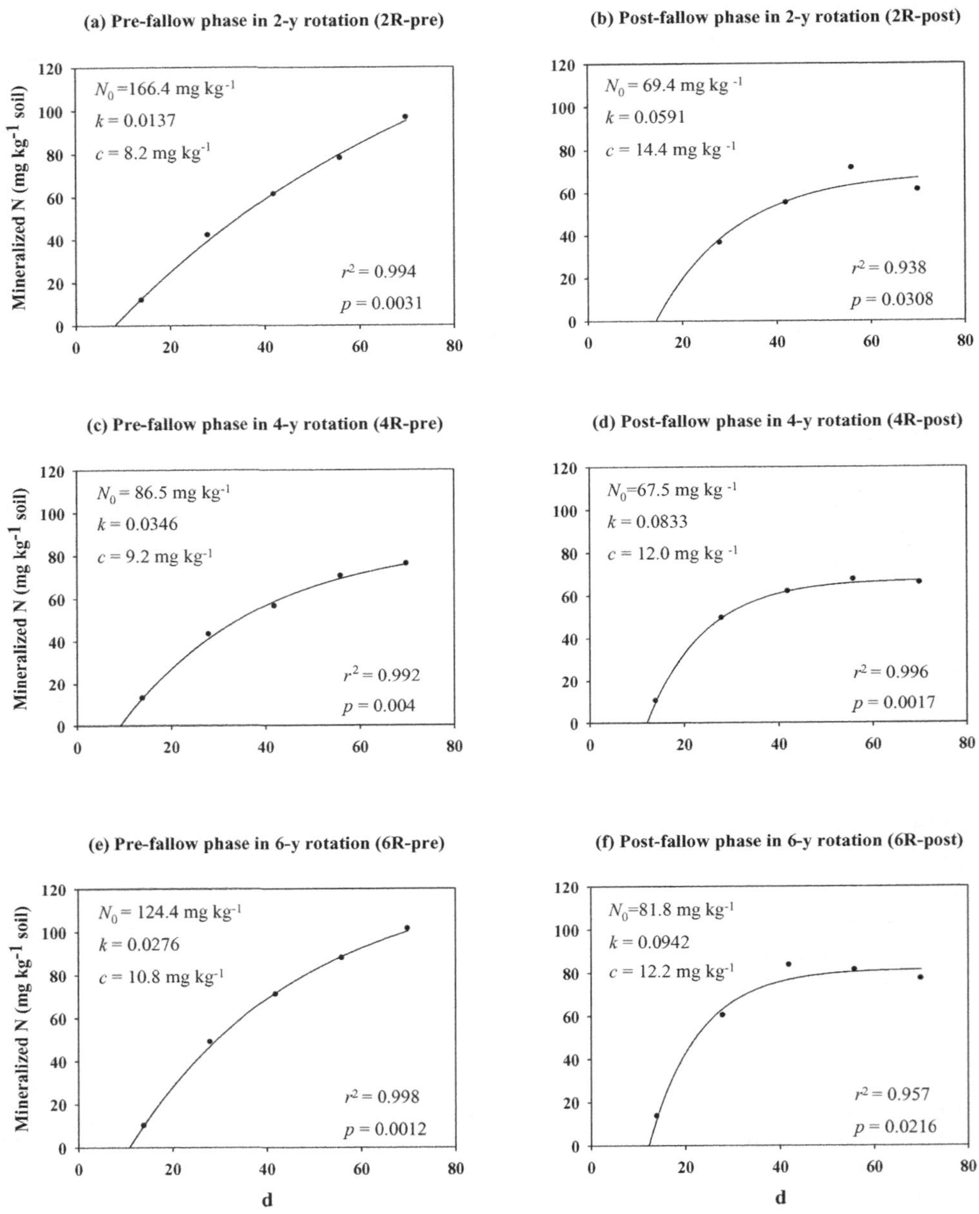

Figure 13. Fitting curves of N mineralization of surface soils from pre- and post-fallow phases of the 2-, 4-, and 6-y wheat-fallow rotations in Southern Chernozem, as described by the first order kinetic model with an initial delay of mineralization ($N\min = N0(1 - e - k(t - c))$, where $N\min$ is mineralized N at time t, $N0$ is potentially mineralizable N (PMN), k is a mineralization rate constant, and c is an initial delay in mineralization).

6-y and 869 kg ha^{-1} per cropping year in CW plots in 1997. In contrast, the grain yields were negatively proportional to the weed biomass and to the duration of rotation. The largest weed infestation was observed in the CW system, where weed biomass (dry matter) exceeded the grain yield by a factor of 1.8.

Grain yield in 1998 was low in all the rotations (510, 520, 540 and 490 kg ha^{-1} y^{-1} for 2-y, 4-y, 6-y and CW plots, respectively (Table 11, K.A. Akhmetov, 1999, local citation). Although low grain yield is common for the region, where lack of water is a main limiting factor for wheat growth, the 1998 season was extremely dry. Lack of application of fertilizer and herbicide also affected grain yields, which was undertaken to exclude all other factors except the summer fallows.

5.2 *Effect of summer fallow on SOM status*

Soil organic C and total N: To protect the fields against weeds and to store more moisture and nutrients in the soil, fallowed field are normally cultivated 4 to 5 times during the growing season. Such intensive mechanical disturbance enhances mineralization of SOM in fallow due to better aeration of surface soil, and exposing organic matter occluded within aggregates to microbial attack upon disruption of aggregates. Additionally, bare fallow does not contribute plant residues for the replenishment of SOM.

In general, distributions of SOC and TN among rotations with different fallow frequencies were comparable to those reported by Collins et al. (1992), Campbell and Zentner (1993) and Biederbeck et al. (1994) for Chernozem soils. Frequently fallowed systems (such as 2R) contained less SOM than less frequently fallowed systems (such as 6R). These results confirmed the findings from North American arable systems that frequent fallowing systems accelerate mineralization of SOM (Janzen, 1987; Campbell and Zentner, 1993; Biederbeck et al., 1994). *Potentially mineralizable C:* Continuous wheat (CW) and 6-y systems (6R) had higher amount of PMC that was inversely proportional to fallowing frequency, and also indicated the long-term effect of fallowing. These results corroborate the study of Campbell et al. (1999) who observed for a silt-loam in southwestern Saskatchewan that mineralized C (measured after 30 days at 21°C) represented 1.06 and 1.45% of SOC in a 2-y fallow-wheat rotation and continuous growing of wheat, respectively. Campbell et al. (1992) observed that C mineralization was not related to the amount of crop residue from the previous year. In the present study, PMC was slightly higher in the pre-fallow (2R-pre, 4R-pre and 6R-pre) than in the post-fallow (2R-post, 4R-post and 6R-post) phases, probably reflecting the input of crop and weed residues during the preceding year (Table 10). *Soil mineral N:* The CF system maintained the highest amount of soil min-N probably due to enhanced mineralization of SOM compared to the other systems. The short-term effect of fallowing on the accumulation of min-N is also evident. During fallow phase, min-N is not subjected to either plant uptake or leaching, thus resulting in a greater accumulation of soil min-N in post-fallow (2R-post, 4R-post and 6R-post) than in pre-fallow phases (2R-pre, 4R-pre and 6R-pre). *Potentially mineralizable N:* Larger amounts of mineralized nitrogen (N_0) in the pre-fallow phases indicate larger storage of PMN in these soils than in post-fallow soils. Differences in the rate constant (k) between pre- and post-fallow phases indicate that fallowing causes changes in the quality of the PMN.

Due to multiple cultivations during fallowing the soil is subjected to alternating wet-dry cycles. The wet period creates better moisture conditions for microorganism activity and produces greater biomass than in cropped fields. In the subsequent dry period, the greater biomass turned into necromass due to drought. This cycle may be repeated several times during a cropping season. And later, during incubation in the laboratory, the microbial necromass and living biomass were rapidly mineralized, showing a higher mineralization rate constant in the post fallow than in the pre-fallow phases (Figure 13).

Soils from the post-fallow phase had a longer initial delay of mineralization, suggesting that higher concentration of min-N compared to pre-fallow phase probably stimulated microbial activity and resulted in immobilization of mineralized N during the initial stages of incubation (Mamilov et al., 1985).

The long-term effect of fallow was not observed for soil min-N or PMN, suggesting that N mineralization is only affected by the substrate added during the previous year or the latest cycle of rotation. Nitrogen in the forms of NH_4^+ and NO_3^- is assimilated by plants and returned into soil whereas C originates from CO_2 in the air and is plowed as organic residue into the soil. Nitrogen transformations are closely related to the processes of mineralization of its organic forms

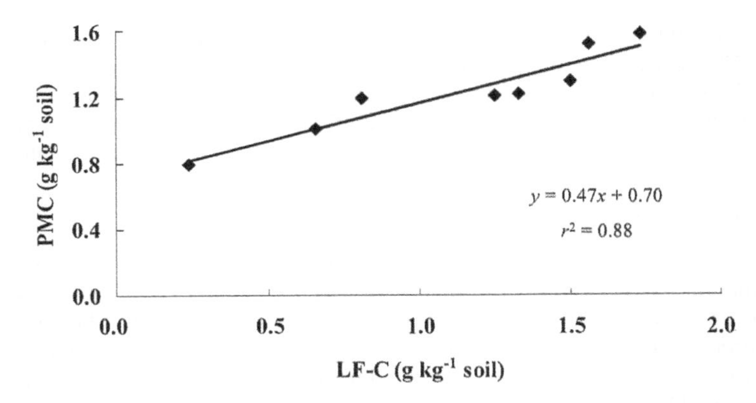

Figure 14. Correlation of potential mineralizable C (PMC) with 'light fraction' C (LF-C).

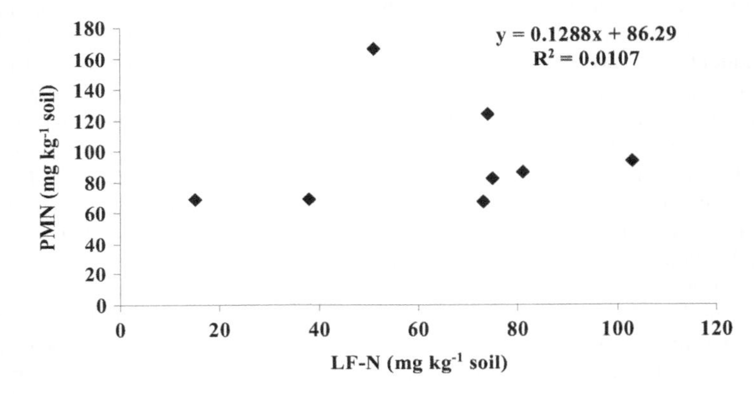

Figure 15. Correlation of potential mineralizable N (PMN) with 'light fraction' N (LF-N).

in plant-soil system. Therefore, in plant-soil systems N cycling is affected over shorter periods than is C cycling. *'Light fraction' organic matter:* The 'light fraction' of SOM (LF-OM) consists mainly of plant residues, small animals and microorganisms adhering to plant-derived particulate matter at various stages of decomposition that serves as a readily decomposable substrate for soil microorganisms and also as a short-term reservoir of plant nutrients (Gregorich et al., 1994).

The 'light fraction' C (LF-C) was positively correlated with PMC (Figure 14), and confirms the hypothesis that the reduced fallowing system has more potential to supply soil with easily mineralizable C. However, there was no linear correlation between LF-N and PMN (Figure 15), presumably because the high C:N ratio of the LF-OM temporarily induced N immobilization (Janzen et al., 1992).

The concentration of labile OM, which is closely related to LF-OM, may be governed by the degree to which temperature and moisture conditions constrain decomposition of accumulated residues (Beiderbeck et al., 1994). Under the CW system decomposition of residues during periods of favorable soil temperature was retarded by the depleted soil moisture reserve (Shields and Paul, 1973; Douglas et al., 1992; Akhmetov, 1999, local citation). Then, when moisture and temperature constraints were alleviated during laboratory incubations, soil respiration rate was high (Janzen et al., 1992). On the contrary, residues in the 2R system during the fallow phase were always exposed to favorable moisture and temperature regimes over an extended period. Therefore, labile SOM was rapidly depleted in the field, and the laboratory respiration rates were much lower in 2R than in CW (Biederbeck et al., 1994).

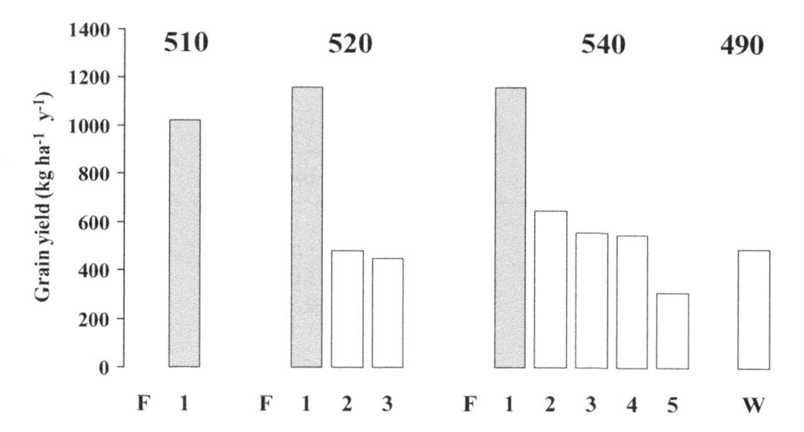

Figure 16. Grain yield (1994–1999) of spring wheat as affected by years after fallow (F is a fallow; 1,2,3...5 are succession of crops after fallow. The values above bars are average yield per rotation including the fallow year).

Grain yields and weed biomass: The highest grain yield and the lowest weed biomass were produced during the first year. However, the yield decreased sharply in the second and successive years after fallowing (Table 11; Figure 16). This trend is, firstly, because plants in a post-fallow phase take advantage of higher soil min-N. Secondly, because when a field is in fallow it provides the only break for weed infestation, the amount of weeds was generally less in the first year after fallow which reduced competition for nutrients.

In contrast to the grain yield, weed infestation was the maximum in the second years after fallow in 4R and 6R treatments. Probably, some of the weeds were not destroyed during the fallow and their seeds remained dormant and germinated during the second year after fallowing (Akhmetov, 1999, local citation).

Correlation between the grain yield and weed infestation, average for 1986–1996, is presented by the following equation of multiple linear regression (Eq. 6):

$$Y = 20.82 - 0.189X, \tag{6}$$

where, X is the total amount of weeds plants m^{-2}.The coefficient of determination was also high ($r^2 = 0.78$) representing 78% of changes of the yield due to weed infestation.

The highest grain yield, considering the whole rotation, was obtained in 6R (540 kg ha^{-1}) and it parallels the distribution of soil labile OM (PMC and LF-C) and supports the hypothesis that the longer rotations with fewer fallows contribute more to the accumulation of SOM than shorter rotations with higher frequency of fallowing.

6 SPATIAL VARIABILITY OF ORGANIC MATTER DYNAMICS IN THE SEMI-ARID CROPLANDS OF NORTHERN KAZAKHSTAN

The present study was conducted in large-scale upland fields located near the Barayev Kazakh Research and Production Centre of Grain Farming in Shortandy, Akmolinsk Oblast, northern Kazakhstan. The data reported here were obtained under on-farm conditions, which were managed in the conventional ways following the collapse of the Soviet Union. At present, most of the farms are managed without the use of fertilizers and herbicides. The land use of the study sites consisted mostly of arable land with the cultivation of spring wheat as major crop and barley and oats as minor crops. Other types of land use included fallow, grassland, and abandoned land. The crop rotation system at this study site consisted of a four-year rotation, i.e., fallow-wheat-wheat-wheat/barley.

Table 11. Grain yield (1986–1996 and 1998) and weed biomass (1986–1996 and 1997) in wheat-based rotation systems with different frequency of fallow in Southern Chernozem.

	Grain yield (Mg ha^{-1})		Weeds' biomass dry matter) (Mg ha^{-1})	
Rotation	1986–1996	1998	1986–1996	1997
Two-year (2R)				
Fallow	–	–	–	–
Wheat after fallow	1.85	1.02	0.12	0.13
Average of rotation	0.93	0.51	0.06	n/a
Four-year (4R)				
Fallow	–	–	–	–
Wheat 1st year	1.84	1.16	0.14	n/a
Wheat 2nd year	1.78	0.48	0.26	n/a
Wheat 3rd year	1.56	0.45	0.12	n/a
Average of cropping year	1.73	0.70	0.17	0.16
Average of rotation	1.30	0.52	0.13	n/a
Six-year (6R)				
Fallow	–	–	–	–
Wheat 1st year	2.09	1.16	0.09	n/a
Wheat 2nd year	1.66	0.55	0.22	n/a
Wheat 3rd year	1.61	0.56	0.20	n/a
Wheat 4th year	1.58	0.65	0.18	n/a
Wheat 5th year	1.55	0.31	0.15	n/a
Average of cropping year	1.70	0.65	0.17	0.16
Average of rotation	1.41	0.54	0.14	n/a
Continuous wheat cropping (CW)				
CW	1.00	0.49	1.13	0.87

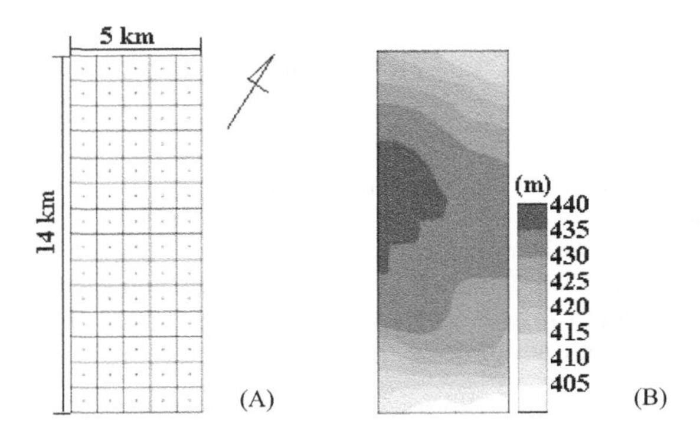

Figure 17. (A) Schematic diagram of the sampling sites indicated as dots in the field and (B) topography of the study site.

The study field (14 × 5 km) was divided into 70 plots (1 × 1 km each), as shown in Figure 17(A), and the SOM dynamics was measured at the center of each plot. Soils were sampled to a depth of 90 cm at 15-cm intervals in June 2001, at the beginning of the growing season. Plant sampling was done between late August and September 2001. In addition to soil and plant sampling, the elevation of the center point of each plot was measured using a differential Global Positioning System (GPS) (Magellan ProMARK X) in May 2001 to assess the micro-topography of the study field. The elevation ranged from 402 m to 437 m with an average of 427 m. The central plateau

Table 12. Descriptive statistics of the soil, plant and water properties.

Field properties	Mean	Maximum	Minimum	CV (%)[a]
Soil				
Organic carbon: 0–15 cm ($g\,kg^{-1}$)	25.6	36.1	16.3	17.1
Organic carbon: 15–30 cm ($g\,kg^{-1}$)	20.4	31.1	6.3	21.6
Organic carbon: 30–45 cm ($g\,kg^{-1}$)	17.0	34.9	3.9	27.5
Organic carbon: 45–60 cm ($g\,kg^{-1}$)	13.5	21.5	4.3	27.5
Organic carbon: 60–75 cm ($g\,kg^{-1}$)	10.1	18.5	1.4	41.4
Organic carbon: 75–90 cm ($g\,kg^{-1}$)	6.1	13.4	0.5	55.2
Organic carbon: 0–15 cm ($Mg\,ha^{-1}$)	39.8	65.1	22.7	19.4
Organic carbon: 15–30 cm ($Mg\,ha^{-1}$)	37.0	56.9	11.2	21.0
Organic carbon: 30–45 cm ($Mg\,ha^{-1}$)	31.1	63.3	7.0	27.1
Organic carbon: 45–60 cm ($Mg\,ha^{-1}$)	26.7	41.4	8.2	27.6
Organic carbon: 60–75 cm ($Mg\,ha^{-1}$)	21.7	41.1	3.2	41.6
Organic carbon: 75–90 cm ($Mg\,ha^{-1}$)	14.6	32.2	1.2	55.2
Organic carbon: 0–90 cm ($Mg\,ha^{-1}$)	170.9	250.3	108.1	17.4
Potentially mineralizable carbon: 0–15 cm ($Mg\,ha^{-1}$)	2.72	6.87	0.69	40.4
Plant				
Yield ($Mg\,ha^{-1}$)[b]	1.38	3.52	0.00	56.4
Ear C: output C ($Mg\,ha^{-1}$)	0.61	1.51	0.00	56.6
Residue C: input C ($Mg\,ha^{-1}$)	1.22	2.33	0.27	42.5
Total C ($Mg\,ha^{-1}$)	1.82	3.72	0.41	42.2
Soil water				
Soil water: 0–15 cm (mm)	32	52	15	18.1
Soil water: 15–30 cm (mm)	39	50	22	15.1
Soil water: 30–45 cm (mm)	40	49	27	11.4
Soil water: 45–60 cm (mm)	41	54	21	14.2
Soil water: 60–75 cm (mm)	42	60	25	18.2
Soil water: 75–90 cm (mm)	43	57	27	17.7
Soil water: 0–90 cm (mm)	237	293	158	12.0
Snow depth (mm)	302	462	135	21.8
Topography				
Altitude (m)	427	437	402	2.1

[a] Coefficient of variation. [b] On a dry weight basis.

had the highest elevation and the north-facing slope and south-facing slope extended gently from the plateau (Figure 17,B). The snow depth was measured at each sampling site in March 2002 to determine the distribution of snow accumulation.

Soil samples (0–90 cm) were analyzed for SOC and TN concentrations, and C:N ratio was computed to assess the SOM status. The amount of PMC was also determined for the fresh surface soil samples (0–15 cm) after aerobic incubation for 19 weeks, controlling the water content at 60% of the maximum WHC at 30°C.

The mean, maximum and minimum values and the coefficient of variation of each property were calculated as descriptive statistics, followed by a geostatistical analysis. A semivariogram was used to evaluate the spatial variability of the properties (Oliver, 1987; Webster and Oliver, 2001; Yanai et al., 2001). Maps were subsequently computed using block kriging to evaluate the regional patterns of variation rather than local details. The geostatistical software, GS+ Version 5.3 for Windows (Gamma Design Software), was used for the analysis (Robertson, 1998).

6.1 *General trend of organic matter dynamics*

Table 12 shows the descriptive statistics of the field properties. The average SOM concentration of 25.6 $g\,kg^{-1}$ for the surface soil decreased with depth to 6.1 $g\,kg^{-1}$ at 75–90 cm depth. In accord

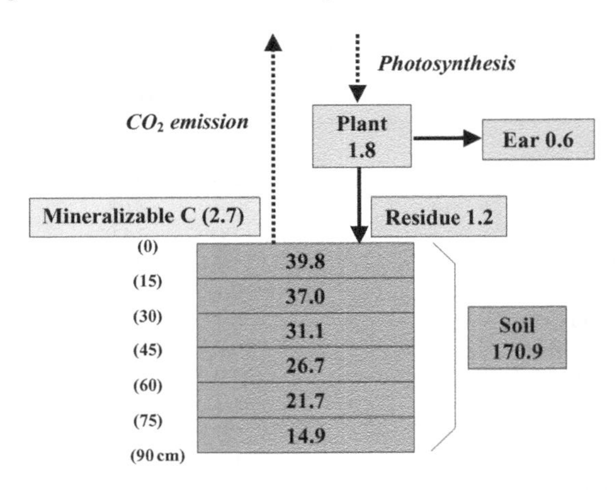

Figure 18. Average C stock and flow of the soil-plant system at the study site (Mg ha^{-1}). Mineralizable C indicates the potential C pool evaluated under optimal conditions for organic matter mineralization.

with the SOC profile, soil bulk density increased with depth; 1.04, 1.21, 1.22, 1.32, 1.43 and 1.60 Mg m^{-3} from 0–15 cm to 75–90 cm depth. Thus, the SOC pool of 39.8 Mg ha^{-1} for the surface soil gradually decreased with increase in depth, i.e. 37.0, 31.1, 26.7, 21.7 and 14.6 Mg ha^{-1} for 15–30, 30–45, 45–60, 60–75 and 75–90 cm, respectively. Even though there was a decreasing trend with depth, the subsoil also contained a considerable amount of SOC pool, which cannot be ignored from the environmental viewpoint. The total amount of SOC pool within the 90-cm depth was 170.9 Mg ha^{-1}, with a coefficient of variation of 17.4%. PMC in the surface soil was 2.7 Mg ha^{-1} or 6.8% of the total SOC pool, suggesting that a considerable part of the SOC pool may be released as CO_2 under favorable conditions of SOM decomposition. The coefficient of variation exceeded 40%, suggesting a high variation compared to the total SOC pool (17.1%), presumably because the amount of PMC depends on both the amount of chemically mineralizable C and the microbiological activity for mineralization.

Plant biomass contained 1.8 Mg ha^{-1} of C, of which 1.2 Mg ha^{-1} was returned to the field as plant residues and 0.6 Mg ha^{-1} was removed as crop (ear). Average crop yield, calculated on the basis of 54 data with crop cover, was 1.38 Mg ha^{-1} on a dry weight basis, which was almost similar to the average crop yield in this area. The coefficients of variation of all the plant properties exceeded 40%, suggesting a large variation in the field. The average SOC pool and flux of this soil-plant system are presented in Figure 18.

The amount of soil water stored at the beginning of the growing season was about 30–40 mm for each 15-cm depth with a tendency for a slight increase with increase in depth. Soil water for the 0–90 cm depth was 237 mm, which was more than one and a half times the amount of the average precipitation during the growing season (about 160 mm). This suggests the importance of stored soil water in springtime for sound growth of wheat/barley in this region, even though not all the soil water would be available to plants. Furthermore the mean snow depth measured during the winter of 2002 was 302 mm corresponding to 75 mm of water assuming snow density of 0.25 Mg m^{-3}. Management of snow during the winter would, therefore, contribute considerably to the storage of available water in the soil profile.

Correlation among field properties: Table 13 shows the correlation matrix of selected field properties. Elevation was positively correlated with the amount of soil water in both surface layer (0–15 cm) and whole profile (0–90 cm), snow depth, SOC concentration in both surface layer (0–15 cm) and whole profile (0–90 cm) ($p < 0.01$), PMC content, plant ear C content and plant yield ($p < 0.05$). The amount of soil water (0–15 cm, 0–90 cm) was positively correlated with the SOC concentrations, but not with the plant properties, even though the soil water content (0–30 cm) had a moderately positive relationship with the yield and ear C content ($p < 0.10$). Snow depth was

Table 13. Correlation matrix of selected field properties.

	Altitude	SW[a] 0–15	SW[a] 0–90	Snow[b]	SOC[c] 0–15	SOC[c] 15–30	SOC[c] 30–45	SOC[c] 0–90	PMC[d]
Soil water: 0–15 cm	0.33**[c]								
Soil water: 0–90 cm	0.46**	0.59**							
Snow depth	0.41**	0.12**	0.30*						
Soil organic carbon: 0–15 cm	0.47**	0.44**	0.33**	0.10					
Soil organic carbon: 15–30 cm	0.06	−0.14	−0.03	0.12	0.19				
Soil organic carbon: 30–45 cm	0.14	−0.14	0.02	0.25*	0.16	0.72**			
Soil organic carbon: 0–90 cm	0.32**	0.17	0.40**	0.25*	0.47**	0.61**	0.66**		
Potentially mineralizable carbon	0.28*	0.19	0.12	0.06	0.43**	−0.14	−0.04	0.11	
Plat ear C	0.29*	0.07	0.05	0.28*	−0.03	0.15	0.06	−0.03	−0.06
Plant residue C	0.05	0.15	0.09	0.32*	−0.04	0.06	−0.01	−0.02	−0.12
Plant total C	0.17	0.14	0.08	0.35*	−0.04	0.10	0.02	−0.03	−0.11
Plant yield	0.31*	0.08	0.05	0.29*	−0.01	0.16	0.08	−0.01	−0.04

[a] Soil water, [b] Snow depth, [c] Soil organic carbon and [d] Potentially mineralizable carbon, [e]* and ** indicate significant level of 0.05 and 0.01, res.

Table 14. Geostatistical parameters of the soil, plant and water properties.

Field properties	Nugget	Sill	Range (km)	Q value	Model[a]
Soil					
Organic carbon: 0–15 cm (Mg ha^{-1})	24.7	79.2	6.9	0.69	S
Organic carbon: 15–30 cm (Mg ha^{-1})	4.3	63.6	1.6	0.93	S
Organic carbon: 30–45 cm (Mg ha^{-1})	13.9	77.4	1.6	0.82	E
Organic carbon: 45–60 cm (Mg ha^{-1})	1.9	54.4	1.0	0.97	S
Organic carbon: 60–75 cm (Mg ha^{-1})	10.6	83.3	1.6	0.87	E
Organic carbon: 75–90 cm (Mg ha^{-1})	1.4	64.7	1.3	0.98	S
Organic carbon: 0–90 cm (Mg ha^{-1})	115	901	2.6	0.87	E
Potentially mineralizable carbon: 0–15 cm (Mg ha^{-1})	1.02	2.04	9.0+	0.50	E
Plant					
Yield (Mg ha^{-1})[b]	0.40	1.04	9.0+	0.38	S
Ear C: output C (Mg ha^{-1})	0.08	0.22	9.0+	0.34	S
Residue C: input C (Mg ha^{-1})	0.24	0.26	1.7	0.95	L
Total C (Mg ha^{-1})	0.5	0.62	9.0+	0.81	L
Soil water					
Soil water: 0–15 cm (mm)	20	37	3.5	0.46	E
Soil water: 15–30 cm (mm)	5	36	2.6	0.85	E
Soil water: 30–45 cm (mm)	10	29	9.0+	0.65	S
Soil water: 45–60 cm (mm)	5	36	1.4	0.86	E
Soil water: 60–75 cm (mm)	35	69	7.7	0.50	S
Soil water: 75–90 cm (mm)	43	98	9.0+	0.56	E
Soil water: 0–90 cm (mm)	575	1259	9.0+	0.54	E
Snow depth (mm)	1512	4021	3.0	0.62	S

[a]S: Spherical, E: Exponential and L: Linear.

positively correlated with the SOC concentration at the 30–45 cm and 0–90 cm depths ($p < 0.01$) and all the plant properties ($p < 0.05$). The amount of PMC was positively correlated with the SOC concentration at the same depth ($p < 0.01$), indicating the presence of a strong link between the C source and the amount of CO_2 emission. As a result, topography, available water content, soil C pool, plant C content and yield were all interrelated. N.K. Azarov, Kazakh Research Institute of Grain Farming in Shortandy, suggested that the geographical characteristics were correlated with the snow depth, humus content, moisture content and cereal productivity.

6.2 *Spatial variability of organic matter dynamics*

Geostatistical parameters of the field properties are shown in Table 14. The Q values of SOC ranged between 0.7 and 1.0, suggesting the existence of a highly developed spatial structure, whereas those of PMC and soil water were about 0.5–0.7, suggesting a considerable development of the spatial structure. For the plant properties, total and residue C displayed a well-developed spatial structure, whereas yield and ear C displayed a poorly developed one. As the spatial structures were moderately to well-developed, the ranges could be interpreted as the limit distances of spatial dependency. The ranges of the soil water content (0–15 cm), snow depth and plant residue C were 3.5, 3.0 and 1.7 km, respectively, suggesting a relatively short spatial dependency. On the contrary, the SOC concentration (0–15 cm), PMC and plant yield were characterized by relatively long ranges of 6.9 or >9 km. These results suggest that most of the field properties displayed a well-developed spatial structure and hence had the potential to be managed spatially or site-specifically based on this spatial dependency.

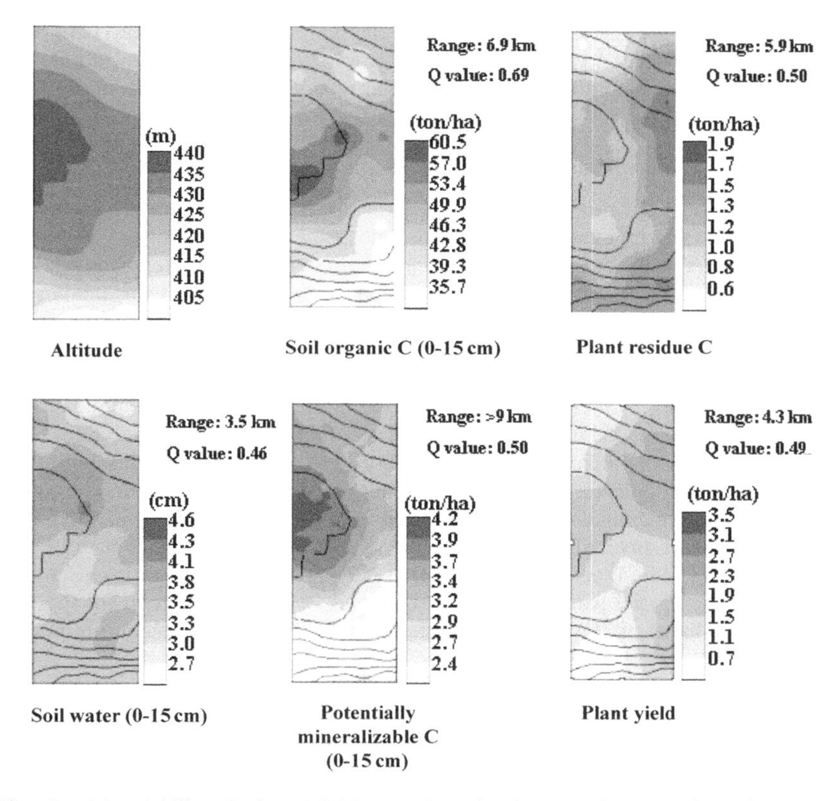

Figure 19. Spatial variability of selected field properties related to organic matter dynamics.

Figure 19 shows the isarithmic maps or spatial patterns of the selected properties in the field, which were obtained from the data of spatial dependency described above. Spatial pattern of elevation is also shown for comparison and contour lines in other maps indicate the isarithm of elevation. Soil water content (0–15 cm) was relatively high in the central plateau and low in the north-facing and south-facing slope areas. This trend confirmed the strong correlation between the elevation and soil water content ($p < 0.01$), presumably because soil water is retained more stably in the central flat plateau than in other slope areas. The SOC concentrations and the amount of PMC followed a trend similar to that of the soil water content (0–15 cm), indicating that the value in the central plateau was the highest, followed by that in the north-facing slope and then the south-facing slope areas. These results suggest that the central plateau had the maximum SOC concentration and accordingly was also the largest source of CO_2. The lowest SOC concentration in the south-facing slope area may be due to the lower plant C input, as mentioned below, reflecting the lower soil water content due to the larger amount of sunshine and hence high evaporation. The spatial pattern of the amount of PMC showed only a general trend of CO_2 emission under optimal conditions for SOM decomposition. Further studies need to be conducted to estimate actual values of CO_2 emission *in situ* under field conditions. The C budget of the field can be properly evaluated, if the amount of CO_2 emission is estimated on the basis of the amount of PMC in soil and annual data of the field conditions (e.g., soil temperature and moisture content).

In contrast to soil, plant properties showed slightly different spatial patterns. Plant residue C was higher in the north-facing slope area, toward slightly north compared with the SOC concentration. This tendency was more pronounced for the plant yield. The yield was the highest in the north-facing slope area followed by the central plateau and south-facing slope area. Ear C concentration, an index of C output from the system, had an almost similar pattern to that of yield. These patterns are mainly attributed to soil water content and available or mineralizable SOM concentration because water and available N are generally among major determinants of plant growth in this region.

6.3 *Possibility of site-specific management for sustainable agriculture*

Spatial patterns of the field properties strongly indicate that SOM dynamics in the field was markedly affected by the topography and that the most favorable area for the SOM sequestration was slightly different from that for the agronomic production. The yield was relatively high and the soil C pool was moderate in the north-facing slope area. The yield was moderate and the SOC pool and release of CO_2 from soil were the highest in the central plateau, whereas the yield and SOC dynamics were relatively low in the south-facing slope area. These results support the conclusion that site-specific management is a viable alternative for sustainable agriculture in this region. A possibility is to intensify management in the north-facing slope area to maximize crop yield without accelerating SOM decomposition. This could be achieved by high rates of fertilizer and seed application in this area. Another possibility is to reduce or even discontinue agricultural management in the south-facing slope area because crop yield is expected to be considerably lower under the current management. Since the most appropriate management varies regionally, depending on both environmental and socio-economic conditions (Paustian et al., 1997), the type of site-specific management must also be carefully selected. Thus, site-specific agricultural management based on the spatial patterns of SOM dynamics is a suitable option for the promotion of sustainable agricultural production and for limiting SOC decomposition and minimizing soil degradation.

7 SOM BUDGET IN DIFFERENT TYPES OF STEPPE SOILS IN NORHTERN KAZAKHSTAN

There are three types of Chernozems in northern Kazakhstan: Dark Chestnut soils (DC; Typic Haplustolls), Southern Chernozems (SC; Typic Haplustolls), and Ordinary Chernozems (OC; Pachic Haplustolls). Soil color is the main difference between DC and SC which is affected by the SOM concentration, which is generally lower in DC than in SC. Farmers use a 4- or 5-year crop rotation system, comprising summer fallowing and 3 or 4 years of continuous cereal cultivation. The present study was conducted in the Thelinogratsukaya district (N51°12′, E71°06′) east of Astana, in the Shortandy district (N51°40′, E71°02′; same as the experimental field in the previous section), 60 km north of Astana, and in the Reonidovka district (N53°57′, E69°32′) 300 km north of Astana, chosen as representative sites of DC, SC, and OC, respectively. The climate of Astana is drier than that of the upper north. In the Reonidovka district, the soil color is progressively darker, presumably due to decreasing evapotranspiration. Monthly climatic data for the three sites during the cropping season (May to September) for the years 2002, 2004 and 2005 are shown in Figure 20. Mean air temperature was the maximum in July, and was higher at the DC than the OC site. Mean air temperature was 16.1, 15.9, and 15.8°C, and total precipitation was 178, 193, and 202 mm at the DC, SC, and OC sites, respectively.

Surface soil samples were taken from cropped fields under different land use stages (i.e., cereal cropping, summer fallow, and pasture) from these sites between 2003 and 2005. All plots were situated within a diameter of 50 km. Plant samples were also taken in 2002, 2004, and 2005 at the soil sampling plots. Total C content of crop residue and of weed samples collected from cereal, pasture, and fallow fields was determined by the dry-combustion method (VarioMax CN; Elementar Inc.). In pasture fields, the C input was calculated as follows (Eq. 7):

$$\text{Input C} = \text{increase in below ground biomass C (Aug–May)} + \text{aboveground biomass C (in August)} \qquad (7)$$

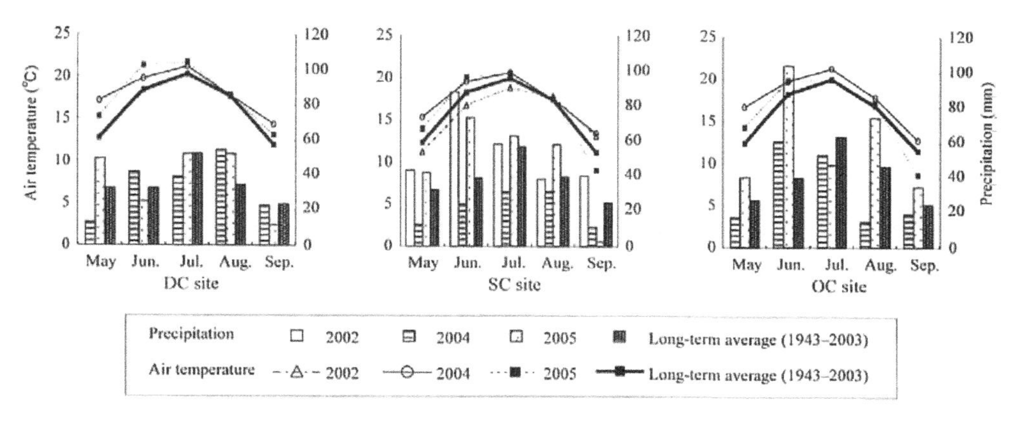

Figure 20. Climatic condition at each site during CO_2 emission monitoring period.

Total SOC concentration was determined by the Tyurin method and PMC was measured according to the method described earlier.

The CO_2 emission was measured during the cropping season (May to September) in 2002 (SC site only), 2004, and 2005 using an alkali absorption method. Maximum and minimum temperatures at 5-cm depth were recorded at the time of measuring CO_2 emission by using a maximum-minimum thermometer (AD–5625; AND, Inc.) installed at each site. In addition, volumetric water content (%) at 0–15 cm depth was measured by using a soil moisture sensor (Hydro Sense; Campbell Scientific, Inc.). Air temperature, soil temperature (5-cm, 15-cm depths), soil moisture (0–15 cm), and precipitation were monitored at representative plots in 2002 (SC site), 2004 (OC and SC sites), and 2005 (OC, SC, and DC sites) using a datalogger (CR–10X; Campbell Scientific, Inc.) at 30-min intervals during the cropping season.

7.1 *CO_2 emission rate and its controlling factors*

The data on monthly precipitation and mean air temperature during the crop-growing season are shown in Figure 20. Precipitation patterns show relatively high annual variability at the SC and OC sites. In 2002, the SC site experienced exceptionally high moisture conditions due to high precipitation since the preceding year throughout the growing season (see also Table 2). In 2004, there was a severe drought from May to July at the SC site and from August to September at the OC site. In contrast, May to August 2005 was a wet period at the SC and OC sites. Only the DC site received normal precipitation in both years. Mean air temperature was slightly higher than normal during 2004 and 2005 at all sites.

The data on PMC and SOC for the CO_2 monitoring plots are shown in Table 15. At all sites, PMC and SOC were higher in pasture than in the cereal or fallow plots, except for PMC at OC site. Differences between PMC and SOC among land use types are associated with long-term differences in substrate addition (Schomberg and Jones, 1999). In cereal plots, PMC and SOC pools ranged from 1.8 (DC) to 2.8 (SC) Mg C ha^{-1} and 29.7 (SC) to 46.5 (OC) Mg C ha^{-1}, respectively. The clay content was about 40%, and was the same for all sites and land use types (data not shown). Daily CO_2 emission, air temperature, and precipitation during the cropping season are shown in Figures 21 to 23. Fluctuations in CO_2 emission from the soils closely followed a pattern similar to that of the temperature, which was the maximum for each site in June or July, and started to decrease in September. These results are similar to those reported by Rochette et al. (1991) and Singh and Shekhar (1986). Rochette et al. (1991) reported that the maximum CO_2 emission occurred in June, corresponding with the maximum growth of the plant. Buyanovsky et al. (1986) reported that maximum CO_2 emission coincided with the maximum air and soil temperatures in winter wheat ecosystems.

Table 15. Potentially mineralizable orgnic carbon content and soil organic content at the CO2 monitoring plots.

	Dark Chestnut			Southern Chernozem			Ordinary Chernozem		
		PMC	SOC		PMC	SOC		PMC	SOC
	No of plots	average (S.E) (Mg C ha^{-1})	average (S.E)	No of plots	average (S.E) (Mg C ha^{-1})	average (S.E)	No of plots	average (S.E) (Mg C ha^{-1})	average (S.E)
2002									
Cereal	–	–	–	9	2.65 (0.30)	39.1 (3.4)	–	–	–
Fallow	–	–	–	–	–	–	–	–	–
Pasture	–	–	–	–	–	–	–	–	–
2004									
Cereal	8	1.82 (0.14)	29.7 (1.8)	9	2.38 (0.22)	35.9 (1.6)	6	1.64 (0.07)	46.9 (2.2)
Fallow	2	1.90 (0.01)	30.3 (3.8)	1	3.27	40.9	4	2.21 (0.16)	45.8 (1.4)
Pasture	2	2.47 (0.12)	36.9 (1.3)	2	3.61 (0.68)	41.8 (3.9)	2	2.41 (0.19)	60.7 (0.8)
2005									
Cereal	11	2.20 (0.13)	32.7 (1.1)	17	2.80 (0.18)	37.2 (1.1)	10	1.87 (0.12)	46.5 (1.4)
Fallow	7	1.86 (0.16)	29.9 (2.0)	5	2.32 (0.12)	33.6 (2.1)	1	0.79	51.5
Pasture	2	2.47 (0.12)	36.9 (1.3)	4	3.42 (0.30)	41.0 (1.9)	2	2.41 (0.19)	60.7 (0.8)

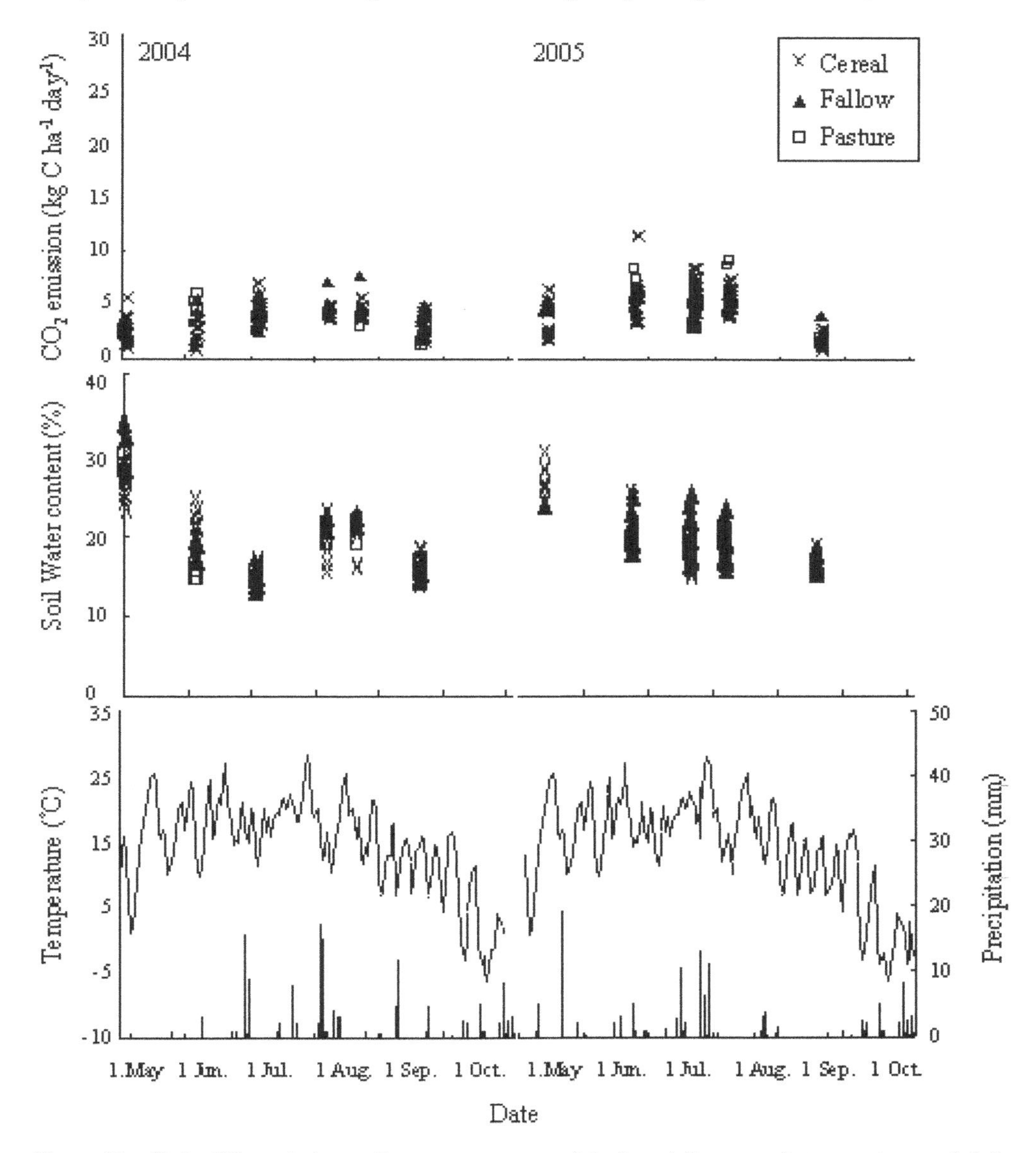

Figure 21. Daily CO_2 emission, soil water content, precipitation, daily mean air temperature, and daily precipitation at DC site. Upper figure shows daily CO_2 emission, middle figure shows soil water content, and lower figure shows mean air temperature (solid line) and precipitation (rods).

Precipitation also markedly influenced the soil water content and CO_2 emission, both were lower in June 2004 than in June 2005 at all sites. Because of a severe drought in June 2004, especially at the SC site, only small amounts of CO_2 were emitted at that time. Despite air temperature being the same in 2004 and 2005, CO_2 emission in June 2004 was 2 to 5 times lower than that in 2005 at the SC site. A large increase in CO_2 emission was observed under laboratory conditions when dry samples were rewetted (Orchard and Cook, 1983), probably due to increases in soil microbial population and activity. Even though the study by Orchard and Cook was conducted under controlled laboratory conditions, the same event may also occur under field conditions.

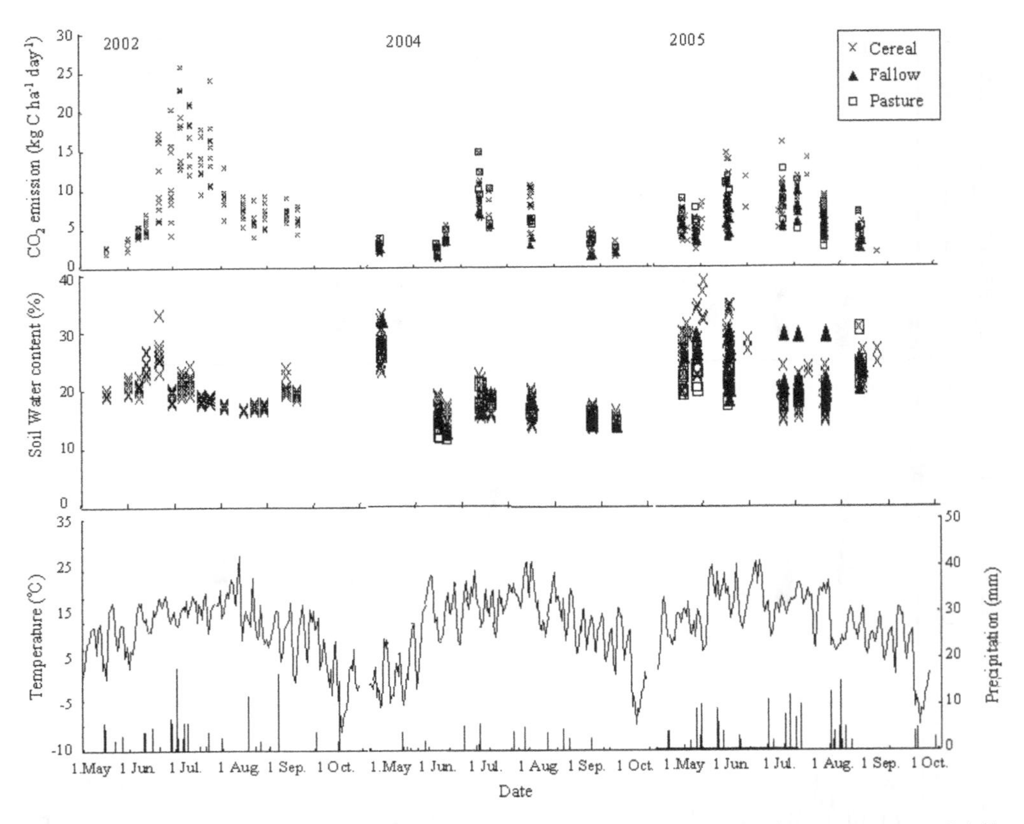

Figure 22. Daily CO_2 emission, soil water content, precipitation, daily mean air temperature, and daily precipitation at SC site. Upper figure shows daily CO_2 emission, middle figure shows soil water content, and lower figure shows mean air temperature (solid line) and precipitation (rods).

Emmerich (2003) suggested that some of the CO_2 emission may originate from inorganic carbon pool after precipitation under the semiarid environment. However, the surface soil of Emmerich's study site contains many rock fragments of limestone and low SOC concentration (8 g kg^{-1}). In general, the contribution of inorganic pool to CO_2 emission in present study is rather limited.

Land use also markedly affected the seasonal variation in the amount of CO_2 emission. The influence of increased CO_2 emission together with the increased temperature was not as clearly defined in pasture fields as was in cereal and summer fallow fields during the drought period of 2004. In contrast, maximum CO_2 emission was observed in pasture fields at the OC site during the wet June 2005. Tang et al. (2003) suggested that temperature sensitivity is relatively low in the dry season, and Xu and Qi (2001) showed that this sensitivity is strongly correlated with the soil moisture content. It is likely that in 2004 the soil profile in pasture fields was generally drier than that under other land-uses due to strong evapotranspiration in pasture fields. Therefore, soil microorganism activity may be lower in pasture fields than under other land-uses during periods of drought. In contrast, the soil profile in June and July 2005 may have contained sufficient moisture for soil microbial activity to be higher in pasture fields, where substrates like PMC accumulated more than under other land-uses at the OC site (Table 15).

7.2 *Prediction equations of CO_2 emission*

Stepwise multiple regression of the Arrhenius model was used to estimate daily CO_2 emission using the following parameters: soil temperature, dryness factor, precipitation, and PMC by soil

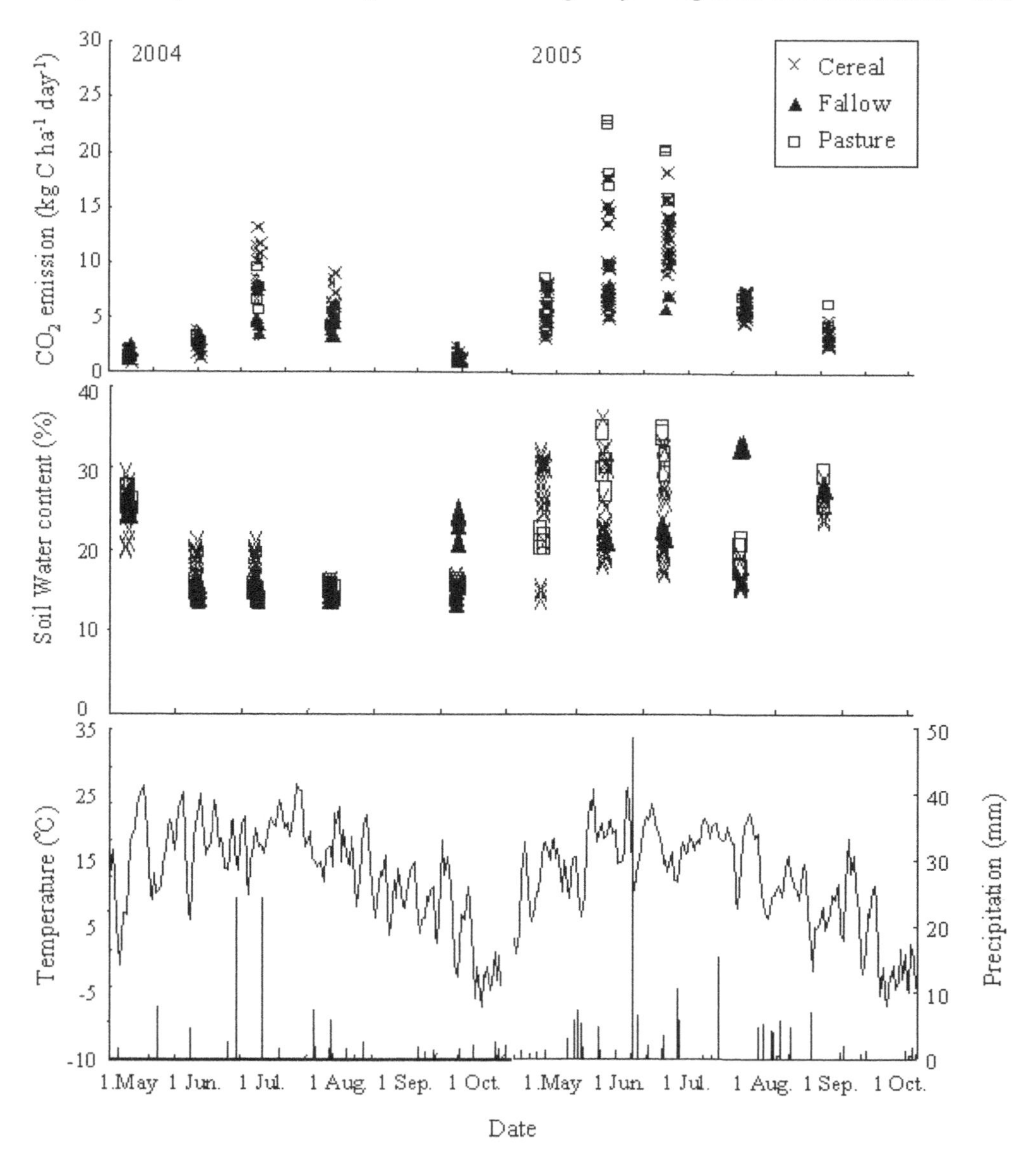

Figure 23. Daily CO_2 emission, soil water content, precipitation, daily mean air temperature, and daily precipitation at DC site. Upper figure shows daily CO_2 emission, middle figure shows soil water content, and lower figure shows mean air temperature (solid line) and precipitation (rods).

type and land use. To avoid multicollinearity, either precipitation or dryness factor was used in the regression analysis, and factors were selected on the basis of high r^2. The Arrhenius equation along with the extensive form of estimating CO_2 emission can be stated as follows (Eqs. 8, 9):

$$C_{\text{em}} = aC_0^b W^c \text{e}^{-E/RT} \tag{8}$$

$$\ln C_{\text{em}} = \ln a + b \ln C_0 + c \ln W - E/RT \tag{9}$$

where C_{em} is daily CO_2 emission, C_0 is PMC, W is precipitation or dryness factor, E is activation energy, R is the gas constant, T is mean soil temperature (K), and a, b, and c are the constants related to decomposition rate, C_0, and W, respectively. The dryness factor is derived from the potential evapotranspiration (PET) and precipitation, in which daily PET is estimated using the Hargreaves-Samani equation (Hargreaves and Samani, 1982; Hargreaves et al., 1985), and the daily dryness factor is computed by dividing precipitation by PET.

The results are given in Table 16. Soil temperature had a high and a positive contribution in all 9 equations, and a significant relationship at 1% level was obtained between the CO_2 emission and the activation energy (E), indicating a significant dependency of CO_2 emission on soil temperature. Such dependency was also reported for Chernozem soils in northern Kazakhstan (this study) and in Russia by Kudeyarov and Kurganova (1998). Based on the value of E, the estimated Q_{10} values for temperatures between 10 and 20°C ranged from 1.55 to 3.18. Raich and Schlesinger (1992) reviewed soil respiration rates of diverse ecosystems and observed that Q_{10} ranged from 1.3 to 3.3, with a mean value of 2.4. In the previous section, the Q_{10} values in cereal fields at an SC site ranged from 1.3 to 2.0. In the present study, Q_{10} values of fallow and cereal fields were higher at the OC than DC site. Schinner (1982) reported higher Q_{10} values at lower temperatures, a finding also supported by the present results.

Cumulative precipitation or the dryness factor also showed a positive contribution in all 9 equations generally at a significance of 1% level. Precipitation was selected in cereal plots at all sites, and the dryness factor was selected in DC and SC fallow plots and DC and OC pasture plots. There was a significant correlation between precipitation and the dryness factor, which drastically affected soil moisture content (data not shown). It is the soil water status that governs the inter-seasonal variation in CO_2 emission (Schlentner and Van Cleve, 1985; Rochette et al., 1991; Kessavalou et al., 1998; Davidson et al., 2000; Tang et al., 2003; Huxman et al., 2004). Cumulative days was selected from 8 to 15 days, and suggested that CO_2 emission *in situ* was controlled by several precipitation events which occurred during that period.

7.3 *Relationship between land use and determining factors of CO_2 emission*

Mean Q_{10} value of the three soil classes for each land use type was 1.78 for cereal, 2.04 for fallow, and 2.62 for pasture. The mean Q_{10} value of fallow was higher than that of cereal, implying that tillage alters the decomposition environment by aerating the soil, breaking up soil aggregates, and incorporating weeds into the soil profile (Beare et al., 1994; Fortin et al., 1996; Lee et al., 1996; Prior et al., 1997; Six et al., 2002). Pasture fields had the highest Q_{10} value at each site, indicating that the sensitivity of CO_2 fluctuation to soil temperature was the highest under a pasture management system. Precipitation factor was selected at all sites under cereal cropping, whereas the dryness factor tended to be selected for fallow and cereal fields. These results suggest that the sensitivity of fluctuation of CO_2 is more strongly influenced by dryness in fallow and pasture fields than in cereal fields.

The contribution of SOM (i.e. C_0) was somewhat uncertain in fallow and pasture fields, where CO_2 sampling plots were fewer than in cereal plots. Prediction of the variance of CO_2 emission using the parameter of SOM may thus not have been possible, due to the limited number of CO_2 sampling data sets in fallow and pasture fields.

7.4 *Annual carbon flow and carbon budget under different crop rotations in the three regions*

Prediction equations were used to estimate annual CO_2 emission along with the climatic data from data loggers or the meteorological station closest to the CO_2 measuring sites. Air temperature monitored by the meteorological observatory was calibrated with the soil temperature at 5-cm depth at each site by correlation analysis using monitored air temperature and soil temperature data from the dataloggers.

The data on annual C flow and C budget of the crop rotation system are summarized in Table 17. Annual CO_2 emission is the summation of daily CO_2 emission from April 1 to October 31. Estimates

Table 16. Coefficients determined by stepwise multiple regression analysis.

		Coefficients								
	Site	ln*a*	*b*	*W*	Cumulative days for *W*	*c*	*E* ($kJ\,mol^{-1}$)	Q_{10} (10–20°C)	r^2	*n*
DC	Cereal	23.05**	0.23*	Precipitation	8	0.09**	42.47**	1.81	0.62**	156
	Fallow	22.87**	–	Dryness	15	0.30**	41.42**	1.79	0.49**	60
	Pasture	28.94**	2.63*	Dryness	8	0.04	62.04**	2.39	0.65**	34
SC	Cereal	25.58**	0.15	Precipitation	15	0.19**	48.31**	1.97	0.40**	417
	Fallow	27.86**	0.68*	Dryness	9	0.08**	54.38**	2.14	0.52**	51
	Pasture	29.95**	–	Precipitation	10	0.17**	58.81**	2.28	0.65**	57
OC	Cereal	17.34**	0.33*	Precipitation	15	0.54**	31.34**	1.55	0.83**	148
	Fallow	28.54**	–	Precipitation	15	0.15	55.84**	2.19	0.72**	45
	Pasture	40.83**	–	Dryness	15	0.36**	82.53**	3.18	0.74**	38

*, **: Significant at 5% and 1% levels, respectively.
$C_{cm} = aC_0^b\, W^e\, e^{E/RT}$
C_{cm}: Daily CO_2 emission ($kg\,C\,ha^{-1}\,d^{-1}$
C_0: Potentially mineralizable C ($kg\,C\,ha^{-1}$)
W: Cumulative precipitation or dryness factor; cumulative days is tested from 2 to 15 days, selected by the highest r^2.
E: Activation energy ($J\,mol^{-1}$)
R: Gas constant ($8.31\,J\,mol^{-1}\,K^{-1}$)
T: Soil Temperature (K)
a, b, c: Coefficients

Table 17. Annual (from Apr. to Oct.) carbon flow and carbon budget of crop rotation system.

Site	Land use	Year	Carbon input as plant residue	Carbon output	Carbon budget	Estimation under different fallow frequencies			
			average (S.E)	average (S.E)		Average carbon budget (2004 and 2005)	3-year rotation (F*1–C*2–C)	4-year rotation (F–C–C–C)	5-year rotation (F–C–C–C–C)
			(Mg C ha^{-1})						
DC	Cereal	2004	0.75 (0.07)	0.75 (0.01)	0.00	0.10	−0.51	−0.42	−0.32
		2005	0.99 (0.11)	0.80 (0.01)	0.19				
	Fallow	2004	0.02	0.72 (0.01)	−0.70	−0.70			
		2005	0.02	0.71 (0.00)	−0.69				
	Pasture	2004	1.56 (0.31)	0.61 (0.06)	0.95	0.81			
		2005	1.39 (0.40)	0.68 (0.08)	0.66				
SC	Cereal	2002	1.82 (0.17)	1.06 (0.02)	0.77				
		2004	1.35 (0.01)	1.01 (0.02)	0.36	0.27	−0.30	−0.03	0.25
		2005	1.32 (0.07)	1.14 (0.02)	0.18				
	Fallow	2004	0.02	0.92	−0.90	−0.84			
		2005	0.02	0.79 (0.03)	−0.77				
	Pasture	2004	2.11 (0.24)	0.66 (0.01)	1.45	1.17			
		2005	1.61 (0.12)	0.73 (0.02)	0.88				
OC	Cereal	2004	1.31 (0.01)	0.92 (0.01)	0.39	0.35	−0.10	0.25	0.60
		2005	1.44 (0.25)	1.13 (0.02)	0.31				
	Fallow	2004	0.02	0.85 (0.02)	−0.83	−0.80			
		2005	0.02	0.79	−0.77				
	Pasture	2004	2.50 (0.35)	1.22 (0.01)	1.28	1.26			
		2005	2.57 (0.37)	1.34 (0.02)	1.23				

* 1: Fallow, * 2: Cereal

of the mean annual CO_2 emission of cereal fields ranged from 0.75 (DC; 2004) to 1.14 (SC; 2005) $Mg\,C\,ha^{-1}$. These emission values are similar to those from the Canadian prairie (de Jong, 1974), but lower than those for Eurasian Chernozems (Vugakov and Popova, 1968; Larionova et al., 1998). In the SC site, the annual soil respiration at the SC site was low (1.01–1.14 $Mg\,C\,ha^{-1}$) compared with that for the adjacent experimental station (2.5–3.2 $Mg\,C\,ha^{-1}$) in the previous section). This discrepancy in results may be due to more intensive management in the experimental field, through snow management, fertilization, herbicide application and deep cultivation, all of which magnify the C dynamics by increasing C input and output.

Estimates of the mean annual CO_2 emission of fallow and pasture fields ranged from 0.71 (DC; 2005) to 0.92 (SC; 2004) $Mg\,C\,ha^{-1}$ and from 0.61 (DC; 2004) to 1.34 (OC; 2004) $Mg\,C\,ha^{-1}$, respectively. Annual CO_2 emission of cereal and pasture fields in 2005 was slightly higher than that in 2004 at each site, because precipitation in 2005 was higher than that in 2004, which greatly affected water availability for soil microbes.

The C input as plant residues in 2004–2005 in cereal fields ranged from 0.75 (DC; 2004) to 1.44 (OC; 2005) $Mg\,C\,ha^{-1}$, with the highest value at the OC site, followed by that at SC and DC sites. There was extremely high C input at SC site in 2002. The high C input as plant residues in 2002 may be from the record precipitation received in June at SC site and, therefore, exceptional. The annual C budget, calculated by subtracting annual CO_2 emission from C input as plant residues, varied between 0.00 (DC; 2004) to 0.77 (SC; 2002) $Mg\,C\,ha^{-1}$. The cereal cropping fields at all sites contributed to C sequestration. Burning plant residue is an occasional practice in this study area, and it has agronomic advantages linked to soil management such as tillage. However, not burning is advantageous to long-term increase in SOM and its impact on aggregation (Crovetto, 1996). Therefore, the benefits of using plant residues as mulch to C sequestration must be carefully considered.

The C budget of summer fallow fields ranged from −0.90 (SC; 2004) to −0.69 (DC; 2005) $Mg\,C\,ha^{-1}$. Based on the average values of the annual C budget for each land use stage in the respective study sites, the C budget is calculated for whole rotation systems with different fallowing frequencies. The average value of 2004 and 2005 was used for this calculation, since the input of plant residue at SC in 2002 was exceptionally high due to better moisture conditions during that year. Thus, estimates of C budget of a 3-year crop rotation system ranged from −0.51 (DC) to −0.10 (OC) $Mg\,C\,ha^{-1}$. The negative C budget at all sites indicates that a 3-year crop rotation system depleted the SOC pool in all sites. In contrast, estimates of C budget of a 5-year crop rotation system ranged from −0.32 (DC) to 0.60 (OC) $Mg\,C\,ha^{-1}$, with only the DC site contributing as a C source. The adverse effect of summer fallow on C budget is exacerbated by the drier climatic condition. Thus, increasing water storage is extremely important to C budget. In pasture fields, on the other hand, the average C budget in 2004 and 2005 ranged from 0.81 (DC) to 1.26 (OC) $Mg\,C\,ha^{-1}$ and pasture management at all sites contributed to C sequestration. Therefore, inclusion of pasture into the crop rotation system in some topographical location is beneficial to improving the overall SOM status.

8 CONCLUSION

Several experiments related to dynamics of water and SOM were conducted in 1998–2000 at Barayev Kazakh Research and Production Center of Grain Farming in northern Kazakhstan. Predominant soils of the study sites are Southern Chernozems or Haplustolls with mean annual temperature of 1.6°C and total annual precipitation of 324 mm. The management treatments comprised of a variety of the frequencies in summer fallow and other water-harvesting management techniques, i.e. snow collection and subsoil cutting.

The data on soil moisture and temperature regimes showed that both summer fallow and snow collection practices accumulated additional ca. 100 mm moisture (0–90 cm soil) when practiced individually. When used together, benefit of 100 mm additional moisture from snow collection was negated by the summer fallow, which induced subsoil freezing even during early spring and reduced percolation and loss of thawed snow through surface runoff and evaporation. Snow collection with subsoil cutting management is more important than summer fallowing for capturing additional moisture for better plant growth.

The data on SOM dynamics in a cropping season showed that the CO_2 emission modeled with Arrhenius relationship based on daily soil respiration rate and its determining factors (i.e. soil temperature, soil moisture, and the activation energy) was 2.5–3.2 Mg C ha^{-1} and the input of plant residues ranged from 0 (fallow) to 4.5 Mg C ha^{-1} (cropping oat). Thus, the SOM budget ranged from −2.9 (fallow) to 1.3 Mg C ha^{-1} (oat) with an average of −0.6 Mg C ha^{-1} under wheat cropping. The SOM concentration is generally stable under wheat cropping, but decreases under summer fallow, resulting in the gradual decrease by 2.9 Mg C ha^{-1} in a conventional rotation cycle for 5 years which may be equivalent to 4% of the total SOM pool in the plow layer (30 cm).

With reference to changes in soil quality, frequent summer fallowing decreased SOC, total N, PMC, and LF-C. The decrease was stronger in PMC and LF-C than in SOC and total N. The PMN, min-N, and LF-N were less affected by the frequency of summer fallowing, but more by the fallowing phase The, pre-fallowing phase resulted in significantly higher PMN and LF-N and lower min-N than post-fallowing phase. The SOM depletion due to summer fallowing can be observed with the changes in soil characteristics related to soil C status, particularly those affecting microbial activity (i.e. PMC and LF-C). During the summer fallow, soil N fertility increases through conversion of PMN into min-N, which is, however, subjected to surface runoff and/or leaching during the snow thawing period in spring before being absorbed by crops.

The SOM dynamics under farmers' practices were also studied in cropped fields in different regions of northern Kazakhstan, i.e. DC (N51°12′, E71°06′), SC (N51°40′, E71°02′; near the Center in Shortandy) and OC (N53°57′, E69°32′), between 2001 and 2005. In the SC study field (14 × 5 km), spatial patterns of the field properties strongly suggested that the SOM dynamics in the field was markedly affected by the topography and that the most favorable area for the storage of SOM was slightly different than that for crop production. The yield was relatively high and the SOC pool was moderate in the north-facing slope area, the yield was moderate and the SOC pool and release of CO_2 from soil were the highest in the central plateau, whereas the yield and soil C dynamics were relatively low in the south-facing slope area. These results suggested that a site-specific agricultural management based on the spatial patterns of SOM dynamics is a viable option for the promotion of sustainable agricultural production and for reducing SOM decomposition or soil degradation.

Then the SOM budget was determined in the three sites with special reference to land use (i.e. cereal cropping, fallow and pasture management). Estimates of annual CO_2 emission in 2004 and 2005 ranged from 0.75 to 1.14 Mg C ha^{-1} in cereal field, 0.71 to 0.92 Mg C ha^{-1} in fallow field, and 0.61 to 1.34 Mg C ha^{-1} in pasture field. Annual C budget was calculated by subtracting the annual CO_2 emission from C input as plant residues. In cereal fields, it ranged from 0.00 to 0.77 Mg C ha^{-1}, indicating that fields under cereal cropping at all sites contributed to C sequestration during the cultivation year. The C budget of summer fallow fields, however, was always negative, ranging from −0.90 to −0.69 Mg C ha^{-1}. Computations of the C budget for the whole rotation systems showed that the adverse effect of summer fallow increases as the climatic becomes drier, and the importance of capturing water increases. On the contrary, the C budget in pasture fields ranged from 0.66 to 1.45 Mg C ha^{-1} and pasture management at all sites contributed to C sequestration. The extensive use of summer fallowing must be reconsidered to reduce the loss of SOM. Based on the data of water dynamics, the snow management is recommended as an alternative approach for improving soil moisture conditions at some topographical locations. In addition, it is recommended that pasture management be introduced as a part of the crop rotation system, especially in drier soils such as the DC sites on some topographical locations. Since the spatial dependency of PMC and C input as plant residue were well developed in this region and such spatial patterns of the SOM dynamics under grain farming are markedly affected by the topography, site-specific management is a viable option. The latter includes appropriate arrangement of fallow and/or pasture management in time and space as well as differentiated water capturing strategies. Site-specific management is an alternative approach for sustainable agriculture in this region.

REFERENCES

Aase, J.K. and F.H. Siddoway. 1980. Stubble height effects on seasonal microclimate, water balance and development of no-till winter wheat. Agricultural Meteorology 21:1–20.

Alvarez, R., O.J. Santanatoglia, and R. Garcia. 1995. Soil respiration and carbon inputs from crops in a wheat-soybean rotation under different tillage systems. Soil Use and Management 11:45–50.

Amelung, W., K.W. Flach, and W. Zech. 1998. Climatic effects on soil organic matter composition in the Great Plains. Soil Science Society America Journal 61:115–123.

Anderson, J.M. 1991. The effects of climate change on decomposition processes in grassland and coniferous forests. Ecological Applications 1:326–347.

Baldock, J.A. and P.N. Nelson. 2000. Soil organic matter. p. 25–84. In: M.E. Summner (ed.), Handbook of Soil Science. CRC Press, Boca Raton, FL.

Batjes, N.H. and W.G. Sombroek. 1997. Possibilities for carbon sequestration in tropical and subtropical soils. Global Change Biology 3:161–173.

Beare M.H. P.F. Hendrix, and D.C. Coleman. 1994. Water-stable aggregates and organic matter fractions in conventional- and no-tillage soils. Soil Science Society America Journal 58:777–786.

Biederbeck, V.O., H.H. Janzen, C.A. Campbell, and R.P. Zentner. 1994. Labile soil organic matter as influence by cropping practices in an arid environment. Soil Biology and Biochemistry 12:1647–1656.

Bird, M.I., A.R. Chivas, and J. Head., 1996. A latitudinal gradient in carbon turnover times in forest soils. Nature 381:143–146.

Black, A.L. and J.F., Power. 1965. Effect of chemical and mechanical fallow methods on moisture storage, wheat yields, and soil fertility. Soil Science Society America Proceedings 29:465–468.

Brookes, P.C., A. Landman, G. Pruden, and D.S. Jenkinson. 1985. Chloroform fumigation and the release of soil nitrogen: A rapid direct extraction method to measure microbial biomass nitrogen in soil. Soil Biology and Biochemistry 17:837–842.

Buyanovsky, G.A., C.L. Kucera, and G.H. Wagner. 1987. Comparative analysis of carbon dynamics in native and cultivated ecosystems. Ecology 68:2023–2031.

Buyanovsky, G.A., G.H. Wangner, and C.H. Gantzer. 1986. Soil respiration in a winter ecosystem. Soil Science Society America Journal 50:338–344.

Campbell, C.A., A.P. Moulin, K.E. Bowren, H.H. Janzen, L. Townly-Smith, and V.O. Biederbeck. 1992. Effect of crop rotation on microbial biomass, specific respiratory activity and mineralizable nitrogen in a Black Chernozemic soil. Canadian Journal Soil Science 72:417–427.

Campbell, C.A. and R.P. Zentner. 1993. Soil organic matter as influenced by crop rotations and fertilization. Soil Science Society America Journal 57:1034–1040.

Campbell, C.A., V.O. Biederbeck, B.G. McConkey, D. Curtin, and R.P. Zentner. 1999. Soil quality-effect of tillage and fallow frequency. Soil organic matter quality as influenced by tillage and fallow frequency in a silt loam in southwestern Saskatchewan. Soil Biology and Biochemistry 31:1–7.

Caprio, J.M., G.K. Grunwald, and R.D. Snyder. 1989. Conservation and storage of snowmelt in stubble land and fallow under alternate fallow-strip cropping management in Montana. Agricultural and Forest Meteorology 45:265–279.

Chepil, W.S., F.H. Siddoway, and D.V. Armbrust. 1964. Prevailing wind erosion direction in the Great Plains. Journal Soil and Water Conservation 19:67–70.

Chuluun, T. and D. Ojima. 2002. Land use change and carbon cycle in arid and semi-arid lands of East and Central Asia. Science in China (Series C) 45: 48–54.

Christensen, B.T., 1992. Physical fractionation of soil and organic matter in primary particles and density separates. Advances in Agriculture 20:2–90.

Christensen B.T. 1996. Carbon in primary and secondary organo-mineral complexes. Advances in Soil Science 24:97–165.

CIMMYT, 2000. World wheat overview and outlook 2000–2001. International Maize and Wheat Improvement Center, Texcoco, Mexico.

Clarke, A.L. and J.S. Russell. 1977. Crop sequential practices. p. 279–300. In: J.S. Russell and E.L. Greacen (eds.), Soil Factors in Crop Production in a Semi-arid Environment. University of Queensland Press, St. Lucia.

Collins, H.P., P.E. Rasmussen, and C.L. Douglas. 1992. Crop rotation and residue management effects on soil carbon and microbial dynamics. Soil Science Society America Journal 56: 783–788.

Crovetto, C. 1996. Stubble over the soil: the vital role of plant residue and soil management to improve soil quality. American Society of Agronomy, Madison, WI, 245 p.

Davidson, E.A., L.V. Verchot, C.J. Henrique, I.L. Ackerman, and J.E.M. Carvalho. 2000. Effects of soil water content on soil respiration in forests and cattle pastures of eastern Amazonia. Biogeochemistry 48:53–69.

de Jong. 1974. Soil aeration as affected by slope position and vegetative cover. Soil Science 131:34–43.

Douglas Jr., C.L., R.W. Rickman, B.L. Klepper, J.F. Zuzel, and D.J. Wysocki. 1992. Agroclimatic zones for dryland winter wheat producing areas of Idaho, Washington, and Oregon. Northwest Science 66:26–34.

Drinkwater, L.E., P. Wagoner, and M. Sarrantonio. 1998. Legume-based cropping systems have reduced carbon and nitrogen losses. Nature:396:262–265.

Duiker, S.W. and R. Lal. 2000. Carbon budget study using CO_2 flux measurements from a no till system in central Ohio. Soil Tillage Research 54:21–30.

Dzhalankuzov, T.D. and V.V. Redkov. 1993. Changes in morphological and agrochemical properties of calcareous Southern Chernozems of North Kazakhstan due to long-term cultivation. p. 53–58. In: Proceedings of Academy of Sciences of Republic of Kazakhstan. Biology Series 1 (in Russian).

Elliott, E.T., K. Paustain, and S.D. Frey. 1996. Modeling the measurable or measuring the modelable: a hierarchical approach to isolating meaningful soil organic matter. p. 407–419. In: E.T. Elliott and C.A. Cambardella (eds.), Physical Separation of Soil Organic Matter. Agriculture, Ecosystems and Environment 34.

FAO, 1998. AQUASTAT – FAO's Information System on Water and Agriculture. http://www.fao.org/ag/AGL/aglw/aquastat/countries/.

FAO, 2004. Carbon sequestration in dryland soils. Food and Agriculutre Organizaton of United Nations, Rome, Italy, pp.108.

FAO, 2005. FAO statistical year book 2005/2006, Food and Agriculutre Organizaton of United Nations, Rome, Italy, http://www.fao.org/statistics/yearbook/vol_1_1/index_en.asp.

Farahani, H.J., G.A. Peterson, D.G. Westfall, L.A. Sherrod, and L.R. Ahuja. 1998. Soil water storage in dryland cropping systems: The significance of cropping intensification. Soil Science Society America Journal 62:984–991.

Ferguson, W.S. and B.J. Gorby. 1971. Effect of various periods of seed-down to alfalfa and bromegrass on soil nitrogen. Canadian Journal Soil Science 51:65–73.

Fortin, M.C., P. Rochette, and E. Pattey. 1996. Soil carbon dioxide fluxes from conventional and no-tillage small-grain cropping systems. Soil Science Society America Journal 60:1541–1547.

Glaser, B., M.B. Turrion, D. Solomon, A. Ni, and W. Zech. 2000. Soil organic matter and quality in mountain soils of the Alay range, Kyrgyzia, affected by land use change. Biology and Fertility of Soils 31:407–413.

Gossen, E. 1998. Agrolandscape agriculture and forestry management as the basis of sustainable grain production in the steppes of Eurasia. p. 44–48. In: Spring Wheat in Kazakhstan: Current Status and Future Directions. Proceedings of the Kazakhstan-CIMMYT Conference, 1997. Shortandy, Akmola, Kazakhstan.

Gregorich, E.G., M.R. Carter, D.A. Angers, C.M. Monreal, and B.H. Ellert. 1994. Towards a minimum data set to access soil organic matter quality in agricultural soils. Canadian Journal Soil Science 74:367–385.

Hardy, J.P., P.M. Groffman, R.D. Fitzhugh, K.S. Henry, A.T. Welman, J.D. Demers, T.J. Fahey, C.T. Driscoll, G.L. Tierney, and S. Nolan. 2001. Snow depth manipulation and its influence on soil frost and water dynamics in a northern hardwood forest. Biogeochemistry 56:151–174.

Hargreaves, G.H. and Z.A. Samani. 1982. Estimation potential evapotranspiration. Journal Irrigation and Drainage Engineering 223–230.

Hargreaves, G.L., G.H. Hargreaves, and J.P. Riley. 1985. Irrigation water requirements for Senegal River Basin. J. Irrig. Drain. Eng., ASCE 111:265–275.

Haynes, R.J. 2000. Labile organic matter as an indicator of organic matter quality in arable and pastoral soils in New Zealand. Soil Biology and Biochemistry 32:211–219.

Houghton, J.T., L.G. Meria Filho, B.A. Callander, N. Harris, A. Kattenberg, and K. Maskell. 1996. Climate change 1995: The science of climate change, contribution of working group 1 to the second assessment report of the Intergovermental Panel on Climatic Change. Cambridge, England: Cambridge University Press.

Huxman, T.E., K.A. Snyder, D. Tissue, A.J. Leffler, K. Ogle, W.T. Pockman, D.R. Sandquist, and S. Schwinning. 2004. Precipitation pulses and carbon fluxes in semiarid and arid ecosystems. Oecologia 141:317–324.

Janzen, H.H. 1987. Soil organic matter characteristics after long-term cropping to various spring wheat rotations. Canadian Journal Soil Science 67:845–856.

Janzen, H.H., C.A. Campbell, S.A. Brandt, G.P. LaFond, and L. Townley-Smith. 1992. Light-fraction organic matter in soils from long-term crop rotations. Soil Science Society America Journal 56:1799–1806.

Johnsson, H. and L-C. Lundin, 1991. Surface runoff and soil water percolation as affected by snow and soil frost. Journal Hydrology 122:141–159.

Kessavalou, A., J.W. Doran, A.R. Mosier, and R.A. Drijber. 1998. Greenhouse gas fluxes following tillage and wetting in a wheat-fallow cropping system. Journal Environmental Quality 27:1105–1116.

Kharin, N.G., R. Tateishi, and I.G. Gringof. 1998. Use of NOAA AVHRR data for assessment of precipitation and land degradation in Central Asia. Arid Ecosystem 4:25–34.

Kokovina, T.P. 1979. Soil processes in a plowed typical thick chernozem. Soviet Soil Science 11:494–503.

Kudeyarov, V.N. and I.N. Kurganova. 1998. Carbon dioxide emissions and net primary production of Russian terrestrial ecosystems. Biology and Fertility of Soils 27:246–250.

Lal, R. 2002. Carbon sequestration in dryland ecosystems of west Asia and north Africa, Land Degradation and Development 13:45–59.

Lal R, 2004. Carbon sequestration in soils of central Asia, Land Degradation and Development, 15, 563–572.

Larionova, A.A., A.M. Yermolayev, S.A. Blagodatsky, L.N. Rozanova, I.V. Yevdokimov, and D.B. Orlinsky. 1998. Soil respiration and carbon balance of gray forest soils as affected by land use. Biology and Fertility of Soils 27:251–257.

Lee, L.J., C.W. Wood, D.W. Reeves, J.A. Entry, and R.L. Raper. 1996. Interactive effects of wheel-traffic and tillage system on soil carbon and nitrogen. Communication Soil Science and Plant Analysis 27:3027–3043.

Leggett, G.E. 1959. Relationships between wheat yield, available moisture and available nitrogen in eastern Washington dry land areas. Bulletin No. 609, Washington Agricultural Experiment Station, Institute of Agricultural Sciences, Washington State University (cited from Fuentes, J.P., M. Flury, D.R. Huggins, and D.F. Bezdicek. 2003. Soil water and nitrogen dynamics in dryland cropping systems of Washington State, USA. Soil and Tillage Research 71:33–47).

Lioubimtseva, E., R. Cole, J.M. Adams, and G. Kapustin. 2005. Impacts of climate and land-cover changes in arid lands of Central Asia. Journal Arid Environment 62:258–308.

Longmire, J. and A. Moldashev. 1999. Changing competitiveness of the wheat sector of Kazakhstan and source of future productivity growth, Economics working paper 99–06, International Maize and Wheat Improvement Center (CIMMYT), Texcoco, Mexico. 45 p.

Lykov, A.M. and A.M. Tulikov. 1976. Practical work on agriculture with basics of soil science. M. Kolos.

Mamilov, Sh.Z., K.O. Beisenova, A.Sh. Mamilov, and M.K. Yanovskaya. 1985. Biological activity and dynamic of nutrients in Chernozem soils with different agronomic use. p. 132–134, In: State and Rational Use of Soils of Kazakhstan, Soil Science Society of Kazakhstan, Alma-Ata (in Russian).

Mathews, O.R. and T.J. Army. 1960. Moisture storage on fallowed wheatland in the Great Plains. Soil Science Society America Proceedings 24:414–418.

Meng, E. and A. Morgounov. 2000. Changing competitiveness of the wheat sector of Kazakhstan, CIMMYT Economics Working Paper, International Maize and Wheat Improvement Center, Texcoco, Mexico. 4 p.

Mikhailova, E.A., R.B. Bryant, I.I. Vaasenev, S.J. Schwager, and C.J. Post. 2000. Cultivation effects on soil organic carbon and nitrogen contents at depth in the Russian Chernozem. Soil Science Society America Journal 64:738–745.

Mizina, S.V., I.B. Eserkepova, and V.R. Sutyushev. 1997. Wheat Voluerability Assessment under possible climate change in Kazakhstan. Hydrometeorology Ecology 3:64–72.

Morgounov, A., M. Karabayev, D. Bedoshvili, and H.J. Braun. 2001. Improving wheat production in Central Asia and the Caucasus. p. 65–68. In: Research highlights of the CIMMYT Wheat Program 1999–2000. International Maize and Wheat Improvement Center, Texcoco, Mexico.

Morgounov, A. and L. Zuidema. 2001. The legacy of the Soviet agricultural research system for the Republics of central Asia and the Caucasus. ISNAR Research Report, 20, 52 p.

Nobe, K. 1998. The agriculture of Kazakhstan during shifting socio-economic systems. Nougyou Sougou Kenkyuu, 54, 1–111 (in Japanese).

Oliver, M.A. 1987. Geostatistics and its application to soil science. Soil Use and Management, 3:8–20.

Orchard, V.A. and F.J. Cook. 1983. Relationship between soil respiration and soil moisture. Soil Biology and Biochemistry 15:447–453.

Paustian, K., O. Andren, H.H. Janzen, R. Lal, P. Smith, G. Tian, H. Tiessen, M. Van Noordwijk, and P.L. Woomer. 1997. Agricultural soils as a sink to mitigate CO_2 emissions. Soil Use and Management 13:230–244.

Poluektov, Y.V. 1983. Protection of soils from erosion through strip cropping. Soviet Soil Science 15:79–84.

Prior, S.A., H.H. Rogers, G.B. Runion, H.A. Torbert and D.C. Reicosky. 1997. Carbon dioxide enriched agroecosystems: Influence of tillage on short-term soil carbon dioxide efflux. Journal Environmental Quality 26: 244–252.

Raich, J.W. and W.H. Schlesinger. 1992. The global carbon dioxide flux in soil respiration and its relationship to vegetation and climate. Tellus 44B:81–99.

Redkov, V.V. 1964. Soils of Tselinograd oblast. Nauka, Alma-Ata, p.325–326. (in Russian).

Robertson, G.P. 1998. GS+: Geostatistics for the environmental sciences. Gamma Design Software, Plainwell, MI.

Rochette, P., R.L. Desjardins, and E. Pattey. 1991. Spatial and temporal variability of soil respiration in agricultural fields. Canadian Journal Soil Science 71:189–196.

Rubinstein, M.I. 1959. Decomposition rate of organic matter of virgin Chernozem in Northern Kazakhstan during their cultivation. Soviet Soil Science 11:1332–1335.

Schinner, F. 1982. Soil microbial activities and litter decomposition related to altitude. Plant and Soil 65:87–94.

Schlentner, R.E. and K. Van Cleve. 1985. Relationships between CO_2 evolution from soil, substrate temperature, and substrate moisture in four mature forest types in interior Alaska. Canadian Journal Forest Research 15:97–106.

Schomberg, H.H. and O.R. Jones. 1999. Carbon and nitrogen conservation in dryland tillage and cropping systems. Soil Science Society America Journal 63:1359–1366.

Shibusawa, S. 1999. Introduction for precision agriculture in U.S. Journal Japanese Society Agricultural Machinery 61:7–12.

Shegebaev, O.S. 1998. Scientific support for spring wheat production in Kazakhstan. p. 24–29. In: A.I. Morgounov, A. Satybaldin, S. Rajaram, and A. McNab (eds.), Spring Wheat in Kazakstan: Current Status and Future Directions. Proceedings of the Kazakstan-CIMMYT Conference; Shortandy, Akmola, Kazakstan; 22–24 Sep 1997. International Maize and Wheat Improvement Center, Texcoco, Mexico.

Shields, J.A. and E.A. Paul. 1973. Decomposition of ^{14}C-labelled plant material under field conditions. Canadian Journal Soil Science 53:297–306.

Shkonde, E.I., M.V. Lola, and D.N. Pryanishnikov. 1982. Effect of fertilizers and plants on the content and composition of humus in Ciscaucasian chernozem. Soviet Soil Science 14:626–634.

Singh, K.P. and C. Shekhar. 1986. Seasonal pattern of total soil respiration, its fractionation and soil carbon balance in a wheat-maize rotation cropland at Varanasi. Pedobiologia 29:305–318.

Six, J, C. Feller, K. Denef, S.M. Ogle, Sa.J.C. De Moraes, and A. Alberecht, 2002. Soil organic matter, biota and aggregation in temperate and tropical soils – Effects of no-tillage. Agronomie 22:755–775.

Smika, D.E. and G.A. Wicks. 1968. Soil water storage during fallow in the Central Great Plains as influenced by tillage and herbicide treatments. Soil Science Society America Proceedings 32:591–595.

Soil Survey Staff. 2003. Keys to Soil Taxonomy. Ninth Edition. U.S. Department of Agriculture and National Resources Conservation Service, Washington. 332 p.

Sollins, P., G. Spycher, and C.A. Glassman. 1984. Net nitrogen mineralization from light and heavy-fraction forest soil organic matter. Soil Biology and Biochemistry 16:31–37.

Sorokina, N.P. and V.M. Kogut. 1997. The dynamics of humus content in arable chernozems and approaches to its study. Eurasian Soil Science 30:146–151.

SPSS Inc. 1998a. SYSTAT version 8.0, Statistics, Chicago, IL. 1086 p.

SPSS Inc. 1998b. SigmaPlot version 5.0, Programming guide, Chicago, IL.

Spycher, G., P. Sollins, and S. Rose. 1981. Carbon and nitrogen in the light fraction of a forest soil: vertical distribution and seasonal patterns. Soil Science 2:79–87.

Srivastava, J. and E. Meyer. 1998. Is conservation tillage a viable option in the CIS? World Bank Report. 31 p.

Suleimenov, M., K.A. Akshalov, Z. Kaskarbayev, A. Kireyev, L. Martynova, and R. Medeubayev. 2005. Role of wheat in diversified cropping systems in dryland agriculture of Central Asia. Turkish Journal Agriculture and Forestry 29:143–150.

Tang, J., D.D. Baldocchi, Y. Qi, and L. Xu. 2003: Assessing soil CO_2 efflux using continuous measurements of CO_2 profiles in soils with small solid-state sensors. Agricultural and Forest Meteorology 118:207–220.

Trumbore, S.E., O.A. Chadwick, and R. Amundson. 1996. Rapid exchange between soil carbon and atmospheric carbon dioxide driven by temperature change. Science 272:393–396.

Vance, E.D., P.C. Brookes, and D.S. Jenkinson. 1987. An extraction method for measuring soil microbial biomass C. Soil Biology and Biochemistry 19:703–707.

Vladimir, A., I. Shutov, and L. Kaljuzny. 2001. Snow management in agricultural landscapes, Hydrosphere. 6 p.

Vugakov, P.S. and Y.E.P. Popova. 1968. Carbon dioxide regime in soils of the Krasnoyarsk forest steppe. Soviet Soil Science 1:795–801.

Webster, R. and M.A. Oliver. 2001. Geostatistics for environmental scientists. John Wiley & Sons, Chichester, UK.

World Bank. 2003. Kazakhstan drylands management project. Project proposal document, 25929-KZ.

Yanai, J., C.K. Lee, T. Kaho, M. Iida, T. Matsui, M. Umeda, and T. Kosaki. 2001: Geostatistical analysis of soil chemical properties and rice yield in a paddy field and application to the analysis of yield-determining factors. Soil Science and Plant Nutrition 47:291–301.
Xu, M. and Y. Qi. 2001. Spatial and seasonal variations of Q_{10} determined by soil respiration measurements at a Sierra Nevadan forest. Global Biogeochemical Cycles 15:687–696.
Yu, F., K.P. Price, J. Ellis, and P. Shi. 2003. Response of seasonal vegetation development to climatic variations in eastern central Asia. Remote Sensing of Environment 87:42–54.

CHAPTER 23

Conservation agriculture in the steppes of Northern Kazakhstan: The potential for adoption and carbon sequestration

P.C. Wall
International Maize and Wheat Improvement Center (CIMMYT), Harare, Zimbabwe

N. Yushenko
Central Kazakh Agricultural Research Institute, Karaganda, Kazakhstan

M. Karabayev
International Maize and Wheat Improvement Center (CIMMYT), Astana, Kazakhstan

A. Morgounov
International Maize and Wheat Improvement Center (CIMMYT), Ankara, Turkey

A. Akramhanov
International Maize and Wheat Improvement Center (CIMMYT), Astana, Kazakhstan

1 THE AGRICULTURAL ENVIRONMENT OF NORTHERN KAZAKHSTAN

1.1 *Climate and soils*

The steppes of northern Kazakhstan are representative of the vast Eurasian grasslands spreading from the foothills of the southern Ural Mountains in the west to the foothills of the Altai Mountains in the east (Gilmanov *et al.*, 2000) (Figure 1). In general there is a gradient in annual precipitation from approximately 350 mm in the northern and northwestern steppes to 200 mm or less in the southern and south-eastern steppes. Some precipitation can be expected in all months, with approximately one third of annual precipitation falling as winter snow. July is the wettest month, although drought periods in July are also common. The soils of the region are Chernozems and Kastanozems (chestnut soils), with Chernozems predominating in the wetter, north and north-western regions and becoming increasingly lighter in the lower rainfall areas. The black chernozem soils have approximately soil organic matter 6.0% (SOM) (1 Mha), the southern black chernozems (e.g. around Kostanay) have approximately 4.0% SOM (1.5 million ha) and the chestnut soils approx. 3.5% SOM (V. Dvurechenski, personal communication).

Results reported in this paper are from the Kazakh Research Institute of Grain Farming, (KRIGF) near the town of Shortandy, about 40 km NNW of Astana (51° 40′ N, 71° 00′ E, 367 m a.s.l.) and the Central Kazakh Agricultural Research Institute (CKARI) in the southern steppe region about 60 km NW of the city of Karaganda (50° 10′ N, 72° 37′ E, 540 m.a.s.l.). The SOM levels of the natural steppe at the KRIGF are approximately 3.5% (Suleimenov et al., 2005). Mean annual temperature at Shortandy is 1.6°C, and mean annual rainfall 323 mm (Figure 2).

1.2 *History of tillage*

Traditionally the steppes were extensively grazed, but from 1954 to 1964 about 18 Mha were opened up to agriculture under a grain production scheme of the USSR (Gossen, 1998). Intensive

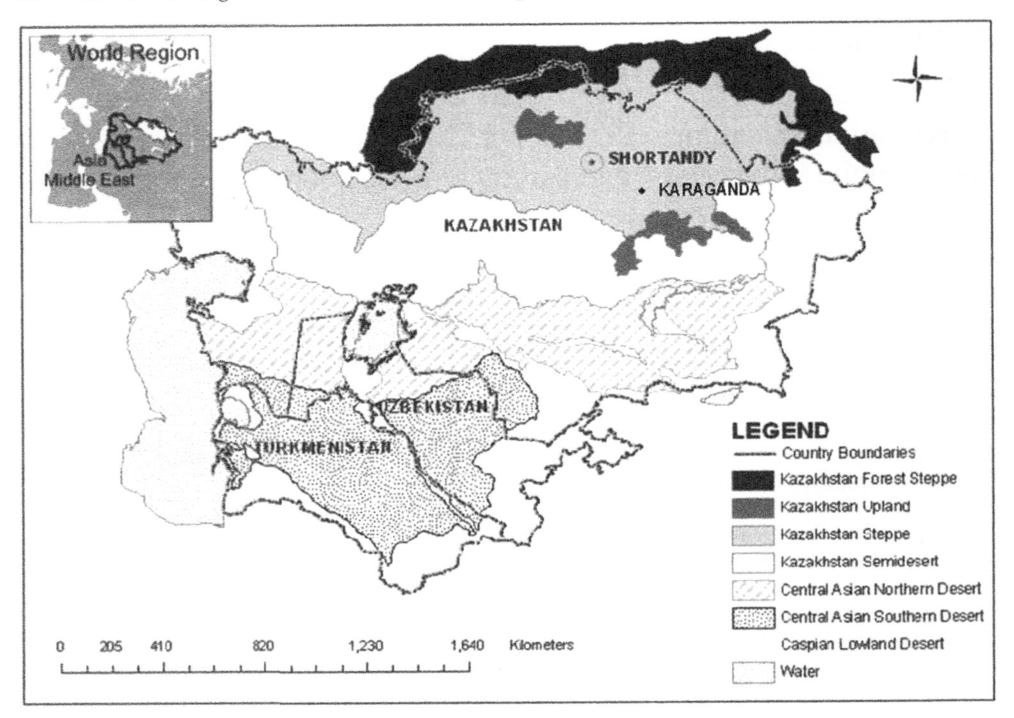

Figure 1. Location of the Shortandy and Karaganda sites on the ecoregion map of Central Asia (map modified from Gilmanov et al., 2000).

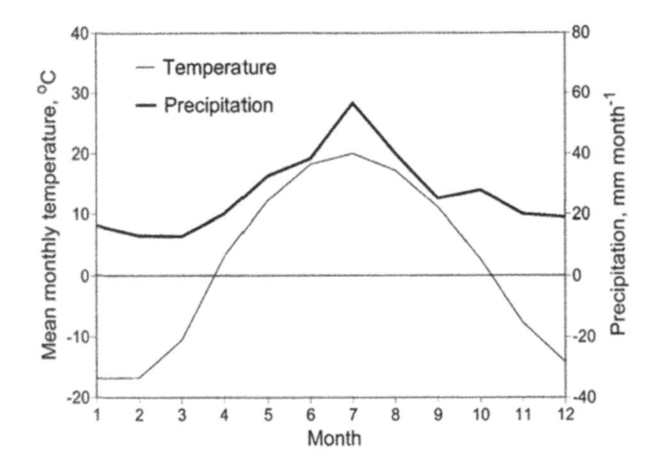

Figure 2. Climatic diagram for the study site near Shortandy, Kazakhstan (long-term means for 1936–2000 for Shortandy meteorological station, Kazakh Research Institute of Grain Farming; graph from Gilmanov et al., 2000).

tillage soon led to severe problems of wind erosion, leading to research on a reduced tillage system at the KRIGF by a research team headed by Dr. A.I. Barayev. Once developed, the system, based on deep tillage with a chisel plow in the fall (to improve the water infiltration with snow melt) followed by a light harrowing in spring prior to seeding became the accepted practice under an

Table 1. Spring wheat yield (Mg ha^{-1}) in a 5-course grain-fallow rotation as affected by methods of main soil tillage on southern calcareous chernozem (A.I. Barayev, 1978, as reported by Suleimenov, 2000).

Soil tillage method	Crop after summer fallow			
	1st	2nd	3rd	4th
Plow	1.59	0.97	0.77	0.76
Sub-tillage	1.65	1.31	1.21	1.26
Gain (%)	+3.8	+35.0	+57.1	+65.8

edict from the Soviet Politiburo and was used on 62 million hectares in the steppe region of the USSR, including 20 million hectares in northern Kazakhstan.

Originally (1960s and 1970s) continuous wheat cropping was enforced in northern Kazakhstan, but researchers observed that the wheat-fallow system led to more stable yields. The common production system in the region today is continuous wheat, interspersed with occasional fallow years. The recommended frequency of the fallow is once every four to five years, but although this was followed up until independence in 1991, since then farmers have decreased the fallow frequency, as a response to low crop prices. Although little tillage is conducted for the wheat crop, tillage of the fallow is intensive, and commonly four to eight passes of a sweep cultivator are made to control weeds, leaving a clean "black fallow" (A.Kurishbayev, personal communication). Suleimenov *et al.* (2005) say that recently, because of the high cost of tillage, the black fallow is generally replaced by a weedy fallow. However, personal observations suggest that intensively tilled fallows are still quite common.

1.3 *Benefits of the fallow year*

Suleimenov (2000) reports results of Barayev showing the benefits of deep fall tillage in a rotation of fallow and four years of continuous wheat (Table 1). It is interesting to note that the benefits to deep fall tillage are small in the first year of wheat after fallow, but increase sequentially in both real and relative terms in the following wheat crops. This data is consistent with increases in profile moisture with deep fall tillage, but suggest that the effect of the fallow in moisture storage is limited: yield in the first crop after fallow was only 0.34 Mg ha^{-1} higher than the yield of the second wheat crop after fallow.

The reason for the benefits of the fallow season may have changed over time. Today it appears that the two major benefits associated with the fallow year are nitrogen (N) mineralization and weed control. The recommended practice for N fertilizer applications to wheat is to fertilize the second and subsequent wheat crops after fallow (generally with about 20 kg ha^{-1} N) but not the first wheat crop after fallow. Even so, declines in grain yield as the succession of cereal crops continues after the fallow year are common.

1.4 *Soil organic matter*

An accumulation of mineralized N during the fallow implies a break-down in SOM. Valentin Duvurushenski (personal communication) quotes data of A.A. Bektemirov on the decline in SOM in the soils of Kostanay Oblast in northern Kazakhstan with tillage (Table 2). Probably based on this data, Duvurushenski, the Director of the North-West Research and Agricultural Production Center in Kostanay, summarizes that tillage in that environment reduces SOM content by 0.5% per year: or by 20% of it's original level with 40 years of tillage. Recent data from Djalankuzov *et al.*, 2004 on five soil types from northern Kazakhstan show reductions in SOM not only in the A (tilled) horizon, but also in the B_1 horizon and in most soils a slight reduction even in the B_2 horizon (Table 3).

Table 2. Data from A.A. Bektemirov on the reduction of soil organic matter through long-term tillage in three soils of Kostanay Oblast, northern Kazakhstan (V. Dvuruchensky, personal communication).

	Soil Organic Matter (%)		
	Virgin	Cultivated	% Reduction
Common Chernozem	8.30	6.30	24
Southern Chernozem	5.30	4.22	20
Dark Chestnut (Dark Kastanozem)	4.10	3.40	17

Table 3. Soil organic matter (SOM) of different soils from northern Kazakhstan with and without a history of tillage (Data from Djalankuzov et al., 2004).

		% SOM in Soil Horizon		
	Land state	A	B_1	B_2
Common Chernozems	Virgin	7.6	4.5	2.6
	Tilled	6.0	3.8	2.6
Common carbonate Chernozems	Virgin	7.2	4.4	2.8
	Tilled	5.6	3.6	2.7
South Chernozems	Virgin	5.6	3.4	2.2
	Tilled	4.3	2.3	1.9
South carbonate Chernozems	Virgin	5.5	3.5	2.3
	Tilled	4.0	2.9	2.2
Dark-Chestnut carbonate	Virgin	3.9	2.6	1.8
	Tilled	2.8	2.2	1.6

In the reigning conservation tillage (CT) system in northern Kazakhstan, it is the fallow year that involves intensive tillage and causes the reduction in SOM, giving the short-term benefit of N mineralization. Suleimenov (2005) quotes data of Akhmetov et al., (1998), from a long-term trial at the KRIGF. The trial, started in 1962, included different frequencies of fallow. After 35 years, the original level of 3.9% SOM was reduced by 10% by a system with fallow every second year, by 11% in a system with fallow one year in three, and by 2% with fallow one year in four. With fallow one year in six or with continuous cropping no reduction in SOM was observed. However, Suleimenov stresses that the relatively small reductions in SOM observed in this trial may be due to the fact that all straw was spread and retained on the fields in this trial, whereas in common farmer practice much of it would have been burned. This possibly explains the difference in SOM reduction between the Kostanay and Akmola experience.

1.5 *Agricultural research and Perestroika*

Under the Soviet system, production, not economic efficiency, was the primary goal and farm managers were rewarded for high yields, not for economic surpluses. Production practices were centrally defined, and there were, therefore, few differences in input levels and production practices across extensive regions. Research tended to be centrally controlled, especially at the institute level, and very conservative, given the personal risks involved in making recommendations which might be judged not to be in the interests of the Party. This resulted in a wealth of long-term studies on agricultural systems, but unfortunately it is still difficult to access much of the accumulated data and information. The Soviet system also affected the way research was managed. As the primary goal was production and not economic efficiency, research was more oriented towards optimizing productivity. While sustainability was the focus of much of the research, the concept of client demand for technology was not prevalent, and the idea of talking to producers to ascertain their problems, and thus orient research programs, was very uncommon. Even today many researchers have difficulty in reorienting their research programs to reflect farmers' problems and economic efficiency.

1.6 *Agricultural machinery*

Before Perestroika, 10% of the agricultural machinery of Kazakhstan was renewed each year. This policy of programmed replacement ended abruptly in 1991, and little machinery has been replaced since then. Thus much of the farm machinery today is at least 14 years old, and some up to 23 years old. Obviously maintaining this equipment is costly, and much of it is beyond repair or extremely inefficient. This offers an opportunity for reduced tillage technologies, especially those technologies such as zero tillage which require less traction force, and therefore can be used with less efficient tractors.

1.7 *Output markets*

After 1991, markets for agricultural products were disorganized or absent. Whereas most of the wheat crop had previously been shipped to Russia, suddenly this demand dried up and there was little outlet for the millions of tons of grain, and prices fell to the equivalent of US$40 per ton. For farmers just becoming accustomed to producing economically this made the challenge even more difficult, and restricted severely the inputs used for crop production. Whereas phosphorus (P) fertilizer was routinely applied before independence, after independence fertilizer was seldom used, and levels of nutrients in the soil fell markedly. This meant that much of the research on fertilizer responses also has to be reassessed.

2 CONSERVATION AGRICULTURE AND POTENTIAL BENEFITS TO NORTHERN KAZAKHSTAN

2.1 *What is conservation agriculture?*

Conservation agriculture (CA) is the name presently given to a system of cropping that involves minimal soil disturbance, and maintains the soil surface covered permanently and to as great an extent as possible with crops and crop residues: it is a crop production system as close as possible to the natural system. Thus zero tillage or minimum tillage (although this latter term is often ill-advisedly used for systems that involve considerable tillage) together with crop residue retention on the soil surface are the principal characteristics of CA. In systems where tillage is obviated and residues retained, biological activity and diversity increases: the residues provide a habitat and food source and the lack of soil tillage ensures a stable habitat. However, certain diseases which survive on the residues of crops may be carried over to the following crop, and so crop rotation becomes an important component of CA systems. Crop rotation also imparts other benefits to a functioning CA system.

2.2 *The benefits of conservation agriculture*

The benefits of CA include the protection of the soil surface from the elements, especially raindrops and solar radiation. Raindrops falling on a bare soil, especially one that has been recently tilled, break down surface aggregates, leading to the soil surface sealing and the formation of a crust when the soil surface dries. With residue cover, surface porosity is maintained, which, when combined with continuous pores in the soil below the surface, aids water infiltration and gas exchange. Eliminating soil tillage can results in an increase of continuous large pores, as a result of increased activity of tunneling soil fauna and root channels of previous crops. The corollary of greater water infiltration is that less water runs off the field, thus reducing the potential for soil erosion. Combined with this, surface residues impede water run-off, reducing the velocity and therefore the erodibility of the run-off water – it is common to observe clean water flowing off conservation agriculture fields and at the same time greater volumes of muddy water flowing off conventionally cultivated fields after a rain storm. Surface residues are also effective in reducing soil erosion by wind, the degree of protection being enhanced by the density of the cover and the height of the residues.

Soil surface cover with crop residues also protects the soil surface from solar radiation and therefore reduces water evaporation from the soil. The combination of increased water infiltration and reduced evaporation leads to a markedly improved crop water balance, especially in drier climates where run-off and evaporation can constitute high proportions of the total water balance. Yield benefits to the introduction of CA systems may be evident in the first year in severely moisture-limited environments and seasons. Over time, because of the buffering of the system against short-term moisture stress, crop yield stability is enhanced and, therefore, the risks of crop or economic loss reduced.

Soil tillage generally leads to reductions in SOM over time. Soil aggregation reduces the mineralization of both N and C. However, soil tillage mechanically breaks down soil aggregates, pulverizing the soil and exposing cloistered C to mineralization. In turn the reduction in soil organic carbon (SOC) levels lead to weaker aggregation and a decline in soil structure. By maintaining surface cover with crop residues, which provides a constant food source for soil fauna and flora, and reducing soil movement and aggregate disturbance by tillage to a minimum, CA practices lead to a gradual increase in SOC, and, associated with this, increased nutrient availability, improved soil structure and increased water-holding capacity. A recent summary of C sequestration by no-tillage agriculture, one of the principal components of CA, suggests that in a wide variety of soils and environments world-wide, no-tillage leads to an increase in SOC levels of 325 ± 123 kg C $ha^{-1}yr^{-1}$ (Six et al., 2002). There has been much discussion on how long the phase of SOC accumulation and sequestration can continue once CA has been adopted, but to date there do not appear to be any definite data on this aspect.

2.3 *Farmer appreciation of conservation agriculture*

Although the benefits of CA on soil health and system sustainability can be very marked, these are not normally the factors that convince farmers to adopt the new system. Often the main characteristic of CA that attracts farmers, both large-scale and small-scale, is the reduction in energy required to produce a crop. In small-scale agriculture this can be in the form of human labor or in animal traction, whereas in larger-scale agriculture it is normally in the form of fuel savings and being able to cover a greater area with the same traction source – "more hectares per horse-power" (Wall, 2002). Other advantages for mechanized farmers are the reduction in wear and tear of machinery and maintenance costs: tractor engines run under lower load for seeding and crop spraying, the main activities under CA, than they do for plowing, sub-soiling and ripping of soil. Older tractors, unable to cope with the load requirements of conventional agriculture, can still be used adequately in CA. Also as there is little soil movement, dust is kept to a minimum, and equipment operating in the relatively dust-free air requires less maintenance.

Another major benefit seen by farmers in some irrigated areas where two or more crops are grown each year, and turn-around time between crops is short, is that without the need to till the land the farmer has a greater capacity to seed crops on time (Hobbs et al., 1997). This factor is also important in many dryland conditions where farmers must wait for the first rains to be able to complete soil tillage, thus losing the opportunity for early seeding. In all dryland conditions the loss of moisture through tillage is an important factor, and can make the difference between good initial crop stand and a mediocre stand: the need to reseed crops is markedly lower in CA due to this factor and the generally better moisture conditions under the mulch.

Due to the increases in SOC, cation exchange capacity (CEC) and the availability of nutrients increase over time, and thus lower levels of applied fertilizers are necessary for the same yield levels. Mr. Franke Dijkstra, one of the pioneer CA farmers in Brazil reports that today he produces double the maize yield that he produced 25 years ago, and does so with only half the fertilizer that he used then. He also reports 50% greater soybean yields with again only half of the fertilizer that he used when he started CA (Derpsch, 2005). This fact highlights one of the major advantages of CA: increased use efficiency of inputs, especially fertilizer, labor and fuel inputs. In some instances the same yields can be attained with conventional agriculture as they are from a stabilized CA system that has acquired a new equilibrium of soil health and quality, but the conventional system generally requires considerably more external inputs to achieve the same yield level. Economic efficiency is thus an important component of CA.

2.4 *Problems and difficulties with conservation agriculture*

There are of course difficulties with the adoption of CA. The principal one of these is that of mind-set, and the culture of the plow. There is often considerable peer-pressure to stop trying to farm without tillage, although luckily this is becoming less important as the information about the successes of CA become more widespread. This problem of mind-set is however, still very important for research workers and extension agents: many professionals who have been preaching the need for soil tillage for decades find it very difficult to accept that tillage is not necessary, and even more difficult to admit this to farmers.

Weed control is one of the principal reasons for soil tillage, and when tillage is reduced or obviated, weed control is one of the major management changes that has to be faced. In most instances, in the first years of CA, the use of chemicals for weed control increases. However, the principal herbicide used for weed control at or prior to crop establishment is glyphosate – a herbicide for the total control of weeds. Luckily glyphosate is relatively environmentally benign: it has very low mammalian and invertebrate toxicity; it is tightly bound to clay particles in the soil and so is not leached; and is broken down by soil microbes, generally in the space of about three months. As soil erosion is reduced by CA, the chance of glyphosate getting into waterways is low, and even then it is so tightly bound to the clay particles that it is not released into the water. However, one concern is the extremely widespread use of glyphosate, and the appearance of resistant weeds: populations of eleven weeds resistant to glyphosate have been reported worldwide (International Survey of Herbicide Resistant Weeds, 2006).

Even though CA generally depends on herbicides for weed control in the first years, weed populations decline rapidly in well-managed CA systems. As weed seed is not incorporated into the soil by tillage, it tends to germinate more uniformly and completely: whereas in conventionally plowed systems weed seed may be buried as deep as 20 cm or more below the surface, only to be returned to the surface to germinate in a subsequent season, this does not happen in CA. Also the level of biological activity in the residue mulch and the surface soil is considerably higher in CA than it is in conventional agriculture, with greater and more diverse populations of both flora and fauna. Weed seed is attacked by both insects and fungi at the surface and weed seed viability is reduced in CA. This, combined with the physical effects of residues in hindering weed germination and the reduction in the soil seed reserve with a few years without soil inversion, results in marked declines of weed populations. Crop rotation often implies a rotation of herbicide products and families – an added avenue in reducing weed populations in well-managed CA systems. However, it is important to manage weeds correctly, and to avoid weeds setting seed.

In some systems, especially smallholder systems, crop residues have alternative uses. Residues may be used as animal feed, albeit generally of low quality, building material, fuel for cooking etc. To be able to spread the use of CA in such circumstances it is important to be able to show farmers that leaving at least part of the residues on the soil surface increases system productivity more than their use for other ends. Sayre et al. (2001) has shown in a dryland situation in central Mexico that the increase in productivity of maize and maize-wheat systems under CA, half of the residues can be removed without affecting productivity. Moreover, because of the yield increase with the surface mulch, half the residues of mulched systems is more than all the residues from the unmulched systems, demonstrating the far greater system productivity of the CA systems.

However, demonstrating increases in system productivity to farmers can be difficult – especially to smallholder farmers who generally manage mixed crop-livestock systems and are more likely to rely on the residues for animal feed.

In irrigated, high-productivity agriculture and in large-scale mechanized agriculture, crop residues may be more of a hindrance than a prized resource. In many of these cases, including most farms in northern Kazakhstan, straw burning has been, or is, a common practice. Burning crop residues, apart from being detrimental to the environment and liberating CO_2 into the atmosphere, also squanders an important resource for sustainable agricultural systems. However, in conditions where residue production is high, it may be costly to remove or manage the residues.

As tillage is reduced there is less mineralization of N, often one of the major "benefits" of intensive tillage. Especially in soils that have low levels of available N, this may result in moderate to severe N deficiency in crops grown without tillage, especially where considerable levels of crop residues remain on the surface. This N shortage is generally overcome with the application of approximately 20–30 kgha^{-1} of N fertilizer for a few years, until SOM levels increase and a new level of SOM turn-over and N mineralization is established.

3 RESEARCH ON CONSERVATION AGRICULTURE SYSTEMS IN NORTHERN KAZAKHSTAN

Given the problems observed with the present system, and based on experiences elsewhere, in 2000 the International Maize and Wheat Improvement Center (CIMMYT), the KRIGF and the CKARI initiated a modest effort to explore the benefits of CA systems under the conditions of the northern steppes.

3.1 *Adequate seed drills*

As is generally the case, the first concern was to obtain equipment for the seeding of wheat into residues without previous tillage: direct seeding. Although in many cases in the present CT system the wheat crop is seeded into previously untilled soil, the seeder shoes commonly used in northern Kazakhstan are fitted with V-shaped sweeps that give a shallow tillage of the whole seed-bed. It was decided that instead of importing seeding units from abroad, it would be better to try and develop seeding units or seeding shoes for direct seeding that could be fitted to existing seeders, given the current economic situation and the cost of imported equipment. In the early days of zero tillage in Brazil and Paraguay it had been found that farmers preferred to first purchase "kits" that could be fitted to their existing conventional seeders, even though these were considerably lighter than dedicated zero tillage seed drills. However, this allowed them to try zero tillage with a functional, albeit not perfect, seeder, before committing themselves to the purchase of a better, but expensive, zero tillage seeder.

Advantages from following this route in Kazakhstan also came from the uniformity of the seed drills used on the farms. Nearly all seed drills had been manufactured during the Soviet period at the Agromash factory in Astana (then Tselinograd), the capital. This meant that only one set of modified seeding units could be fitted to practically all the seeders in the region. Also the equipment is heavier than common seed drills for conventionally tilled conditions elsewhere, possibly because of the widespread use of CT in the steppes.

Prototype seeding shoes were made by Dr. M. Matchuskov of the Department of Agricultural Engineering of KRIGF, and after several rounds of testing and modification, were fitted to seeders at the two institutes for trials the following spring. At the same time a seeder was modified with an extra beam so that coulters could be fitted ahead of the seeding shoes in case these accumulated too much straw during seeding. However, it was found that with the shoes staggered over the three tool-bars of the common seeders, straw accumulation was not a problem, at least at the present wheat yield levels.

The "Matchuskov" seeding shoes still cause considerable soil movement and longer (front to back), narrower shoes which cause less lateral soil velocity would be an improvement. Recently, chisel points from India have been taken to Kazakhstan, and a new modification to the standard seeding shoes made by Dr. V. Duvurushenski. Manufacture of these was tried at both the Agromash factory in Astana, Kazakhstan, and a factory in Omsk, Siberia, with the latter giving better results because of the hardness of the steel used, and therefore the extended life of the shoes.

3.2 *Field trials*

Trials were established at both the KRIGF and the CKARI in the spring of 2001. As the fallow year is the weak point of the present system, treatments concentrated on modifications to the fallow year in a four-course rotation: one year of fallow followed by three years of wheat. In these trials the conventional mechanical (black) fallow is compared with a chemical fallow, a combined fallow that has only two tillage passes and one or two chemical applications, and two treatments with oat/pea green manure cover crops (GMCCs) combined with chemical weed control. The first of these was planted in the spring and terminated before seed set, while the second was seeded in mid-summer and left to be killed by cold. Four replications of this trial were initiated each year in 2001, 2002 and 2003 at both institutes, but were, unfortunately, suspended at the KRIGF in 2004. All of the trials continue at CKARI, but treatments will be modified in the 2006 season. At this site the trials are repeated on three landscape positions (plateau, north slope and south slope) because of the effect of this on snow accumulation, rainfall and soil temperature. The plateau generally captures the most snow and the south slope the least as prevailing winter winds are from the north. Snow also melts faster on the south slope because of the sun angle.

The following sections summarize some of the principal results of these trials to date:

Snow and moisture capture: The effect of the different treatments on snow capture measured at 10 points in each plot during the month of February, 2004, is shown in Table 4. In the conventionally tilled "mechanical" fallow and the combined fallow that received two tillage passes during the fallow year, there was no standing stubble, and so snow capture was considerably lower than in the treatments that were not tilled during the fallow year and had standing stubble and dry weed or GMCC plants. As approximately one third of the total annual precipitation is in the form of snow, the extra snow capture in the untilled fallow treatments could be very important.

Effects on soil moisture at seeding: Increased snow capture, however, does not necessarily translate directly into increased moisture in the soil profile at seeding time, as this also depends on the amount of moisture in the profile in the autumn, the speed of the spring snow thaw, spring rainfall and the timing of soil thaw with respect to snow melt. If the soil contains appreciable levels of soil moisture before the onset of winter, or if enough snow is accumulated even in treatments with least snow accumulation, or if the snow melts before there is appreciable soil thaw, or there

Table 4. Snow accumulation (cm) by five fallow treatments on three landscape positions at the Central Kazkah Agricultural Research Institute in February, 2004.

	Landscape position		
Fallow treatments[1]	Plateau	South	North
Mechanical	5.0	6.0	8.0
Reduced	8.0	7.0	8.0
Chemical	11.1	15.3	19.0
Early GMCC[2]	15.3	11.2	19.1
Late GMCC	14.1	12.3	15.2

[1] See text for full treatment descriptions.
[2] Green manure cover crop.

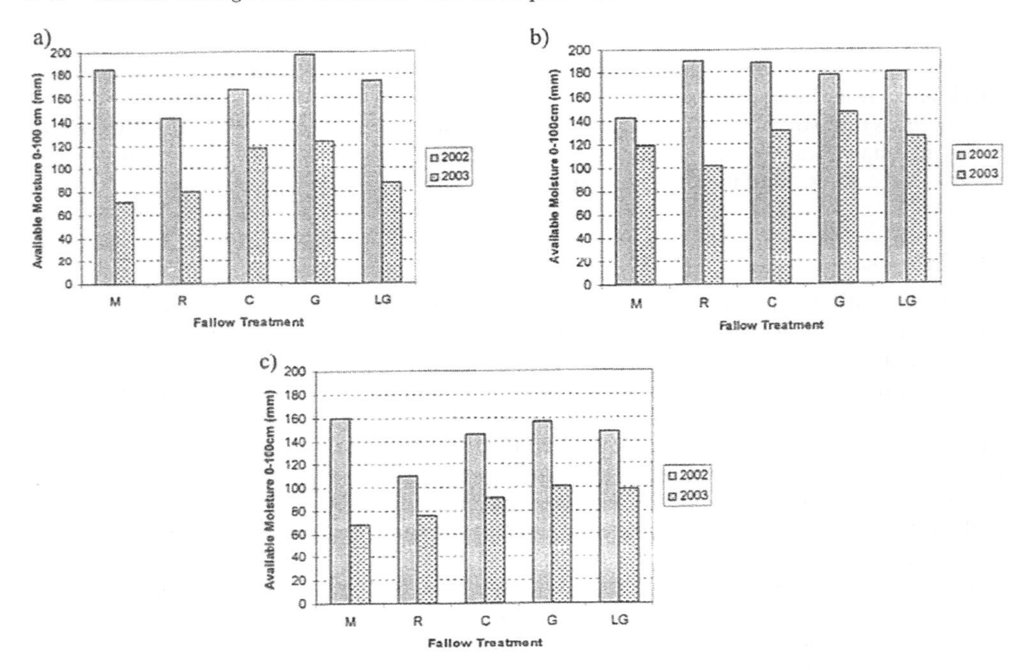

Figure 3. Effect of fallow treatments on soil available moisture (0–100 cm depth) at seeding of the first wheat crop after fallow on three landscape positions in two seasons at the Central Kazakh Agricultural Research Institute. a) plateau; b) north slope; c) south slope. M = conventional mechanical fallow; R = reduced tillage fallow; C = chemical fallow; G = chemical fallow + early green manure cover crop; LG = chemical fallow + Late green manure cover crop.

is good spring rainfall, differences between different practices in spring soil moisture content may be small. The effect that these factors may have on spring soil moisture on the three landscape positions at the CKARI can be seen in Figure 3. The dry spring of 2003 contrasted with a wet start to the season in 2002 and marked differences in profile soil moisture were evident in the different fallow tillage treatments which were not evident in the 2002 season.

Fallow treatments provoked few differences in soil moisture at seeding at CKARI in the wet spring of 2002, but much bigger differences in the drier 2003 spring. Similarly, at KRIGF in 2002 differences between treatments in soil moisture accumulation in the profile were small. However, at CKARI there were still some important differences between treatments in the 2002 season, e.g. the reduced fallow treatment (two tillage passes plus chemical weed control) had considerably less moisture on the plateau and south slope, but was one of the moistest treatment on the north slope. We hypothesize that this is due to a combination of the snow capture, snow melt and soil thaw factors mentioned earlier, and which will require far more intensive studies to unravel. In the 2003 season however, results were very consistent. All three treatments without fallow tillage had more soil moisture the following spring than did the treatments with fallow tillage. There was little difference between the chemical fallow treatment and the early green manure cover crop (GMCC) treatment, although the latter consistently resulted in slightly higher profile moisture levels. The effect of the late GMCC differed between landscape positions depending on the growth of the GMCC in the autumn: on the plateau where growth of the GMCC was relatively abundant, soil moisture levels were reduced in the following spring, but on the south slope where there was little GMCC growth, there was no effect on spring soil moisture levels.

Effects on spring soil erosion: The wet spring of 2002 resulted in severe soil erosion of fallow fields in much of northern Kazakhstan. Although slopes are not marked, they are very long and water accumulation and velocity can be great. The difference in soil erosion between areas with

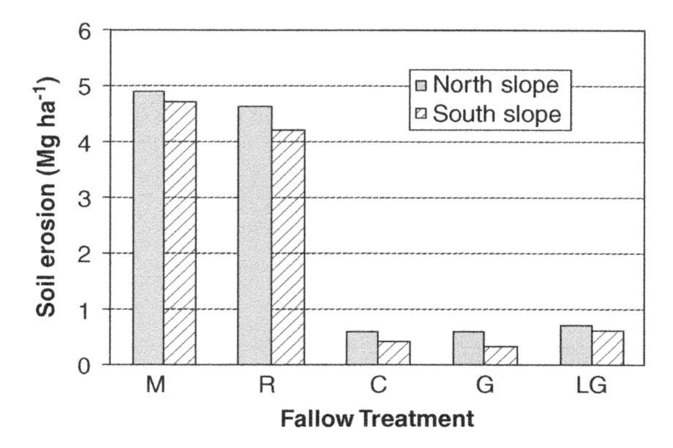

Figure 4. Effect of fallow treatments on spring soil erosion at two landscape positions at the CKARI, 2002. M = conventional mechanical fallow; R = reduced tillage fallow; C = chemical fallow; G = chemical fallow + early green manure cover crop; LG = chemical fallow + Late green manure cover crop.

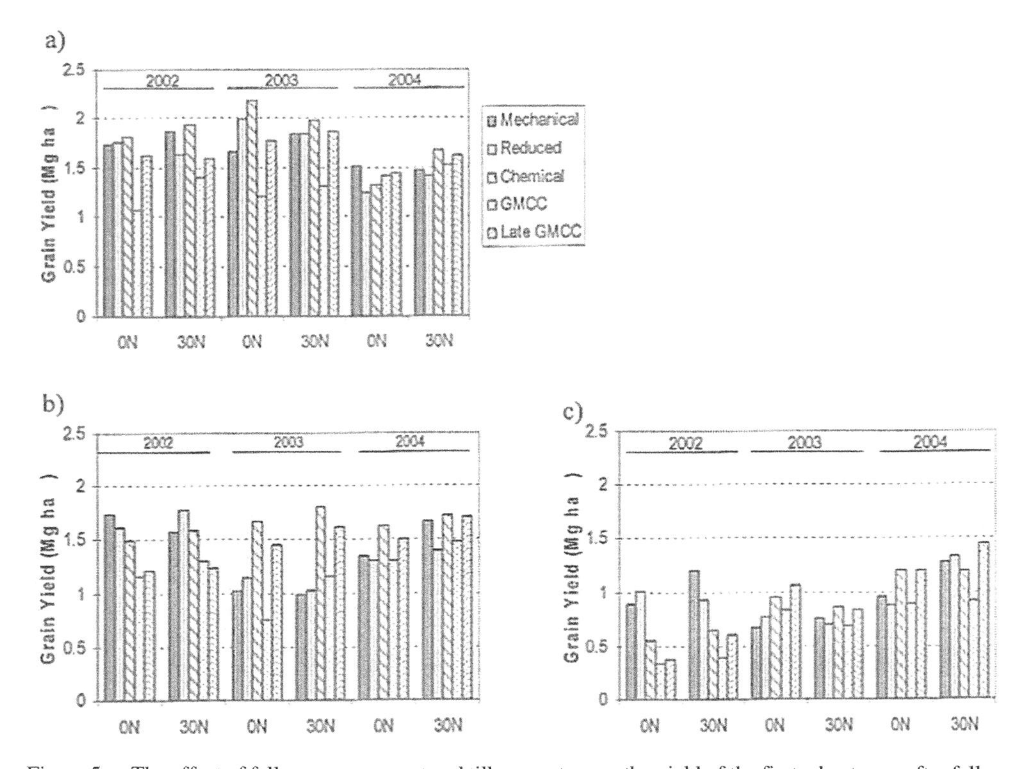

Figure 5. The effect of fallow management and tillage system on the yield of the first wheat crop after fallow on three landscape positions. CKARI, 2002–2004. a) Plateau; b) north slope; c) south slope. See text for full treatment descriptions.

crop residues and the fallow lands was extremely marked: there was seldom any erosion on the cropped areas with retained residues. These effects were also evident on the relatively small plots of the tillage trials at both the KRIGF and the CKARI. Erosion was estimated on all plots at CKARI by the volume of the gullies formed in each plot, and data are shown in Figure 4. The difference in erosion between the tilled fallow plots and the untilled plots is extremely marked: erosion without fallow tillage was only 12% of that measured on the tilled plots.

Table 5. Tissue N levels (N% of whole shoot at flowering) of first year wheat after fallow in five systems with different fallow and tillage management on the plateau landscape position (CKARI, 2003).

	Nitrogen applied to wheat (kg ha^{-1})	
Fallow treatment[1]	0	30
Mechanical	1.81	2.11
Reduced	1.77	2.10
Chemical	1.66	1.86
Early GMCC [2]	1.63	1.89
Late GMCC	1.71	2.10

[1] See text for treatment descriptions.
[2] Green manure cover crop.

Effects on crop yield: Apart from the direct effects of the seeder efficiency, including seeding depth and uniformity, the major effects of a change from the conventional CT practices to conservation agriculture practices would be expected in the first wheat crop after the fallow, as little tillage is carried out between wheat crops in the conventional (conservation tillage) system. In Figure 5, the grain yield data from the first wheat crop after fallow in three consecutive seasons (adjacent blocks of land) on three landscape positions at CKARI are shown for the five tillage treatments described earlier. The generally greater yield of the plateau site (average yield 1.62 Mg ha^{-1}) compared to the north slope (1.41 Mg ha^{-1}) and the south slope (0.88 Mg ha^{-1}) is evident in the Figure. In the 2002 season, some problems with the newly developed prototype seeder resulted in lower plant populations in the direct seeded treatments, which resulted in lower grain yields than the conventional practices on the north and south slopes. However, after this initial problem, the first wheat crop after a chemical fallow yielded at least as well as that after the conventionally tilled fallow, and often gave a slight yield advantage. However, the yield of the wheat crop following the early GMCC fallow was considerably lower than the other direct seeded treatments, especially in the 2003 season (drier spring) and on the drier north and south slope landscape positions. This is consistent with the hypothesis that the GMCC reduced moisture storage in the profile. However, this was not the case, as was shown in Figure 3. Tissue analysis conducted on whole shoot at flowering from each treatment on the plateau showed, however, that tissue N levels were lower in all the untilled treatments than in the treatments with a tilled fallow, and the lowest levels were found in the treatment with the early GMCC (Table 5), presumably because the GMCC had taken up N which remained in the undecomposed residues. The tissue N levels in the treatments with untilled fallow and no N fertilizer application to the wheat crop were below the critical level for N in whole shoot tissue at this stage of approximately 1.7% (Reuter, 1986). The lower N status of the crops following an untilled fallow are also evident in Plate 1, a composite of photographs of the wheat crop after a conventionally tilled fallow (left) and after an untilled chemical fallow (right). The difference in N status between treatments is undoubtedly due to the breakdown of SOM as a result of tillage during the fallow season, and subsequent mineralization of N from this organic matter.

The problem of different levels of crop nutrition with different fallow management treatments and tillage practices was also evident in soil analysis results from the KRIGF in the first season of trials in 2002 (Figure 6). Data are not available for subsequent seasons. Grain yield closely mirrored soil nutrient levels, which were the highest in the treatment with the tilled fallow. Critical levels established by past research at the Institute are 13–15 mg kg^{-1} for NO_3–N and 25–30 mg kg^{-1} for P_2O_5. As in the tissue data from CKARI reported earlier, soil N levels were lower in all treatments following an untilled fallow, especially in the treatment with an early-planted GMCC in the fallow year, again presumably because the GMCC took up N which was still incorporated in the undecomposed residues.

While NO_3–N levels were adequate in the tilled treatments, P levels were very low in all treatments. Phosphorus fertilizer (60 kg ha^{-1} P_2O_5) was applied to the conventionally tilled fallow and

Plate 1. First year wheat after the recommended tilled fallow (left) and a chemical (untilled) fallow (right). After a chemical fallow even a 20 kg ha^{-1} application of N did not give the same crop color and development as the crop after a tilled fallow and without applied N (CKIGF, 2004; photo by P.C. Wall).

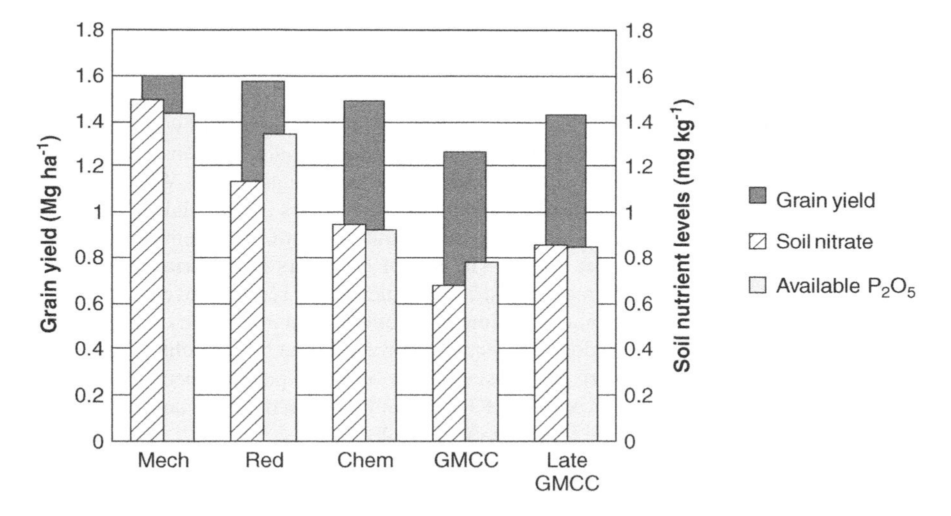

Figure 6. Grain yield of the first wheat crop after fallow with five fallow management treatments, and soil NO_3 and P levels before seeding of the wheat crop (KRIGF, 2002).

the reduced tillage fallow in the autumn of the fallow season, but even so P levels were low in these treatments. In the other three treatments with untilled fallows, P fertilizer was applied at a rate of 20 kg ha^{-1} P_2O_5 at seeding, after the samples had been taken for the above analyses. Previous research had shown that 60 kg ha^{-1} P_2O_5 in the fallow year was sufficient for the whole of the four-year rotation, and so this common practice and recommendation continued. However, after Perestroika, the unfavorable economic conditions resulted in little fertilizer use, even on the research institute farms, and this in turn resulted in low soil P levels. This fact underscores one of the big problems of research in Kazakhstan. The country has a rich history of very good and meticulous agricultural research. However, the focus of this research, and the recommendations that emanated from it, were oriented towards a very different philosophy of agricultural production and economic circumstances. This will require considerable effort in revisiting past research results to reinterpret these, or repetition of much research to develop new recommendations.

Effects on soil carbon: The focus of this workshop is on C sequestration. To date we have not seen significant differences in SOM levels between treatments at the CKARI. However, over the four years of the trial on average there is approximately 0.2% more SOM in the treatments without fallow tillage and with direct seeding of the wheat crop with narrow seeder shoes (as opposed to the sweeps used on the conventional equipment). This translates into approximately 400 kg C ha^{-1} yr^{-1}, a level which is consistent with the review of Six *et al.* (2002) who found a general increase in SOC under conservation agriculture practices of 325 ± 113 kg C ha^{-1} yr^{-1} under a wide range of temperate and tropical conditions.

3.3 *What next?*

In 2003, FAO initiated an 18-month Technical Cooperation Project in northern Kazakhstan. This project installed large-scale (100 ha) conservation agriculture demonstration plots on four farms in the region and conducted numerous farmer awareness meetings and field days. It was evident that there was much farmer interest in the new system, but that there were still numerous issues to be addressed before widespread adoption of CA will occur.

Training and awareness: There is still a need for much effort in farmer and researcher awareness and training. CA involves a change in mind-set – without this it becomes too easy to resort to tillage when any problem occurs in a newly implanted CA system.

Weed control: Weed control is one aspect that needs further research. Herbicides are costly in Kazakhstan, especially when compared to depressed grain prices. Also strategies for weed control in the initial years of CA need to be developed. If the spring is cool and moist, weed populations are low at seeding and only a residual herbicide will be required, whereas in warmer springs and later plantings, glyphosate and a residual herbicide may be necessary. Options for weed control with different weed spectra and these different conditions must be available. Weed pressure will decline over time with CA, but good weed control in the early years is essential.

Fertilization strategies: Information on fertilization strategies in CA systems still needs to be developed, as indicated by the results above. The use of fertilizers is important not only to get optimum grain yields, but also to increase the straw production for residue cover.

Crop rotations: The development and adoption of functional and economic crop rotations, and the general diversification of the production system, is probably the biggest challenge to the successful development of CA in northern Kazakhstan. Work on this aspect had been envisaged in the CIMMYT research program with CKARI and KRIGF but had been deferred until a second phase because of the limited funds available (Karabayev et al., 2003).

Residues pose a potential disease threat to the following wheat crop, as necrotrophic diseases may harbor on the residues and infect the young plants of the following crop. Residue retention from one wheat crop to the next has been common in the CT system of the region, and a major disease outbreak is probably unlikely. However, Dr. Julie Nicol found higher levels of necrotrophic diseases in the CA plots of farmers managing the FAO demonstration plots.

Crop rotation should have other benefits, including weed control, nutrient cycling, and, if legume species can be incorporated into the rotation, nutrient provision. However, markets for alternative crops to wheat will need to be developed. Suleimenov et al. (2005) show that other crops, especially peas and lentils have higher profit margins than bread wheat, but do not explain whether the profit margin was due to higher output market prices, and whether these prices are likely to change as production volumes increase. However these studies do show the potential for more diversified systems in northern Kazakhstan which bodes well for the future. Policy emphasis should be placed on market development for alternative crops.

Equipment: Equipment is still likely to be a significant factor in the spread of CA in northern Kazakhstan. Although functional modifications to the common seed drills have been made, there is still scope for considerable improvement: the new chisel points developed by V. Duvuruchenski based on the Indian model is an important step in this direction. However, other equipment is also necessary for the successful management of CA, especially sprayers for uniform herbicide application and straw spreaders for the combine harvesters so that more uniform residue cover can be obtained.

Policy: Recently CA has been receiving support at high levels of government. Advances to date have been a result of broad partnerships between many institutions, including the Ministry of

Agriculture, the Kazakh National Farmers Union, international research and development agencies and US universities, and considerable effort has been invested in increasing the awareness of CA principles and practice among technical staff and farmers. The Ministry of Agriculture has recently initiated a development plan to stimulate the production of 1 Mha of canola on the northern steppes, including subsidized credit for machinery and equipment. This support for the diversification of agriculture, and the support for new equipment, should provide incentives for the switch to CA.

4 CONCLUSION

The steppes of northern Kazakhstan were opened for agriculture in the 1950's and soon suffered from severe erosion both from wind and water. A system was developed at the Kazakh Research Institute of Grain Faming which is still widely used in northern Kazakhstan, Ukraine and Siberia. The common cropping system in northern Kazakhstan is a wheat-fallow system, with 3–5 years of continuous spring wheat followed by a year of tilled fallow. The SOM content continues to decline in the system, largely due to the fallow tillage. Wheat after fallow benefits from mineralized N and the weed control achieved in the fallow year, and possibly also from moisture storage in drier seasons. Recently, CA systems have been proposed and tested, with a focus on reducing tillage in the cropped years, and replacing the fallow with a cash crop or green manure cover crop. The principal predicted benefits of CA are a reduction in soil erosion, an increase in soil available water and SOM and, ultimately, increased productivity and profitability of agricultural production in the region. The factors that favor adoption are the reductions in tractor time and fuel use, and the availability of zero tillage kits to adapt the local seeders, and the possibility of using the ageing machinery pool, whereas the factors that prejudice adoption are fuel subsidies, the relatively high price of herbicides, and lack of information on technological options. Although there are trends showing increases in SOC levels after four years of continuous CA, differences are not yet statistically significant. However, it appears that carbon sequestration in SOC with CA in northern Kazakhstan is in the order of 400 kg ha^{-1} yr^{-1}.

The biggest challenge to the widespread adoption of CA in northern Kazakhstan is one of mind-set. This has been the case in all other countries where CA has spread, and we believe Kazakhstan will be no different. However, there is clear evidence that the system works under the conditions of the region and there are some hard-working enlightened individuals who see that the principles of CA are not only functional, but important to halt the marked, albeit slow, soil and land degradation in the region. There are currently over 12 million hectares of zero tillage, mostly with residue cover, in Canada (Derpsch, 2005), much of it under conditions very similar to the conditions of northern Kazakhstan. Thus Kazakh farmers do not have to learn all the lessons themselves but can benefit from both the positive experiences and lessons learned by their Canadian counterparts.

REFERENCES

Derpsch, R. 2005. The extent of Conservation Agriculture adoption worldwide: Implications and Impact. Proceedings on CD, III World Congress on Conservation Agriculture, 3–7 October, 2005. Nairobi, Kenya.

Djalankuzov, T.D., S.D. Abdykhalykov, V.V. Redkov, M.I. Rubinstein, and B.U. Suleymenov. 2004. Main results of the long-term soil monitoring in northern Kazakhstan, p. 45–56. In: A.S. Saparov et al. (eds.) Essential Problems of Soil Science. Tetis, Almaty.

Gilmanov, T.G., D.S. Johnson, N.Z. Saliendra, K. Akshalov, and B.K. Wylie. 2004. Gross Primary Productivity of the True Steppe in Central Asia in Relation to NDVI: Scaling Up CO_2 Fluxes. Environmental Management Volume 33, Supplement 1, S492–S508.

Gossen, E. 1998. Agrolandscape agriculture and forestry management as the basis of sustainable grain production in the steppes of Eurasia. p. 44–48. In: A. Morgounov, A. Satibaldin, S. Rajaram and A. McNab (eds.), Spring Wheat in Kazakhstan: Current Status and Future Directions. International Center for Maize and Wheat Improvement, Texcoco, Mexico.

Hobbs, P.R., G.S. Giri, and P. Grace. 1997. Reduced and zero tillage options for the establishment of wheat after rice in South Asia. RWC Paper No. 2. Mexico, D.F.: Rice-Wheat Consortium for the Indo-Gangetic Plains and CIMMYT.

International Survey of Herbicide Resistant Weeds. Glyphosate resistant weeds – 2006. http://www. weedscience. org/glphosate.gif.

Karabayev, M., P.C. Wall, A. Kenzhebekov, N. Yushenko, and A. Bektemirov. 2003. Initial Experiences with Zero Tillage in Northern Kazakhstan. Proceedings on CD of the 2nd World Congress on Conservation Agriculture. Iguassu Falls, Paraná, Brazil, August 11–15, 2003.

Reuter, D.J. 1986. Temperate and sub-tropical crops. p. 38–99. In: D.J. Reuter and B.J. Robinson (eds.), Plant Analysis: An Interpretation Manual. Inkata Press, Melbourne, Australia.

Sayre K.D., M. Mezzalama, and M. Martinez. 2001. Tillage, crop rotation and crop residue management effects on maize and wheat production for rainfed conditions in the altiplano of central Mexico. p. 575–580. In: L. Garcia-Torres, J. Benites, and A. Martinez-Vilela (eds.), Vol. II, Conservation Agriculture: A Worldwide Challenge. ECAF/FAO, Córdoba, Spain.

Six, J., C. Feller, K. Denef, F.M. Ogle, J.C. de Moraes Sa, and A. Albrecht. 2002. Soil organic matter, biota and aggregation in temperate and tropical soils – effects of no tillage. Agronomie 22:755–775.

Suleimenov, M.K. 2000. Development of soil tillage systems in northern Kazakhstan. p. 94–100. In: M. Karabayev, A. Satybaldin, J. Benites, T. Friedrich, M. Pala and T. Payne (eds.), Conservation Tillage: A Viabler Option for Sustainable Agriculture in Eurasia. Almaty, Kazakhstan: CIMMYT; Aleppo, Syria: ICARDA.

Suleimenov, M., K. Akhmetov, Z. Kaskarbayev, A. Kireyev, L. Martynova, and R. Medeu-bayev. 2005. Role of wheat in diversified cropping systems in dryland agriculture of Central Asia. Turkish J. Agric. For. 29:143–150.

Wall, P.C. 2002. Extending the Use of Zero Tillage Agriculture: The Case of Bolivia. Paper presented at the International Workshop on Conservation Agriculture for Sustainable Wheat Production in Rotation with Cotton in Limited Water Resource Areas, Tashkent, Uzbekistan, October 13–18, 2002.

CHAPTER 24

Cover crops impacts on irrigated soil quality and potato production in Uzbekistan

A.X. Hamzaev, T.E. Astanakulov, I.M. Ganiev, & G.A. Ibragimov
Samarkand Agricultural Institute, Samarkand, Uzbekistan

M.A. Oripov
Bukhara State University, Bukhara, Uzbekistan

K.R. Islam
Crops, Soil and Water Resources, Ohio State University South Centers, Piketon, OH, USA

1 INTRODUCTION

Farmers in Uzbekistan are challenged with the maintenance of irrigated soil quality for economic crop production under a continental climate. Irrigated soils produce over 96% of the gross agricultural outputs of Uzbekistan; however, its soils are low in organic matter and available nutrient content. They cannot maintain and/or support production agriculture without substantial fertilizer input. The loss of organic matter and nutrients is due in great measure to continued deep plowing of soils which accelerates chemical oxidation of organic matter and greater mining of nutrients, thus affecting soil quality over time (Lal, 2004). Crops are also affected by water stress which is considered one of the most limiting factors for crop production under dry climates. The progressive decline in irrigated soil quality is becoming a major agricultural problem in Uzbekistan.

Soil quality management for profitable crop production is the foundation of sustainable agriculture. A progressive decline in soil quality which is caused by the practice of conventional agriculture has prompted management practices that enhance soil biology and nutrient recycling through organic amendments (Tester, 1990; Nahar et al., 1995; Islam and Weil, 2000). Cover crops are one of the important components of conservation management systems. They provide organic matter as mulch which reduces evaporation and surface runoff, increases biological activity, improves the status of saline soils, decreases soil compaction by increasing porosity and water infiltration, supplies N to succeeding crops, recycles subsoil nutrients, and enhances soil quality (Alekseev, 1936 and 1957; MacRae and Mehuys, 1985; Hargrove et al., 1989; Tester, 1990; Hussain et al., 1992; Decker et al., 1994; Mahmood and Aslam ,1999).

Non-legume cover crops can fix tropospheric CO_2, however, legumes fix both CO_2 and N in the plant biomass, and subsequently provide N for succeeding crops by recycling of N-rich crop residues in soil (Decker et al., 1994). Oripov (1991) reported that plow down of cover crops residues significantly increased earthworm populations. There were six to seven times higher than fallow plowed soil without any cover crops. Growth of cover crops has also been considered as a potential strategy to control soil pathogens and diseases (Rouatt and Atkinson, 1950). Several scientists have reported that the planting of rape, pea, oil radish, and mustard as cover crops caused a significant reduction of soil-borne diseases in crops, including potatoes (Sanford, 1946; Weinhold et al., 1964; Berdnikov and Kosyanchuk, 1999).

Potato (*Solanum tuberosum* L.) is one of the important food crops in Uzbekistan (Astanaqulov, 1991). At present, the average potato production in Uzbekistan is only 16 Mg ha^{-1} (Anonymous,

2005). Within the next few years, the potato crop is expected to achieve 1 to 1.2 M Mg yr^{-1} on 1 M ha of land under irrigation with the application of chemical fertilizers (Anonymous 2005). Short-term nutrient replenishment benefits can be achieved through the use of chemical fertilizers; however, the use of cover crops together with chemical fertilizers may enhance the long-term soil productivity of agroecosystems due to its impact on soil biological, chemical, and physical properties (Khan et al., 1996; Berdnikov and Kosyanchuk, 1999; Malik et al., 2002). In addition, the use of cover crops to increase organic matter, such as mulch, could enhance soil moisture retention for early establishment of crops – thereby reducing a common problem faced by the arid farming systems of Uzbekistan.

Objectives of the study reported in this paper were to (i) determine the biomass contribution of barley, peas, radish and rape as cover crops in pure stand or in mixed cropping, (ii) compare the effects of successive cover crop biomass contributions on soil biological, chemical, and physical properties, as well as on soil quality, and (iii) evaluate the impacts of cover crops and soil quality on yields and the profitability of irrigated potato production in Uzbekistan.

2 MATERIALS AND METHODS

2.1 *Site and experiment*

The field study was conducted from 1997 to 2000 on an irrigated silt loam soil (typical gray sierozems) in the Kashka Darya region of Uzbekistan. The region has a continental climate with an annual average rainfall of less than 360 mm. More than 95% of the rainfall occurs from October to May. The average monthly minimum temperature in January is 3°C and the maximum temperature is 29°C in July. Mean monthly relative humidity is highest in February (78%) and drops to 35% in June and July. The predominant farming system in the region is based on irrigated cotton, wheat, maize, potato, fodder, and pulses.

The field experiment had six cover crop treatments (pure stands and mixed crops) in a randomized complete block design. Each treatment was replicated three times. Cover crops included barley (*Hordeum vulgare*, L. cv. Temur) field peas (*Pisum sativum*, L, cv. Vostok-55), oilseed radish (*Raphanus olipheria* L. cv. Raduga), rape (*Brassica napus* L. cv. Nemerchansky-2268), mixtures of peas and barley and peas and radish. They were planted in mid-July 1997. Barley at 160, peas at 70, radish at 18, and rape at 16 kg ha^{-1}, respectively were planted in 2 m wide × 7 m long replicated plots. In the case of mixed planting, half of the seeding rates were used for each crop. Rape and radish seeds were planted at 1.5 to 2 cm; barley at 5 to 6 cm; and peas at 6 to 7 cm soil depth. Before planting the cover crops, all field plots were conventionally tilled and fertilized with ammonium nitrate at 30 kg ha^{-1}, ammonium phosphate at 100 kg ha^{-1}, and potash 60 kg ha^{-1}. Control treatments were maintained when fertilizers were applied without any cover crops. All cover crop plots were surface irrigated nine times at 55.6 to 66.7 m^3 ha^{-1}, starting in July, 1997. Cover crops were plowed down at their maximum vegetative growth in late October, 1997.

Potato (cv. Sante) tubers (30 to 100 g weight) were planted using 70 cm × 20 cm spacing in March, 1998 at 3.3 Mg ha^{-1}. Potatoes were surface irrigated nine times. Manual weeding was performed twice during the growing season. The potatoes were harvested in early July, 1998 after random selection of 20 plants from each replicated plot. Total marketable yield, production cost, market price, and cost efficiency of growing potatoes were assessed. The experiment was repeated during the 1998 to 2000 growing seasons in the same manner as described above.

2.2 *Sampling, processing and analysis of cover crops and soil*

At maximum vegetative growth, the above- and below ground biomass of different cover crops were sampled from three randomly selected 50 cm × 50 cm subplots within each replicated plot to measured their growth and biomass production. A sub-sample of fresh biomass was oven-dried at 55°C until a constant weight was obtained to calculate the dry-matter biomass contribution of cover crops to soil. After harvesting the cover crop biomass, soils within randomly selected subplots (50 cm×50 cm) were dug out to a 50 cm depth in order to collect earthworms. Data on earthworms were presented on a per ha basis.

Soil samples from randomly selected subplots within each replicated plot were collected at 0 to 30 cm depth before harvesting potatoes in June, 2000. The field-moist soil samples were air-dried, 2-mm sieved, and analyzed for selected chemical and physical properties. Total organic carbon (TOC) content of soil was measured by the modified Tyurin spectrophotometric method (Arinushkin, 1970). Nitrate (NO_3), phosphorus (P_2O_5), and potassium (K_2O) were measured by following standard methods (Arinushkin, 1970; Volodin et al., 1989). Soil physical properties, such as bulk density (ρb), porosity (ft), macroaggregate (MA) and microaggregates (MI), and moisture content, were also measured (Kachinskii, 1965). The pool of MA, MI, TOC, NO_3, P_2O_5, and K_2O were calculated by multiplying their concentration by soil depth (30 cm) and concurrently measured ρb. The amount of soil within 30 cm depth before establishment of the experiment was used to calculate the equivalent TOC pool (ETOC) and rate of C sequestration in soil over time. All the results were expressed on the basis of oven-dried (~105°C) weight of soil.

2.3 *Calculation of soil quality index*

A soil quality index (SQ_{index}) was calculated by modifying the concept that "higher values of selected soil properties are better indicators of soil fertility" (Islam and Weil, 2000b). Values (X) for individual soil properties, such as the number of earthworms, ft, MA, AR, TOC, NO_3, P_2O_5 and K_2O were normalized between 0 to 1 relative to the highest value (X_{max}) of each parameter in the dataset ($X_n = X\, X_{max}{}^{-1}$). The values of pH, sand, silt, silt, clay, MI, and ρb were not included in the calculation because the criteria of "more is better" is not true or may be uncertain for the range of values of the study. Equal weight was assigned to normalized values of each property. The normalized values of soil properties were then averaged to calculate the $SQ_{index} = (\Sigma X_n = X\, X_{max}{}^{-1})\, n^{-1}$.

2.4 *Statistical analyses*

Data were combined (1997 to 2000) and analyzed to separate the effects of cover crops on soil properties and potato yields compared to control treatment using the SAS analysis of variance program (SAS 2001). They were combined because of the lack of consistent differences in the dependent variable over time. Differences in mean values at the $p \leq 0.05$ were considered significant. A single degree-of-freedom orthogonal linear contrast between group means in response to cover crops vs. control on soil properties, SQ_{index}, and potato yield and economics was performed. Regression and correlation analyses were used to establish relationships among SQ_{index}, potato yield, and economic data for the crops.

3 RESULTS AND DISCUSSION

3.1 *Biomass production of cover crops*

As shown in Figure 1, total fresh and dry biomass production of cover crops varied consistently over time. Among the cover crops, radish produced the greatest amount of biomass (>32 Mg ha^{-1}) while barley produced the lowest (~20 Mg ha^{-1}). Peas and radish as mixed cover crops produced more than 30 Mg ha^{-1} of fresh biomass. Significantly radish crops produced the greatest amount of dry biomass (>6 Mg ha^{-1}) followed by peas and barley as mixed cover crops (>5 Mg ha^{-1}), followed by barley (4 Mg ha^{-1}). Both fresh and dry biomass of cover crops was higher when planting peas in combination with radish or barley than peas and barley as individual cover crops.

The pronounced variation in biomass production among cover crops may be related to their differences in genetic potential, plant architecture, adaptability, rooting pattern, nutrients and water

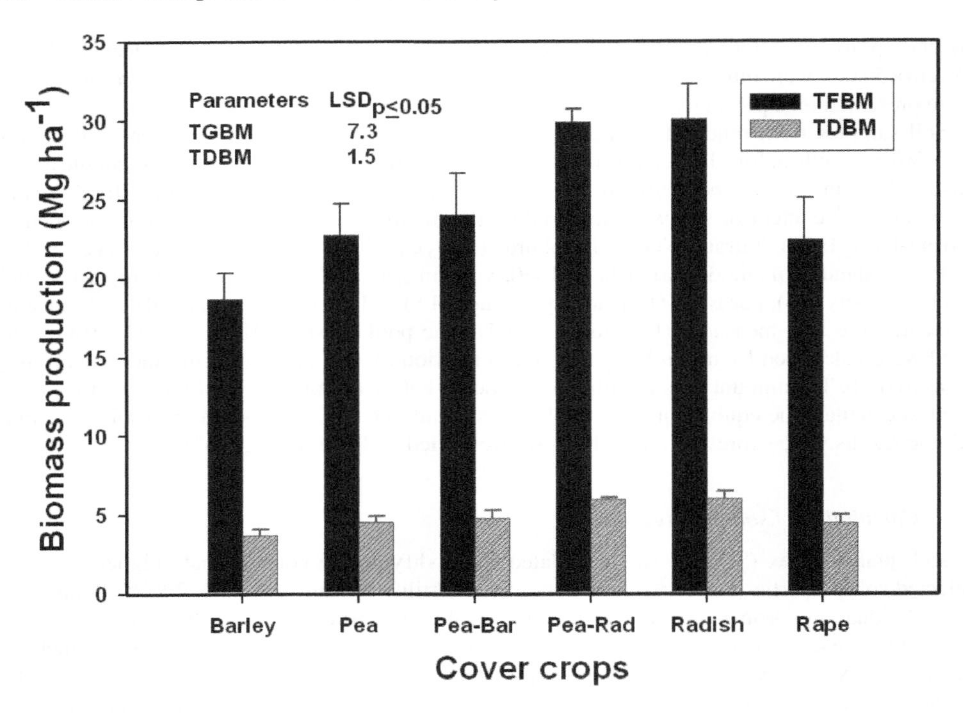

Figure 1. Total fresh biomass (TFBM) and total dry biomass (TDBM) production of various cover crops (1997–2000 average data).

uptake, and N-fixing capacity (Hussain et al., 1992; Clark, 1998; Berdnikov and Kosyanchuk, 1999; Malik et al., 2002). Greater biomass production of cover crops, especially radish alone or in combination with N-enriched peas, may facilitate greater deposition of N-enriched residues to enhance soil quality.

3.2 *Effect of cover crops on soil biological, chemical and physical properties*

Soil properties were found to vary significantly in response to the effects of cover crops over time. As shown in Table 1, consistent increase in number of earthworm populations (134 to 256%) was observed in soils under fertilized cover crops compared to soils that were fertilized but with no cover crops (control). The effect of cover crops on earthworms was more pronounced when peas and radish were used as mixed cover crops. Among the cover crops, barley and rape exerted less pronounced effects on earthworm populations.

Soil aggregate structures (i.e., MA and MI) were changed in response to cover crops as show in Table 1. Cover crops significantly increased MA (32 to 82%), AR (42 to 115%), and ft with an associated decrease in ρb and MI (6 to 16%) compared to the control plot. The effect of cover crops on ρb, ft, MA, MI, and AR was most pronounced for pea and radish mixtures followed by peas alone, and peas and barley mixtures. Among the cover crops, barley and rape crops had relatively less effect on soil physical properties. As illustrated in Table 1, an orthogonal linear contrast between group means of soil properties indicated that, on average, cover crops increased earthworm numbers by 200%, MA by 62%, and AR by >80% than control.

Table 2 contains data on concentration and pool of TOC, and rate of C sequestration in soil at 30 cm depth. Plots with cover crops had significantly higher concentrations of TOC than control plots. TOC concentration was highest (6.64 g kg^{-1}) in soils with peas and radish cover crop mixtures, intermediate in soils with pea, rape, and pea and barley cover crop mixtures (6.61 g kg^{-1}),

Table 1. Cover crop effects on earthworm populations, bulk density, porosity, and macroaggregate and microaggregate stability at 0–30 cm soil depth.

Cover crops	ρb g cm^{-3}	*ft* m^3 m^{-3}	MA g 100 g^{-1}	MA Mg ha^{-1}	MI g 100 g^{-1}	MI Mg ha^{-1}	AR	Earthworms 10^3 ha^{-1}
Control	1.29	0.52	8.4	32.5	32.4	125.1	0.26	86
Barley	1.25	0.53	11.5	43.0	31.2	117.0	0.37	201
Pea	1.23	0.54	15.6	57.4	30.1	110.9	0.52	255
Pea + barley	1.23	0.54	14.9	55.0	29.7	109.9	0.50	273
Pea + radish	1.22	0.54	16.2	59.3	28.8	105.5	0.56	306
Radish	1.24	0.53	13.2	49.1	31.1	115.4	0.43	285
Rape	1.24	0.53	14.1	52.5	30.6	113.9	0.46	223
LSD$_{p \leq 0.05}$	0.02	0.01	1.8	6.1	0.88	4.8	0.08	24
Orthogonal linear contrast								
Ccrop vs. cont.	−0.04	0.02	5.8	20.3	−2.2	−13	0.21	171
T-test (p)	0.001	0.003	0.006	0.006	Ns	0.005	0.001	0.001

ρb = Soil bulk density, *ft* = Total porosity, MA = Macroaggregate stability, MI = Microaggregate stability, and AR = Macroaggregate and microaggregate ratios.

Table 2. Cover crop effects on total organic C concentration and pool, and rate of total organic C sequestration at 0–30 cm soil depth (1997 vs. 2003).

Cover crops	TOC (g kg^{-1})	TOC (Mg ha^{-1})	ETOC Mg ha^{-1}	C sequestration rates (kg ha^{-1})
Initial (1997)	6.38	24.4	24.4	–
Control	6.41	24.8	24.5	38.1
Barley	6.53	24.5	25.0	186.0
Pea	6.58	24.3	25.2	260.0
Pea + barley	6.64	24.6	25.4	333.9
Pea + radish	6.64	24.4	25.4	333.9
Radish	6.53	24.2	25.0	186.0
Rape	6.61	24.6	25.3	297.0
LSD$_{p \leq 0.05}$	0.08	0.25	0.3	103.8
Orthogonal linear contrast				
Ccrop vs. cont.	0.18	−0.4	0.7	228.1
T-test (p)	0.011	0.01	0.012	0.001

TOC = soil total organic C; ETOC = total organic C content in equivalent amount of soil at 30 cm depth; and Ccrop = cover crops.

and lowest (6.38 g kg^{-1}) in soils found in the control plots. A linear comparison between group means of cover crop plots and control plots suggests that cover crops have significantly increased TOC concentration over time. Data in Table 2 show that TOC pool did not increase significantly for cover crop plots compared to control plots but the TOC content in equivalent amount of soil (ETOC) at 30 cm depth increased significantly (~3%) over time. The highest increase (>4%) was recorded in soils under peas with barley or radish as mixed cover crops. Soils under peas with barley or radish had significantly greater C sequestration (334 kg ha^{-1} yr^{-1}) than in soils (38.1 kg ha^{-1} yr^{-1}) under without cover crops over time (Table 2). Soils under peas and rape in pure stands have sequestered C at 260 kg ha^{-1} yr^{-1} than only 186 kg ha^{-1} yr^{-1} in soils under barley and radish in pure stands.

Table 3. Cover crop effects on nitrogen, phosphorus and potassium concentration and pool at 0–30 cm soil depth.

	NO_3		P_2O_5		K_2O	
Cover crops	($g\,kg^{-1}$)	($kg\,ha^{-1}$)	($mg\,kg^{-1}$)	($kg\,ha^{-1}$)	($mg\,kg^{-1}$)	($kg\,ha^{-1}$)
Control	7.39	28.6	20.3	57.9	294.9	440.3
Barley	7.62	28.6	23.0	67.1	251.5	513.4
Pea	17.03	62.8	32.2	95.7	303.7	839.9
Pea + barley	13.27	49.0	31.0	91.9	308.5	814.8
Pea + radish	14.94	54.8	33.9	101.0	314.9	904.5
Radish	9.64	35.8	30.8	90.9	306.2	718.9
Rape	10.30	38.3	34.1	100.4	313.7	786.6
$LSD_{p \leq 0.05}$	4.22	15.4	ns	24	ns	183.4
Orthogonal linear contrast						
Ccrop vs. cont.	4.74	16.5	10.5	33.2	4.8	317.1
T-test (p)	0.01	0.01	0.01	0.008	ns	0.009

NO_3 = nitrate; P_2O_5 = phosphorus; and K_2O = potassium.

Table 3 contains data on nutrient content in soils. Plots with cover crops had significantly greater concentration NO_3, and pools of P and K in soil than control plots. While the difference in P and K contents between control plots cover crop plots was highest in soils under peas and radish mixtures, NO_3 content was highest in soils under peas alone (see Table 3). An orthogonal linear comparison between group means of cover crop plots and control plots suggests that cover crops have significantly increased concentration and pools of N (greater than 60%), P (greater than 50%), and K (greater than 70%) over time.

Use of cover crops as soil amendments has resulted in significantly better scores on SQ_{index} compared to those for the control treatment (see Figure 2). The beneficial effects of application of successive cover crops to improve SQ was more pronounced for plots with pea and radish mixture cover crops (~55%) followed by peas along cover crops (50%) as compared to control plots. Barley had the lowest difference in effect (~15%) to improve SQ. Radish and rape cover crops results in similar differences (>35%) on SQ. The effects of peas and radish as mixed cover crops or pea alone on SQ were more pronounced than any other cover crop alone or mixed cover crop. As shown in Figure 2, linear contrast between group means of control and cover crops plots suggests that, on average, cover crops significantly (~40%) improve SQ_{index} over time.

Cover crop biomass has resulted in a consistent increase in earthworm populations, ft, MA, AR, TOC, C sequestration, NO_3, P and K content and SQ_{Index} with an associated decrease in ρb and MI after three years of successive use of cover crops. They have acted as soil amendments because of the amount of organic matter deposited in soils by them. Larger amounts of cover crops biomass with a lower ρb have probably explains the significantly lower ρb and associated increases in ft, MA, and AR of soil over time. Significant decreases in ρb imply greater pore space and improved aeration which result in a suitable environment for enhanced biological activity (Hargrove et al., 1989; Tester, 1990; Mahmood and Aslam, 1999). Increasing biological activity associated with greater amounts of organic matter may have produced more organic binding agents (e.g., polysaccharides) for soil macroaggregation (Tester, 1990; Islam and Weil, 2000a) when cover crops are used. Since crop residues are the main source of C and N in soil, a greater deposition of organic matter from cover crops especially peas and radish should have increased the TOC concentration and N content. But the temporal effects of greater amount of low density biomass deposition from cover crops had significantly reduced the soil ρb with an associated decrease in TOC pool in soils under cover crops compared to control plots. When using equivalent amount of soil at 30 cm depth, the TOC pool and C sequestration rates were found significantly greater in soils under peas with radish or barley compared to control plots. However, the rate of N increase in soil was more pronounced than the

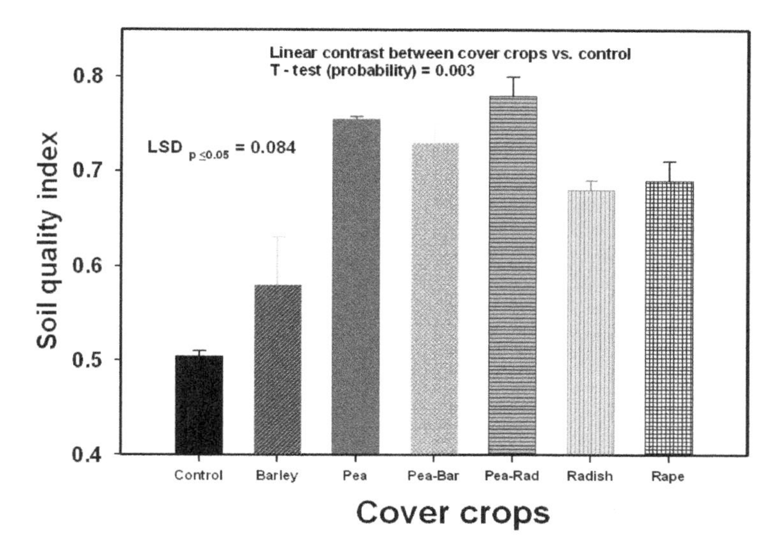

Figure 2. Cover crops effects on soil quality index (Data presented were average across 1997 to 2000). Bar indicates standard error of mean.

N-enriched residue from when only peas are used as a cover crop as opposed to used of peas and radishes as a mixed cover crop. Greater biomass production and N-fixing potential of cover crops, especially peas and radishes, may have had a positive effect on biomass N, P and K concentration which result in increased N, P and K content in soil. This may be explained by the fast rate of growth of peas and radishes which have the ability to quickly scavenger nutrients from the subsoil, and recycle biomass nutrients to the surface soil upon decomposition (Clark, 1998; Khan et al., 2001). Furthermore, an excess of H^+ produced during symbiotic N-fixation by legumes such as peas is often exchanged for greater uptake of P and K^+ by the roots (Liu et al., 1989). This process of H^+ release and exchange for basic cations by the legume roots may cause rhizosphere acidity over time (Liu et al., 1989) which helps legumes to uptake occluded P from soil while exuding organic acids (Gardner et al., 1981). Legumes often facilitate the release of K^+ from K bearing minerals through root contact exchange and rhizosphere acidity (Bajwa, 1987). As a result, legumes tend to accumulate a greater concentration of P and K in biomass, when used as cover crops, and they subsequently make these nutrients available in the soil for the succeeding crops.

Greater biomass incorporation may have improved the SQ over time which followed by enhanced biological activity, macroaggregate formations, and nutrient recycling under successive use of peas and radishes as mixed cover cropping. Hence, the ability of peas and radishes in mixed cropping for three successive years provide a more conducive biological, physical and chemical environment to improve SQ over control plots and those using other cover crops. In contrast, a significant decline in soil properties and SQ under continuous chemical fertilization is most probably related to fallow and the limited application of organic matter to soil over time.

3.3 *Cover crops and soil quality effects on potato production*

Potato yields were significantly higher for plots that used cover crop amendments as compared to control treatment plots. Peas had the highest beneficial effects on potato yields followed by peas and radish and barley in mixed cropping. All have greater effects than use of chemical fertilizers as shown in Table 4. Compared to control plots, the greatest potato yields (31 Mg ha^{-1}) were obtained when using peas as a cover crop and the lowest yield difference (23.8 Mg ha^{-1}) was obtained when barley was used as a cover crop. Peas in combination with radishes or barley significantly increased (29 to 30 Mg ha^{-1}) potato yields; however, the use of radish as a cover

Table 4. Economic efficiency of cover crops as soil amendments for potato production.

Cover crops	Potato yield ($Mg\,ha^{-1}$)	Production cost Average ($\$\,Mg^{-1}$)	Production cost Total ($\$\,ha^{-1}$)	Market price Average ($\$\,Mg^{-1}$)	Market price Total ($\$\,ha^{-1}$)	Profit ($\$\,ha^{-1}$)	Economic efficiency (%)*	Net efficiency (%)**
Control	21.8	38.7	842.7	63.0	1382.9	540.1	64.1	0
Barley	23.8	39.2	932.7	63.0	1502.6	569.9	61.1	−3
Pea	31.1	32.4	1007.9	63.0	1959.3	951.4	94.3	30.2
Pea + barley	28.8	34.2	984.3	63.0	1814.4	830.1	84.3	20.2
Pea + radish	30.1	33.2	998.8	63.0	1896.3	897.5	89.8	25.7
Radish	25.8	36.9	953.2	63.0	1628.6	675.3	70.8	6.7
Rape	27.1	37.3	1011.1	63.0	1710.5	699.4	69.4	5.3
$LSD_{p \leq 0.05}$***	2.62	ns	ns	63.0	ns	353.9	18.4	24.1
Orthogonal linear contrast								
Ccrop vs. cont.	6.1	−3.2	138.7	–	36.9	230.5	14.2	14.2
T-test (p)	0.013	ns	0.004	–	0.013	0.045	0.017	0.015

*Efficiency = (total market price – total production cost) × 100/total production cost;
**Net efficiency = (efficiency of individual cover crops – efficiency of control treatment);
***LSD = least significant difference.

crop resulted in minor increases in yields over those of the control plot. There were no significant differences in potato yields in response to peas alone or to peas used in combination with other cover crops.

The total and average costs of potato production did not vary significantly among the treatments. However, the total market price of potatoes was higher when only peas were used as a cover crop or when peas were used in combination with radish or barley as opposed to control groups as shown in Table 4. The higher returns were generated by successive use of peas as cover crops in pure stands or in combination with radishes and barley. Economic efficiencies were 20–30% higher when potatoes were grown as a cash crop after only peas were used to amendment soil or when used in combination with radishes or barley to improve SQ. A linear contrast between the effects of chemical fertilization and cover crops shows that cover crops as amendments have significantly higher increases in potato yields (~28%), profit levels (~43%), and economic efficiency levels (22%) which compared to control treatment plots (Table 4).

Significant positive effects of cover crops on potato yields are possibly due to the incorporation of large amounts of labile and nutrient enriched organic matter which resulted in an improvement in SQ. A number of studies have reported that legume and *Brassica* cover crops (e.g., peas and radishes), when incorporated into the soil, break-down and release labile organic matter and nutrients, such as N, and antimicrobial compounds that improve SQ and reduce root associated pathogen populations and diseases. This facilitates greater nutrient and water uptake by roots which results in higher growth and yield (Clark, 1998; Abdallahi and N'Dayegamiye, 2000; Islam and Weil, 2000a). Others have reported that long-term use of cover crops confers benefits in terms of biological, chemical and physical properties, such as efficient biological activity, increased organic matter and nutrient content, improved soil structure, and greater soil quality (Al-Khatib et al., 1997; Clark, 1998; Abdallahi and N'Dayegamiye, 2000; Islam and Weil, 2000a).

Improvements in soil quality resulting from cover crop use are significantly correlated with potato yields as shown in Figure 3. A significant linear relationship ($R^2 = 0.906$***) exists between SQ_{index} and potato yields which reflects the substantial contribution of cover crops – especially peas alone or peas in combination with radishes or barley – to the improvement of SQ. Improved SQ results in greater quantities of and profits from potato production. Variance related to the successive use of cover crops as organic amendments helped to explain increased profits related to potato production ($R^2 = 0.894$***) as shown in Figure 4.

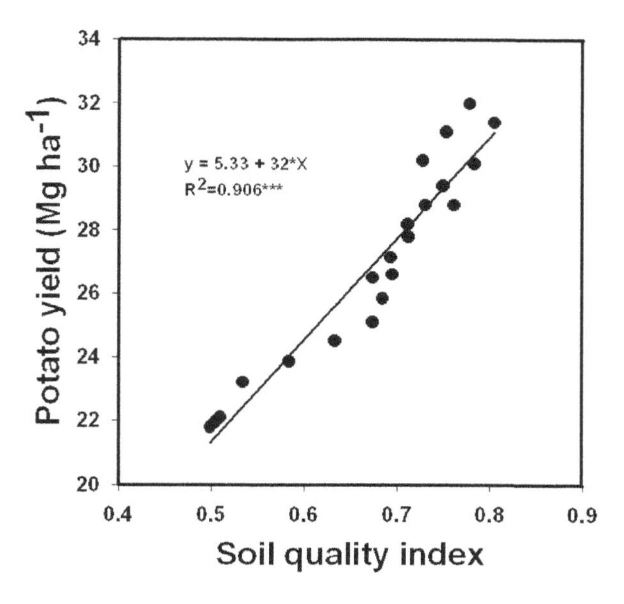

Figure 3. Linear relationship between the Soil Quality Index and differences in potato yields in response to cover crops as opposed to only chemical fertilizers (Data presented were averaged across 1997 to 2000).

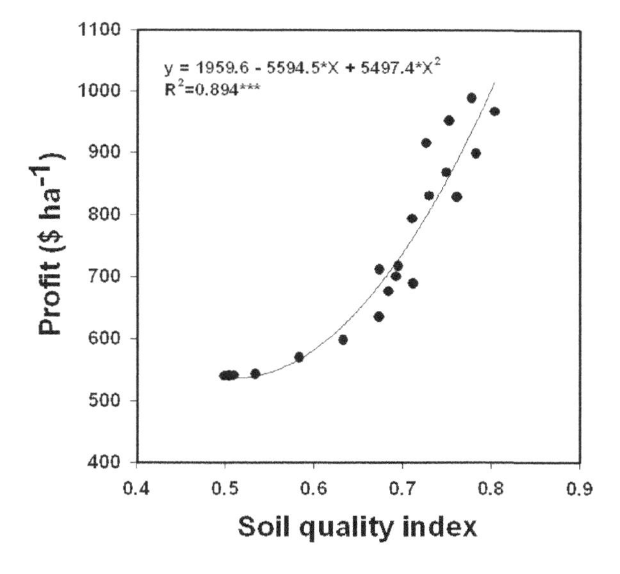

Figure 4. Non-linear relationship between the Soil Quality Index and profit differences in potato yields from plots using cover crops as opposed to those using chemical fertilizers alone (Data presented were average across 1997 to 2000).

4 CONCLUSIONS

The use of cover crops in combination with chemical fertilizers resulted in significantly improved irrigated SQ and potato yields through greater incorporation of labile and nutrient enriched biomass, enhanced biological activity, greater macroaggregate stability and aeration, C sequestration, and nutrient recycling. The 3 year time period represented by this study may be deemed insufficient to induce consistent measurable changes in C sequestration by some. Although accelerated chemical oxidation of organic matter takes place rapidly under hot and dry conditions, however, the trend to sequester C as soil organic matter in response to successive use of cover crops under the continental climate conditions encountered in Uzbekistan is evident. Field peas alone or in combination with radish or barley were found to be the most effective preceding cover crop to improve soil biological, chemical and physical properties. Incorporation of greater amounts of N-enriched and labile biomass from peas and radishes resulted in significant improvements in irrigated soil quality. This resulted in greater potato yields that those obtained when only using chemical fertilizers. Thus, the use of peas alone or peas in combination with radish or barley as cover crops is recommended in order to improve SQ and thus greater potato yields for typical sierozems soils under irrigation in Uzbekistan.

REFERENCES

Abdallahi, M.M. and A. N'Dayegamiye. 2000. Effects of green manures on soil physical and biological properties and on wheat yields and N uptake. Can. J. Soil Sci. 80:81–89.

Al-Khatib, K., C. Libbey, and R. Boydston. 1997. Weed suppression with *Brassica* green manure crops in green pea. Weed Sci. 45:439–445.

Anonymous. 2005. Agricultural Statistics of Uzbekistan (2004). Uzbekistan Statistics Committee, Tashkent, Uzbekistan.

Astanaqulov, T.E. 1991. Potato selection, seed production and agrotechnics in condition of Zarafshan valley. Ph.D. dissertation, University of Leningrad, Leningrad, Russia.

Alelseev, Y.K. 1936. Theory and practice of green manure (in Russian). p. 21. Agriculture Publishing House, Moscow, Russia.

Alekseev, Y.K. 1957. Green manure in irrigated lands (in Russian). p. 23. Agriculture Publishing House, Moscow, Russia.

Arinushkin, E.V. 1970. Manual of soil chemical analysis. MGU., Moscow, Russia. p. 487 (in Russian).

Bajwa, M.I. 1987. Potassium mineralogy of Pakistan soils and its effects on potassium response. p. 203–16. In: Proc. Potassium and Fertilizer Use Efficiency. Islamabad, Pakistan.

Berdnikov, A.M. and V.P. Kosyanchuk. 1999. Potato growing with siderite usage. Journal of Agriculture 4: 26. (In Russian).

Clark, A. 1998. Managing cover crops profitably. 2nd ed., Sustainable Agriculture Network, SARE/CSREES, USDA, USA.

Decker, A.M., A.J. Clark, J.J. Meisinger, F.R. Mulford, and M.S. McLntosh. 1994. Legume cover crops contributions to no-till corn production. Agron. J. 86, 126–135.

Gardner, W.K., D.G. Parbery, and D.A. Barber. 1983. The acquisition of phosphorus by *Lupinus albus* L. III. The probable mechanism by which phosphorus movement in the soil/root Interface is enhanced. Plant Soil. 70:107–24.

Hargrove, W.L., K.A. McVay, and D.F. Radcliff. 1989. Winter legume effects on soil properties and nitrogen fertilizer requirements. Soil Sci. Soc. Amer. J. 53:1856–1862.

Hussain, T., A.A. Sheikh, M.A. Abbas, G. Jilani, M. Yaseen. 1992. Efficiency of various green manures for N fertilizer substitution and residual effect on the following wheat crop. Pak. J. Agri. Sci. 29, 263–67.

Islam K.R. and R.R. Weil. 2000a. Soil quality indicator properties in mid-Atlantic soils as influenced by conservation management. J. Soil Water Cons. 55:69–78.

Islam, K.R. and R.R. Weil. 2000b. Land use effects on soil quality in a tropical forest ecosystem of Bangladesh. Agriculture, Ecosystems and Environment 79:9–16

Johnson, A.M., A.M. Golden, D.L. Auld, and D.R. Sumner. 1992. Effects of Rapeseed and vetch as green manure crops and fallow on nematodes and soil-borne pathogens. J. Nemat. 24:117–126.

Kachinskii, N.A. 1965. Soil Physics, Part-I. Vissahyaskhola, Moscow.

Khan, P.K., M. Musa, M.A. Shahzad, N.A. Adal, and M. Nasim. 1996. Wheat performance on green manured and fertilized fields. Pak. J. Soil Sci. 2:84–86.

Lal, R. 2004. Carbon sequestration in soils of Central Asia. Land Degradation & Development 15: 563–572.

Liu, W.C., L.J. Lund, and A.L. Page. 1989. Acidity produced by leguminous plants through symbiotic dinitrogen fixation. J. Environ. Qual. 18:529–534.

MacRae, R.J. and G.R. Mehuys. 1985. The effect of green manuring on the physical properties of temperate area soils. Adv. Soil Sci. 3:71–94.

Mahmood, I.A. and M. Aslam. 1999. Salient characteristics of cultivated soil influenced by legume and cereal cropping system. Pak. J. Biol. Sci. 2:95–97.

Malik, M.A., M.F. Saleem, and M.A. Cheema. 2002. Substitution of nitrogen requirement of wheat (*Triticum aestivum* L.) through green manuring. Intern. J. Agric. Biol. 4:145–147.

Nahar, K., J. Haider, and A.J.M.S. Karim. 1995. Residual effect of organic manures and influence of N fertilizer on soil properties and performance of wheat. Annals Bang. Agric. 5:73–78.

Oripov, R.O., N. Kholmanov, and S. Shonazarov. 1991. Rape in cotton-lucerne crop rotation. Proc. of the Intensification problems of feed-processing in irrigated farming. p. 25. Tashkent, Uzbekistan.

Rouatt, J.W. and R.G. Atkinson. 1950. The effect of the incorporation of certain cover crops on the microbiological balance of potato scab infested soil. Can. J. Bot. 28:140–152.

Sanford, G.B. 1946. Soil-borne diseases in relation to the microflora associated with various crops and soil amendments. Soil Sci. 61: 9–21.

SAS Institute. 2001. *SAS Users' Guide*. R 8.2; SAS Institute, Inc., Cary, NC.

Tester, C.F. 1990. Organic amendment effects on physical and chemical properties of a sandy soil. Soil Sci. Soc. Am J. 54:827–831.

Volodin, V.M. et al. 1989. Methods of soil fertility evaluation. Inventor's Certificate 1528142, Bulletin of Inventions # 11.

Weinhold, A.R., J.W. Oswald, T. Bowman, J. Bishop, and D. Wright. 1964. Influence of green manures and crop rotation on common scab of potato. Am. Potato J. 41:265–273.

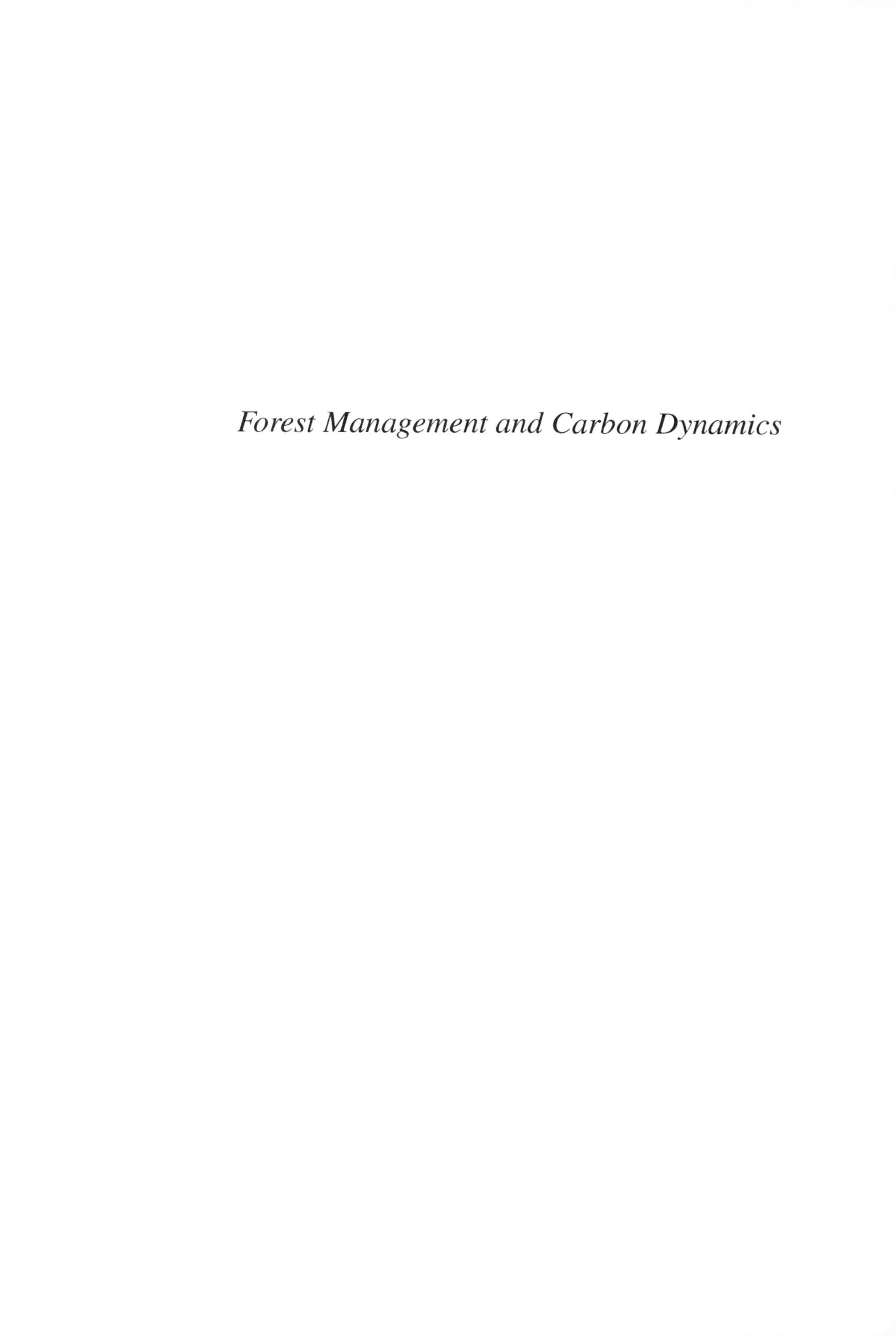

Forest Management and Carbon Dynamics

CHAPTER 25

Forest carbon sequestration and storage of the Kargasoksky Leshoz of the Tomsk Oblast, Russia – Current status and the investment potential

Roger A. Williams
School of Natural Resources, The Ohio State University, Columbus, OH, USA

Sarah E. Schafer
International Programs, The Ohio State University, Columbus, OH, USA

1 INTRODUCTION

The Kyoto Protocol lays out a possible framework for the creation of transferable emission reductions through investment in mitigation projects operated under Article 6 (Joint Implementation) or Article 12, the Clean Development Mechanism (CDM). The criteria for CDM project are real, measurable and long-term benefits related to the mitigation of climate change (Article 12.5b) and Reduction in emissions that are additional to any that would occur in the absence of the certified project activity (Article 12.5c). The CDM governs project investments in developing countries that generate certified emission reductions for industrialized countries undertaking specific commitments under the United Nations Framework Convention on Climate Change (UNFCCC) and the Kyoto Protocol.

The CDM allows 38 industrialized countries to invest in projects that reduce greenhouse gas (GHG) emissions in developing countries. The developed countries can use the carbon (C) credits from the reduced emissions, in order to comply with the protocol's requirement that they reduce their own emissions. Basically, C credits are GHG emission reductions that are created when a CDM project reduces or even eliminates the emission GHG's, such as (CO_2) or methane (CH_4), as compared to what would be released in absence of such a project. For purposes of CDM qualification, the project must lessen emissions of greenhouse gases (GHGs). These CDM projects are grouped into four basic categories: energy, transportation, waste management, and forestry. However, the rules and regulations for each category have still not been fully developed.

Also included are Joint Implementation (JI) projects, which involve the investment by an industrialized country, or entities in the country, in projects in another industrialized country to reduce emissions or sequester C. Both project types allow for International Emissions Trading (IET), which allows for the trading of emission reductions among industrialized countries that have ratified the Protocol. JI projects result in Emission Reduction Units (ERUs) and CDM projects result in Certified Emission Reductions (CERs).

Forests are a major focus area of the Kyoto Protocol, which includes afforestation (establishment of plantations on land that was bare of trees in 1990), the managed forest, and C credits as a commodity. Because the vegetation of forests contains more than 75% of all C accumulated in the vegetation found in terrestrial ecosystems (Olson et al., 1983), the role of forests in global climate change mitigation is critical. A study released by the International Institute for Applied Systems Analysis (Nilsson et al., 2000) notes that since reductions of the magnitude sought by the Kyoto Protocol would be difficult to achieve, the Protocol offers countries an alternative: reducing a country's CO_2 burden by planting more forests or creating or improving other C sinks.

The forestry sector is viewed as potentially providing economically viable opportunities for C offsets through the conservation of existing C pools, sequestered C in managed and new forests, through the substitution of forest products for more energy intensive materials in construction, and through substitution of biomass fuels for fossil fuels. The Kyoto Protocol recognizes that forests can be a C sink and, therefore, accounted in terms of the net achievement of an emissions target, with a target for the first national accounting to be between 2008 and 2012. Accordingly, forests, while only one component in the overall GHG emissions equation, are measurable C sinks with potential C offset use.

Of particular interest are the vast boreal forests of the Northern Hemisphere. Globally, the boreal forest comprises almost 25% of the world's closed canopy forest as well as vast expanses of open transitional forest. Russia holds over half of the world's northern boreal forest and 20% of all global forests. Most of these contiguous forests occur in Siberia, where human populations are sparse and access is very limited. Russia signed onto the Kyoto Treaty in October 2004, and thereby makes it an eligible country for establishing, trading, and selling C credits. Accordingly, Russia will need to more accurately document and quantify its C storage to determine its sink in relation to its GHG emissions. Russia is now in a position to sell-off any credits it determines to have in surplus to offset the country's GHG emissions. In addition, outside investors from Kyoto ratified countries could have the opportunity to develop CDM projects in Russia for the purpose of establishing C credits.

The purpose of this paper is to exam the current C storage and sequestration in Siberia, and exam the potential for outside investors in CDM projects. Specifically data for the Tomsk Oblast, located in southwestern Siberia, is examined for the purpose of example.

2 METHODS

The Tomsk Oblast is a province located in southwestern Siberia (Figure 1). It is divided into 26 leshozes, or forest management units. The Kargasoksky Leshoz was chosen for this paper as it represents one of the largest units in Tomsk and is heavily forested. Russian forests are subdivided into three types according to the main goods and services they provide. Type I forests are for ecosystem services, which include all forests providing for water regulation, environmental protection, and for important sanitary or health-improving functions as well as scientific, historical, social, or cultural value. Type II forests are to be used for light extensive uses and include all forests within densely populated areas that provide both environmental protection and goods from limited exploitation. Type III forests are forest designated for intensive uses, which include all forests of richly wooded regions that are mainly managed for exploitation.Type III forests are expected to sustainably provide the national demand for timber, while preserving some of their broader ecosystem services. The type III forests were selected for the analysis because these forests permit more latitude in forest management activities, and have fewer restrictions as compared to Type I and II forests.

The most recent forest inventory of the 2004 of the State Forest Fund Account (SFFA) gathered from the Ministry of Natural Resources for the Kargasoksky Leshoz. This included data on species growth rates; volume estimates including the crown, branches, roots and trunk for dominate species, and age structure of the forested region. Type III forests in the Kargasoksky Leshoz are comprised of six forest types, as defined by the dominant species (Table 1 and 2). These include Scotch pine (*Pinus sylvestris* L.), Siberian larch (*Larix sibirica* Ledeb.), Siberian spruce (*Picea obovata* Ledeb.), Siberian fir (*Abies sibirica* Ledeb.), birch (*Betula pendula* Roth., *B. pubescens* Ehrh.), and aspen (*Populus tremula* L.). For purposes of this study, however, larch and willow were not included, as the total area for larch comprises only 0.01% of the forests in the Kargasoksky Leshoz, all of which reside in the Mature and Overmature category, and willow has no commercial value and contributes even less than larch to the total forest composition.

Information about the area of distribution of the forest types by specific age is not presented in the SFFA. However, this information is kept in the SFFA database. Forest stands are grouped into the age groups based on the age of final harvest and on the age classes. Age class as defined by the SFFA is an interval, characterizing the age of trees and shrubs as follows: (1) 20 years

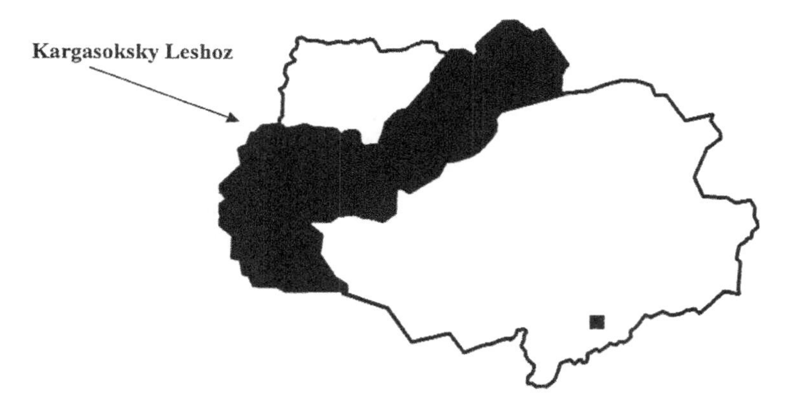

Figure 1. Map displaying the location of the Tomsk Oblast in southwestern Siberia. The City of Tomsk is the capital. The Kargasoksky Leshoz and the closest major cities (Novosibirsk and Kemerovo, in neighboring regions) are also shown.

gradation was set for hardwoods and coniferous of seed origin; (2) 10 years gradation was set for softwoods and hardwoods of sprout origin; and (3) 5 years or one year gradation was set for shrubs. Siberian pine (*Pinus sibirica* Du Tour.) forests are an exception as the age class gradation is 40 years. Therefore, depending on the age, forest stands are divided into five groups by the SFFA: (1) young growth of the 1st age group, (2) young growth of the 2nd age group, (3) middle aged forests, (4) maturing forests, (5) mature and overmature forests.

There is a direct correlation between wood volume and C, which was used rather than biomass to determine C values, as data for biomass was not complete. Oven dry wood densities were gathered for each species and applied to the volume of each of the age classes for each species (Table 3). A value of 0.5 was than applied to the total dry weight of the respective forests to convert volume to C. When specific C content is unknown, different researchers have estimated C content as 50% of the absolutely dry mass of the stem, roots, and leafless branches (Alexeyev and Birdsey, 1998; Koch,1989).

Table 1. Forest area (ha) by forest type and age class in the group III forests of the Kargasoksky Leshoz, Tomk Oblast, Russia.

		Age class				
		Young growth				
Species group	Total	Class 1	Class 2	Middle aged	Approaching maturity	Mature and Overmature
Pine	890751	83633	60883	149530	100643	496062
Spruce	17162	1874	931	1491	4253	8613
Fir	24980	2266	3179	2513	986	16036
Larch	153	Pine	0	0	0	153
Total conifer	933046	87773	64993	153534	105882	520864
Birch	599185	47314	16237	57184	18740	459710
Aspen	163842	5288	2886	1307	583	153778
Willow	31	0	0	10	0	21
Total hardwood	763058	52602	19123	58501	19323	613509
Total	1696104	140375	84116	212035	125205	1134373

Table 2. Forest volume (thousand m^3) by forest type and age class in the group III forests of the Kargasoksky Leshoz, Tomsk Oblast, Russia.

		Age class				
		Young growth				
Species group	Total	Class 1	Class 2	Middle aged	Approaching maturity	Mature and Overmature
Pine	105422.1	1610.5	3435.7	15817.9	12935.2	71622.8
Spruce	2488.4	63.4	41.0	179.8	730.7	1473.5
Fir	4112.1	65.0	124.8	190.1	154.5	3577.7
Larch	26.8	0.0	0.0	0.0	0.0	26.8
Total conifer	112049.4	1738.9	3601.5	16187.8	13820.4	76700.8
Birch	85302.3	508.8	457.4	4188.8	2153.2	77994.1
Aspen	35349.1	95.8	128.5	145.7	85.7	34893.4
Willow	2.2	0.0	0.0	0.3	0.0	1.9
Total hardwood	120653.6	604.6	585.9	4334.8	2238.9	112889.4
Total	232703	2343.5	4187.4	20522.6	16059.3	189590.2

Table 3. Ovendry wood density values used for converting cubic meter volume of trees to ovendry phytomass in the Type III forests of the Kargasoksky Leshoz in the Tomsk Oblast, Russia.

Species	Ovendry wood density (kg mg^{-3})	Source
Aspen	336	Tt Timber Group, 2004
Birch	485	Heräjärvi et al., 2004
Scotch pine	512	Christie Timber Company, 2004
Siberian larch	595	Timberwise, 2004
Siberian fir	401	Alden, 1997[a]
Sigerian spruce	449	Alden, 1997[b]

[a] Values for balsam fir (*Abies balsamea*) used.
[b] Values for white spruce (*Picea glauca*) used.

Based on the volume to biomass conversions, total C and subsequently C ha^{-1} were determined for each forest type. Ratios of bark mass to timber, crown mass to timber and root mass to timber, and understory phytomass in tons per hectare for each of the age classes were than applied to timber values to determine the amount of total standing C in t ha^{-1}. The ratios used were reported by Alexeyev and Birdsey (1998) in their report on C storage in forests throughout Russia.

Since specific ages were not available for the different forest types but rather age groupings, numbers 1 through 5 were assigned to each of the age groups, with 1 representing the youngest and 5 representing the oldest age classes. The age group and C ha^{-1} data pairs were used to develop a model to estimate the C sequestration rates. A non-linear form of the Richards growth function model was used to fit the data:

$$C = b1[1 - e(^{-b2e})^{(b3A)}]$$

Where C = t C ha^{-1}, A = the age group, e = base of the natural logarithm, and b1, b2, and b3 are coefficient estimates. The Gauss-Newton method of non-linear regression was used to estimate the model coefficients. In this model form, the b1 coefficient represents the asymptote, or the maximum accumulated carbon/ha; the b2 coefficient indicates the rate at which the asymptote is reached, and the b3 coefficient represents the initial growth rate. The C/ha includes the total forest C minus the soil carbon. Accordingly, it includes the bole, crown, roots and understory.

3 RESULTS AND DISCUSSION

The coefficient estimates for each forest type is provided in Table 4. According to the model coefficients, pine and spruce forests achieve their maximum C storage/ha earlier than the other forest types, attaining 55.3 and 66.0 t C ha^{-1} respectively. It would appear from the plot of the models (Figure 2) that the other forest types have yet to peak in their storage of carbon. This is substantiated by the b2 coefficient, which reflects the rate at which this asymptote is reached. Pine and spruce have values of 0.12 and 0.9, respectively, which is higher than values for the other forest types. The fact that pine and spruce forests reach their asymptotes earlier than the other forests may be explained by the b3 coefficient of the model. The b3 coefficient, which describes the initial growth rate, is higher for both pine and spruce forests as compared to the other forest types.

Currently 67% of all type III forests (based on area) within the Kargasoksky Leshoz are in the mature-overmature age class. Birch and aspen by far contain the greatest amount in this age class, comprising 77% and 94%, respectively, by forest type. Fir, pine and spruce have lesser amounts, representing 64%, 56%, and 50%, respectively, within each forest type. Old-growth or old forests contain the greatest amounts of stored aboveground C/ha than any other defined land cover in the biosphere (Clausen and Gholz 2001). However, these mature forests sequester C at slower annual rates than do young forests or even-aged natural stands.

Table 4. Coefficients and statistics of models[a] to estimate the forest carbon sequestration rate (t ha^{-1}) for type III forests in the Kargasoksky Leshoz, Tomsk Oblast, Russia. Carbon includes tree boles, crowns, roots, and forest understory.

Species	b1	b2	bB3	R^2	MSE	N
Aspen	69.9694	0.0746	0.6211	0.9939	14.5553	5
Birch	79.2158	0.0585	0.6779	0.9967	9.9070	5
Fir	110.7000	0.0607	0.6058	0.9950	22.1675	5
Pine	55.3258	0.1207	0.8089	0.9989	3.5945	5
Spruce	66.0256	0.0941	0.8993	0.9857	69.0797	5

[a] Model: $C = b1(1 - e^{(-b2e)^{(b33A)}})$, where C = total carbon (t/ha) and A = age class category.

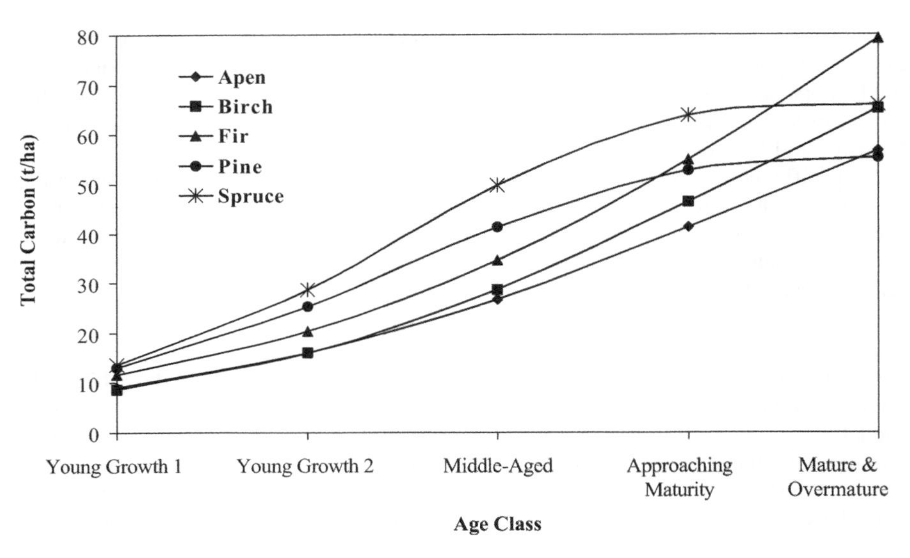

Figure 2. Plot of the carbon sequestration models by forest type for the type III forests of the Kargasoksky Leshoz, Tomsk Oblast, Russia.

Even though their total sequestered C is less compared to old forests, young forests and particularly plantations, sequester C at much higher rates than old forests. Approximately 26% of all type III forests in the Kargasoksky Leshoz occur in the young growth through middle-aged age classes. The pine forest type has the largest amount of young forest with 33% of the pine forest occurring within this young age-class range. This is followed in decreasing order by fir, spruce and aspen with 32%, 25%, and 20%, respectively, occurring within this young age class range by forest type. Aspen forests have the smallest amount of forests existing in this age class range at 6% within this forest type.

Pine forests account for 53%, or a little over half of the total forest area of the type III forests within this leshoz. This, plus the fact that it displays a more equal distribution across the age classes and displays one of the highest sequestration rates, makes these forests the most attractive in terms of potential for C mitigation management and investment.

However, some institutions must be in place before this possibility can exist. First there is the definition of Kyoto forests that is debated, and what constitutes eligible forests for C credits. Currently, Kyoto forests, as defined in article 3.3, must arise from a change in land use, and planted not before 1990. Growers must have evidence to prove their forests meet these qualifications. Note also that C sequestered by the forests during 2008–2012 alone is tradable. Decision is pending about the period after 2012.

Under the protocol, C sequestered in trees (i.e., C credit) must come from Kyoto forests, which are new forests that are (1) planted on land, which historically has not been covered by forest (i.e., afforestation); (2) planted on land which historically has contained forest but which has been used for another purpose since last being covered by forests (i.e., reforestation); and (3) additional to those that would otherwise have been planted. Accordingly, industrialized countries must include lands subject to afforestation, reforestation, and deforestation since 1990 in their GHG accounting for the 2008–2012 accounting period.

Forest management activities applied to natural stands established prior to 1990 is currently being discussed and debated. At present, only about 11% of the world's forests are being actively managed with long-term plans for goods and services (Clausen and Gholz, 2001), and most of the management activities are occurring in countries where the temperate and boreal forests exist. Accordingly, there is the potential for management activities to increase the global C stores.

Most of the analysis performed on C sequestration in the forest sector has focused on activities covered under Article 3.3 – afforestation, reforestation, and deforestation. The distinction between articles 3.3 and 3.4 is that article 3.3 deals with 'new' forests created since 1990, while article 3.4 deals with management practices for forests established before January 1990.

Article 3.4 of the Kyoto Protocol also provides a potential basis for claiming emission units from 'additional' sink activities, where the amount of C absorbed and stored as a result of the management of forests, croplands, and grazing lands, and of re-vegetation activities, has increased since 1990. Additional sink activities include the management of forests established before January 1990 (so called non-Kyoto forests). Management in this context is defined broadly. The current negotiating text defines forest management as a system of practices for stewardship and use of forestland aimed at fulfilling relevant ecological (including biodiversity), economic and social functions of the forest in a sustainable manner. It is not mandatory for countries to account for Article 3.4 activities in the first commitment period of 2008–2012. Countries can choose to account for them, but need to do so no later than 2006.

Accordingly, forest management activities for the purpose of adding C to Russia's mitigation accounts would have to be distinguished from business as usual forest management and demonstrate net increases in carbon. Forest management activities considered for C sequestration include the addition of amendments such as fertilizer, longer rotations, less invasive selective cutting, soil conservation, forest protection (from pests, fire, etc.) recycling forest products, genetic engineering, and lengthening the residence time of forest products.

Based on our preliminary findings, the pine forests in the Kargasoksky Leshoz should become the focus of any management activities. They occupy over half of the forest area, and demonstrate one of the highest sequestration rates as compared to the other forests. While mature forests contain more stored C than young forests, the rate at which they sequester additional C has been diminished as compared to young forests. Also, any thorough C accounting system needs to include products that are removed from these forests before all C impacts are determined. Manriquez (2002) modeled a series of forest rotations to look at C stored and removed, including that removed and stored in products. The model revealed that forests placed on shorter rotations do not lead to greater C emissions, which is an implicit assumption in the Kyoto Protocol. And because managed forests remove dead, dying and decayed material, the risk of wildfire, which also releases C into the atmosphere, is reduced. Accordingly, managed forests maintained in younger age classes with products included in the accounting sequestered more C and maintained a C surplus than unmanaged forests allowed to attain mature status.

4 CONCLUSIONS

The 2004 forest inventory data from the Tomsk Oblast was acquired and analyzed to estimate stored forest C in 5 age class categories: young growth 1, young growth 2, middle-aged, approaching maturity, and mature-overmature. The Kargasoksky Leshoz, one of 26 forest management units in the Tomsk Oblast, was chosen for this analysis as it is one of the largest management units in the oblast and is very sparsely populated. The category III forests in this unit were the basis for analysis because these forests allow more flexibility in utilization and management, as determined by the Russian Forest Code. Birch (*Betula* sp.) and pine (*Pinus sylvestris*) account for nearly 90% of forests within this category. Fir (*Abies sibirica*), spruce (*Picea sibirica*), larch (*Larix sibirica*), and aspen (*Populus tremula*) comprise the remainder of these forests. Approximately 68% of all forests exist in the mature-overmature age class, which presents both challenges and opportunities for C management and investments. C was estimated in these forests using standard and developed conversions based on volume and biomass. Sequestration models were developed for all forest types, and the pine and spruce forests were found to sequester more C at earlier ages compared to other forests. However, both of these forest types reach asymptotes earlier and at lower tons of C per hectare than the other forest types. The total forest area occupied by a forest type and the sequestration rate potential will be critical in determining which forests display the greatest

C credit investment potential. The potential use of these forests for mitigation purposes and the opportunities for outside investment in these forests for C credits is discussed.

Scotch pine forests in the type III category display faster C sequestration rates than most other forest types of the Kargasoksky Leshoz. Since these forests occupy over half of the type III forests, they hold the greatest promise for C mitigation management, including the establishment of CDM and JI projects. The investment potential of these forests will only be realized after a full C account has been established for Russia. Russia will be able to sell and market the C credits in the EIT system once it has been determined that excess C exists in Kyoto recognized forests to offset C emissions. This is a likely scenario as Russia agreed to stabilize their emissions at 1990 levels in the first commitment period, and is projected to arrive at emissions that are 250 Mt C yr^{-1} below 1990 levels within this first commitment period (Gurney and Neff 2000). Thus, it is expected that Russia will ultimately show a surplus of carbon.

REFERENCES

Alden, H.A. 1997. Softwoods of North America. Gen. Tech. Rep. FPL–GTR–102. Madison, WI: U.S. Department of Agriculture, Forest Service, Forest Products Laboratory. 151 p.

Alexeyev, V.A. and R.A. Birdsey (eds.). 1998. Carbon storage in forests and peatlands of Russia. Gen. Tech. Rep. NE-244. Radnor, PA: U.S. Department of Agriculture, Forest Service, Northeastern Forest Experiment Station. 137 p.

Christie Timber Company. 2004. http://www.christie-timber.co.uk/timber_species/softwoods/Pinus_Sylvestris.htm

Clausen, R.M. and H.L. Gholz. 2001. Carbon and forest management. USAID Development Experience Clearinghouse (DEC)/Development Information Services (DIS) Clearinghouse/CD-DIS/Development Experience System (DEXS). 73 p.

Gurney, K. and J. Neff. 2000. Carbon sequestration potential in Canada, Russia and the United States under Article 3.4 of the Kyoto Protocol. WWF Report, July 2000, 28 p.

Heräjärvi, H., A. Jouhiaho, V. Tammiruusu, and E. Verkasalo. 2004. Small-diameter Scots pine and birch timber as raw materials for engineered wood products. International Journal Forest Engineering. 15(2):1–13.

Koch, P. 1989. Estimates by species group and region in the USA of: I. Below-ground root weight as a percentage of ovendry complete-tree weight; and II. Carbon content of tree portions. Consulting report. 23 p.

Manriquez, Carolina. 2002. Carbon sequestration in the Pacific Northwest: a model.

Master's thesis. University of Washington, Seattle, Washington, USA. 158 p.

Nilsson, S., A. Shvidenko, V. Stolbovoi, M. Gluck, M. Jonas, and M. Obersteiner. 2000. Full Carbon Account for Russia. International Institute for Applied Systems Analysis Interim Report IR-00-021. Laxenburg, Austria. 191 p.

Olson, J.S., J.A. Watts, and L.J. Allison. 1983. Carbon in live vegetation of major world ecosystems. Env. Sci. Div. Pub. 1997. Oak Ridge, TN: U.S. Department of Energy. 164 p.

Timberwise. 2004. http://www.timberwise.fi/products_larch.html

Tt Timber Group. 2004. http://www.tt-timber.com/servlet/control/wtk_page/SpeciePopup/pk/22021

CHAPTER 26

Soil and vegetation management strategies for improved carbon sequestration in Pamir mountain ecosystems

S. Sanginov & U. Akramov
Soil Science Institute of Tajikistan, Dushanbe

1 INTRODUCTION

Globally, 75% of terrestrial carbon is found in soil organic matter. Soils are the basic resources of mountainous areas. They provide the medium in which plants and forests grow and their properties determine water availability for vegetation. Low organic matter content and high degradation of soil in the mountain slopes are the main causes of the low vegetation productivity in the Pamir ranges. The soil organic matter consists of a mixture of plant and animal residues in various stages of decomposition, of substances formed either chemically or biologically from various breakdown products, and of a biomass of soil organisms and its decomposing remains. Recent concern for increasing atmospheric CO_2 levels, and also the losses of nonhuman carbon stored in arable and forest soils, has resulted in increased interest in the mineralization of plant carbon its sequestration by humic substances.

Because of its influence on many important soil factors, the mineralization and humification of plant biomass carbon added to soil could represent an important indicator of soil quality. In mountainous areas of Tajikistan, the strategy for irrigation of high eroded dry slopes with the construction of small channels, private land ownership, community pasture and forest management, the introduction of double crop system, and the mulching of orchard plantation, are directly connected with the restoration of degraded soils. These practices enhance soil quality, improve water use efficiency, and sequester carbon in soil and biomass.

Tajikistan is a mountainous country with an area of 143.1 km^2. The relief is extremely diversified and ranges from 300 m to 7,495 m above sea level. The highest mountain of Central Asia is the Peak Somoni (7,495 m above sea level), which is located in the northern Pamirs (Tajikistan). The climate of Tajikistan is continental, with average winter temperatures varying between −3°C and −20°C, and average summer temperatures varying between 19°C and 32°C. Average annual precipitation rate of this region is 400 mm, ranging from less than 200 mm in the plains and deserts to 2,400 mm in the mountains.

Historically in ancient times, Tajikistan was one of the main agricultural centers of the continent. Tajikistan is an agrarian country with a national economy that is based mainly on the exploitation of its primary natural resource; i.e., its: soils which are mainly serozems (80%). Cotton has been continuously produced for almost a century, which has caused soil productivity to become very low. During the Soviet time, agriculture was characterized by a high rate of mechanization, heavy use of chemical fertilizers and pesticides, development of monoculture in the regions, and use of excessive quantities of water for irrigation. As a result, soil deterioration has become dramatic. Although substantial resources were allotted for agricultural intensification from 1970 to1990, farm yields have continued to decrease during the last decade.

As shown in Table 1, about 2.2% of the total area in Tajikistan is cropland, about 25% is used as rangeland, about 3.8% is forested, with the remaining 64.8% utilized for other purposes. About 75% of the total cropland is irrigated. The arable land per capita in Tajikistan amounts to 0.08 ha.

Table 1. Land use in Tajikistan.

Land use	Hectares	%
Rainfed cropland	142,000	0.6
Rangelands	3,550,000	25.2
Forests	537,000	3.8
Other area	9,113,000	64.8
Irrigated cropland	718,000	1.6
Total area	14,060,000	100

Table 2. Soil parameters of natural fertility of main type of soils in Tajikistan by vertical zones.

Number	Soil types	Altitudes (m)	Depth of organic matter horizon (cm)	Depth of soil profile (cm)	Organic matter humus (%)
1	Sandy desert	250–300	0–10	–	less 0.5
2	Grey brown	300–450	10–15	15–25	0.5–0.7
3	Light sierozem	450–500	20–35	30–35	0.7–1.2
4	Typical serozem	500–700	25–30	30–40	1.0–1.5
5	Dark serozem	700–1000	30–45	40–50	1.5–2.5
6	Mountain soils	1000–2500	45–50	50–60	2.0–3.0
7	High mountain soils	2500–4800	70–80	100–120	6.0–9.0

Table 3. Distribution of carbon in irrigated soils (0–50 cm).

	Quantity of total C in soil (t ha^{-1})					
Regions	15	30	60	90	120	Total
Badakshan	30,000	90,000	120,000	180,000	120,000	540,000
Central	48,720	1,163,790	1,980,900	237,420	41,400	3,472,230
Hatlon	717,720	6,417,510	2,375,640	783,360	73,680	10,367,910
Sugd	119,235	2,798,100	354,960	63,450	7,800	3,343,545
Total	915,675	10,469,400	4,831,500	1,264,230	242,880	17,723,685

There are more than 25 different types of soil in the Central Asia, of which some characteristics are shown in Table 2. They are common to the Central Asia region.

2 CARBON STOCKS IN THE IRRIGATED SOILS

Soils are the largest carbon reservoir of carbon in the crop production cycle. According to Batches and Sam Brock (1997) soils contain a significant quantity of carbon, about three times more than that found in vegetation and twice as much as that found in the atmosphere. Most agricultural production occurs on irrigated land in Tajikistan. According to the most recent soil survey data, the total quantity of carbon stock found in irrigated lands is 17.2 m tons. Around 5.2% of the irrigated area contains less than 15 tons/ha of carbon in the 0–50 cm of soil layer. Approximately 86.3% of irrigated areas contain 30–60 tons/ha of carbon; approximately 9.5% contain 90 t ha^{-1} and the remaining area contains more than 120 t ha^{-1} of carbon. Table 3 indicates that more than 58.5% of the total carbon stocks is located in the Hatlon region of the country, around 19% in the Sugd region, and about 20% in the central part and Badakshan regions.

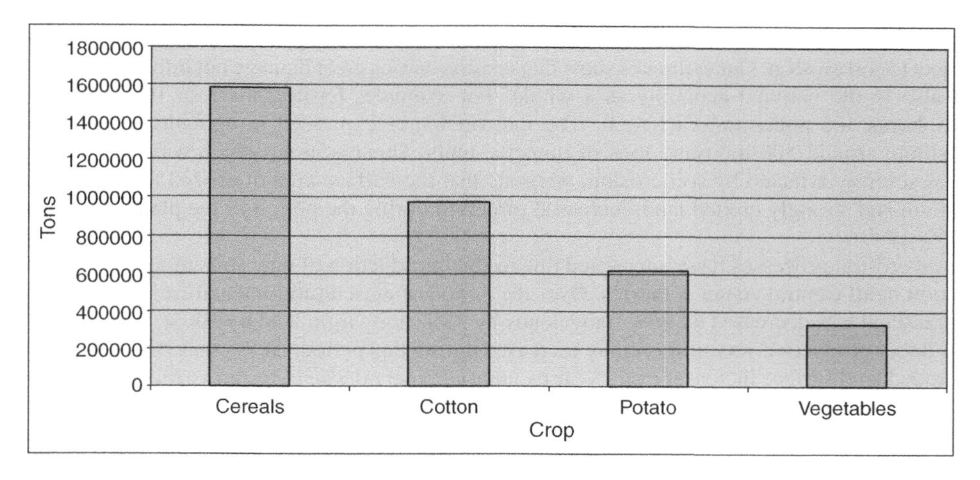

Figure 1. Input of crops to carbon stocks.

Table 4. Slope land in Tajikistan (%).

Slope (in degrees)	Total	Arable land	Grassland and badland
2	9.1	24.1	6.4
2–5	15.4	40.6	11.0
5–10	12.5	17.3	12.2
10–15	11.4	8.7	11.9
15–20	12.7	6.0	13.9
20–30	21.5	3.3	24.6
30–45	13.8	–	16.3
>45	3.2	–	3.7

3 TOTAL CARBON STOCKS FOR PRESENT LAND UTILIZATION TYPE IN TAJIKISTAN

An assessment of the utilization of carbon stocks in under current land use regimes in Tajikistan is an important aspect of evaluation of production systems. This is important for food, fodder, and fiber production. It is also important to assess the potential of the system for carbon stocks and sequestration. Evaluation of carbon stocks an indicator of the extent of land degradation as well as the economic and biophysical suitability of land use types.

According to our calculations for agricultural crop vegetation, cereals are the principal contributor to carbon stocks taking into account both above ground and belowground biomass. As shown in Figure 1, they contribute approximately 44.6% of the total for crop vegetation. Cotton contributes approximately 27.9%; potatoes contribute approximately 17.5%; and the remainder is association with the production of vegetables and flax. About 53% of total carbon stocks resulting from vegetation is found in the Hatlon region.

4 SOIL DEGRADATION AND LOSS OF CARBON

The main causes of soil degradation in Central Asia are soil erosion, landslides, soil compaction, and secondary salinization caused by irrigation. As indicated in Table 4, more than 90% of the territory in Tajikistan is situated on the slopes greater than 5°, Thus, millions of hectares of soil are subject to soil erosion. Our estimates show that erosion causes great damage not only to agriculture, but also to the national economy as a whole. For example, fertile soils lose 100,000 tons of phosphorus and potassium each year. The indirect losses expressed in agricultural production constitute around 500 thousand tons of nutritive units. Our research, which was carried out in zones severely affected by soil erosion, suggests that the surface area of eroded land, especially medium and strongly eroded land, increased threefold during the period of the planned economy (1950–1990).

Soil erosion processes have intensified due to the introduction of new soils to agricultural production in all Central Asian countries. Over the period of agricultural intensification, the surface of grassland was decreased by 96%, rangelands by 35%, and virgin land by 100%. The amount of land under production was substantially increased during this period. At the same time the amount of eroded land surface increases every year by approximately 0.8%. In recent years more and more damage has been caused by wind. Unfortunately, this form of erosion is rarely considered in Central Asia, although its importance is growing due to deforestation and practices that do not consider natural balances in the natural resource base. The main natural way to protect against erosion is to diversity techniques of land management, taking into account the susceptibility of soils to wind erosion. An important response to most soil degradation problems is the application of certain soil conservation technologies. They can be sued to address adaptive and/or preventive strategies which are now being considered. Adaptive strategies include policies and actions to diminish soil degradation rates within the framework of current territory organization and farming. Preventive strategies include policies and actions to transform traditional farming into soil conserving, soil protecting and soil reclaiming farming.

5 SOIL CONSERVATION TECHNOLOGIES FOR AREAS WITH PRONOUNCED SLOPES

The current system of agriculture being practiced on sloped lands should be replaced by erosion-protection agriculture in order to prevent soil erosion. Without the adoption of these practices it is difficult to ensure high yields and to decrease production costs. Agriculture in erosion prone regions should envisage a specialization of farms and specific agricultural cropping practice that increase the fertility of these erosive soils. Ideally, soil erosion control is practiced taking into account the particular conditions of the territory to which the land pertains. Given variations in climate, relief, soil and vegetative cover in different regions, requires that various approaches be used to control erosion.

Data reported in this study are from the ICARDA "On-farm soil and water management for sustainable agriculture in Central Asia" project. As a first step the soil degradation rate for the hydrographic basin was identified which enable us to identify whether erosion is intensifying due to natural conditions or to an anthropogenic factor. We identified soil degradation rates for specific regions and for all of Central Asia, using geographic-comparative, cartographic, laboratory-analytical and mathematical-statistical methods.

5.1 *Mulching*

Mulching is an important technology for various reasons. It improves soil microclimate, enhances soil micro organisms, improves soil structure and fertility, conserves soil moisture, reduces weed growth, prevents soil erosion caused by rainfall, and reduces tillage rates. Mulching is also a very important technology for water harvesting as illustrate in Figure 2. In the areas of rain-fed agriculture, effective management of water is of great importance, especially when rainfalls are very irregular and thus less favorable for crop production. Mulching with plastics and plant residues was found to help improve the water harvesting capacity of soil and the growth of trees in southern Tajikistan. In general, adequate vegetative soil cover, erosion reduction practices, organic matter

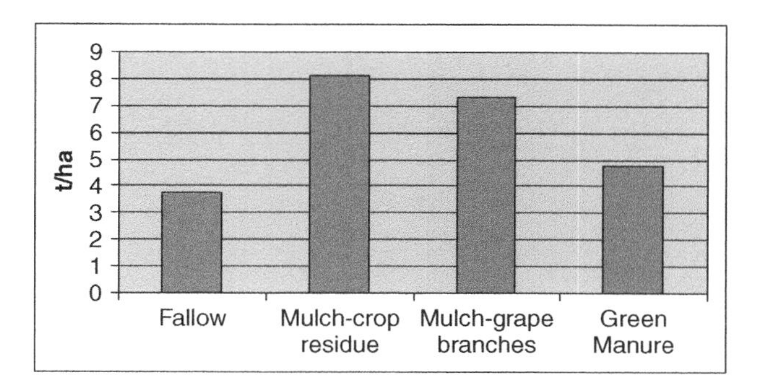

Figure 2. Effect of mulch on the yield of grapes.

Table 5. Changes of organic matter and total nitrogen in typical serozem soils during last 15 years ($t\,ha^{-1}$).

		Carbon		Nitrogen	
Treatments	Depth (cm)	1986	2001	1986	2001
Fallow	0–30	47.3	34.3	3.74	3.29
	30–50	18.0	18.2	1.53	1.50
	0–50	65.3	52.5	5.27	4.79
Mulching	0–30	48.1	117.6	3.87	4.70
	30–50	24.2	57.3	1.39	3.96
	0–50	72.3	174.9	5.26	8.66

management, and guiding the flow of excess water are the most important ways to conserve soil and to increase farming productivity. But as shown in Table 5 mulching has tripled soil organic matter and total nitrogen in soil over the past 15 years.

5.1.1 *Influence of mulching on soil moisture regime*

Different mulching methods were tested as water-saving technologies on moderately eroded brown soil terrains. Mulching methods used were: 10 cm thick hay, grape shoots and plastic films. Manure containing 9 kg of N, 3 kg of P, and 6 kg of K was incorporated into the top 10–15 cm of soil. As shown in Figure 3, the mulching increased soil moisture content.

The year 2000 was a year of pronounced drought. Measurements taken on August 8 of that year indicate that the soil moisture content was reduced by 54% using plowing, by 74% using green manure, by 87% using manure, by 67% using mulching with plastic films. The year 2002 was a year of substantial rain. Measurements taken on August 8 of that year indicate that soil moisture content was reduced by 46% using plowing, by 57% using green manure, by 49% using manure, by 63% using plastic films. Mulching of the soil surface resulted in an accumulation of soil moisture and increased the yield of grapevines. Using mulching, soil moisture content was two times greater in dry year 2000 and 8% greater in wet year 2002 as compared to plowing. Using green manure and manure, soil moisture content was 12–21% higher than if it had been plowed. Despite the dry weather in 2000 and 2001, the use of manure resulted in an increase of 2.09 t/ha in the yield of the grapevines. This compared to lower increases resulting from traditional practices such as the use of green manure (0.8–1.4 $t\,ha^{-1}$) and use of mulching (0.4–1.3 $t\,ha^{-1}$). The yield of the grapevines was lower in 2002, despite the abundant rains because of a heavy hailstrom in May of that year compared to the previous years, but the same tendency was observed. Compared to a control plot,

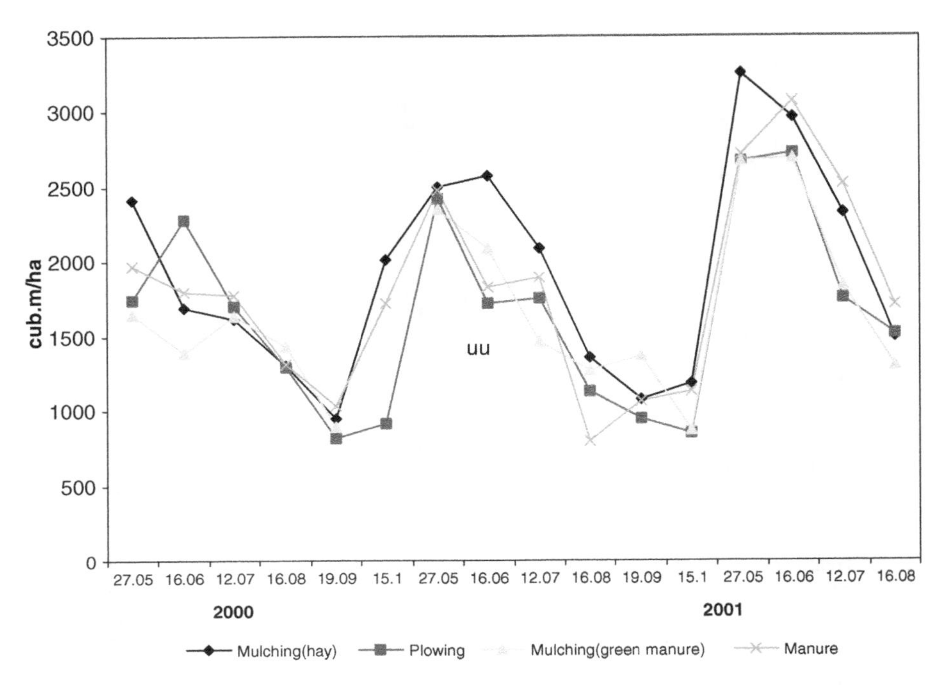

Figure 3. The influence of mulching on soil moisture content.

manure increased the yield of the grapevines by 2.4 t ha^{-1}; mulching with hay increased yields by 0.8 t ha^{-1}; and mulching using shoots of grapevines increased yields by 1.8 t ha^{-1}.

5.2 *Mulching and supplemental irrigation*

As shown in Figure 4, mulching with plastic films and plant residue increased soil moisture storage by 20–30% compared to fallow. Although soil moisture was higher under black films, plant residues were found to be economical compared to expensive plastic films.

Within the control option, soil moisture storages on 20 July of 2000 amounted to 50% of that recorded on 27 April of that year. Using black plastic films, soil moisture storages increased by 84%. They increased by 54% using white thick films; by 57% using white thin films; by 46% using white thin film and nitrogen, and by 59% using plant residues. On 7 July of that year soil moistures storages were 80% higher when using black film; 15% higher when using white thick films; 20% higher when using white thin films; and 26% higher when using plant residues. Within the control option, soil moisture storages on 20 July of 2000 amounted to 50% of that recorded on 27 April of that year. Using black plastic films, soil moisture storages increased by 84%. They increased by 54% using white thick films; by 57% using white thin films; by 46% using white thin film and nitrogen, and by 59% using plant residues. On 7 July of that year soil moistures storages were 80% higher when using black film; 15% higher when using white thick films; 20% higher when using white thin films; and 26% higher when using plant residues.

The period stretching from February to May 2001 was very dry, which led to supplemental irrigation (SI). Walnut and pine tree saplings were irrigated twice a month at a rate of 3l per sapling from July to October to prevent withering. SI in addition to mulching improved soil moisture. Soil moisture concentration associated with mulching using black plastic film was higher by 27–45% as compared to control plots. Survival rates varied from 60–70% as opposed to 32% for the control plot.

Lack of rain in July and August of 2001 sharply reduced soil moisture storage and survival rate of saplings. This led to a decision to increase SI to a rate of 3.0 liters per sapling three times a month

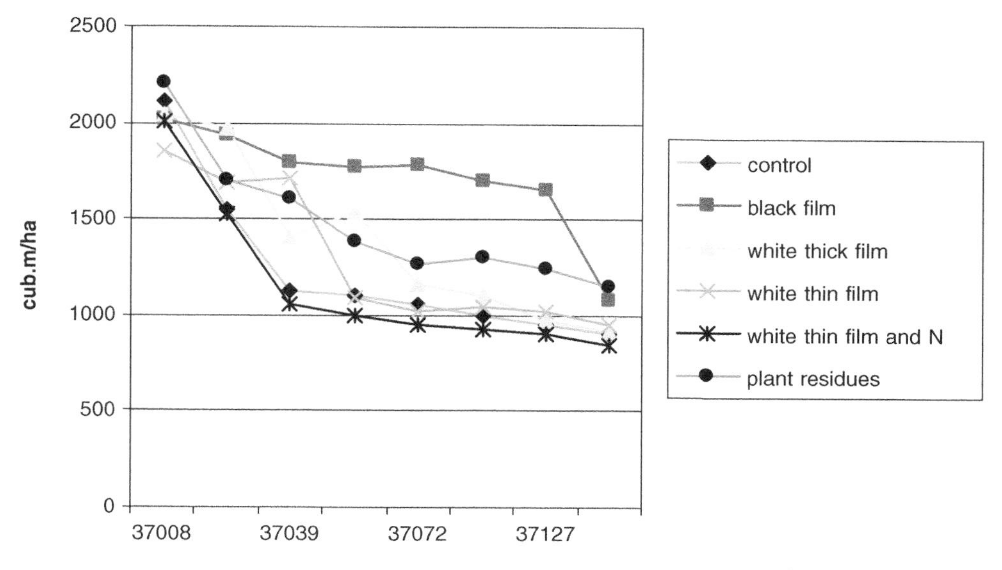

Figure 4. Influence of mulch materials on soil moisture content under grapevines (m^3 ha^{-1}).

Table 6. Dynamic of soil moisture and survival rate of walnut saplings using mulching and supplemental irrigation, 2001.

		July		October	
	Number saplings	Soil moisture (m^3 ha^{-1})	Survival rate (%)	Soil moisture (m^3 ha^{-1})	Survival rate (%)
Control	17	109	43	58	32
Black plastic film	30	139	86	84	70
White plastic film	30	126	71	88	65
Natural grasses	30	118	66	83	60

along with testing mulch materials and manure after June, 2002. Under this regime soil moisture concentration levels were 25% higher using black plastic films, 16% higher using white plastic films, and 8% higher using natural grasses. The soil moisture concentration using mulch and in for plots using the SI options was 118–139 m^3 ha^{-1} in July and 83–88 m^3 ha^{-1} in October as opposed to 109 and 58 m^3/ha respectively using the control option. The survival rate from 60–70% as opposed to only 32% for theat control option. In December 2001, fields were plowed at 30–35 cm depth at the experimental site; and 50 × 60 cm holes were dug for cherry saplings and grapevines and 60 × 80 cm holes for walnut and pine saplings. In February, snow was retained to save moisture at the holes. In the beginning of March, cherry saplings, grapevines and walnut were planted in the holes. Pine saplings were planted in April. Before sowing, 5.0 kg of manure was applied in each hole. Between the 25th of May and the 10th of June, land around the saplings was cultivated and then mulched with plant residues. In early July, the mulching material was renewed after further cultivation. Starting from July, additional irrigation of the saplings was done three times per month. Three liters of water were applied every time to each sapling. Irrigation was done after 16:00. As shown in Table 6, this irrigation increased the survival rate of the saplings.

In April 2002, an experiment was conducted to assess the influence of mulching and SI on soil moisture content for terraces. The experiment was laid out at a site with plum and walnut saplings.

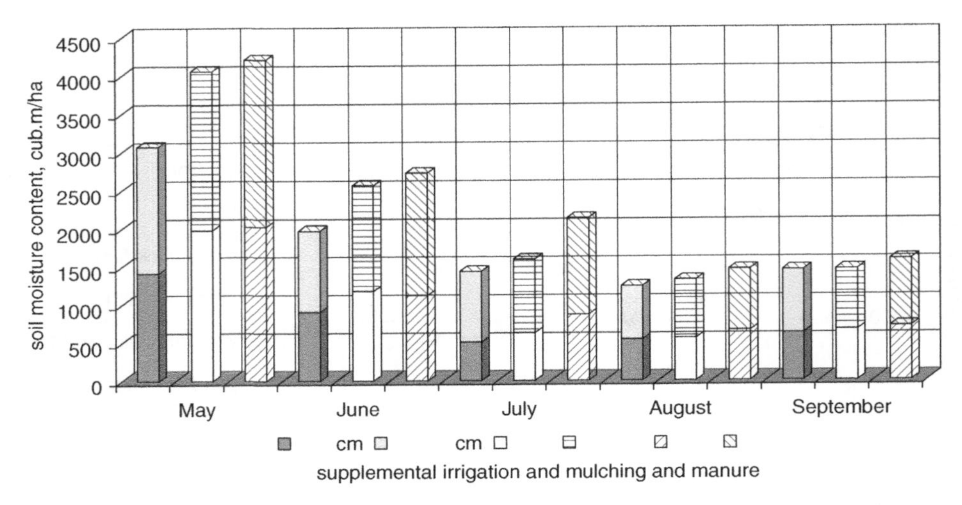

Figure 5. Soil moisture storages ($m^3\ ha^{-1}$) under plum trees under mulching and supplemental irrigation regimes.

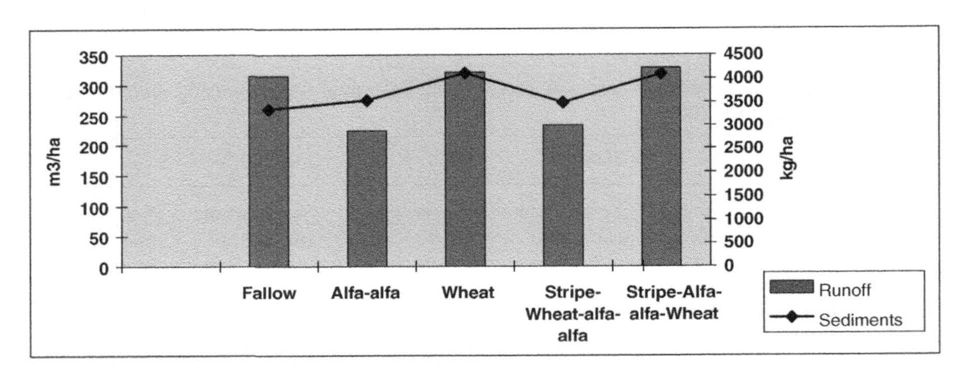

Figure 6. The relationship between runoff and sedimentation for different soil cover regimes.

Mulching was applied under trees for a 50 cm × 50 cm area. SI was done in July 2002. As indicated in Figure 5, the influence of the mulching and SI became apparent, especially in topsoil, beginning in May.

Difference in soil moisture storage at the 0–50 cm layer in treatments, for mulching and SI and for SI only, was considerable during May–July. Soil moisture concentrations were 569 and 993 $m^3\ ha^{-1}$ more with mulching and SI compared to SI treatment only at 50 cm and 1.0 m layers, respectively. These differences disappeared in August. Manure saved an additional 40–241 $m^3\ ha^{-1}$ of soil moisture in the 0–50 cm soil layer. Therefore SI in combination with mulching and manure provided adequate minimum soil moisture for growing tree saplings and increased their survival rate. However, water flows may cause considerable soil erosion. For farmers, this means losses of both production and loss of the farm soil. As shown in Figure 6, use of strip cropping technology can minimize the risks and losses of soil and water resources.

5.3 *Strip cropping*

Traditional wheat production on the high slopes results in greater soil and water loss compared strip cropping systems for wheat and grasses, or for grasses and wheat, or for legumes and grasses.

Table 7. Saline soils in Tajikistan (thousand ha).

Years	Total irrigated area	Lightly saline	Medium saline	Heavy saline	Total saline areas
1980	633.8	78.3	20.4	12.1	110.8
2000	703.0	80.0	28.0	8.2	116.2

Strip cropping on slopes increases the quantity of organic matter in the soil. According to our data from Obi-kiik site for an ICARDA project, the quantity of humus for strip cropping treatments increased from 1.34 to 1.85%.

As depicted in Figure 6, integrating strip cropping into land management processes can help control soil erosion, improve farm productivity and conserve natural resources.

6 RECLAMATION OF SALINE AND WATERLOGGED SOILS

Table 7 indicates that Tajikistan has more than 116,000 ha of saline soils and more than 30,000 ha of waterlogged areas. Irrigation and drainage practices in the higher zones of valleys have raised the groundwater level in the lower lying areas. These groundwater levels have risen to only 1–2 meters below the soil surface. The installation of drainage facilities can help prevent this water logging and salinization, especially for arid and semi-arid areas. Drainage, in combination with adequate irrigation scheduling, enables excess salts to be leached from the plant root zone. Information about salinization of soils resulting from irrigation was provided for only seven countries, including all five Central Asian countries. Saline areas varied from 5 to 50% of total irrigated area.

Development of an irrigation system in Tajikistan had been accompanied by an increase in water use. In the 1980s, water usage was 5.5 $km^3\ yr^{-1}$ and now it is 11.2 $km^3\ yr^{-1}$, with consumption exceeding 20,000 $m^3\ ha^{-1}\ yr^{-1}$. Great losses of fresh water have become a negative consequence of ineffective use of irrigations waters. As a result, 30,000 ha of irrigated land are being impacted by secondary salinization and water logging. Salinization and water logging are the main challenges to sustainable agriculture. Degradation processes and their various combinations cause different types of damage throughout Tajikistan. Only several attempts have been made to evaluate the influence of soil salinization on the total cotton production area. According to estimates, the country annually looses 70–100 thousand t of cotton, 100–150 thousand t of grain, 400 million m^3 of irrigation water, and 2.5 thousand t of nitrogen-potassium fertilizers in drain water. Irrigation of 100–110 thousand ha of upper slope lands resulted in a rise in ground water levels and greater salinization of soils areas closer to sea level.

7 CARBON STOCKS AND SOIL CONSERVATION TECHNOLOGIES

Data found in Table 8 suggests that implementation of soil conservation technologies will increase carbon sequestration and increase annual profits from farming by out $4,000,000.

8 CONCLUSIONS

1. Mulching, SI, strip cropping reclamation of waterlogged and saline soils, among other conservation practices, have an important impact on the flow of carbon into the soil and vegetation. With proper husbandry, these practices can dramatically improve soil quality, the amount of stored in the soil, and the productivity of the soil.

Table 8. Impact of implementation of different conservation technologies on soil carbon sequestration and farm income.

	Technology	Area, annual ha	Sequestration of C, $t\,ha^{-1}\,yr^{-1}$	Total C $t\,ha^{-1}\,yr^{-1}$	Annual gains US$*
1	Conservation tillage	50,000	0.15	7,500	112,500
2	Mulching with plastic	15,000	0.05	750	11,250
3	Mulching with crop residue	1,000	0.05	50	750
4	Application of organic manure and compost	150,000	0.20	30,000	450,000
5	Plant nutrient management	200,000	0.15	30,000	450,000
6	Rehabilitation of eroded soil	10,000	0.20	2,000	30,000
7	Reclamation of saline soils	5,000	0.1	500	7,500
8	Reclamation of waterlogged soils	2,000	0.1	200	3,000
9	Water conservation and irrigation	750,000	0.2	150,000	2,250,000
10	Afforestation	10,000	0.1	1,000	15,000
11	Pasture improvements	150,000	0.1	15,000	225,000
12	Fertilization and intensification	300,000	0.1	30,000	450,000
Total				267,000	4,050,000

*Based on C value of $\$15\,t^{-1}$.

2. Use of the FAO methodology to assess carbon stocks can be of great help in evaluating the soil degradation rates and the suitability of present production systems.
3. Enhancement of carbon sequestration in the mountainous dry area has direct environmental, economic, and social benefits. This is particularly the case if it is related to retaining biomass on the soil surface by practices such as strip cropping that can increase soil fertility and water retention. It can also increase productivity of farming practices and provide farmers and farm families with improved livelihoods.

REFERENCES

Agro Climatic Reference Book of Tajikistan. 1989. Dushanbe 179 p.

Batches, N.H. and W.G. Sombrock. 1977. Possibilities for carbon sequestration in tropical and subtropical soils. Glob. Change Biol. 3:161–173.

FAO. 2004. Carbon sequestration in dryland soils. Soil Recourses Report No.102. Rome.

Hernandez, R.P., P. Koohafkan, and J. Antoine. 2004. Assessing carbon stocks and modeling win-win scenarios of carbon sequestration through land-use changes. Food and Agriculture Organization, Rome.

IUSS. 2003. International Workshop on Practical Solutions for managing optimum C and N content in Agricultural soils. Prague, 120 p.

Lal, R. 2002. Carbon sequestration in dryland ecosystems of west Asia and North Africa. Land Deg. Dev. 13:45–59.

Tajik. Statistical Agency. 2005. Yearbook. Dushanbe, 255 p.

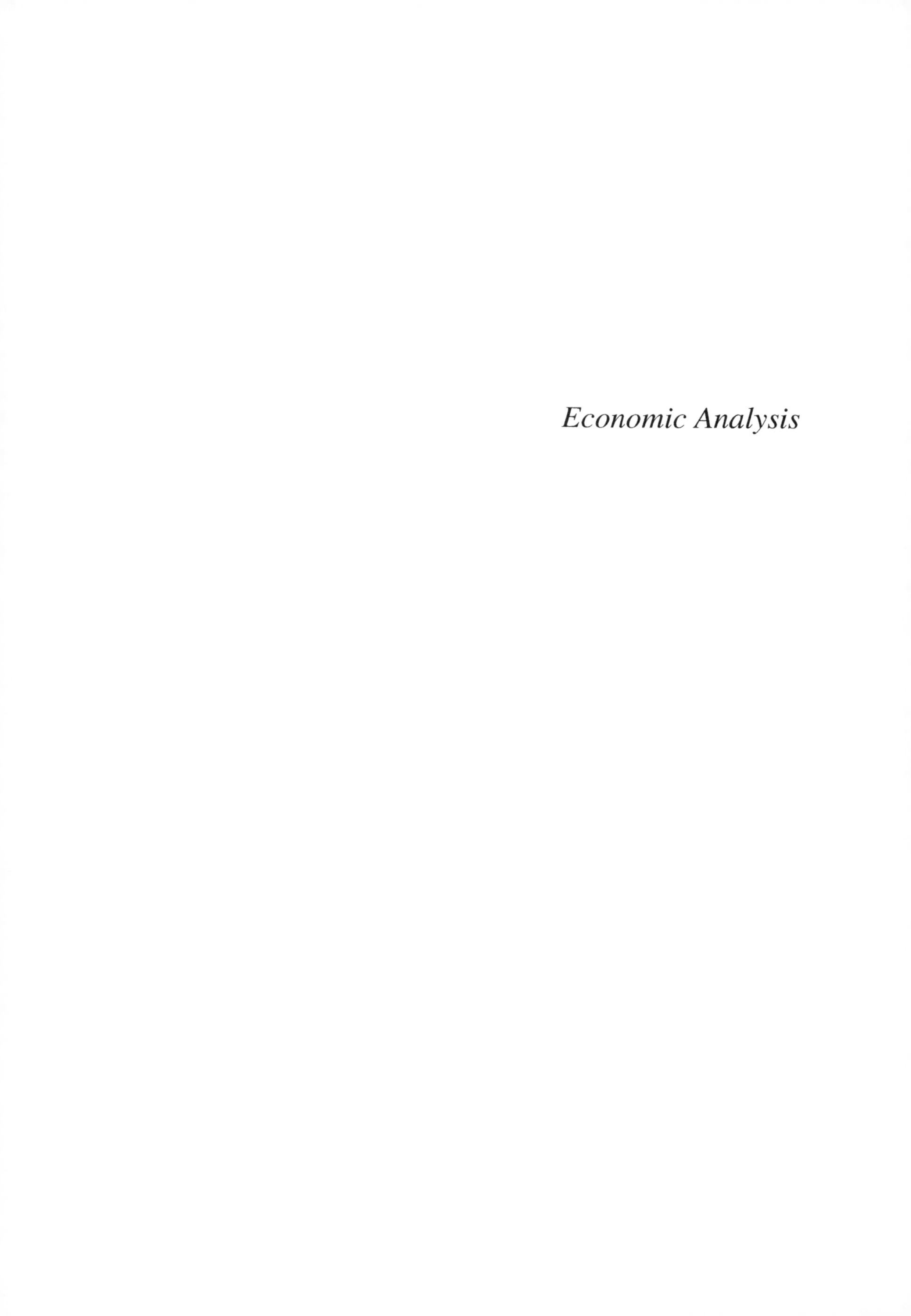

Economic Analysis

CHAPTER 27

An economic comparison of conventional tillage and conservation tillage for spring wheat production in Northern Kazakhstan

Paul E. Patterson & Larry D. Makus
Agricultural Economics, University of Idaho, USA

1 INTRODUCTION

Agriculture in Kazakhstan and other Central Asian Republics is undergoing a transformation as the former Soviet Republics evolve from centrally planned economies with state owned farms into market-based economies with privately owned farms. The rainfed grain production region of northern Kazakhstan exemplifies a region in transition. These recently privatized farms tend to be large, heavily mechanized, and in comparison to similar size grain farms in the United States and Canada, they employ a large number of workers. But labor in Kazakhstan is relatively inexpensive and currently it is readily available. The equipment and cropping systems used on these farms are similar, which facilitates the development of standardized costs and returns estimates that can be used to analyze the economics of alternative cropping systems. However, the equipment on these farms, much of it dating back to Soviet times, is obsolete and badly deteriorated.

A crucial question both for farmers in Kazakhstan and for the Kazak government is, what should Kazak agriculture look like in the future? This is a particularly important question because current production systems may not be economically viable or sustainable in the future. Kazak agriculture needs a large infusion of cash or credit to help farmers replace and modernize equipment. Should farmers simply purchase a machinery compliment needed to continue the conventional cropping system? Or, should farmers use this opportunity to adopt a different cropping system? And if they change, which system should they adopt? Are there economically viable alternatives to the conventional tillage system that can also provide environmental benefits?

Grain farmers in many parts of the world have switched or are switching from intensive tillage systems to conservation tillage (CT) systems. Conservation tillage can refer to a range of strategies and techniques used to establish a crop. The term is applied to systems that decrease the number or the intensity of tillage operations between crops. The term conservation tillage can also be applied to a no-till or direct seeding system where tillage is not used to establish a seedbed. The rationale for changing tillage systems can be economic, environmental, or both. A need to achieve government mandated environmental benefits, primarily reduced soil erosion and improved water conservation, drove the initial changes in the United States and Canada. More recently, farmers are changing because of economic benefits of CT. Economic benefits accrue to a farmer in three ways. First, machinery operating expenses (fuel, lubricants, repairs and labor) are lower when field operations are eliminated. Second, machinery ownership costs (depreciation, interest, taxes and insurance) can be lower with a CT system because CT generally requires less equipment when compared to CVT, or existing equipment is used less, thus extending its useful life and lowering the annual ownership cost. In either case, there are lower ownership costs per hectare. A third potential economic benefit results if farmers are paid to adopt or continue using environmentally beneficial production practices. While these "green payments" are not universally available, they are becoming more common and they represent an important new policy tool in many countries. Governments currently provide most of the funding. But the role of the private sector is growing as trading of carbon credits, for example, becomes more widely accepted. Paying farmers for their carbon © sequestration rights can provide important economic benefit to them while providing environmental benefits for society.

The purpose of this project was to develop detailed costs and returns estimates for the conventional tillage (CVT) system in the rainfed grain production region of northern Kazakhstan, as well as for a proposed conservation tillage (CT) system and to compare the economics of these systems. The crop rotation was two years of wheat following a fallow year for both systems.

2 DATA

Data for this analysis were collected in July 2004 from four farms cooperating in a three-year project co-sponsored by the Food and Agricultural Organization of the United Nations (FAO), the Ministry of Agriculture of the Republic of Kazakhstan (MoA), International Maize and Wheat Improvement Center (CIMMYT) and the Union of Farmers of the Republic of Kazakhstan (UFK). Data on all field operations were collected by the authors using an interpreter to conduct personal interviews and included the machinery implements and tractors used, fuel consumption for tractors and combines, labor requirements and inputs applied. Data collected from the cooperating farms along with data provided by Dr. Ivan A. Vasco from the Research and Production Center of Grain Farming in Shortandy, Kazakhstan were used to develop a model wheat farm for northern Kazakhstan. The model farm was used as a basis to compare productions costs of the two cropping systems. Dr. Muratbek Karabayev, CIMMYT-CAC office in Almaty and the project's director, provided two years of yield data on CVT and CT wheat production collected from these same cooperating farms. Costs and returns estimates (enterprise budgets) were developed for each crop year of the 3-year rotation for conventional tillage (CVT) and conservation tillage (CT) spring wheat systems based on this data.

Because Kazakhstan has not completed the transition to a market economy, market values were not available for all resources used on these farms. Existing equipment on the cooperating farms, although useable, is approaching both physical and technological obsolescence, and does not reflect an optimum equipment compliment for CVT or CT. Since most of the Soviet era equipment was acquired with the farm, it had no distinct and separate value. For these reasons, ownership costs for most existing equipment (primarily depreciation and interest costs) are difficult to accurately assess. Ownership costs based on existing equipment compliments are quite low. In contrast, the operating cost for equipment is high given the advanced age and general poor condition of most equipment currently in use. Breakdowns are frequent and repair costs are high.

Conventional farming practices on large grain farms in northern Kazakhstan are similar in terms of field operations and equipment used. This uniformity is likely a function of the Soviet centrally planned collective farm system that preceded the current situation. This high degree of uniformity made it relatively easy to develop a representative sequence of field operations for both CVT and CT.

3 PROCEDURES

The first objective of the project was to develop a model farm based data collected from farmers and researchers. The second objective was to construct costs and returns estimates for each grain crop year and fallow year. The third objective was to make an economic comparison between the CVT and CT systems.

3.1 *Model farm*

The model farm is 1,000 hectares (ha) with 990 tillable ha divided into three fields of equal size. The CVT system follows a traditional 3-year rotation: mechanical summer fallow – spring wheat – spring wheat. Conventional summer fallow uses four mechanical tillage operations to control weeds and preserve moisture (as noted later, one tillage operation overlaps with conventional post-harvest fall tillage operations). The annual cost of land used in the analysis is based on a rental rate of $8.50 per ha charged to all 1,000 has for a total farm cost of $8,500. Land rent is only assigned to the 660 cropped has, not to summer fallow. This results in a land rent cost of $12.88 per cropped ha shown in the individual costs and returns estimates found in the appendix. The model farm has seven permanent salaried employees and an owner/manager who does not receive a salary. One full time laborer is eliminated in the CT budget due to lower labor requirements from fewer tillage operations. Total permanent worker wages amount to $11,400 annually under the CVT system and $9,000 under the CT system. In addition, two seasonal workers are hired for 14 days in the spring at $22 per day to help with planting and three seasonal workers are hired for 20 days during the fall to help with harvest, also at $22 per day. Total seasonal worker wages amount to $1,936 annually for both conventional and CT. Food and transportation costs during planting and harvest for seasonal and permanent workers add an additional $642 to the CVT labor costs and $568 to the CT labor costs. This is computed at $2.20 per day while the seasonal workers are employed.

The machinery compliment is composed of both new and used equipment. Both CVT and CT model farms have two (one new and one used) large 4-wheel drive tractors (K700), two (one new and one used) crawler tractors (DT75), and three (one new and two used) smaller wheeled tractors (MTZ80). One small tractor is eliminated from the CT model farm. There are three Russian built combines, one new and two used. There are ten SZS (2.1 meters) grain seeders, and five seeders are pulled as an operational unit. The cost of each CVT seeder is $2,800, while the cost of the CT seeders is $2,440. Each model farm also has an OPU 2000 sprayer, six wagons for hauling grain, and a variety of tillage equipment. Machinery shed and grain storage buildings are the only buildings assigned a specific value. Equipment for grain cleaning (although commonly used) is not included due to the lack of complete information on investment and operational costs. None of the ownership costs from equipment and buildings are assigned to the fallow budgets.

3.2 *Conventional tillage field operations*

Eight field operations are used in the CVT system. Each operation is described below, indicating the month it takes place, the equipment used, the speed, fuel consumption, as well as any inputs applied and the cost of the inputs. Fuel is valued at the government-subsidized rate of $0.20 l^{-1}$.

Field Operations for Wheat Following Wheat or Wheat Following Mechanical Fallow

1. Sept–Oct Plowing
   KPG 3.5 — Speed: 8 km hr^{-1}.
   K700 Tractor — Fuel: 12 l ha^{-1}.

2. Dec–Feb Snow Retention (expected is 1–3 times, though only 1 operation is budgeted)
   SVU 2.6 — Speed: 5 km hr^{-1}.
   DT75 Crawler Tractor — Fuel: 5 l ha^{-1}.

   SVU 2.6 — Speed: 5 km hr^{-1}.
   DT75 Crawler Tractor — Fuel: 5 l ha^{-1}.

3. Apr–May Harrow for Water Retention and Weed Control
   10 BEG 3.0 pulled as one unit — Speed: 5 km hr^{-1}.
   DT75 Crawler Tractor — Fuel: 6 l ha^{-1}.

   10 BEG 3.0 pulled as one unit — Speed: 5 km hr^{-1}.
   DT75 Crawler Tractor — Fuel: 6 l ha^{-1}.

4. May Heavy Disk for Planting Preparation
   LDG 15 Speed: 8 km hr^{-1}
   K700 Tractor Fuel: 8 l ha^{-1}.

5. May Plant and Fertilize
   5 SZS 2.1 pulled as one unit Speed: 8 km hr^{-1}.
   K700 Tractor Fuel: 9 l ha^{-1}.

   5 SZS 2.1 pulled as one unit Speed: 8 km hr^{-1}.
   K700 Tractor Fuel: 9 l ha^{-1}.

Materials:

Seed: 120 kilograms/ha. @ \$175 t^{-1} (\$0.175/kilogram)
Fertilizer: 100 kg ha^{-1} of AmmoPhos @ \$100 t^{-1} (\$0.10 km^{-1})

6. June Herbicide Application
   OPU 2000 Sprayer Speed: 10 km hr^{-1}.
   MTZ80 Tractor Fuel: 2.0 l ha^{-1}.

Materials:

Herbicide: 1 l broadleaf herbicide @ \$5.50 ha^{-1}.

7. Aug–Sep Harvest
   1 New Yenesi Combine Fuel: 6.5 l ha^{-1}. Speed: 6 km hr^{-1}.
   2 Used Yenesi Combines Fuel: 9.5 l ha^{-1}. Speed: 5 km hr^{-1}.

8. Aug–Sep Hauling: 20 km per roundtrip 41 trips
   2 Kazak Wagons (10 t capacity per wagon)
   K700 Tractor Fuel: 32 l per trip

Conventional Mechanical Fallow Field Operations

Note: The last summer fallow operation (fall plowing) is the same as the first field operation after wheat harvest. Therefore, only three field operations are assigned to summer fallow for budgeting purposes.

1. June Cultivation for Weed Control (at 8 cm depth)
   KPS 9 Field Cultivator Speed: 10 km hr^{-1}.
   K700 Tractor Fuel: 8 liters/ha.

2. July Cultivation for Weed Control (at 10 cm depth)
   KPS 9 Field Cultivator Speed: 10 km hr^{-1}.
   K700 Tractor Fuel: 9 l ha^{-1}.

3. August Cultivation for Weed Control (at 10 cm depth)
   KPS 9 Field Cultivator Speed: 10 km hr^{-1}.
   K700 Tractor Fuel: 6 l ha^{-1}.

4. Sept–Oct Plowing 25 cm
   KPG 3.5 Speed: 8 km hr^{-1}.
   K700 Tractor Fuel: 12 l ha^{-1}.

Wheat yields used in the budgets represent the average for the two years of the project on all four farms. Wheat following mechanical fallow has a yield of 1.35 t, while wheat following wheat using CVT has a yield of 1.12 t ha^{-1}.

3.3 *Conservation tillage field operations*

Six separate operations are used in the conservation tillage system. These are described below, indicating the month, the equipment used, the speed and fuel consumption, as well as any inputs applied and the cost of the inputs. Fuel is valued at the government-subsidized rate of $\$0.20\,l^{-1}$. The chemical fallow operation is also described.

Field Operations for Wheat Following Wheat or Wheat Following Chemical Fallow

1. Dec–Feb Snow Retention (expected is 1–3 times, though only 1 operation is budgeted)
   SVU 2.6 Speed: $5\,km\,hr^{-1}$.
   DT75 Crawler Tractor Fuel: $5\,l\,ha^{-1}$.

   SVU 2.6 Speed: $5\,km\,hr^{-1}$.
   DT75 Crawler Tractor Fuel: $5\,l\,ha^{-1}$.

2. May Glyphosate Application
   OPU 2000 Sprayer Speed: $10\,km\,hr^{-1}$.
   MTZ80 Tractor Fuel: $2.0\,l\,ha^{-1}$.

Materials:

Herbicide: $2.5\,l\,ha^{-1}$ glyphosate @ $\$6.50\,l^{-1}$, or $\$16.25\,ha^{-1}$.

3. May Plant and Fertilize
   5 SZS 2.1 pulled as one unit Speed: $8\,km\,hr^{-1}$.
   K700 Tractor Fuel: $9\,l\,ha^{-1}$.

   5 SZS 2.1 pulled as one unit Speed: $8\,km\,hr^{-1}$.
   K700 Tractor Fuel: $9\,l\,ha^{-1}$.

Materials:

Seed: 120 kilograms/ha. @ $\$175\,t^{-1}$ ($\$0.175\,kg^{-1}$)
Fertilizer: $100\,kg\,ha^{-1}$ of AmmoPhos @ $\$100\,t^{-1}$ ($\$0.10\,kg^{-1}$)

4. June Herbicide Application
   OPU 2000 Sprayer Speed: $10\,km\,hr^{-1}$.
   MTZ80 Tractor Fuel: $2.0\,l\,ha^{-1}$.

Materials:

Herbicide: $1\,l\,ha^{-1}$ broadleaf herbicide @ $\$5.50\,ha^{-1}$.

5. Aug-Sep Harvest
   1 New Yenesi Combine Fuel: $6.5\,l\,ha^{-1}$. Speed: $6\,km\,hr^{-1}$.
   2 Used Yenesi Combines Fuel: $9.5\,l\,ha^{-1}$. Speed: $5\,km\,hr^{-1}$.

6. Aug–Sep Hauling: 20-kilometer roundtrip 45 trips
   2 Kazak Wagons (10 t capacity per wagon)
   K700 Tractor Fuel: 32 l per trip

The yield for wheat following chemical fallow seeded with a modified grain seeder is $1.49\,t\,ha^{-1}$. Yield for the second year conservation-tilled wheat (wheat following wheat) is $1.24\,t\,ha^{-1}$.

Wheat yields used in the budgets represent the average for two years of the project on all four farms.

Chemical Fallow Operations

1. June Herbicide Application
   OPU 2000 Sprayer Speed: 10 km hr^{-1}.
   MTZ80 Tractor Fuel: 2.0 l ha^{-1}.

Materials:
Herbicide: 3 l ha^{-1} glyphosate @ \$6.50/l or \$19.50 ha^{-1}.

4 ANALYSIS AND RESULTS

Costs and returns estimates (or enterprise budgets) represent a systematic method for calculating, organizing and reporting revenues and costs for a production enterprise and were used as the basis of this study. Crop production enterprises are typically defined by commodity, the field production practices, or a combination of the two. There are three enterprise budgets for each of the two tillage systems: 1) fallow, 2) wheat following fallow and 3) wheat following wheat.

Three separate enterprise budgets are presented for each tillage system for a total of six budgets (Appendices I through VI): CVT wheat following mechanical fallow (Appendix I); CVT wheat following wheat (Appendix II); mechanical fallow for CVT wheat (Appendix III); CT wheat following chemical fallow (Appendix IV); CT wheat following wheat (Appendix V); and chemical fallow for CT wheat (Appendix VI). Only the direct operating costs for the fallow operations are assigned to the two fallow budgets. The cropped acres are assigned all ownership costs (machinery, land, and buildings).

A comparison of revenue, costs and net returns for the two tillage systems on a whole-farm basis is shown in Table 1. Conservation tillage shows some clear economic advantages compared to the traditional tillage system based on this analysis. Some of these are most apparent only on a whole farm basis. For the whole farm, CT generated added revenue of \$11,583 (\$121,622 versus \$110,039) because of higher yields. The CT system also had \$5,094 in higher costs (\$105,726 versus \$100,632), mostly due to glyphosate, an expensive non-selective herbicide used in both years of wheat production and in the fallow year. Net returns for CT were \$6,489 higher than for CVT on a whole-farm basis (\$15,896 versus \$9,407) as higher revenues more than offset higher costs.

The analysis also compared the costs per ha and per ton of wheat produced under the two tillage systems. Since there is no revenue generated in the fallow year, costs for fallow year must be covered by has in production in order for the cost per ton of wheat to reflect all costs. Wheat-1 refers to the wheat crop following fallow and wheat-2 refers to the wheat crop following wheat. Since the wheat crop following fallow gets the higher yield benefit from additional moisture, this enterprise is also assigned the fallow costs. Rather than developing a wheat-fallow costs and returns estimate by including the fallow operations in the wheat-1 budgets, separate fallow budgets were prepared for both conventional and CT. This provides more flexibility in analyzing costs and allowed for a direct comparison of fallow costs between the two tillage systems. Table 2 shows the cost per ha for the

Table 1. Whole farm revenue, total costs, and net returns comparison by tillage system.

Tillage system	Revenue	Total cost	Net returns
Conventional tillage	\$110,039	\$100,632	\$9,407
Conservation tillage	\$121,622	\$105,726	\$15,896
Change	+\$11,583	+\$5,094	+\$6,489

fallow year, the first year of wheat following fallow (wheat-1) and the combined wheat-1 + fallow both per ha and per t. Table 3 shows the cost per ha and per t for the second year of wheat (wheat-2).

The cost per fallowed ha for the CT (Table 2) was $12.33 higher than with CVT ($23.76 versus $11.43). While the CT fallow used less fuel and labor than the CVT fallow, the cost of glyphosate used in the CT fallow year more than offset these savings. Conventional tillage fallow used 27 l of fuel and $5.40 of labor ha^{-1}, while CT fallow used only 2 l and $2.73 of labor per ha. The cost per ha for wheat following fallow was $3.19 higher under CT ($149.99 versus $146.80). For wheat + fallow, the cost ha^{-1} for CT was $15.52 higher than for CVT ($173.75 versus $158.23). But when these costs are converted to a per ton basis, CT and CVT are nearly the same, with a negligible $0.60 t^{-1} lower cost for CT.

Tables 4 and 5 compare revenues, costs, and net returns for wheat-1 + fallow and wheat-2, respectively. In Table 4, CT shows higher revenue of $18.90 per ha compared to CVT ($201.15 versus $182.25), higher costs of $15.52 ha^{-1} ($173.75 versus $158.23) and higher net returns of $3.38 ha^{-1} ($27.40 versus $24.02). In Table 5, CT shows higher revenue of $16.20 ha^{-1} ($167.40 versus $151.20), essentially the same costs ha^{-1} ($146.65 versus $146.72) and a higher net return of $16.27 ($20.75 versus $4.48).

The reduction in field operations under conservation tillage saves on labor, fuel and repairs as shown in Table 6. Fertilizer and seed costs remain unchanged. Lower machinery ownership costs result from a reduction in equipment usage and help cover part of the increase in operating costs

Table 2. Cost per hectare for fallow, wheat-1 and wheat-1 + fallow, and cost per ton for wheat-1 + fallow under conventional and conservation tillage.

Tillage system	Fallow (Cost ha^{-1})	Wheat-1 (Cost ha^{-1})	Wheat + fallow (Cost ha^{-1})	Wheat + fallow (Cost t^{-1})
Conventional tillage	$11.43	$146.80	$158.23	$117.21
Conservation tillage	$23.76	$149.99	$173.75	$116.61
Change	+$12.33	+$3.19	+$15.52	−$0.60

Table 3. Cost per hectare and per ton for wheat-2 under conventional and conservation tillage.

Tillage system	Wheat 2 (Cost ha^{-1})	Per Unit (Cost t^{-1})
Conventional tillage	$146.72	$131.00
Conservation tillage	$146.65	$118.27
Change	−$0.07	−$12.73

Table 4. Wheat 1 + fallow comparison of revenue, costs, and net return per hectare.

Tillage system	Revenue	Total cost	Net returns ha^{-1}
Conventional Tillage	$182.25	$158.23	$24.02
Conservation Tillage	$201.15	$173.75	$27.40
Change	+$18.90	+$15.52	+$3.38

Table 5. Wheat 2 comparison of revenue, costs and net returns per hectare.

Tillage system	Revenue	Total cost	Net returns ha^{-1}
Conventional Tillage	$151.20	$146.72	$4.48
Conservation Tillage	$167.40	$146.65	$20.75
Change	+$16.20	−$0.07	+$16.27

Table 6. Whole farm operating costs for conventional and conservation tillage.

Operating cost	Conventional	Conservation	Difference	% Change
Seed	$13,860	$13,860	$0	
Fertilizer	$6,600	$6,600	$0	
Fuel	$10,331	$4,821	−$5,510	−53.3%
Labor	$13,978	$11,504	−$2,474	−17.7%
Pesticides	$3,630	$21,863	+$18,233	+502.3%
Repairs	$7,176	$6,100	−$1,076	−15.0%
All	$55,575	$64,748	+$9,173	+16.5%

Table 7. Machinery ownership costs for cropped hectares.

	Per ha	Total farm
Conventional tillage	$52.95	$34,947
Conservation tillage	$46.77	$30,868
Difference	−$6.18	−$4,079

(see Table 7). Some equipment is no longer needed and other equipment will last longer. Overall, the machinery ownership cost savings are small, in part because of the older equipment values used in the assumed machinery compliment. Ultimately the additional revenue associated with the higher yields experienced for CT more than compensates for the higher production costs.

5 DISCUSSION AND CONCLUSION

An increase in fuel price would increase the cost savings for CT relative to CVT practices. The CVT system uses more fuel, but since subsidized fuel costs ($\$0.20\,l^{-1}$) are used in the budgets the advantage of reducing fuel consumption is not fully captured. Even with the subsidized fuel cost, switching to CT saves the farm $5,510 in annual fuel costs (Table 6).

The capital investment needed to fully apply the CT system should be less than what will eventually be required to maintain the traditional cropping system. At some point, additional capital investment in equipment will be required. Lower potential investment costs for the CT system is a significant potential savings in a region where access to capital is a significant challenge. The budgets presented here assume limited investment in new equipment, so ownership costs associated with such equipment are relatively small. Two new tractors, one new grain combine and 10 new seeders are assumed in the equipment compliment. All other equipment is used. New seeders are assumed primarily because limited information was available to assess the value of existing or modified seeders. Overall, one of the more significant weaknesses of the analysis has to do with the relatively old, and therefore inexpensive, tillage equipment. Little is saved in terms of ownership costs when this equipment is eliminated. Additionally, total equipment repair cost savings shown in the analysis are only about $1,076 for CT (Table 6). This amount likely under quantifies potential repair cost savings, but limited information was available to adequately address repair costs on tillage equipment. Once equipment issues regarding the type and size are resolved, the analysis should be repeated using new equipment values for optimum machinery compliments.

The CT system also presents an issue related to increased operating costs associated with the high cost of glyphosate (one annual application is assumed in the CT budgets, including fallow). Less expensive alternatives may be available for a non-selective herbicide, but such alternatives are not explicitly evaluated. Additionally, a single application of broad leaf herbicide is assumed for both conservation and CVT. There was some debate during the interviews about the need for any additional herbicide beyond glyphosate in the CT practices, and the need for an herbicide to control wild oats under both conventional and CT. Reductions in the use of a broadleaf herbicide for CT, or the additional use of a wild oat herbicide exclusively for CVT will tend to place additional cost advantages on CT practices.

The assumptions used in setting up the model farm may not show all other potential cost savings. With a high proportion of the labor costs defined as permanent labor, only $2,474 was saved in labor costs for CT. Again, labor is a relatively inexpensive resource, and permanent labor reductions may not always occur when switching to CT practices.

Additional gains associated with potential long-term soil productivity enhancements from CT cannot be adequately evaluated given the short time period associated with this study. Studies in other regions have suggested such enhancements are possible due to increases in organic matter, better soil profiles, and reductions in soil erosion.

ACKNOWLEDGMENTS

The authors wish to thank Dr. Muratbek Karabayev and Dr. Alexi Morgounov from the CIMMYT-CAC office in Almaty, Dr. Ivan A. Vasco from the Research and Production Center of Grain Farming in Shortandy, Kazakhstan, and Bakhyt Mussiraliev who served as our interpreter. We would also like to acknowledge Winrock International for their assistance in providing travel, per diem and lodging funds under the Farmer to Farmer Program.

REFERENCES

Commodity Costs and Returns Estimation Handbook. 1998. A Report of the AAEA Task Force on Commodity Costs and Returns. Ames, Iowa.

De Vuyst, E.A., and A.D. Halvorson. 2004. Economics of annual cropping versus crop-fallow in the Northern Great Plains as Influenced by tillage and nitrogen. Agron. J. 96:148–153.

Halvorson, A.D., G.A. Peterson, and C.A. Reule. 2002. Tillage system and crop rotation effects on dryland crop yields and soil carbon in the Central Great Plains. Agron. J. 94:1429–1436.

Janosky, J.S., D.L. Young, and W.F. Schillinger. 2002. Economics of conservation tillage in a wheat-fallow rotation. Agron. J. 94:527–531.

Katsvairo, T.W., and W.J. Cox. 2000. Economics of cropping systems featuring different rotations, tillage and management. Agron. J. 92:485–493.

Ribera, L.A., F.M. Hons, and J. Richardson. 2004. An economic comparison between conventional and no-tillage farming systems in Burleson County, Texas. Agron. J. 96:415–424.

Appendix I. Conventional tillage budget for wheat* following fallow in north Kazakhstan.

Item	Quantity per ha	Unit of measurement	Price/cost per unit	Value per hectare	Total value
VALUE OF PRODUCTION					
Wheat	1.35	ton	$135.00	$182.25	$60142.50
Gross returns				$182.25	$60142.50
OPERATING COSTS					
Seed:				$21.00	$6930.00
Wheat seed	120.00	kilo	$0.17	$21.00	$6930.00
Fertilizer:				$10.00	$3300.00
Ammonium phosphate	100.00	kilo	$0.10	$10.00	$3300.00
Pesticides:				$5.50	$1815.00
Herbicide: broadleaf	1.00	ha	$5.50	$5.50	$1815.00
Fuel:				$12.59	$4154.20
Labor:				$19.45	$6419.00
Repairs:				$9.99	$3295.80
Total operating costs				$78.53	$25914.00
Operating cost per ton			$58.17		
Net returns above operating costs				$103.72	$34228.50
OWNERSHIP COSTS					
Land				$12.88	$4250.00
Machinery depreciation				$37.05	$12225.00
Machinery interest				$15.90	$5248.50
Building depreciation				$1.39	$460.00
Building Interest				$1.05	$345.00
Total ownership costs				$68.27	$22528.50
Ownership cost per ton			$50.57		
Total costs				$146.80	$48442.50
Total cost per ton			$108.74		
NET RETURNS				$35.45	$11700.00

*3-year rotation: conventional wheat following mechanical fallow for 1,000 ha farm.
Number of ha: 330.
Labor costs include spring and fall food and transportation costs.
Fuel costs include lubricants.

Appendix II. Conventional tillage budget for wheat* following wheat in north Kazakhstan.

Item	Quantity per ha	Unit of measurement	Price/cost per unit	Value per ha	Total value
VALUE OF PRODUCTION					
Wheat	1.12	ton	$135.00	$151.20	$49896.00
Gross returns				$151.20	$49896.00
OPERATING COSTS					
Seed:				$21.00	$6930.00
Wheat seed	120.00	kilo	$0.17	$21.00	$6930.00
Fertilizer:				$10.00	$3300.00
Ammonium phosphate	100.00	kilo	$0.10	$10.00	$3300.00
Pesticides:				$5.50	$1815.00
Herbicide: broadleaf	1.00	ha	$5.50	$5.50	$1815.00
Fuel:				$12.51	$4127.80
Labor:				$19.45	$6419.00
Repairs:				$9.99	$3295.80
Total operating costs				$78.45	$25887.60
Operating cost per ton			$70.04		
Net returns above operating costs				$72.75	$24008.40
OWNERSHIP COSTS					
Land				$12.88	$4250.00
Machinery depreciation				$37.05	$12225.00
Machinery interest				$15.90	$5248.50
Building depreciation				$1.39	$460.00
Building Interest				$1.05	$345.00
Total ownership costs				$68.27	$22528.50
Ownership cost per ton			$60.96		
Total costs				$146.72	$48416.10
Total cost per ton			$131.00		
NET RETURNS				$4.48	$1479.90

*3-year rotation: conventional wheat following wheat for 1,000 ha farm.
Number of Hectares: 330.
Labor costs include spring and fall food and transportation costs.
Fuel costs include lubricants.

Appendix III. Conventional tillage budget for mechanical fallow in north Kazakhstan.

Item	Quantity per ha	Unit of measurement	Price/cost per unit	Value per ha	Total value
VALUE OF PRODUCTION					
Gross Returns				$0.00	$0.00
OPERATING COSTS					
Fuel				$6.21	$2049.00
Labor				$3.45	$1140.00
Repairs				$1.77	$584.40
Total operating costs				$11.43	$3773.40
Operating cost per unit			NA		
Net Returns above operating costs				$−11.43	$−3773.40
OWNERSHIP COSTS					
Machinery depreciation				$0.00	$0.00
Machinery interest				$0.00	$0.00
Total ownership costs				$0.00	$0.00
Ownership cost per unit			NA		
Total costs				$11.43	$3773.40
Total cost per unit			NA		
NET RETURNS				$−11.43	$−3773.40

*3-year rotation: mechanical fallow following second year of wheat.
Number of Hectares: 330.
No machinery ownership costs were assigned to the fallow budget.
Fuel costs include lubricants.

Appendix IV. Conservation tillage budget for wheat* following fallow in north Kazakhstan.

Item	Quantity per ha	Unit of measurement	Price/cost per unit	Value per ha	Total value
VALUE OF PRODUCTION					
Wheat	1.49	ton	$135.00	$201.15	$66379.50
Gross returns				$201.15	$66379.50
OPERATING COSTS					
Seed:				$21.00	$6930.00
Wheat seed	120.00	kilo	$0.17	$21.00	$6930.00
Fertilizer:				$10.00	$3300.00
Ammonium phosphate	100.00	kilo	$0.10	$10.00	$3300.00
Pesticides:				$25.00	$8250.00
Herbicide: broadleaf	1.00	ha	$5.50	$5.50	$1815.00
Herbicide: glyphosate	3.00	liters	$6.50	$19.50	$6435.00
Fuel:				$7.12	$2348.90
Labor:				$16.07	$5302.00
Repairs:				$8.71	$2874.00
Total operating costs				$87.90	$29004.90
Operating cost per ton			$58.99		
Net Returns above operating costs				$113.25	$37374.60
OWNERSHIP COSTS					
Land				$12.88	$4250.00
Machinery depreciation				$31.90	$10527.17
Machinery interest				$14.87	$4907.10
Building depreciation				$1.39	$460.00
Building interest				$1.05	$345.00
Total ownership costs				$62.09	$20489.27
Ownershipcost per ton			$41.67		
Total costs				$149.99	$49494.17
Total cost per ton			$100.66		
NET RETURNS				$51.16	$16885.33

*3-year rotation: conservation tillage wheat following chemical fallow.
Number of ha: 330.
Labor costs include spring and fall food and transportation costs.
Fuel costs include lubricants.

Appendix V. Conservation tillage budget for wheat* following wheat in north Kazakhstan.

Item	Quantity per ha	Unit of measurement	Price/cost per unit	Value per ha	Total value
VALUE OF PRODUCTION					
Wheat	1.24	ton	$135.00	$167.40	$55242.00
Gross Returns				$167.40	$55242.00
OPERATING COSTS					
Seed:				$21.00	$6930.00
Wheat seed	120.00	kilo	$0.17	$21.00	$6930.00
Fertilizer:				$10.00	$3300.00
Ammonium phosphate	100.00	kilo	$0.10	$10.00	$3300.00
Pesticides:				$21.75	$7177.50
Herbicide: broadleaf	1.00	ha	$5.50	$5.50	$1815.00
Herbicide: glyphosate	2.50	liters	$6.50	$16.25	$5362.50
Fuel:				$7.03	$2320.10
Labor:				$16.07	$5302.00
Repairs:				$8.71	$2874.00
Total operating costs				$84.56	$27903.60
Operating cost per ton			$68.19		
Net Returns above operating costs				$82.84	$27338.40
OWNERSHIP COSTS					
Land				$12.88	$4250.00
Machinery depreciation				$31.90	$10527.17
Machinery interest				$14.87	$4907.10
Building depreciation				$1.39	$460.00
Building Interest				$1.05	$345.00
Total ownership costs				$62.09	$20489.27
Ownership Cost per ton			$50.07		
Total costs				$146.65	$48392.87
Total cost per ton			$118.27		
NET RETURNS				$20.75	$6849.13

*3-year rotation: conservation tillage wheat following chemical fallow
Number of Hectares: 330
Labor costs include spring and fall food and transportation costs.
Fuel costs include lubricants.

Appendix VI. Conservation tillage budget for chemical fallow in north Kazakhstan.

Item	Quantity per ha	Unit of measurement	Price/cost per unit	Value per ha	Total value
VALUE OF PRODUCTION					
Gross returns				$0.00	$0.00
OPERATING COSTS					
Pesticides:				$19.50	$6435.00
Herbicide: glyphosate	3.00	liter	$6.50	$19.50	$6435.00
Labor				$2.73	$900.00
Repairs				$1.07	$352.00
Fuel				$0.46	$152.00
Total operating costs				$23.76	$7839.00
Operating cost per unit			NA		
Net Returns above operating costs				$−23.76	$−7839.00
OWNERSHIP COSTS					
Machinery depreciation				$0.00	$0.00
Machinery interest				$0.00	$0.00
Building depreciation				$0.00	$0.00
Building interest				$0.00	$0.00
Total ownership costs				$0.00	$0.00
Ownership cost per unit			NA		
Total costs				$23.76	$7839.00
Total cost per unit			NA		
NET RETURNS				$−23.76	$−7,839.00

*3-year conservation tillage: Chemical fallow with glyphosate following second year of wheat.
Number of ha: 330.
No machinery ownership costs were assigned to the fallow budget.
Fuel costs include lubricants.

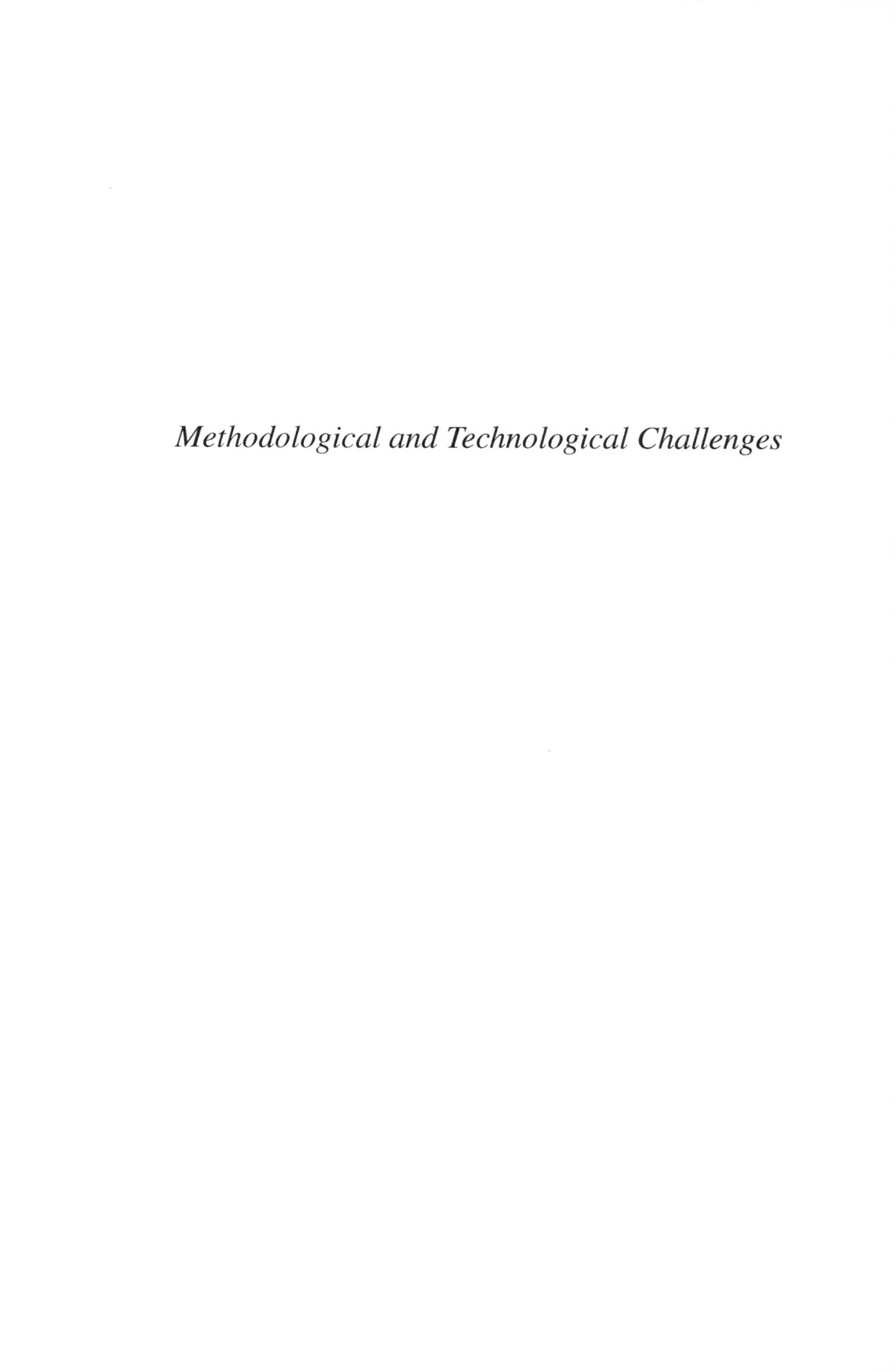

Methodological and Technological Challenges

CHAPTER 28

An assessment of the potential use of SRTM DEMs in terrain analysis for the efficient mapping of soils in the drylands region of Kazakhstan

Erik R. Venteris
Ohio Division of Geological Survey, Columbus, OH OH, USA

Konstantine M. Pachikin
Department of Soils, Institute of Kazakhstan Ministry of Education and Science, Almaty, Kazakhstan

Greg W. McCarty & Paul C. Doraiswamy
USDA-ARS Hydrology & Remote Sensing Laboratory, Beltsville, MD, USA

1 INTRODUCTION

The association between soil type and terrain is a fundamental concept in pedology. A soil's position relative to landform has been a recognized factor in soil formation from the very earliest scientific work on the distribution of soils (e.g., see Dokuchaev's works dating back to 1879, Jenny, 1941). The shape of the land affects how water and sediment are transported, deposited and stored as well as the aspect of the landform, which affects insolation. These landform processes in turn affect moisture balance, plant growth, and geochemistry, leading to great diversity in soils even over relatively short distances. An understanding of these processes and how they lead to soil diversity is the main basis for the qualitative mapping of soils by traditional soil survey techniques (Hudson, 1992). Such approaches typically involve field investigations and the application of a conceptual model to delineate like soils on the landscape. Worldwide, such soil surveys have greatly enhanced our understanding of soil diversity, and have provided an essential tool for land use management and soil conservation. However, in unmapped or coarsely mapped areas (such as the dry lands region of central Asia) traditional soil survey techniques may not be the best approach to increase the resolution of information, particularly when an inventory of a specific soil property is sought such as soil organic carbon (SOC). Maps made by these techniques are labor intensive and typically take up to decades to complete for large areas. Moreover, the traditional soil survey is a model of general soil diversity, not of individual soil properties. The soil survey is often not the optimal way to predict individual soil physical and geochemical properties (Webster and Beckett, 1968).

Maps of soil properties created through GIS techniques may provide a viable alternative to the traditional soil mapping. Qualitative modeling techniques (McBratney, et al., 2000) based on terrain analysis from digital elevation models (DEMs) (Wilson and Gallant, 2000) and other GIS data sets can be used to efficiently and accurately map soil properties over large areas (Gessler, 1996, McKenzie and Ryan, 1999). The general concept is to link discrete soil sample data to continuous GIS datasets in statistical, geostatistical or physical models to generate spatially continuous maps. Such methods not only allow for efficient generation of spatial models, but they also increase scientific rigor due to the ease in quantifying error through standard statistical validation techniques and reproducibility of the results.

The Shuttle Radar Topography Mission (SRTM) flown in 2000 AD has greatly increased the viability of quantitative soil mapping worldwide. A key difficulty in the past was obtaining a useful DEM for terrain analysis at the scale appropriate for modeling at the toposequence (hill slope) scale.

Figure 1. Image of the study area from Google Earth showing the town of Aksu Ayuly with the SRTM DEM (shaded relief) used for terrain analysis.

For remote regions, digital elevation data was often only available from 1 km resolution DEMs as compiled by the United States Department of Defense (GTOPO30). The resolutions of these datasets were not appropriate for soil modeling at the landscape scale. An alternative is the digital capture of legacy topographic maps and subsequent conversion to DEMs using software such as ANUDEM (Hutchinson and Gallant, 2000). The DEMs from such procedures have proven useful for soil modeling, but the digitization of contour maps is labor intensive.

The SRTM data set holds much promise to increase the use of terrain data in statistical and process-based soil models in areas that previously lacked quality terrain data at the toposequence scale. The SRTM data covers the range of latitudes from 60 degrees north to 56 degrees south at a horizontal resolution of 90 m. The data can also be used to make 30 m resolution DEMs, but these are currently not available worldwide. The nominal absolute vertical error of this DEM is stated as 16 meters (Smith and Sandwell, 2003), but local (relative) errors are generally less than this amount. For the purposes of quantitative soil modeling, errors in capturing the shape of the land at the hill slope scale are the most important. An investigation of the accuracy of SRTM data in Kazakhstan using precision elevation data such as available from phase tracking global positioning systems could provide useful information. However, the value of the DEM can be evaluated directly by simply studying whether it contains predictive information useful for the modeling of soil organic carbon content.

The critical question for application of these techniques in the dry lands region of Kazakhstan is the strength of the relationship between landform and SOC distribution. Field investigations conducted in 2003 to observe soils, geomorphic setting, and present agricultural practices suggest that quantitative soil mapping using DEMs holds promise for the region. The area is characterized by moderate relief (150 m) hills with broad alluvial valleys. There is sufficient relief to drive soil forming processes and the magnitude of relief is such that SRTM DEMs can capture it. Soil pits were dug in the area of Aksu Ayuly (Figure 1, 48° 42′ 38″N, 73° 47′ 07″E) in 2003. A soil pit in an area

Figure 2. Photographs of soil profiles from the middle of an alluvial terrace (left) and from an area of concentrated water flow from surrounding hill slopes (right). The soil profile on the left is darker with a thicker A horizon, likely indicating a greater amount of soil organic carbon.

of concentrated hill slope drainage showed a much darker and thicker A horizon (30 cm vs 10 cm) than the soil on an alluvial terrace (Figure 2). While far from a definitive analysis, such examples suggest that topography and geomorphology may have important roles in the spatial prediction and inventory of SOC. However, it is important to remember the full range of soil forming factors (applicable at the scale of interest) when mapping soils using spatial modeling techniques. Maps of vegetation, land use and other parameters available from the analysis of remote sensing data will also likely prove useful to the mapping of SOC. This report concentrates on the use of the newly available SRTM DEMs and illustrates the calculation of a key terrain derivative that has proven useful for SOC prediction at other locations.

2 AN ILLUSTRATED EXAMPLE OF USING SRTM DATA AND TERRAIN ANALYSIS

The full workflow of obtaining SRTM DEMs and the calculation of terrain derivatives useful to soil prediction is presented as an example to guide future research. The first step is to download the SRTM DEM from the Internet. The United States Geological Survey (USGS) provides the dataset through its Seamless Data Distribution System (http://seamless.usgs.gov/). The SRTM data is found under the "*View and Download International Data*" link. Upon clicking this area, a web-mapping application is launched (Figure 3). The user can use the zoom tools to focus on an area of interest. Next, the user should click the "*Download*" tab to the right of the map and open the layer "*Elevation*". Within the "*Elevation*" layer is a box labeled "*SRTM Finished 3 arc second*" which should be checked. On the right is a set of tools labeled "*Downloads*". These tools allow an area of interest to be selected for download by various methods, a selection box being the simplest. Once and area is selected, the user is lead through a serious of prompts to download the data. The data is provided in ArcGIS (ESRI, 2007) GRID format (compressed as a .zip file) in a geographic coordinate system. Complete instructions for downloading DEM and other spatial data are provided on the USGS website.

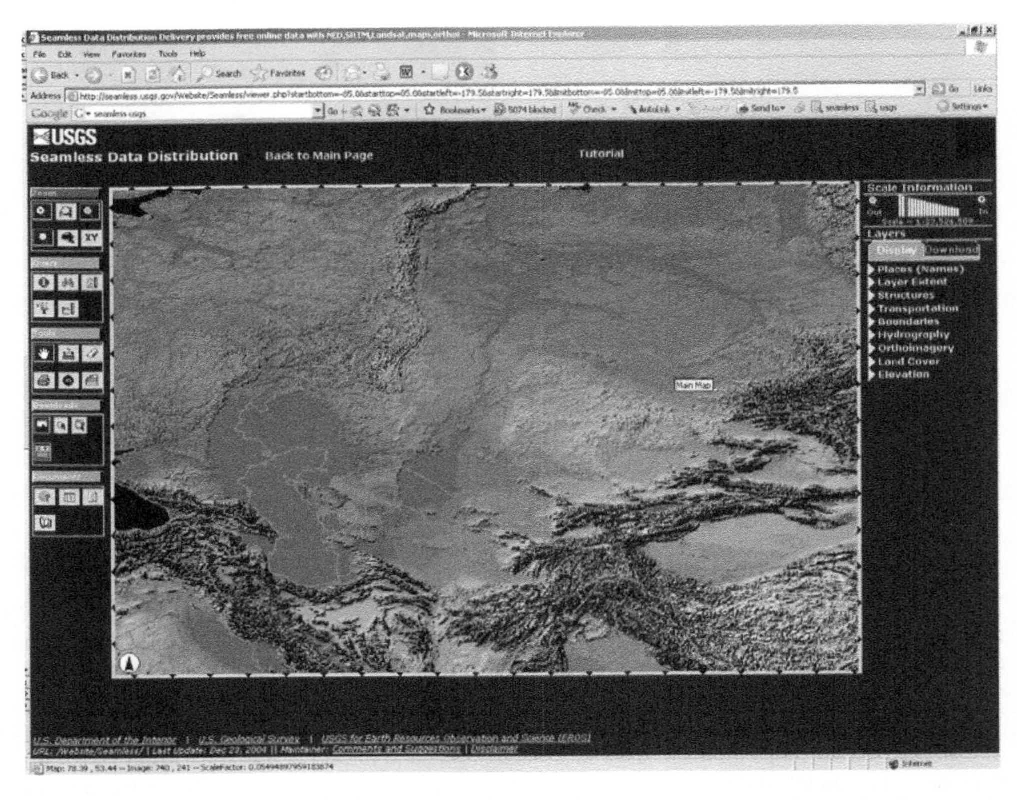

Figure 3. Screen shot showing the United States Geological Survey's Internet mapping application, used to download SRTM DEMs.

Table 1. Key terrain analysis software freely available for download from the Internet.

Software	Hyperlink	Notes
TauDEM	http://hydrology.neng.usu.edu/taudem/	Conducts flow routing analysis as toolbar within ArcGIS, a convenient, one software system approach
Terrain Analysis System (TAS)	http://www.sed.manchester.ac.uk/ geography/research/tas/	Comprehensive tool set for calculation of terrain derivatives, many of the latest advanced algorithms included. Must convert grid format to use from and to ArcGIS
SAGA	http://www.saga-gis.uni-goettingen.de/ html/index.php	GIS system that has many terrain analysis tools.
GRASS GIS	http://grass.itc.it/	The only way to calculate flow line density for very large grids (r.terraflow). Currently only works on LINUX

An important issue in conducting GIS based soil analysis is the choice of software. Experience has shown that using ArcGIS for data storage and cartography, in conjunction with specialty software for terrain analysis is the most effective approach (Venteris and Slater, 2005). There are many specialty packages for terrain analysis available, and all are freeware (Table 1). For this

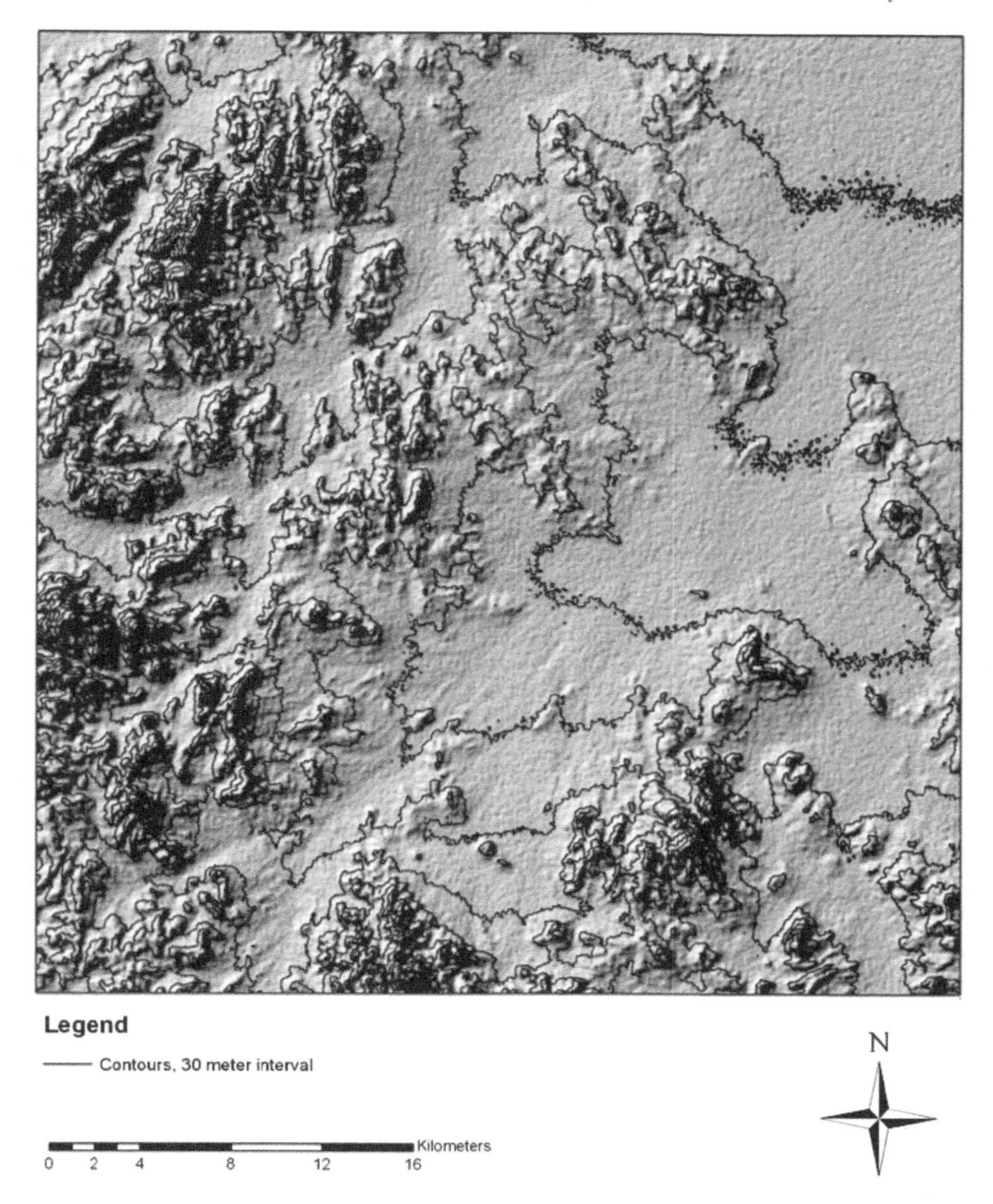

Figure 4. Shaded relief and contour map of the raw DEM (projected to UTM coordinate system). Each contour line is heavily crenulated with many surrounding isolated contours. The raw DEM is clearly noisy.

example, terrain analysis will be conducted in TAS (Terrain Analysis System), with display and cartography conducted within ArcGIS.

A DEM from just south west of the Aksu Ayuly (Figure 1) was downloaded to illustrate the procedure for terrain analysis. The ultimate goal was to obtain a terrain model that approximated the effect of terrain on the distribution of annual wetness (water flow and storage). The first step was to construct a map to visualize the strengths and weaknesses of the SRTM DEM. The 90 meter resolution DEM was converted to the Universal Transverse Mercator projection and a combined shaded relief and contour model was created using Spatial Analyst in ArcGIS (Figure 4).

The raw DEM contained a significant amount of noise over short distances, as seen in the 'bumpiness' of the surface and the excessive amount of crenulations (kinks) in the derived contours (Figure 4).

Such errors will affect all terrain derivatives, but the issue is the most critical when using the DEM to calculate how water flows on the surface. Algorithms that calculate flow accumulation require DEMs that give an uninterrupted sequence of lower elevations in the directions of down slope flow and stream coalescing. Spurious "pits" (as opposed to real pits in the landscape due to geomorphic processes) in the DEM will interrupt the flow accumulation algorithm, giving an unrealistic estimate of the amount of water reaching downstream locations. Smoothing and/or pit-filling routines are typically applied to raw DEMs to improve the modeling of surface drainage.

There were several options for processing DEMs to obtain realistic models of surface flow. For this study, the results from using a simple low pass filter (smoothing) were compared with a pit filling routine in TAS. The DEM was smoothed using a 7 × 7 cell low-pass filter (all cells assigned equal weight). The reduction in the small-scale noise was immediately apparent (Figure 5), as the crenulations in the contour lines were much reduced and the DEM appears smoother overall. However, there was a faint striping effect in the Northeast-Southwest direction that remained. This striping could be removed with the use of specialized filters. For comparison, a pit filling routine contained in TAS was applied as an alternate approach to smoothing (i.e. choosing the *Remove Topographic Depressions* option, using the Planchon and Darboux (2001) method, with drainage enforced on flats). The overall appearance of the DEM has changed little from the original (Figure 6). The critical question was how much of the granularity seen in the original DEM was reflecting the shape of the terrain and how much was due to noise. This was impossible to determine without an independent DEM at a higher resolution. It is likely that much of the surface roughness was noise, as the pattern carried across to the alluvial plains, which are typically very smooth due to the depositional processes involved.

The effect of pit filling or smoothing on the calculation terrain derivatives is the most critical issue when using the DEM for soil-landscape modeling. There is a wide range of topographic parameters available for use in quantitative soil mapping (Wilson and Gallant, 2000). The topographic wetness index (TWI) (Beven and Kirby, 1979) is a commonly used terrain derivative in soil-landscape modeling. It is particularly applicable to the modeling of soil organic carbon, as TWI is a proxy for many processes important to SOC such as water balance, plant growth, microbial action, and sediment transport processes. TWI is defined as

$$TWI = \ln\left(\frac{A_s}{\tan B}\right) \tag{1}$$

where A_s is the specific catchment area (area of upslope contributing cells as determined by the flow algorithm, divided by the cell width) and tan B is the slope. There are many different methods available for calculating A_s, and the differences in results are significant. Past work (Wilson and Gallant, 2000, Venteris and Slater, 2005) has shown that algorithms that allow for multiple flow directions give more realistic results than the original D8 method of O'Callahan and Mark, 1984. For this study, the deterministic infinity approach (D-inf) of Tarboton et al. (1991) was arbitrarily chosen. The final choice of algorithm for A_s calculation should be done empirically, based on which is the most predictive of the soil property to be modeled.

TWI was calculated for the different DEMs to make qualitative comparisons between the affects of DEM smoothing and pit filling. There were marked differences in TWI results between the pit filling and low-pass filtering approaches. Using pit filling only (Figure 7) resulted in low connectivity between cells on the hill slope. Overall, this approach left the uplands "dry", with little accumulation between cells. TWI patterns on the hill slopes were irregular, and areas of converging flow at the bases of hill slopes were not well delineated. This pit-filling algorithm also forced drainage in flat areas so a clear channel network (not channels precisely, but dendritic patterns of high TWI) was developed in the valley bottoms (it is possible to force the position

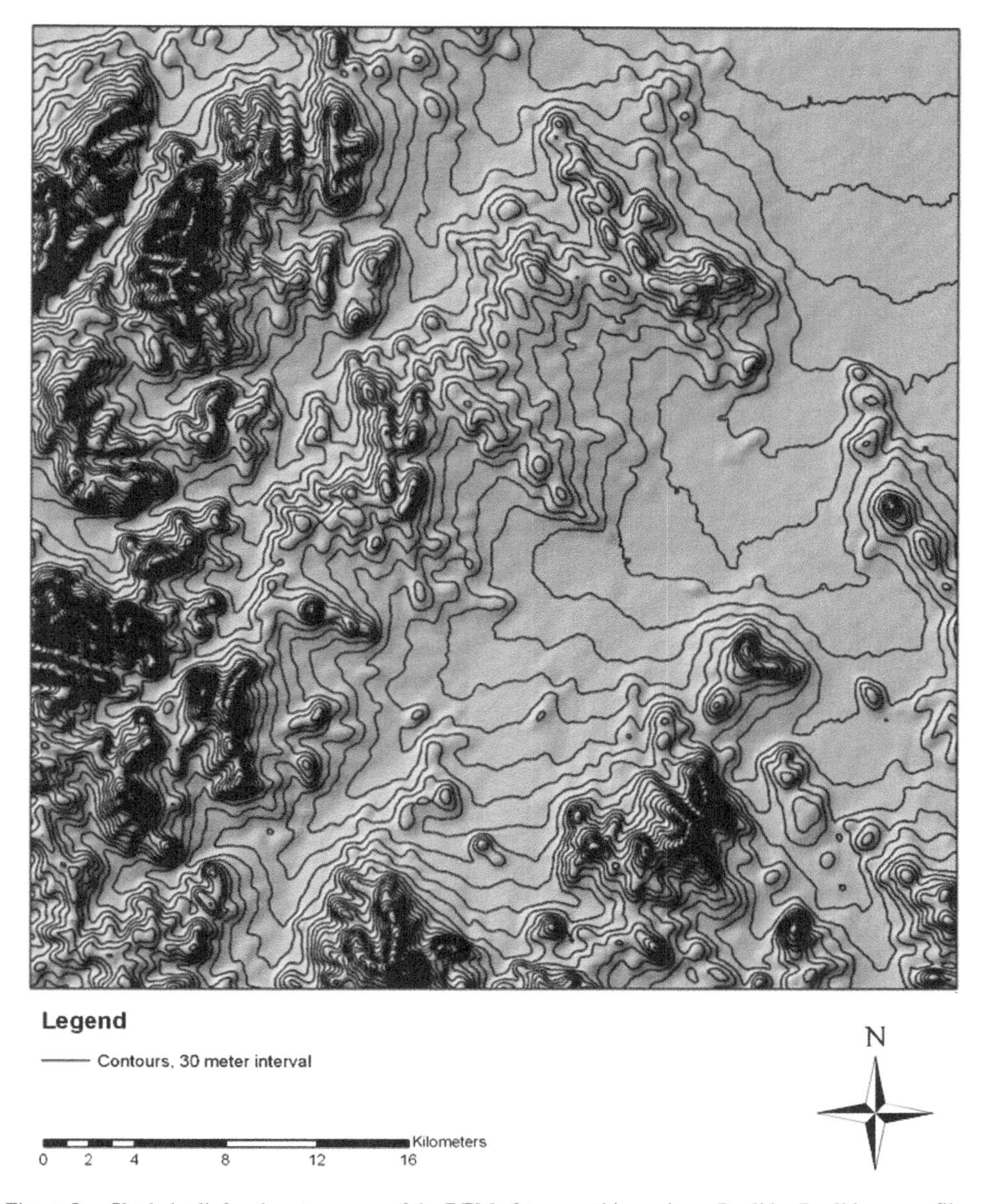

Figure 5. Shaded relief and contour map of the DEM after smoothing using a 7 cell by 7 cell low pass filter. The contours are much smoother, and the isolated contours have been eliminated.

of these channels using maps of the actual stream positions in TAS). TWI from the smoothed DEM provided a much different pattern for the distribution of TWI this landscape (Figure 8). The flow between cells on the hill slope was more connected, resulting in smooth increases in TWI down slope. Areas of convergent flow at the base of hill slopes were clearly delineated. However, the pattern of channels was unsatisfactory, with several unrealistic terminations in the bottom of the valleys (see center of Figure 8). An alternate combined approach was to use the pit filling routine on the smoothed DEM, which provided both a good resolution of potential

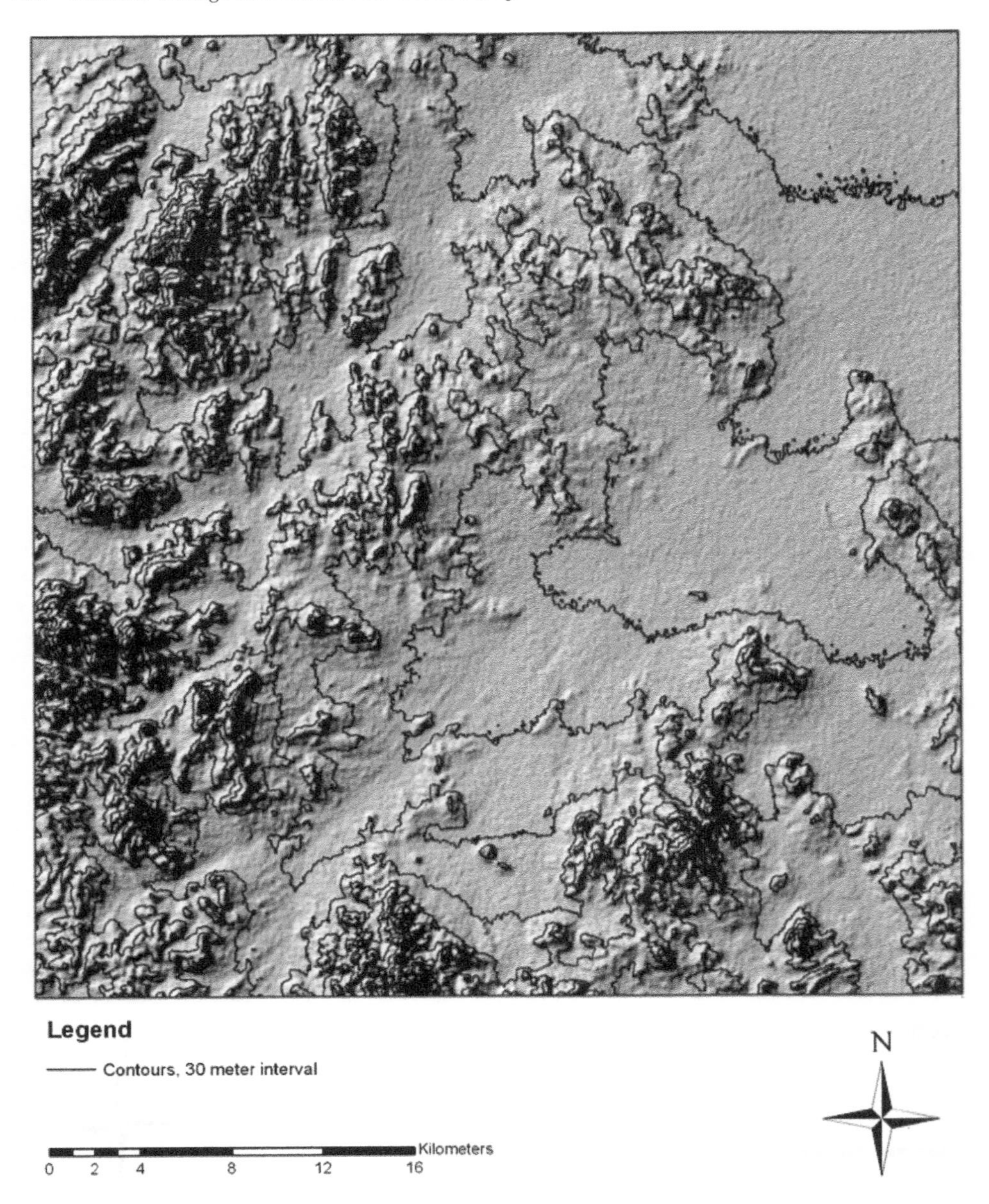

Figure 6. Shaded relief and contour map of the DEM after the pit-filling routine was applied. The overall shape of the DEM only shows subtle differences from the unprocessed version.

moisture differences between hill slope positions, and realistic channel patterns in the valley bottoms (Figure 9).

It is not possible to determine which approach to calculating TWI is best without independent data on soil moisture, water drainage, and/or soil properties (SOC). The approaches described above only provide some examples of the myriad of combinations of DEM processing and TWI algorithms possible. Ultimately, the choice should be made using a comprehensive database of SOC values for soil samples, and thorough statistical validation to determine which approach has the most predictive value.

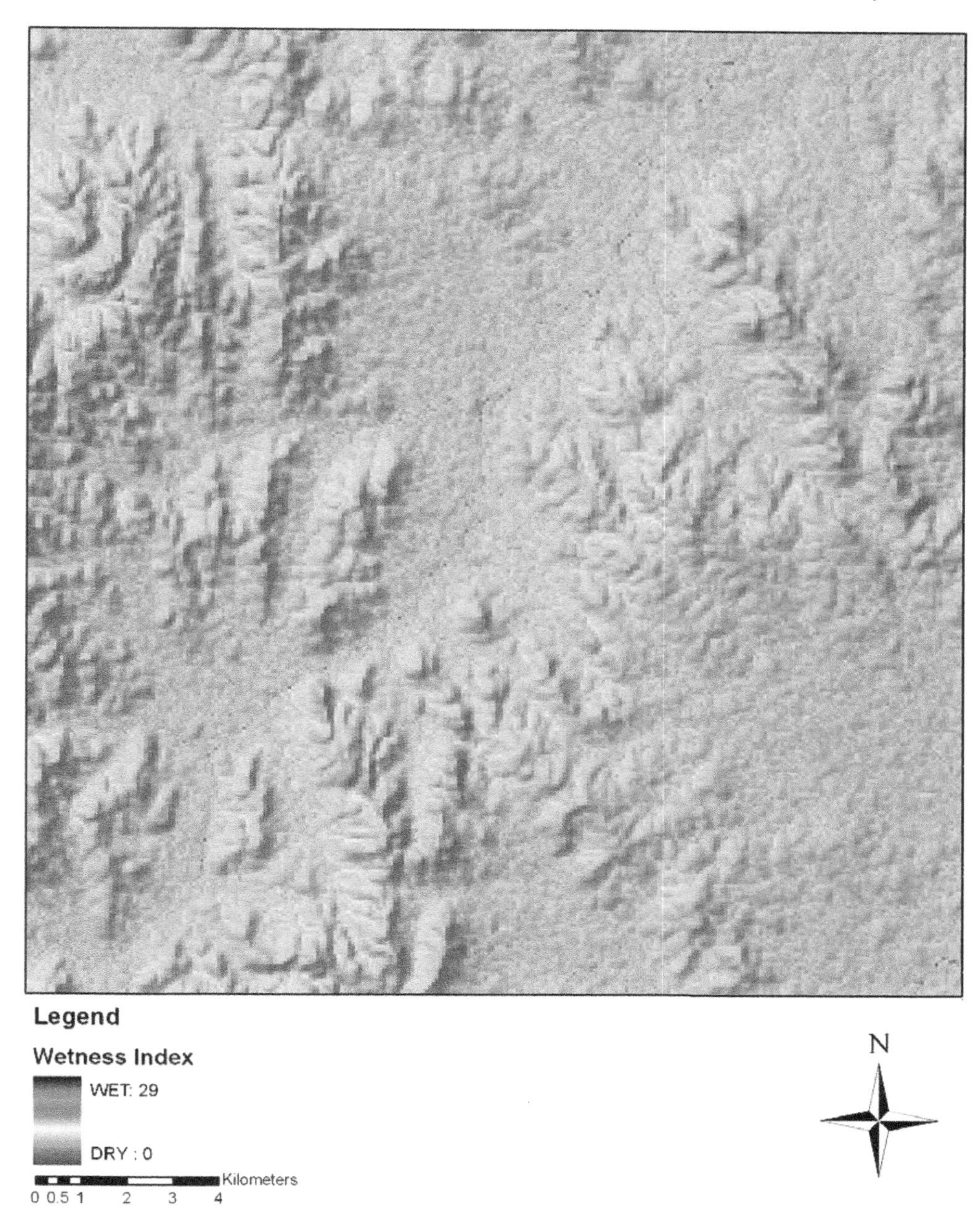

Figure 7. TWI based on DEM with pit-filling only. There is less connection between the cells overall. Areas of potential concentrated or channelized flow are delineated, but there is little apparent potential for modeling TWI differences within the hill slopes.

3 CONCLUSIONS

The relationship between soil properties and landscape position is well established and needs to be incorporated into SOC inventory and process modeling studies. This contribution only points to the

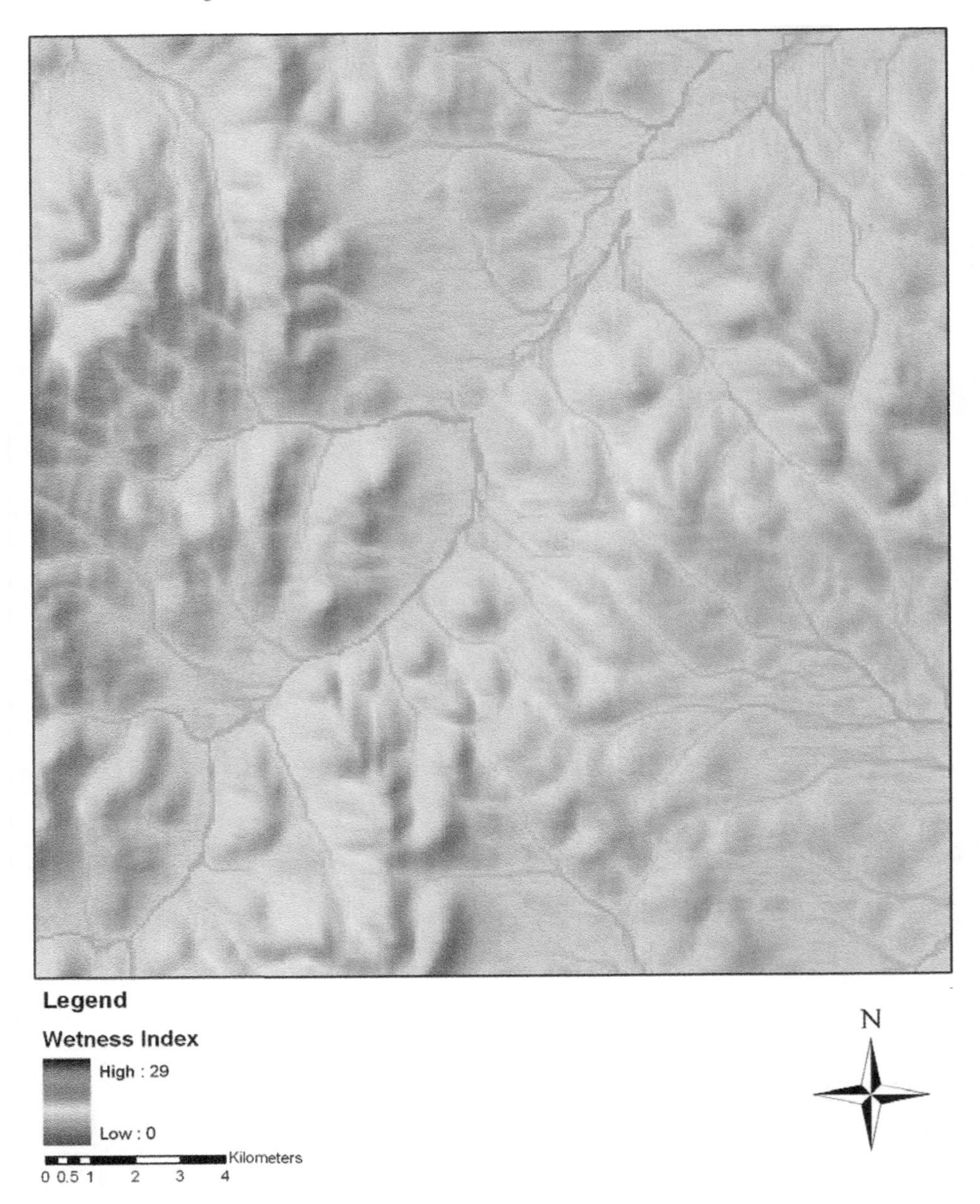

Figure 8. TWI based on the DEM smoothed with a low pass filter. There are smooth connections between cells on the hill slope and areas of convergent flow are clearly delineated. Because a pit filling routine was not used, channels in the valley bottoms terminate prematurely.

feasibility of using terrain information; much work remains to prove the hypotheses that SOC can be mapped quantitatively using spatial data sets. Key items needed to further this study include-

1. A comprehensive data set of soil cores or pits that are stratified by landscape position, vegetation type and/ or agricultural practice is needed. The number of sampling sites needs careful thought, but 30 locations for each combination of topography and vegetation type should be considered a minimum. Nested sampling over a range of scales is useful for interpreting variability over

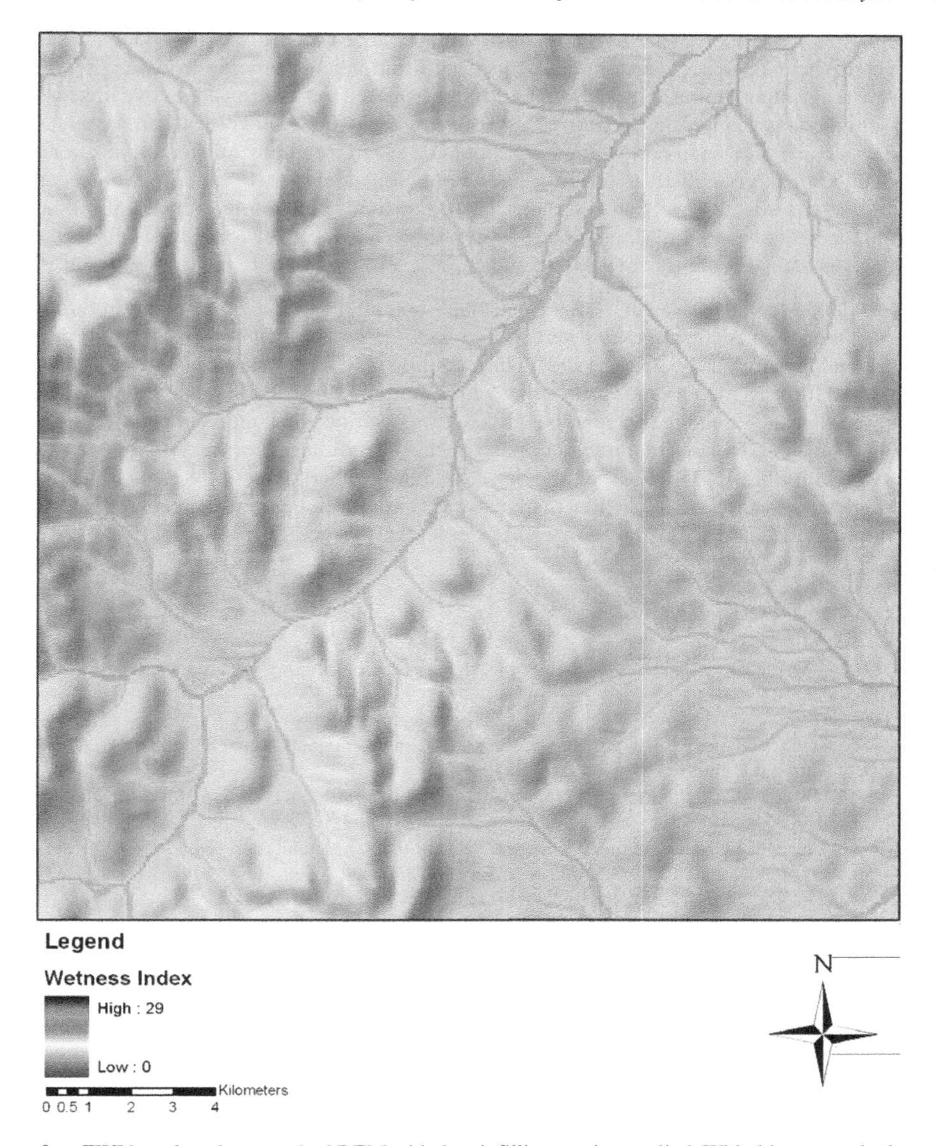

Figure 9. TWI based on the smoothed DEM with the pit filling routine applied. With this approach, there are smooth transitions in TWI on the hill slopes and realistic representation of flow concentrations in the alluvial valleys.

short distances. SOC should be measured along with a full suite of physical and chemical soil properties to give insight into processes. The use of Cs 137 for erosion and deposition modeling (Walling and He, 1999) should be given careful consideration, as sedimentary processes are critical to measuring the SOC budget.

2. Ancillary data on topography and soil moisture should be collected to guide terrain analysis. High resolution DEMs should be created for the test location that can aid in the separation of signal from noise in the SRTM DEMs. Approaches using precision GPS or LIDAR are favored for such studies. In addition, detailed grids of soil moisture using periodic sampling by time domain reflectometry or soil core collection would help in determining the best approach to

calculating TWI. Moreover, TAS contains many advanced algorithms to model surface flow that could be used to account for the effects of infiltration and evaporative losses.

3. A wide range of modeling approaches should be considered for creating predictive links between the sample data and terrain derivatives. In addition, spatial datasets other than terrain such as land use from multi-band remote imagery should be tested for improved predictions. For spatial modeling, both geostatistical (kriging and simulation) and statistical regression techniques should be considered (McBratney et al., 2000). For SOC forecasting, numerical models that include the full range of hill slope processes should provide the most accurate results. Numerical models that can account for soil erosion and deposition are especially favored (Liu et al., 2003).

The dry lands region of Kazakhstan has excellent potential for SOC accumulation through the adoption of management practices to reverse losses caused by Khrushchev's Virgin Lands programs of the 1950s (McCauley 1976, Grote 1998). The modeling of SOC in space is needed to set baselines for numerical forecast models and verify soil C sequestration. Modeling in space and time is needed to forecast gains and compare potential benefits between various management practices. All these modeling efforts need to consider interaction between soil forming processes and topography. The recent availability of SRTM DEMs for the dry lands regions facilitates study of such interactions. DEM-based studies have great potential to increase the understanding of soil processes in the region and to improve the accuracy of inventory and forecasts involving SOC.

REFERENCES

Beven, K.J. and M.J. Kirkby. 1979. A physically based, variable contributing area model of basin hydrology. Hydrol. Sci. Bull. 24:43–69.

ESRI. 2007. http://www.esri.com

Gessler, P.E. 1996. Statistical soil-landscape modeling for environmental management. Ph.D. Thesis, The Australian National University.

Grote U. 1998. Central Asian environments in transition. Manila, Philippines: Asian Development Bank.

Hudson, B.D. 1992. The soil survey as paradigm-based science. Soil Sci. Soc. Am. J. 62:836–841.

Hutchinson, M.F. and J.C. Gallant. 2000. Digital elevation models and representation of terrain shape. p. 29–50. In: J.P. Wilson and J.C. Gallant (eds), Terrain Analysis. Wiley, New York.

Jenny, H. 1941. Factors of Soil Formation. McGraw Hill, New York. 269 p.

Liu, S., N.B. Bliss, E. Sundquist, and T.G. Huntington. 2003. Modeling carbon dynamics in vegetation and soil under impact of soil erosion and deposition. Global Biogeochemical Cycles 17:1074.

McBratney, A.B., I.O. Odeh, T.F. Bishop, M.S. Dunbar, and T.M. Shatar. 2000. An overview of pedometric techniques for use in soil survey. Geoderma 97:293–327.

McCauley, M. 1976. Khrushchev and the development of Soviet Agriculture, The Virgin Land Programme 1953–1964. Plymouth: The Bowering Press

McKenzie, N.J. and P.J. Ryan. 1999. Spatial prediction of soil properties using environmental correlation. Geoderma 89:67–94.

O'Callaghan J.F. and D.M. Mark. 1984. The extraction of drainage networks from digital elevation data. Computer Vision, Graphics, and Image Processing 28:323–344

Planchon, O. and F. Darboux. 2001. A fast, simple algorithm to fill the depressions of digital elevation models. Catena 46: 159–176.

Smith, B. and D. Sandwell. 2003. Accuracy and resolution of shuttle-radar topography-mission data. Geophysical Research Letters 30:1467.

Tarboton, D.G., R.L. Bras, and I. Rodrigues-Iturbe. 1991. On the extraction of channel networks from digital elevation data. Hydrological Processes 5:81–100

Venteris, E.R. and B.K. Slater. 2005. A comparison between contour elevation data sources for DEM creation and soil carbon prediction, Coshocton, Ohio. Transactions in GIS 9:179–198.

Walling, D.E. and Q. He. 1999. Improved models for estimating soil erosion rates from cesium-137 measurements. J. Environ. Qual. 28:611–622.

Webster, R. and P.H. Beckett. 1968. Quality and usefulness of soil maps. Nature 219:680–682.

Wilson, J.P. and J.C. Gallant. 2000. Terrain Analysis, Principles and Applications. John Wiley and Sons, New York.

CHAPTER 29

Potential for soil carbon sequestration in Central Kazakhstan

Greg McCarty & Paul Doraiswamy
USDA-ARS Hydrology and Remote Sensing Laboratory, Beltsville, MD, USA

Bakhyt Akhmedou
Science Systems and Applications, Inc., Lanham, MD, USA

Konstantin Pachikin
Department of Soils, Institute of Kazakhstan Ministry of Education and Science, Almaty, Kazakhstan

1 INTRODUCTION

As a result of the "Virgin Lands" program begun by the Soviet Union in the 1950's, over 30 million hectares (Mha) of native steppe in Kazakhstan were cultivated to wheat production (McCauley, 1976). This transformed an ecosystem that previously supported nomadic grazing into cropland that supplied over 25% of wheat demand within the Soviet Union (Kaser, 1997). As a result of this activity large areas of marginal land were brought into crop production and this led to wide-scale degradation of soils in addition to excess use of fertilizers and large irrigation projects that caused formation of saline soils and water pollution. Overall effect was increased desertification and deterioration of the surrounding ecosystems (Grote, 1998). With the collapse of the Soviet Union in 1991, Kazakhstan became independent and became the second largest country formed out of the disintegration. The gain of independence destroyed the highly regulated market system established in the Soviet era and generally resulted in increased poverty, high inflation, scarcity of food and other products, and large declines in grain production (Alaolmolki, 2001). This in turn has led to wide scale abandonment of marginal croplands which has left the landscape populated with abandoned production fields with degraded soils.

Rehabilitation of degraded lands in Kazakhstan is consistent with the goals of sequestering atmospheric CO_2 within the terrestrial ecosystem. Long-term storage of carbon (C) within the soil C pool is seen as a viable mechanism to offset increases in atmospheric C due to fossil fuel consumption. The Clean Development Mechanism (CDM) of the Kyoto Protocol would be one means to establish C sequestration with developing countries within the region although currently soil C sequestration (agricultural sinks) is not eligible for CDM funding before 2012 at the earliest (Robbins, 2004). If soil C sequestration were implemented under international mechanisms to promote development such as CDM, reductions in atmospheric CO_2 (sequestration) would be expected to be real, measurable, and long term and, C sequestration (additionally) would be measured against baseline emissions from business as usual.

Before implementation of C sequestration projects, it is necessary to establish estimates of the potential capacity to sequester C within the region of interest and assessment of economic considerations (Antle and Mooney, 1999). Investment of international markets in C sequestration projects will require robust estimates of capacity to store C in soils in the region of interest. We conducted this study to determine current and potential soil C sequestration for the Shetsky area in central Kazakhstan based on land-use changes that are being implemented within the *Kazakhstan Dryland Management Project*, sponsored by the World Bank and the Global Environmental Facility

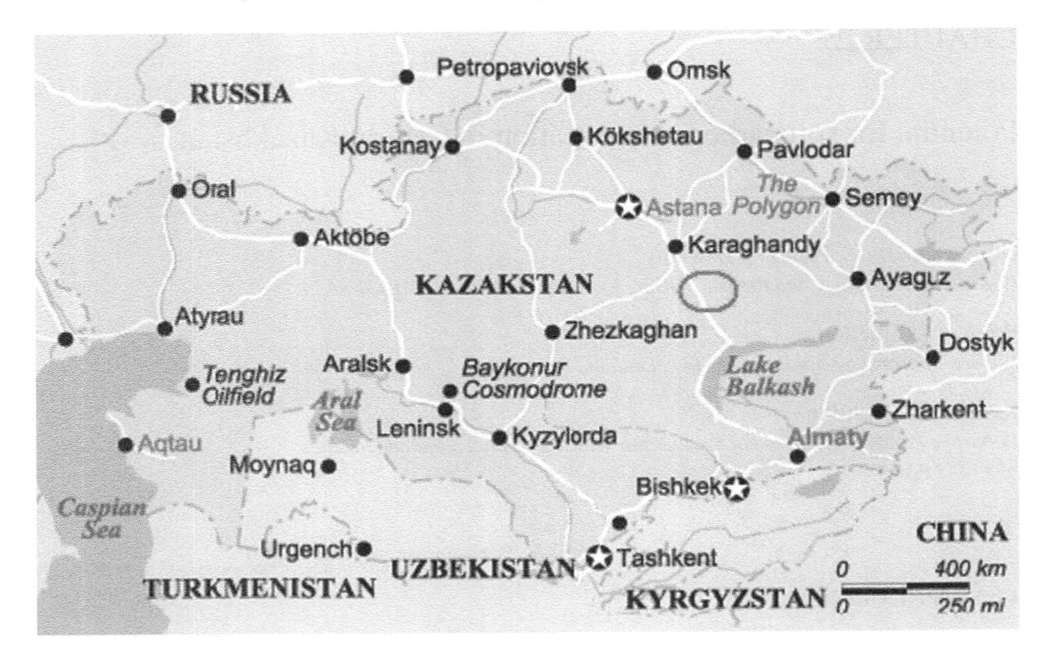

Figure 1. Map of Kazakhstan with study area designed with red circle.

(GEF). By use of this project as a prototype, we evaluated the feasibility of sequestering C in the soils of degraded lands in Kazakhstan through the restoration of vegetation adapted to grazing land use.

2 DESCRIPTION OF PROJECT AREA

The *Kazakhstan Drylands Management Project* was implemented in the northern zone of Shetsky Rayon located in central Kazakhstan within an area where acute rural poverty prevails (see Figure 1). This region of Central Kazakhstan is a part of so-called "risky agriculture" zone distinguished by unstable climatic conditions. In addition to the high precipitation years that can produce grain yields as high as 1.4–1.5 t ha^{-1}, years of drought repeat every 3 to 4 years and typically produce yields of 0.3–0.4 t ha^{-1} and sometimes plantings that burn out completely. Extreme soil, climatic, social, economic and technological conditions restrict the economic development opportunities. Two-thirds of the cereal production areas from the Soviet era are unprofitable for grain production (less than 0.6 t ha^{-1}) and have largely been abandoned. The soils and landscapes are suitable for introduction of sustainable land use practices such as sown fodder/pastures/rangeland or rehabilitation of natural vegetation. Annual rainfall in the region is below 300 mm yr^{-1}. The study site is easily accessible for demonstration purposes making it will suited for the *Drylands Management Project*. The project zone is representative of other areas in the region which will allow it to be replicated in other areas of interest.

The region that encompasses the study area contains over one M ha (11 sub-districts in the Shetsky rayon). Based on local statistics for this land area, 60% has been allocated to private farmers and 53,000 ha currently cultivated in wheat, potatoes, vegetables and forage-crops. And 143,000 ha are abandoned wheat or perennial grasses. The Drylands Management Project will provide 70,000 to 100,000 ha of abandoned lands with planting of perennial grass (principally, *Agropyron*). In collaboration with farmer groups, demonstration plots at selected sites in the project area were established to facilitate the transfer of the rehabilitation technology.

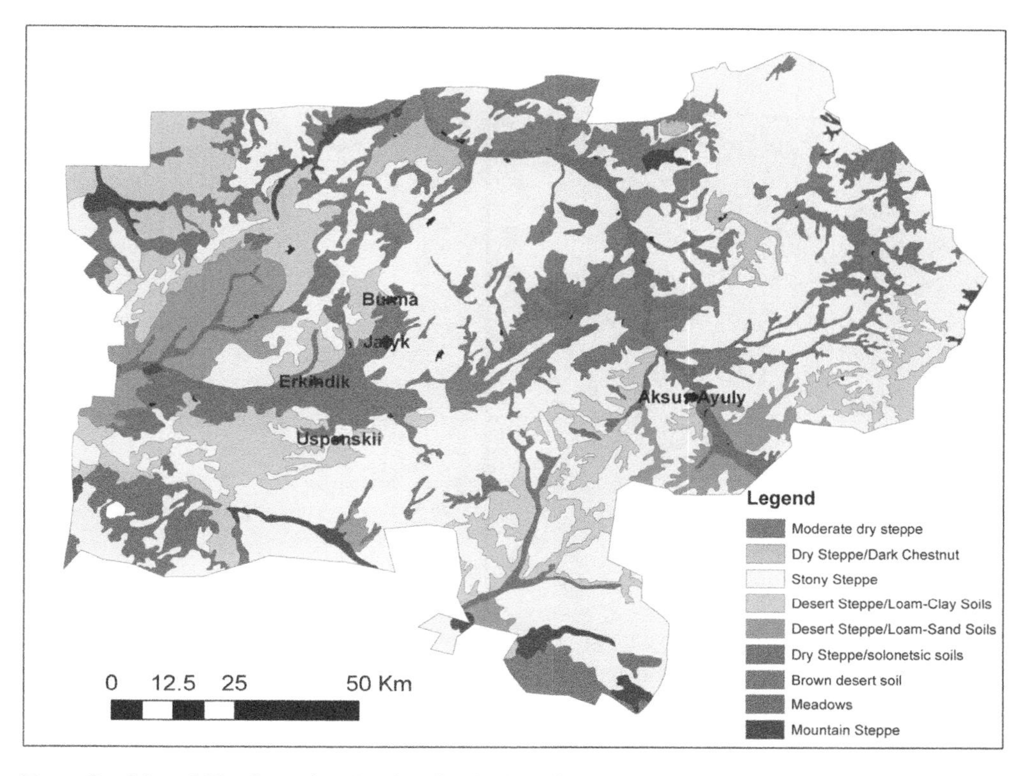

Figure 2. Map of Shetsky region showing distribution of the nine ecotype classifications (see Table 1 for ecoregion descriptions).

Geographically, the project's territory is situated in the central part of Kazakh melkosopochnik (a highly eroded plateau) and is characterized by complex structure of the surface. Melkosopochnik and low mountain massifs with many dense layer outcrops. The soil cover of the project territory is formed in the conditions of arid with a gradation from steppe to desert-steppe from south to north. Parent material, heterogeneity of relief and moisture regime of landscape position are all important parameters dictating soil properties such as C content. The region contains primarily chestnut and light chestnut soils.

3 ESTIMATING THE POTENTIAL FOR CARBON SEQUESTRATION IN THE STUDY AREA

An unsupervised classification of land use was created using multi-temporal Landsat 7 Thematic Mapper (TM) data (Figure 2). This analysis showed that 15% of the land area of the study site was in crop production and 30% as rangeland. A map of nine common ecoregions was generated for the study area from information on soil type, topography and vegetation (Figure 3). Based on expert knowledge recommendations were developed concerning best sustainable land use for these ecoregions with recommendation that some cropland be taken out of production (Table 1). The current land use data was combined with ecoregion data to establish the aerial extent of active or abandoned cropland in unsustainable environments (Table 2).

The different ecoregions were assigned different capacity to sequester and storage C based on current and projected land-use and land-use changes. Two methods were used to estimate C sequestration potential. Both methods are based on the C information derived from the soils map

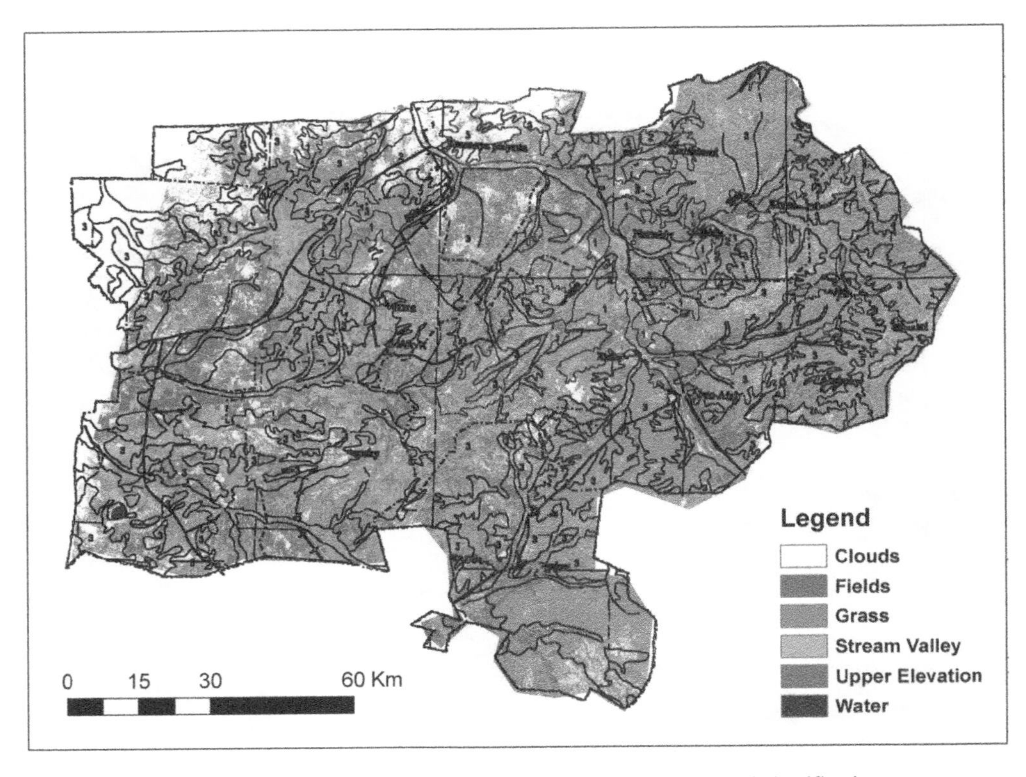

Figure 3. Land-use classification for the Shetsky region based on unsupervised classification.

Table 1. Landuse change recommendations for croplands in different ecoregions based on present objectives.

Ecoregion	Soil types	Ecosystems	Description of sustainable landuse
1	Dark chestnut loamy, heavy loamy clay and carbonate soils	Moderate dry steppes	Keep as fields
2	Dark chestnut light loamy, loamy sandy soils	Dry steppes	Keep as fields
3	Dark and light slightly developed soils and mountain soils	Stony steppes	Convert to pastures
4	Light chestnut loamy, heavy loamy clayey and carbonate soils	Desert steppes (semi deserts)	Convert to pastures
5	Light chestnut light loamy, loamy sandy soils	Semi deserts desert steppes	Convert to pastures
6	Dark chestnut and light chestnut solonetzic soils	Dry steppes	Convert to pastures
7	Brown desert normal, brown solonetzic soils	Steppe deserts	Convert to pastures
8	Meadow chestnut, meadow boggy, alluvial soils	Meadows	Do not plow
9	Steppe, meadow steppe, desert solonetz	Steppe and desert Solonetz	Do not plow

Table 2. Estimation of carbon sequestration potential for use by two different methods.

		Method 1[a]		Method 2[b]		
Ecoregion	Fields (ha)	C sequestration potential ($t\ ha^{-1}$)	Soil class 3–9 Converted (t)	Virgin carbon	0.3 Virgin carbon	Soil class 3–9 converted (t)
1	35700	4	not converted	88.5	26.5	not converted
2	3300	5.5	not converted	71	21.3	not converted
3	85000	1.4	119100	77	23.1	1964800
4	27600	6.5	179700	69.5	20.8	576600
5	17700	11	194900	98	29.4	521100
6	21500	limited information		limited information		
7	3500	limited information		limited information		
8	21000	18	377700	140	42	881500
9	6000	15	90400	69	20.7	124800
	157400	Total if all fields are converted	962000			4068700
		Method 1 Total C	$6.10\ t\ ha^{-1}$		Method 2 Total C	$25.8\ t\ ha^{-1}$

[a] Method of estimation based on estimates of C in virgin and cultivated soils from the limit soil surveys within the region.
[b] Method of estimation based on assumption that cultivated soils contain 70% of C in virgin soils.

for the study area in Shetsky. The map has nine ecoregions and information on land use within each environment. Two C levels are provided for each soil class, first level is the C (C) for never-tilled soils under natural ecosystem vegetation and the second is the C content of cultivated soils (see table associated with map). An extensive survey of soil C contents in soils in the region was conducted during the summer of 2004 and 2005 which included paired sites under virgin vegetation and cultivation (the basis for Method 1). It is well established that natural ecosystem conversion to agricultural production results in substantial soil C loss (Lal et al., 1998). Davidson and Ackerman (1993) estimated that between 20 to 40% of soil C is lost with conversion of lands into agricultural production. For the purposes of the feasibility study it is assumed that 30% of soil C was lost during the period of agriculture associated with the Virgin Lands program (the basis for Method 2). Reestablishment of vegetation similar to those found on virgin lands should permit soils to return to pre-cultivation contents of soil C.

Calculation of potential C sequestration within the region by way of landuse change excluded cropland in Ecoregions 1 and 2 because it was recommended that these two classes, which are primarily in river valleys, should remain in crop production. There was insufficient information to provide recommendation for Ecoregions 6 and 7 and, therefore, not used in this computation. The remainder of cropland area was assumed to be converted to pasture and rangelands in this computation. The total of 157,400 ha shown at the bottom of column 2 represents the total cropland areas for Ecoregions 3, 4, 5, 8, and 9.

Both methods of analysis demonstrate substantial capacity to sequester C within the Shetsky region based on the land use changes to be implemented within the Kazakstan Dryland Management Project. However there is substantial divergence in estimates of capacity for sequestration. Method 1 estimated an average regional C sequestration potential of approximately $6\ t\ C\ ha^{-1}$, whereas Method 2 estimated a much higher potential of $26\ t\ C\ ha^{-1}$. For Method 1 the field data available was still very limited concerning differences between soil C in natural ecosystem versus production fields. Use of literature estimates for C loss resulting from typical implementation of agriculture are likely more reliable. As a result, Method 1 is very conservative and probably underestimates true potential for sequestration based on region adoption of landuse changes. The C market payments based on these estimates of C storage would result in substantial investment of foreign capital within the Shetsky region.

This analysis indicates that the region has substantial capacity for increased soil carbon storage with conversion of production fields with marginal and very low productivity. Activities of the Dryland Management Project have good likelihood of resulting in substantial carbon sequestration within the Shetsky region and can provide a good prototype for projects implemented to sequester carbon in soil. Ability to translate increased carbon storage into verifiable carbon credits will require increased capacity for measurement and modeling of carbon dynamics within ecosystems of the region.

4 CONCLUSIONS

The World Bank Kazakhstan Drylands Management Project has the goal of restoring degraded soils associated with abandoned croplands in Kazakhstan. Global markets for carbon sequestration are likely to grow with continued implementation of international agreements such as Kyoto as well as those expected to follow. The aim of this study was to assess the feasibility carbon sequestration projects within the Shetsky region of Kazakhstan. To perform this analysis, we overlaid detailed maps of landuse and soil ecotypes to estimate amounts of land area which should remain in crop production or converted to range/pasture land. This analysis used estimates of carbon sequestration potential for the converted lands to assess impact landuse conversion activities under the Drylands Management Project. The analysis indicated that the region has substantial capacity for increased soil carbon storage with conversion of production fields that have degraded soils. Activities of this project will result in substantial carbon sequestration within the Shetsky region. The project can serve as a prototype for other carbon sequestration projects implemented in the Central Asia region.

REFERENCES

Alaolmolki, N. 2001. Life after the Soviet Union, the newly independent Republics of the Transcaucasus and Central Asia. Albany, NY: University of New York Press.

Antle, J. and S. Mooney. 1999. Economics and policy design for soil carbon sequestration in agriculture, Montana State University-Bozeman, research discussion paper No. 36.

Davidson, E.A. and I.L. Ackerman 1993. Changes in soil carbon inventories following cultivation of previously untilled soils. Biogeochemistry 20:161–193.

De Beurs, K.M. and G.M. Henebry. 2004. Land surface phenology, climatic variation, and institutional change: Analyzing agricultural land cover change in Kazakhstan. Remote Sensing of Environment 89:497–509.

Grote U. 1998. Central Asian environments in transition. Manila, Philippines: Asian Development Bank.

Kaser, M. 1997. The economics of Kazakhstan and Uzbekistan. London: The Royal Institute of International Affairs.

Lal, R., J.M. Kimble, R.F. Follett and C.V. Cole. 1998. The Potential of U.S. Cropland to Sequester Carbon and Mitigate the Greenhouse Effect. Ann Arbor Press, Chelsea, MI.

McCauley, M. 1976. Khrushchev and the development of Soviet Agriculture, The Virgin Land Programme 1953–1964. Plymouth: The Bowering Press.

Robbins, M. 2004. Carbon Trading, Agriculture and Poverty. Special Publication No. 2, World Association of Soil and Water Conservation. Beijing China, p. 48.

CHAPTER 30

Application of GIS technology for water quality control in the Zarafshan river basin

T.M. Khujanazarov
Department of Desert Ecology and Water Resources Research, Samarkland, Uzbekistan

T. Tsukatani
Institute of Economic Research, Kyoto University, Japan

1 INTRODUCTION

At the beginning of the 21st century one of the crucial problems of the Central Asian countries Uzbekistan, Kazakhstan, Turkmenistan, Tajikistan and Afghanistan are water resources. The irrational use of water resources for intensive agriculture development, as well as global climate change that are currently taking place in this region could result in changes of the established equilibrations "climate/water" system (Glantz, 2002; Agaltseva, 2002; Khujanazarov et al., 2005).

Historically the Aral Sea has varied a lot in size, depending on the climate, since its origin birth approximately ten thousands years ago, when it was very small and only the Syr Darya was discharged into it. Amu Darya, four to five thousand years ago changed its course into the Aral Sea, which at that time was four to five times larger than in the middle of the 20th century. Thus, the one of the most severe manmade environmental and ecological disasters of all time was created during the last decades of the 20th century mainly due to the lack of the integrated effects of water withdrawal for large-scale irrigation. Such phenomenon reduced the inflow of water from the Amu Darya and the Syr Darya Rivers and their tributaries to the Aral Sea by 90% compared to the middle of the 20th century levels.

Today due to the intensification of aridity, very little of the water reaching the Sea, (only 20%) is originated from the countries downstream of Tajikistan and the Kyrgyzstan. Long- term extending of production of cotton monoculture demands of water use was beyond any sustainable limits. In 1989 4.3 million hectares in the Amu Darya Basin and 3.3 million hectares in the Syr Darya basin were irrigated. Almost 30% of irrigated land is severely salinized and the crop yields are reduced by 20 to 50%. And the salt-concentration of the sealake almost tripled between 1960 and 1990. The increasingly less productive condition of the soils resulted in an overuse of fertilizers, herbicides and pesticides. All this has made the water in the rivers as well as in the shrinking Aral Sea unfit for any type of human consumption and the soils have become polluted and unproductive. Additionally the altered land use and massive irrigation caused basin-wide soil and water salinization, loss of biodiversity, poor quality of drinking water and chronic health problems for the population. Due to the heavy pollution, there are high incidence rates of water-borne diseases such as typhoid fever, dysentery, viral hepatitis A, and intestinal inflections occur in the provinces along the river corridor (Fayzieva, 1999; Fayzieva et al., 2001). The increasing environmental degradation and its effects on the living conditions and human health in this region have been known for decades. In spite of several initiatives designed to reverse the trend, however, very little improvement has been achieved and the impoverished people in the area continue to suffer.

The Amu Darya and the Syr Darya are the main sources of surface water in the Aral Sea Basin, but there are also several smaller rivers of importance. They include the Zerafshan, Sherabad, Kashkadarya and others, which flow west out of the mountains in Uzbekistan. They originally ended in terminal lakes in the desert or overflowed into the Amu Darya, but now are totally diverted for irrigation except in wet years when some flow reaches the terminal lakes. Zarafshan River once contributed its waters to Aral Sea, but in last decades due to intensive irrigated agricultural development, the flow of the river has decreased drastically. This phenomenon arised as result of poor watershed management, increased salinity levels and water consumption for irrigation as well as misuse of water for leaching of arable lands. The impact of some industrial sources on the quality and ecological state of natural water is not well known, because of a lack of a control system and poor observance by those who pollute water. Moreover, the constant application of existing agricultural practices is contributing to the increase of accumulation of toxic salts and other types of pollutants, in soil, plants, surface and drinking water (Toderich et al., 2001; Tsukatani and Katayama, 2001; Ko et al., 2006; Toderich et al., 2006). Therefore, Zarafshan River Valley located in the center of Zarafshan inter-mountainous depression of Nuratau and Zarafshan ridges (Tian-Shan mountain system), we have chosen as an object for biomonitoring to understand the impact of intensive irrigated agriculture and cattle/goat/sheep farming on environmental state of arid and sandy desert ecosystems.

Since there is currently very little data on the quality of Zarafshan river water, the objectives of the present paper were formulated as follows:

- to assess the current ecological state of Zarafshan River Basin;
- to identify the water quality of main streams and canals which are providing irrigation and collector-drainage system;
- to incorporate the static databank into dynamic database through GIS mapping to visualize water quality parameters by using GIS that will allow to identify the temporal distribution of different pollutants, including heavy metals.

2 METHODS AND MATERIAL

Data collection and creation of data bank on water resources use and its water quality in Zarafshan river basin were conducted during 1999–2006 within framework of several Uzbek-Japanese projects.

The target area for water samples as shown on Figure 1 was along the watercourse of Zarafshan River and its tributaries (irrigated canals, drainage collectors, lakes, water reservoirs) from the settlement of Rahvatabad (Tajikistan border) to Samarkand, Dargom canal, Karadarya and Akdarya rivers, some mountainous streams at the footsteps of Nuratau mountains, surface streams of Navoi industrial zone, collector-and-drainage network of Gijduvan and Shafirkan area, drainage system and Zarafshan river throughout Bukhara oasis and Karakul plateau (south-east part of Kyzylkum desert), including big transboundary Amu-Karakul canal, near Uzbek-Turkmenian frontier.

At each sampling site the global position by SONY GPS; pH, electronic conductivity (Ec), turbidity (Tur), dissolved oxygen (Do), water temperature (T, °C), total dissolved salts (TDS) and salinity by Horiba multiple water quality monitoring system as U-10; LMO or U-20, Meter TOLEDO, ATAGO were obtained. Ion concentrations were determined by Shimadzu ion chromatographic analyzer HIC-6A: Na^+, K^+, Mg^{2+}, Ca^{2+}, Cl^- and SO_4^{2-}. Measurement of dissolved inorganic nitrogen (DIN), and dissolved inorganic phosphorus (DIP) was done by Technician Instruments Corporation Auto Analyzer II. Standards for NH_4, NO_2 and NO_3 were special grade reagents of $(NH_4)_2SO_4$, $NaNO_2$ and $KNO_3$from Wako Pure Chemical Industries, Ltd. These reagents were first dried for one hour at a temperature of 110°C, and adjusted to each concentration after getting cooled in desiccators. Standard PC'4 from Wako Pure Chemical Industries, Ltd. was also used. At the laboratory of Chemistry and Radioactive at the University of Kyoto, metal concentration analyses were performed using a Hewlett Packard HP 4500 inductively coupled plasma mass spectrometer (ICP-MS) for Pb^{208}, Sb^{121}, Cd^{111}, Se^{82}, As^{75}, Zn^{66}, Cu^{63}, Ni^{60}, Fe^{57}, Fe^{56}, Mn^{55}, Cr^{53}. They also measured Al^{27} and B^{11} were detected. Standard solutions were blank, 2.50, 5.00, 10.0, 25.0 and 50.0 ppb for Zn^{66}, Fe^{57}, Fe^{56}, Mn^{55}, Al^{27}, and B^{11}. Standards for U^{238} and Sb^{121}

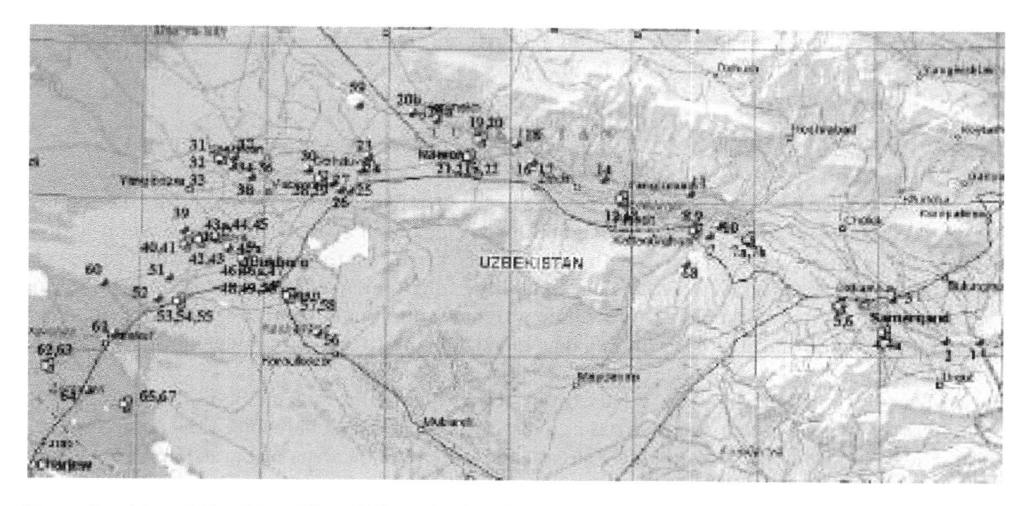

Figure 1. Map of Zarafshan River Valley with location of the sampling points and target area.

were blank, 0.05, 0.10, 0.20, 0.05 and 1.00 ppb. Soil and vegetation analysis was conducted by X-ray-spectral, instrumental-activation, absorption and chromatography methods in Central Laboratory of Samarkand Geology. Using original programs, we conducted statistic processing and built computer maps based on Arch Gis 3.1 (ARC Hydro, 2002).

3 SIMULATION OF SPATIAL-TEMPORAL DISTRIBUTION OF KEY HABITAT FACTORS IN THE GIS

A coupling of the Zarafshan River basin model with the ZerGIS was developed for the spatial evaluation of the effect of long-term water availability on the Environmental state. The river network model was georeferenced by manually assigning the arcs of the main tributaries along the Zarafshan River. The indexes used for arcs in the ZerGIS model network were transferred as attributes to the polylines. This enables direct linking of the modeled water flow values for the Zarafshan River and its main tributaries into a simulation model with the GIS. The length of the river stretches in its southern part involved huge areas of Bukhara oasis and Karakul Plateau where Zarafshan River disappears in the sands. So, our water quality assessment will be performed mostly for these areas. Therefore all monitored water samples points are summed and withdrawn at three large parts in the river network, as upper stream points (from 1 to 22); middle stream (from 22–38), and downstream areas (from 38 to 67). Thus, the resulting water distribution in the GIS is only an approximation of the spatially explicit water availability. The ecological evaluation has been performed by using the environmental models from where such resolution is mostly efficient (Lauver et al., 2002).

4 DISCRIPTION AND CRITARIA OF SELECTION

Representativeness. The chosen geographical area gives us unique opportunity to analyze all statistical data of water distribution. Selected points are located in the divergence area of Zarafshan River that allows to fully utilizing unique capabilities of GIS. Allocating data to each point, we can quickly visualize the perspective of the overall current ecological status.

Each set of points is on the same altitude, which eliminates the effects due to difference in altitudes, and lets us analyze solely the factors in focus.

5 METHOLOGICAL WORK ON USING GIS FOR ZARAFSHAN RIVER VALLEY

Many attempts have been undertaken to integrate distributed hydrological models with GIS, mainly for spatial data retrieval and preprocessing and for visualization of model results. Several hydrological software tools have interfaces with GIS. GIS can provide information on channel geometry, terrain and channel slope, soil type, land use, spatially explicit micro climate, etc. that are important inputs, e.g., precipitation models. To facilitate a tighter coupling beyond data preprocessing and visualization, hydrological tools have been developed that allow for hydrological modeling directly in the GIS (e.g., ArcHydro, 2000).

However, for our investigation, we used only three main Blocks Data Bank, collected by us during last years. They were summarized as average daily and seasonal water flow data; agrometeorological data for upper stream, down stream and lower stream areas of Zarafshan River Valley and multi-annual data of physical and chemical characteristics of water samples, analyzed across Zarafshan River course. Target area was vectorised by using maps at 1:100 000 scale. Such approach allowed us to produce a digital mapping of the irrigated agriculture area of the Zarafshan River Basin (ZerGIS map), which we use as a matrix for further mapping of physical and chemical water parameters (Figure 2).

6 RESULTS AND DISCUSSION

6.1 *Current assessment of water resources management in the Zarafshan River Valley*

Zarafshan River Basin with total area of 17 700 km^2 covers the territory of Tajikistan and Uzbekistan. Only 31% of river's total flow of the river has an underground source, the rest comes from glacier and mountain snow. The river basin is transitional, with the upper 300 km of the river located in Tajikistan. Although most of the annual river discharge originates in Tadjikistan, only 8% of the water is used there with the rest of water fully used by Uzbekistan. The Zarafshan River Basin is densely populated, second to Fergana and North-Eastern Uzbekistan. Population density of the basin is 80–100 km^{-2}. Zarafshan River basin is the most ancient heart of agricultural and urban civilization of Central Asia. The largest cities are Samarkand, Bukhara and Navoi. The basin is located in the center of Zarafshan inter-mountainous depression bordering with Nuratau Mountains in the North-East, Zarafshan ridges in the South-West. The length of depression is up to 180 km; width is from 10 to 60 km inclined to North-West. Absolute elevation points from 720 to 280 m.

The river is the sole source of potable water for much of the region, yet it receives municipal, industrial and agricultural wastes as it passes through the Samarkand, Navoi and Bukhara main industrial provinces in Uzbekistan. From Navoi to the Bukhara and Karakul oasis, the stream of Zarafshan becomes of collector-drainage system type. Surface of water there is 4–6 m deeper from surface of soils formed by fine-fractured material.

As is shown on Figure 3, Zarafshan is a terminal river basin, running toward Bukhara, it reduces its water volume because of the irrigation and runs through this vast irrigation land with increasing minerals and metal content, and diminishes in the neighborhood of Karakul, trying in vain to reach Amu Darya. Irrigated area of Zarafshan basin was 557 000 ha. Annual water volume with multi-aqueous years from 1953 to 1998 measured at Dupuli (upper stream) is much higher than water volume flowing into the lower reaches of the main stream of the river (Ziaddin, Kazalinsk and Navoi). Data were summarized and shown in Figure 4.

This feature determines the favor for irrigation within annual flow distribution where 80–90% of the annual flow is generated from April to October; the maximum flood falls from June to August and within last years the maximum daily was noted in 1998, in the upper stream of the Zarafshan River (Figure 5).

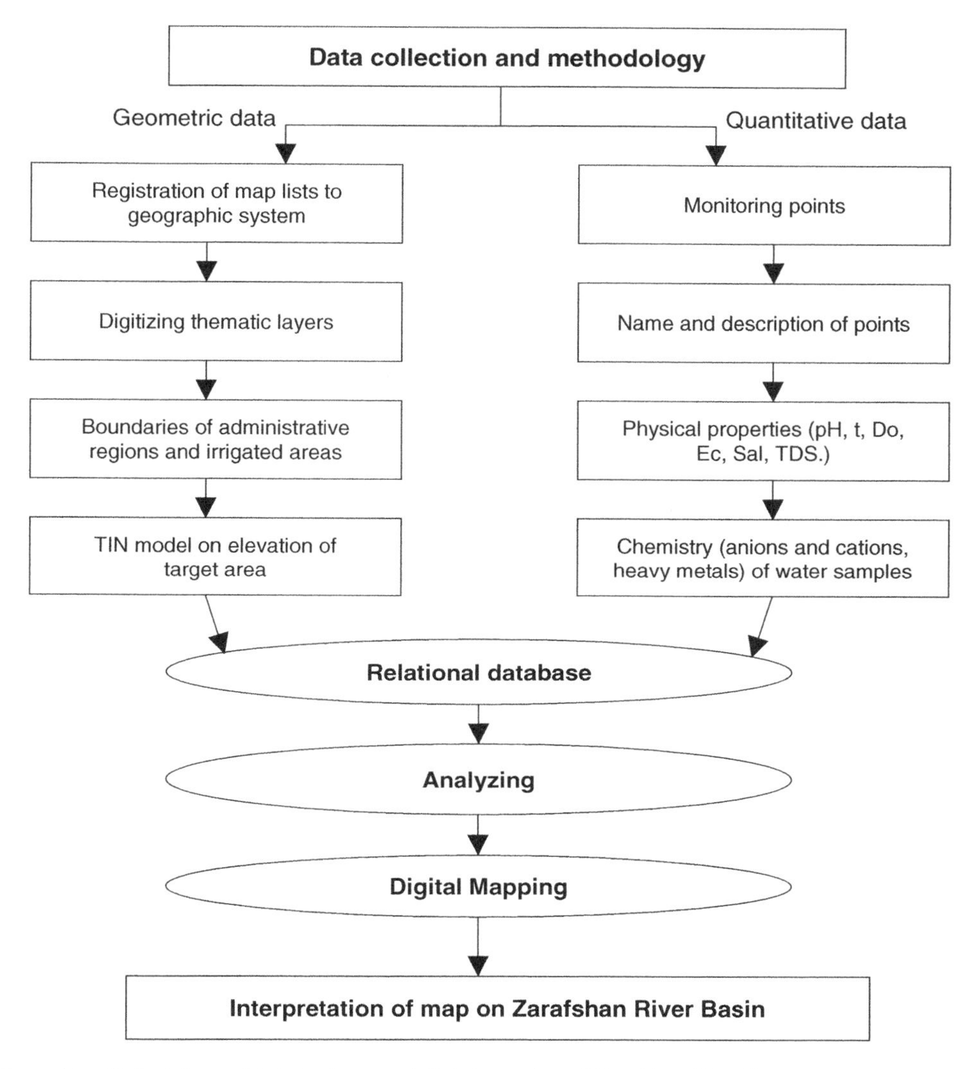

Figure 2. Methodology of data bank use and digital mapping of Zarafshan River Basin.

6.2 *Water Quality Assessment and Monitoring*

Results of reduced chemical analysis of water samples (67 points) analyzed by us during 2000–2004, indicate that mineralization of river's water changes within surveyed area from 0.3 to 2.7 $g\,l^{-1}$. Down the stream from mountains to Navoi meridian mineralization increases from 0.3 to 1 g/l and then up to Bukhara oasis it reaches 2.6 $g\,l^{-1}$. In the same direction the chemical composition of water changes – hydrocarbonate ion decreases and sulphate ion increases. Mineralization level of collector-and-drainage water, broadly used within Bukhara oasis, is higher and ranges from 2.5 to 4.9 $g\,l^{-1}$. Lower mineralization (0.6–0.7 $g\,l^{-1}$) occurs in canals water taken from Amudarya River and used for irrigation and partially for potable water supply. It was concluded that the HCO_3^- content is decreasing and the Cl^- and SO_4^{2-} are generally increasing from Navoi to Bukhara. For example, Figure 6 shows a gradual increase of cations and anions (Na^+ as a particular case) as water moves downstream in the Zarafshan River with a maximum in Buchara oasis being distinctly visible.

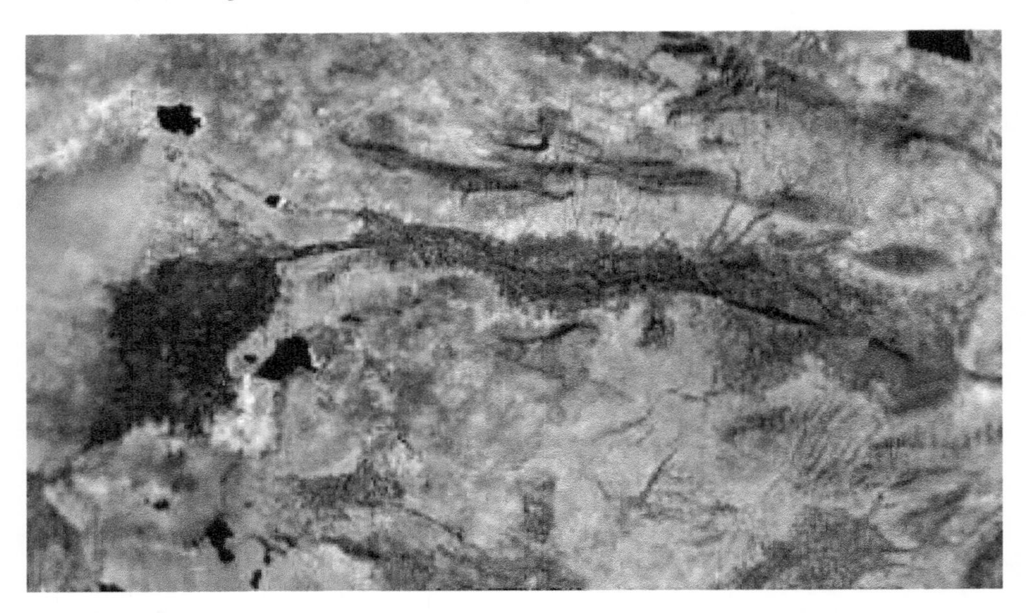

Figure 3. View to the Zarafshan River Valley from space (taken from Google maps).

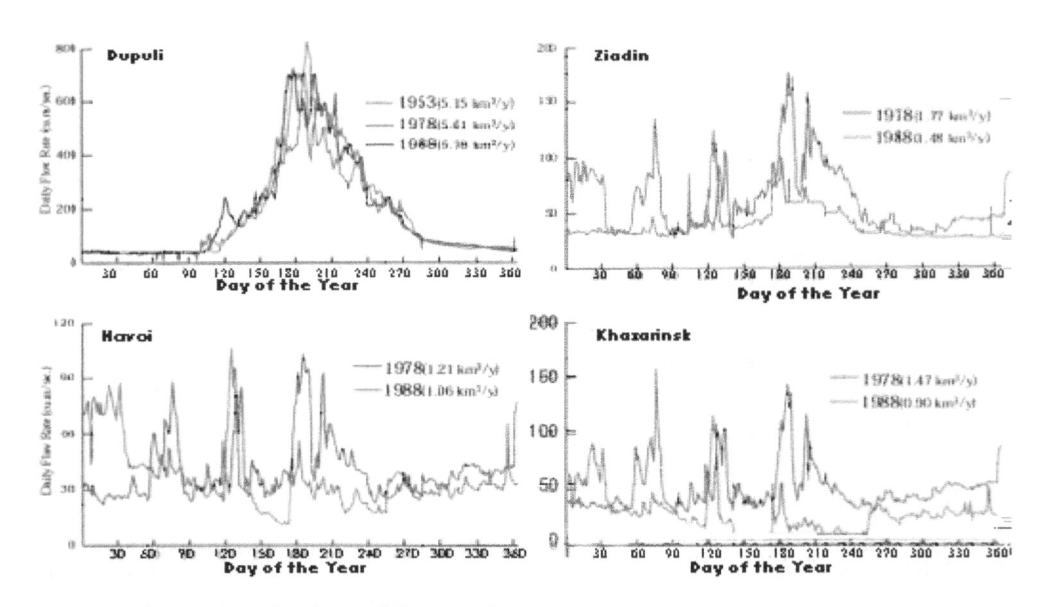

Figure 4. Water volume flowing on different stations.

Salinity also reflects the water quality in such arid areas as Zarafshan basin. The river keeps the lowest salinity up to Kattakurgan; almost same lower salinity is seen between Kattakurgan and Navoi except an inflow from Karatau Mountain. The remarkable feature of saline distribution at many tributaries after passing Navoi towards Bukhara into Karakul plateau is the existence of sandy-clayey conglomerates, sandy hills and saline lakes. Maximum value of salinity was of 5.57 g l^{-1} in sampled water from various collectors in the Karakul plateau and Kuimazar canal (near

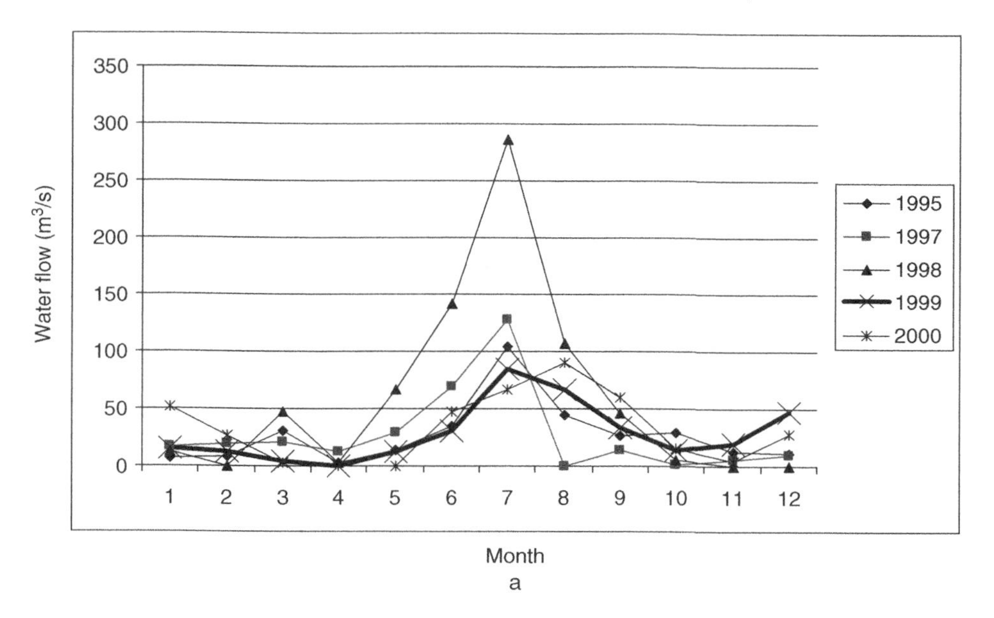

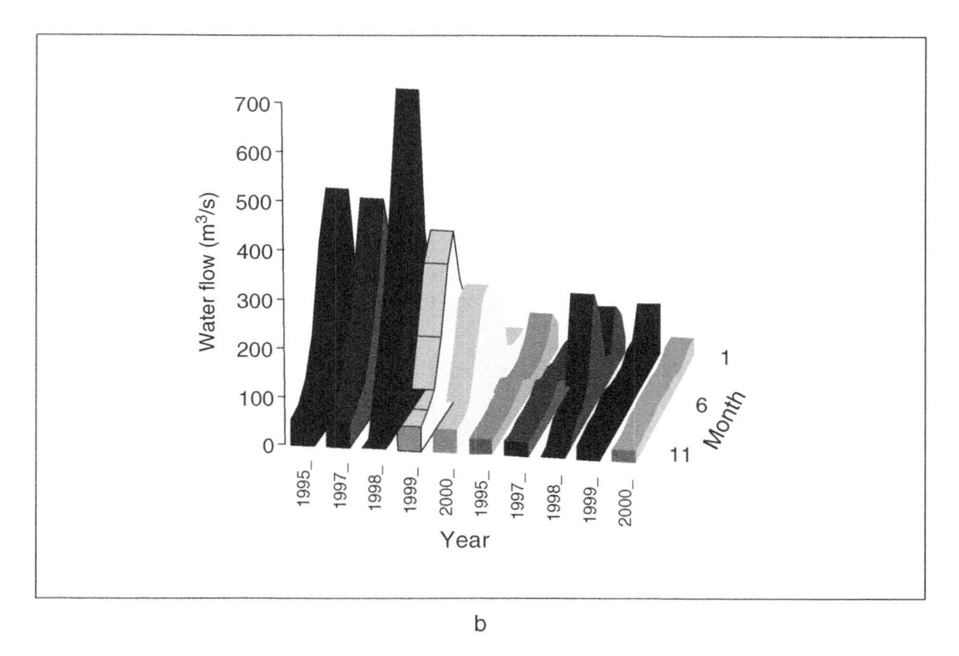

Figure 5. Annual water flow through years of observation.

Bukhara) to which there was a discharge from cotton field. At sampling sites such as Kanimekh, Kyzyltepa, Galasiya, Kuymazar canal, Karaulbazar, Yamanzhar, Aktepa, Shaikhan, as well as some waterways that flow through Shafircan district and some sites of Karakul plateau where salinity was comparatively high, Ec and other anions and cations correspond with their high concentration. This might be caused both by the usage of some chemicals at the agricultural field in harvest season and/or by the discharge of some technogenic chemicals from industry, which are especially

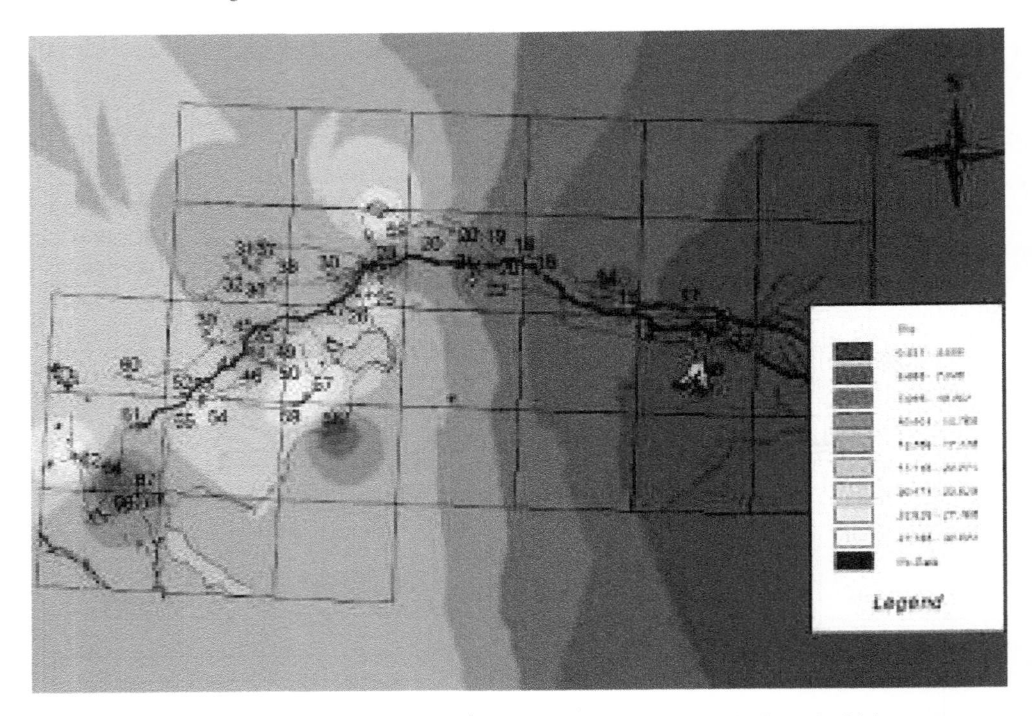

Figure 6. Spatial – temporal distribution of Na^{+} cations; light-brown colour indicate the highest concentration in the lowest stream of Zarafshan River (taken from GIS model).

concentrated in the lower reaches of Zarafshan River Basin. Our results also indicate that alkalinity of water samples changes within surveyed area from 2.00 up to 6.80 meq l^{-1}. Downstream from the mountains to the Navoi meridian, alkalinity increases from 3.52 meq l^{-1} reaching maximum in sandy Kyzylkum desert at values of 38, 43, 57 and 65.

6.3 *Dynamics of salinity level of surface waters of Zarafshan River basin for the period of 2001–2005*

The river waters generated in the mountain areas are of high quality, with salinity levels generally in the range 0.15 to 0.25 g l^{-1}. Salinity levels generally increase with progression downstream, as a result mainly of the salt load in the return flows from irrigated areas discharged via the collector drains. Thus, in the lower reaches there have been significant increases in salinity over time with the expansion of irrigation. Salinity levels have now stabilized, and in fact over the last decade (1991–2000) there has been a drop in mean annual values of salinity in the middle and lower reaches of both rivers (Figure 7). This is attributable to a decrease in drainage flows related to the changes in water management and economic conditions in the region.

Sulfate ion well correlates with salinity and with other cations and chloride ions. The gradual accumulation of NH_4^+ within the investigated area of the basin from Tajikistan border through Bukhara oasis into Kyzylkum sandy desert are well correlated with the reduction of water volume, low water flow rate and decreasing of oxygen marked at the same point. It was also found out that at increasing of water volume in water reservoirs or river accelerate the accumulation of biogenic elements due to the leaching of remains from agricultural fields (Nikolaenko, 1975).

Remarkable feature of saline distribution is at many tributaries after passing Navoi. Maximum value of salinity was of 6.4 g l^{-1} at site where we sampled water from Kuimazar Canal to which there was a discharge from cotton fields. Ec also showed the highest value at the same site. At

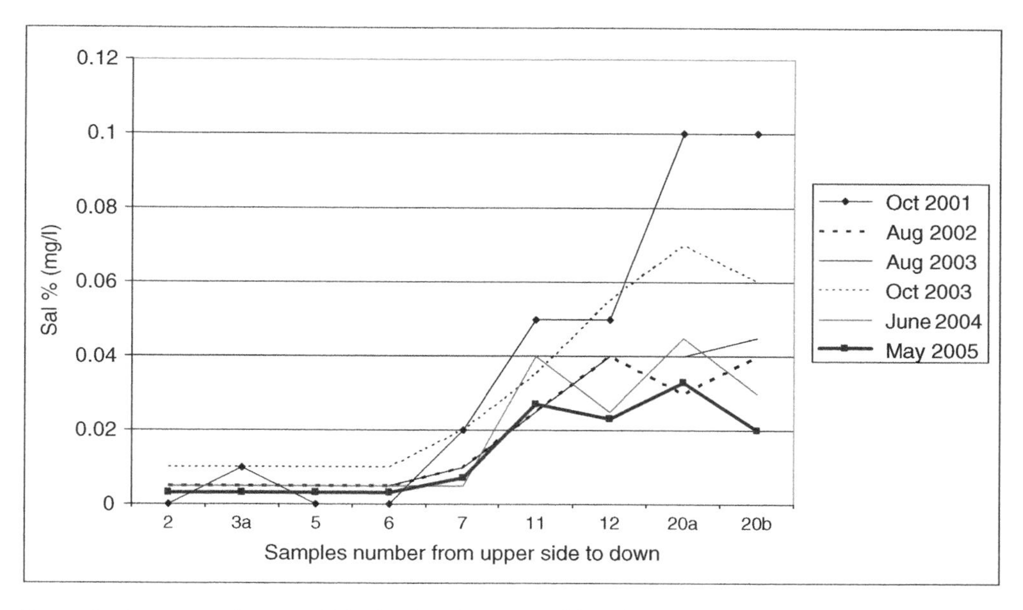

Figure 7. Dynamics of water salinity level on the course of Zarafshan River Valley (2002–2005).

Table 1. Correlation coefficients between each item.

Item	NH_4^+	K^+	Mg^+	Ca^{2+}	Cl^-	$N\text{-}NO_3^-$	SO_4^{2-}	pH	EC	Salinity	Alkalinity
Na^+	−0.11	0.79	0.84	0.94	0.96	0.20	0.93	−0.38	0.96	0.97	0.28
NH_4^+		0.01	0.01	−0.05	−0.08	0.23	0.03	−0.18	−0.10	−0.11	0.13
K^+			0.94	0.84	0.66	0.30	0.82	−0.59	0.84	0.83	0.60
Mg^+				0.90	0.71	0.46	0.90	−0.68	0.91	0.89	0.67
Ca^{2+}					0.85	0.36	0.96	−0.50	0.95	0.94	0.38
Cl^-						0.09	0.88	−0.26	0.90	0.91	0.11
$N\text{-}NO_3^-$							0.41	−0.50	0.32	0.29	0.69
SO_4^{2-}								−0.54	0.95	0.95	0.42
pH									−0.51	−0.49	−0.69
Ec										0.99	0.45
Salinity											0.40

sampling sites 25, 35, 43, 47, 53, 54, 57, 61, 65, and 67, where salinity was comparatively high, Ec and anions and cations corresponded with their high concentration. This is the main reason of high correlation between each ion concentration that is shown in Table 1. This might be caused both by the usage of some chemicals at the agricultural field in harvest season and by the discharge of some technogenic chemicals from industry.

Micro component content of surface water in Zarafshan basin includes higher level of iron, manganese, potassium, cadmium, copper, nickel, selenium in unfiltered samples. When filtered, content of most of the heavy metals in water does not exceed various guidelines from WHO, EEC and others. That means heavy metal content in water is directly related to mechanical suspension. Electronic microscope study of mechanical suspension of Zarafshan River indicated contrast size of fractured material. Prevailing size of fractures is from 5 to 10–15 μm. Mechanical suspension near Samarkand has significant sizes, poor roundness, and rich content of quartz, feldspar, and miscellaneous minerals. In ore and sulfate minerals some indications of pyrite, magnetite, hematite, barite, celestine, sphene, antimonite are found. Size of ore minerals varies from 0.5–1.3 μm.

Water samples taken from the eastern part of Zarafshan showed higher concentrations of antimony up to 0.5 LPC. This probably resulted from the mining operation in Ayni town (39°23′N, 68°32′ E: 80 km east of Pendjikent), Tadjikistan that mines the Jijakrut mercury-antimony deposit. Microprobe analysis of suspension on filters detected frequent traces of antimonite of 1–3 μm in size. Sampling for pesticides and organic compounds showed presence of butiphos exceeding LPC.

7 CONCLUSION

GIS couple with hydrological modeling assists in delineation of the river network. Decision Support Systems have benefited from the incorporation of GIS in the aspects of database, interface and model integration. Database created by us it not only brings the spatial dimension into traditional water resources use and water quality multi-annual dynamic Data Base, but also has the ability to integrate various social, economic and environmental factors related to water resources planning and management. A critical analysis of the adequacy of the existing data is required, and an accurate assessment of all available water resources. Precipitation records and patterns and other climatic factors are to be studied, and consideration is to be given to long-term changes in both local and general climatic conditions. Later phases are to finalize the water and salt balances and prepare draft regional policy, strategy and action programs.

REFERENCES

Agaltseva, N. 2002. The assessment of climate changes impact on the existing water resources in the Aral Sea Basin. In the book: Dialoque about water and climate: Aral Sea as a particular case. Glavgidromet, Tashkent: 3–59.

Arc Hydro. 2002. GIS for Water: Tutorial and Guide. CD-ROM Copyright © 2002 ESRI.

Fayzieva, D. 1999. Some aspects of the integrated information system for health impact assessment in Zarafshan River Basin. J. Questions of Cybernetics, Tashkent, V. 158:154–161.

Fayzieva, D., L.A. Lutfullaev, and I.A. Usmanov. 2001. Influence of Anthropogenic Pollution of the river Zarafshan on Intestinal Infections morbidity of the population. p. 227–235. In: Water Pollution – YI. WIT Press, Southampton, Boston.

Glantz, M.H. 2002. Water shortages and climate change collide in the Amudarya basin in Give and Take, ISAR (Initiative for Social Action and Renewal in Eurasia). Washington, DC, USA, 5(2/3):27–28.

Khujanazarov, T.M., T. Tsukatani, K.N. Toderich, and M. Abdusamatov. 2005. Introductory assessment of transboundary water resources in Central Asian Riparian Countries. Journal Agricultural Metereology 60:621–625.

Ko, S., V. Aparin, Y. Kawabata, K. Shiraishi, M. Yammamoto, M. Nagai, and Y. Katayama. 2006. Application of ICP-MS on analysis of water quality in Zarafshan River. Journal Arid Land Studies 15–4:375–378.

Lauver, C.L., W.H. Busby, and J.L. Whistler. 2002. Testing a GIS model of habitat suitability for a declining grassland bird. Environmental Management 30(1):88–97.

Nikolaenko, V.A. 1975. Content of biogenic elements and organic substances in the water of Charvak reservoir (Uzbekistan). Bulletin of Glavgidromet 2(83):17–27.

Toderich, K.N., R.I. Goldshtein, V.B. Aparin, K. Idzikowska, and G.Sh. Rashidova. 2001. Environmental state and an analysis of phytogenetic resources of halophytic plants for rehabilitation and livestock feeding in arid and sandy deserts of Uzbekistan. p. 154–165. In: Sustainable Land Use in Deserts. Spriger Publisher, Berlin.

Toderich, K.N., N.P. Yensen, Y. Katayama, Y. Kawabata, V.A. Grutsinov, G.K. Mardonova and L.G. Gismatullina. 2006. Phytoremediation technologies: using plants to clean up the metal/salts contaminated desert environments. Journal Arid Land Studies 15S:183–186.

Tsukatani, T., and Y. Katayama. 2001. Water Quality of Zarafshan River Basins. Discussion paper No 527, Kier, Kyoto University Publisher, Japan. 28 p.

CHAPTER 31

Remote sensing application for mapping terrestrial carbon sequestration in Kazakhstan

U. Sultangazin, N. Muratova & A. Terekhov
Kazakh Space Research Institute, Almaty, Kazakhstan

1 INTRODUCTION

Typical landscapes of northern Kazakhstan are steppe and forest-steppe. Total area of such regions exceeds 1 m km^2, about 30% of the total area of the Republic. Carbon sequestration in vegetation in these regions can be achieved by sequestration in natural grasses and in cultivated crops. The most fertile parts of this region (more than 10 m ha) are used for cultivation of spring crops. So an estimation of CO_2 sequestration requires development of methods of estimating productivity of local vegetation, and control of the steppe fires which are a common occurrence in this zone. The satellite data are an objective source of the information for monitoring vegetation growth and fires in the steppe regions of Kazakhstan, and are useful to the decision-making process relevant to CO_2 sequestration.

The remotely sensed estimation of vegetative cover is based on spectral responses registered on satellites, and depends on the canopy cover transparency and color, and also plant architecture. For example, the plane, large leaves located parallel to the ground give different responses than the system of vertically located narrow leaves. It is assumed that for an identical soil type and humidity and the specific vegetation type, there is a unique relation between leaf cover area, evaluated by remote sensing data, and biometric parameters of vegetation (biomass, density, etc.).

Cultivation of cereals is rather effective and most widespread for the drought-prone climate in Northern Kazakhstan. There exists a strong relationship between weight of the grain and the remainder of the biomass (the harvest index) for local cereals. Therefore, the general approach of using the satellite information for estimating the productivity of crops in Northern Kazakhstan is relevant to assessing the above-ground dry vegetative biomass estimation in this region.

This chapter describes the basic results of application of the satellite data to the identification of arable lands in Northern Kazakhstan, planted areas and to estimation of productivity. Assessment of the amount of CO_2 sequestration by crops and nature grasses is also discussed.

2 IDENTIFICATION OF CROP LANDS DETECTION IN NORTHERN KAZAKHSTAN

More than 10 m ha of arable lands are under spring crops in Northern Kazakhstan. Other types of croplands are of minor importance. The estimation of the amount of grain production is based on two basic factors: area planted and crop yields.

Large sized cereal fields (400 ha) and their close proximity provide an opportunity to use TERRA/MODIS satellite data with middle resolution (1st and 2nd channels, 250 m) for plant state control during vegetation season. Swath width of about 2200 km of TERRA or AQUA /MODIS and daily monitoring of the agriculture territories by two satellite systems can provide detailed

Figure 1. Land use map of the Shortandy district of Akmolinskaya oblast in 2001 (built on the base of satellite data).

information about vegetation spectral dynamics. Identification of cultivated crop fields is considerably facilitated by the predominant land area under spring crops which occupy more than 90% of arable lands.

The kowledge of the planted area is important to obtaining estimates of grain production and CO_2 balance. The quantitative estimation of the area under spring crops is based on the results of joint analysis of EOS MODIS thematic processed images and GIS data of administrative districts. The type of fields (cropland, pasture) is identified by the satellite data and their areas are estimated from the land use map (Figure 1).

Recognition of spring crops in the steppe zone is based on satellite data analyses during two periods – from May to the beginning of June (planting and seedling stage), and from second half of July to beginning of August (flowering stage). At this time, the maximum spectral differences are observed between spring crops, and other land use and natural grasses (Figure 2).

The widely practiced crop-fallow rotation in Northern Kazakhstan leads to thee classes of the basic land uses in the region: spring crops, fallow and abandoned land. The separation of spring crop fields and natural grasses can be made when a grain crop is fully developed. Identification of the fallow fields is based on significant differences in the spectral characteristics of these land uses during the season of vegetative growth (Figure 3).

Water deficit, caused by climatic features of Kazakhstan steppe zone, is the major factor determining crop productivity. The second important factor is weed infestation. High cost of herbicides, and low price of crop grains, necessitates the use of mechanical systems and fallowing of crop fields to control weeds and increase soil nitrogen stock. Thus, fallowing is widely practiced in the region.

Identification of the basic types of land uses within the 4-year period (2002–2005) allowed restoration of adequate number of cultivation cycles after fallowing of separate fields (Figure 4). These cultivation-fallow cycles play an important role in forecasting crop productivity.

3 ESTIMATION OF CROP PRODUCTIVITY

Estimates of crop yields are made during the period of flowering/heading one month before harvest, as the expected amount of grain production. Algorithm of the forecast is based on the information received during the annual ground routing survey of croplands and analysis of the synchronous

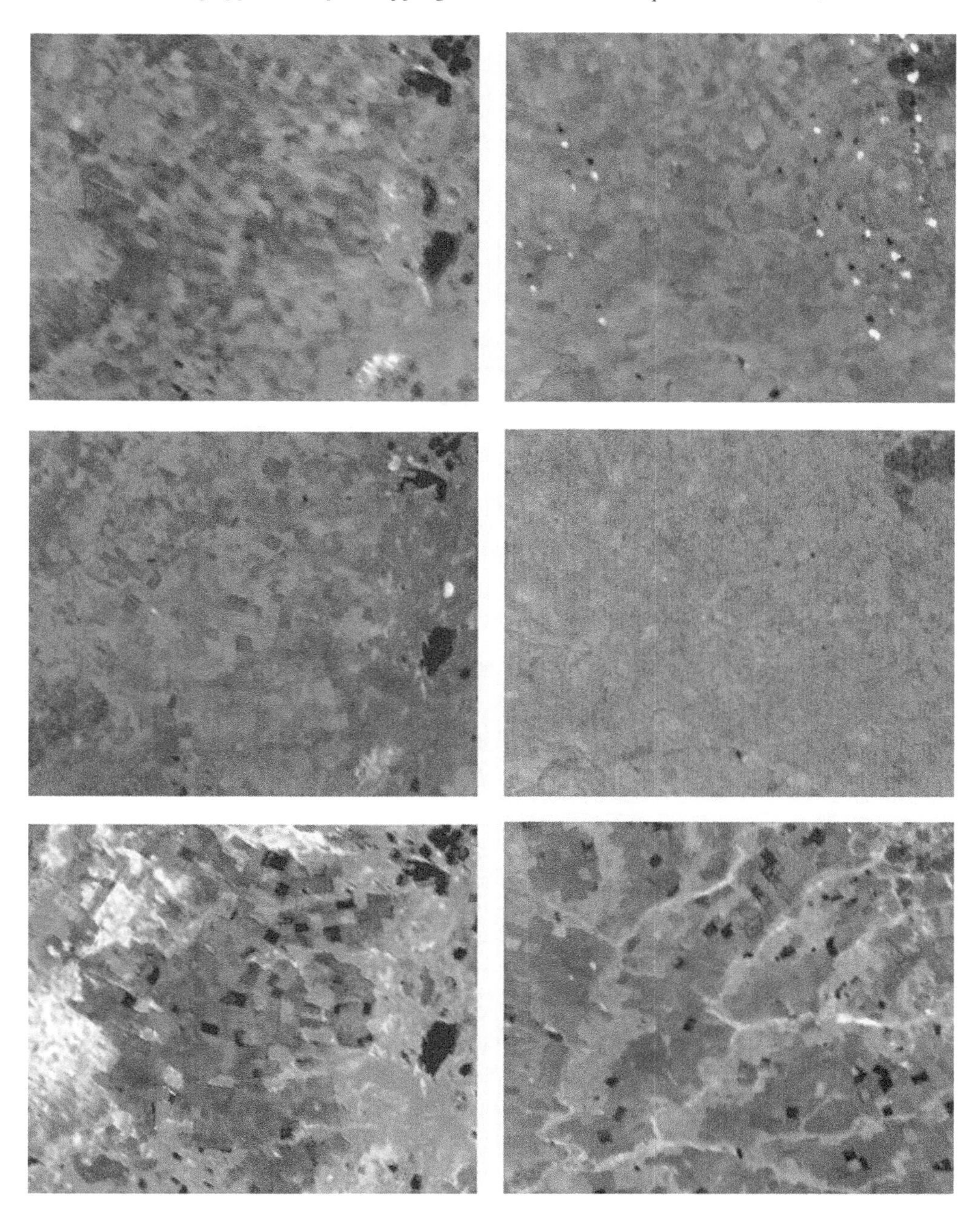

Figure 2. Typical MODIS data (channel – 1,2; resolution 250 m) for agricultural territory of Northern Kazakhstan during more informative time periods. Upper left – RGB-211 composite, May 23, 2001; Upper right – RGB-211 composite, July 29, 2001; Middle left – correspondently composite RGB-222; (July: May: July), territory of Kostanai oblast; Middle right – RGB-211, composite, June 5, 2002; Bottom left – RGB-211, composite, August 3, 2002; Bottom right – correspondently composite RGB-222; (August: June; August), territory of Akmola obalast.

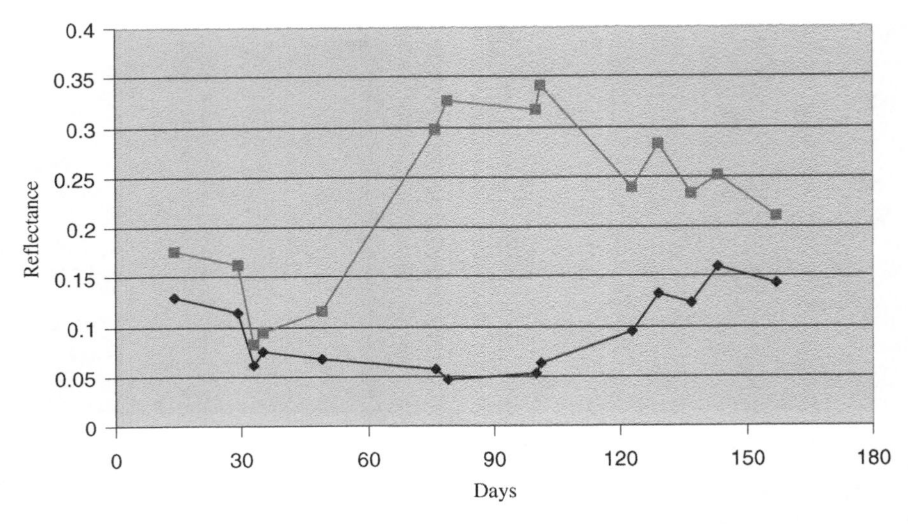

Figure 3. Reflectance values of typical wheat (A) and fallow (A) fields in Akmola oblast during 2004 vegetation season (zero point is 1st May). EOS MODIS data, 2nd channel: 841–876 nm.

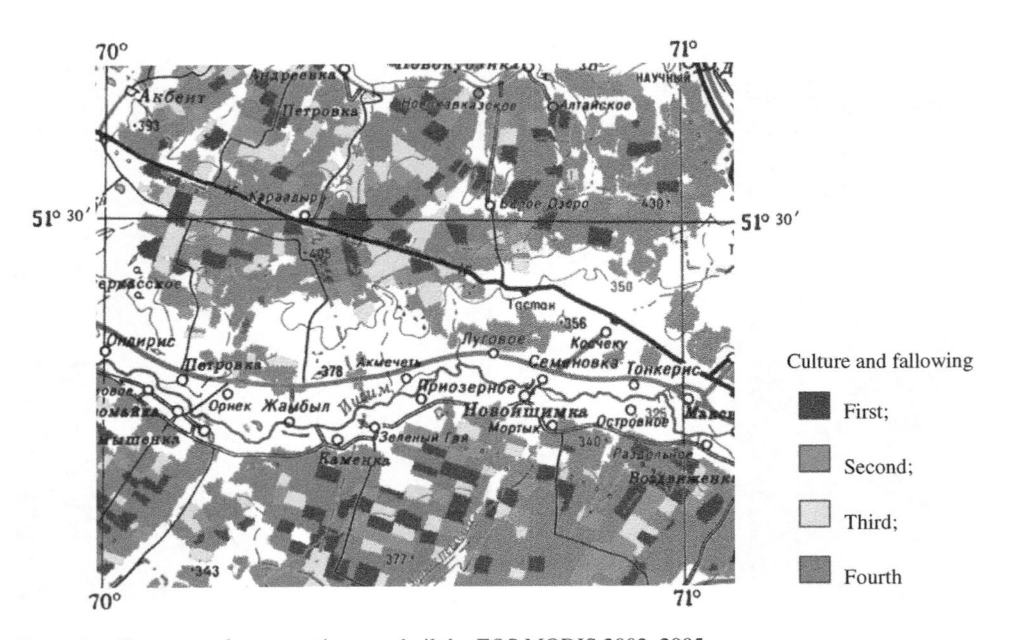

Figure 4. Fragment of crop rotation map built by EOS MODIS 2002–2005.

EOS/MODIS satellite information: 1 (620–670 nm), 2 (841–876 nm) channels with resolution of 250 m. Wheat biometrical parameters and expected yield on surveyed fields are collected during ground observations. The EOS/MODIS monitoring information is processed by obtaining the cloudless images covering all territory during earing-flowering stage. It can be one space image (cloudless weather), or mosaic from several images when there are problems with the cloud cover.

Grain production and productivity is estimated by a statistical linear model with use of satellite vegetation index. A wide range of indices are used depending on weather conditions, state of vegetation, soil cover and the climatic/weather factors. Soil moisture, vegetative cover, dust content, and atmospheric transparency are dynamic parameters. Thus, different types of vegetation indices are used to estimate different crop growth scenarios. For example, soil contribution to a total signal from crop fields is more in dry years of low canopy than during the wet years of 100% canopy cover. If the contribution of soil is high, the condition of vegetation is described more precisely by the specialized index such as Soil Adjusted Vegetation Index. So the choice of optimum vegetation index (VI) occurs among seven basic types: Near Infrared Red (NIR), Ratio Vegetation Index (RVI = NIR/Red), Normalized Difference VI (NDVI = (NIR−Red)/(NIR−Red)), Infrared Percentage VI (IPVI = NIR/(NIR + Red)), Difference VI (DVI = NIR−Red), Soil Adjusted VI (SAVI = (1 + L)*(NIR − RED)/(NIR + RED + L)), Global Environmental Monitoring Index (GEMI = v*(1 − 0.25*v) − (RED − 0.125)/(1−RED), where v = $[2*(NIR^2 - RED^2) + 1.5*NIR + 0.5*RED]/(NIR + RED + 0.5)$).

On the basis of the available ground and satellite information, the set of calibration curves is developed relating the expected productivity to satellite vegetation index. The index that has the best correlation coefficient with the ground data is used for estimating crop productivity. For example, in 2005 correlation of indexes with the ground data was as follows: DVI ($R^2 = 0.64$); NIR ($R^2 = 0.63$); GEMI ($R^2 = 0.62$); SAVI ($R^2 = 0.61$); RVI ($R^2 = 0.60$); IPVI ($R^2 = 0.55$); NDVI ($R^2 = 0.55$). It is important to note that procedure of estimating the total grain production evaluation includes additional correction for the weed infestation factor. However, for estimating the CO_2 sequestration total vegetative biomass is important which can be estimated through basic calibration equation without any additional correction.

4 HARVEST OF SPRING CROPS AND ITS SPECTRAL PATTERN

There are more than 10 m ha of area under spring crops in Northern Kazakhstan, and it is a significant part of vegetative cover of this region. Cultivation of spring grain crops plays an important role in the CO_2balance. The dry vegetative remains or crop residues are exposed to a range of various transformations during harvest, which causes additional flux of CO_2 emission or sequestration. There are three modes of crop biomass transformations after its harvest: (i) cutting and scattering on a field; (ii) collection and removal from a field; and (iii) burning at the time of field clearing. The satellite imagery is necessary for the account of actual harvest practice and estimation of CO_2balance as affected by agricultural production.

Harvesting of spring crops in Northern Kazakhstan takes place 90–95 days after planting. Depending on the weather conditions during the vegetation period, the harvest can begin from the third week of August (hot droughty years) and finish in middle of October (cold wet years). The harvest is usually completed over 30–40 day period. Harvesting is done by two ways: direct or separate combine harvest. Lodging into swaths at separate harvest allows harvesting to begin earlier. Faster grain drying in swaths leads to ending the harvest earlier, which reduces the risk of grain losses because of early frosts. Direct combine harvest causes lower losses of grain output and is more economical, since it is a one-pass field operation.

The spectral pattern of crop field before harvest is primarily determined by plant population, stalk height and degree of weed infestation. Crop field after harvest can have a range of various spectral images. After felling down into swaths or direct combine harvesting with mechanical straw scattering, the field striped structure is formed with the characteristic spatial scale corresponding to width of a harvester (usual 6–9 m). Remote sensing data are capable to fix similar structures if the spatial resolution of an image is less or equivalent to width of a harvester. For example, IRS-PAN data with 6 м resolution is suitable for registration of this structure (Figure 5A). The field characteristic structure after harvest is completely lost in IRS-LISS data with

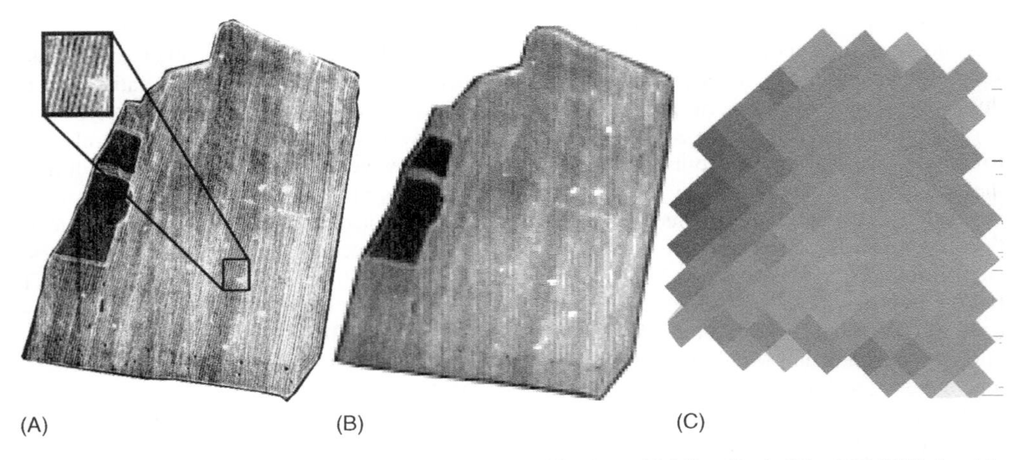

Figure 5. Space images of crop field in September, 17 2004: (A) – IRS-Pan (6 m); (B) – IRS-LISS, band 2: 620–680 nm (23 m); (C) – MODIS, band 1: 620670 нм (250 m).

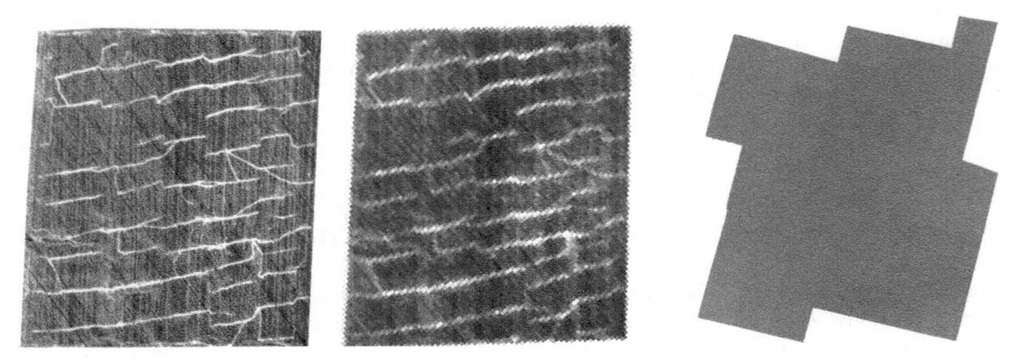

Figure 6. Space images of crop field in September, 17 2004: Left – IRS-Pan (6 m); Middle – IRS-LISS, band 2: 620–680 nm (23 m); Right – MODIS, band 1: 620–670 нм (250 m).

23 m resolution (Figure 5B) and especially MODIS with 250 m resolution (Figure 5C) due to spatial generalization.

The collection of straw on a field after harvest (direct combine harvesting with shocker or swaths thrashing by combine with shocker) forms another structure connected with system of shocks or traces after their collection and removal. Spatial scale of structure is determined by distance between shock sets, which depends on quantity of straw (height of plants and their density or number m^{-2}). The distance usually varies from 100 to 500 m with shocker's volume of about 2–3 m^3. Similar structures are fixed in details on IRS-Pan or IRS-LISS. However, these are not visible on MODIS images (Figure 6).

Recently, there has been an increase in use of direct combine harvesting with cutting and straw scattering by ventilator use. Straw scattering is recommended as a necessary procedure for maintenance of soil fertility and improvement of its structure. Scattered straw forms a high surface cover by dry vegetative biomass and accordingly has a high reflectance in visible and near infra-red parts of spectrum. Significant increase of reflectance on TERRA/MODIS images specifies with scattering of straw by harvesting (Figure 7).

In general, a wet growing season promotes luxurious crop growth, development of above-ground biomass, and production of a large amount of straw (Figure 8). After crop maturity, a significant

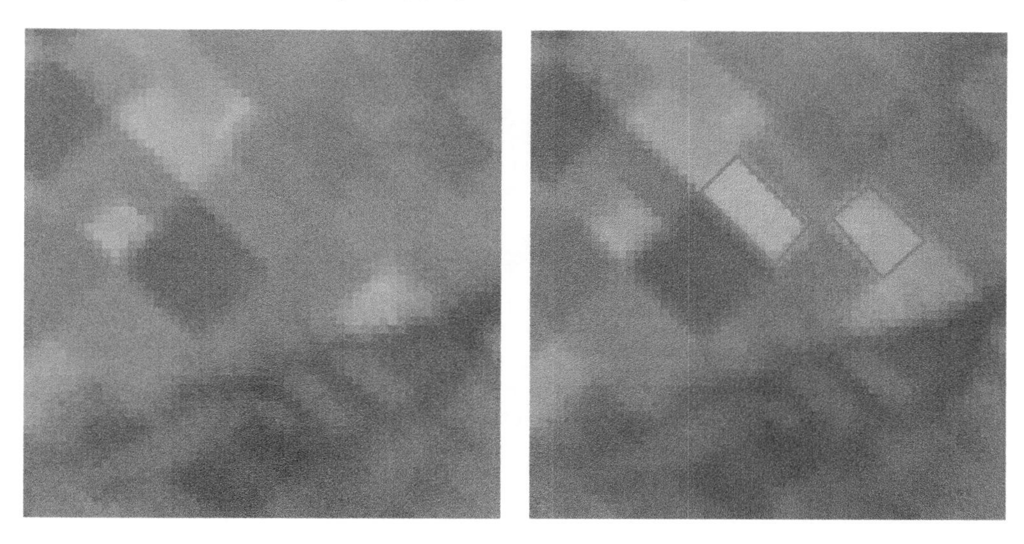

Figure 7. Example of spectral characteristics changes connected with crop harvest. EOS/MODIS composite (211) images in August, 26 (left) and August, 27 (right).

Figure 8. Typical crop fields in Northern Kazakhstan grown under deficit and adequate soil moisture content.

volume of dry vegetative material is formed on fields which create problems for the next sowing operation. One of the traditional ways of addressing this problem is to burn the crop residues. Biomass burning can be controlled or managed by altering of the biomass of natural grasses. Freely burnt land has specific spectral signature (Figure 9).

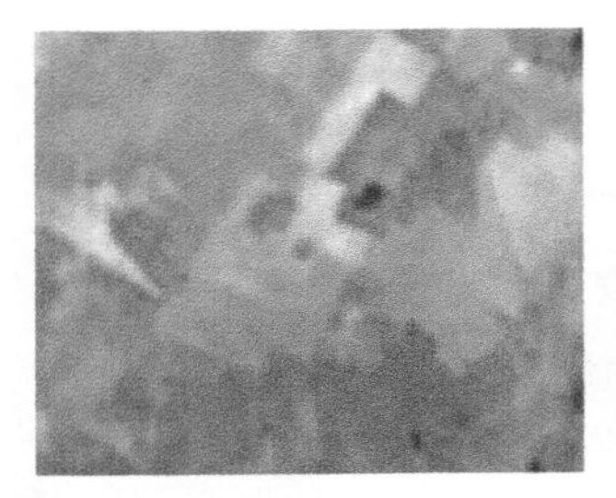  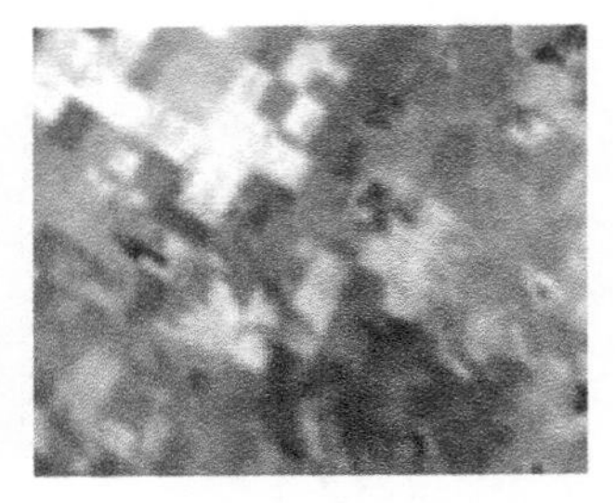

Figure 9. TERRA/MODIS images (composite RGB-211 channels, 250 resolution) during the crop maturing and harvest stages in 2002. Black regions represent freshly burnt areas.

Thus, the satellite data can be a source of the information about scales of application of various harvesting technologies which permit a precise estimation of CO_2 balance under a range of crop production ecosystems in Kazakhstan.

5 ESTIMATES OF CO_2 SEQUESTRATION UNDER NATURAL GRASSES

The natural vegetation of Kazakhstan steppe zone primarily comprises grasses alternated with low shrubs. Trees grow only in the wettest sites on rather small areas. Highly diverse vegetation with a range of canopy architecture make it difficult to precisely estimate the volume of dry vegetative biomass in the context of species composition, growth stage, and other conditions. Therefore, simplification of structure of vegetative cover is necessary.

The basic simplification is based on the assumption that cereals constitute the predominant vegetative cover in the steppe zone (Figure 10). All cereals are characterized by thin, vertical stem of 10–90 cm height, with rather small leaves, located on an acute angle on the stem. Similar structure of grassy plants is typical under droughty conditions.

Relatively equal distribution of the biomass between grains and other dry above-ground components of a plant is approximately observed for the spring crops in Northern Kazakhstan. This implies that the harvest index of most cereals is about 1. Therefore, strategies of estimating crop productivity are based on satellite vegetation indices calculated during periods of the maximal volume of green biomass. These strategies can also be used for estimation of the natural vegetation dry biomass. Adoption of this approach to steppes natural vegetation facilitates estimating the dry biomass (primary product) on the basis of seasonal maximum of satellite vegetative index. Regional crop yield data for a set of Kazakhstan steppe territories (West-Kazakhstan, Aktyubinsk, Kostanaiskaya and Akmolinskaya oblasts) for 1997 and 1998 was used to prepare a calibration curve relating seasonal NOAA/NDVI maximum with crop productivity and with the dry weight of above-ground parts of the steppe vegetation (Figure 11).

Weight of vegetation biomass of a steppe zone in dry autumn season is influenced by weather conditions during the year. The most important factor is moisture availability. Adequate precipitation can considerably increase the volume of the vegetative biomass (Figure 12). The second important factor in determining the volume of dry vegetative biomass in the autumn is steppe fires. During the wet years, the grassy vegetation forms well-developed close cover, which is very prone to fire after summer drying. Fires can burn to hundreds of thousands of hectares (Figure 13). Common causes of fires, as a general rule, are the activities connected with agricultural production, transport, careless handling of fire and natural factors. Fires are the cheapest way of pasture cleaning and eliminating unwanted grasses. However, together with pastures, millions of hectares of natural vegetation are also burnt. For example, more than 30% of all territories of Kazakhstan steppe zone were burnt during 2002.

The burnt areas in Kazakhstan steppe zone and the volume of burnt out organic material were estimated by using the EOS MODIS data (Figure 14). Vegetation on more than 20 m ha was destroyed by fire in 2002, and the amount of CO_2 emission was about 30 m t.

Figure 10. Steppe typical landscape (left) and different crop fields (right).

6 CONCLUSIONS

Kazakhstan has the 9th largest land area in the world. Grasses and cereals grown in steppes, natural and cultivated, have a high potential of CO_2 sequestration. The potential for enhancing carbon (C)

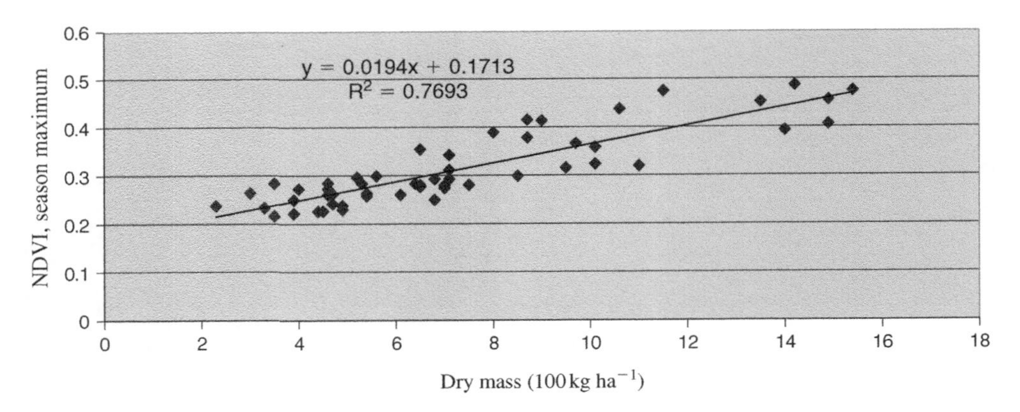

Figure 11. Relationship between biomass yield and the NOAA/NDVI.

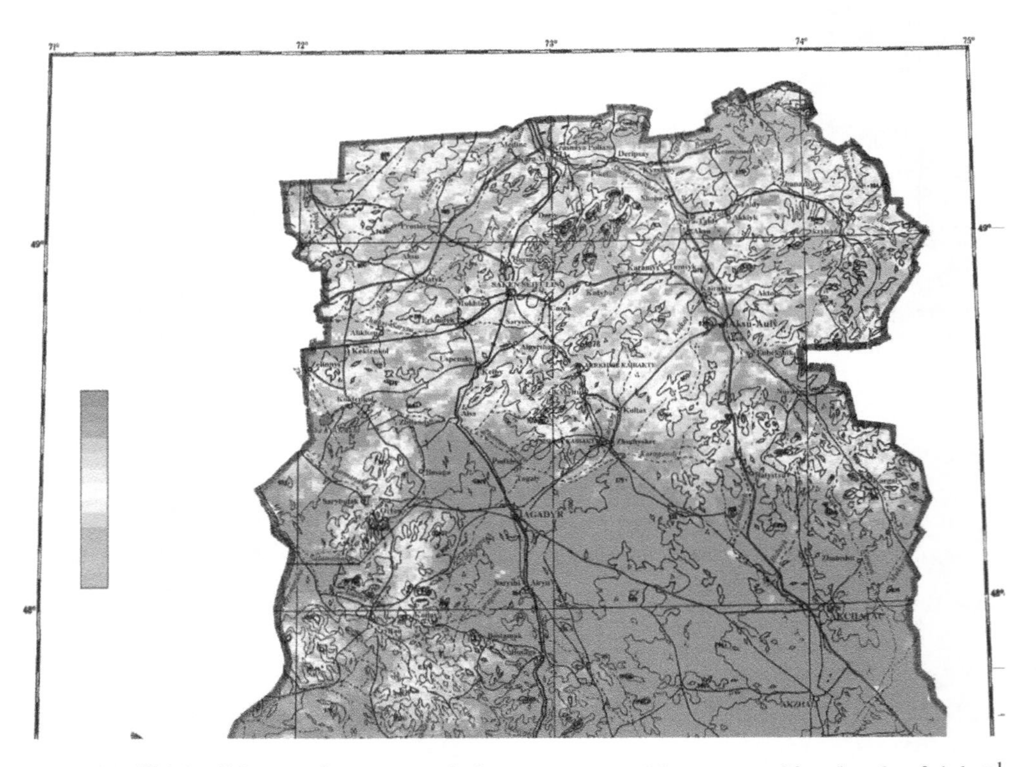

Figure 12. Weight of biomass of a steppe zone in dry autumn season (biomass ranged from less than 0.4 t ha^{-1} for the dark brown area to more than 15 t ha^{-1} for the dark blue area).

sequestration is especially high while considering all components including soils, and terrestrial biomass in agricultural croplands, forestry and other natural vegetation. The IPCC, in its book "Land Use, Land-Use change and Forestry (LULUCF)" refers to articles 3.3 and 3.4 of the Kyoto Protocol (IPCC 2000). Country reports can include such activities as forest, arable lands and pastures management, and forest restoration and secondary growth. Agricultural production is one

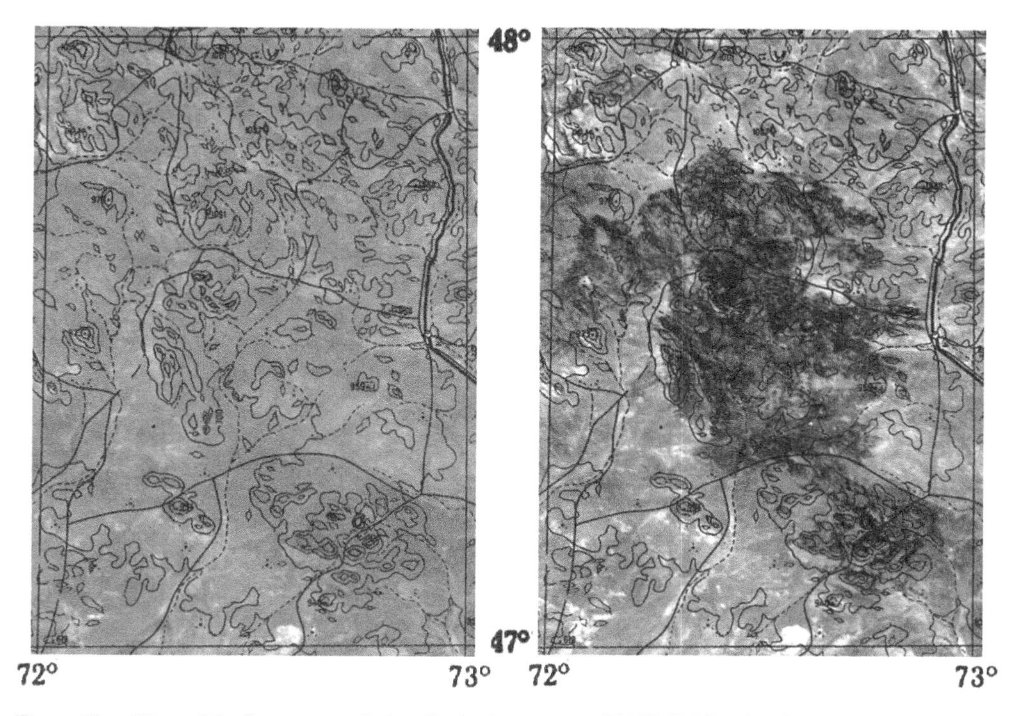

Figure 13. View of the Steppe zone during fire in the autumn of 2002 (EOS MODIS images).

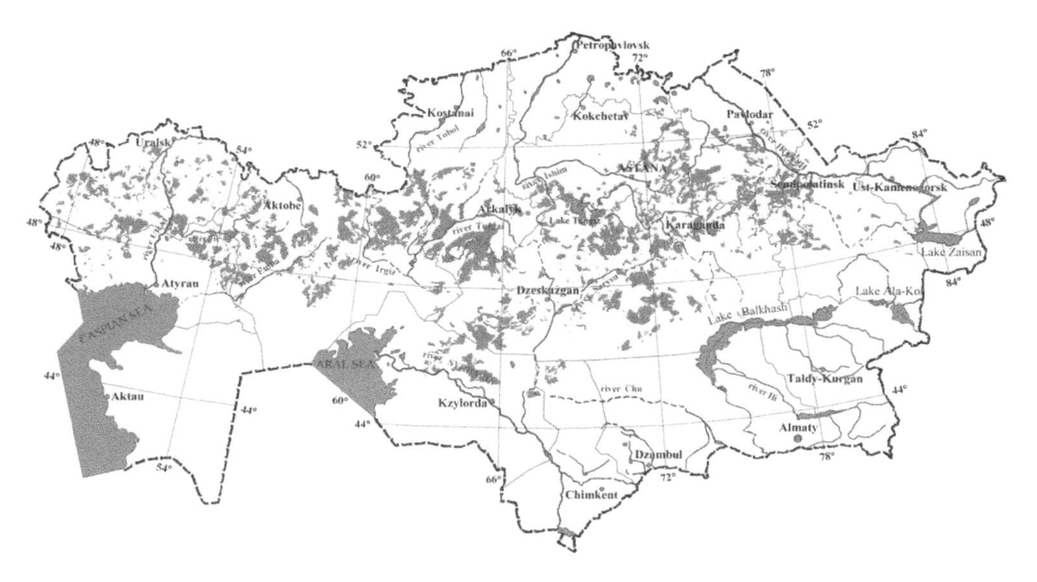

Figure 14. Kazakhstan steppe burned areas in 2002.

of the basic economic sectors in Kazakhstan. Agricultural lands can be a source of CO_2 emission and also of sequestration depending on their management. The proposed algorithms based on the remote sensing data and ground surveys are useful to estimating CO_2 balance including sequestration by Kazakhstan steppe grassy vegetation (cultural and natural) during the growing season.

CHAPTER 32

Possible changes in the carbon budget of arid and semi-arid Central Asia inferred from landuse/landcover analyses during 1981 to 2001

Elena Lioubimtseva
Department of Geography and Planning, Grand Valley State University, Allendale, MI, USA

1 INTRODUCTION

Climate records indicate that the Central Asian region has experienced a very significant warming trend during the past century (Chub, 2000; Lioubimtseva et al., 2005; Lioubimtseva, 2005). Regional land-use/landcover changes, together with responses to global climate change, result in changes in the hydrological cycle, ecosystem dynamics and the boundary layer of the atmosphere. The arid and semi-arid zones of Central Asia are very vulnerable to human disturbances and regional climate changes, because their ecosystems may be the first to reach tipping points under current human disturbances and climate change (Lioubimtseva and Adams, 2004).

To fully understand the impact of human activities, it is also necessary to consider the extent to which anthropogenic effects have modified the background level of carbon storage, and whether change in the intensity of either process has any evident potential to take up or release carbon from the desert-zone carbon reservoir. Much uncertainty exists about the possible impacts of global climate change on the sequestration of carbon in the vegetation and soils of the arid zones in general, and in those of Central Asia in particular. It is possible that regional land-use/landcover change and global climate change could result in significant changes in carbon reservoirs in these areas. Estimates of the carbon pools in the desert soils are still very uncertain (Lioubimtseva and Adams, 2002; Lioubimtseva et al., 2005).

Landcover changes during the past twenty years are assessed in this paper. Satellite imagery from the Pathfinder Advanced Very High Resolution Radiometer (AVHRR) Land dataset is used for Central Asia. Key factors related to recent changes in vegetation cover and the carbon budget of this vast arid area are also discussed.

2 CLIMATE, PHYSIOGRAPHY, AND CARBON BUDGET OF CENTRAL ASIA

The Central Asian region comprises the Iran-Turaninan Lowlands and the southern margin of the Kazakh Hills. It is bounded by Kopet-Dagh, of the Tian-Shan and Pamir-Alaï mountain ranges on its southern and southeastern edges (Figure 1). The continental climate of this region implies that summers are hot, cloudless and dry, and that winters are moist and relatively warm in the south and cold with severe frosts in the north. Precipitation has a distinctive spring maximum, which is associated with the northward migration of the Iranian branch of the Polar front. Detailed discussion of physical geography of Central Asia can be found in Lioubimtseva (2002).

The major types of landscapes in this vast continental region vary according to climate, lithology of parental rocks, and types of soils (Petrov, 1976; Glazovskaya, 1996; Babaev, 1996; Lioubimtseva, 2002).They include several types of desert ecosystems. Among them are: (a) sand desert with loose-sandy serozems and greyish-brown soils of ancient alluvial plains; (b) pebble-sand desert on

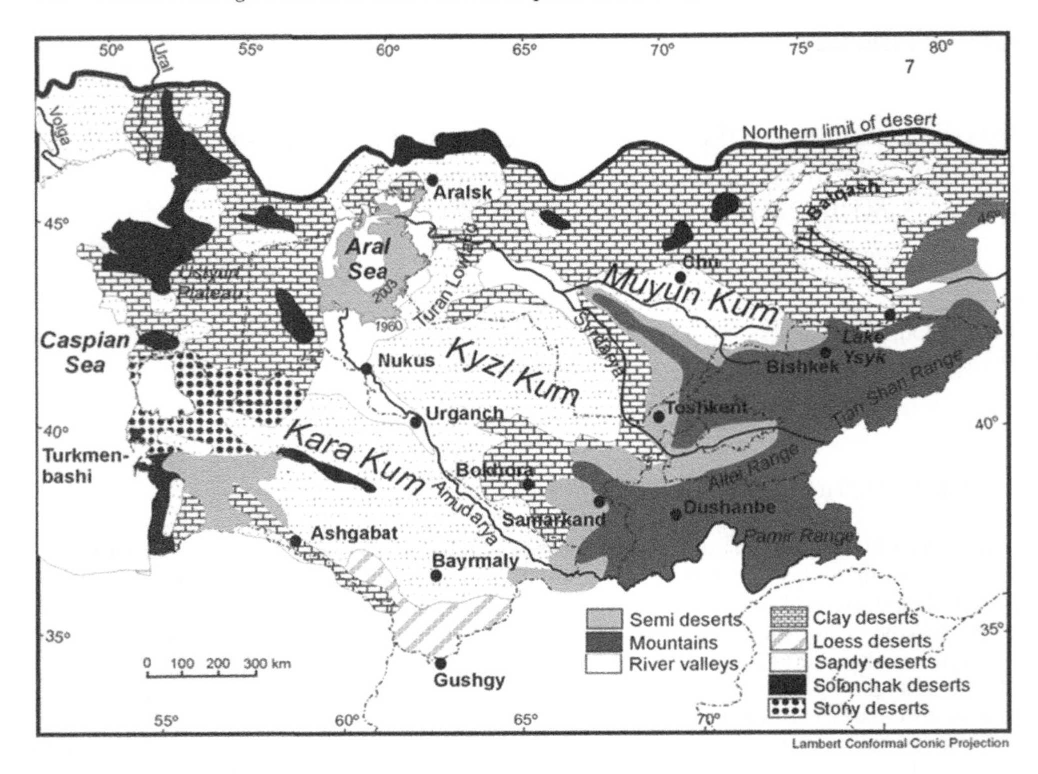

Figure 1. Major landcover types of Central Asia.

gypsiferous, greyish-brown soils, (c) gravelly, gypsiferous deserts; (d) stony deserts on mountain slopes and hills; (e) loamy deserts on slightly carbonate, greyish-brown soils of the elevated plains of the Aral Sea area; (f) loess semideserts of piedmont plains; (g) clayey *takyrs* on plains and ancient river deltas; (h) badlands on the Palaeogene low-mountain deserts and semideserts composed of saliferous marls and clays; and (i) solonchak deserts and semi-deserts in saline depressions and along the coasts of the inland seas (Table 1).

The seasonal dynamics of CO_2 fluxes on desert and semi-desert vegetation of Central Asia have not been measured. No data are currently available on their possible and long-term changes. Even more uncertain is the soil carbon budget of the region. Earlier publications (Adams and Lioubimtseva, 2002; Lioubimtseva and Adams, 2002) have suggested that currently existing global databases (e.g., Zinke et al., 1984; Batjes and Sombroeck, 1997) tend to overestimate the carbon content in temperate arid ecosystems of Central Eurasia. Russian sources indicate between 0.5 and 1.5 kg C m^{-2} in sierozems and sandy soils and up to 5–6 kg C m^{-2} in the chernozem-like meadow soils of Kazakhstan (Table 1). A detailed discussion of the key contradictions between global and regional carbon soil and vegetation data in Central Asia can be found in the articles by Lioubimtseva and Adams (2002) and Lioubimtseva et al. (2005).

3 RECENT LANDCOVER CHANGES RECORDED BY REMOTE SENSING

Long temporal series of remote sensing data offer a very useful and often the only possible approach to climate change for regions with limited amount of ground observations. Such monitoring techniques are based on the estimation of the statistical relationships between climate aridity and vegetation phytomass estimations from satellite imagery (Kogan, 1995; Lambin, 1997; Qi et al.,

Table 1. A summary of the range of types of soils found in the Central Asian desert region, together with typical organic carbon storage and land use.

Ecosystem/ landcover	Soil group	Vegetation C density (kg C m^{-2})	Soil organic carbon (kg C m^{-2})	Depth of human horizon (cm)	Dominant land use	Anthropogenic degradation processes
Meadow steppe	Chernozem-like meadow soils	2–3	5–6	30–35	Pastoral, arable	Degradation of vegetation, salinization
Semi-desert	Alkanized chestnut soils	0.6–1.5	0.9–4.0	35–40	Pastoral, arable irrigated	Degradation of vegetation, soil compactness increase
Rock desert (hammada)	Burozem	0.1–1	0.01–0.5	1–2	Pastoral	Degradation of vegetation
Rock desert (hammada)	Siero-burozem	0.1–1	0.01–0.5	0.5–1.5	Negligible (extensive pastoral)	Degradation of vegetation
Loess desert	Light sierozem	0.2–1.5	1–1.5	3–5	Pastoral, arable irrigated	Trampling and compactness, fertilization, salinization
Saline desert	Solonchak	0.1–1.0	0.5–2	1–3	Pastoral	Degradation of vegetation, trampling and compactness, fertilization
Semi-desert	Solonetz	0.5–2	1.5–2.8	5–10	Seasonal pastoral	Salinization, degradation of vegetation, trampling and compactness, fertilization
Sandy desert	Sandy desert soils	0.6–1.5	0.5–1.5	1–2	Seasonal Pastoral	Soil loosening, wind erosion, fertilization
Clayey desert	Takyr	0.1–0.5	Negligible	Negligible	–	–
Floodplains and delta	Alluvial meadow and swampy soils	2.5–5.0	3–4	10–20	Arable or Pastoral	Salinization, water erosion

2000). Zolotokrylin (2002) has developed a simple empirical aridity index for Central Asian deserts and semideserts, which can be defined by the duration of the period with a normalised difference vegetation index (NDVI) less than 0.07. This indicator reflects the zonality of heat exchange between arid land and atmosphere as the relationship between radiation and evapotranspiration mechanisms in the regulation of thermal conditions of soil surface and the lower layer of atmosphere (Zolotokrylin, 2002).

3.1 *Image data*

The Pathfinder Advanced Very High Resolution Radiometer Land dataset is an excellent tool for analysis of land use and landcover change at the regional or global scale. It provides non-interrupted series of preprocessed medium-resolution data since 1981. Parameters produced as a part of this dataset include reflectance and brightness temperatures derived from the five-channel cross-track scanning AVHRR aboard the NOAA Polar Orbiter 'afternoon' satellites (NOAA-07: Jul 81 to Jan 85, NOAA-09: Feb 85 to Oct 88, NOAA-11: Nov 88 to Sep 94, and NOAA-14: Jan 85 to Oct 01), along with a derived Normalized Difference Vegetation Index (NDVI), cloud and quality control indicators, and ancillary data. These data are derived from the NOAA Global Area Coverage (GAC) Level 1B data spanning a period of more than 20-years.

Pigments in green leaves, such as chlorophyll absorb strongly at red and blue wavelengths. Lack of such absorption at near-infrared wavelengths results in strong scatter from leaves. The contrast between red and near-infrared reflectance of vegetation is captured by NDVI. A commonly used greenness index [(near infrared-RED)/(near infrared + RED)] is often used as a proxy for biomass, net primary productivity, and leaf area index (Kogan, 1995; Qi et al., 2000). Although more advanced indices and NDVI modifications have been developed during the past decade, NDVI still represents a very useful tool to monitor landcover/land-use and to study vegetation phenology and biomass changes. The NDVI is robust, accessible from the AVHRR and other sensors, it can be easily calibrated with terrestrial observations and is highly correlated with the amount of photosynthesizing vegetation. Advantages and disadvantages of NDVI applications for monitoring desert vegetation have been thoroughly discussed in the international remote sensing literature during the past decade (Kogan, 1995; Qi et al., 2000; De Beurs and Henebry, 2004).

Six hundred ninety eight (698) 10-day maximum value composites were available in grid form at a resolution 8 km by of 8 km from July 10, 1981 to September 30 2001. They were used to establish the NDVI temporal trends in arid and semi-arid Central Asia. A permanent data gap exists from the middle of September to the end of December of 1994 due to satellite failure. September 30, 2001 is currently the most recent day of available data.

Simple statistical analyses of these data series involved three major steps: (1) a review of temporal variations revealed in the 10-day and monthly NDVI composites with a particular focus on annual trends in spring NDVI (associated with precipitation peaks); (2) the analysis of statistical relationships between NDVI and precipitation in four key areas (Eastern Kara Kum, Plateau Usturt; Central Kyzyl Kum, and an area around the eastward side of the Aral Sea); and (3) the computation and analysis of spatial and temporal patterns of an empirical aridity index derived from the NDVI.

Statistical analyses of NDVI series revealed a substantial increase in NDVI between 1986 and 1994 with a very prominent peak in 1993–1994, followed by a return to more arid conditions (Figure 2). Interestingly, this trend is observable in all parts of Central Asia and adjacent areas, from the Middle East to Mongolia and China, despite the great variability in local landscapes and meteorological conditions. A greenness peak in 1993–1994 has also been reported in other NDVI-based studies in the adjacent arid and semi-arid regions of Eurasia (DeBeurs and Henebry, 2004; Gunin et al., 2005; Kharin and Tateishi, 2004; Bajargardal and Karnieli, 2005).

Previous studies have found a fairly strong correlation between NDVI and precipitation for most of the region during the 1980s and 1990s. This is consistent with data reported in the earlier NOAA-based studies in Uzbekistan and Kazakhstan (Kharin et al., 1998; Zolotokrylin, 2002). Socio-economic and institutional changes in this region after 1991 and significant decreases for agricultural subsidies from Russia may be important cause for recent land-use and landcover changes in Central Asia.

Changes in biogeophysical processes are reflected in NDVI trends. They include changes in vegetation biomass and productivity, as well as onset and timing of land surface phenology. They link the ecological dynamics of the vegetated surface with the atmospheric dynamics of the boundary layer and hydrological cycle. A simple empirical index of aridity is based on the assumption of a steady relationship between NDVI and precipitation (1).

$$A = N_{NDVI<\alpha}/T \quad (1)$$

Where N is number of 10-day intervals with NDVI less than a threshold value α (climate-related variable that was set to 0.07 for arid Central Asia) and T is a number of 10-day intervals between

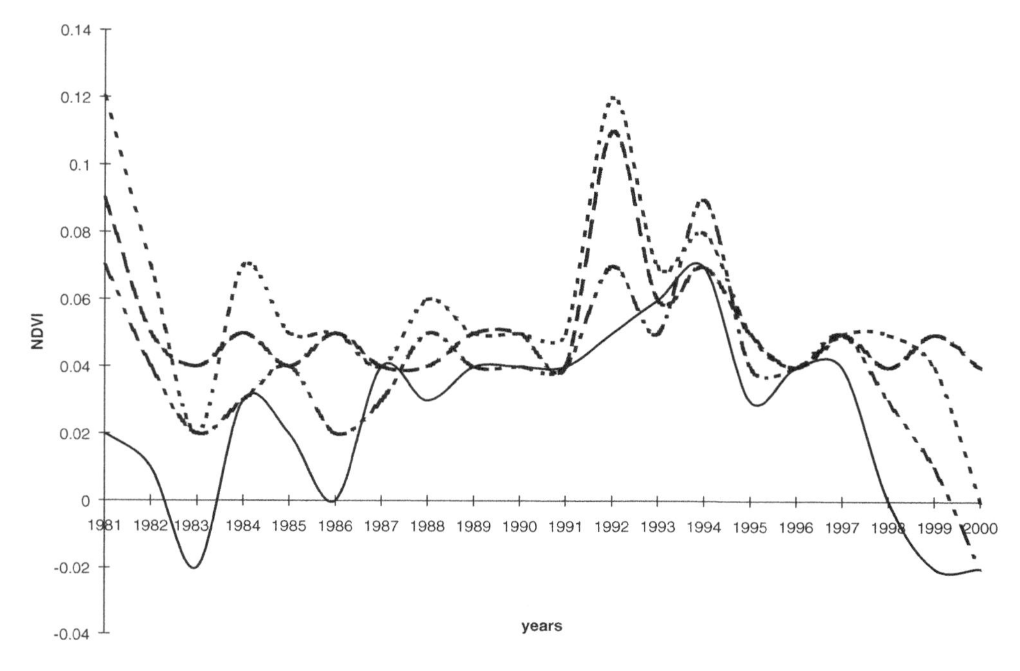

Figure 2. Temporal variation of the Normal Difference Vegetation Index (NDVI) values.

Table 2. Changes in the NDVI-based aridity index.

Years/location	East Karakum	Usturt	KyzylKum	East Aral
1981–1990	0.033854	0.035024	0.029463	0.023627
1991–2000	0.028701	0.025141	0.038037	0.044112

April 1 and October 31 during the period of observation. The index was derived from NDVI and used to assess the spatial and temporal climate variability. This index of aridity was computed both for the 1980s and 1990s. The data reveal a decrease of aridity in most of Central Asian region except for the Aral Sea area (Table 2). The increase of the number of 10-day intervals with arid NDVI in a vicinity of the Aral Sea is apparently caused by the severe human-induced desertification of this area, a combination which has been confirmed by ground observations. Temporal variations of 10-day periods with NDVI greater than 0.07 between April 1 and October 31 are illustrated in Figure 3.

3.2 *Interpretation of NDVI trends*

Reasons for the observed greening trend in Central Asian deserts are not completely clear. They might signal a globally significant shift in the carbon budget. Our analysis of precipitation data during the same period did not reveal consistent changes in the amounts of precipitation during NDVI peak times, except for the major irrigated areas, for which precipitation increase from 15 to 25% have been recorded during the past twenty years. Some recent studies suggest this trend might have been caused by rapid and very prominent institutional and land-use policy changes in the countries of the former Soviet Union and its client states. They may have caused "deintensification"

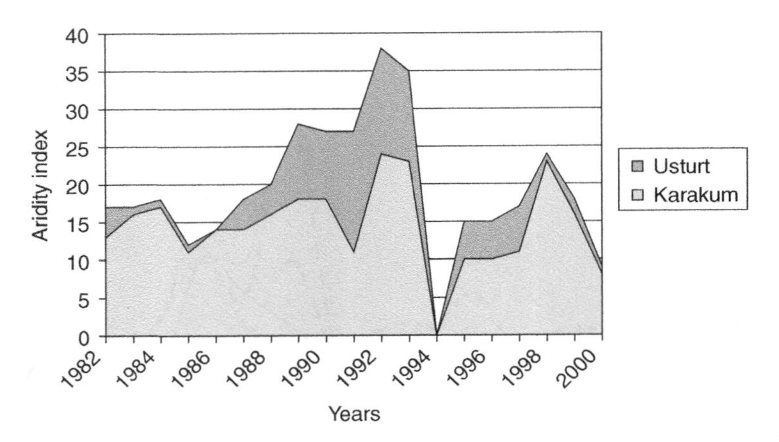

Figure 3. Temporal variation in 10 day periods with NDVI > 0.07.

of the region's agriculture (DeBeurs and Henebry, 2004). Such dramatic changes in agriculture could have caused profound changes in hydrometeorology, phenology, and biogeochemistry at the macro-regional scale.

Land-use changes have been caused by rapid economic transformation of this area. They are the most prominent drivers of change in landcover, climate and the carbon budget of the region. Irrigated arable lands have increased by 60% from 1962 to 2002 and the total irrigated area in five "Stans" of Central Asia increased by over half a million hectares (5.2%) just from 1992 to 2002 (Lioubimtseva et al., 2005). Turkmenistan alone accounted for 59% of the change. It increased its irrigated area by 300,000 hectares (20%). At the same time the grazing impact on vegetation has constantly decreased during the past decade. Cattle, camel, horse, and sheep populations have decreased by 27, 18, 28, 53 percent respectively over the last 10 years in the Central Asian countries (FAO, 2005) Overall, the numbers of small ruminates have decreased over the past ten years by over 30 million head. Undergrazing rather than overgrazing has been the main reason for vegetation changes during the past two decades in arid rangelands of Central Asia (Kharin et al., 1998; 2004).

Desert biomes cover a considerable part of arid Central Asia. It is important to realize that arid zones are featured by a very sparse vegetation cover. More than 65% of the surface reflectance signal in this very coarse-resolution AVHRR pixel data is caused by soil rather than vegetation. Assuming that the barren soil signal is fairly constant, the revealed NDVI difference can be attributed to vegetation changes. On the other hand, the soil surface of the desert might be far from being literally "barren" and it is important to estimate the proportion of the observed NDVI signal coming from the biogenic crusts on the soil surface compared to the signal coming from vegetation (Karnieli et al., 1999). Biogenic crusts on the soil surface vary from a few millimeters to several centimeters in thickness. They play a significant role in the desert ecosystems that controls processes as water retention and carbon and nitrogen fixation in soils. The accelerated growth of such biogenic crusts has been observed in the Kara Kum desert of Turkmenistan during the past 40–50 years. It has usually been attributed to the undergrazing caused by decreases in wild fauna and insufficient pressure on the desert rangelands. Accelerated growth of "black mosses" in the Kara Kum may also be a response of microphytes to increasing concentrations of CO_2 in the atmosphere (Lioubimtseva et al., 2005). Remote sensing data series including the blue channel of electromagnetic spectrum (such as Landsat TM, MODIS, and VEGETATION-SPOT), are absent in AVHRR NOAA. They allow for separation of the signal produced by the biogenic crusts that produced by higher vegetation (Karnieli et al., 1999; Lioubimtseva, 2005). Further research is needed to derive there two separate signals from remote sensing data. Unfortunately, image series including blue channel such as MODIS and VEGETATION data cover much shorter observation periods.

Theoretically CO_2 fertilization can be partly responsible for the greening trend revealed by NDVI, not only in biogenic crusts but also in the higher vegetation. Increased atmospheric CO_2 concentration has direct and, relatively immediate effects on two important physiological processes in plants. It increases the photosynthetic rate. However, it decreases stomatal opening and therefore the rate at which plants lose water. The combination of these two factors, increased photosynthesis and decreased water loss, implies a significant increase of water efficiency (the ratio of carbon gain per unit water loss) and productivity and a reduction in the sensitivity to drought stress in desert vegetation because of elevated atmospheric CO_2 (Smith et al., 2000). Temperate deserts and semi-deserts of Central Asia are dominated by vegetation with the C_3 photosynthetic pathway. It is expected that plants that use the C_3 photosynthetic pathway will respond more strongly to raised CO_2 than species with the more water-efficient and CO_2-efficient C_4 photosynthetic system (Graybill and Idso, 1993; Grünzweig, and Körner, 2000; Smith, 2000; Huxman et al., 2000).

4 REGIONAL CLIMATE CHANGE

To gain better understanding of environmental changes discussed in the previous sections of this paper, it is useful to see them in the context of a larger picture of climate change.

The IPCC Third Assessment Report provides very limited insights into present and future climate change in arid Central Asia and does not address spatial variability of temperature or precipitation trends within the region (IPCC, 2001).

In order to evaluate several climate change scenarios generated by the leading Global Circulation Models (GCMs) and to assess their credibility at the regional scale we used MAGICC-SCENGEN 4.1 model developed by Dr. Tom Wigley at the National Center for Atmospheric Research (2004). MAGICC stands for a Model for the Assessment of Greenhouse-Gas Induced Climate Change. This climate model has been used in all IPCC assessments to project global-mean temperature and precipitation. SCENGEN is a climate Scenario. This software uses regionalization algorithm to generate regional climate change scenarios based on the output from MAGICC and the library of GCM scenarios.

Model outputs are summarized in Tables 3 and 4. They indicate that the current trend of increase in arid Central Asia is likely to continue. Precipitation projections vary from one model to another and are highly controversial. Almost all models project a slight decrease of precipitation in western part of the region, and a 5 to 24% increase in central and east central Asia. However, a temperature increase is projected for the entire area. This suggests that the "greening" trend in central and eastern parts of central Asia may continue, while the western sector of the region (Turkmenistan and western Uzbekistan) can become even more arid. However, given the low absolute amounts of precipitation and extremely high annual and interannual variability of precipitation in arid regions, the task of reliable climate change modeling remains problematic (Hulme et al., 2001; Lioubimtseva, 2004). Atmospheric dynamics in desert regions are known to be very sensitive to natural climate variability at short temporal scales and the effects of such variability on longer temporal trends are not clear.

5 GENERAL DISCUSSION AND CONCLUSION

Coarse-resolution remote sensing data suggest that a greening trend has resumed in the arid and semi-arid zones of Central Asia between 1982 and 1995 with a prominent peak in 1994, followed by a decline between 1996 and 2001. Since the early 1980s more than 70% of the arid zones of Central Asia have become greener by about 10%, which may imply an increased carbon sequestration capacity. These landcover changes may indicate of very significant changes in regional and global carbon budgets. Similar trends have been reported in several other arid regions of the world.

This "greening" trend in Central Asia, as recorded by remote sensing data, may stem from several factors, such as climate change and dramatic land-use changes that in turn could trigger hydrometeorological changes of the regional scale, as well as a CO_2 fertilization effect on arid vegetation

Table 3. Temperature change scenarios.

Model	Institution	Annual temperature change ranges (compared to 1961–1990), degrees C			
		2025	2050	2075	2100
HadCM2	Hadley Centre Unified Model 2 (UK)	0.8–1.2	1.2–2.1	2.3–3.1	3.0–4.3
UKTR	UK Met. Office/Hadley Centre Transient Model (UK)	1.2–1.9	2.3–3.2	2.9–4.7	3.8–6.0
CSIRO-TR	Commonwealth Scientific and Ind. Research Organization, Transient Model (Australia)	1.0–1.3	1.9–2.3	2.5–3.3	3.3–4.3
ECHAM 4	European Centre/Hamburg Model 4 Transient (Germany)	1.4–1.7	2.1–2.9	3.3–4.2	4.3–5.5
UKHI-EQ	UK Met. Office High Resolution (UK)	1.2–1.7	1.9–3.0	3.0–4.4	4.2–5.7
CSIRO2–EQ	Commonwealth Scientific and Ind. Research Organization mode., Mark 2 (Australia)	0.8–1.1	1.5–1.9	1.9–2.7	2.7–3.5
ECHAM 3	European Centre/Hamburg Model 3 Transient model (Germany)	1.3–1.6	2.3–2.7	3.3–4.0	4.3–4.9
UIUC-EQ	University of Illinois at Urbana-Champaign model (USA)	0.9–1.2	1.5–2.0	2.2–2.8	2.8–3.3
ECHAM 1	Europen Centre/Hamburg Model 1 Transient (Germany)	0.7–1.5	1.7–2.6	2.3–3.1	2.6–4.8
CSIRO1–EQ	Commonwealth Scientific and Ind. Research Organization model, Mark 1 (Australia)	0.9–1.3	1.7–2.2	2.4–3.1	2.8–3.9
CCC-EQ	Canadian Climate Centre model (Canada)	1.1–1.6	2.1–3.0	2.8–4.1	3.9–5.3
GFDK-TR	Geophysical fluid Dynamics Laboratory Transient model (USA)	1.1–1.3	1.9–2.2	2.7–3.2	3.5–4.0
BMRC-EQ	Bureau of Meteorology Research Centre model (Australia)	1.2–1.5	2.1–2.6	3.1–3.7	3.8–4.2
CGCM1–TR	Canadian Climate Centre for Modeling and Analysis 1 Transient model (Canada)	0.9–1.8	1.7–3.2	2.5–4.6	3.1–5.9
NCAR-DOE	National Center for Atmospheric Research (DoE) Transient Model (USA)	0.9–1.1	1.5–1.9	2.2–2.9	2.9–3.7
CCSR-NIES	Center for Climate Research Studies/NIES (Japan)	1.2–1.4	2.1–2.5	3.0–3.5	3.9–4.8
Average scenario based on all available GCM runs		0.9–1.0	1.6–2.0	2.4–2.8	3.2–3.9

and biogenic crusts. The complexities of climate change, vegetation-climate feedbacks, and the direct physiological effects of CO_2 on ecosystems present particular challenges for understanding possible changes in the carbon budget of Central Asia.

Rapid land-use changes of the past two decades are associated with profound political, social and economic changes in Central Asia. They include increased irrigation of land and a decline in grazing pressure on desert rangelands. Regional NDVI trends, examined in this study, confirm earlier conclusions by Kharin et al. (1998; 2004) who observed that desertification has not been a major issue for arid rangelands of central Asia during the past decades. The area adjacent to the Aral Sea and irrigated arable lands may be an exception. However, further research is necessary to answer the question "Could such dramatic deintensification of agriculture in the arid and semi-arid

Table 4. Precipitation change scenarios.

Model	Institution	Annual precipitation change ranges (compared to 1961–1990), degrees C			
		2025	2050	2075	2100
HadCM2	Hadley Centre Unified Model 2 (UK)	0.6–20.6	−3.2–26.9	−4.5–61.4	−5.9–66.4
UKTR	UK Met. Office/Hadley Centre Transient Model (UK)	−1.8–5.3	−3.2–11.6	−4.6–13.1	−5.9–21.2
CSIRO-TR	Commonwealth Scientific and Ind. Research Organization, Transient Model (Australia)	−5.6–7.9	−12.3–15.7	−22.0–22.4	−22.7–32.4
ECHAM 4	European Centre/Hamburg Model 4 Transient (Germany)	−3.0–0.9	−7.1–1.1	−6.8–1.6	−7.5–2.8
UKHI-EQ	UK Met. Office High Resolution (UK)	−0.9–0.6	0.0–4.6	−1.3–6.6	−10.1–12.4
CSIRO2-EQ	Commonwealth Scientific and Ind. Research Organization mode., Mark 2 (Australia)	−9.5–5.2	−16.3–11.0	−23.3–15.7	−30.0–11.5
ECHAM 3	European Centre/Hamburg Model 3 Transient model (Germany)	−10.4–(−)0.9	−17.9–(−)6.0	−24.0–(−)8.6	−38.1–(−)2.9
UIUC-EQ	University of Illinois at Urbana-Champaign model (USA)	−6.5–(−)1.1	−12.8–(−)1.6	−18.2–(−)2.3	−23.5–2.9
ECHAM 1	Europen Centre/Hamburg Model 1 Transient (Germany)	−1.3–14.6	−7.7–25.2	−10.9–37.2	−18.1–46.9
CSIRO1-EQ	Commonwealth Scientific and Ind. Research Organization model, Mark 1 (Australia)	−.5–2.8	1.6–11.6	2.3–16.5	2.2–21.3
CCC-EQ	Canadian Climate Centre model (Canada)	−4.3–0.4	−4.4–0.7	−10.0–1.0	−12.9–1.3
GFDK-TR	Geophysical fluid Dynamics Laboratory Transient model (USA)	−2.5–7.1	−6.8–9.2	−5.7–10.8	−12.5–13.4
BMRC-EQ	Bureau of Meteorology Research Centre model (Australia)	−8.3–(−)3.8	−13.9–(−)6.7	−24.0–(−)6.5	−31.0–(−)10.4
CGCM1-TR	Canadian Climate Centre for Modeling and Analysis 1 Transient model (Canada)	−1.0–8.6	−1.8–15.0	−2.6–11.2	−3.3–27.6
NCAR-DOE	National Center for Atmospheric Research (DoE) Transient Model (USA)	−0.7–2.7	−1.7–2.3	−2.4–6.8	−3.2–5.7
CCSR-NIES	Center for Climate Research Studies/NIES (Japan)	−3.6–9.0	−6.5–46.5	−5.2–74.8	−11.9–53.0
Average scenario based on all available GCM runs		−0.3–0.9	−0.3–8.9	−3.0–9.4	−3.0–9.1

zones of Central Asia have caused profound changes in hydrometeorology, phenology, and carbon cycle of this vast region?"

Considerable changes in desert and semi-desert vegetation cover, as indicated by NDVI trends, may be also result from a combination of greenhouse-related climate change and direct physiological CO_2 effects on vegetation. Changes in photosynthesis and water-use-efficiency may become even more pronounced over the coming century. They could have implications for crop growth in desert-marginal areas, favor greater productivity, and perhaps increase the productivity and biomass of natural desert vegetation and soil organic matter. Field research in the desert rangelands of Turkmenistan and Uzbekistan suggest that at least part of the observed NDVI signal in this area is likely to be caused by the accelerated growth of biogenic crusts, rather than increased vegetation. While this trend of accelerated growth of biogenic crusts has been generally attributed to decreased grazing pressure on the desert rangelands, the accelerated growth of "desert mosses" may also be a response of microphytes to increased concentrations of CO_2 in the atmosphere.

Remote sensing imagery including the blue channel of electromagnetic spectrum, associated with the recently available MODIS, and VEGETATION-SPOT, can further separate the signal produced by the biogenic crusts from that produced by higher vegetation, thus helping to quantify carbon budget of Central Asia.

Temperate deserts and semi-deserts of Central Asia represent a relatively significant, dynamic, and still poorly quantified carbon sink, that is likely to play an important role in global and regional climate change. The carbon budget of the deserts and semi-deserts of newly independent states of Central Asia has been changing as a result of dramatic land-use/landcover changes that occurred during the past decades. Analyses of temporal and spatial variations of the normalized difference vegetation index (NDVI), derived from NOAA meteorological satellites images from 1981 to 2001, and showed significant changes in the land cover of Central Asia during this period. Despite its limitations, NDVI is a useful indicator of leaf area and vegetation net primary production. It is closely related to carbon content in vegetation cover. Preliminary estimations suggest that a greening trend occurred in arid Central Asia between 1982 and 1996, but that it was followed by on opposite trend between 1996 and 2001. Since the early 1980s more than two thirds of arid zones of Central Asia have become 'greener' by about 8 to 11%. That indicates a significant potential change in the carbon budget. This greening trend can be partly attributed to rapid land-use change in this region during the past two decades, and partly, to biotic responses to precipitation and temperature changes.

ACKNOWLEDGEMENTS

This research was funded by NASA grant NNH04ZYS005N as a part of the research project "Evaluating the effects of institutional change on regional hydrometeorology". I would like to thank Tom Wigley (National Center for Atmospheric Research) who provided MAGIC/SENGEN 4.1 software. I am also thankful to Geoff Henebry and Jiaguo Qi for insightful conversations about the data discussed in this article. Special thanks to Rattan Lal and Bob Stewart for their generous editorial help.

REFERENCES

Adams, J.M. and E. Lioubimtseva. (2002). Some key uncertainties in the global distribution of soil and peat carbon. p. 459-469 In: J.M. Kimble, R. Lal, and R.F. Follett (eds.), Agricultural Practices and Policies for Carbon Sequestration in Soil, pp. 459–469, CRC Press Lewis Publishers, Boca Raton, London, New York, Washington DC, 512 pp.

Babaev, A.G. (1996) Problemi osvienija aridnij zemel (Problems of Arid Land Development). Moscow University Press, Moscow. 282 p. In Russian

Batjes, N.H. and W.G. Sombroek, (1997). Possibilities for carbon sequestration in tropical and subtropical soils. Global Change Biology 3:161–173.

Bajargardal and Karnieli, 2005. NOAA-AVHRR derived NDVI and LST for detecting droughts in Mongolia. Arid Ecosystems 11(26–27):73–78.

Chub, V.E. (2000). *Climate change and its impact on the natural resources potential of the Republic of Uzbekistan*. (Izmenenija klimata I ego vlijanije na prirodno-resursnij potensial respubliki Uzbekistan), Gimet, Tashkent, 253 p. (in Russian)

De Beurs, K.M. and G.M. Henebry. 2004. Land surface phenology, climatic variation, and institutional change: Analyzing agricultural land cover change in Kazakhstan. Remote Sensing Environment 89:497–509.

FAO (2005) *FAO Agricultural Data*. Accessed at the following URL: http://apps.fao.org/, United Nations Food and Agriculture Organization, Rome (accessed December 2005).

Glazovskaja, M.A. (1996) Soils of Northern Eurasia, Map 1:5 M with explicatory note (in Russian). Map of Soil Degradation processes (1:20 M) Moscow State University Press, Moscow.

Graybill, D.A. and S.B. Idso. 1993. Detecting the aerial fertilization effect of atmospheric CO_2 enrichment in tree-ring chronologies. Global Biogeochemical Cycles 7:81–95.

Grünzweig, J. and C. Körner. 2000. Growth and reproductive responses to elevated CO_2 in wild cereals of the northern Negev of Israel. Global Change Biology 6:631–638.

Gunin, P.D., A.N. Zolotokrylin, A.A. Vinogradova, and S.N. Bazha. 2004. The study of vegetation dynamics in Southern Mongolia using NDVI data. Arid Ecosystems 10:(24–25) 29–34.

Hulme, M., R.M. Doherty, T. Ngara, M.G. New, and D. Lister. 2001. African climate change: 1900-2100. Climate Research 17, 145–168.

Huxman, T.E., R.S. Nowak, S. Redar, M.E. Loik, D.N. Jordan, S.F. Zitzer, J.S. Coleman, J.R. Seeman, and S.D. Smith. 2000. Photosynthetic responses of *Larrea tridentata* to a step-increase in atmospheric CO_2 at the Nevada Desert FACE Facility. Journal Arid Environments 44:425–436.

IPCC. 2001. Climate Change: The Scientific Basis Contribution of Working Group I to the Third Assessment Report of the Intergovernmental Panel on Climate Change (IPCC) J.T. Houghton, Y. Ding, D.J. Griggs, M. Noguer, P.J. van der Linden and D. Xiaosu (eds.). Cambridge University Press, UK. 944 p.

Karnieli, A., G.J. Kidron, C. Glaesser, and E. Ben-Dor. 1999. Spectral characteristics of cyanobacteria soil crust in semiarid environment. Remote Sensing Environment 69:67–75.

Kharin N.G., R. Tateishi, and I.G. Gringof. 1998. Use of NOAA AVHRR data for assessment of precipitation and land degradation in Central Asia. Arid Ecosystems 4(8): 25–34.

Kharin, N.G. and R. Tateishi. 2004. Map of degradation of the drylands of Asia. Arid Ecosystems 10(24–25): 17–28

Kogan, F.N. 1995. Application of vegetation index and brightness temperature for drought detection. Advances Space Research 15:91–100.

Lambin, E. 1997. Modelling and monitoring land-cover change processes in tropical regions, Progress Physical Geography 21(3):375–393.

Lioubimtseva, E. 2002. Arid Environments. p. 267–283. In: M. Shahgedanova (ed.). Physical Geography of Northern Eurasia. Oxford University Press, Oxford.

Lioubimtseva, E. and J.M. Adams. 2002. Carbon Content in Desert and Semidesert Soils in Central Asia. p. 209–256. In: J.M. Kimble, R. Lal, and R.F. Follett (eds.). Agricultural Practices and Policies for Carbon Sequestration in Soil. CRC Press Lewis Publishers, Boca Raton, London, New York, Washington, DC.

Lioubimtseva, E. and J.M. Adams. 2004. Possible implications of increased carbon dioxide levels and climate change for desert ecosystems. Environmental Management 33(S1):S388–S404.

Lioubimtseva, E. Climate Change in Arid Environments: revisiting the past to understand the future. Progress Physical Geography 28(14):502–530.

Lioubimtseva, E., R. Cole, J.M. Adams, and G. Kapustin. 2005. Impacts of Climate and Land-Cover Changes in Arid Lands of Central Asia. Journal Arid Environments. 62(2):285–308

Lioubimtseva, E. Environmental Changes in Arid Central Asia Inferred from Remote Sensing Data and Ground Observations. 2005. Arid Ecosystems 11(26–27):12–23.

Petrov, M.P. (1976) Deserts of the World. Chichester, John Wiley & Sons. 447 p.

Qi, J., Y. Kerr, M. Moran, M. Weltz, A. Huete, S. Sorooshian, and R. Bryant. 2000. Leaf area index estimates using remotely sensed data and BRDF models in a semi-arid region. Remote Sens. Environ. 73:18–30.

Smith, S.D., T.E. Huxman, S.F. Zitzer, T.N. Charlet, D.C. Housman, J.S. Coleman, L.K. Fenstermaker, J.R. Seemann, and R.S. Nowak. 2000. Elevated CO_2 increases productivity and invasive species success in an arid ecosystem. Nature 408:79–82.

Wigley, T.M.L., Raper, S.C.B., Hulme, M. and Smith, S. (2000) The MAGICC/SCENGEN Climate Scenario Generator: Version 2.4, Technical Manual, Climatic Research Unit, UEA, Norwich, UK. 48 p.

Zinke P.J., A.G. Stangenburger, W.M. Post, W.R. Emmanuel, and J.S. Olson. 1984. Worldwide organic soil carbon and nitrogen data. Environmental Sciences Division, Publication No. 2212. Oak Ridge National Laboratory, US Department of Energy.

Zolotokrylin, A.N. 2002. The indicator of climate aridity (Indicator klimnaticheskoj aridnnosti). Arid Ecosystems 8(16):49–57 (in Russian with summary in English).

CHAPTER 33

Western Siberian peatlands: Indicators of climate change and their role in global carbon balance

S.N. Kirpotin
Tomsk State University, Tomsk, Russia

A.V. Naumov
Institute of Soil Science and Agrochemistry, Novosibirsk, Russia

S.N. Vorobiov
Tomsk Regional Environmental Administrative Unit, Tomsk, Russia

N.P. Mironycheva-Tokareva & N.P. Kosych
Institute of Soil Science and Agrochemistry, Novosibirsk, Russia

E.D. Lapshina
Yugorskiy State Univeristy, Khanty-Mansiysk, Russia

J. Marquand
University of Oxford, UK

S.P. Kulizhski
Tomsk State University, Tomsk, Russia

W. Bleuten
University of Utrecht, Utrecht, The Netherlands

1 INTRODUCTION

The United Nations Millennium Declaration expresses the intention "to make every effort to ensure the entry into force of the Kyoto Protocol ... and to embark on the required reduction in emissions of greenhouse gases." To implement the Millennium Declaration the UN indicated that it would be necessary to reverse the loss of environmental resources. In this context, it emphasized that carbon dioxide emissions are the largest source of the greenhouse gas effect.

Global warming is a major environmental issue and is expected to be greatest at high latitudes. Arctic and sub-arctic landscapes are particularly sensitive to temperature change because of permafrost thawing (Callaghan and Jonasson, 1995). Some areas of the arctic have already experienced warming of up to 0.75°C per decade. The vast region of Siberia is one of them and some have noted that the observed warming over the last 50 years is probably the result of increased greenhouse gas concentrations (Dlugokencky et al., 1998; IPCC, 2001).

Asian Russia has been estimated as a big terrestrial sink of 0.58 Gt atmospheric carbon per year (Kudeyarov, 2004; Zavarzin and Kudeyarov, 2006). However, the precise functional role of pristine peatlands in the global and regional carbon cycle has not yet been evaluated. In particular carbon exchanges in sub-arctic peatlands, such as the lightly vulnerable ones that are prevalent in Western Siberia, have not been adequately researched. It is very important to study them in depth in order to assess the reaction of peatlands to future climate change.

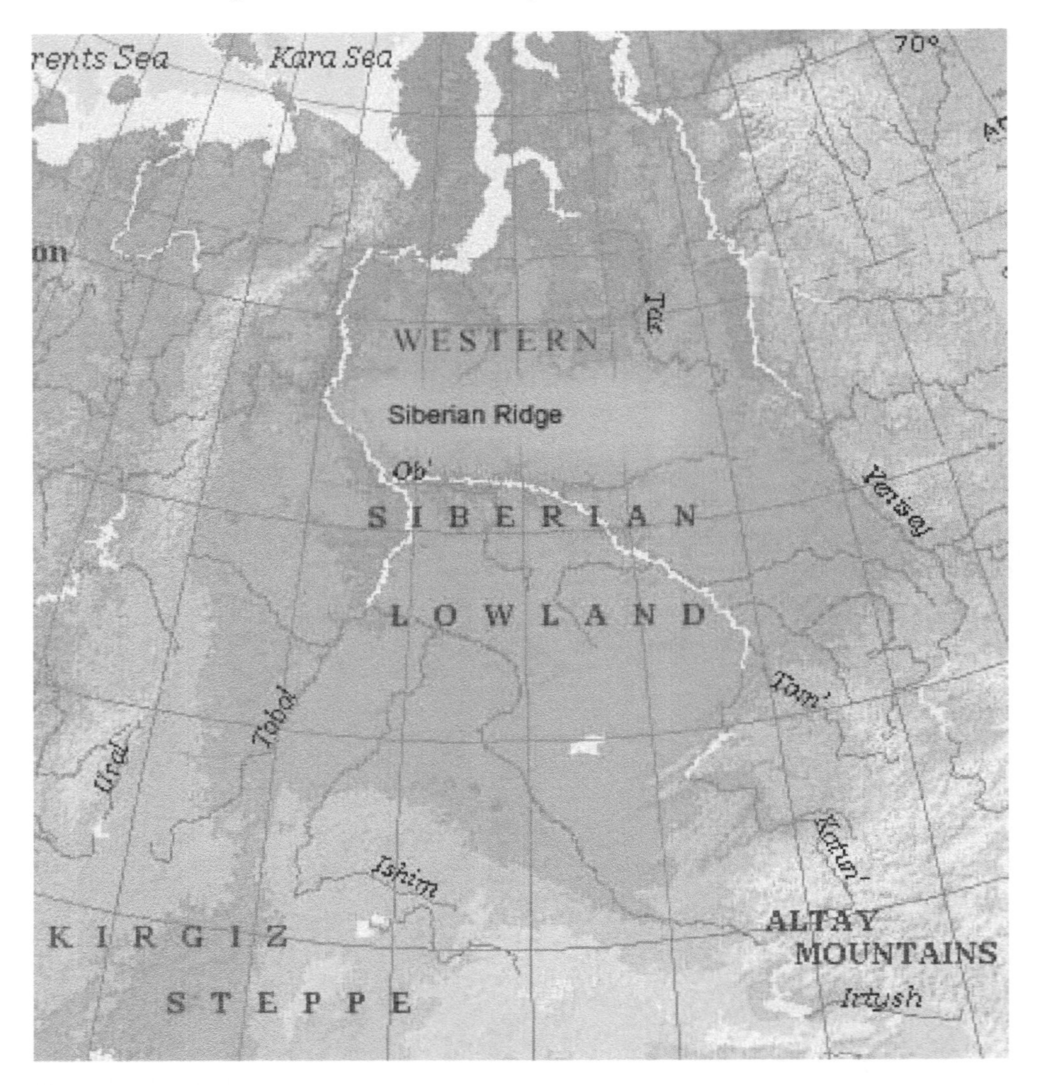

Figure 1. Map of Western Siberian Lowland.

2 PHYSIOGRAPHIC FEATURES OF THE REGION

The Western Siberian plain extends from the Ural Mountains in the west to the central Siberian plateau in the east. The South to North extension is almost 2500 km. A clear landscape sequence of bioclimatic zones has developed on the low, flat relief of the area – from southern steppes to northern tundra. This surface area is about 3 M(million) km^2 in area. Of this total, peatland fens and bogs have been estimated to make up a little more than 1 M km^2 (Vaganov et al., 2005).

The Western Siberian region is divided into two parts by the Siberian Ridge (*Sibirskie uvaly*) which extends in an east-west direction (See Figure 1). The southern part of the Western Siberian Plain is characterized by bioclimatic sub-zones, such as taiga and sub-taiga, forest steppes and steppes with different types of mires. Permafrost reliefs are only present in isolated frozen peat "islands" near slopes of the Siberian Ridge. These small permafrost islands are very unstable and sensitive to climate-driven changes in the landscape (Lapshina and Kirpotin, 2003).

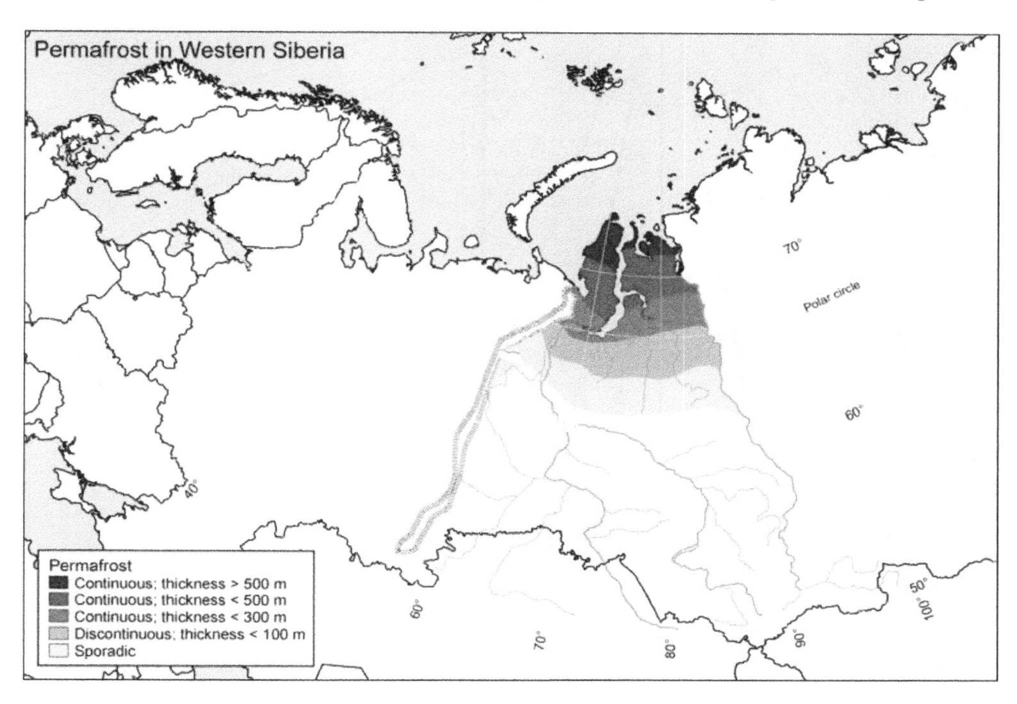

Figure 2. Permafrost distribution in Western Siberia.

North of the Siberian Ridge, the forest-tundra landscapes appear which gradually change into typical tundra in more northern regions. The forest-tundra zone, especially that part found between the Nadym and Pur rivers, is characterised by vast peatlands. The frozen peatlands form flat "palsas" (Matthews et al., 1997). They are found in up to 70% of the surface of the watershed (See Figure 2). Lakes and fens are present over deeper permafrost or where permafrost is absent between the flat *palsas*. Woodlands, open Larch (*Larix sibirica*), and Pine (*Pinus sibirica*) forests cover up to 20% of the total surface. They are found where permafrost is absent, such as on the flood plains, on terraces of large rivers, on narrow strips around some lake shores, and on high mounds between bogs (Kirpotin et al., 1995; Kirpotin et al., 2003).

Contrasting processes are occurring in the southern and northern parts of Western Siberia Bogs are expanding in the southern and middle taiga sub-zones of this region. Bogs in the northern taiga and forest-tundra zones (Kirpotin et al., 2004; Lapshina et al., 2006) are experiencing thermokarst and are being colonized by trees (Kirpotin et al., 1995; Callaghan et al., 1999; Kirpotin, 2003). These processes are probably connected to recent climate changes that have led to global warming.

3 INDICATORS OF CLIMATE CHANGE IN SUB-ARCTIC PEATLANDS

Large areas of arctic and sub-arctic wetlands are very vulnerable to climate change. Changes in them can have an important feedback mechanism in the process of global warming due to their large carbon stocks and the presence of permafrost in them (Gorham, 1991; IPCC, 2001). Extensive mire complexes are present north of latitude N57° in Western Siberia. Shallow permafrost occurs north of latitude N60° and is continuous above the Polar circle. The zone that is most vulnerable to climate change in Western Siberia is about 600 km wide. Large pristine wetlands with discontinuous permafrost are present within this zone. Lapshina et al. (2001) found evidence of climate change in these wetlands, namely, recent peat accumulations on top of partly frozen peat.

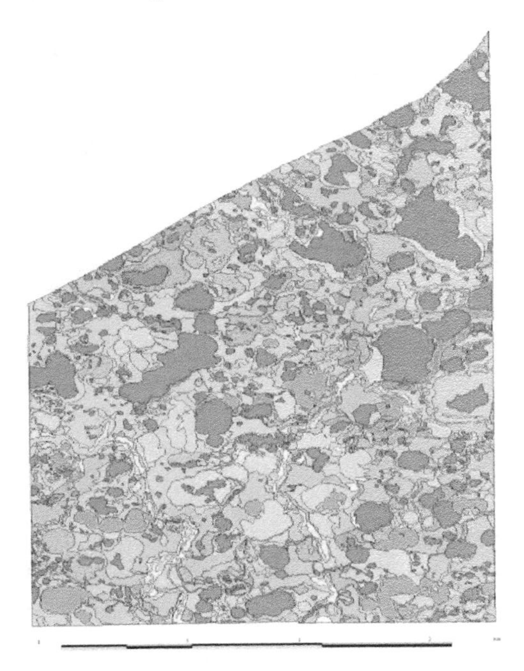

Figure 3. Map of landscape patterns for Puritey-Malto site.

P*alsa* mires are especially sensitive to warming trends due to thawing of frozen peat (Kirpotin et al., 1995; Callaghan et al., 1999; Kirpotin and Vorobiov, 1999; Muldiyarov et al., 2001; Kirpotin et al., 2003; Nelson and Anisimov, 1993; Matthews et al., 1997; Sollid and Sorbel, 1998). Thermokarst features are spreading over extensive areas of the sub-arctic region of Siberia as a consequence.

The authors have made several expeditions to the sub-arctic region of Western Siberia during the past 15 years. They have observed first hand the change dynamics in the landscapes of this region. The flat *palsas* of the Pur-Pe–Tanlova interfluves (64°–65° NL, 75°–76° EL) from 1989 to 1991 were studied in the field. These analyses were supported with aerial photographs (Scale 1: 10 000 and 1: 25 000) (See Figure 3). Cyclic successions in the development of the *palsa* complexes in this region were described in detail. These analyses showed that landscape units experience continuous transformation from one type into other ones (Kirpotin et al., 1995; Kirpotin et al., 2003).

The permafrost, which is 60–100 meters thick, is almost continually distributed here with the only exception being the valleys of large rivers. Thawing of the upper part of the permafrost was observed in the flood-plains of small rivers, on the higher mineral soils under ridge-hollow complex bogs, under the thermokarst lakes, and in thawed *palsa* hollows.

The area between the Nadym and Pur rivers is flat lowland of marine and lacustrine-alluvial genesis and with a prevalence of sandy, loamy soils. Although many rivulets and brooks exist in it, drainage is hampered by its low relief and a subsurface that consists of impervious, frosted soil layers. The presence of many drained thermokarst lakes (these are dry lake basins and are called "Khasyrei" in Russian) has been interpreted by Zemtsov (1976) as indicative of an active neotectonic rise of this area. Khasyrei occupy 20% of the total lake area in the center of the watershed as shown in Figure 4.

Another powerful relief-maker factor that influences the process of swamp development in the sub-arctic conditions is the presence of continuous strong northerly winds. Winds have caused coastal abrasions of the thermokarst lakes as well as their shallowing and swamping from the southern lee side. This process has been facilitated by the accumulation of peat materials in these areas that have been washed away by waves (See Figure 5).

Figure 4. Khasyreis" – Drained lakes (Kirpotin, 1999, 2004).

Figure 5. Coastal abrasion of the thermokarst lake with shallows and swamps on the Southern Lee Side.

4 CURRENT DEVELOPMENT OF PALSAS

Based on their own research, Scandinavian scientists have suggested that flat *palsas* undergo cyclic development. This perspective is based on detailed long-term observations of these *palsas* and has been substantiated with photographic data on the individual stages of this cycle (Matthews et al., 1997; Sollid and Sorbel, 1998). Scandinavian scientists only put forward the idea of the cyclic development of palsas after careful, long-term observation of the formation of separate frozen mounds and inter-palsas thawed hollows. However, palsas are not common to Scandinavia and these limited observations may not be applicable to other regions where similar palsas exist.

(a) (b)

Figure 6. The first stage of permafrost melting (Thermokarst) on the "Palsa" surface: a – ground view; b – porous surface of palsa (air photo, Kirpotin et al., 2004).

Although a completely different situation exists in the sub-Arctic region of western Siberia, the cyclic succession is shown very clearly over the extensive space occupied by flat palsas. All of the cyclic stages of this process are visible and landscape boundaries precisely reflect their processes of development (Kirpotin et al., 2003). The cross over pattern of succeeding stages khasyrei lake palsas are even recognizable from satellite images.

During the first stage of this process, small (0.5–3 m) saucer-shaped round closed dwarf shrub-sedge-sphagnum thermokarst depressions are formed (See Figure 6a). Thermokarst areas are formed by thawing of the upper part of the permafrost which enlarges the "active layer". This process is supported by relatively warm summer rains. In the aerial photographs such palsas are shown to have a characteristic "porous" surface. The surface appears to have been corroded forming numerous round shaped pits (See Figure 6b).

Cracks in the lichen cover and drying of the underlying peat during rainless periods are conditions that lead to the formation of some thermokarst areas (See Figure 7). Moisture remains in the cracks, and some of them increase in size. They burst when the newly added moisture fast freezes. In such cases the affected areas are large and they develop so quickly that sphagnum mosses and/or sedges do not have sufficient time to settle. Bare soil or attenuated wet peat covered by a thin sheet of *Drepanocladus exannulatus* or *Warnstorphia fluitans* can be observed. Sometimes it is actually submerged below open water. Linear areas provide the basis for the formation of inter-palsa hollows and water-tracks with cotton-grass-sedge-sphagnum and sedge-sphagnum vegetation. The hollows and water tracks provide a system for draining the remaining from melted soil-ice, snow and added precipitation water from the flat palsas.

The frozen peat found in the mounds gradually thaws during the summer season and the moisture formed as a result of its thawing flows to inter-palsa hollows, streams and lakes. Therefore, once initiated, thermokarst areas can increase in size even during relatively dry periods. If the area is not intercepted by a water flow, it will gradually increase in size and will normally turn into a small round shaped thermokarst lake (See Figures 8 and 9).

Lakes beds steadily increase in area due to lake shore erosion that is induced by wind born waves as depicted in Figure 4. Shore materials are transported by compensation currents along the lake bottom to upwind shores where new eutrophic sedge fens can develop as indicated in Figure 5.

Numerous thermokarst lakes are spread over the area. Quite often they appear close to each other. Higher located lakes can change into khasyrei when they are drained by water channels and lower situated lakes eventually fill with water as illustrated in Figure 10. Drainage activity can develop are a result of collapses of permafrost flat palsa areas between lakes or due to groundwater flows below thicker areas in the active layers found above permafrost.

Figure 7. Cracks in the Lichen cover and underlying peat (Kirpotin et al., 2004).

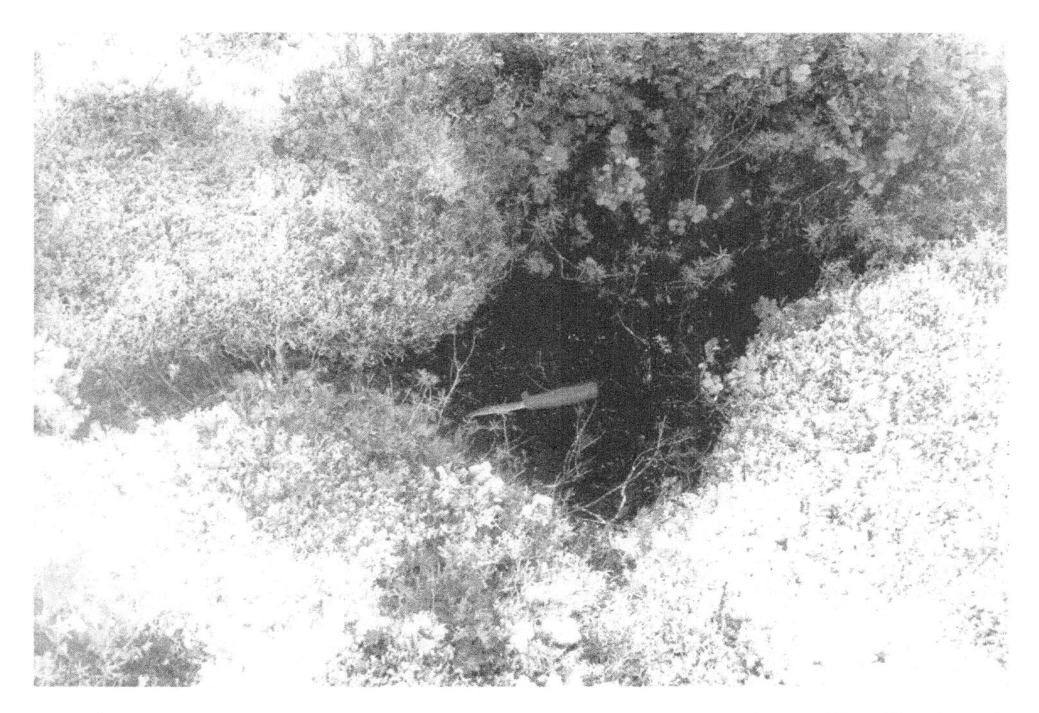

Figure 8. Origin of an initial thermokarst lake during the third stage of permafrost melting (Kirpotin et al., 2004).

Figure 9. A round mature lake during the fourth stage of permafrost degradation (Kirpotin et al., 2004).

In summary, during the first stage of the cyclic degradation of flat palsa complexes, thermokarst lakes may appear as a result of the appearance of different sized hollows. These lakes can increase in size due to shore erosion since lake water acts as a heat source which induces further thawing of permafrost layers. These thermkarst lakes can also turn into a khasyrei.

Cotton-grass-sedge-sphagnum swamps develop on the sandy or peat bottoms of drained lake basins. Their khasyrei floors are one to three meters lower than the surrounding flat palsas and are therefore susceptible to late summer frost cause by inflows of cold air. Ice lenses can develop at low, moist places in the mineral or peat soil by such cooling process in periods before a permanent snow cover is formed. The presence of permafrost below the khasyrei floor may enhance this new permafrost formation. The new ice lenses may grow and push up the soil above them. This results in the formation of small two to five meter tall dome-shaped mounds of rounded or oval form. Lichens and dwarf shrubs typically associated with palsa settle on the surface of these small mounds as illustrated in Figures 11 and 12. This process is verified by the presence of a thin layer of sedge-moss peat typically found on *khasyrei* floors, but not on dry tops of mounds formed by ice-heaving. The dry tops of new mounds allow for the settlement of birch shrubs (*Betula nana*) and even of birch trees (*Betula pubescens*). Two to ten year old birch trees have been found which substantiates the presence of ice-heaving activity.

As the heaving by renewed permafrost goes on, the isolated small mounds merge into a uniform system and, depending on the capacity of the peat deposit, they turn into typical flat palsa plateaus as shown in Figure 13. Edges of the drained lake basin can still be seen in aerial photographs at this stage of the cycle.

In summary, it appears that there is a steady cycling of cryogenic processes. The thermokarst and permafrost heaving have been peculiar to Western-Siberia's sub-arctic region for a long time.

When these processes were studied in the early 1990s, it seemed that thermokarst was starting to prevail over new permafrost heaving. However, changes in the sub-arctic region still did not have a dominant character then and it was difficult to say with full confidence where the cryogenic

Figure 10. "Khasyrei" drained lake which lost water to another reservoir or river as fifth stage of permafrost degradation (Kirpotin et al., 2004).

Figure 11. A mature *Khasyrei* with regenerated frozen peat mounds (aerial photo, 1989).

pendulum would swing. Nevertheless, the prevailing view was put forward at that time (Kirpotin et al., 2003), namely, that in the near future the landscape could change significantly if the surface of flat mounds and palsa plateaus covered with lichens were to be steadily reduced in area and giving way to inter-palsa hollows.

During August, 2004 expeditions to New-Urengoy and Pangody (N 66° E 74°) were organized by the EU-INTAS project to study "The effect of climate change on the pristine peatland ecosystems and (sub) actual carbon balance of the permafrost boundary zone in sub-arctic western Siberia." This field research facilitated comparisons with earlier findings. Findings from this research suggested that the thermokarst has indeed expanded and that it now dominates the area being studied. As observed through aerial photographs, the white surfaces, typical for flat palsas covered with whitish

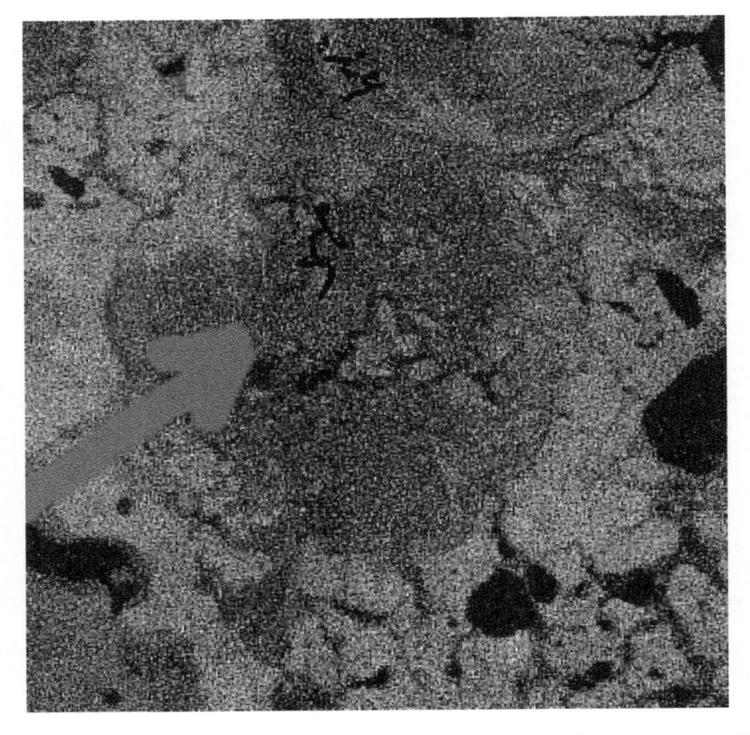

Figure 12. Mature *Khasyrei* with regenerated frozen peat mounds viewed from the ground (Kirpotin et al., 2004).

Figure 13. Old "Khasyrei' with restored plateau mounds in final stage of the "Palsa" development cycle (aerial photo, 1989).

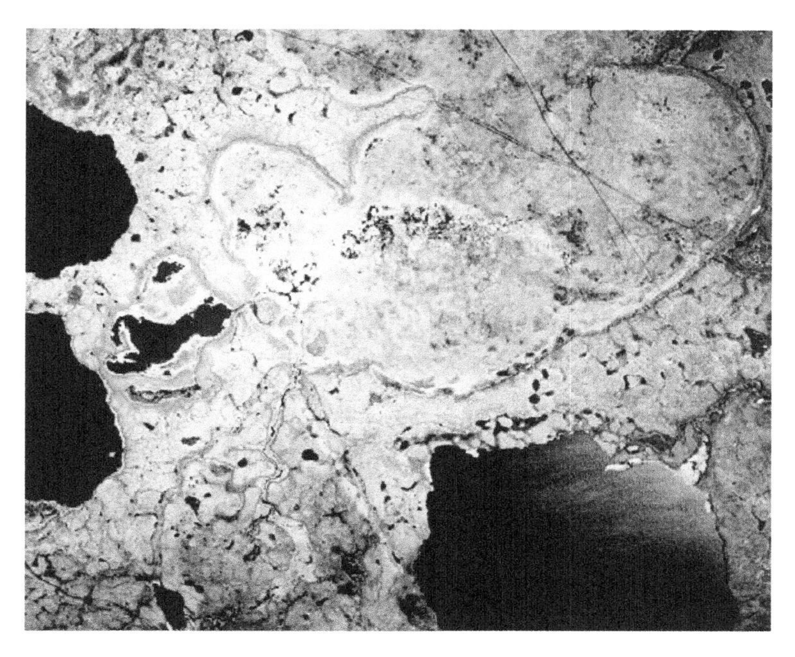

Figure 14. Fresh thermokarst area with drowned dwarf shrubs (Kirpotin et al., 2004).

lichens, decreased in area while green or brown colored hollows increased in area. Thermokarst depressions on the surface of flat palsa develop so swiftly that lichens and dwarf shrubs drawned (Figure 14), and sphagnum mosses in most cases have not had time to settle in them, or are only starting to occupy these fresh water-bearing sites (Figure 15).

Shores of the big thermokarst lakes are one kilometre or more in diameter and are evidently retreating (See Figure 16). However it was not possible to determine the rate of shore retreat by comparing LANDSAT images of 1987 with those of 2002. This suggests that the growth of thermokarst lakes could not have been more than 30 meters over this period. It may be that the rates of shore retreat and thawing of permafrost are even more recent developments caused by warming trends.

Small and middle-sized lakes have also appreciably increased in area as shown in Figure 17. The strip of freshly-submerged dwarf shrubs that are from one to three meters in width – primarily *Ledum palustre* and *Betula rotundifolia* – is clearly visible on the shores of these lakes. The presence of dwarf shrubs leads to the conclusion that these changes are recent, having occurred during the last 3–4 years. Most of the dwarf shrubs still have not had time to decompose.

These data agree with the latest observations of degradation of Arctic sea ice. "Arctic specialists at the US National Snow and Ice Data Centre at Colorado University, who have documented the gradual loss of polar sea ice since 1978, believe that a more dramatic melt began about four years ago…" (Connor, 2005).

The peat is thicker at the Puritey/Malto Key Site (N 64° 40–45′, E 75°24–29′) than at the Northern Key Sit. Prior to undertaking another expedition to it during summer, 2005, we expected to find that frozen bogs – flat palsas – would be more sensitive to climactic warming at the southern edge of the permafrost area than at that further north. However, we found the opposite to be true. The southern palsas were more stable than northern palsas. The changes that we observed at the New Urengoy-Pangody Key Site in 2004 are much greater than those we observed in 2005 at this site. Apparently, this is explained by the level of thermokarst activity, which depends directly on the thickness of the peat layer of palsas. The thick layer of frozen peat protected the palsas at the southern site and did not allow for deep melting. In contrast, as one moves further north, the annual growth of mosses becomes progressively less and the peat layer of palsas becomes correspondingly

Figure 15. Sphagnum mosses occupy the thermokarst area (Kirpotin, 2004).

Figure 16. Shore of big lake with diameter greater than one kilometer (Kirpotin et al., 2004).

thinner. The thinner layer of peat melts more easily in the summer, leading to thermokarst in the more northern areas. Thus, very active thermokarst is apparent in areas where the peat layer of the frozen bogs is thin – about 20–30 cm and 50 cm maximum – and it is almost absent on the southern edge of the permafrost zone where the thickness of frozen peat varies from 1.5–2 m (Figure 18). Thus our rough estimate of the area of active thermokarst activity within the West-Siberian Plain is located between 65°–68° latitude N.

Figure 17. Shore of small lake of about one hundred meters in diameter (Kirpotin et al., 2004).

Figure 18. Graphic description of key sites.

Data from the 2004 and 2005 expeditions are still being processed. Thus, it is not yet possible to quantify the scale, pattern sizes and speed of the thermokarst processes which were observed. However preliminary observations suggest a recent increase in thawing of thermokarst. If so, it may be the start of an irreversible change in the thermokarst landscape of the sub-arctic regions of Western Siberia.

5 MATHEMATICAL MODELLING OF CRYOGENIC PROCESSES

Flat palsas are characterized by ice formations which tend to segregate them. These peat bogs have peat layers with various levels of thickness and are spread over sandy, clay and loamy grounds. Clay soils have the potential to totally collapse up to 6 m in depth. Sandy soils, on the other hand, collapse less than 0.5 m (Geocryological Forecast, 1983). According to processes to estimate freezing and thawing, the basic variables influencing the intensity of their presence on peat grounds are (1) heat conductivity, (2) heat from phase transitions, and (3) thermal resistance to isolation from snow cover, vegetation, etc. Features of permafrost degradation are illustrated by the application of equations to estimate freezing and thawing that were developed by Gosstroy (Recommendations ..., 1988).

Freezing and thawing depths are described respectively by the following equations:

$$H_{freez} = 1,2\left(\sqrt{\left(2\lambda_{fr}\sum t_{fr}^{*}720{*}3,6/Q_{phas} + R_{win}^{2}\lambda_{fr}^{2}\right)} - R_{win}\lambda_{fr}\right)$$

$$H_{taw} = \sqrt{\left(2\lambda_{warm}\sum t_{warm}^{*}720{*}3,6/Q_{phas} + R_{summer}^{2}\lambda_{warm}^{2}\right)} - R_{summer}\lambda_{warm}$$

where λ_{fr} and λ_{warm} = the heat conductivity of frozen and thawed ground;
$\sum t_{fr}$ and $\sum t_{warm}$ = –the sum of monthly average surface temperatures during cold and warm periods;
Q_{phas} = the heat of phase transitions;
R_{win} and R_{summer} = the total thermal resistance of isolation; and
$R = h/\lambda$, h = the thickness of isolation from snow, peat, etc.

These equations indicate that the basic condition of permafrost existence is the excess of freezing depth over thawing depth. Analyses of data from these equations allow for the determination of some of the main soil properties that influence freezing and thawing processes. The most essential factors are soil humidity, peat layer thickness, and snow cover thickness. The presence of permafrost in soils depends on thickness of organic horizons, thickness of peat layers and thickness of a snow covers at constant mid-annual temperature and humidity levels.

Figure 19 illustrates the dynamics of freezing and thawing of peat lands according to the given equations and using air temperature, humidity soil humidity and now cover thickness data from the Tarko-Sale meteorological station.

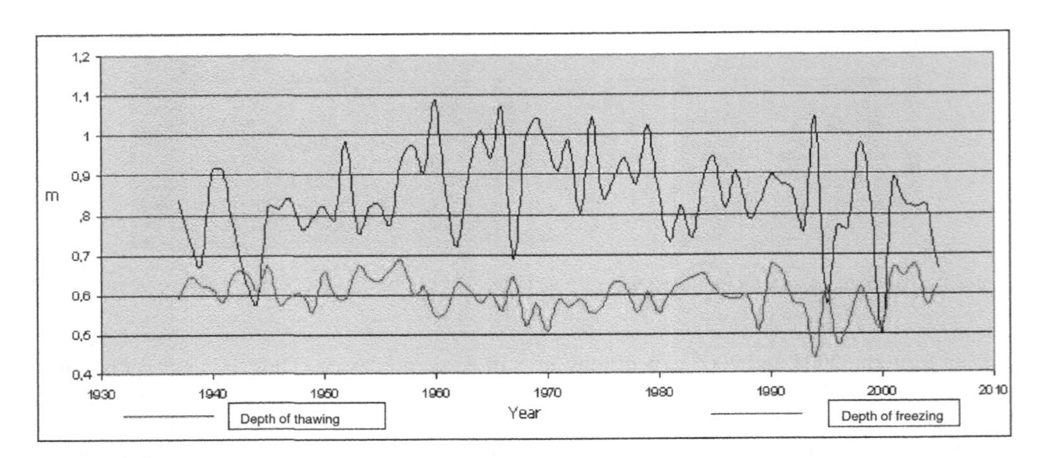

Figure 19. Changes in freezing and thawing depths of peat grounds.

In general, freezing depths of peat are greater than thawing depths in high latitudes. However, climatic conditions, that favor development of thermokarst processes, can appear at any time.

Data analyses using the freezing-thawing equations suggest that intensification of thermokarst processes can be caused by heterogeneity of microclimatic conditions and change processes related to one another. For example, increases in the depth of seasonal thawing and the formation of thermokarst areas are made possible by reductions in or damage to the mosses-lichen layer caused by fire, etc. Although sub-arctic ecosystems allow for restoration of minor damage resulting from bogging processes and peat accumulation, the development of thermokarst is more probable when thermokarst areas are constantly fed by moisture as shown in Figures 20 and 21. The depth of thawing increases as a result of snow accumulation thermokarst craters. Thickness of snow cover is one of major factors that influence freezing and thawing processes in regions that have been studied.

Theory related to cryogenic processes suggests that sites with a low-thickness of peat deposit are the most vulnerable to thermokarst activity because thawing of adjoining grounds will result in greater deformations. It is connected to a high level of ice content and potential collapse of substratum grounds as well as to the greater heat conductivity of mineral grounds compared to peat. Thick layers of peat act as thermo-insulation and keep underlying grounds from thawing.

This feature explains the rather unusual phenomenon observed at the Puritey-Malto key site in 2005 which was described in the previous section. Active thermokarst is apparent in areas where the peat layer of the frozen bogs is less than 50 cm. It is almost absent on the southern edge of the permafrost zone where the thickness of frozen peat varies from 1.5–2 m.

Changes in hydrological conditions can activate permafrost heaving processes. The application of quantitative methods in forecasting the cryogenic heaving processes of freezing peat is rather difficult because researchers have generally focused on studying mineral grounds. Use of developed methods to forecast cryogenic processes of freezing peat is very difficult because no experimental values for hydraulic conductivity of peat exist for unfrozen conditions once thawing has occurred. Furthermore, design procedures are based on characteristics of mineral grounds, such as plasticity and lamination limit, which do not apply to peat. Finally, no consensus exists regarding processes of heaving and segregation related to ice formation. Most scientists recognize heaving, but consider it to be insignificant in the process of segregation of ice formations (Vtjurin and Vtjurina, 1980; Ershov, 1982).

Locations of frost mounds in river valleys, at the bottoms of drained lakes – khasyreis – and at the mouths of brooks is evidence of the important role of moisture in their formation. It is obvious, that

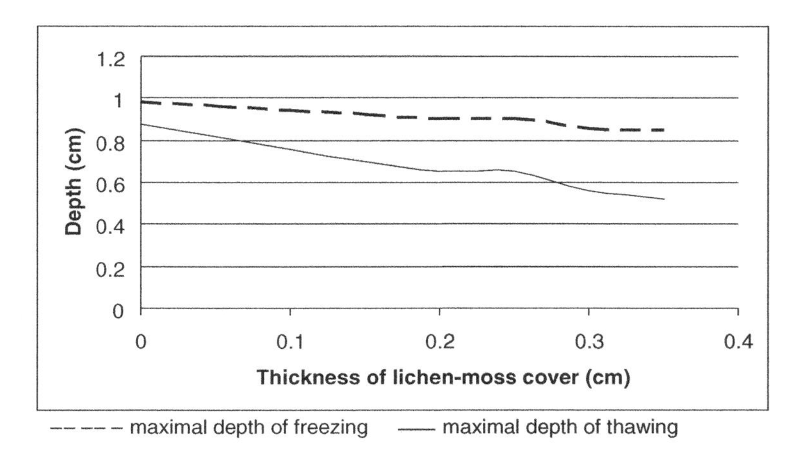

Figure 20. Dependence of the maximal depths of freezing and thawing at different thickness of lichen-moss layer.

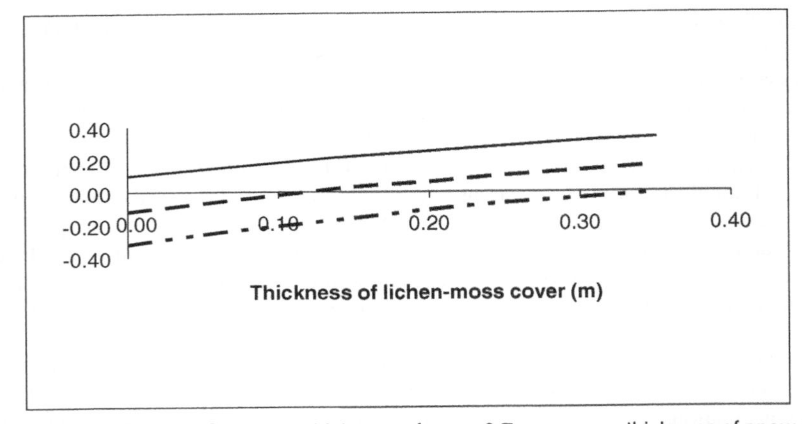

Figure 21. Dependence of a difference of depth of the freezing-thawing border of peat grounds at thickness of snow cover and lichen-moss layer.

palsa mound formation processes are connected to spatial variations in freezing-thawing processes which occur because of variations in vegetation and ground moisture. The general tendency of heaving is probably connected to increases in freezing which are in turn related to increase in moisture supply (Grechischev et al., 1980).

6 WESTERN-SIBERIAN PEATLANDS AND THE GLOBAL CARBON CYCLE

Previous discussion highlights the highly sensitive and dynamics processes of change in occurring in northern bog landscapes. For this reason, it is important to analyze probable future changes in the carbon cycle and their impact on peatlands, particularly those north of the Sibirsky Uvala ridge in Western Siberia. There is no clear evidence that the area is either a net sink or source of emission into the atmosphere. Furthermore, no data exist on extremely high gas fluxes of carbon dioxide or methane that result from the thawing of pristine northern bogs. Thus, analyses of stabilization mechanisms related to the carbon cycle and carbon balance of peatlands of Western-Siberia are important to predicting climatic changes.

Many authors believe that peat accumulation in Western Siberia began simultaneously during the early Holocene period in several locations that were favorable for swamping (Blyakharchuk et al., 2001; Lapshina et al, 2001; Liss, 2001; MacDonald et al., 2001; Muldiyarov et al., 2001; Velichko et al., 2001; Zelikson et al., 2001; Lapshina et al., 2001; Preis and Antropova, 2001; Turunen et al., 2001). The most typical features of this process are reflected in Table 1. The principal bio-climatic parameters change, from South to North and accordingly, the carbon stock in peat deposits and peatland areas decreases along the specified direction. However, peat deposits and carbon accumulation occurred irregularly in connection with climatic changes during the entire Holcene period.

The average annual temperature varies from −1.1°C in the south to −6.6°C in the north. These variations clearly influence the functional conditions and appearances of the Western Siberia bogs. The precipitation gradient is not so well defined for the territory under consideration.

The South–North temperature gradient and variations in the occurrence of permafrost may limit peatland distribution in the North.

The biological productivity of bogs and fens is in general agreement with variations along the South to North gradient. No significant differences among net primary production values (NPP) of the middle and northern taiga bog ecosystems have been revealed. Wide ranges of NPP have

Table 1. Climatic and carbon accumulation features of bio-geographical sub-zones in western Siberia.

Parameter	Forest tundra	Northern taiga	Middle taiga	Southern taiga
Mean annual temperature, °C	−6.6	−5.3	−3.1	−1.1
Precipitation, mm	400–430	410–500	500–550	450–500
Peatlands area, 10^3 km^2 [1]	108	226	254	316
Peat accumulation in Holocene, mm C yr^{-1} [2,3,4,5]	0.2	0.1–0.3	0.3–0.8	0.8–1.4
Carbon accumulation rate in Holocene, g C m^{-2} yr^{-1} [3,4,5]	n.d.	7–11	12–35	25–60
Peat carbon stock, Pg C [1]	1.9	10.2	15.5	24.2
Net primary production, g dry phytomass, m^{-2} yr^{-1} (min-max) average [6,7]	(300–600)/462	(350–960)/608	(500–890)/570	(240–2400)/813

Note: [1] Yefremov and Yefemova, 2001; [2] Liss, 2001; [3] Lapshina and Polgova, 2001; [4] Bleuten and Lapshina, 2001; [5] Turunen et al., 2001; [6] Kosykh et al., 2003; [7] Kosykh, Mironycheva-Tokareva, 2005; n.d., no data.

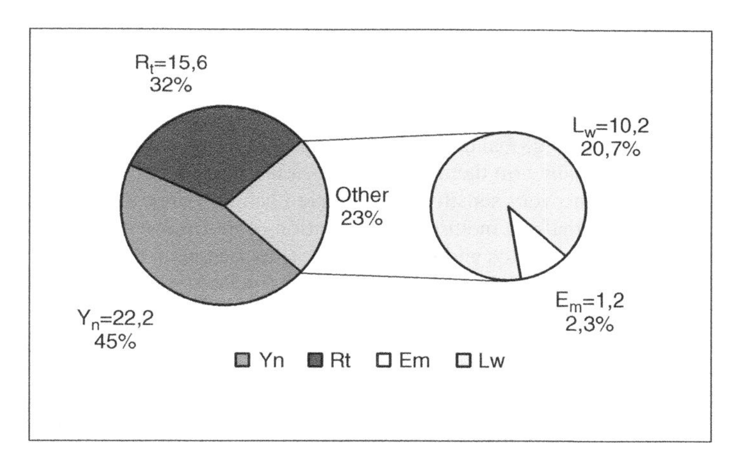

Figure 22. Quantity proportions between carbon cycle components of oligotrophic and mesotrophic mires of West Siberia by (Naumov, 2004), in Tg C yr^{-1} (R_t – total C loss through autotrophic and heterotrophic respiration; E_m – methane-C emission; L_w – dissolved carbon output with mire water flow; Y_n – net carbon sequestration).

been established and they are related to the variety of peatland types which are found in sub-zones. Existing estimations do not take into account the net primary production of marshes and swamps.

The total NPP, including that of oligotrophic and mesotrophic bog ecosystems within the boundaries of the taiga zone, was compared with balance characteristics of the carbon cycle based on direct gas flux measurements in the field (See Figure 22). In the North about 160 TgC is fixated per year by peatland vegetation, assuming a C-content of dry phytomass of 0, 45, which corresponds to 18–32% of the net carbon fixation by terrestrial ecosystems of Russia (Nilson et al., 2003; Zavarzin and Kudeyarov, 2006). Total net carbon sequestration by Western-Siberia forest-bogs excluding the contribution of marshes and swamps, is estimated to be 22.2 TgC yr^{-1} (See Figure 22). The average net Holocene carbon accumulation was estimated to be 11.8 Tg yr^{-1} for raised string bogs that represent 68 M ha in Western Siberia (Turunen et al., 2001). This value is apparently underestimated because researched lands were repeatedly impacted by fires. In addition the extrapolation of a data set from the middle taiga to all Western Siberia probably is not valid because of the wide range of climatic variations in the region. However, a consensus exists that present-day sequestration of atmospheric carbon by northern peatlands is higher than it previously was.

Total emissions through autotrophic and heterotrophic respiration make up 32% of net assimilation flux or 49 Tg C yr^{-1}. Methane emissions represent only 2.3% of total emissions or 1.2 Tg C yr^{-1}. Net carbon assimilation is only about 31% of the NPP calculated using direct counts of shoot and root growth. This discrepancy is explained by a conceptual model of the carbon cycle offered recently by Naumov (2004). According to this model more than 60% of NPP is related to internal ecosystem resources. The internal cycle provides rather independent functioning and stability to the mire ecosystem.

Measurements of carbon gas and methane fluxes from the surface of frozen and thawing bogs that are situated in a permafrost zone, fail to confirm the assumed large emissions of greenhouse gases into the atmosphere as a result of thermokarst degradation (Naumov, 2001). The optimum temperature for the total respiration of northern peatlands is about 12–13°C and is associated with a narrow temperature range of biota activity. Thus, climate warming will inevitably result in the occurrence of more productive ecosystems across diverse ecological systems which will in turn lead to greater carbon accumulation.

7 CONCLUSIONS

Western Siberia contains many peatlands that represent a largue unique bog sinc on earth. They have been a sinc for atmospheric carbon since the last deglaciation period. In this paper, the contribution of Western Siberia peatlands to global carbon balance is assessed as well as possible influences of climatic and environment change on them. Northern peatlands, such as "palsas" in the sub-arctic region of western Siberia, are very sensitive to climatic change. Large areas are subject to rapid the sequential processes permafrost melting and activation of thermokarst drops. These processes are clearly indicated by maps, photos and models of these processes in the region.

Numerous data exist about the rapid change dynamics associated with northern bog landscapes. However, none of them suggest that they will lead to gross infringements in the carbon balance. Thermokarst is a natural phenomenon which has occurred for a long time. Changes in relief and vegetation, occurring as a result of this process, can apparently be considered as "natural" for northern landscapes, but they are accelerated by human activity. An intermediate stage of this evolution may be characterized by progressive swamping and increases in mire ecosystem productivity. Mires have a unique internal mechanism of carbon balance regulation. Due to their non-saturated cycle, they have the capacity to buffer global warming by reducing carbon gases in the atmosphere.

ACKNOWLEDGEMENTS

This research is financially supported by EU-INTAS project 34.35.25.

REFERENCES

Blyakharchuk, T.A., S.N. Kirpotin, S.N. Vorob'ev, E. Muldiyarov. 2001. Preliminary results of spore-pollen analysis of a peat section in the northern ecotone of the forest zone in Western Siberia. p. 19–21. In: S.V. Vasiliev, A.A. Titlyanova, and A.A. Velichko (eds.), West Siberian Peatlands and Carbon Cycle: Past and Present. Proc. Int. Field Simpos. Noyabrsk.

Callaghan, T.V., and S. Jonasson. 1995. Arctic ecosystems and environmental change. Phil. Trans. Roy. Soc. London A. 352:259–276.

Callaghan, T.V, S.N. Kirpotin, B. Werkman, S.N. Vorobyev, I. Brown, and V.V. Lukyantsev. 1999. Investigation of contrasting climate impact on vegetation and landscape processes in forest tundra and taiga of the Western

Siberian plain as a basis for the opening up of the North. Exploration of the North: Traditions and challenge of time. p. 62–63. In: Proc. Russian-Canadian Workshop, Tomsk State University, Tomsk, Russia.
Bleuten, W. and E.D. Lapshina (eds.). 2001. Carbon storage and atmospheric exchange by West Siberian peatlands. FGUU Scientific Reports 2001–1, Utrecht (NL), ISBN 90-806594-1.X, 165 p.
Connor, S. 2005. Global warming "past the point of no return". The Independent. Friday, 16 Sept. http://news.independent.co.uk/world/science_technology/article312997.ece
Dlugokencky, E.J., K.A. Masarie, P.M. Lang, and P.P. Tans. 1998. Continuing decline in the growth rate of the atmospheric methane burden. Nature 393: 447–450 [in Russian].
Ershov, E.D. 1982. Cryolithogenesis. Moscow: Nedra, 211 pp.
Geocryological forecast for West-Siberian gas-bearing provinces. 1983. Novosibirsk: Science: 182 p [in Russian].
Gorham, E. 1991. Northern peatlands: role in the carbon cycle and probable responses to climatic warming. Ecol. Appl. 1:182–195.
Grechischev, S.E., L.V. Chistotinov, and J.U.L. Schur. 1980. Cryogenic physical-geological processes and their forecast. Moscow: Nedra, 382 p [in Russian].
IPCC 2001. Climate Change 2001: The Scientific Basis. The carbon cycle and atmospheric carbon dioxide. p. 183–227. In: J.T. Houghton, Y. Ding, D.J. Griggs, M. Noguer, P.J. van der Linden, X. Dai, K. Maskell, and C.A. Johnson (eds.). Cambridge Univ. Press. 183–237.
Kirpotin, S., S. Vorobev, V. Chmyz, T. Guzynin, S. Skoblikov, and A. Yakovlev. 1995. Structure and dynamics of the vegetative cover of palsas in Nadym-Pur interfluve of West Siberia Plain. Botanichesky Zhurnal 80 (8): 29–39. [in Russian].
Kirpotin, S.N., T.A. Blyakharchuk, and S.N. Vorobyev. 2003. Dynamics of subarctic West Siberian palsas as indicator of global climatic changes. Bulletin of Tomsk State University. TSU Press, Appendix 7: 122–134 [in Russian].
Kirpotin, S.N. and S.N. Vorobiov. 1999. The natural dynamics of sub-arctic landscapes in the West Siberian Plain as indicator of global changes of climate. In: Proc. 42nd Annual Symposium of the IAVS "Vegetation and Climate" – 1st ed. Vitoria-Gazteiz: Servico Central de Publicaciones del Gobierno Vasco: 74.
Kirpotin S.N., E.D. Lapshina, N.P. Mironycheva-Tokareva, and W. Bleuten. 2004. "Landslide" thermokarst in the West-Siberian sub-arctic region and the tendency of global climatic changes // Ecological, humanitarian and sports aspects of underwater activity. Proceedings of 3rd International Scientific-Practical Conference, October, 21–23, 2004. Tomsk: Tomsk State University Press: 163–169 [in Russian].
Kosykh, N.P., N.P. Mirinycheva-Tokareva, and W. Bleuten. 2003. Productivity of southern taiga bogs of West Siberia. Bulletin of Tomsk State University. TSU Press, Appendix 7: 142–152 [in Russian].
Kosykh, N.P. and N.P. Mironicheva-Tokareva. Net primary production of mires in the north of Western Siberia //Peat bogs and biosphere: Proceedings of IV Scientific School (Tomsk, 12–15 September). Tomsk, 2005. – p. 228–231. [in Russian].
Kudeyarov, V.N. 2004. Present-day estimates of carbon cycling at global scale and for territory of Russia. p. 17–26. In: Emission and sink of greenhouse gases on the Northern Eurasia Territory. Pushchino: PSC RAS. [in Russian].
Lapshina, E.D. and S.N. Kirpotin. 2003. Natural dynamics of sub-arctic landscapes in the West Siberian Plain as indicator of global changes of climate. p. 39–45. In: Proc. 7th Korea-Russia Int. Symp. of Science and Technology KORUS 2003, Univ. of Ulsan.
Lapshina, E.D., A.A. Velichko, O.K. Borisova, K.V. Kremenettsky, and N.N. Pologova. 2001. Holocene dynamics of peat accumulation. p. 47–72. In: W.Bleuten and E.D. Lapshina (eds.), Carbon storage and atmospheric exchange by West Siberian peatlands. FGUU Scientific Reports 2001-1, Utrecht (NL), ISBN 90-806594-1-X.
Liss, O.L. 2001. The evolution of West Siberian bogs since the Pleistocene. p. 33–36. In: S.V. Vasiliev, A.A. Titlvanova, and A.A. Velichkn (eds.), est Siberian Peatlands and Carbon Cycle: Past and Present, Proc. Int. Field Simpos. Noyabrsk.
MacDonald, G.A., L.C. Smith, K.E. Frey, S. Peugh, A. Velichko, K. Kremenetski, O. Borisova, E.D. Lapshina, E.E. Muldiyarov, and P.A. Dubinin. 2001. History of carbon accumulation in the northern part of West Siberia. p. 37. In: S.V. Vasiliev, A.A. Titlyanova, and A.A. Velichko (eds.), West Siberian Peatlands and Carbon Cycle: Past and Present, Proc. Int. Field Simpos, Noyabrsk.
Matthews, J.A., S.O. Dahl, M.S. Berrisford, and A. Nesje. 1997. Cyclic development and thermokarstic degradation in the mid-alpine zone at Leirpullan, Dovrefjell, Southern Norway. Permafrost and Periglacial Processes. 8: 107–122.
Muldiyarov, E.Y., E.D. Lapshina, K. Kremenetskiy, and E.V. Perevodchikov. 2001. History of development and structure of the peat layer of bogs of northern taiga of west Siberia. p. 41–44. In: S.V. Vasiliev,

A.A. Titlyanova, and A.A. Velichko (eds.), West Siberian Peatlands and Carbon Cycle: Past and Present. Proc. Int. Field Simpos. Noyabrsk. [in Russian].

Naumov, A.V. 2001. Emission of CH_4 and CO_2 in connection with temperature conditions of peat bog soils in the northern taiga subzone. p. 110–112. In: S.V. Vasiliev, A.A. Titlyanova, A.A. Velichko (eds.), West Siberian Peatlands and Carbon Cycle: Past and Present. Proc. Int. Field Simpos. Noyabrsk. [in Russian].

Naumov, A.V. 2004. Carbon Budget and Emission of Greenhouse Gases in Bog Ecosystems of Western Siberia. Eurasian Soil Science 31(1):58–64.

Nelson, F.E. and O.A. Anisimov. 1993. Permafrost zonation in Russia under fnthropogenic climatic change. Permafrost and Periglacial Processes 4:137–148.

Nilson, S., A. Shvidenko, V. Stolbovoi, I. McCallun, and M. Jonas. 2003. Full carbon account of the terrestrial ecosystem of Russia in 1988–2002. p. 88–89. In: Emission and sink of greenhouse gases on the northern Eurasia territory. Proc. Second Int. Conf. Pushchino, June 16–20, 2003. Abstracts. Pushchino.

Preis, Yu. and N. Antropova. 2001. Permafrost as a main factor of formation of Westwrn Siberia peatlands. p. 198–201. In: S.V. Vasiliev, A.A. Titlyanova, A.A. Velichko (eds.), West Siberian Peatlands and Carbon Cycle: Past and Present. Proc. Int. Field Simpos. Noyabrsk.

Recommendations for a technique of regulation of seasonal freezing and thawing of grounds and development of thermokarst at settling of Western Siberia. 1988. Moscow: Stroyizdat: 72.

Sollid, J.L. and L. Sorbel. 1998. Palsa dogs as a climate indicator – examples from Doverfjell, Southern Norway. AMBIO. 27(4):287–291.

Turunen, J., A. Pitkänen, T. Tahvanainen, and Tolonen. 2001. Carbon accumulation in West Siberian mires. Russia. Global Biogeochem. Cycles 15:285–296.

United Nations Millenium Declaration. http://www.un.org/millennium/declaration/ares552e.pdf

Vaganov, E.A., E.F. Vedrova, S.V. Verkhovets, S.P. Efremov, T.T. Efremova, V.B. Kruglov, A.A. Onuchin, A.I. Sukhinin, and O.B. Shibistova. 2005. Forests and swamps of Siberia in the global carbon cycle. Siberian ecological journal. 4: 631–649. [in Russian].

Velichko, A.A., W. Bleuten, E.D. Lapshina, K.V. Kremenetski, O.K. Borisova, E.M. Zelikson, E.Yu Novenko, Yu M. Kononov, V.A. Klimanov, V.P. Nechaev, E.E. Muldiyarov, O.A. Chichagova, V.V. Pisareva, and P.A. Dubinin. 2001. Carbon cycle and Holocene history of bog in West Siberia taiga zone. p. 61–63. In: S.V. Vasiliev, A.A. Titlyanova, A.A. Velichko (eds.), West Siberian Peatlands and Carbon Cycle: Past and Present. Proc. Int. Field Simpos. Noyabrsk.

Vtjurin, B.I. and E. Vtjurina. 1980. Principles of classification lithocryogenous processes and phenomena. Geomorphology 13–22 [in Russian].

Yefremov, S.P. and T.T. Yefremova. 2001. Stocks and forms of deposited carbon and nitrogen in bog ecosystems of West Siberia. p. 148–151. In: S.V. Vasiliev, A.A. Titlyanova, A.A. Velichko (eds.), West Siberian Peatlands and Carbon Cycle: Past and Present. Proc. Int. Field Simpos. Noyabrsk.

Zavarzin, G.A. and V.N. Kudeyarov. 2006. Soil as a main source of carbonic acid and organic carbon reservoir on territory of Russia. Bulletin Russian Academy Sciences 76(1):14–29. [in Russian].

Zelikson, E.M., O.K. Borisova, O.A. Chichagova, S.P. Efremov, V.A. Klimanov, K.V. Kremenetski, and E.Yu Novenko. 2001. New data on vegetation and climate history in West Siberian taiga zone. p. 64–67. In: S.V. Vasiliev, A.A. Titlyanova, and A.A. Velichko (eds.), West Siberian Peatlands and Carbon Cycle: Past and Present. Proc. Int. Field Simpos. Noyabrsk.

Zemtsov, A.A. 1976. Geomorphology of West-Siberia plain (northern and central parts). Tomsk. Izd. Tomskogo Universiteta: 344.

Research and Development Priorities

CHAPTER 34

Researchable priorities in terrestrial carbon sequestration in Central Asia

R. Lal

The Ohio State University, Carbon Management and Sequestration Center, Columbus, Ohio, USA

1 INTRODUCTION

Strategies of mitigation of severe problems of soil and environmental degradation in Central Asia can be identified and implemented through a strong data base on properties, processes and practices affecting carbon (C) sequestration in soil and vegetation (e.g., trees, shrubs). The available research information collected in this book is an excellent starting point. However, there are numerous knowledge gaps that must be filled in order to develop a comprehensive development-oriented program. Being a large area comprising of several distinct ecosystem/biomes, soil specific data are needed from well designed research plots and on-farm conditions. Requirements of the site-specific information are also justified by unique land use and farming systems which are characteristics of the region. Important among these systems are extensive grazing, irrigated agriculture, and sparsely vegetated and barren or desertified lands. In addition, there are also mountain systems, rainfed cropland and forest lands (see Chapter 1 by Eddy de Pauw). Therefore, long-term objectives of a research program are to strengthen the database on C pool and flux with regard to specific land use systems, soil properties, eco-regional characteristics, and cultural/ethnic niches.

2 INFORMATION AVAILABLE

The data presented in this book support the following principal conclusions:

1. Soil degradation and desertification are severe problems throughout the Central Asia region. The problem is exacerbated by excessive and inappropriate irrigation, intensive tillage, summer fallowing, and uncontrolled grazing.
2. No-till farming with residue mulch and elimination of summer fallowing are viable options for cropland management in Central Asia.
3. Precision farming or soil-specific management is a useful strategy for enhancing nutrient use efficiency, and reducing risks of water contamination.
4. Controlled grazing, sowing of improved species and improvement of soil fertility are important to enhancing terrestrial C sequestration in grazing land.
5. Restoration of degraded soils and ecosystems and adoption of recommended management practices can lead to C sequestration in soil and biomass. Preliminary estimates of C sequestration are given in Table 1. Total potential for the region ranges from 37 to 68 Mt C yr^{-1} (Table 1).

3 FUTURE RESEARCH NEEDS

There are numerous important knowledge gaps which must be filled to identify restorative land uses and conservation-effective soil/crop management practices. Important among these are described below.

Table 1. Preliminary estimates of the rates and total potential of terrestrial C sequestration.

				Accumulation rates (kg ha^{-1} yr^{-1})						Sequest. Pot.[a] Mt yr^{-1} OC	
				Organic C				Inorganic C			
System	Area (Mha)	Pixels	Area (Mha)	Veget. Low	Veget. High	Soil Low	Soil High	Low	High	Soil Low	Soil High
Extensive Mountain systems											
Rangelands	5.32	356,869	21.3	100	400	50	200	0		1.06	4.25
Agroforestry only	2.13	142,895	8.5			100	200	0		0.85	1.70
Forests only	0.72	48,552	2.9			150	300	0		0.43	0.87
Irrigated systems	6.22	417,278	24.9			200	400			4.97	9.94
Rangelands											
a. Deserts (<150 mm)	18.11	1,214,371	72.3			100	100	<10		7.23	7.23
i. with cool winters	2.74	184,007	11.0								
ii. with cold winters	15.36	1,030,364	61.4								
b. Steppe (>150 mm)	53.51	3,588,899	213.7	150	250	50	150	10	30	10.69	32.06
Rainfed grain production zone, plains	93.21		372.0							36.74	67.55

[a] Sequestration potential in million tonnes per year of organic carbon.
CASANet, CIRIT-OXU (prepared by the working group of the workshop participants).

3.1 *Basic research in terrestrial C pool and dynamics*

Research information is needed in processes affecting both soil and biomass C (Figure 1). Soil-specific research is needed in understanding processes governing dynamics of soil organic carbon (SOC) and soil inorganic carbon (SIC) pools. With regards to SOC, it is important to now the historic C loss, total SOC sink capacity, fate of SOC transported by erosional processes, rate of SOC sequestration by different management practices, and the effect of SOC on soil quality and agronomic productivity. Similar to SOC, understanding dynamics of SIC is also important especially in irrigated soils. The magnitude of leaching of biocarbonates is affected by the quality of irrigation water and soil properties. Rate of formation of secondary carbonates (pedogenic carbonates) is influenced by biogenic processes, availability of cations (Ca^{+2}, Mg^{+2}, K^{+}), and management practices (e.g., liming, application of biosolids) must be determined for different soil types and ecoregions of Central Asia.

Two distinct but inter-related components of biomass C are the above-ground and below-ground components. With regards to the above-ground biomass, research information needed is the rate of net primary productivity (NPP) for species suited to different eco-regional niches, effect of management on NPP, and of the possibility of establishing energy plantations to produce biofuel feedstock. Little, if any, research information is known about the rate and turnover time of the below-ground C pool. Research data needed for different species include the following: depth distribution of biomass C, the nature of recalcitrant compounds present in the root biomass, the rate of turnover of fine roots, relationship between the belowground biomass including leaf litter and

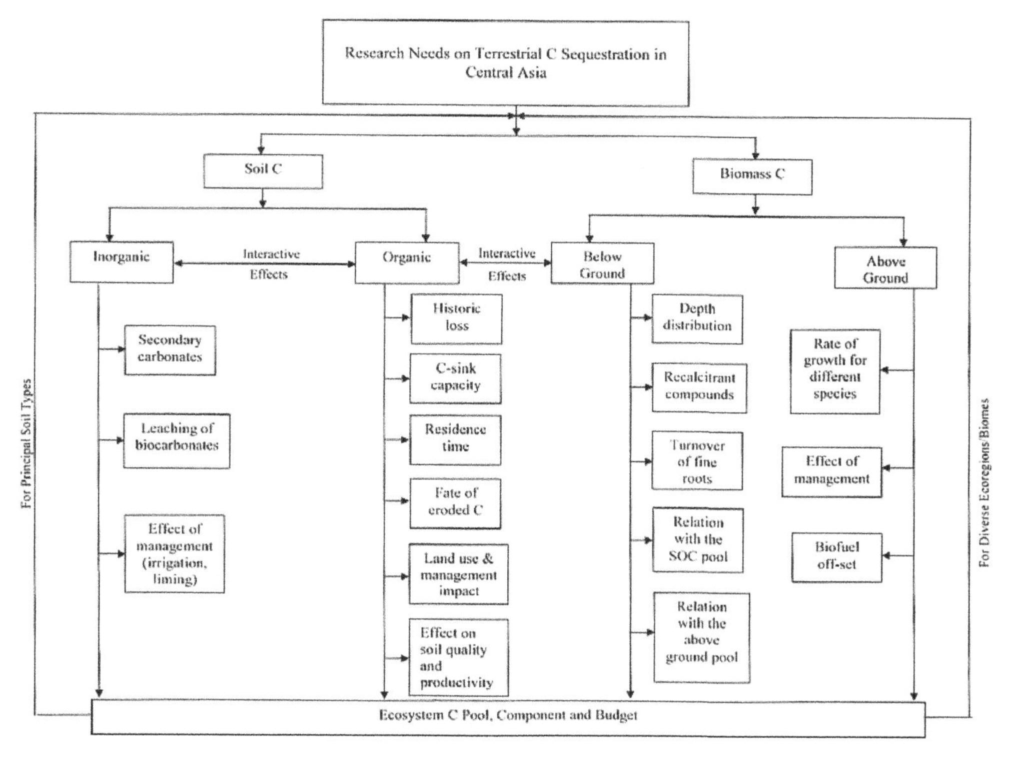

Figure 1. Needs for basic research on the terrestrial C pool and dynamics.

the SOC pool and dynamics, and the relationship between the above-ground and the below-ground biomass for different species grown under a wide rage of management practice.

3.2 *Strategies for sustainable management of soil and water resources*

Terrestrial C sequestration, in soil and biomass, as outlined in Figure 1, requires identification of appropriate strategies to realize the potential sink capacity of the diverse ecosystems. Some strategies of achieving the potential, once it has been determined, are outlined in Figure 2. The goal is to identify systems for sustainable management of soil resources. With regards to soils of the arid and semi-arid regions, soil-water management is crucial to enhancing biomass production. Closely interacting with soil-water management is the irrigation management. Soil and water management techniques must be identified to enhance the water use efficiency (WUE) so that more biomass is produced per unit amount of the water consumed. There exists a strong link between WUE and soil fertility or the availability of essential plant nutrients. Thus, WUE can be enhanced only through increasing the nutrient use efficiency (NUE), and vice versa. In addition to managing cropland and grazing land, restoring degraded/desertified soils and ecosystems are extremely important. The goal is to enhance ecosystem services, especially with regards to increasing terrestrial C pool, improving water quantity and quality, and increasing biodiversity.

3.3 *Strategies for terrestrial carbon sequestration*

Soil and vegetation management have strong impact on the terrestrial C sequestration. Two components of terrestrial C sequestration (total C pool) involve management of: soil C pool in cropland

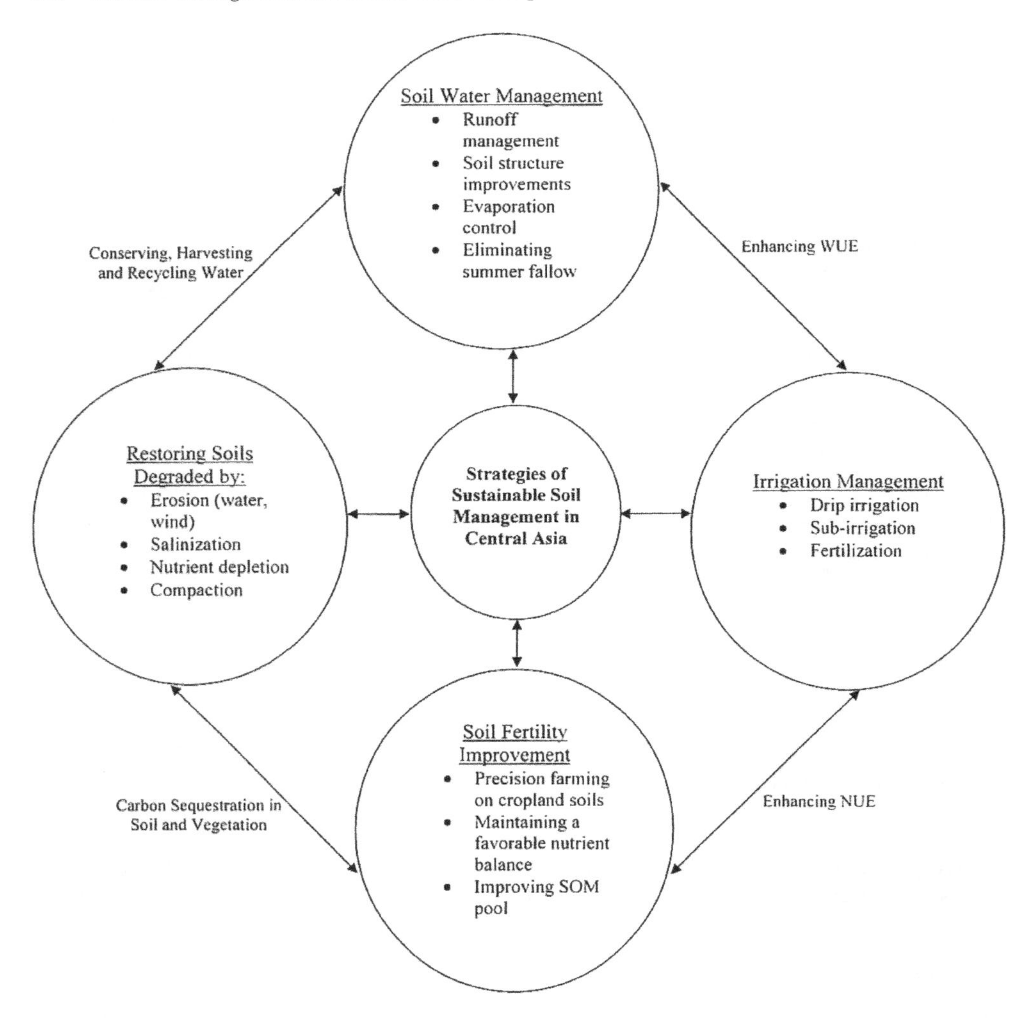

Figure 2. Strategies for sustainable management of soils in Central Asia.

and rangeland, and biotic (biomass) C pool in the forestland (Figure 3). Numerous chapters in this book outline impact of no-till farming, eliminating summer fallow and enhancing soil fertility on agronomic/biomass productivity. Similarly, adopting controlled grazing and conserving water while improving soil fertility are important to rangeland management. In addition to the choice of appropriate species, stand management (e.g., thinning, weeding) conserving/recycling water and improving soil fertility are important strategies of managing forestland. Establishing biofuel/energy plantations is another topic of relevance to C sequestration in the biomass. Halophytes, irrigated with brackish water, can be grown to produce biofuel feedstock. In addition, drought tolerant woody species and perennial grasses may be identified for producing biomass which can be used to produce ethanol.

The concept of "carbon farming" must be introduced to land managers, policy makers and the farming community. Terrestrial C is a marketable commodity and must be grown on land similar to other farm produce (e.g., cotton, wheat, meat, milk).

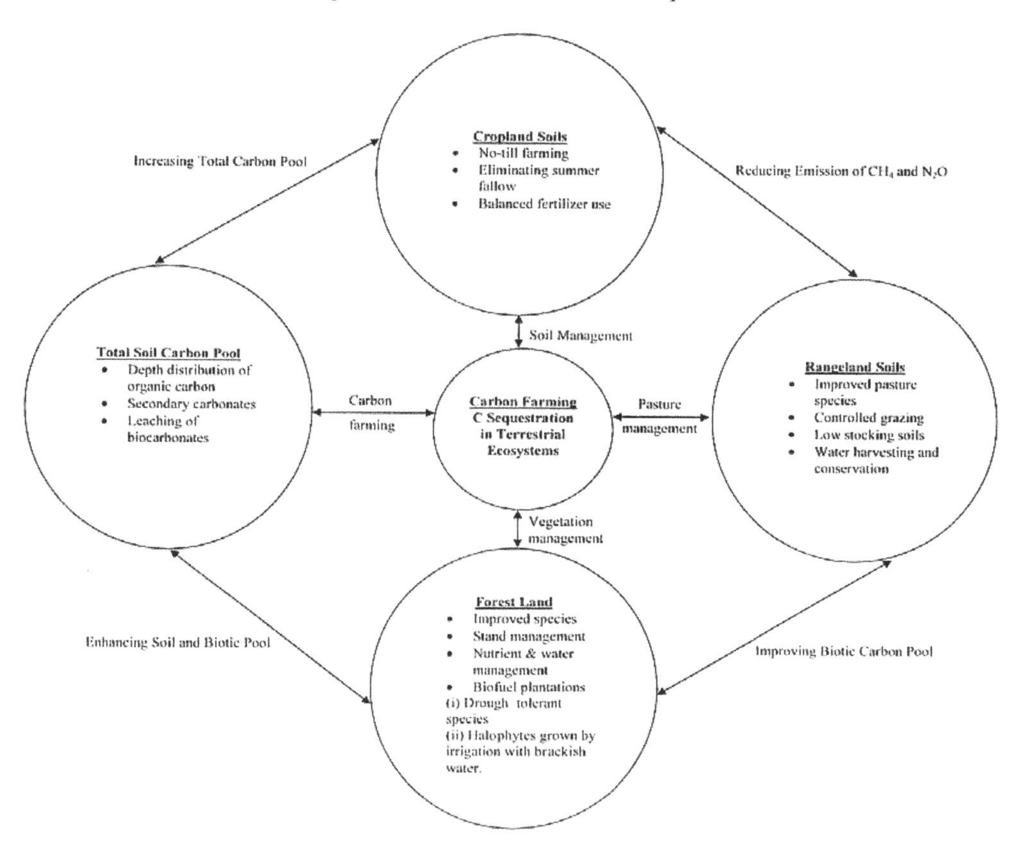

Figure 3. Strategies for terrestrial C sequestration and carbon farming.

3.4 *Trading carbon credits*

Operationalization of C farming involves trading (buying and selling) of C credits. There are numerous steps in commodification of C (Figure 4). Important among these are (i) measurement of the rate of C sequestration in soil and biomass at community or regional scale with reference to a specific baseline, (ii) evaluating permanence or the residence time of C sequestered in relation to the recommended land use and soil/plant management and the current/future risks of soil degradation/desertification, (iii) determining the price of terrestrial C in relation to the market (demand) and societal value of terrestrial C pool in relation to the ecosystem services, and (iv) establishing a market either through the Kyoto Protocol (CDM), World Bank or the private institution (The Chicago Climate Exchange). At present, there is no procedure available to trade terrestrial C credits in Central Asia, and its operationalization is required to set the process in motion.

3.5 *Methodology for assessing the terrestrial C pool*

A standard protocol must be established for soil sampling, sample preparation, and data analysis (Figure 5). Rather than using the present techniques of laboratory-based analyses, several attempts are being made to develop in-situ methods of measuring C under field conditions (e.g., LIBS, INS). Scaling and data aggregation are also important to obtaining estimates of C pool at watershed, landscape or regional scale. In this regards the importance of GIS and remote sensing techniques cannot be over emphasized (Figure 5).

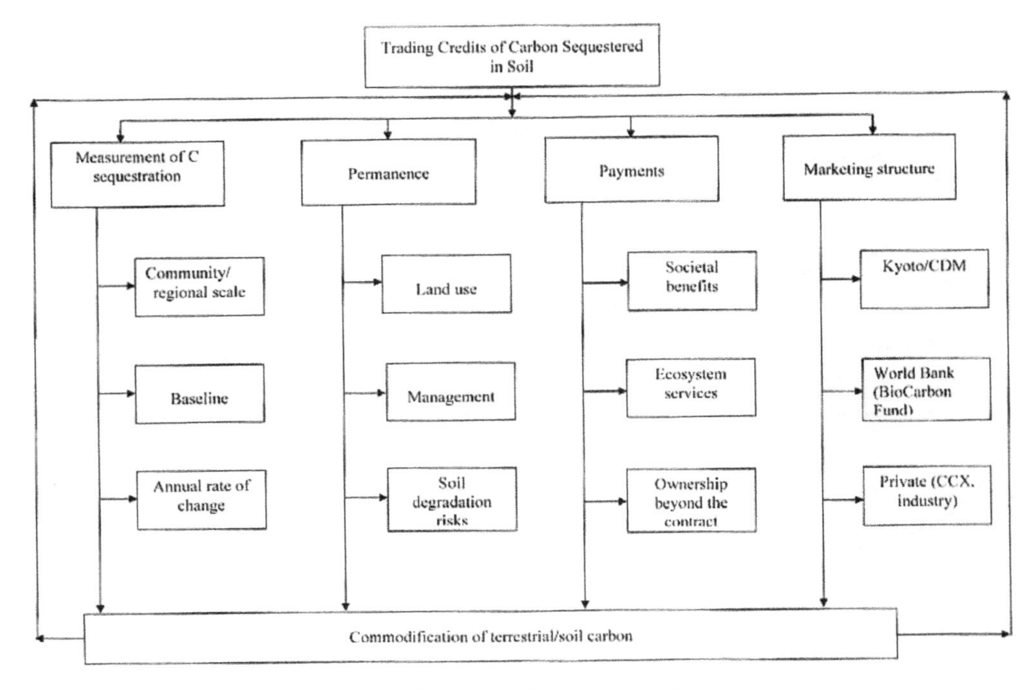

Figure 4. Operationalizing "carbon farming" and trading carbon credits.

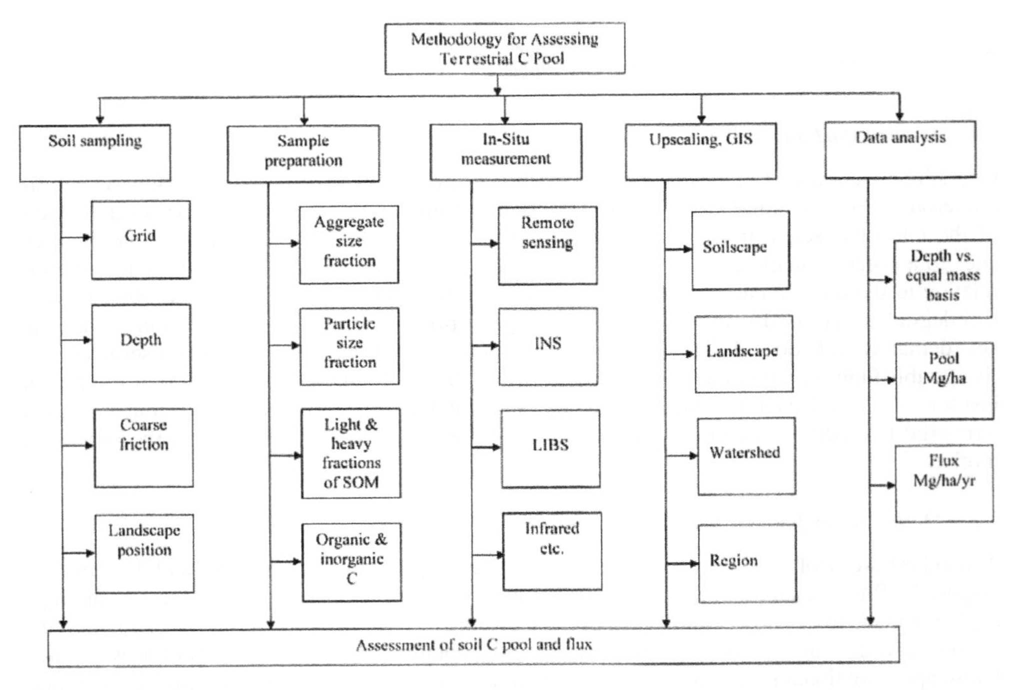

Figure 5. Research priorities for developing methodology of assessing soil C pool and flux.

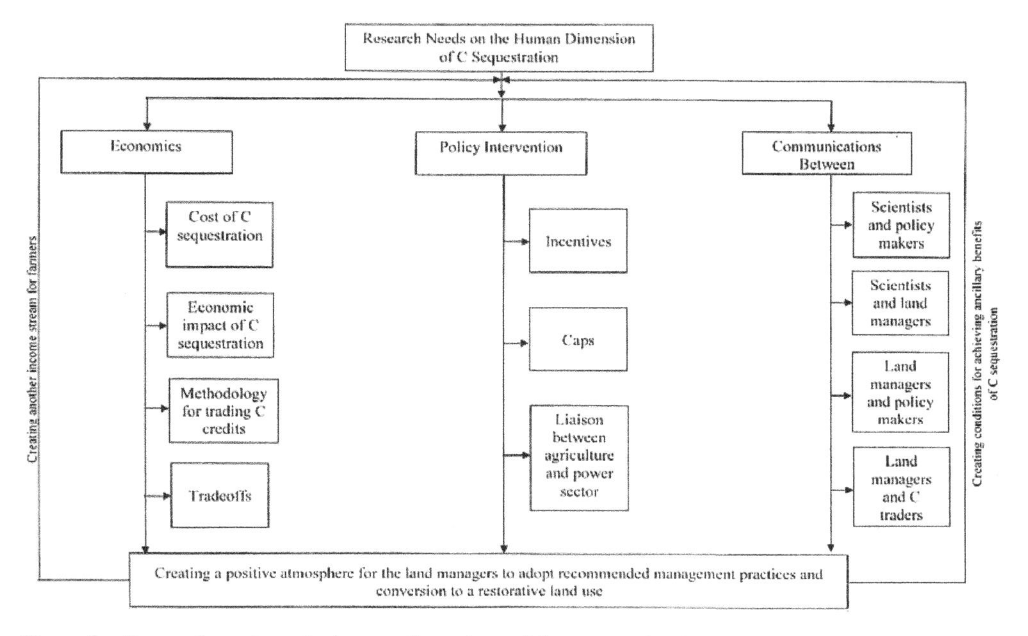

Figure 6. Research needs on the human dimensions of C sequestration in terrestrial ecosystems.

3.6 *The human dimension of C sequestration*

Commodification of terrestrial C requires several economic and social issues which need to be addressed (Figure 6). Important among these issues were discussed by a workgroup and summarized in the CASCANet brochure. These are briefly described below:

a. Land Tenure Policy: Land users do not become land improvers unless they feel that they and theirs have some exclusive access to the benefits accruing from those land improvements. In Central Asia, while irrigated plots may be held in marginally secure leaseholds, much of the remainder of the agricultural land remains at the disposition of the State. Land privatization policies are allegedly most advanced in Turkmenistan even though that country has the least sequestration potential. A homesteading policy, through which some kind of secure title could be passed over to the land improver on condition that those improvements were maintained, would seem to be in order. Particularly important would also be a range of tenure policy encouraging pastoral nomads to protect perennial grass and shelterbelt plantings in return for some sort of privileged access. Here a water point exclusivity protocol, if designed to fit the ecologically optimal transhumance patterns, might prove very beneficial for the landscape.
b. Water Conservation Policy: Central Asia is world renowned for its uneconomical use of irrigation water. Not only does this exacerbate the salt logging of too much bottomland, it draws water away from the plant life that could be protecting the hillsides from erosion. National water policies featuring a water use fee structure are long overdue.
c. Commodity Marketing: Deliberate steps should be taken to favor the sustainable production and marketing of field and range produce from perennial, soil and water conserving plants. An inventory of such possibilities and their market forecasts need to be developed. Once the comparative advantage of certain landscape friendly commodities has been identified, then international consumers and corporate investors should be sought out.
d. Research to Land-Use Planning Linkages: Central Asia appears to have relatively strong agricultural research and environmental science capacity. This development of National land-use plans maximizes each ecological zone's sustainable agriculture, commodity production and

carbon sequestering potential. While these plans should incorporate full input from local land users, once they are implemented, farmers ignoring their conservation covenants should be subject to a higher land tax penalty.

e. Education and Public Awareness: These National Plans should also be lined to a public media effort underscoring the significant carbon sequestration revenues ($ billions) that could accrue to the region if they were effectively implemented. Civic awareness can also be heightened if this same education outreach shows how quickly Central Asia's land resource endowments will be completely depleted if drastic conservation, carbon sequestering, steps are not immediately taken.

3.7 *Inter-disciplinary team approach*

Terrestrial C sequestration is an inter-disciplinary issue, and must be addressed through an inter-disciplinary team approach (Figure 7). Such a team must be based on a close interaction between biophysical scientists (e.g., pedologists, geologists, hydrologists, foresters and agronomists, remote sensing and GIS specialists) with economists and political scientists. Physical scientists can determine C processes at the landscape/soilscape scales, modelers can develop predictive models using GIS and scaling techniques, and economists and social scientists can identify policy imperatives and develop decision support systems that land managers and policy makers can use (Figure 7).

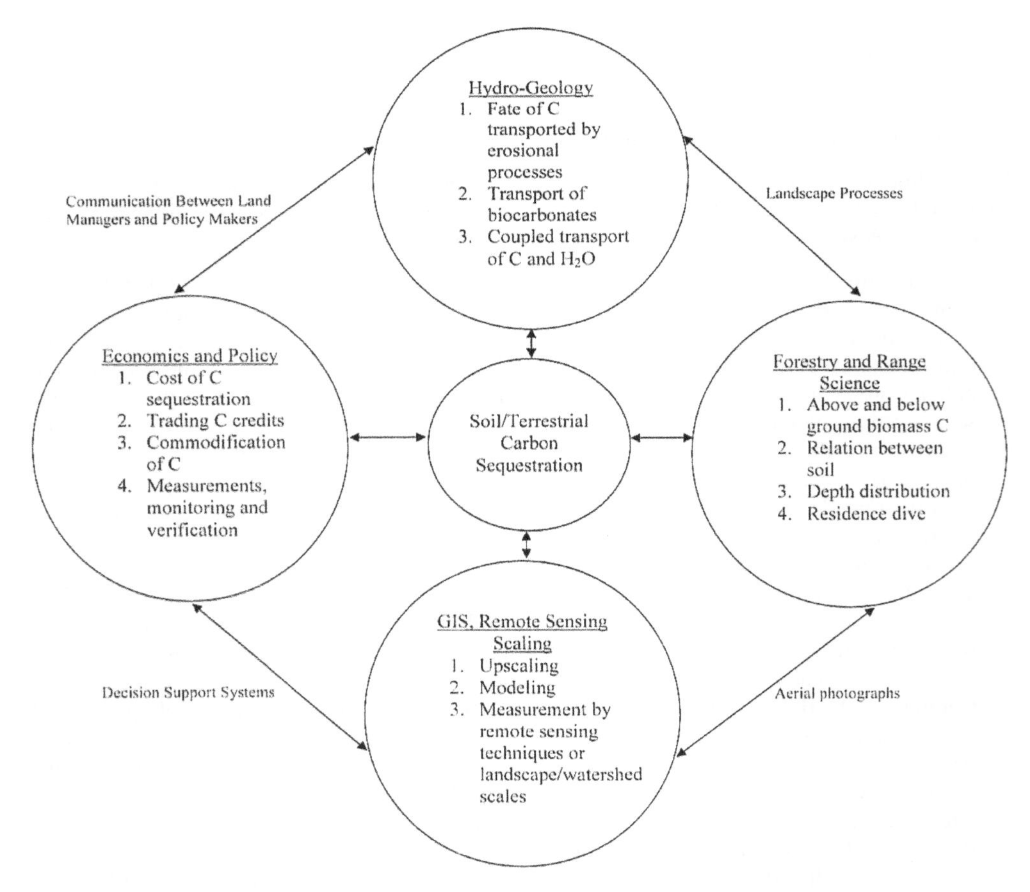

Figure 7. Interdisciplinary team approach to soil-terrestrial carbon sequestration and climate change.

4 CONCLUSIONS

Carbon sequestration in soil and biomass is an important and emerging field of science. There exists a strong need to develop a comprehensive research program in understanding processes affecting C dynamics, ecosystem properties which influence the rate of C sequestration and magnitude of the sink capacity, and management practices which upon adoption can lead to realization of the attainable C sink capacity. Ancillary benefits of terrestrial C sequestration must also be assessed in terms of the improvements in crop/biomass yield, reduction in non-point source pollution, and off-set of CO_2 emissions by industrial activities. Operationalization of the process of trading C credits is a priority issue which requires an inter-disciplinary team approach. Biomass C in forests, plantations, and woody shrubs must also be systematically assessed. Developing relation between above-ground and below-ground biomass, and evaluating the impact of below-ground biomass on soil organic C pool are important researchable priorities. Establishing energy plantations, of perennial grasses and short rotation woody perennials, in appropriate ecoregions where soil moisture is adequate or halophytes can be grown by irrigation with brackish water, is an emerging issue which must also be linked to the climate change while producing C-neutral or C-negative fuel sources.

Subject index

Author index

For Product Safety Concerns and Information please contact our EU representative GPSR@taylorandfrancis.com Taylor & Francis Verlag GmbH, Kaufingerstraße 24, 80331 München, Germany

Printed and bound by CPI Group (UK) Ltd, Croydon, CR0 4YY
08/06/2025
01897007-0015